Selected Problems in Fluid Flow and Heat Transfer

Selected Problems in Fluid Flow and Heat Transfer

Special Issue Editor

Artur J. Jaworski

MDPI • Basel • Beijing • Wuhan • Barcelona • Belgrade

Special Issue Editor
Artur J. Jaworski
School of Computing and Engineering,
University of Huddersfield
UK

Editorial Office
MDPI
St. Alban-Anlage 66
4052 Basel, Switzerland

This is a reprint of articles from the Special Issue published online in the open access journal *Energies* (ISSN 1996-1073) from 2018 to 2019 (available at: https://www.mdpi.com/journal/energies/special_issues/fluid_flow)

For citation purposes, cite each article independently as indicated on the article page online and as indicated below:

LastName, A.A.; LastName, B.B.; LastName, C.C. Article Title. *Journal Name* **Year**, *Article Number*, Page Range.

ISBN 978-3-03921-427-3 (Pbk)
ISBN 978-3-03921-428-0 (PDF)

Contents

About the Special Issue Editor

Artur J. Jaworski, PhD; MSc(Eng); DIC; CEng; FRAeS; FHEA, is currently Chair in Mechanical Engineering and Head of Department of Engineering and Technology at the School of Computing and Engineering, University of Huddersfield, UK. He has previously served as Chair in Energy Technology and Environment at the University of Leeds (2013–2017), Chair in Engineering and Head of the Thermofluids Research Group at the University of Leicester (2011–2013), and Lecturer and Senior Lecturer at the University of Manchester (2000–2011). In 2015, he held a Visiting Professor appointment at the Faculty of Engineering, University of Cagliari, Italy, and in 2017, a Visiting Professor appointment at the School of Computing and Engineering, University of Huddersfield, UK.

Professor Jaworski received his MSc(Eng) from the Faculty of Power and Aeronautical Engineering of the Warsaw University of Technology (1986–1991), and PhD and DIC from the Department of Aeronautics, Imperial College of Science, Technology and Medicine, London (1992–1996). This was followed by a position as Postdoctoral Research Associate at the Department of Chemical Engineering, University of Manchester Institute of Science and Technology (1996–2000).

Professor Jaworski's research track record includes studies in the theoretical and numerical analysis of heat and mass transfer processes in thermal solar systems, experimental fluid dynamics and aerodynamics related to vortex dynamics and coherent structures generated by vortex breakdown flow field, aerodynamic flow control using synthetic jet actuators, sensor design, measurement and instrumentation for multiphase processes, non-invasive imaging techniques such as industrial process tomography and, most recently, thermoacoustic technologies which utilize thermal fluid interactions between acoustic fields and compressible fluids to engineer thermodynamic machines (engines and coolers) with no moving parts. He has held two prestigious Fellowships in Thermoacoustic Technologies: EPSRC Advanced Research Fellowship (2004–2009) and Royal Society Industry Fellowship (2012–2015). This research gave rise to his interest in the thermal fluid processes which underpin this Special Issue of Energies. He has co-authored over 140 publications, of which over 60 are in refereed scientific journals.

Preface to "Selected Problems in Fluid Flow and Heat Transfer"

Development of modern engineering applications demand the heat transfer enhancement. At the same time efficient heat transfer equipments are necessary to reduce energy consumptions and improve the energy saving. Such techniques allow also to reduce emissions and pollution and to obtain more control in carbon dioxide increase. Heat transfer enhancement is both very attractive and challenging issues in the research and industry fields. It has a fundamental role to improve the energy efficiency as well as to develop thermal systems with high performances.

Heat transfer enhancement techniques can be found in different engineering applications such as solar energy system, thermal control, electronics cooling, nuclear reactor, heat exchanger, automotive cooling, refrigeration and chemical process, etc. They are classified in passive methods or in active methods. In the first methods no direct application of external power is required, whereas in the second methods an external power is necessary. The effectiveness of the heat transfer enhancement techniques is strongly related to the heat transfer mechanisms. It can be due to from single-phase free convection to dispersed-flow film boiling. Plate fins, nanofluids, porous insertions can be considered as examples of passive techniques, while magnetic fields, induced vibrations, rotations are examples of active techniques.

Therefore, this book includes original papers on the most recent research activities on the topic to provide useful guidelines for future research directions and engineering applications.

Artur J. Jaworski
Special Issue Editor

Editorial

Special Issue "Selected Problems in Fluid Flow and Heat Transfer"

Artur J. Jaworski

School of Computing and Engineering, University of Huddersfield, Queensgate, Huddersfield HD1 3DH, UK;
a.jaworski@hud.ac.uk

Received: 8 July 2019; Accepted: 31 July 2019; Published: 7 August 2019

Fluid flow and heat transfer processes play an important role in many areas of science and engineering from the planetary scale (e.g., influencing weather and climate) to microscopic scales of enhancing heat transfer by the use of nanofluids; they also underpin the performance of many energy systems understood in the broadest possible sense. This topical special issue is dedicated to the recent advances in this very broad field. The main criteria for paper acceptance were academic excellence, originality and novelty of applications, methods or fundamental findings. All types of research approaches were equally acceptable: experimental, theoretical, computational, and their mixtures; the papers could be both of fundamental or applied nature, including industrial case studies. With such a wide brief, it was naturally very difficult to define a finite list of relevant disciplines. However, it was broadly anticipated that the authorship and ultimate readership would come from the fields of mechanical, aerospace, chemical, process and petroleum, energy, earth, civil and flow instrumentation engineering, but equally biological and medical sciences, as well as physics and mathematics; that is, everywhere where "fluid flow and heat transfer" phenomena may play an important role or be a subject of worthy research pursuits. Cross disciplinary research and development studies were also most welcomed.

The response to this special issue invitation was very impressive. We received 85 manuscripts. A strict refereeing process adopted by the editorial team of *Energies* meant that only 25 papers made it to the final special issue. Also, in the process, some papers were judged not suitable for the special issue, but generally of suitable standard for the regular issues of *Energies* and those were transferred over to alternative editors and subsequently published elsewhere. Publishing this special issue was of course a team effort, and thanks are due to all involved, in particular a group of reviewers who have to remain anonymous, my colleagues from the editorial board of *Energies* and the tireless editorial team led by Ms. Julyn Li, without whom the special issue would have never succeeded.

For obvious reasons, the range of topics covered by the papers is very diverse. However, to help bring an order to this motley collection of works, I have attempted to identify a few main themes in which the contributions were made. These are listed below, the ultimate one being named "Miscellaneous Problems" as a reflection of my futile attempts to associate these papers with other themes. This is by no means a reflection on the academic excellence present in these papers.

The first theme can be arbitrarily defined as "**Turbomachinery and boundary layer flow**" to include the following works:

- "Analysis of Different POD Processing Methods for SPIV-Measurements in Compressor Cascade Tip Leakage Flow" by Shi et al. [1];
- "Unsteadiness of Tip Leakage Flow in the Detached-Eddy Simulation on a Transonic Rotor with Vortex Breakdown Phenomenon" by Su et al. [2];
- "Effect of Rotor Thrust on the Average Tower Drag of Downwind Turbines" by Yoshida et al. [3];
- "POD Analysis of Entropy Generation in a Laminar Separation Boundary Layer" by Jin and Ma [4];

- "Experimental Investigation of Flow-Induced Motion and Energy Conversion of a T-Section Prism" by Shao et al. [5].

Here, the traditional "turbomachinery" topic has been widened to include the aerodynamics of wind turbines [3]. Usually turbomachinery incorporates boundary layer flows and boundary layer separation processes which was the logic behind grouping paper [4] under this heading. The last paper deals with flow-induced motion for harvesting of flow energy [5] which is related to vortex induced vibrations and hence its loose association with boundary layer flows and separation.

The second theme is defined as **"Heat transfer and heat exchangers"**. This is relatively self-explanatory and includes a mix of numerical and experimental works of both fundamental and applied nature in heat transfer and design of heat exchangers. The contributions are as follows:

- "Flow Structure and Heat Transfer of Jet Impingement on a Rib-Roughened Flat Plate" by Alenezi et al. [6];
- "Numerical Study on Thermal Hydraulic Performance of Supercritical LNG in Zigzag-Type Channel PCHEs" by Zhao et al. [7];
- "A Machine Learning Approach to Correlation Development Applied to Fin-Tube Bundle Heat Exchangers" by Lindqvist et al. [8];
- "A Numerical Study on the Light-Weight Design of PTC Heater for an Electric Vehicle Heating System" by Kang et al. [9];
- "Unsteady Simulation of a Full-Scale CANDU-6 Moderator with OpenFOAM" by Kim et al. [10].

The third group of contributions fits rather well the traditional field of multiphase flows, specifically **"Two-phase flow"** (gas–liquid), due to the type of problems considered in the papers. The contributions included:

- "Visualization Study on Thermo-Hydrodynamic Behaviors of a Flat Two-Phase Thermosyphon" by Wang et al. [11];
- "Gas–Liquid Two-Phase Upward Flow through a Vertical Pipe: Influence of Pressure Drop on the Measurement of Fluid Flow Rate" by Ganat et al. [12];
- "Investigation on the Handling Ability of Centrifugal Pumps under Air–Water Two-Phase Inflow: Model and Experimental Validation" by Si et al. [13].

The fourth theme of research that emerged can be referred to as **"Flow with micro- and nano-scale features"**. This includes the following papers:

- "Resonant Pulsing Frequency Effect for Much Smaller Bubble Formation with Fluidic Oscillation" by Desai et al. [14];
- "Bubble Size and Bubble Concentration of a Microbubble Pump with Respect to Operating Conditions" by Jeon et al. [15];
- "Spherical Shaped (Ag–Fe$_3$O$_4$/H$_2$O) Hybrid Nanofluid Flow Squeezed between Two Riga Plates with Nonlinear Thermal Radiation and Chemical Reaction Effects" by Ahmed et al. [16]
- "Numerical Study of the Magnetic Field Effect on Ferromagnetic Fluid Flow and Heat Transfer in a Square Porous Cavity" by El-Amin et al. [17];
- "Experimental Study of Particle Deposition on Surface at Different Mainstream Velocity and Temperature" by Zhang et al. [18].

There are two papers dealing with microbubbles present in the flow [14,15], a nanofluids application [16], ferromagnetic fluid behaviour [17] and a particle deposition problem [18].

The fifth grouping includes only two papers. It is referred to as **"Waste heat recovery"**. It is clearly an important and growing field in energy research, especially in the context of climate change and efficiency drives in manufacturing industries to reduce carbon emissions. The two contributions to this special issue include:

- "Development and Assessment of Two-Stage Thermoacoustic Electricity Generator" by Hamood et al. [19];
- "Investigation of the Concepts to Increase the Dew Point Temperature for Thermal Energy Recovery from Flue Gas, Using Aspen®" by Fedorova et al. [20].

Finally, the last grouping brings together **"Miscellaneous problems"**, purely on the basis that the works submitted are within relatively niche fields that include a single paper each in this special issue. The papers are devoted to fire spreading [21], pyrolysis [22], lubrication [23], water hammer effect [24] and a refined method of calculating friction losses in pipes [25].

- "Experimental Study on the Fire-Spreading Characteristics and Heat Release Rates of Burning Vehicles Using a Large-Scale Calorimeter" by Park et al. [21];
- "Macro and Meso Characteristics of In-Situ Oil Shale Pyrolysis Using Superheated Steam" by Wang et al. [22];
- "A Numerical Study on Influence of Temperature on Lubricant Film Characteristics of the Piston/Cylinder Interface in Axial Piston Pumps" by Song et al. [23]
- "Investigation on Water Hammer Control of Centrifugal Pumps in Water Supply Pipeline Systems" by Wan et al. [24];
- "One-Log Call Iterative Solution of the Colebrook Equation for Flow Friction Based on Padé Polynomials" by Praks and Brkić [25].

The guest editor and the editorial team of *Energies* hope that the readership will find the selection of articles presented here a useful contribution to the broad field of fluid flow and heat transfer in the context of energy systems.

Conflicts of Interest: The author declares no conflict of interest.

References

1. Shi, L.; Ma, H.; Lixiang, W.L. Analysis of Different POD Processing Methods for SPIV-Measurements in Compressor Cascade Tip Leakage Flow. *Energies* **2019**, *12*, 1021. [CrossRef]
2. Su, X.; Ren, X.; Li, X.; Gu, C. Unsteadiness of Tip Leakage Flow in the Detached-Eddy Simulation on a Transonic Rotor with Vortex Breakdown Phenomenon. *Energies* **2019**, *12*, 954. [CrossRef]
3. Yoshida, S.; Fujii, K.; Hamasaki, M.; Takada, A. Effect of Rotor Thrust on the Average Tower Drag of Downwind Turbines. *Energies* **2019**, *12*, 227. [CrossRef]
4. Jin, C.; Ma, H. POD Analysis of Entropy Generation in a Laminar Separation Boundary Layer. *Energies* **2018**, *11*, 3003. [CrossRef]
5. Shao, N.; Lian, J.; Xu, G.; Liu, F.; Deng, H.; Ren, Q.; Yan, X. Experimental Investigation of Flow-Induced Motion and Energy Conversion of a T-Section Prism. *Energies* **2018**, *11*, 2035. [CrossRef]
6. Alenezi, A.H.; Almutairi, A.; Alhajeri, H.M.; Addali, A.; Gamil, A.A.A. Flow Structure and Heat Transfer of Jet Impingement on a Rib-Roughened Flat Plate. *Energies* **2018**, *11*, 1550. [CrossRef]
7. Zhao, Z.; Zhou, Y.; Ma, X.; Chen, X.; Li, S.; Yang, S. Numerical Study on Thermal Hydraulic Performance of Supercritical LNG in Zigzag-Type Channel PCHEs. *Energies* **2019**, *12*, 548. [CrossRef]
8. Lindqvist, K.; Wilson, Z.T.; Næss, N.; Sahinidis, N.V. A Machine Learning Approach to Correlation Development Applied to Fin-Tube Bundle Heat Exchangers. *Energies* **2018**, *11*, 3450. [CrossRef]
9. Kang, H.S.; Sim, S.; Shin, Y.H. A Numerical Study on the Light-Weight Design of PTC Heater for an Electric Vehicle Heating System. *Energies* **2018**, *11*, 1276. [CrossRef]
10. Kim, H.T.; Chang, S.-M.; Son, Y.W. Unsteady Simulation of a Full-Scale CANDU-6 Moderator with OpenFOAM. *Energies* **2019**, *12*, 330. [CrossRef]
11. Wang, C.; Yao, F.; Shi, J.; Wu, L.; Zhang, M. Visualization Study on Thermo-Hydrodynamic Behaviors of a Flat Two-Phase Thermosyphon. *Energies* **2018**, *11*, 2295. [CrossRef]
12. Ganat, T.A.; Hrairi, M. Gas–Liquid Two-Phase Upward Flow through a Vertical Pipe: Influence of Pressure Drop on the Measurement of Fluid Flow Rate. *Energies* **2018**, *11*, 2937. [CrossRef]
13. Si, Q.; Bois, G.; Jiang, Q.; He, W.; Ali, A.; Yuan, S. Investigation on the Handling Ability of Centrifugal Pumps under Air–Water Two-Phase Inflow: Model and Experimental Validation. *Energies* **2018**, *11*, 3048. [CrossRef]

14. Desai, P.D.; Hines, M.J.; Riaz, Y.; Zimmerman, W.B. Resonant Pulsing Frequency Effect for Much Smaller Bubble Formation with Fluidic Oscillation. *Energies* **2018**, *11*, 2680. [CrossRef]

15. Jeon, S.-Y.; Yoon, J.-Y.; Jang, C.-M. Bubble Size and Bubble Concentration of a Microbubble Pump with Respect to Operating Conditions. *Energies* **2018**, *11*, 1864. [CrossRef]

16. Ahmed, N.; Saba, F.; Khan, U.; Khan, I.; Alkanhal, T.A.; Faisal, I.; Mohyud-Din, S.T. Spherical Shaped (Ag—Fe_3O_4/H_2O) Hybrid Nanofluid Flow Squeezed between Two Riga Plates with Nonlinear Thermal Radiation and Chemical Reaction Effects. *Energies* **2019**, *12*, 76. [CrossRef]

17. El-Amin, M.F.; Khaled, U.; Beroual, A. Numerical Study of the Magnetic Field Effect on Ferromagnetic Fluid Flow and Heat Transfer in a Square Porous Cavity. *Energies* **2018**, *11*, 3235. [CrossRef]

18. Zhang, F.; Liu, Z.; Liu, Z.; Liu, Y. Experimental Study of Particle Deposition on Surface at Different Mainstream Velocity and Temperature. *Energies* **2019**, *12*, 747. [CrossRef]

19. Hamood, A.; Jaworski, A.J.; Mao, X. Development and Assessment of Two-Stage Thermoacoustic Electricity Generator. *Energies* **2019**, *12*, 1790. [CrossRef]

20. Fedorova, N.; Aziziyanesfahani, P.; Jovicic, V.; Zbogar-Rasic, A.; Khan, M.J.; Delgado, A. Investigation of the Concepts to Increase the Dew Point Temperature for Thermal Energy Recovery from Flue Gas, Using Aspen®. *Energies* **2019**, *12*, 1585. [CrossRef]

21. Park, Y.; Ryu, J.; Ryou, H.S. Experimental Study on the Fire-Spreading Characteristics and Heat Release Rates of Burning Vehicles Using a Large-Scale Calorimeter. *Energies* **2019**, *12*, 1465. [CrossRef]

22. Wang, L.; Yang, D.; Li, X.; Zhao, J.; Wang, G.; Zhao, Y. Macro and Meso Characteristics of In-Situ Oil Shale Pyrolysis Using Superheated Steam. *Energies* **2018**, *11*, 2297. [CrossRef]

23. Song, Y.; Ma, J.; Zeng, S. A Numerical Study on Influence of Temperature on Lubricant Film Characteristics of the Piston/Cylinder Interface in Axial Piston Pumps. *Energies* **2018**, *11*, 1842. [CrossRef]

24. Wan, W.; Zhang, B.; Chen, X. Investigation on Water Hammer Control of Centrifugal Pumps in Water Supply Pipeline Systems. *Energies* **2019**, *12*, 108. [CrossRef]

25. Praks, P.; Brkić, D. One-Log Call Iterative Solution of the Colebrook Equation for Flow Friction Based on Padé Polynomials. *Energies* **2018**, *11*, 1825. [CrossRef]

Article

Analysis of Different POD Processing Methods for SPIV-Measurements in Compressor Cascade Tip Leakage Flow

Lei Shi [1] **, Hongwei Ma** [1,*] **and Lixiang Wang** [2]

[1] School of Energy and Power Engineering, Beihang University, Beijing 100191, China; sladeshi@126.com
[2] Midea Corporate Research Center, Foshan 528300, China; wanglixiang1106@163.com
* Correspondence: mahw@buaa.edu.cn; Tel.: +86-131-2188-5794

Received: 15 January 2019; Accepted: 11 March 2019; Published: 15 March 2019

Abstract: Though the proper orthogonal decomposition (POD) method has been widely adopted in flow analysis, few publications have systematically studied the influence of different POD processing methods on the POD results. This paper investigates the effects of different decomposition regions and decomposition dimensionalities on POD decomposition and reconstruction concerning the tip flow in the compressor cascade. Stereoscopic particle image velocimetry (SPIV) measurements in the blade channel are addressed to obtain the original flow field. Through vortex core identification, development of the tip leakage vortex along the chord is described. Afterwards, each plane is energetically decomposed by POD. Using the identified vortex core center as the geometric center, the effects of different decomposition regions with respect to the vortex core are analyzed. Furthermore, the effects of different single velocity-components as well as their combination are compared. The effect of different decomposition regions on the mode 1 energy fraction mainly impacts the streamwise velocity component. Though the addition of W velocity component in the decomposition does change the spatial structures of high-order modes, it does not change the dynamic results of reconstruction using a finite number of POD modes. UV global analysis is better for capturing the kinetic physics of the tip leakage vortex (TLV) wandering.

Keywords: POD; tip leakage flow; decomposition region; decomposition dimensionalities; vortex identification; SPIV

---

1. Introduction

The efficiency, the pressure rise and the stable operating range are three prominent parameters for energy and power machinery, such as axial fans and compressors. The tip leakage flow (TLF), driven by the pressure difference between the pressure side and suction side of the blade in the tip region, plays a significant role in all three parameters mentioned above [1–3]. In recent years, the TLF, especially the tip leakage vortex (TLV), has been found to be an inherently unsteady flow phenomenon, with several distinct unsteady behaviors, including vortex wandering [4], vortex splitting [5,6] and vortex breakdown [7,8]. Previous studies showed that these unsteady behaviors not only have profound effects on the TLF mean characteristics, but also a close connection with some important flow phenomena such as rotating stall [9,10], rotating instability [11], oscillations of the TLV [12] and vortex shedding. Therefore, it is necessary to study the unsteady behaviors of the TLV to obtain a thorough knowledge of it.

Stereoscopic particle image velocimetry (SPIV) has been extensively employed to investigate the characteristics of the TLV for its ability to capture an instantaneous snapshot of tip flow structures at various scales [5,8,13,14]. Additionally, multiple averaging methods are also applied, among which the time-averaged method is frequently used. Several stochastic and deterministic characteristics of the TLV have been obtained through this method and statistics analysis [8,13,15]. Besides the simple averaging (ensemble averaging), some researchers employed a triple decomposition to extract the coherent wandering motion from the SPIV data [16]. Through this method, the TLV cores are collocated by linearly shifting the instantaneous core positions to the mean vortex core. This method was concluded to be superior to the simple averaging and recommended to analyze the characteristics of a concentrated vortex [17]. Oweis and Ceccio [14] used a vortex reconstruction method in which ideal Gaussian vortices were fitted to multiple vortices. As a result, they could reduce the effect of primary vortex wandering and reveal changes in flow variability that are masked by the vortex wandering process. However, the TLF is characterized by large-scale vortex and small-scale turbulence motion together. Many important information of unsteady flow behaviors associated with scale may be ignored using these averaging methods as a result.

The proper orthogonal decomposition (POD) method can decompose the examined flow field into orthogonal modes in space and identify the dominant mode based on energy rank [18,19]. It has been widely applied to both experimental and numerical data to identify some coherent structures, such as impeller flow [20], cylinder engine flow [21], wind turbine wake [22] and Jet and vortex actuator-induced flow [23]. Based on optimizing the mean square of the field variable, the POD method can extract flow structures of different scales. It is a promising key technique to analyze the flow structure, the kinematic and dynamic characteristics of the TLV in different length scales and energy levels. Unfortunately, there is a lack of use of the POD method to analyze the unsteady tip leakage flow. Recently, Li [24] introduced the POD method in turbine tip leakage flow analysis. Using the POD method, dominant flow modes governing the unsteady evolution in the tip region are successfully obtained. It is concluded that the POD method with advanced simulation approach can bring an interesting view of what can be done to better understand the leakage flow. Since the rapid development of numerical simulation and experimental techniques, the POD method is providing a new perspective about the tip flow physics.

However, spatial modes and energy distribution can in practice be obtained by different POD processing methods, such as using different decomposition dimensionalities [25,26] or different decomposition regions [27]. In the tip region, the streamwise velocity component is much stronger than the velocity components in the secondary flow direction. The TLV and wall boundary layer would introduce velocity deficits as well. Thus, the velocity and kinetic energy deviation would be quite diverse in different regions and different directions. This inevitably leads to different POD decomposition and reconstruction results using different decomposition regions and decomposition dimensionalities. Though the effects of many factors (POD algorithms [19], number of snapshots [25], analyzed flow variable [28]) have been assessed extensively, few publications have systematically studied the effects of decomposition regions and decomposition dimensionalities on the POD results. A clarification of the impacts of these two factors on POD decomposition and reconstruction results will help with a better and more efficient application of POD method in the TLF analysis.

In this paper, the effects of different decomposition regions and decomposition dimensionalities on POD decomposition and reconstruction of the tip flow are clarified. First, in the blade channel along a full chord length of a single compressor cascade, a SPIV measurement is addressed to obtain the original flow field for the POD decomposition and reconstruction. Combined with vortex core identification, development of the TLV along the chord is described. Then each plane is energetically decomposed and reconstructed using POD. In order to avoid the influence of main flow disturbance and extract more flow field structures associated with the TLV unsteady characteristics, POD decomposition regions are selected with caution. Using the identified vortex core as geometric center, the effects of different decomposition region with respect to the vortex core on POD decomposition are analyzed. Furthermore, the influences of different single velocity-components as well as their combination in POD decomposition are compared, so that to carry out a systematical studying on the effect of these two factors on the POD decomposition. Finally, combined with vortex statistical analysis, the impacts of decomposition dimensionalities in the reconstructed flow field are discussed. It is expected that findings presented in this paper will inspire other researchers who use POD to analyze the TLF.

This paper is organized as follows: firstly, the experimental apparatus and layout are introduced as follows; secondly, an overview of POD method and vortex core identification is given; in the following parts, the detailed comparison results are analyzed; the last section concludes this work.

2. Apparatus and Techniques

2.1. Experimental Facility and Test Conditions

The experiment is carried out in a low speed wind tunnel. The wind tunnel, as schematically shown in Figure 1, has a rectangular exit section of 250 mm by 120 mm (width × height). The test section of wind tunnel consists of a compressor cascade with seven blades, tip wall, hub wall and profile side walls as indicated in Figure 2. The parameters of the cascade and test conditions are listed in Table 1. During the experiment, the inlet velocity of the compressor cascade is kept at 30 m/s. At this velocity, the turbulence intensity in the mainstream is about 2.6%. We also measured the inlet boundary layer since it has a significant impact on the TLF roll-up procedure by interacting with the incoming main flow (Figure 3). The thickness of the boundary layer on the tip wall is about 3 mm.

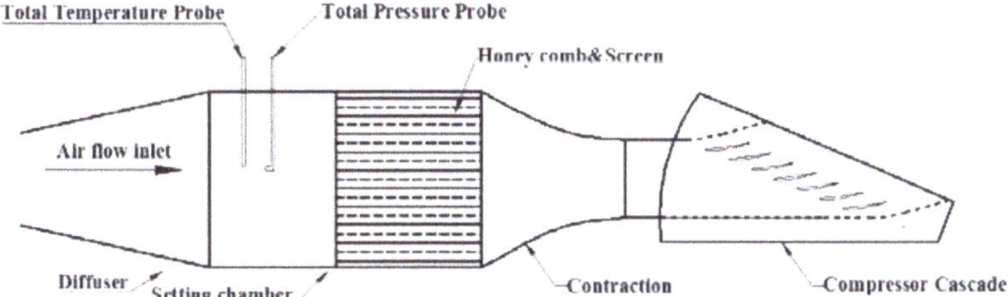

Figure 1. Schematic of the low-speed cascade wind tunnel.

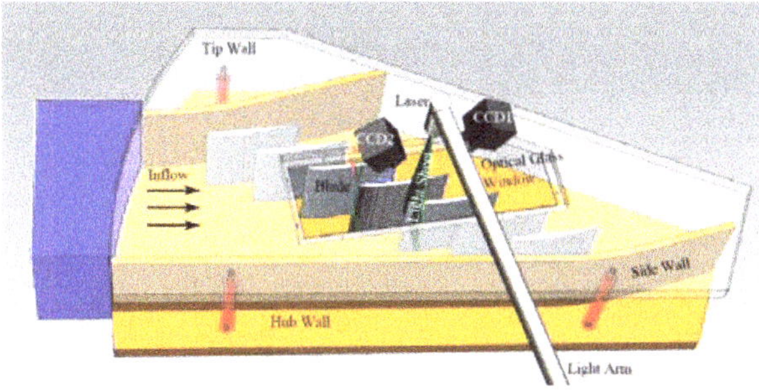

Figure 2. Schematic of compressor cascade and the SPIV measurement configuration.

Table 1. Cascade parameters and test condition.

Items	Details
Number of blades, N	7
Chord length, c	126.8 mm
Span, H	120 mm
Pitch, p	72 mm
Reynolds No., Re	2.81×10^5
Incidence angle	0 degree
Tip clearance/c	5%

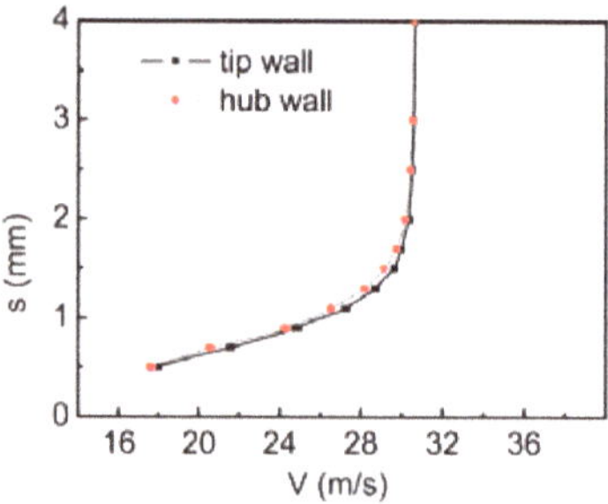

Figure 3. Velocity profiles at the hub and tip walls. Note that the y-axis is the vertical distance of test points to the wall.

2.2. SPIV Technique

An advanced commercial SPIV system is employed to measure the TLF field in the blade passage. As shown in Figure 2, the SPIV system is mainly composed of two 2048 × 2048 pixels, 12 bit frame-straddling based CCD cameras, a dual cavity Nd: YAG laser (200 mJ/pulse at a 15 Hz repetition rate) and a laser arm. The system is 2D-3C (two dimensional, three components) with two cameras located at different sides of the light sheet, which enables users to conduct the measurement of all three velocity components of a 3D flow.

The measurement cross sections are set to be perpendicular to the chordwise direction in the passage. 12 measurement locations along the chordwise direction (z-direction) are set, from $L/c = 0$ to $L/c = 1.1$ with an interval of 10% chord length, as shown in Figure 4. At each measurement location, at least 800 effective instantaneous images are recorded. The valid size of the cross section is about of $30 \times 60 \ \text{mm}^2$, which covers nearly half spanwise range of the whole passage. The spatial resolution of the calibrated image is about 0.039 mm/pixel.

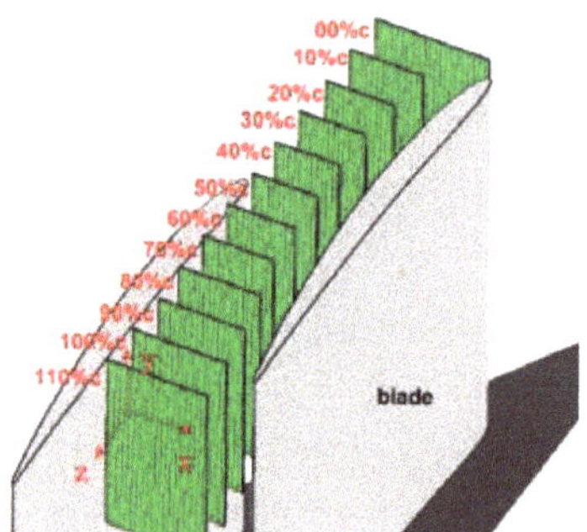

Figure 4. Layout of SPIV measurement cross-sections.

For the data processing, the inter-frame time is set to 10 µs based on the velocity of inlet flow and yields a maximum particle-image displacement of less than 8 pixels. A median-subtraction filter algorithm is applied to all the images to remove non-uniformities in the background light intensity. Subsequently, the particle-image is calculated by the software MicroVec V3 with the two-passes PIV interrogation algorithm. The dimensions of the interrogation region are 64×64 pixels in the first pass, and 32×32 pixels in the second pass. As a result, the spatial resolution of a single velocity vector is 1.26 mm × 1.26 mm. Once a vector field is calculated, vector validation algorithms must be used to eliminate spurious vectors. In the present work, the relative bias from the average vector is used as a post-processing criterion for eliminating questionable vectors, and the relative bias is set to be 50%.

2.3. Discussion of Uncertainties

The measurement error of the SPIV technique, including both random errors and bias errors, have been thoroughly discussed by many researchers [29–36]. Treated with caution during both the experiment and the data processing, random errors can be effectively eliminated. Bias errors depend crucially on the accuracy of particles displacement concerning the SPIV measurement in the compressor cascade. Among bias errors, the one caused by peak locking [30,32] should be dominated. In order to reduce this error, the least squares Gaussian sub-pixel fit for peak detection [5] is employed. Figure 5 illustrates the histogram of the displacement of the particles at the L/c = 1.0 measurement plane, which shows a well control of the peak locking effect during the SPIV experiments.

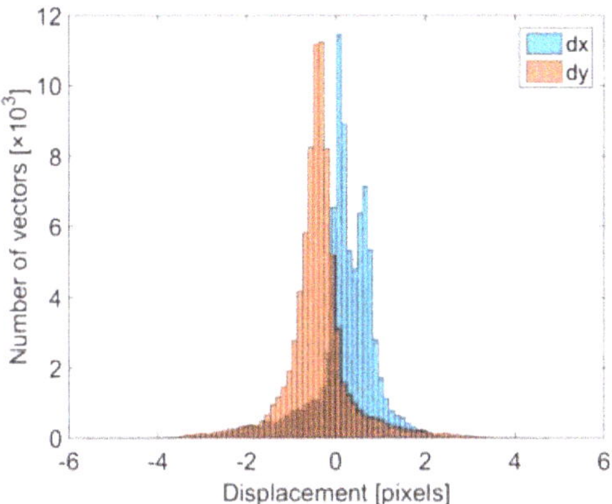

Figure 5. Histograms of the measured particle-image displacement at L/c = 1.0. (79,680 vectors in total, dx and dy are the particle-image displacements in the x and y directions).

According to the analyses of Westerweel [34,36] and Raffle [30], the displacement errors of the in-plane component in the SPIV measurement is about 0.05 pixel. Taking other uncertainties (background noise, acceleration, velocity gradient) into consideration, the displacement errors of the in-plane component are estimated as 0.1 pixels.

Based on the theoretical analysis by Zang and Prasad [37], as shown in Equation (1) the error ratio between the out-of-plane and in-plane components depends on the camera included half angle:

$$\frac{U_z}{U_x} = \frac{U_z}{U_y} = \tan \alpha \tag{1}$$

where α is the camera included half angle; U_x, U_y and U_z are the uncertainties of the x, y and z velocity components. In this paper, this angle is set to $45°$. Therefore, the displacement error of the out-of-plane component is 0.1 pixels as well.

The relative uncertainty of instantaneous velocity in the u-component can be estimated from the equation:

$$U_{B,u_{ij}^k} = \frac{\Delta/K}{u_{ij}^k} \tag{2}$$

where K is converge factor related to measured velocity distribution characteristics. It is set to 2.576 for a 99% confidence interval. u_{ij}^k is the corresponding instantaneous velocity at the i, j grid node in the kth snapshot. Δ is error bound calculated by Equation (3). Likewise, the relative uncertainties of instantaneous velocity field in the v- and w-components can also be deduced:

$$\Delta = \frac{d}{m \cdot dt} \tag{3}$$

where d is estimated displacement error, m is image magnification and dt is the inter-frame time.

The measurement uncertainties of the time-averaged velocity, contain both the statistical factor (type A uncertainty) and the factors unrelated with the statistical analysis (type B uncertainty).

Type A relative uncertainty of the time-averaged velocity can be estimated as [38]:

$$U_{A,\overline{u_{ij}}} = \frac{\sqrt{\frac{1}{N(N-1)} \sum_{k=1}^{N} (u_{ij}^k - \overline{u_{ij}})^2}}{\overline{u_{ij}}} \tag{4}$$

where u_{ij} is the time-averaged velocity in the u-component at the i, j grid node, N is the number of snapshots acquired, u_{ij}^k is the corresponding instantaneous velocity in the kth snapshot. Similarly, type A relative uncertainty of the time-averaged mean velocity in the other two components can also be deduced.

In the compressor cascade, the main factor contributing to the type B uncertainty of the time-averaged velocity($U_{B,\overline{u_{ij}}}$) is the time-averaged of the instantaneous velocity deviation shown in Equation (2).

Thus, the combined standard relative uncertainty of the time-averaged velocity in the u-component can be given as:

$$U_{\overline{u_{ij}}} = \sqrt{(U_{A,\overline{u_{ij}}})^2 + (U_{B,\overline{u_{ij}}})^2} \tag{5}$$

The uncertainties of the time-averaged velocity can be calculated by the propagation of error formula:

$$U_{velocity} = \sqrt{\frac{(\overline{U} \cdot \sigma_U)^2 + (\overline{V} \cdot \sigma_V)^2 + (\overline{W} \cdot \sigma_W)^2}{(\overline{U}^2 + \overline{V}^2 + \overline{W}^2)^2}} \tag{6}$$

where $\sigma_U, \sigma_V, \sigma_W$ are the uncertainties of the time-averaged velocity in u-, v- and w-component.

The uncertainties of instantaneous velocity and time-averaged velocity are shown in Figure 6. Since the introduction of type A uncertainty, the uncertainty in the time-averaged velocity field is elevated. It can be seen that the uncertainty of time-averaged velocity is about 0.5–2% in the mainstream region, and it can achieve 6% near the tip wall. Moreover, it should be pointed out that in the tip region the accuracy of the in-plane components is about 0.5–3% and for the out-of-plane component is about 5–10%, and these results are not shown here for simplicity. Though the uncertainty of the out-of-plane component is higher, the statistical results are worthful for qualitative analyses of the unsteady characteristics in the compressor cascade, and constructive conclusions can be drawn from these results.

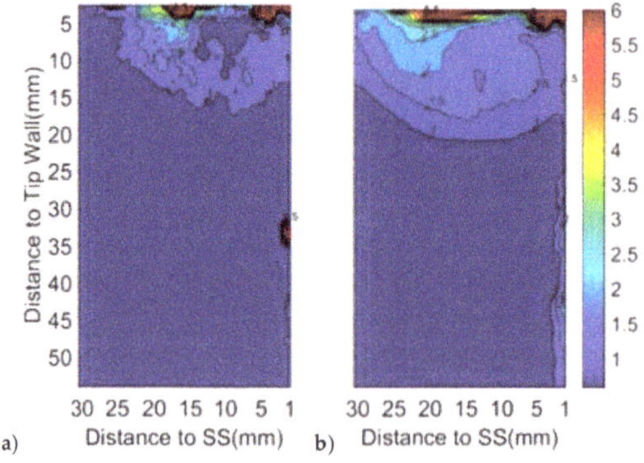

Figure 6. Distribution of measurement relative uncertainties in percentage terms at L/c = 1.0, (**a**) the instantaneous velocity, (**b**) the time-averaged velocity. The velocity is the resultant velocity of components in the x-, y- and z- direction.

2.4. PODMethod

In this paper, a set of instantaneous SPIV measured velocity field $V^{(k)}$ (snapshots) with total number of K is decomposed into a linear combination of K spatial orthonormal basis function (POD modes, φ_m) and the corresponding coefficients $C_m^{(k)}$:

$$V^{(k)} = \sum_{m=1}^{K} C_m^{(k)} \varphi_m \tag{7}$$

with the constraints that the following function is minimized [39]:

$$\sum_{k=1}^{K} \left\| V^{(k)} - \sum_{m=1}^{M} C_m^{(k)} \varphi_m \right\|^2 \rightarrow \min \tag{8}$$

Considering the instantaneous snapshots $V^{(k)}$ made available from the SPIV measurement:

$$V^{(k)} = \left(u_{i,j}, v_{i,j}, w_{i,j} \right)^{(k)} \tag{9}$$

where k is the snapshot index, u, v, w are the velocity components in x, y and z direction respectively, i, j are the index of spatial position coordinates in the velocity distributions for a collocated setup, as illustrated in Figure 7.

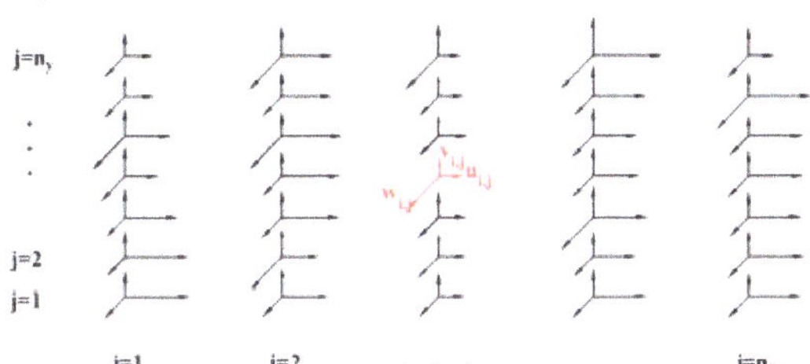

Figure 7. Example of discrete SPIV data on a uniform collocated grid.

Then the velocity component of every single snapshot $V^{(k)}$ is collated as following:

$$U = \begin{bmatrix} U^{(1)} \\ U^{(2)} \\ \vdots \\ U^{(K)} \end{bmatrix} = \begin{bmatrix} u_{i=1,j=1}^{(1)} & u_{i=1,j=2}^{(1)} & \cdots & u_{i=1,j=n_y}^{(1)} & u_{i=2,j=1}^{(1)} & \cdots & u_{i=n_x,j=n_y}^{(1)} \\ u_{i=1,j=1}^{(2)} & u_{i=1,j=2}^{(2)} & \cdots & u_{i=1,j=n_y}^{(2)} & u_{i=2,j=1}^{(2)} & \cdots & u_{i=n_x,j=n_y}^{(2)} \\ \cdots & \cdots & \cdots & \cdots & \cdots & & \cdots \\ u_{i=1,j=1}^{(K)} & u_{i=1,j=2}^{(K)} & \cdots & u_{i=1,j=n_y}^{(K)} & u_{i=2,j=1}^{(K)} & \cdots & u_{i=n_x,j=n_y}^{(K)} \end{bmatrix} \tag{10}$$

where $n_x \times n_y$ is the total number of the spatial positions.

The other two component-wise velocities are processed the same way and matrix V and matrix W are obtained, respectively. Then, simultaneously using both u and v velocity components and neglecting w, most researchers tend to define the spatial correlation matrix as follows [21,40]:

$$C = \frac{1}{K}\left(UU^T + VV^T \right) \tag{11}$$

However, according to the works of Arányi [25] and Lengani [26], the results would be noticeably different if a multi-dimensional field is analyzed component-wise instead of globally. In this paper, we make a deeper quantitatively comparison of the decomposition results using different analysis dimensionalities to clarify whether decomposition dimensionalities impact the whole modes or just specific modes. Moreover, as the aim of most POD analyses is to reconstruct approximately the vector field, we also compare the reconstruction results based on two reconstruction criteria. Therefore, the spatial correlation matrices (C) are defined according to different analysis dimensionalities (AD) as shown in Table 2.

Table 2. Five spatial correlation matrices for different decomposition dimensionalities.

Processing Methods	Global Analysis		Component-Wise Analysis		
AD	UVW-simultaneously	UV-simultaneously	U	V	W
C	$\frac{1}{K}(UU^T + VV^T + WW^T)$	$\frac{1}{K}(UU^T + VV^T)$	$\frac{1}{K}UU^T$	$\frac{1}{K}VV^T$	$\frac{1}{K}WW^T$

The minimization problem described in Equation (8) is realized by solving the eigenvalue problem of correlation matrix C:

$$C\beta_m = \lambda_m \beta_m \tag{12}$$

where Eigenvalues λ_m are arranged in descending order according to the magnitude of eigenvalues, the basis function φ_m are obtained by projecting the snapshots onto the eigenvector β_m.

The POD coefficients $c_m^{(k)}$ are obtained by projecting the snapshots onto the basis function φ_m and $\frac{1}{2}(c_m^{(k)})^2$ represents the energy contribution by the mth mode to the kth snapshot.

The K $\times$ K coefficient matrix $c_m^{(k)}$:

$$c_m^{(k)} = \begin{bmatrix} c_1^{(1)} & c_2^{(1)} & \cdots & c_K^{(1)} \\ c_1^{(2)} & c_2^{(2)} & \cdots & c_K^{(2)} \\ \vdots & \vdots & \vdots & \vdots \\ c_K^{(K)} & c_K^{(K)} & \cdots & c_K^{(K)} \end{bmatrix} \tag{13}$$

The whole kinetic energy from all the snapshots captured by the mth mode is:

$$KE = \frac{1}{2}\sum_{m=1}^{K}\left(\sum_{k=1}^{K}(C_m^{(k)})^2\right) \tag{14}$$

The fraction of the energy captured by the mth mode is:

$$ke_m = \frac{1}{2}\sum_{k=1}^{K}(C_m^{(k)})^2/KE \tag{15}$$

By setting the amplitude of higher-order modes to zero, a low order flow field contains the most energetic modes is reconstructed:

$$V^{(k)} = \sum_{m=1}^{M} c_m^{(k)}\varphi_m \tag{16}$$

where M is the reconstruction order less than K.

2.5. TLV Vortex Core Identification

The TLV is a concentrated vortex before splitting into several small vortices or breakdown. In order to identify the TLV core correctly, some criteria, such as Q criterion, λ_2 criterion and Δ criterion, have been developed. The ability of the three criteria is similar to each other and none of them can be applied without error to all situations. Since the reduced $-\lambda_2$ criterion has been employed in compressor TLV many times and can separates the TLV from the background high shear layer, here the reduced $-\lambda_2$

criterion is employed [41]. This reduced criterion is effective to identify the TLV core and is given by Equation (17):

$$\lambda_2 = \left(\frac{\partial V_x}{\partial x}\right)^2 + \left(\frac{\partial V_y}{\partial y}\right)^2 + 2 \times \frac{\partial V_y}{\partial x}\frac{\partial V_x}{\partial y} \tag{17}$$

where x and y are coordinate axes in the measured cross section as demonstrated in Figure 4, and V_x and V_y are the velocity components in the corresponding directions. In this paper, zero threshold of λ_2 is selected to identify the border of the TLV core.

After the identification, the circulation and radius of the TLV core are quantitatively analyzed. The circulation, which represents the strength of the vortex, is obtained by calculating the line integral of velocity along the border. As shown in Figure 10, a red ellipse is used to fit the TLV core and the center and the radius of the TLV core are calculated from this fitting ellipse. The five-pointed star represents the center of the TLV core identified. The x-axis in the figure is the distance of the TLV core center to the blade suction surface. The y-axis is the distance of the center to the cascade tip wall. Detailed information about the identification program may be found in our previous study [42].

3. Results

3.1. Tip Flow Characteristics

Figure 8 lays out the time-averaged SPIV measurement results of the normalized streamwise velocity and λ_2. The trajectory of the TLV cores can be seen clearly from the λ_2 distribution. The concentrated TLV can be found after L/c = 0.5, which is an ambiguous result for the deficiency of SPIV spatial resolution in the z-direction.

As the TLV propagates downstream, the TLV core radius and the absolute value of vortex circulation expands rapidly (Quantitatively shown in Figure 9a). Figure 9a also quantitatively illustrates that the TLV moves slowly away from the blade tip suction surface in the x-direction (Schematically shown in Figure 9b from a bird's eye view). In the y-direction, the TLV moves slowly away from the tip wall. The variance tendencies of the positions of the TLV in the time-averaged flow fields conform well with the SPIV results obtained in a laboratory-scale compressor at the design condition [5,6,42].

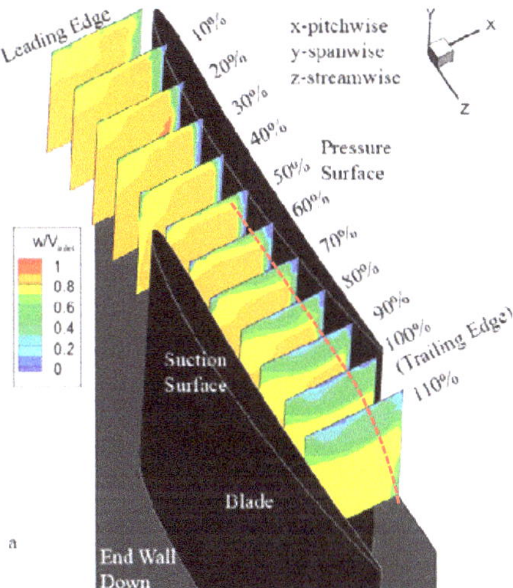

Figure 8. *Cont.*

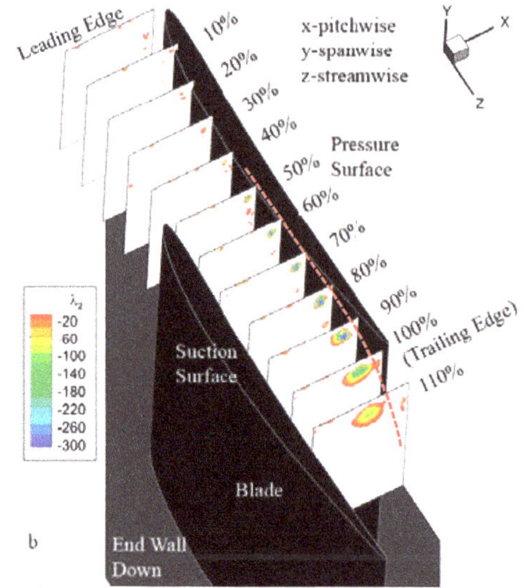

Figure 8. Combined maps of time-averaged SPIV measurement results in the blade passage. (**a**) Normalized Streamwise Velocity; (**b**) λ_2. The red dash lines on both maps sketch the trajectory of the TLV cores.

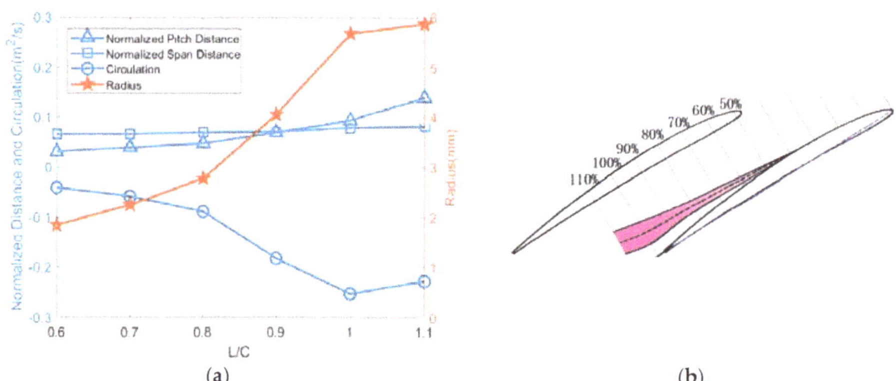

(**a**)   (**b**)

Figure 9. (**a**) Statistics of identified TLV time-averaged parameters; (**b**) Schematic of TLV core trajectory. Normalized pitch distance is the distance of the TLV core center to the blade suction surface in the x-direction normalized by the span, normalized span distance is the distance of the TLV core center to the tip wall in the y-direction normalized by the span.

In Figure 10 the instantaneous vortex cores from $L/c = 0.7$ to $L/c = 1.1$ in all the SPIV snapshots are illustrated by single points. The vortex core center, border and border fitting ellipse of the time-averaged vortex are also shown in Figure 10 for comparison. At $L/c = 0.7$ most instantaneous TLV are near the core center of the time-averaged TLV. The position and circulation distribution of the TLV is in a relatively concentrated form. With the development of the TLV core to the trailing edge, the pitchwise and spanwise distribution range of the vortex expands. From $L/c = 0.9$ to $L/c = 1.1$, when the TLV is shedding form the blade suction side, wandering of the TLV cores becomes significantly stronger and the TLV motions cover a much wider area.

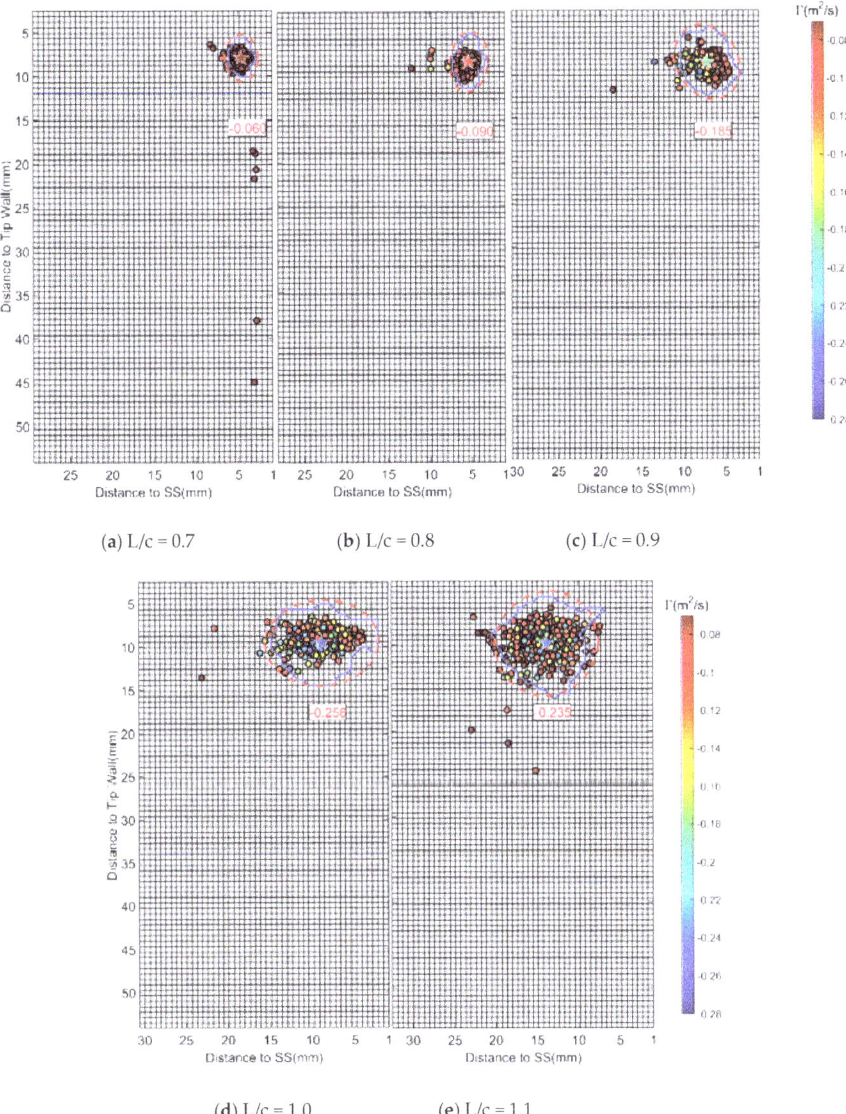

(**a**) L/c = 0.7   (**b**) L/c = 0.8   (**c**) L/c = 0.9

(**d**) L/c = 1.0   (**e**) L/c = 1.1

Figure 10. Statistical distribution of the locations, area and circulation of identified TLV cores. The color of the points shows the vortex circulation.

Figure 11 shows the standard deviation of normalized SPIV measured velocity. The flow velocity in every single snapshot is normalized by the inlet average velocity. The standard deviation is calculated as Equation (18) to show the unsteadiness of the TLV:

$$\sigma = \sqrt{\frac{1}{K-1}\sum_{i=1}^{K}\left(\frac{Vs_i - \overline{V}s}{V_{inlet}}\right)^2} \tag{18}$$

where Vs_i is the instantaneous velocity in every single snapshot, $\overline{V}s$ is the time-averaged velocity.

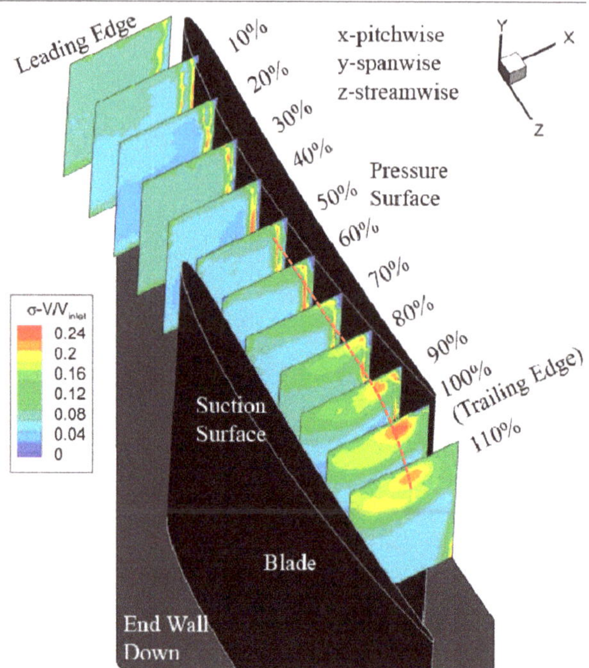

Figure 11. Standard deviation of normalized SPIV measured velocity in the blade passage. The velocity is the resultant velocity of components in the x-, y- and z- direction.

As the vortex propagates downstream, the standard deviation of the normalized SPIV measured velocity near the TLV core increases. In this paper, the unsteadiness of the TLV is mainly originated from the TLV wandering as the vortex maintains a concentrate vortex and does not show any evidence of vortex splitting or breakdown.

The aforementioned experimental results demonstrated the main characteristics of the TLV. In the tip region, streamwise velocity component is much higher than velocity components in secondary flow directions. The TLV and wall boundary layers would introduce velocity deficits as well. Thus, in the compressor cascade, we should keep in mind that the decomposition regions and decomposition dimensionalities would have significant impacts while use the POD method to further analyze the TLV wandering characteristics. Therefore, the two factors should be investigated with caution to get reasonable POD results.

3.2. Effect of Decomposition Region on the Energy Fraction of Mode 1

Since the TLV is a concentrated vortex, to focus on the unsteady characteristics caused by the TLV wandering, circular decomposition regions are deliberately chosen. The decomposition regions are concentric circles with the time-averaged TLV core center as their common center. The radius R_{dr} of the decomposition regions are multiples of the time-averaged TLV core radius R_v. The borders of decomposition regions are shown as dotted line in Figure 12.

Figure 12 b shows the instantaneous TLV cores center, the time-averaged TLV core center and the time-averaged TLV core border. Figure 12c,d shows two instantaneous velocity snapshots chosen randomly. The minimum decomposition region ($R_{dr}/R_v = 2$) covers most instantaneous vortex cores centers. This selection of decomposition regions would keep most information of the TLV wandering and turbulence fluctuation around the TLV while removes the interference of the main flow. While the decomposition region converge to the TLV core center with R_{dr}/R_v reducing from 7 to 2, the effect of main flow on the POD results would be eliminated. Therefore, it is beneficial to extract more flow field structure associated with the unsteady wandering of the TLV and turbulence fluctuation around it.

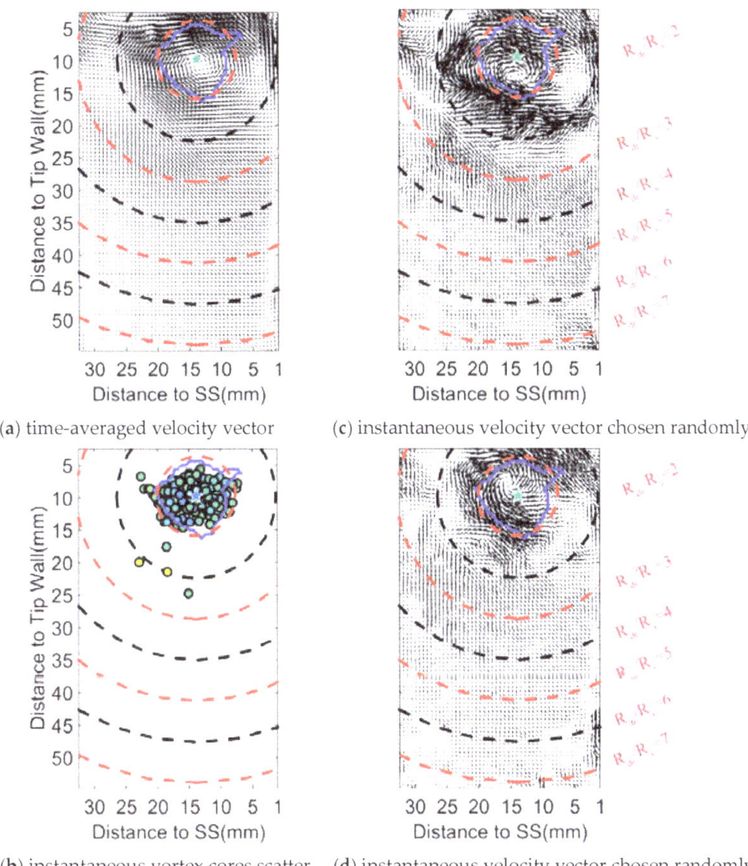

(**a**) time-averaged velocity vector   (**c**) instantaneous velocity vector chosen randomly

(**b**) instantaneous vortex cores scatter   (**d**) instantaneous velocity vector chosen randomly

Figure 12. Decomposition regions at $L/c = 1.1$.

Before the POD decomposition is performed, many researchers have addressed the difference in POD results between subtracting the time-averaged flow field and not subtracting [25,43,44]. As will be confirmed in the following, while the average is not removed, mode 1 is representative of the time-averaged velocity field. Since the POD modes are all orthogonal, the higher-order modes would be well estimates of the fluctuation. In this paper, the POD analysis is conducted without subtracting the average in the snapshots.

The energy rank is the first information provided by the POD eigenvalues λ_m [45]. Among the energy rank, the energy fraction of mode 1 is the most important. The total energy fraction of all other higher-order modes can be obtained by Equation (19):

$$\frac{\sum\limits_{m=2}^{K} \lambda_m}{\sum\limits_{m=1}^{K} \lambda_m} = 1 - \frac{\lambda_1}{\sum\limits_{m=1}^{K} \lambda_m} \tag{19}$$

Therefore, the energy fraction of mode 1 reflects the relative contribution of fluctuation to the whole energy of the flow field. The energy fraction of mode 1 can represent the level to which extent the instantaneous fields approach the time-averaged result.

From L/c = 0.6 to L/c = 1.1, as the fluid propagates downstream, the energy fraction of mode 1 decreases as shown in Figure 13. It should be noted that the results are from POD decompostion using three velocity vector components U, V and W simultaneously. The POD highlights that the flow posses a higher energy in fluctuation as the vortex propagates downstream, consistent with the results of the standard deviation previously observed in Figure 11. Figure 13 also shows that the deviation of mode 1 energy fraction between L/c = 0.6 and L/c = 1.1 becomes evident as the decomposition region converge to the TLV core. Indeed, the TLV experiences the process of vortex shedding from the blade suction side and wanders much stronger at the downstream near the trailing edge.

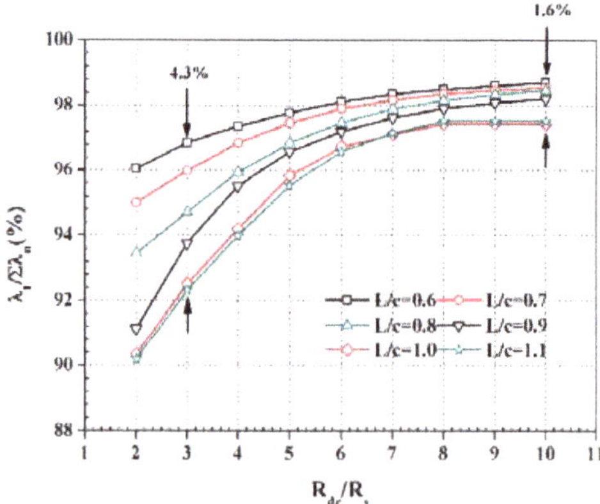

Figure 13. Mode 1 energy fraction at different decomposition region sizes and different chord position.

At the same chordwise position, as the decomposition region size decreases, the energy fraction of mode 1 decreases as well. The energy fraction of fine scale fluid structure, normally contained in the higher-order modes, is weakened by the introduction of main flow. It proves that the decomposition region chosen is beneficial to focus on the unsteady characteristics analysis of the TLV and turbulence fluctuation around it.

To further investigate the effect of decomposition regions on energy distribution, four more different analysis dimensionalities, such as UV global analysis, U component-wise analysis, V component-wise analysis and W component-wise analysis, are used.

3.3. Effect of Decompostion Dimensionalities on the Energy Fraction of Mode 1

Figure 14 demonstrates the energy fraction of mode 1 at different decomposition region sizes and different chord positions using four different analysis dimensionalities.

Note that the decomposition region sizes have marginal impact on the results associated with U component-wise analysis, V component-wise analysis and UV global analysis. However, the impact become considerable on the results associated with W component-wise and UVW global analysis while both of the results have the same variability tendency. This corresponding to the fact that the dominant flow direction is z-direction (streamwise) in this cascade while x-direction and y-direction are the secondary flow directions where velocity amplitude is relatively low.

As the flow travels downstream, the energy fraction of mode 1 decreases dramatically. From L/c = 0.6 to L/c = 1.1, the deviation of the energy fraction of mode 1 is nearly 10% in the U component-wise analysis, and higher than 20% in V component-wise analysis and UV global analysis as shown in Figure 14. While the W component is included, the deviation reduced to less than 5%. Therefore, the POD without the W component highlights the development of unsteady characteristics along the streamwise direction. The figure also shows that, using POD without the W component, mode 1 energy fraction at L/c = 0.9 deviate from that at L/c = 1.1 by twice as much as the deviation from L/c = 0.6 to L/c = 0.9. Near the trailing edge, the TLV is shedding from the blade suction surface, the instantaneous fields diverge considerably from the time-averaged result. UV global analysis captures this physics as the energy transferred from mode 1 expands significantly from L/c = 0.9 to L/c = 1.1.

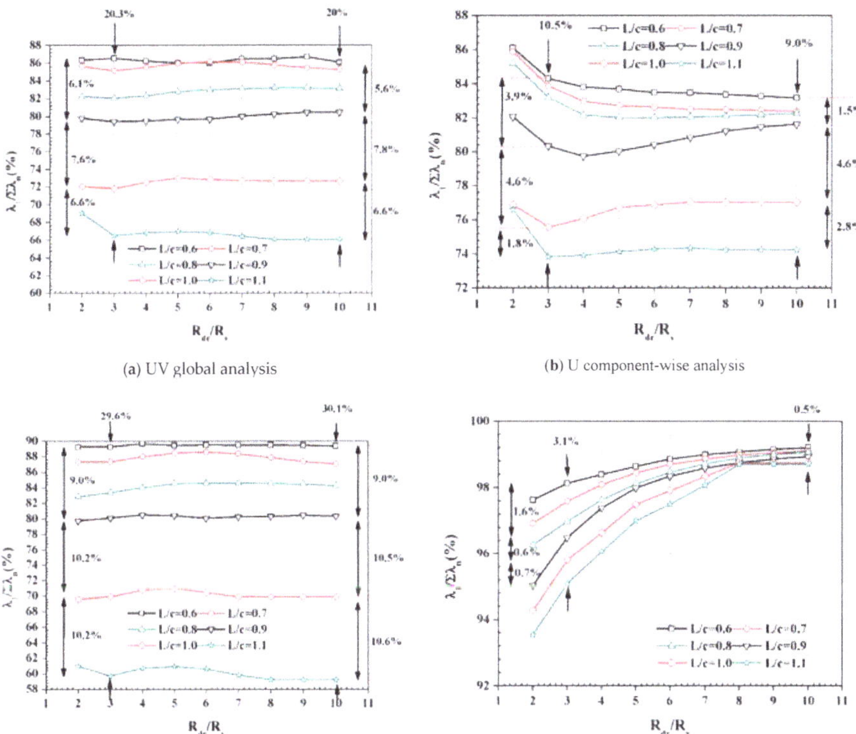

(a) UV global analysis   (b) U component-wise analysis

Figure 14. Mode 1 energy fraction at different decomposition region sizes and different chord position.

Therefore, the effect of decomposition region on the mode 1 energy fraction mainly impacts the streamwise velocity component. When the decomposition region converges to the TLV core, the fraction of streamwise velocity deficit region increases, the impact of main flow with uniform streamwise velocity reduces, the unsteadiness caused by vortex wandering and turbulence fluctuation becomes evident.

Thus, the decomposition region size in the following work is all $R_{dr}/R_v = 3$, in order to extract more flow field structure associated with vortex wandering and turbulence fluctuation and to avoid the effect of main flow disturbance on POD. The results of five decomposition dimensionalities in the selected decomposition region are shown in Figure 15. It is interesting to note that the introduction of W velocity component also changes the absolute amplitude of mode 1 energy fraction. The amplitudes of mode 1 energy fraction are nearly equal in the results of UVW global analysis and W component-wise analysis and they are much higher than the result in the other three analyses.

Based on the aforementioned discussions, we can draw a conclusion that there is a marked difference among mode 1 energy fractions using different decomposition dimensionalities. If the aim of the analysis is to approximately reconstruct the flow field, the effects of decomposition dimensionalities on the spatial structures of POD modes and energy distribution should be urgent to clarify.

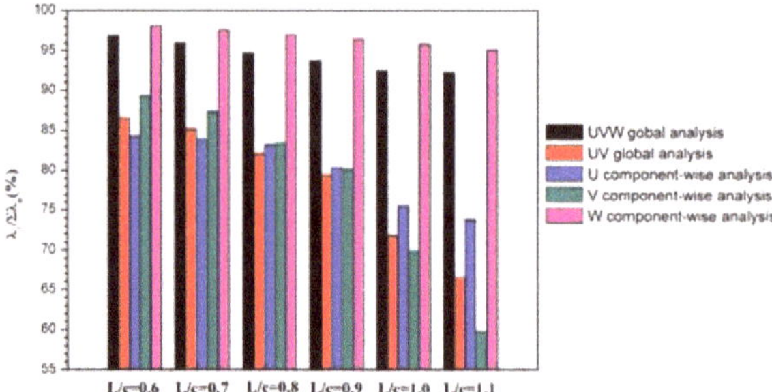

Figure 15. Mode 1 energy fraction at different retained dimensionalities and different chord position when the decomposition region size is Rdr/Rv = 3.

3.4. Effect of Decomposition Dimensionalities on POD Modes

3.4.1. Mode 1 and Its Relationship with the Time-Averaged Flow Field

To quantify the degree of similarity about two velocity vector fields, the relevance index R_p is used. R_p is obtained by projecting one velocity vector field onto another velocity vector field [43]:

$$R_p = \frac{(M1, M2)}{\|M1\| \cdot \|M2\|} \tag{20}$$

where the denominator denotes the product of the L2 norm and the numerator is the inner product of two velocity fields. Rp = 0 means the two velocity field are orthogonal, Rp = 1 if the direction of two velocity vector field are identical, and Rp = −1 if the direction of two velocity vector field are exactly opposite.

The flow patterns of mode 1 from all three decomposition dimensionalities are excellent estimates of the time-averaged velocity field (shown in Figure 16). Moreover, the relevance indices R_p of all three mode 1 and the time-averaged velocity field are nearly 1, which also conforms this conclusion quantitatively. The results from other chordwise positions show the same correlation (not shown).

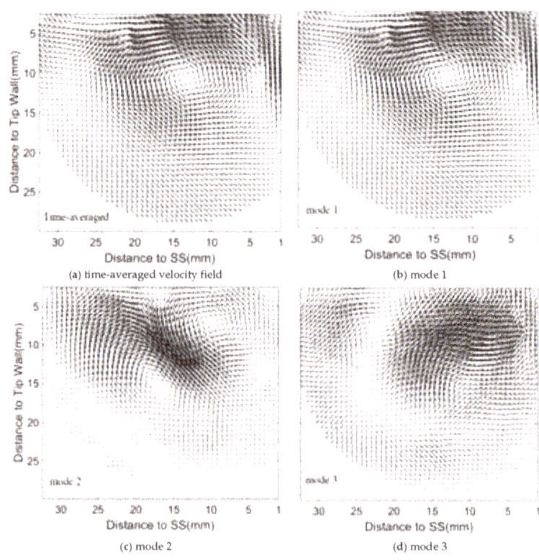

Figure 16. Time-averaged velocity vector field and spatial structures of mode 1, 2 and 3 in UV global analysis at L/c = 1.1.For simplicity, the spatial structures of UVW global analysis and UV component analysis have not shown.

3.4.2. Higher-order Modes

Figure 17 reflects the relative contribution of the modes to the fluctuation energy of the flow field. The modes from W component-wise analysis show a distribution that the energy of the 3rd mode is similar to the 2nd mode. A completely different scenario characterized the POD energy distribution of other four decomposition dimensionalities, especially UVW global analysis. The energy content of the 2nd POD mode appears sensibly higher than that characterizing the 3rd mode.

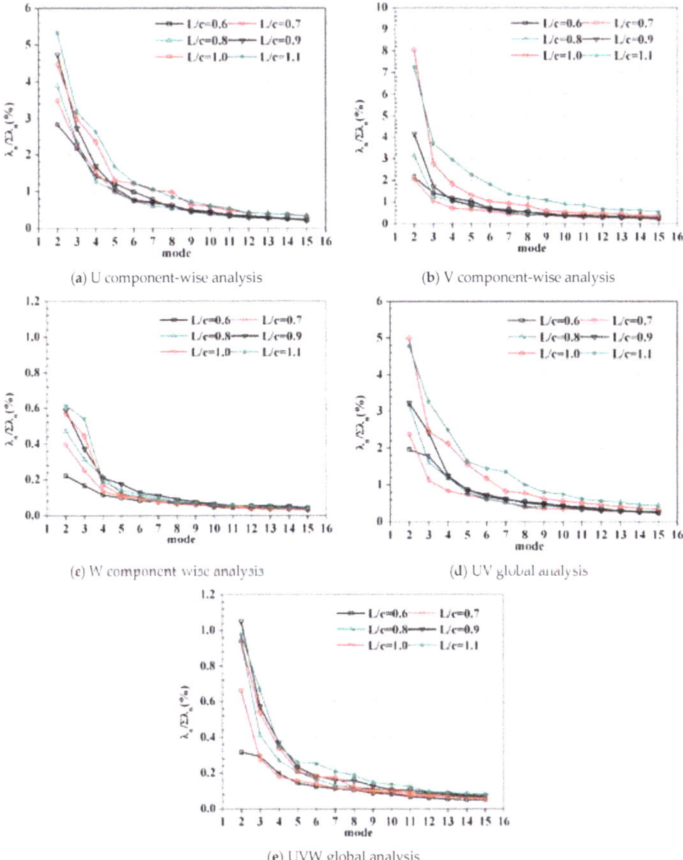

(a) U component-wise analysis

(b) V component-wise analysis

(c) W component-wise analysis

(d) UV global analysis

(e) UVW global analysis

Figure 17. Relative energy of modes higher than 2 from all five different decomposition dimensionalities.

Though the impact of W velocity component in mode 1 energy fraction is dominant as mentioned above, the addition of W velocity component in POD decomposition does not change the relative energy distribution of high-order modes. The 2nd and 3rd POD spatial modes from different decomposition dimensionalities have been checked. No matter which decomposition dimensionality used, two large well-defined vortex-like structures can be observed in the 2nd POD spatial modes (shown in Figure 16). However, the 3rd POD spatial modes are totally different from each other.

To further study the impact of decomposition dimensionalities in the high-order modes, the velocity correlation coefficients account for correlation of each mode from different decomposition dimensionalities are shown in Figure 18. The correlation coefficients of the 2nd POD mode are nearly 1, which verified that the 2nd POD mode is unaffected. The correlation coefficients of modes order higher than 3 are weak. In all, the three different decomposition dimensionalities lead to almost the same 2nd mode while significant difference can be observed in modes which order are higher than 2.

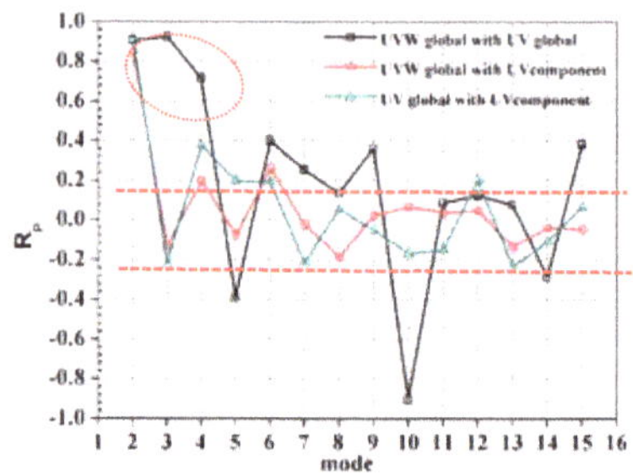

Figure 18. The velocity correlation of high-order modes from different decomposition dimensionalities at L/c = 1.1.

3.5. Effect of Decomposition Dimensionalities on POD Reconstruction

In the tip region, the flow is inherently unsteady. A single POD mode usually cannot bear a significant amount of energy and is not able to completely describe the original flow structures. An ensemble of POD modes is demanded to describe a particular dynamics. To study the dynamics of the TLV wandering, it is of great interest to analyze the vortex distribution characteristics with random noise being removed. In this section, to study the effect of different decomposition dimensionalities on the reconstructed flow field, the modes required to represent 95% energy of the original flow field and vortex distribution characteristics in the reconstructed flow field are compared.

3.5.1. Modes Required to Represent 95% Energy of the Original Flow Field

At the same chordwise position, modes required to represent 95% energy of the original flow field from UVW global analysis and W component-wise analysis are much less than that from other three processing methods. It also proves that the addition of W velocity component in the decomposition changes the energy fraction of every single mode and results in more energy gathering in the low-order modes.

From L/c = 0.6 to L/c = 1.1, an interesting feature in the POD reconstruction is that the modes required are dramatically increased using the U and V velocity components while maintain almost the same using the W velocity component (Figure 19). As the vortex propagates downstream, more and more modes are required to adequately represent the flow field in the x-direction and y-direction. Combining with the discussion in Section 3.3, UV global analysis is better for capturing the kinetic physics of the TLV because it is expected that more high-order modes will be present as the flow travels downstream due to vortex wandering and vortex shedding.

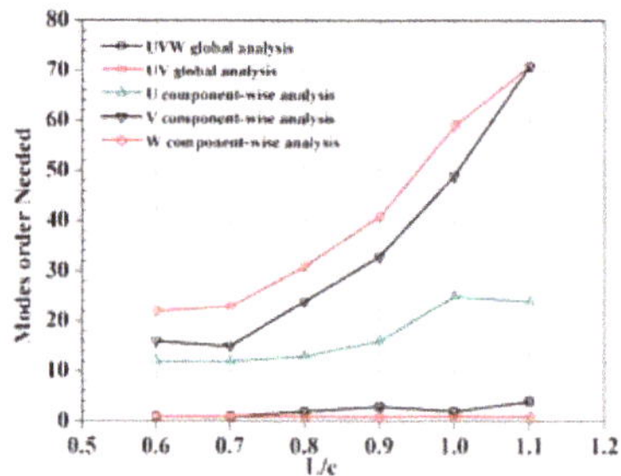

Figure 19. The modes required to represent 95% energy of the original flow field using five different POD processing methods.

3.5.2. Effect of Decomposition Dimensionalities on the Vortex Distribution Characteristics

Two criteria of reconstructing flow field using POD modes is studied in this paper. Reconstruction using a finite number of modes [28,46,47] and reconstruction using modes that represent specific energy portion of the original flow field [45,48,49]. To compare the effect of decomposition dimensionalities on the vortex distribution characteristics in the reconstructed flow field, every single snapshot is reconstructed according to above two criteria.

Results from the reconstruction relying on the same retained energy in the reconstructed flow field are shown in Figure 20. Figure 21 quantitatively shows the proportion of the TLV core parameters. They together demonstrate the geometric and kinematic characteristics of the TLV in the reconstructed flow field. Since the modes required to reconstruct 95% energy of the original flow field is much less in UVW global analysis, the TLV shows more concentrated distribution characteristics in Figure 20. In the reconstructed flow field from UV global analysis and UV component analysis, the TLV cores are decentralized and have the similar distribution characteristics with the original flow field. The histogram of vortex distribution from UVW global analysis is symmetric that the mean value provides a good estimate for the center of the data. The histograms from UV global analysis and UV component analysis have the same shape, and are for a distribution that are skewed left. The reconstruction relying on the retained energy in the reconstructed flow field is sensitive to different decomposition dimensionalities.

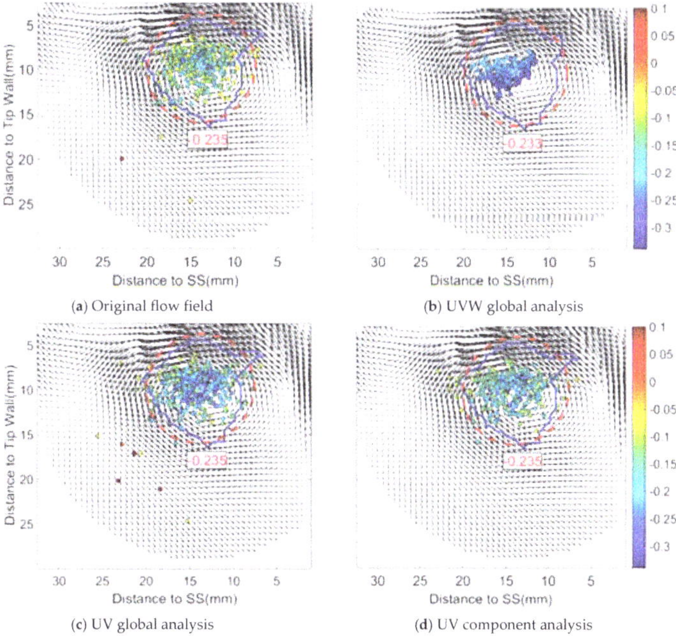

Figure 20. Vortex distribution in the reconstructed flow field that represents 95% energy of the original flow field at L/c = 1.1. The color of the points shows the vortex circulation.

Figures 22 and 23 show the corresponding reconstruction results using the first ten modes. The vortex distributions in the reconstructed flow field using different decomposition dimensionalities show the same concentrated characteristics. The histograms of these three analyses have the same shape as well. Though the addition of W velocity component does change the spatial structures of high-order modes, it does not change the dynamic results of reconstruction using the same number of POD modes.

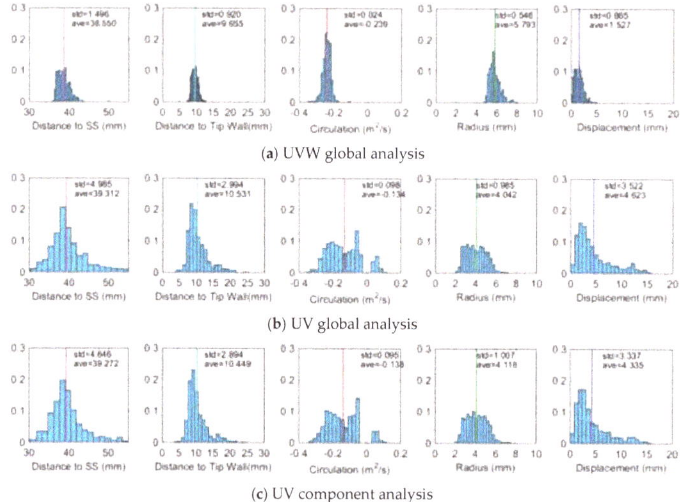

(a) UVW global analysis

(b) UV global analysis

(c) UV component analysis

Figure 21. Statistics of instantaneous reconstructed flow field that represents 95% energy fraction of the original flow field at L/c = 1.1. Note that the parameters from left side to the right are the distance of the TLV core center to the suction side, the distance of the TLV core center to the cascade tip wall, the TLV vortex circulation, the TLV vortex core radius and the distance of instantaneous TLV core center to the time-averaged TLV core center respectively. 'std' and 'ave' in the figure show the standard deviation and time-averaged value respectively.

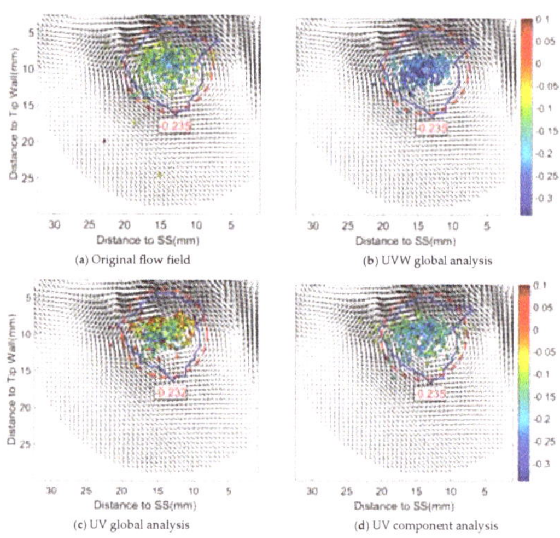

(a) Original flow field

(b) UVW global analysis

(c) UV global analysis

(d) UV component analysis

Figure 22. Vortex distribution from reconstructed flow field using first 10 modes at L/c = 1.1.

To further validate the conclusions drawn above, we checked the reconstruction results using different number of modes and different retained energy portions of the original flow field. The standard deviation of the TLV radius and displacement are chosen to quantitatively analysis the difference between the two reconstruction criteria. The results in Figure 24 confirm that reconstructions using the same modes number can obtain nearly the same vortex distribution characteristics while reconstructions relying on the same energy fraction would lead to totally different vortex distribution characteristics.

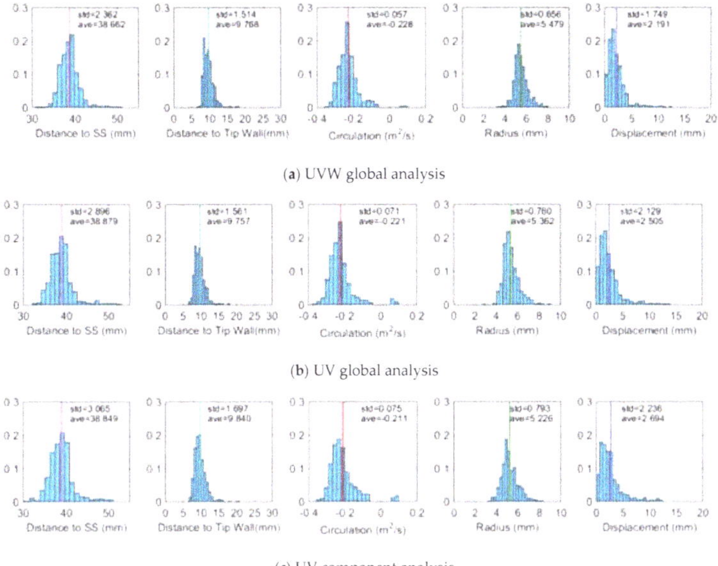

(a) UVW global analysis

(b) UV global analysis

(c) UV component analysis

Figure 23. Statistics of instantaneous reconstructed flow field using first 10 modes at L/c = 1.1.

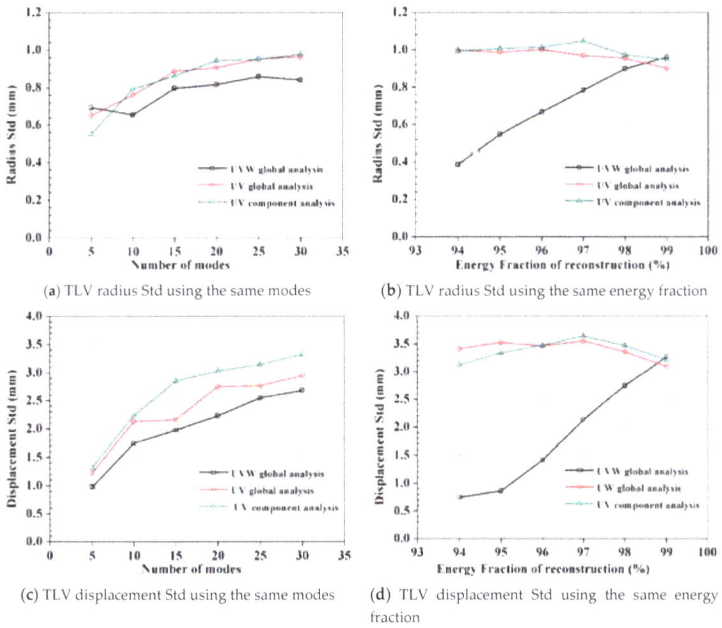

(a) TLV radius Std using the same modes

(b) TLV radius Std using the same energy fraction

(c) TLV displacement Std using the same modes

(d) TLV displacement Std using the same energy fraction

Figure 24. Comparison of reconstruction results based on the two reconstruction criteria at L/c = 1.1. The displacement is the distance of instantaneous TLV core center to the time-averaged TLV core center.

4. Conclusions

In this paper, based on the original tip flow field in the compressor cascade obtained by a SPIV measurement, the effects of different decomposition regions and decomposition dimensionalities on POD decomposition and reconstruction have been clarified. Several conclusions can be made as listed below:

(1) From $L/c = 0.9$ to $L/c = 1.1$, when the TLV is shedding from the blade suction side, wandering of the TLV core becomes significantly stronger and the TLV motions cover a much wider area. As the TLV propagates downstream, the energy fraction of mode 1 decreases.

(2) The decomposition region sizes have marginal impact on POD decomposition of secondary flow velocity. The impact become considerable on the results associated with W component-wise velocity.

(3) For POD decomposition, using different dimensionalities, energy distributions and modes higher than 2 would be totally different. The dominant POD modes, such as the 1st and 2nd POD modes, stay unaffected. Using POD without the W component, mode 1 energy fraction at $L/c = 0.9$ deviate from that at $L/c = 1.1$ by twice as much as the deviation from $L/c = 0.6$ to $L/c = 0.9$.

(4) The reconstruction relying on the retained energy in the reconstructed flow field is sensitive to different decomposition dimensionalities. The reconstruction using a finite number of POD modes is unaffected.

(5) UV global analysis is better for capturing the kinetic physics of the TLV. As the flow travels downstream, more high-order modes with higher energy fraction will be present in UV global analysis, which is corresponding to the kinetic physics of the TLV that the unsteadiness of vortex wandering and vortex shedding becomes stronger.

Author Contributions: Conceptualization, L.S. methodology, L.S.; software, L.S.; validation, L.S.; formal analysis, L.S.; investigation, L.S.; resources, L.S.; data curation, L.S.; writing—original draft preparation, L.S.; writing—review and editing, H.M., L.W.; supervision, H.M.; funding acquisition, H.M.

Funding: This research was funded by the National Natural Science Foundation of China (Grant No. 51776011).

Conflicts of Interest: The authors declare no conflict of interest.

Nomenclature

i,j	The index of the grid points in the velocity distributions
k	Snapshot index
L	SPIV measurement cross section chordwise position
c	Chord length
H	Span
p	Pitch length
Re	Reynolds number of the inlet flow
s	The vertical distance of test points to the wall.
Γ	Circulation
ave	Time-averaged value
Std	Standard deviation
KE	Whole kinetic energy
ke	Energy of specific mode
R_p	Relevance index
M1	Velocity vector fields
R_{dr}	Radius of the decomposition region
R_v	Radius of the TLV core
λ_m	POD eigenvalues
U_x, U_y, U_z	The uncertainties of the x, y and z velocity components
Δ	Error bound
K	Converge factor
$\sigma_U, \sigma_V, \sigma_W$	The uncertainties of the time-averaged velocity in u-, v- and w-component
φ_m	POD mode
AD	Analysis dimensionalities
C	Correlation matrices

References

1. Denton, J.D. *Loss Mechanisms in Turbomachines*; ASME 1993 International Gas Turbine and Aeroengine Congress and Exposition, 1993; American Society of Mechanical Engineers: New York, NY, USA, 1993.
2. Storer, J.; Cumpsty, N. An approximate analysis and prediction method for tip clearance loss in axial compressors. *J. Turbomach.* **1994**, *116*, 648–656. [CrossRef]
3. Adamczyk, J.; Celestina, M.; Greitzer, E. *The Role of Tip Clearance in High-Speed Fan Stall*; ASME 1991 International Gas Turbine and Aeroengine Congress and Exposition, 1991; American Society of Mechanical Engineers: New York, NY, USA, 1991.
4. Straka, W.A.; Farrell, K.J. The effect of spatial wandering on experimental laser velocimeter measurements of the end-wall vortices in an axial-flow pump. *Exp. Fluids* **1992**, *13*, 163–170. [CrossRef]
5. Liu, B.; Yu, X.; Liu, H.; Jiang, H.; Yuan, H.; Xu, Y. Application of SPIV in turbomachinery. *Exp. Fluids* **2006**, *40*, 621–642. [CrossRef]
6. Xianjun, Y.; Baojie, L.; Haokang, J. Characteristics of the tip leakage vortex in a low-speed axial compressor. *AIAA J.* **2007**, *45*, 870–878. [CrossRef]
7. Furukawa, M.; Inoue, M.; Saiki, K.; Yamada, K. The Role of Tip Leakage Vortex Breakdown in Compressor Rotor Aerodynamics. *J. Turbomach.* **1999**, *121*, 469–480. [CrossRef]
8. Brandstetter, C.; Jüngst, M.; Schiffer, H.-P. Measurements of Radial Vortices, Spill Forward, and Vortex Breakdown in a Transonic Compressor. *J. Turbomach.* **2018**, *140*, 061004. [CrossRef]
9. Day, I. *The Fundamentals of Stall and Surge*; Lecture series-van Kareman Institute for fluid dynamics; Von Kareman Institute for fluid dynamics: Rhode-Saint-Genèse, Belgium, 1996; pp. B1–B27.
10. Inoue, M.; Kuroumaru, M.; Tanino, T.; Yoshida, S.; Furukawa, M. Comparative studies on short and long length-scale stall cell propagating in an axial compressor rotor. *J. Turbomach.* **2001**, *123*, 24–30. [CrossRef]
11. Mailach, R.; Lehmann, I.; Vogeler, K. Rotating instabilities in an axial compressor originating from the fluctuating blade tip vortex. *J. Turbomach.* **2001**, *123*, 453–460. [CrossRef]
12. Zierke, W.; Farrell, K.; Straka, W. Measurements of the tip clearance flow for a high-reynolds-number axial-flow rotor. *J. Turbomach.* **1995**, *117*, 522–532. [CrossRef]
13. Liu, B.; An, G.; Yu, X.; Zhang, Z. Quantitative Evaluation of the Unsteady Behaviors of the Tip Leakage Vortex in a Subsonic Axial Compressor Rotor. *Exp. Therm. Fluid Sci.* **2016**, *79*, 154–167. [CrossRef]
14. Oweis, G.F.; Ceccio, S.L. Instantaneous and time-averaged flow fields of multiple vortices in the tip region of a ducted propulsor. *Exp. Fluids* **2005**, *38*, 615–636. [CrossRef]
15. Wu, H.; Tan, D.; Miorini, R.L.; Katz, J. Three-dimensional flow structures and associated turbulence in the tip region of a waterjet pump rotor blade. *Exp. Fluids* **2011**, *51*, 1721–1737. [CrossRef]
16. Reynolds, W.; Hussain, A. The mechanics of an organized wave in turbulent shear flow. Part 3. Theoretical models and comparisons with experiments. *J. Fluid Mech.* **1972**, *54*, 263–288. [CrossRef]
17. Van der Wall, B.G.; Richard, H. Analysis methodology for 3C-PIV data of rotary wing vortices. *Exp. Fluids* **2006**, *40*, 798–812. [CrossRef]
18. Cordier, L. *Proper Orthogonal Decomposition: An Overview*; Lecture series 2002-04, 2003-03 and 2008-01 on post-processing of experimental and numerical data; Von Karman Institute for Fluid Dynamics: Rhode-Saint-Genèse, Belgium, 2008.
19. Taira, K.; Brunton, S.L.; Dawson, S.T.M.; Rowley, C.W.; Colonius, T.; McKeon, B.J.; Schmidt, O.T.; Gordeyev, S.; Theofilis, V.; Ukeiley, L.S. Modal Analysis of Fluid Flows: An Overview. *AIAA J.* **2017**, *55*, 4013–4041. [CrossRef]
20. Guo, R.; Li, R.; Zhang, R. Reconstruction and Prediction of Flow Field Fluctuation Intensity and Flow-Induced Noise in Impeller Domain of Jet Centrifugal Pump Using Gappy POD Method. *Energies* **2018**, *12*, 111. [CrossRef]
21. El-Adawy, M.; Heikal, M.; A Aziz, A.; Adam, I.; Ismael, M.; Babiker, M.; Baharom, M.; Firmansyah; Abidin, E. On the Application of Proper Orthogonal Decomposition (POD) for In-Cylinder Flow Analysis. *Energies* **2018**, *11*, 2261. [CrossRef]
22. Bastine, D.; Witha, B.; Wächter, M.; Peinke, J. Towards a Simplified DynamicWake Model Using POD Analysis. *Energies* **2015**, *8*, 895. [CrossRef]

23. Cadirci, S.; Gunes, H. *Proper Orthogonal Decomposition Analysis of JaVA Flows for Cross-Flow Conditions*; ASME 2011 International Mechanical Engineering Congress and Exposition, 2011; American Society of Mechanical Engineers: New York, NY, USA, 2011; pp. 871–878.

24. Li, H.; Su, X.; Yuan, X. Entropy Analysis of the Flat Tip Leakage Flow with Delayed Detached Eddy Simulation. *Entropy* **2018**, *21*, 21. [CrossRef]

25. Arányi, P.; Janiga, G.; Zähringer, K.; Thévenin, D. Analysis of different POD methods for PIV-measurements in complex unsteady flows. *Int. J. Heat Fluid Flow* **2013**, *43*, 204–211. [CrossRef]

26. Lengani, D.; Simoni, D.; Ubaldi, M.; Zunino, P. POD analysis of the unsteady behavior of a laminar separation bubble. *Exp. Therm. Fluid Sci.* **2014**, *58*, 70–79. [CrossRef]

27. Maurel1, S.; Borée, J.; Lumley, J.L. Extended Proper Orthogonal Decomposition: Application to Jet/Vortex Interaction. *Flow Turbul. Combust.* **2001**, *67*, 125–136. [CrossRef]

28. Kostas, J.; Soria, J.; Chong, M.S. A comparison between snapshot POD analysis of PIV velocity and vorticity data. *Exp. Fluids* **2005**, *38*, 146–160. [CrossRef]

29. Perret, L.; Braud, P.; Fourment, C.; David, L.; Delville, J. 3-Component acceleration field measurement by dual-time stereoscopic particle image velocimetry. *Exp. Fluids* **2006**, *40*, 813–824. [CrossRef]

30. Raffel, M.; Willert, C.E.; Scarano, F.; Kähler, C.J.; Wereley, S.T.; Kompenhans, J. *Particle Image Velocimetry: A Practical Guide*; Springer: Berlin, Germnay, 2018.

31. Keane, R.D.; Adrian, R.J. Optimization of particle image velocimeters. I. Double pulsed systems. *Meas. Sci. Technol.* **1990**, *1*, 1202. [CrossRef]

32. Huang, H.; Dabiri, D.; Gharib, M. On errors of digital particle image velocimetry. *Meas. Sci. Technol.* **1997**, *8*, 1427. [CrossRef]

33. Hain, R.; Kähler, C. Fundamentals of multiframe particle image velocimetry (PIV). *Exp. Fluids* **2007**, *42*, 575–587. [CrossRef]

34. Westerweel, J. Theoretical analysis of the measurement precision in particle image velocimetry. *Exp. Fluids* **2000**, *29*, S003–S012. [CrossRef]

35. Boillot, A.; Prasad, A. Optimization procedure for pulse separation in cross-correlation PIV. *Exp. Fluids* **1996**, *21*, 87–93. [CrossRef]

36. Westerweel, J. Fundamentals of digital particle image velocimetry. *Meas. Sci. Technol.* **1997**, *8*, 1379. [CrossRef]

37. Zang, W.; Prasad, A.K. Performance evaluation of a Scheimpflug stereocamera for particle image velocimetry. *Appl. Opt.* **1997**, *36*, 8738–8744. [CrossRef] [PubMed]

38. Cantrell, C.D. *Modern Mathematical Methods for Physicists and Engineers*; Cambridge University Press: Cambridge, UK, 2000.

39. Lumley, J. *The Structure of Inhomogeneous Turbulence*; Atmospheric Turbulence and Wave Propagation, Nauka: Moscow, Russia, 1967; pp. 166–178.

40. El-Adawy, M.; Heikal, M.R.; A Aziz, A.R.; Siddiqui, M.I.; Munir, S. Characterization of the Inlet Port Flow under Steady-State Conditions Using PIV and POD. *Energies* **2017**, *10*, 1950. [CrossRef]

41. Jeong, J.; Hussain, F. On the identification of a vortex. *J. Fluid Mech.* **2006**, *285*, 69–94. [CrossRef]

42. Ma, H.; Wei, W.; Ottavy, X. Experimental investigation of flow field in a laboratory-scale compressor. *Chin. J. Aeronaut.* **2017**, *30*, 31–46. [CrossRef]

43. Hao, C.; David, L.R.; Volker, S. On the use and interpretation of proper orthogonal decomposition of in-cylinder engine flows. *Meas. Sci. Technol.* **2012**, *23*, 085302.

44. Liu, K.; Haworth, D.C. *Development and Assessment of POD for Analysis of Turbulent Flow in Piston Engines*; SAE International: Warrendale, PA, USA, 2011.

45. Lengani, D.; Simoni, D.; Pichler, R.; Sandberg, R.; Michelassi, V.; Bertini, F. *On the Identification and Decomposition of the Unsteady Losses in a Turbine Cascade*; ASME Turbo Expo 2018: Turbomachinery Technical Conference and Exposition, 2018; American Society of Mechanical Engineers: New York, NY, USA, 2018; p. V02AT45A013.

46. Hossain, M.S.; Bergstrom, D.J.; Chen, X.B. Visualisation and analysis of large-scale vortex structures in three-dimensional turbulent lid-driven cavity flow. *J. Turbul.* **2015**, *16*, 901–924. [CrossRef]

47. Hellström, L.H.O.; Ganapathisubramani, B.; Smits, A.J. The evolution of large-scale motions in turbulent pipe flow. *J. Fluid Mech.* **2015**, *779*, 701–715. [CrossRef]

48. Raiola, M.; Discetti, S.; Ianiro, A. On PIV random error minimization with optimal POD-based low-order reconstruction. *Exp. Fluids* **2015**, *56*, 75. [CrossRef]

49. Chen, H.; Reuss, D.L.; Hung, D.L.S.; Sick, V. A practical guide for using proper orthogonal decomposition in engine research. *Int. J. Engine Res.* **2012**, *14*, 307–319. [CrossRef]

Article

Unsteadiness of Tip Leakage Flow in the Detached-Eddy Simulation on a Transonic Rotor with Vortex Breakdown Phenomenon

Xiangyu Su , Xiaodong Ren , Xuesong Li * and Chunwei Gu

Institute of Gas Turbine, Department of Energy and Power Engineering, Tsinghua University, Beijing 100084, China; su-xy17@mails.tsinghua.edu.cn (X.S.); rxd@mail.tsinghua.edu.cn (X.R.); gcw@mail.tsinghua.edu.cn (C.G.)
* Correspondence: xs-li@mail.tsinghua.edu.cn; Tel.: +86-10-6278-1739

Received: 22 January 2019; Accepted: 7 March 2019; Published: 12 March 2019

Abstract: Tip leakage vortex (TLV) in a transonic compressor rotor was investigated numerically using detached-eddy simulation (DES) method at different working conditions. Strong unsteadiness was found at the tip region, causing a considerable fluctuation in total pressure distribution and flow angle distribution above 80% span. The unsteadiness at near choke point and peak efficiency point is not obvious. DES method can resolve more detailed flow patterns than RANS (Reynolds-averaged Navier–Stokes) results, and detailed structures of the tip leakage flow were captured. A spiral-type breakdown structure of the TLV was successfully observed at the near stall point when the TLV passed through the bow shock. The breakdown of TLV contributed to the unsteadiness and the blockage effect at the tip region.

Keywords: tip leakage flow; detached-eddy simulation; vortex breakdown; transonic compressor

---

1. Introduction

Driven by the pressure gradient inside the clearance of the rotor and casing, tip leakage is an unavoidable phenomenon in the field of turbomachinery, which scholars have studied for a long time in both compressible [1–4] and incompressible [5–8] fields. As for axial compressors, tip leakage flow plays an even more significant role due to its close relationship with loss and stall characteristics [9,10], which are highly valued in the design or analyzing processes. Aiming at reducing the impact of tip leakage flow, plenty of flow control methods (such as air injection [11], bleeding [12], casing treatment [13] and plasma actuation [14]) have been studied in recent years.

As for low-speed compressors, flow structures as well as unsteadiness of tip leakage vortex (TLV) have been widely studied and certain achievements were obtained through numerical efforts and experiments. As the flow coefficient decreases, the interface between the TLV and the incoming flow moves upstream [15], and the trajectory of which will be aligned with the leading edge when the rotor finally encounters a spike-type stall inception [16]. The criterion for spike-initiated numerical stall that leading-edge spillage and trailing-edge backflow are both essential was proved effective in low-speed compressor experiments [17]. Leading-edge spillage was later found to be an accompanying phenomenon, whose fundamental cause is probably the tornado-like vortex, resulted from the interaction between TLV and leading-edge separation [18]. In rotors with a large gap, attention was also paid to the effects of double-leakage tip clearance flow, which generates a vortex rope and subsequent extra mixing loss in the adjacent blade passage [19]. With the increase of stage loading, the importance of tip leakage flow in high-speed or transonic compressor has been increasingly emphasized. There are indeed similarities in the basic structures and mechanisms of the TLV in low-speed compressors and transonic ones. Nevertheless, conclusions for tip leakage flow in low-speed compressors are not entirely suitable for transonic ones, due to further larger pressure gradient and the existence of shock wave. When the TLV crosses the shock, it interacts both with the shock wave and with the pressure-side secondary flows generating a leakage-interaction-region of low speed, high entropy and high turbulence [2]. The interaction results in extra complexity and less stability in the tip flow field, which is considered a hot spot. Strong self-induced unsteadiness was found in the TLV with a characteristic frequency near 60% BPF (blade passing frequency) [4,20,21] and the oscillation of passage shock was revealed as well [22]. Moreover, the casing boundary layer along with the blade surface boundary layer may participate in the interacting process [23], especially at near stall point where a shock-induced separation inside the boundary layer is likely to happen in most cases. Under these circumstances, the TLV and its interaction can make a great impact on the overall performance and eventually lead to a spike-type stall inception in transonic compressors [24].

Despite the considerable efforts made by scholars in turbomachinery community, the complete mechanism of TLV and its influence on the tip flow field are still not fully understood, especially in transonic compressors. Previous investigations have shown that the interaction between the shock and TLV contributes a lot to the unsteadiness in the tip flow field, but failed to reveal this interacting process in detail. On the other hand, Reynolds-averaged Navier–Stokes (RANS) method is routinely adopted in most simulations among previous studies; however, traditional turbulence models have native defects in predicting the unsteady and vortical flows such as the TLV [25]. As a result, DES (detached-eddy simulation) method is thought to be an alternative in capturing separated or vortical flow with bearable cost [26]. Up to now, many scholars [18,27–30] have applied DES methods to the turbomachinery field and achieved satisfactory results.

In this paper, we carried out DES investigations of a transonic compressor rotor, focused on the structure of the TLV, the interaction with shock wave as well as the unsteady characteristics. Due to the limits of computing resources, a single-row DES calculation was adopted. This paper is organized as follows: the compressor and the numerical method chosen in the present study are demonstrated in Section 2, with the validation results shown in Section 3.1. Section 3.2 mainly deals with the unsteadiness related to the TLV. Detailed structures of the TLV at different working conditions are shown in Section 3.3, focused on the vortex breakdown phenomenon at near stall point. Section 4 mainly deals with the mechanism of leakage vortex breakdown. Finally, a short conclusion is drawn in Section 5.

2. Methodology

2.1. Testing Case

The compressor investigated in the present study is an in-house 1.5-stage transonic axial compressor with 22 rotor blades and a tip clearance of 0.82% chord length, which is modeled from the first stage of an F-class gas turbine. Its schematic structure and design parameters are shown in Figure 1 and Table 1, respectively.

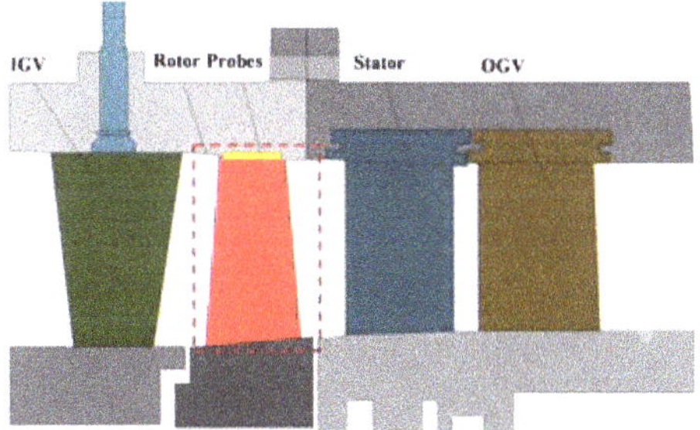

Figure 1. Schematic structure of the 1.5-stage compressor.

Table 1. Design parameters of the compressor.

Parameter	Value	Unit
Rotor blade number	22	-
Rotating speed	24,840	rpm
Mass flow	12	kg/s
Total pressure ratio	1.3	-
Tip Mach number	1.25	-
Tip clearance	0.82%	chord length

2.2. Mesh for DES Calculation

The computational domain is the rotor part (indicated by red dashed line in Figure 1) with inlet and outlet boundaries extended for one-axial chord and two-axial chord, respectively. An unstructured hex-dominant mesh is employed in the DES calculation. As shown in Figure 2, a relatively coarse grid with a refined tip region is adopted to make a trade-off between the flow field resolution and the calculating resources required. To capture the detailed structure of the TLV, the local grid scale at the tip region needs to be at least an order of magnitude smaller than the tip clearance scale in three dimensions, which would be a great challenge if we use a conventional structured mesh. The final mesh is an hybrid grid containing hexahedrons, tetrahedrons, and prisms, with a total element number of 12.3 million, 10 nodes applied in boundary layers to ensure $y^+ < 1$ and 27 nodes applied in the tip clearance region with $\Delta x^+, \Delta y^+, \Delta z^+ < 20$.

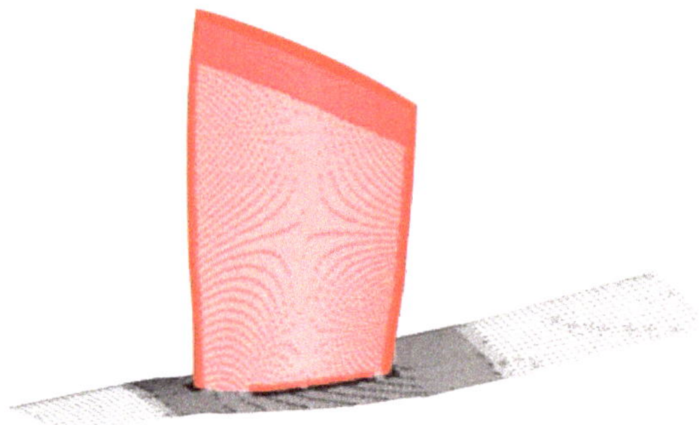

Figure 2. Overview of the detached-eddy simulation (DES) Computational Grid.

2.3. Solver Theory and Calculation Settings

A commercial solver package, FLUENT (14.0, ANSYS, Inc., Canonsburg, PA, USA), was used in the present work, which is a three-dimensional, time-accurate code with implicit second-order scheme, long applied to the field of axial compressors and tip leakage flow [31–37]. The compressible forms of the Reynolds-averaged Navier–Stokes equations were solved in the fluid domain with gravity and volumetric heat source neglected:

$$\frac{\partial \bar{\rho}}{\partial t} + \frac{\partial}{\partial x_j}(\bar{\rho} \tilde{u}_j) = 0 \tag{1}$$

$$\frac{\partial}{\partial t}(\bar{\rho} \tilde{u}_i) + \frac{\partial}{\partial x_j}(\bar{\rho} \tilde{u}_i \tilde{u}_j) = -\frac{\partial \bar{p}}{\partial x_i} + \frac{\partial}{\partial x_j}(\tau_{ij} + \tau'_{ij}) \tag{2}$$

$$\frac{\partial}{\partial t}(\bar{\rho} \tilde{E}) + \frac{\partial}{\partial x_j}[(\bar{\rho} \tilde{E} + \bar{p}) \tilde{u}_j] = \frac{\partial}{\partial x_j}[(\tau_{ij} + \tau'_{ij}) \tilde{u}_i] - \frac{\partial}{\partial x_j}[(\lambda + \lambda') \frac{\partial \tilde{T}}{\partial x_j}] \tag{3}$$

where τ_{ij} is the viscous stress tensor, τ'_{ij} is the Reynolds stress tensor, and $\lambda' = -C_p(\mu_t/Pr_t)$ is the turbulent thermal conductivity.

The ideal air was chosen as the fluid material, which follows the equation of state:

$$\bar{p} = \bar{\rho} \frac{R_m}{M} \bar{T} \tag{4}$$

The properties of air are: molecular weight M = 28.966 g/mol, specific heat capacity at the constant pressure C_p = 1004.4 kJ/(kg · K), thermal conductivity λ = 0.0261 W/(m · K), and dynamic viscosity μ is determined by the Sutherland's formula [38].

The constitutive equations of Newtonian fluid were adopted to model the viscous stress term:

$$\tau_{ij} = 2\mu\bar{S}_{ij} - \frac{2}{3}\mu\bar{S}_{kk}\delta_{ij} \tag{5}$$

where $\bar{S}_{ij} = [(\partial\bar{u}_i/\partial x_j + \partial\bar{u}_j/\partial x_i)/2]$ is the deformation rate tensor. The Reynolds stress term was modeled using the Boussinesq hypothesis, as follows:

$$\tau'_{ij} = 2\mu_t\bar{S}_{ij} - \frac{2}{3}(\rho k + \mu_t\bar{S}_{kk})\delta_{ij} \tag{6}$$

For the detached-eddy simulation, the DES97 model, first developed by Spalart et al. [39], was adopted in the present study with a default DES coefficient C_{DES} of 0.65 [40]. In the DES97 model, the near-wall distance d in the original S-A turbulence model [41] has been replaced by the DES length scale $\tilde{d}$, as shown in Equation (7):

$$\frac{D\tilde{v}}{Dt} = c_{b1}\tilde{S}\tilde{v} + \frac{1}{\sigma}[\nabla \cdot ((v + \tilde{v})\nabla\tilde{v}) + c_{b2}(\nabla\tilde{v})^2] - c_{w1}f_w(\frac{\tilde{v}}{\tilde{d}})^2 \tag{7}$$

where $\tilde{v}$ is the working viscosity and v is the kinematic viscosity. $\tilde{S}$ is a function of another scalar $\bar{S}$ which can be chosen from the vorticity magnitude or the deformation rate [41]. c_{b1}, c_{b2}, c_{w1}, σ and f_w are coefficients of the S-A turbulence model, whose definitions can all be found in [41]. The DES length scale $\tilde{d}$ is defined as follows:

$$\tilde{d} \equiv \min(d, C_{DES}\Delta) \tag{8}$$

where Δ is the local grid scale and C_{DES} is a coefficient in this model. According to Equation (8), DES length scale $\tilde{d}$ will recover to the near-wall distance d when $d \ll \Delta$. This always happens inside a boundary layer so that the original S-A turbulence model is activated. Nevertheless, in the mainstream with high Reynolds number, the production term will balance with the destruction term [39] so that Equation (7) becomes:

$$v_t = (\frac{c_{b1}C_{DES}^2}{c_{w1}f_w})\Delta^2\bar{S} \propto \Delta^2\bar{S} \tag{9}$$

If $\bar{S}$ is defined as the deformation rate and we choose C_{DES} properly, Equation (9) can be the Smagorinsky-Lilly model [42] for LES:

$$v_{SGS} = 2C_s\Delta^2\bar{S} \tag{10}$$

where $\bar{S}$ is the deformation rate. It is worth noting that the RANS equation and the LES equation are formally identical at some time. Taking Equation (2) for example, once a suitable turbulence model is introduced into the momentum equations, the equation itself will no longer carry any information concerning their derivation (averaging). This is true if we always adopt eddy viscosity models in RANS or LES. The tensor τ'_{ij} can be the Reynolds stress for RANS when we consider the superscript "~" as "time-averaging". Whereas, τ'_{ij} can also be the sub-grid-scale stress for LES when the superscript was treated as "spatial-filtering". In general, the DES length scale $\tilde{d}$ acts as a switch for RANS and LES. Therefore, the DES97 model can use LES method in the mainstream and activating RANS method (with S-A turbulence model) inside the boundary layer.

We conduct DES calculations at three working conditions, namely near choke point (NC), peak efficiency point (PE) and near stall point (NS), as shown in Table 2. The physical time of each time step is 1×10^{-6} s, which is small enough to include at least 110 time steps in one blade passing period. Mass flow rate and static pressure were monitored during the calculation to ensure a good convergence. Pressure monitors are located on the tip region of the rotor, with eight points on the casing and one on the blade tip, as shown in Figure 3. As for boundary conditions, adiabatic nonslip-wall conditions were adopted for all solid walls. Radial distributions of total pressure, total temperature, and flow angles were given at the inlet using UDF (user-defined function) files. Static pressure distribution was specified at the outlet.

Table 2. Calculation settings.

Parameter	Setting
Computational domain	One R1 passage
Rotating speed	22,000 rpm, 24,840 rpm
Working condition	NC, PE, NS
Solver	FLUENT
Model	DES97
Time step	1×10^{-6} s
Inner iteration	15

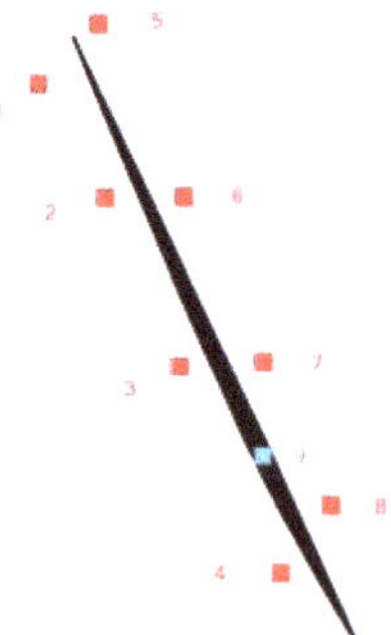

Figure 3. Monitor points in DES calculation (red: on casing, blue: on blade tip).

2.4. Grid Independence Study

The grid independence study was based on RANS calculations with 7 sets of grids at PE, shown in Table 3. The current mesh for DES is the NO.7 grid, which is confirmed to be grid-independent according to Figure 4. We may conclude that the current mesh for DES can provide us a grid-independent result in RANS region. However, the LES region is naturally grid-dependent because the cut-off scale is related to local grid scale. A finer mesh always means a better resolution of the flow field. So an appropriate mesh for LES should meet the requirements of $\Delta x^{+}, \Delta y^{+}, \Delta z^{+}$, which has already been checked in Section 2.2.

Table 3. Grids in the grid independence study.

Mesh No.	1	2	3	4	5	6	7
Elements/million	0.19	0.47	0.80	1.16	2.87	5.83	12.30

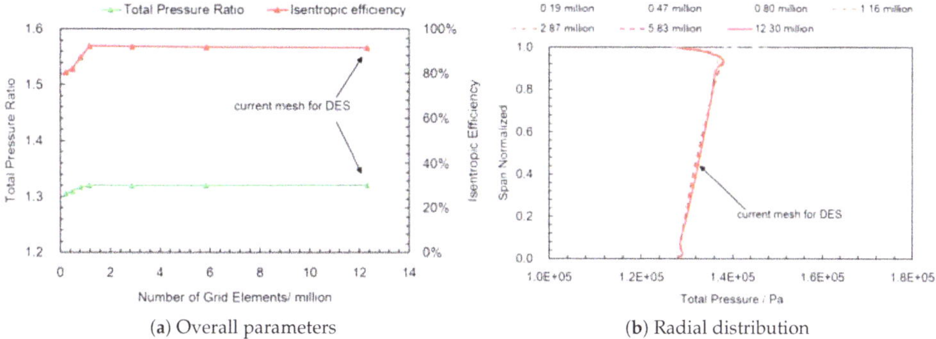

(**a**) Overall parameters

(**b**) Radial distribution

Figure 4. Grid independence study.

3. DES Results

3.1. Validation

DES results are compared with the experimental data as well as multistage RANS results to ensure a reasonable prediction of the overall flow field, as shown in Figure 5. Please note that the point NS1 and point NS2 were both at near stall conditions. These two points were at same working conditions except for the outlet static pressure. We slightly raised the static pressure (1kPa on average) at the outlet boundary for NS1 then we got NS2, aimed at further approaching the stall limit. Numerical calculations have a good prediction for the performance trend of the compressor at different rotating speeds, especially for the total pressure ratio. As for DES results, the maximum deviations of averaged mass flow rate and total pressure ratio at three working conditions are 0.25% and 0.94%, respectively. Other flow details were compared with corresponding RANS results. Figures 6 and 7 shows radial distributions of total pressure ratio and relative flow angle. Parameter distributions of the DES results were consistent with RANS results, indicating the predictions of averaged flow field are not worse than those of RANS. In addition, Figure 8 shows the comparison of relative Mach contour at 99.3% span (slightly below the blade tip) at near stall point at design speed. These two results near the top region have no conflicts in the shock location or the leakage flow behavior, indicating tip leakage flow was correctly captured in DES calculations. In general, DES results are relatively reliable in the present simulation and can be used for following analysis of the tip leakage flow. Besides, there is no experimental data at the design speed. So we conducted the DES calculation at 22,000 rpm instead and compared it to the experimental data aiming to validate the numerical method.

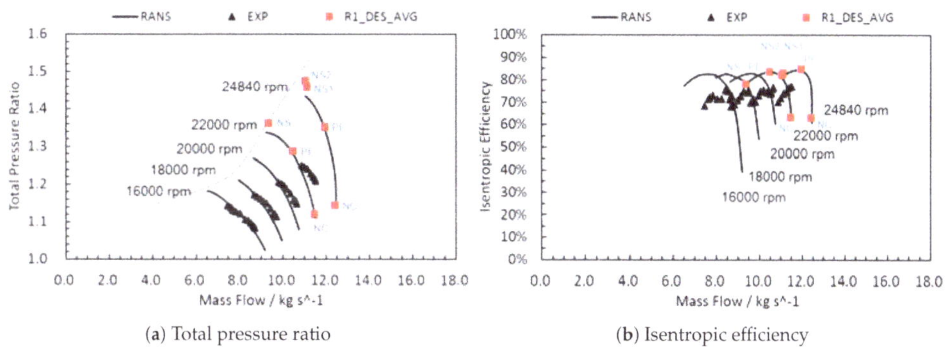

(**a**) Total pressure ratio

(**b**) Isentropic efficiency

Figure 5. Comparison of overall performance.

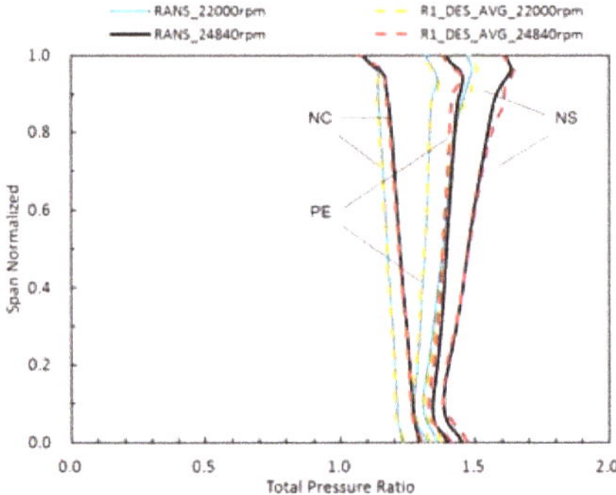

Figure 6. Comparison of total pressure ratio distribution.

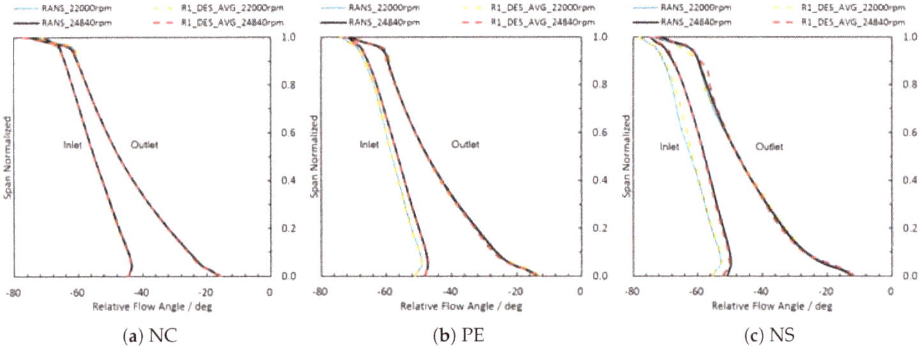

(a) NC   (b) PE   (c) NS

Figure 7. Comparison of relative flow angle distribution. NC: near choke point; PE: peak efficiency point; NS: near stall point.

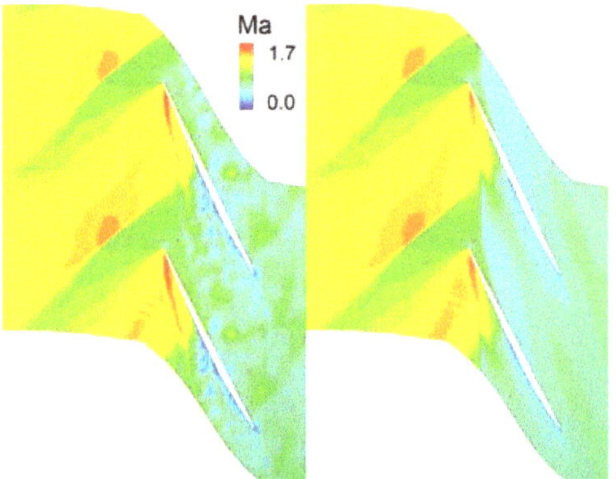

Figure 8. Relative Mach contour at 99.3% span at NS (**left**: DES, **right**: RANS).

Moreover, the correct switch of LES and RANS is critical in DES calculations and needs to be checked in the present study to reduce the impact of grid-induced separation(GIS) and grey area problems of the model itself. A criterion to distinguish LES region from RANS region is as follows:

$$\tilde{\zeta}_{DES} = \frac{\tilde{d} - d}{C_{DES}\Delta - d} \in [0, 1] \tag{11}$$

where $\zeta_{DES} = 1$ indicates LES method is switched on and $\zeta_{DES} = 0$ for RANS. Figure 9 shows LES and RANS regions in DES calculations at 90% span, with LES switched on in the mainstream and RANS used inside the boundary layer as expected. Other locations such as 10%, 50% spans and axial cuts were also checked at different working conditions. The switch is appropriate, and no considerable separation was found in the flow field, which contributes to the credibility of present calculations.

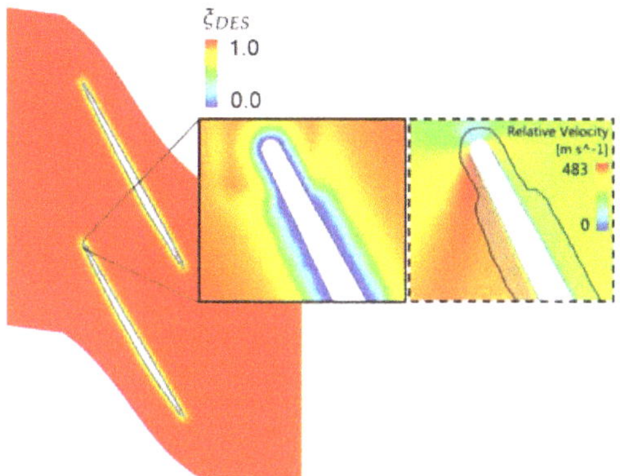

Figure 9. LES and RANS regions in DES results at 90% span.

3.2. Unsteadiness at NS Point

The tip leakage flow in high-speed compressors has self-induced unsteadiness features [4,43]. In present calculation, by monitoring nine static pressure points, it is found that the fluctuations at NS point is much stronger than PE point and NC point. Therefore, the following part will mainly focus on the unsteadiness at NS condition.

Figure 10 shows the static pressure convergence history of different monitoring points. Point 1, 2 and 5 experienced weaker pressure fluctuation than the rest. It is worth noting that point 1 and 2 are exactly in the initial trajectory of the TLV and in front of the shock near suction side, while point 5 is at the leading edge and after the bow shock. For transonic rotors, it is typical that the TLV starts from the leading edge of tip blade, traveling towards the adjacent pressure side. It passes through and interacts with the bow shock, then imprints on the adjacent pressure side, and finally develops downstream along the blade surface. In other words, these three points mentioned above are far away from the interaction region of TLV and the shock, indicating that neither the TLV nor the shock alone is the root cause of the unsteadiness. In addition, points 6 to 9, which experienced strong fluctuations in static pressure, are also located along the trajectory of TLV. However, what makes a difference is that these points are all located after the shock wave. From this aspect, it is exactly the interaction between TLV and the shock that leads to the unsteadiness of the tip region. The detailed reasons will be explained in next section.

Figure 11 shows the spectrum of some representative monitor points at near stall point, which is the result of the fast Fourier transform of the time series data. The frequency characteristics of each point are not the same. Wide though the frequency bands are, they do share some similarities, that is, peak values appeared near two specific frequencies 0.64 BPF and 1.80 BPF. It is worth noting that the former is close to the characteristic frequency of tip leakage flow in typical transonic compressors, for instance, 0.6 BPF [44] for NASA Rotor 37 and 0.57 BPF [4,20] for Darmstadt Rotor 1. Other monitoring points with large fluctuation, such as point 8 and point 9, have similar results as well.

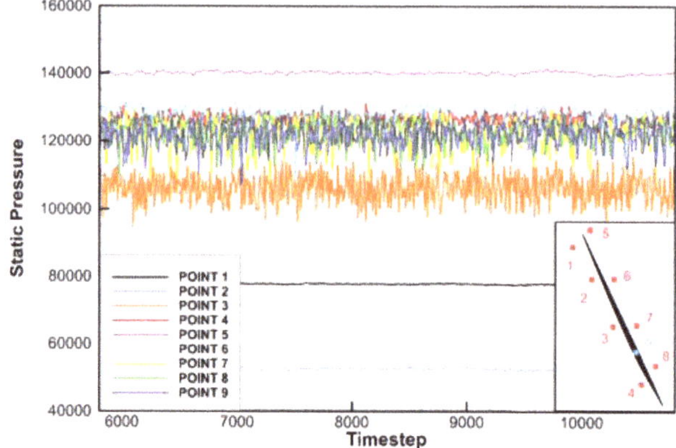

Figure 10. Static pressure history at NS in DES calculation.

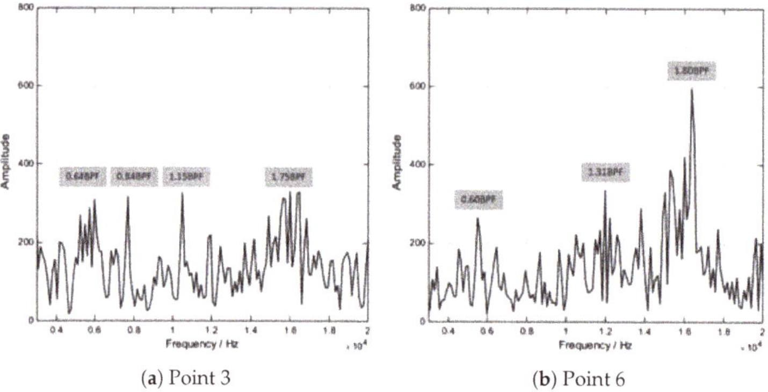

(**a**) Point 3

(**b**) Point 6

Figure 11. Amplitude spectrum of certain monitor points at NS.

The unsteadiness near the tip region inevitably affected the performance of the compressor and is the critical reason for the fluctuation in overall parameters at near stall point, as shown in Figure 12. T1 to T6 are six different moments with the same time interval 5×10^{-5} s, whose corresponding frequency is about 2.2 BPF. Obvious unsteady feature exists above 80% span with a fluctuation range of about 10 degrees for the outlet flow angle and 0.1 (about 6.5% of the time-averaged total pressure ratio) for the total pressure ratio, while the unsteadiness is weak in the middle and root span, indicating that tip leakage flow is an essential factor causing instability at NS point. Nevertheless, there is no obvious fluctuation of the static pressure rise coefficient among the whole blade height range, which means the unsteadiness of the top flow field is mainly caused by the fluctuation of kinetic energy and can be a proof for the wake-like nature of the TLV when passing through the shock [45].

Figure 13 is the top view of R1 shroud with blade profile imprinted. We chose a certain axial location (red line) and obtained its circumferential distribution of static pressure in one period, as shown in Figure 14, taking circumferential angle as the abscissa. There are two low pressure regions, one is at about −1 degrees (indicated by the black arrow) and the other is from −4.5 to −7 degrees (indicated by the red arrow). The former region is at the tip clearance region, very little influenced by the unsteadiness, while the latter experienced a strong oscillation at different time steps, with the valley traveling from the left side to the right side then returning to the left side. According to Figure 13, the latter region is near the pressure side of the tip blade, exactly located in the TLV trajectory after the shock, on its way to hit the adjacent pressure side and to develop downstream. The oscillation of pressure valley indicated that the TLV was no longer stable at NS point and was oscillating in the blade passage.

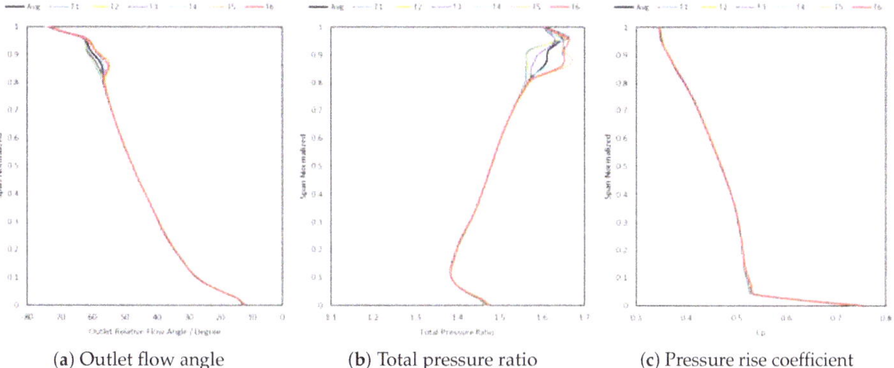

(**a**) Outlet flow angle   (**b**) Total pressure ratio   (**c**) Pressure rise coefficient

Figure 12. Parameter distributions at NS.

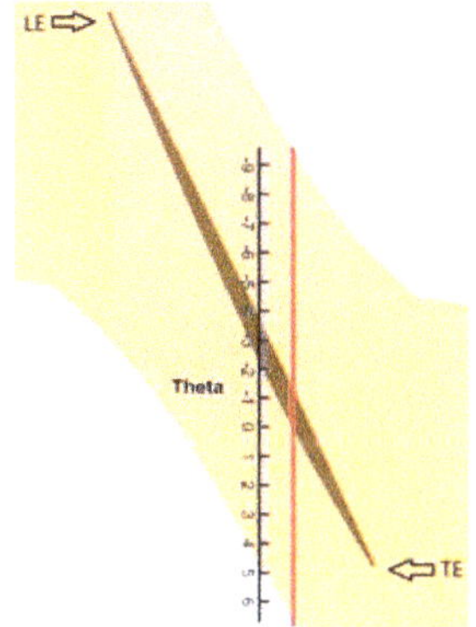

Figure 13. Circumferential line location on the shroud surface.

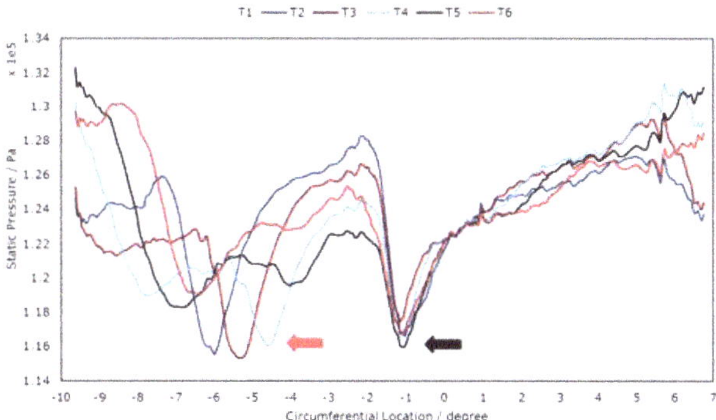

Figure 14. Circumferential distribution of casing pressure at a certain axial location.

Figure 15 can illustrate the above phenomena more clearly, with black dash-dotted line indicated the same axial location in Figure 13. Interacted with the shock and the parallel small vortex, TLV started to swing tangentially at a distance after the shock. This resulted in a small oscillation, developing downstream with an increasing amplitude. As is known, TLV has a lower static pressure value and a lower axial velocity than the mainstream. Every time the trajectory of TLV swept over a certain point, that location would experience a drop in static pressure as well as axial velocity, which is the cause of the oscillation in Figure 14. Eventually, it caused the oscillation of overall parameters such as outlet flow angle, pressure ratio, or efficiency.

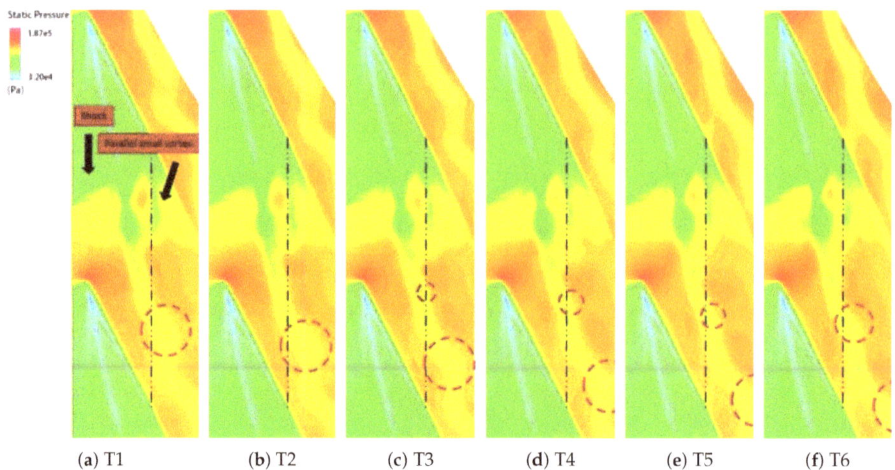

Figure 15. Casing Pressure Contour in One Period at NS.

In general, the unsteadiness of the rotor at NS point is mainly manifested at the tip region, closely related to the TLV. The unsteady characteristics in middle span or root span can be totally ignored in comparison. Neither the TLV nor the shock alone is the origin of the unsteadiness. It was found that the TLV became unstable after the interaction with the shock and experienced a stronger oscillation when developing downstream. The most unstable region in this transonic rotor is near the pressure side of the tip blade, which is similar to the conclusions [4] for Darmstadt Rotor 1.

3.3. Detailed Structure of Tip Leakage Flow

DES calculation can obtain detailed information of the top flow field than RANS and may provide some clues for the unsteady characteristic mentioned above. The general structure of tip leakage flow can be clearly observed in Figure 16, with shock position indicated by dashed line. The vortex structures are in three different forms as the operating condition changes.

At the NC point, there are two shocks, one is the bow shock near the leading edge and the other is the passage shock near the trailing edge. Under this condition, the pressure gradient in the tip clearance is relatively small and the TLV started from the leading edge is weak as well. As a result, the vortex core was slenderer in Figure 16a and vanished directly after the bow shock. Meanwhile, the TLV was too weak to draw downstream leakage fluid into, which led to another vortex system after the middle chord of the blade. At PE point, however, the blade load increased and the leading-edge TLV became stronger, so that most leakage fluid can be drawn into the main leakage vortex. The characteristics of the TLV were kept so well in the developing process that the vortex core remained continuous when passing through the shock with a slight expansion in volume, which indicates a stable flow state at PE point. At NS point, the entire top flow field had a very rich flow pattern, and there were many small vortex structures after the shock. Some of the small vortices were developed from the main leakage vortex. The other parts were from the secondary leakage, that is, after imprinting the pressure surface of the adjacent blade, the TLV leaked to the next blade passage again through the tip clearance of the adjacent blade.

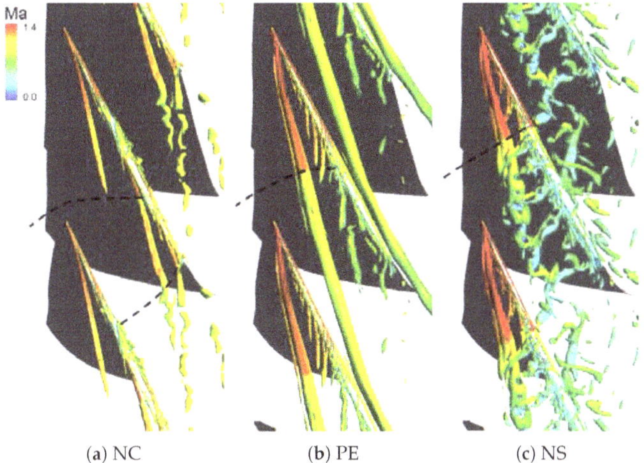

(a) NC   (b) PE   (c) NS

Figure 16. λ_2 Isosurface at tip region of R1 (colored by relative Mach number).

The complex flow field at top region is consistent with the strong unsteadiness mentioned above at NS point. The static pressure oscillation in Figure 15 was caused by the "shedding" of small vortices after the shock. The small vortices did not exist at NC point or PE point and mainly came from the main leakage vortex, which is probably caused by the interaction with the shock. When passing though the bow shock, the TLV at NS point changed from a single solid core to many separate vortices.

Figure 17 is the helicity distribution of different crossflow planes at near stall point, which is used to characterize the intertwining degree of the fluid around TLV core, with the blade surface colored by static pressure, the shock position indicated by the black dotted line, and the beginning of the TLV indicated by the black arrow. The vortex core before the shock was concentrated and slowly increasing in volume. After the shock, the helicity distribution was dispersed and not concentrated anymore, which indicated that the fluid no longer moved around the vortex core tightly after the shock. The distance between the four crossflow planes can be approximated as isometric, but the helicity distribution before and after the shock surface is quite different, which means that the vortex core experienced an abrupt change in its internal structure. Combined with the vortex structure in Figure 16c, we can conclude that a vortex breakdown took place after the interaction between TLV and the shock.

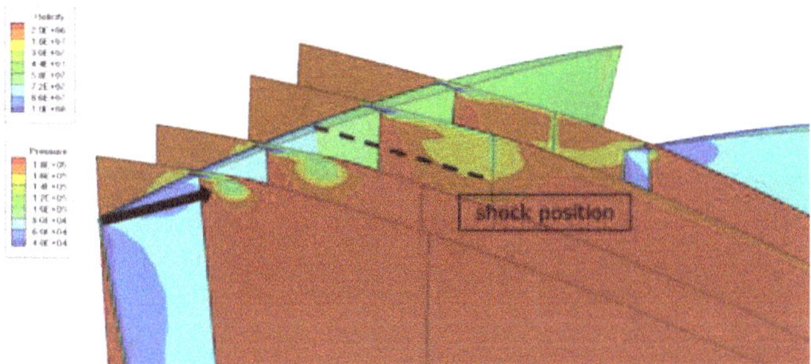

Figure 17. Helicity distribution on crossflow planes at NS (with blade surface colored by static pressure).

The breakdown process can be clearly illustrated in Figure 18, which is a transient λ_2 isosurface colored by relative Mach number. It is confirmed that the TLV changed to a three-dimensional structure and was not axisymmetric anymore when passing through the shock. The vortex core changed its direction to a perpendicular path and started to rotate around the original one when developing downstream, which is the typical structure of a spiral-type breakdown. The structure after the onset of the breakdown process is not stable, with rotating phase related to the flow time, thus causing strong unsteadiness. This contributes a lot to our understanding of the oscillation after the shock in Figure 15 that the previous view on the unsteady flow behavior of the leakage vortex may be incomplete. The unsteadiness is not a two-dimensional phenomenon inside the S1 plane, but a three-dimensional structure in nature. The underlying reason behind is the spiral-type vortex breakdown after the interaction with the shock.

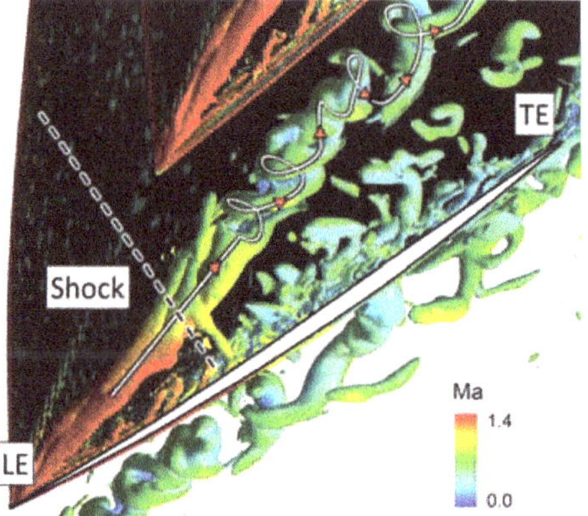

Figure 18. Sketch for spiral-type breakdown of TLV at NS.

To the author's knowledge, this is probably the first time that a detailed structure of the breakdown process for the TLV in a transonic compressor is obtained in numerical investigations. In addition, PIV (particle image velocimetry) measurements of another transonic compressor near stability limit were conducted [46], whose results are in great agreement with the present calculation results, as shown in Figure 19.

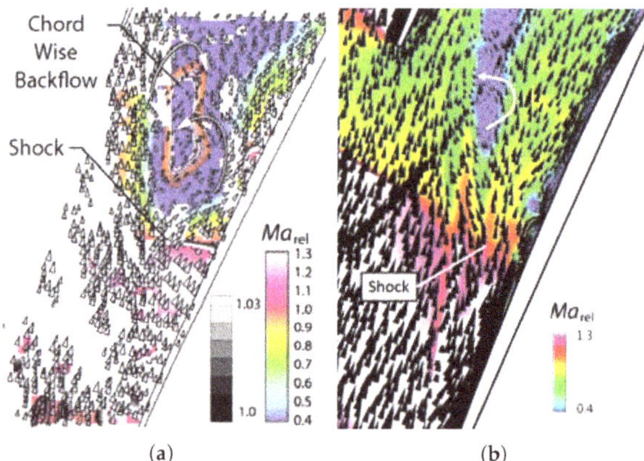

Figure 19. Spiral vortex breakdown at NS.. (**a**) PIV results [46]; (**b**) DES results in the present study.

4. Discussion

Vortex breakdown is a very complex flow phenomenon and is an independent branch in fluid mechanics. The interaction of a streamwise vortex and a normal or oblique shock under supersonic conditions could lead to a vortex breakdown phenomenon. The interacting process and mechanisms were widely studied [47–50]. Recently, lots of efforts [51–55] have been made extending the state of knowledge regarding onset, internal structure and mode selection of vortex breakdown. In the present study, a shock-induced spiral-type vortex breakdown was found at NS. We will deal with the related structure of the breakdown process in this section.

In transonic compressors, the shock wave in the rotor passage can provide a large adverse pressure gradient, with a great influence on the streamwise velocity in the vortex core. As shown in Figure 20, subfigure (a) is the λ_2 isosurface, which illustrates the breakdown process. Subfigure (b) is streamwise velocity contour at the same viewpoint, which indicates the component of relative velocity in the vortex trajectory direction, and subfigure (c) demonstrates the surface streamline. The corresponding cut plane is represented by the red point line in the top view and the location of the bow shock near the suction surface is characterized by the black dashed line. The vortex breakdown is clearly observed in the upper figure with its initial location indicated by the red circle. The breakdown location was not at the shock surface exactly, but at a certain distance downstream, which is in qualitative agreement with experimental results in Figure 19a. Leibovich [56] pointed out that this axial interval is several vortex-core diameters in length. It is worth noting that the initial location of vortex breakdown in (a) coincided with the location S1 in (b) where the streamwise velocity of the vortex core was zero, which indicates that the breakdown of the leakage vortex is closely related to the stagnation point in the center of vortex core. Please note that there were not only one stagnation point in (b). Along with subfigure (c), we could observe a recirculation zone between the two stagnation points S1 and S2. This is consistent with the direct numerical simulation of vortex breakdown in swirling jets and wakes by Ruith et al. [52].

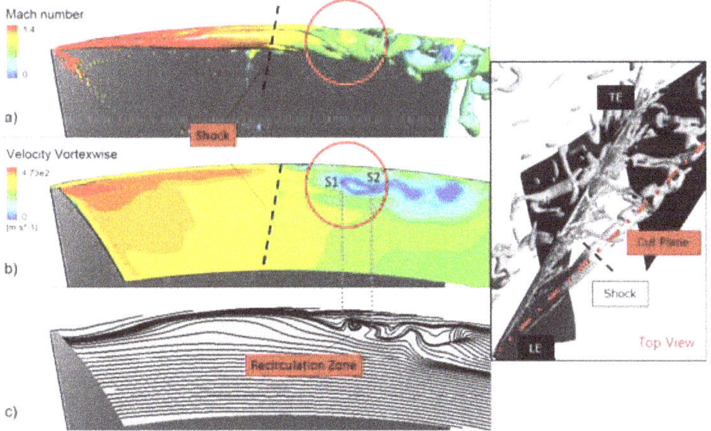

Figure 20. Structures Related to the Breakdown of TLV at NS.

Figure 21 is the swirl velocity vector of the TLV, demonstrating the rotating direction of the TLV before and after the breakdown. For clarity, the adjacent blade was hidden, and the larger images of the dashed zones might have a different view point. Whether before or after the breakdown point, the vortices were temporally rotating in the same direction as the initial TLV, except for the induced vortex in (b). However, the helices after breakdown were coiling in the opposite direction (more clear in Figure 18). This is similar to the findings of Pasche et al. [57] who conducted experiments on obstacle-induced spiral vortex breakdown. The breakdown helices originating at a locally wake-like profile have negative winding sense [52]. Of course, the TLV had a wake-like profile in nature and the breakdown structure in the present study contributed to this view.

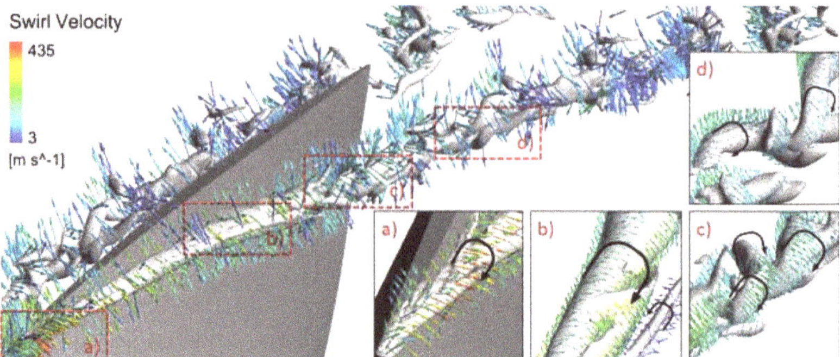

Figure 21. Swirl velocity vector on the λ_2 isosurface.

The causes of vortex breakdown are very complicated. The generation of the recirculation bubble of the vortex breakdown remains unclear but the spiraling motion of the flow behind the recirculation bubble comes from a global unstable mode of the flow [53]. The same mechanism is also observed without recirculation bubble [57], the spontaneous spiraling motion is due to a self-sustained instability. At present, it is consistent to conclude that the reverse pressure gradient and the strength of the vortex itself are crucial factors in the vortex breakdown process [48]. The adverse pressure gradient characterizes the deceleration of the vortex core along the streamwise direction. Criteria for vortex breakdown based on the interaction between the Rankine vortex and the normal shock waves were proposed by Mahesh [58], Smart and Kalkhoran [49] and other scholars. As shown in Figure 22, the abscissa is the freestream Mach number, reflecting adverse pressure gradient in the streamwise direction. The ordinate is the swirl ratio before the shock, which reflects the intensity of the vortex. The swirl ratio is defined as follows:

$$\tau = \frac{\Lambda_{max}}{V_a} \tag{12}$$

where Λ_{max} is the maximum swirl velocity component and V_a is the streamwise velocity component in the vortex core.

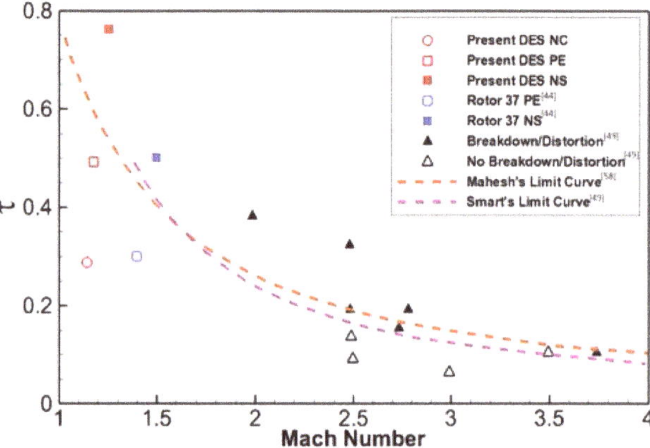

Figure 22. Limit curve for normal shock wave/vortex interaction.

The interaction between the TLV and the bow shock in the transonic compressor could be simplified to the interaction of normal shock and vortex. Both the compressor in the present paper (red marks) and Rotor 37 (blue marks, [44]) exceeded the breakdown limit at NS conditions, and the breakdown of leakage vortex did occur in the detailed flow field. In addition, at PE point and NC point, the vortex breakdown phenomenon was not observed in the present calculation, which is still in good agreement with the breakdown criterion. As the operating point moves to the NS point, the swirl ratio will increase sharply with the increase of the leakage vortex intensity, while the change of Mach number in the axial direction of the vortex core is not obvious. Therefore, transonic rotors with heavy blade loading are likely to experience leakage vortex breakdown at NS point.

We could use the mechanism for the shock-induced breakdown to illustrate the interacting process between the TLV and the bow shock or the breakdown process of the TLV in transonic compressors. When the vortex passes through the shock, the streamwise velocity decreases rapidly due to the strong adverse pressure gradient produced by the shock. The swirl component, however, is orthogonal to the shock surface and is not much affected. In other words, the attenuation of the streamwise component is far greater than that of the swirl component [48]. When the shock is strong enough, the streamwise velocity will decay to zero after the interaction. Under this circumstance, the breakdown phenomenon of the TLV is likely to occur.

The small vortices produced by the TLV vortex breakdown are distributed in the tip region, causing a large blockage effect in the rotor passage. The low energy fluid might subsequently result in leading-edge spillage or induce a spike-type stall inception. Therefore, how to postpone or eliminate the breakdown of TLV may become one of the important ways to enhance the stability at NS point and enlarge the surge margin of transonic compressors.

5. Conclusions

In this paper, a numerical investigation of a transonic compressor rotor using DES method is conducted at three working conditions, focused on the structure of tip leakage flow, the interaction with shock wave and the unsteady characteristics.

Strong unsteadiness was found at NS point with the characteristic frequencies of 0.64 BPF and 1.80 BPF. The most unstable region for this transonic rotor is in the rotor passage near the pressure side of the tip blade, which is the result of the interaction between the TLV and the shock. The affected area is mainly located at the tip region, causing a considerable fluctuation in total pressure distribution and flow angle distribution above 80% span. The unsteadiness at NC point and PE point is not obvious.

Detailed structures of the tip leakage flow were captured, closely related to the working condition. At the NS point, a vortex breakdown can be observed downstream the bow shock. While, at NC point or PE point, the TLV breakdown did not take place. The breakdown process at NS point was confirmed in a spiral-type form. The vortex core changed its direction to a perpendicular path and started to rotate around the original one when developing downstream. Whether before or after the breakdown point, the vortices were temporally rotating in the same direction as the initial leakage vortex. Whereas, the helices after breakdown were coiling in the opposite direction. The breakdown of the TLV generated unstable small vortices and contributed greatly to the unsteadiness at NS point.

Author Contributions: X.S. acquired the numerical data and wrote the paper; and X.R., X.L. and C.G. revised the paper and offered useful suggestions to write the paper.

Funding: This paper is supported by National Science and Technology Major Project (No: 2017-II-0007), National Natural Science Foundation of China (No. 51806118) and by a grant from the Science & Technology on Reliability & Environmental Engineering Laboratory (No. 6142A0501020317).

Acknowledgments: The author thanks Senior Engineer JianBai Li for his critical comments and helpful advice. This work is supported by the "Explorer 100" HPC Platform, State Laboratory for Information Science and Technology, Tsinghua University.

Nomenclature

C_{DES}	the coefficient in DES97 model
C_p	specific heat capacity at the constant pressure
C_s	the coefficient in Smagorinsky-Lilly model
$\tilde{d}$	DES length scale

e	inner energy
E	total energy, $E = e + u_i u_i / 2$
H	total enthalpy, $H = E + p/\rho$
k	turbulent kinetic energy
M	molecular weight
Ma	Mach number
p	pressure
R_m	gas constant, $R_m = 8.314 \, \text{J}/(\text{mol} \cdot \text{K})$
S	deformation rate, $\bar{S} = \sqrt{2\bar{S}_{ij}\bar{S}_{ij}}$
S_{ij}	deformation rate tensor
T	temperature
u	velocity
x	Cartesian coordinates
y^+	nondimensional wall distance
Δ	local grid scale
Δx^+	nondimensional grid scale in x direction
Δy^+	nondimensional grid scale in y direction
Δz^+	nondimensional grid scale in z direction
λ	thermal conductivity
λ'	turbulent thermal conductivity
μ	dynamic viscosity
μ_t	turbulent eddy viscosity
ν	kinematic viscosity
$\tilde{\nu}$	the working viscosity in S-A model
ν_{SGS}	sub-grid-scale kinematic viscosity
ν_t	turbulent kinematic viscosity
ξ_{DES}	a criterion to distinguish LES region from RANS region
ρ	density
τ_{ij}	viscous stress tensor
τ'_{ij}	Reynolds stress tensor or sub-grid-scale stress tensor

Abbreviations

BPF	Blade passing frequency, $1\,BPF = n \cdot NB/60$ (Hz), where n is the rotation speed of the axis with the unit rpm
DES	Detached-eddy simulation
LES	Large eddy simulation
NB	Number of rotor blades
NC	Near choke point
NS	Near stall point
PE	Peak efficiency point
PIV	Particle image velocimetry
RANS	Reynolds-averaged Navier–Stokes
SGS	sub-grid-scale stress
TLV	Tip leakage vortex
UDF	User-defined function

References

1. Storer, J.A.; Cumpsty, N.A. Tip leakage flow in axial compressors. *J. Turbomach.* **1991**, *113*, 252. [CrossRef]
2. Gerolymos, G.A.; Vallet, I. Tip-clearance and secondary flows in a transonic compressor rotor. *J. Turbomach.* **1999**, *121*, 751. [CrossRef]
3. Tan, C.; Day, I.; Morris, S.; Wadia, A. Spike-type compressor stall inception, detection, and control. *Ann. Rev. Fluid Mech.* **2010**, *42*, 275–300.
4. Du, J.; Lin, F.; Chen, J.; Nie, C.; Biela, C. Flow structures in the tip region for a transonic compressor rotor. *J. Turbomach.* **2013**, *135*, 031012. [CrossRef]

5. Liu, Y.; Tan, L.; Wang, B. A review of tip clearance in propeller, pump and turbine. *Energies* **2018**, *11*, 2202. [CrossRef]

6. Tab, L.; Xie, Z.; Liu, Y.; Hao, Y.; Xu, Y. Influence of T-shape tip clearance on performance of a mixed-flow pump. *Proc. Inst. Mech. Eng. Part A J. Power Energy* **2018**, *232*, 386–396. [CrossRef]

7. Liu, M.; Tan, L.; Cao, S. Cavitation-vortex–turbulence interaction and one-dimensional model prediction of pressure for hydrofoil ALE15 by large eddy simulation. *J. Fluids Eng.* **2018**, *141*, 021103. [CrossRef]

8. Hao, Y.; Tan, L. Symmetrical and unsymmetrical tip clearances on cavitation performance and radial force of a mixed flow pump as turbine at pump mode. *Renew. Energy* **2018**, *127*, 368–376. [CrossRef]

9. Denton, J.D. Loss mechanisms in turbomachines. *J. Turbomach.* **1993**, *115*, 621–656. [CrossRef]

10. Vo, H.D. Role of Tip Clearance Flow on Axial Compressor Stability. Ph.D. Thesis, Massachusetts Institute of Technology, Cambridge, MA, USA, 2001.

11. Suder, K.L.; Hathaway, M.D.; Thorp, S.A.; Strazisar, A.J.; Bright, M.B. Compressor stability enhancement using discrete tip injection. *J. Turbomach.* **2001**, *123*, 14. [CrossRef]

12. Eveker, K.M.; Gysling, D.L.; Nett, C.N.; Sharma, O.P. Integrated control of rotating stall and surge in high-speed multistage compression systems. *J. Turbomach.* **1998**, *120*, 440–445. [CrossRef]

13. Müller, M.W.; Schiffer, H.P.; Hah, C. Effect of circumferential grooves on the aerodynamic performance of an axial single-stage transonic compressor. In *Proceedings of the ASME Turbo Expo 2007: Power for Land, Sea, and Air, Montreal, QC, Canada, 14–17 May 2007*; ASME: New York, NY, USA; Volume 6, pp. 115–124. [CrossRef]

14. Vo, H.D. Rotating stall suppression in axial Compressors with casing plasma actuation. *J. Propuls. Power* **2010**, *26*, 808–818. [CrossRef]

15. Hoying, D.A.; Tan, C.S.; Vo, H.D.; Greitzer, E.M. Role of blade passage flow structurs in axial compressor rotating stall inception. *J. Turbomach.* **1999**, *121*, 735. [CrossRef]

16. Vo, H.D.; Tan, C.S.; Greitzer, E.M. Criteria for spike initiated rotating stall. In *Proceedings of the ASME Turbo Expo 2005: Power for Land, Sea, and Air, Peno, NV, USA, 6–9 June 2005*; ASME: New York, NY, USA, 2005; Volume 6, pp. 155–165. [CrossRef]

17. Deppe, A.; Saathoff, H.; Stark, U. Discussion: "Criteria for spike initiated rotating stall" (Vo, H. D., Tan, C. S., Greitzer, E. M., 2008, ASME J. Turbomach., 130, p. 011023). *J. Turbomach.* **2008**, *130*, 015501. [CrossRef]

18. Yamada, K.; Kikuta, H.; Iwakiri, K.i.; Furukawa, M.; Gunjishima, S. An explanation for flow features of spike-type stall inception in an axial compressor rotor. *J. Turbomach.* **2012**, *135*, 021023. [CrossRef]

19. Hah, C. Effects of double-leakage tip clearance flow on the performance of a compressor stage with a large rotor tip gap. *J. Turbomach.* **2017**, *139*, 061006. [CrossRef]

20. Biela, C.; Müller, M.W.; Schiffer, H.P.; Zscherp, C. Unsteady pressure measurement in a single stage axial transonic compressor near the stability limit. In *ASME Turbo Expo 2008: Power for Land, Sea, and Air*; American Society of Mechanical Engineers: New York City, NY, USA, 2008; pp. 157–165.

21. Zhang, Y.; Lu, X.; Chu, W.; Zhu, J. Numerical investigation of the unsteady tip leakage flow and rotating stall inception in a transonic compressor. *J. Therm. Sci.* **2010**, *19*, 310–317. [CrossRef]

22. Hah, C.; Rabe, D.C.; Wadia, A.R. Role of tip-leakage vortices and passage shock in stall inception in a swept transonic compressor rotor. In *Proceedings of the ASME Turbo Expo 2004: Power for Land, Sea, and Air, Vienna, Austria, 14–17 June 2004*; ASME: New York, NY, USA, 2004; Volume 5, pp. 545–555. [CrossRef]

23. Hoeger, M.; Fritsch, G.; Bauer, D. Numerical simulation of the shock-tip leakage vortex interaction in a HPC front stage. *J. Turbomach.* **1999**, *121*, 456–468. [CrossRef]

24. Bergner, J.; Kinzel, M.; Schiffer, H.P.; Hah, C. Short length-scale rotating stall inception in a transonic axial compressor: Experimental investigation. In *Proceedings of the ASME Turbo Expo 2006: Power for Land, Sea, and Air, Barcelona, Spain, 8–11 May 2006*; ASME: New York, NY, USA, 2006; Volume 6, pp. 131–140. [CrossRef]

25. Iim, H.; Chen, X.Y.; Zha, G. Detached-eddy simulation of rotating stall inception for a full-annulus transonic rotor. *J. Propuls. Power* **2012**, *28*, 782–798. [CrossRef]

26. Spalart, P.R. Detached-eddy simulation. *Ann. Rev. Fluid Mech.* **2009**, *41*, 181–202. [CrossRef]

27. Riéra, W.; Castillon, L.; Marty, J.; Leboeuf, F. Inlet condition effects on the tip clearance flow with zonal detached eddy simulation. *J. Turbomach.* **2013**, *136*, 041018. [CrossRef]

28. Riéra, W.; Marty, J.; Castillon, L.; Deck, S. Zonal detached-eddy simulation applied to the tip-clearance flow in an axial compressor. *AIAA J.* **2016**, *54*, 2377–2391. [CrossRef]

29. Yamada, K.; Furukawa, M.; Tamura, Y.; Saito, S.; Matsuoka, A.; Nakayama, K. Large-scale detached-eddy simulation analysis of stall inception process in a multistage axial flow compressor. *J. Turbomach.* **2017**, *139*, 071002. [CrossRef]

30. Lin, D.; Su, X.; Yuan, X. DDES analysis of the wake vortex related unsteadiness and losses in the environment of a high-pressure turbine stage. *J. Turbomach.* **2018**, *140*, 041001. [CrossRef]

31. Nie, C.; Xu, G.; Cheng, X.; Chen, J. Micro air injection and its unsteady response in a low-speed axial compressor. In *Proceedings of the ASME Turbo Expo 2002: Power for Land, Sea, and Air, Amsterdam, The Netherlands, 3–6 June 2002*; ASME: New York, NY, USA, 2002; Volume 5, pp. 343–352. [CrossRef]

32. Shah, P.N.; Tan, C.S. Effect of blade passage surface heat extraction on axial compressor performance. In *Proceedings of the ASME Turbo Expo 2005: Power for Land, Sea, and Air, Reno, NV, USA, 6–9 June 2005*; ASME: New York, NY, USA, 2005; Volume 6, pp. 327–341. [CrossRef]

33. Zhang, H.; Deng, X.; Lin, F.; Chen, J.; Huang, W. A study on the mechanism of tip leakage flow unsteadiness in an isolated compressor rotor. In *Proceedings of the ASME Turbo Expo 2005: Power for Land, Sea, and Air, Barcelona, Spain, 8–11 May 2006*; ASME: New York, NY, USA, 2005; Volume 6, pp. 435–445. [CrossRef]

34. Liu, Y.; Yu, X.; Liu, B. Turbulence models assessment for large-scale tip vortices in an axial compressor rotor. *J. Propuls. Power* **2008**, *24*, 15–25. [CrossRef]

35. Sun, L.; Zheng, Q.; Li, Y.; Bhargava, R. Understanding effects of wet compression on separated flow behavior in an axial compressor stage using CFD analysis. *J. Turbomach.* **2011**, *133*, 031026. [CrossRef]

36. Suman, A.; Kurz, R.; Aldi, N.; Morini, M.; Brun, K.; Pinelli, M.; Ruggero Spina, P. Quantitative computational fluid dynamics analyses of particle deposition on a transonic axial compressor blade—Part I: Particle zones impact. *J. Turbomach.* **2014**, *137*, 021009. [CrossRef]

37. Cameron, J.D.; Bennington, M.A.; Ross, M.H.; Morris, S.C.; Du, J.; Lin, F.; Chen, J. The influence of tip clearance momentum flux on stall inception in a high-speed axial compressor. *J. Turbomach.* **2013**, *135*, 051005. [CrossRef]

38. Sutherland, W. The viscosity of gases and molecular force. *Philos. Mag. Ser. 5* **1893**, *36*, 507–531. [CrossRef]

39. Spalart, P.; Jou, W.; Strelets, M.; Allmaras, S. Comments of feasibility of LES for wings, and on a hybrid RANS/LES approach. In *Advances in DNS/LES*; Greyden Press: Columbus, OH, USA, 1997; p. 1.

40. Shur, M.; Spalart, P.; Strelets, M.; Travin, A. Detached-eddy simulation of an airfoil at high angle of attack. In *Engineering Turbulence Modelling and Experiments 4*; Elsevier: Amsterdam, The Netherlands, 1999; pp. 669–678. [CrossRef]

41. Spalart, P.; Allmaras, S. A one-equation turbulence model for aerodynamic flows. In Proceedings of the 30th Aerospace Sciences Meeting and Exhibit, Reno, NV, USA, 6–9 January 1992; pp. 5–21. [CrossRef]

42. Lilly, D.K. A proposed modification of the Germano subgrid-scale closure method. *Phys. Fluids Fluid Dyn.* **1992**, *4*, 633–635. [CrossRef]

43. Du, J.; Lin, F.; Zhang, H.; Chen, J. Numerical investigation on the self-induced unsteadiness in tip leakage flow for a transonic fan rotor. *J. Turbomach.* **2010**, *132*, 021017. [CrossRef]

44. Yamada, K.; Furukawa, M.; Nakano, T.; Inoue, M.; Funazaki, K. Unsteady three-dimensional flow phenomena due to breakdown of tip leakage vortex in a transonic axial compressor rotor. In *Proceedings of the ASME Turbo Expo 2004: Power for Land, Sea, and Air, Vienna, Austria, 14–17 June 2004*; ASME: New York, NY, USA, 2004; Volume 5, pp. 515–526. [CrossRef]

45. Puterbaugh, S.L.; Brendel, M. Tipclearance flow-shock interaction in a transonic compressor rotor. *J. Propuls. Power* **1997**, *13*, 24–30. [CrossRef]

46. Brandstetter, C.; Schiffer, H.P. PIV measurements of the transient flow structure in the tip region of a transonic compressor near stability limit. *J. Global Power Propuls. Soc.* **2018**, *2*, 303–316. [CrossRef]

47. Cattafesta, L.; Settles, G. Experiments on shock/vortex interactions. In Proceedings of the 30th Aerospace Sciences Meeting and Exhibit, Reno, NV, USA, 6–9 January 1992. [CrossRef]

48. Delery, J.M. Aspects of vortex breakdown. *Prog. Aerosp. Sci.* **1994**, *30*, 1–59. [CrossRef]

49. Smart, M.K.; Kalkhoran, I.M. Flow model for predicting normal shock wave induced vortex breakdown. *AIAA J.* **1997**, *35*, 1589–1596. [CrossRef]

50. Kalkhoran, I.M.; Smart, M.K. Aspects of shock wave-induced vortex breakdown. *Progress Aerosp. Sci.* **2000**, *36*, 63–95. [CrossRef]

51. Delbende, I.; Chomaz, J.M.; Huerre, P. Absolute/convective instabilities in the Batchelor vortex: A numerical study of the linear impulse response. *J. Fluid Mech.* **1998**, *355*, 229–254. [CrossRef]

52. Ruith, M.R.; Chen, P.; Meiburg, E.; Maxworthy, T. Three-dimensional vortex breakdown in swirling jets and wakes: Direct numerical simulation. *J. Fluid Mech.* **2003**, *486*, 331–378. [CrossRef]
53. Gallaire, F.; Ruith, M.; Meiburg, E.; Chomaz, J.M.; Huerre, P. Spiral vortex breakdown as a global mode. *J. Fluid Mech.* **2006**, *549*, 71–80. [CrossRef]
54. Ortega-Casanova, J.; Fernandez-Feria, R. Three-dimensional transitions in a swirling jet impinging against a solid wall at moderate Reynolds numbers. *Phys. Fluids* **2009**, *21*. [CrossRef]
55. Meliga, P.; Gallaire, F.; Chomaz, J.M. A weakly nonlinear mechanism for mode selection in swirling jets. *J. Fluid Mech.* **2012**, *699*, 216–262. [CrossRef]
56. Leibovich, S. The structure of vortex breakdown. *Ann. Rev. Fluid Mech.* **1978**, *10*, 221–246. [CrossRef]
57. Pasche, S.; Gallaire, F.; Dreyer, M.; Farhat, M. Obstacle-induced spiral vortex breakdown. *Exp. Fluids* **2014**, *55*. [CrossRef]
58. Mahesh, K. A model for the onset of breakdown in an axisymmetric compressible vortex. *Phys. Fluids* **1996**, *8*, 3338–3345. [CrossRef]

Article

Effect of Rotor Thrust on the Average Tower Drag of Downwind Turbines

Shigeo Yoshida [1],* [ID], Kazuyuki Fujii [2], Masahiro Hamasaki [1] and Ao Takada [1]

[1] Research Institute for Applied Mechanics, Kyushu University, 6-1 Kasugakoen, Kasuga, Fukuoka 816-8580, Japan; hamasaki@riam.kyushu-u.ac.jp (M.H.); takada@riam.kyushu-u.ac.jp (A.T.)

[2] Interdisciplinary Graduate School of Engineering and Science, Kyushu University, 6-1 Kasugakoen, Kasuga, Fukuoka 816-8580, Japan; fujiik@riam.kyushu-u.ac.jp

* Correspondence: yoshidas@riam.kyushu-u.ac.jp; Tel. +81-92-583-7747

Received: 28 November 2018; Accepted: 8 January 2019; Published: 12 January 2019

Abstract: A new analysis method to calculate the rotor-induced average tower drag of downwind turbines in the blade element momentum (BEM) method was developed in this study. The method involves two parts: calculation of the wind speed distribution using computational fluid dynamics, with the rotor modeled as a uniform loaded actuator disc, and calculation of the tower drag via the strip theory. The latter calculation considers two parameters, that is, the decrease in wind speed and the pressure gradient caused by the rotor thrust. The present method was validated by a wind tunnel test. Unlike the former BEM, which assumes the tower drag to be constant, the results obtained by the proposed method demonstrate much better agreement with the results of the wind tunnel test, with an accuracy of 30%.

Keywords: wind turbine; downwind; tower shadow; load; tower; BEM; actuator disc

---

1. Introduction

Horizontal axis wind turbines are categorized as upwind and downwind turbines according to the position of the rotor relative to the tower. Among these, upwind turbines have been predominant throughout the 30 year history of commercial wind turbines. The most essential reason for the unpopularity of the downwind turbines is the presence of a strong aerodynamic interaction between the rotor and the tower, which is commonly known as the "tower shadow effect". This phenomenon generates impulsive loads and infrasound when one of the blades passes through the wake of the tower [1]. The modeling of the tower shadow effect is a critical technical problem of downwind turbines.

Design loads are calculated on the basis of the international design standard IEC61400-1 [2] or the guidelines for certification bodies DNV GL [3]. In a large number of cases in which the design load is combined with the wind model, the wind turbines can experience failure conditions, as well as various types of wind and marine conditions. The flexibilities of the structure and the controls, in addition to the aerodynamics, hydrodynamics, aero-elastics, and control of the wind turbines, considerably influence the load. The blade element momentum (BEM) method is the most popular tool for determining these characteristics [4,5]. Therefore, the modeling of the tower shadow effect in the BEM is one of the most important technical challenges in the design and development of downwind wind turbines. However, most of the tower shadow models are too simple to express such phenomena, as they consider only the wind speed profile behind an isolated tower [5,6] and not the aerodynamic interaction between the rotor and the tower.

Several studies have been conducted until now that have focused on the variable loads of the downwind turbines caused by the tower shadow effect. Zhao et al. [7] studied the loads on a tower by performing a comparison between two- and three-bladed upwind and downwind rotors in two different rotor speed conditions using computational fluid dynamics (CFD). The peak to average values of the hub bending moment and the tower base bending moment of the downwind turbine were shown to be 5–6 times and 2.5 times larger, respectively, than those of the upwind turbine. Zahle et al. [8] conducted 2-dimensional CFD analysis for three typical tower concepts of downwind turbines as well as for an upwind turbine; a steep load variation was noted for the downwind turbine, and the implementation of streamlined and four-leg towers was shown to be an effective technique to reduce the load variation caused by the tower shadow effects. Matiz-Chicacausa and Lopez [9] investigated the possibilities of the application of the actuator line model (ALM) to the CFD analysis of the tower shadow effects. The ALM was shown to be useful for load estimation of a downwind turbine under specific conditions.

Although the abovementioned research contributed to the expansion of the understanding of the phenomenon, the authors did not aim to develop models for extensive load calculations based on the design standards and guidelines. Wang and Coton [10] developed a high-resolution tower shadow model for downwind turbines based on the prescribed wake vortex model and an efficient near wake dynamics model of the vorticity trailing the blades. The model demonstrated reasonable agreement with the experimental results, with the exception of the blades passing through the tower wake. However, the load of the tower was not discussed in this research. Yoshida and Kiyoki [11,12] developed the load equivalent tower shadow modeling method. The method considered a bell-shaped wind speed profile behind the tower, which helped obtain the equivalent rotor load variation with the wind turbine CFD analysis. This model was applied to the development of the first multimegawatt-scale commercial downwind turbines, including the SUBARU 80/20 [13] and, later, the Hitachi 2 MW, to realize high performance and safety in complex terrains [14,15]. The technology was extended to the later produced Hitachi 5 MW [16] offshore wind turbines. The model has been successfully applied in engineering applications so far; however, two major problems remained unsolved. The first concern was that the CFD modeling of the rotor–tower configuration required for each operating condition was time-consuming, which made the technique inconvenient for use in the design optimization process. The second issue was that no model was established for the tower shadow effects on the tower drag. Considering the situation, the tower load variation caused by the tower shadow effect was modeled by Yoshida by combining the BEM and the Lifting Line Theory [17]. However, the average tower drag was still not discussed in that research.

Considering these situations, this paper proposes a BEM model for the tower average drag. The formulation is presented in Section 2, and the validation of the method by a wind tunnel test is discussed in Sections 3–5.

2. Methodology

The formulation of the average loads of the towers of the downwind turbine by using the BEM is discussed in this chapter. The following two calculations are considered:

(1)   Wind speed distribution by CFD
(2)   Rotor thrust-induced tower average drag

2.1. Wind Speed Distribution by CFD

The wind speed distribution in front of the rotor is calculated using CFD. Figure 1 shows the top view of the rotor and the tower of a downwind turbine. The tower, which has the diameter D_T, is located in front of the rotor. The free stream wind speed U_0 decreases to $U_0(1 - a)$ at the rotor plane as a result of the rotor thrust. Here, a is the axial induction factor. Figure 2 is the schematic of the CFD involved in the present study. The tower is not included in the model, but the position is termed "virtual tower position" in this research. Further, u_T is the wind speed at the virtual tower position, which is between U_0 and $U_0(1 - a)$, depending on the condition. The rotor is modeled by an actuator disc model (ADM). The load distribution is assumed to be uniform to realize a convenient application to the BEM. The thrust per unit volume σ of the ADM is as shown in Equation (1):

$$\sigma = \frac{1}{2\pi r w}\frac{dT}{dr} \tag{1}$$

where w is the thickness of the ADM.

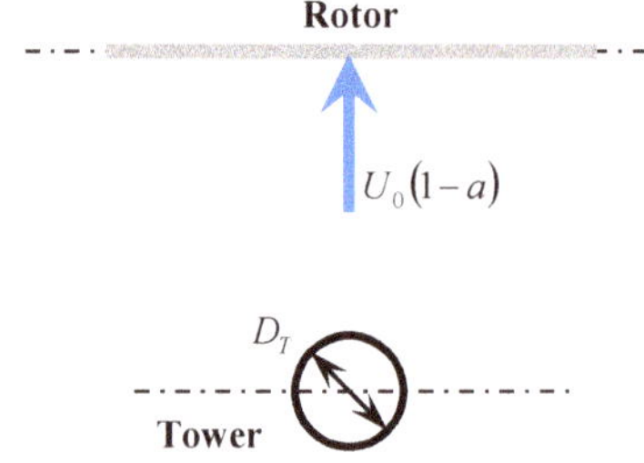

Figure 1. Top view of a downwind turbine. U_0, free stream wind speed, D_T, tower diameter.

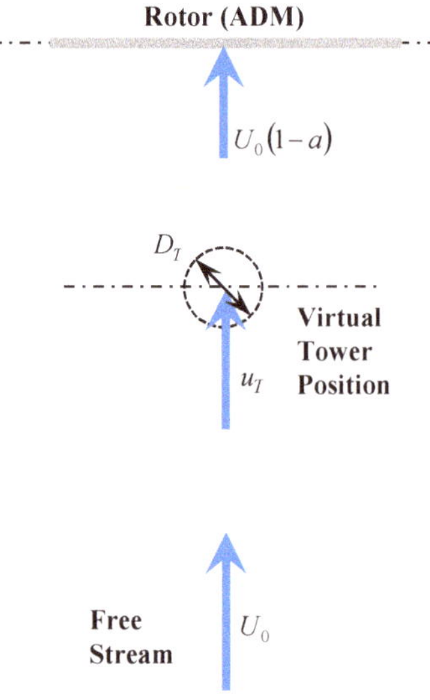

Figure 2. Computational fluid dynamics (CFD) with actuator disc model (ADM). u_T, wind speed at the virtual tower position.

Energies **2019**, 12, 227

2.2. Rotor Thrust-Induced Tower Average Drag

The wind speed at the tower and the tower drag decrease as the rotor thrust increases due to the aerodynamic interaction. Two factors, that is, the wind speed and the ambient pressure gradient caused by the rotor thrust, are considered in this research. The formulation is based on the assumption of the strip theory, according to which the flow between the sections normal to the tower axis is not taken into account.

(1) Wind Speed Effect

The tower drag decreases proportionally to the square of the wind speed when the tower section drag coefficient remains constant. This influence is modeled herein.

The section drag f_{XT0} of the isolated tower with no rotor interaction is as shown in Equation (2):

$$f_{XT0} = \frac{1}{2} \rho U_0^2 D_T C_{dT0} \tag{2}$$

where C_{dT0} is the drag coefficient at the tower section.

In case the effect of the rotor thrust comes into play, the wind speed at the tower position decreases. Furthermore, the tower section drag also decreases from f_{XT0} to f_{XT}. This is expressed in the following two ways, based on the wind speeds at the free stream and at the virtual tower position, as shown in Equation (3):

$$f_{XT} = \frac{1}{2} \rho U_0^2 D_T C_{dT} = \frac{1}{2} \rho u_T^2 D_T C_{dT0} \tag{3}$$

where C_{dT} is the tower section drag based on the free stream wind speed, and u_T is the wind speed at the virtual tower position, which is calculated by CFD without incorporating a tower model, as explained in the previous section. Here, the drag coefficient C_{dT0} is assumed to be constant as in Equation (2).

Therefore, the tower section drag deviation induced by the rotor thrust Δf_{XT} is calculated as in Equation (4):

$$\Delta f_{XTV} - f_{XT} - f_{XT0} - -\frac{1}{2} \rho U_0^2 D_T C_{dT0} \left(1 - \mu_T^2 \right) \tag{2}$$

where the normalized wind speed at the virtual tower is

$$\mu_T = \frac{u_T}{U_0}$$

Therefore, the change in the relevant drag coefficient caused by the wind speed change ΔC_{dTV} is calculated by Equation (4), as given in Equation (5):

$$\Delta C_{dTV} = -C_{dT0} \left(1 - \mu_T^2 \right) \tag{3}$$

This indicates that the term of the tower section drag decreases as the normalized wind speed at the virtual tower, induced by the rotor thrust, decreases.

(2) Effect of the Ambient Pressure Gradient

The tower drag is also dependent on the ambient pressure gradient around the tower. This influence is modeled herein.

The pressure at the virtual tower position p_T is calculated by Bernoulli's law, as in Equation (4):

$$p_0 + \frac{1}{2} \rho U_0^2 = p_T + \frac{1}{2} \rho u_T^2 \tag{4}$$

where p_0 and p_T are the pressures at the free stream and the virtual tower center, respectively. Therefore, the windward pressure gradient $\partial p_T / \partial x_T$ at the virtual tower center is calculated as shown in Equation (5) by the differential of Equation (4):

$$\frac{\partial p_T}{\partial x_T} = -\rho u_T \frac{\partial u_T}{\partial x_T} \tag{5}$$

The section drag by the pressure gradient at the tower section Δf_{XTP} is calculated as in Equation (6) by employing the pressure deviation from the front and back sides of the tower and the tower diameter. The pressure deviation is calculated by the pressure gradient and the windward reference distance Δx_T:

$$\Delta f_{XTP} = \frac{\partial p_T}{\partial x_T} \Delta x_T D_T \tag{6}$$

Assuming a uniform pressure gradient, the tower section drag is expressed as in Equation (7):

$$\Delta f_{XTP} = \int_{-\pi/2}^{\pi/2} \left(\frac{\partial p_T}{\partial x_T} D_T \cos \phi_T \right) \left(\frac{D_T}{2} \cos \phi_T \right) d\phi_T \tag{7}$$

The reference distance is calculated as in Equation (8) using Equations (6) and (7):

$$\Delta x_T = \frac{\pi}{4} D_T \tag{8}$$

Therefore, the change in the tower section drag owing to the pressure gradient is calculated as in Equation (9):

$$\Delta f_{XTP} = \frac{\pi D_T{}^2 \rho u_T}{4} \frac{\partial u_T}{\partial x_T} = \frac{\pi D_T \rho U_0{}^2 \mu_T}{4} \frac{\partial \mu_T}{\partial \zeta_T} \tag{9}$$

where ζ_T is the normalized distance and can be written as

$$\zeta_T = \frac{x_T}{D_T}$$

Therefore, the change in the relevant drag coefficient caused by the pressure gradient ΔC_{dTP} can be written as in Equation (10):

$$\Delta C_{dTP} = \frac{\pi}{2} \mu_T \frac{\partial \mu_T}{\partial \zeta_T} \tag{10}$$

This indicates that the term of the tower section drag decreases as the negative pressure gradient caused by the rotor thrust increases.

(3) Total Average Tower Drag

From (1) and (2), the deviation of the drag Δf_{XT} and the drag coefficient ΔC_{dT} induced by the rotor thrust are derived as Equations (11) and (12), respectively:

$$\Delta f_{XT} = \Delta f_{XTV} + \Delta f_{XTP} = \frac{1}{2} \rho U_0{}^2 D_T \left[-C_{dT0} \left(1 - \mu_T{}^2 \right) + \frac{\pi}{2} \mu_T \frac{\partial \mu_T}{\partial \zeta_T} \right] \tag{11}$$

$$\Delta C_{dT} = \Delta C_{dTV} + \Delta C_{dTP} = -C_{dT0} \left(1 - \mu_T{}^2 \right) + \frac{\pi}{2} \mu_T \frac{\partial \mu_T}{\partial \zeta_T} \tag{12}$$

3. Wind Tunnel Test

3.1. Outline

A wind tunnel test for a wind turbine was conducted to validate the theory described in the previous chapter. The rotor–tower interaction was simulated by a dummy tower placed in front of the rotor of the upwind turbine.

3.2. Facility

The Boundary Layer Wind Tunnel at the Research Institute for Applied Mechanics, Kyushu University [18], was used for the test. The wind tunnel has a test section with a width of 3.6 m, a height of 2.0 m, and a length of 15 m. The maximum wind speed is 30 m/s.

3.3. Test Model

The general specifications and the outline of the model and the tower installed in the wind tunnel are presented in Table 1. The blockage ratio, which is the ratio of the rotor area to the cross section of the wind tunnel, is as small as 5.3%. The rotor speed is controlled by the variable dump load. Figure 3 shows the schematic of the test model and the dummy tower. Although the model is an upwind turbine supported by a tower with a diameter of 50.8 mm, a dummy tower is installed in front of the model to mimic the downwind turbine tower. The diameter of the dummy tower is 64 mm, which is 200% the value of the blade chord length. The dummy tower was placed at two positions; $-6D_T$ and $-4D_T$. Figure 4 shows an image of the test model. The dummy tower is shown as located in front of the turbine.

Table 1. Specifications of the test model and the dummy tower.

Item	Specification
Number of blades	2
Rotor diameter D	700 mm
Rotor radius R	350 mm
Tilt angle	0 deg
Coning angles	0 deg
Airfoil	NACA0013
Blade chord length	32 mm
Twist angle	0 deg
Tower diameter D_T	64 mm
Tower position	-338 mm $(-6D_T)$ -225 mm $(-4D_T)$

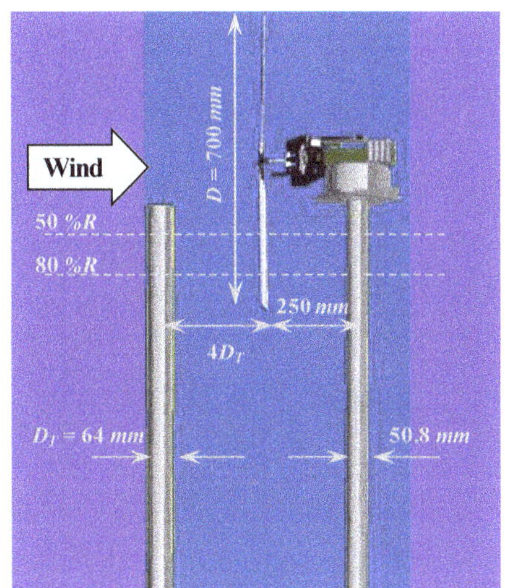

Figure 3. Schematic of the test model (dummy tower at $-4D_T$).

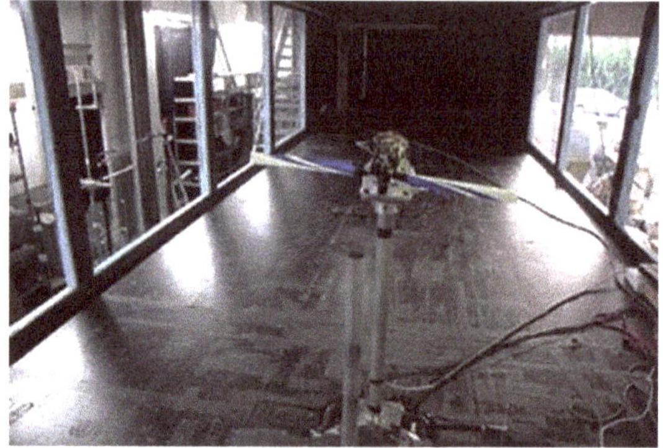

Figure 4. Test model and dummy tower installed in the wind tunnel.

3.4. Measurement

- Rotor speed: Magnetic encoder
- Load (top of the tower): 6-axis Load Cell (Nissho)
- Pressure on the dummy tower (50%R, 80%R, 8 points each): Pressure sensor (Otegiken)

3.5. Test Conditions

- Wind speed (U_0): 6 m/s (uniform, steady)
- Yaw angle: 0 deg
- Blade pitch angle (θ): 4, 6, 8 deg
- Tip speed ratio (λ): 6.6–9.2

3.6. Test Results

The measurements for the power coefficient C_P and the thrust coefficient C_T for the isolated rotor model are shown in Figure 5. The rotor torque was calculated from the rolling moment at the top of the tower. The value of thrust force at the top of the tower was used as the rotor thrust. The maximum value of C_P was noted at approximately $\lambda = 8$ and $\theta = 6$ deg. However, C_T tended to increase with an increase in the tip speed ratio or a decrease in the pitch angle.

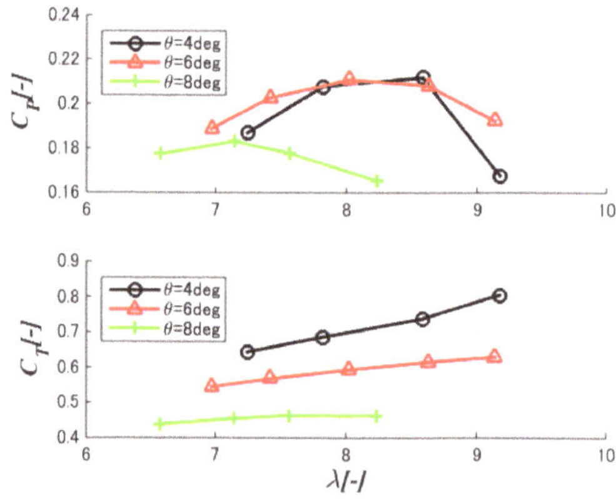

Figure 5. Relationship between the power (C_P) and thrust (C_T) coefficients and the tip speed ratio. θ, blade pitch angle.

The pressure distributions on the dummy tower in typical cases are shown in Figure 6. Here, 0 deg indicates the upwind side of the tower. The pressure on the 50%*R* was relatively higher than that at 80%*R*. In general, the pressures around the downwind side of the tower (180 deg) tended to be higher as the rotor thrust was larger.

The details of other characteristics are discussed in Section 5.

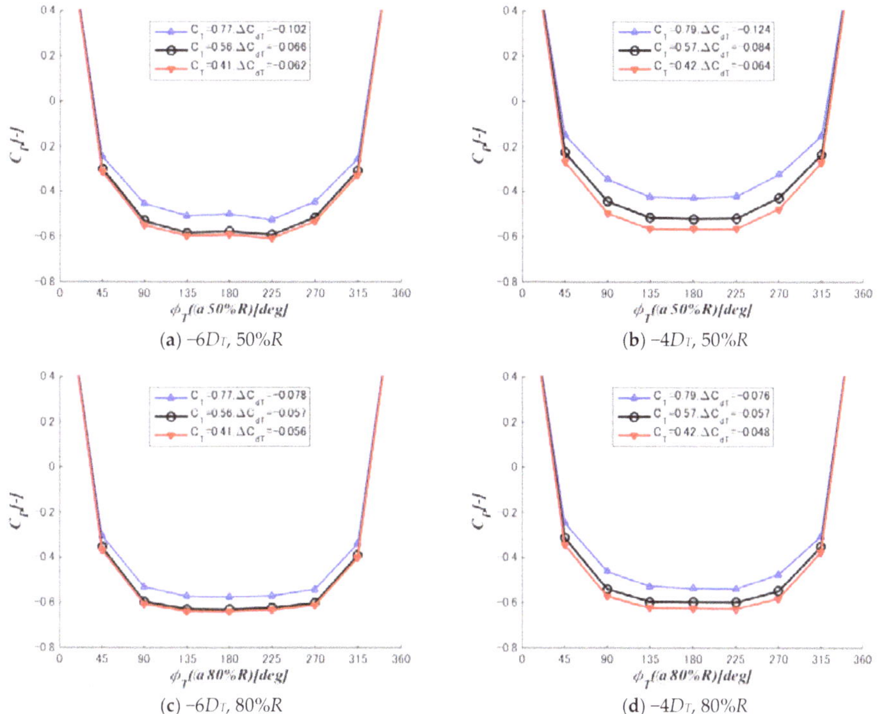

Figure 6. Distribution of the pressure coefficient on the dummy tower in typical cases.

4. Analysis

4.1. Outline

The relationship between the rotor thrust and the tower drag coefficient was determined by the proposed method described in Section 2. The model is generally the same as the one described in Section 3.3; however, the influence of the support was neglected as the support was located far downwind of the rotor, and this distance was fairly large compared to the support's diameter. In addition, the influence of the nacelle was neglected, as it does not affect the outboard sections, as discussed in this study.

4.2. CFD Outline

A CFD analysis was conducted for the rotor using ANSYS CFX [19] considering the k-ω SST turbulence model. The rotor was modeled by an ADM with uniform load distribution.

The simulation domain and the boundary conditions are summarized in Table 2. The model consisted of structured cells around the ADM with $y+$ is about 1 and unstructured ones otherwise. The total number of cells was approximately 50 million. The typical cells are shown in Figures 7 and 8.

Table 2. Simulation domain and boundary conditions.

Boundary	Distance from the Rotor Center	Boundary Condition
Rotor	$-D/140{\sim}+D/140$	ADM
Tower	Not available	Not available
Inlet	$-4D$	Wind speed (uniform)
Outlet	$+5D$	Ambient pressure
Bottom	$-2D$	Slip
Top	$+2D$	Slip
Side	$-2D, +2D$	Slip

"D": Rotor diameter.

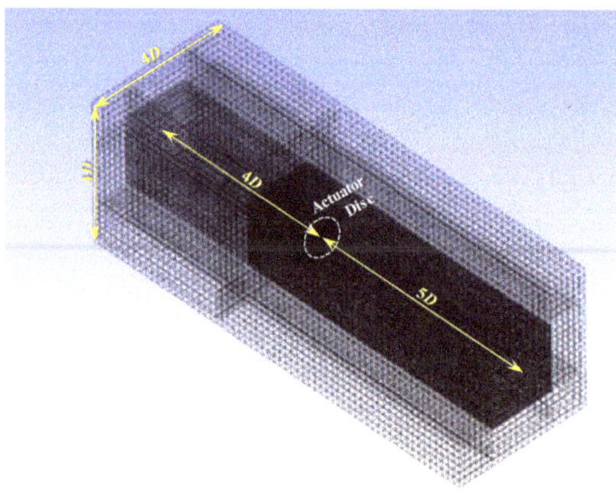

Figure 7. CFD domain.

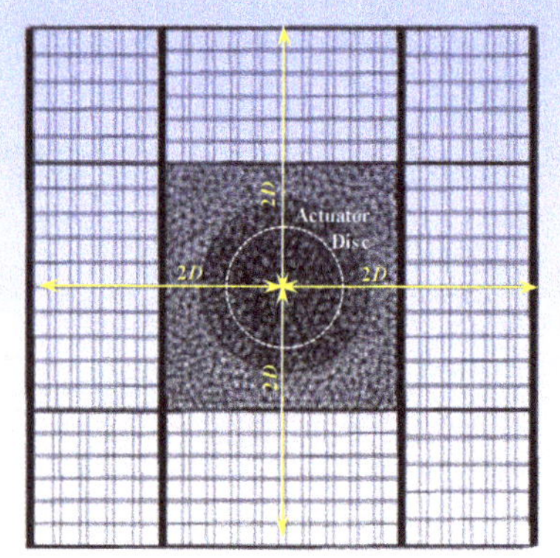

Figure 8. Cells around the rotor.

4.3. Analysis Conditions

- Wind speed (U_0): 6 m/s (uniform, steady)
- Yaw angle: 0 deg
- Thrust coefficient (C_T): 0.4–0.9 (each 1.0)

4.4. Wind Speed Distributions in Front of the Rotor

The wind speed distributions in typical conditions, $C_T = 0.4$ and 0.9, are shown in Figures 9 and 10. The circles in these figures denote the positions of the dummy towers at $-6D_T$ and $-4D_T$. The wind speed tended to decrease in front of the tower in general, and the wind speed was lower in the vicinity of the rotor. A comparison between the two conditions shows that the wind speed around the dummy tower decreased as C_T increased. A comparison between the two sections shows that the wind speed in front of the rotor was lower at 50%R compared to that at 80%R. The load was uniform on the rotor, and the inboard sections were considerably affected by the rotor.

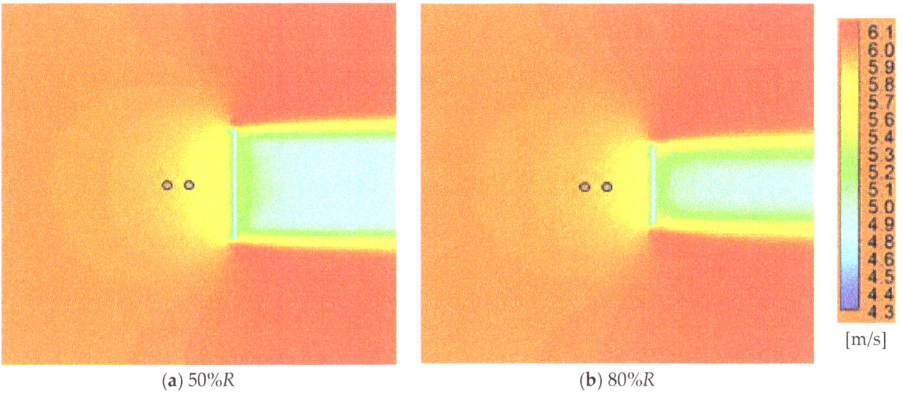

(a) 50%R   (b) 80%R

Figure 9. Wind speed distribution in front of the rotor at $C_T = 0.4$, $U_0 = 6$ m/s. (The circles indicate the dummy tower positions at $-6D_T$ and $-4D_T$).

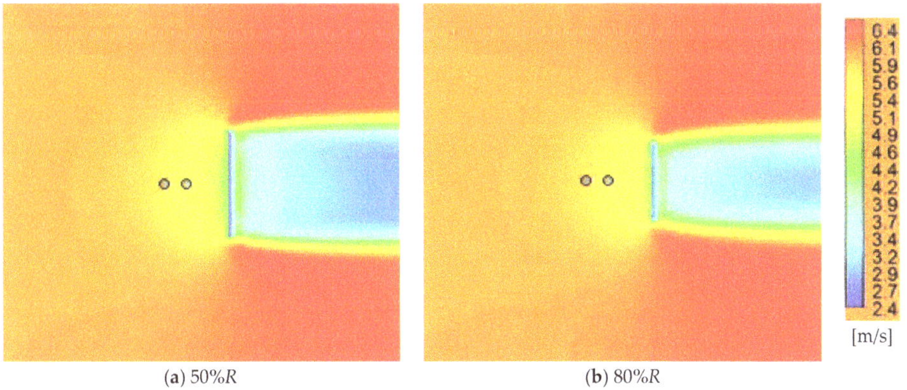

(a) 50%R   (b) 80%R

Figure 10. Wind speed distribution in front of the rotor at $C_T = 0.9$, $U_0 = 6$ m/s. (The circles indicate the dummy tower positions at $-6D_T$ and $-4D_T$).

The distributions of the normalized wind speed μ_T and its differential $\partial\mu_T/\partial\xi_T$ with respect to the normalized distance ξ_T in front of the rotor are shown in Figures 11 and 12. The distributions at $C_T = 0.4$ and 0.9 were identical to those in Figures 9 and 10 in the symmetrical plane. The top subplots correspond to the distributions at 50%R, and the bottom ones correspond to those at 80%R. The wind speed was lower in the vicinity of the rotor and decreased as the thrust coefficient increased, as mentioned above.

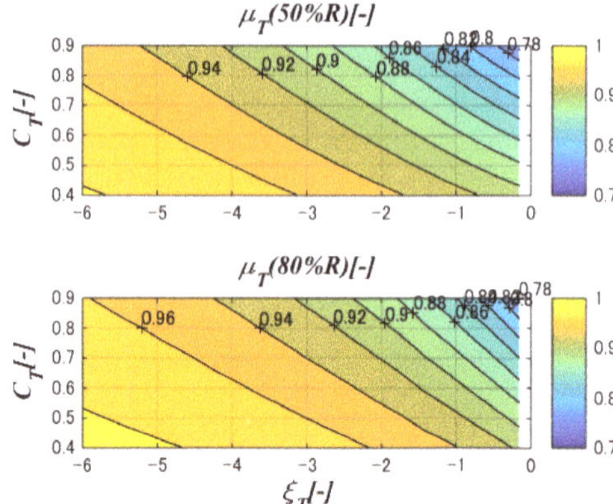

Figure 11. Wind speed in front of the rotor as obtained by CFD. μ_T, normalized wind speed, ξ_T, normalized distance.

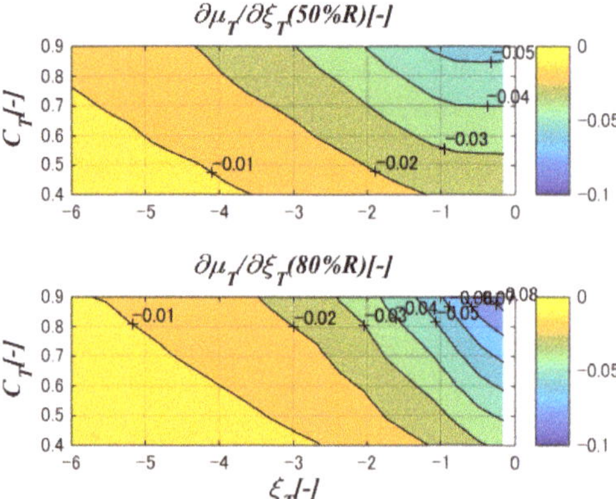

Figure 12. Spatial differential of wind speed in front of the rotor ($\partial\mu_T/\partial\xi_T$) as obtained by CFD.

4.5. Drag Coefficients of the Virtual Tower

The deviation of the drag coefficient of the virtual tower calculated by the present theory is shown in Figure 13. The x-axis denotes the virtual tower position, and the y-axis denotes the thrust coefficient. The diameter of the tower D_T was assumed to be 64 mm, as in the experiment. As indicated by Figures 11 and 12, the drag coefficient tended to be smaller as the rotor–tower clearance decreased and the thrust coefficient increased. The drag tended to be smaller at 50%R than at 80%R.

As shown in Equation (14), the rotor thrust-induced tower drag deviation consisted of the wind speed term and the pressure gradient term. Figure 14 shows the share of the wind speed term to the total deviation. It indicates that approximately 80% of the change in the drag wss caused by the first term of Equation (12), i.e., the decrease in the wind speed, rather than the pressure gradient.

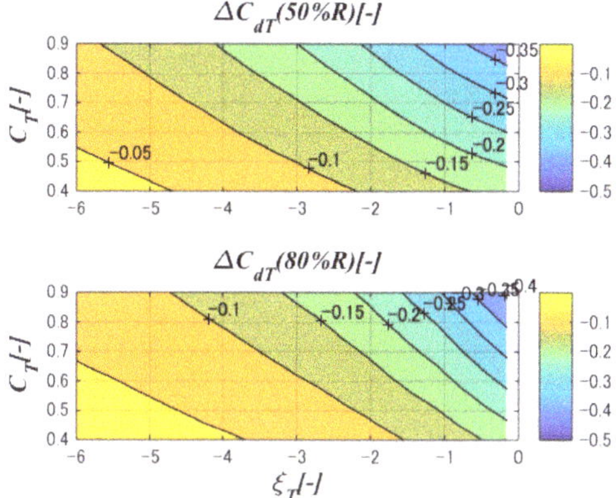

Figure 13. Relationship between the tower section drag and rotor thrust and the virtual tower position. ΔC_{dT}, drag coefficient.

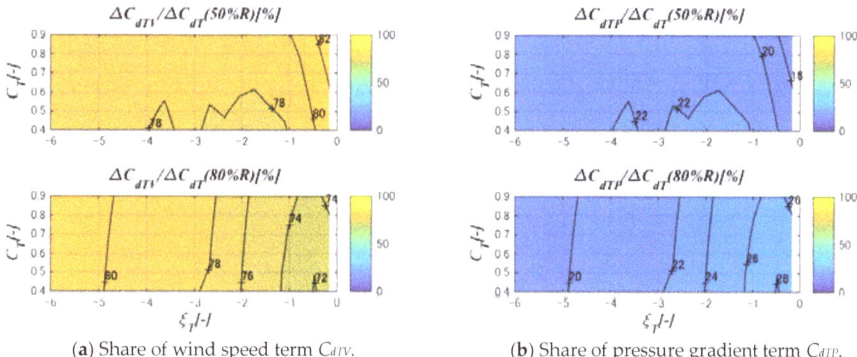

(**a**) Share of wind speed term C_{dTV}.

(**b**) Share of pressure gradient term C_{dTP}.

Figure 14. Share of the wind speed term C_{dTV} in the total C_{dT} tower section drag coefficient.

5. Validation

The thrust-induced drag deviations ΔC_{dT} to the thrust coefficient at $-6D_T$ and $-4D_T$ are shown in Figure 15. The lines are approximations for which the intercepts are at $\Delta C_{dT} = 0$ and $C_T = 0$ by definition. The ΔC_{dT} values for all the cases are shown to be almost proportional to C_T. The proportion factors are summarized in Table 3. The factors of the former BEM are zero, as the method does not consider the influence of the rotor thrust on the tower drag. On the other hand, the proposed method shows good agreement with the test. The drag coefficients tended to decrease as the rotor thrust increased. The drag coefficient at the 50%R section was smaller than at the 80%R section. Further, the drag coefficient decreased as the tower was placed closer to the rotor.

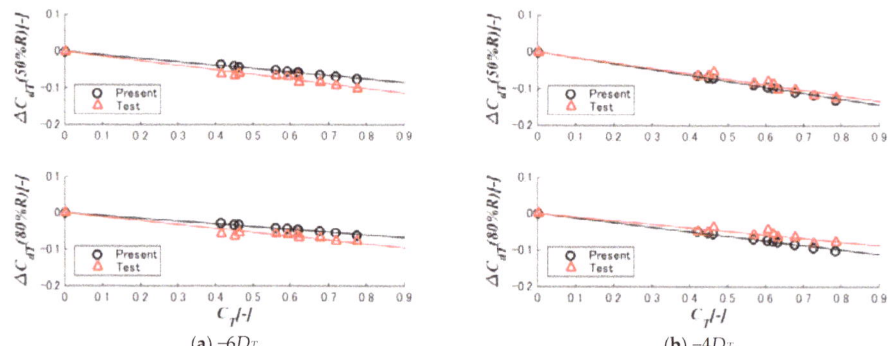

Figure 15. Relationship between tower drag and rotor thrust.

Table 3. Fraction of the deviation of the tower drag coefficient to the thrust coefficient.

Dummy Tower Position	Tower Section	Test	Present	
		$\Delta C_{dT}/C_T$ [-]	$\Delta C_{dT}/C_T$ [-]	Deviation from the Test
$-6D_T$	50%R	−0.1270	−0.0937	−26.2%
	80%R	−0.1058	−0.0741	−30.0%
$-4D_T$	50%R	−0.1484	−0.1594	+7.4%
	80%R	−0.0961	−0.1225	+27.5%

6. Conclusions

A novel analysis method to calculate the average tower drag of downwind turbines, which considers the rotor-induced average tower drag coefficient, was developed. It consists of two terms, that is, the decrease in the wind speed and the pressure gradient caused by the rotor thrust. The rotor thrust distribution is assumed to be uniform. The method was validated by a wind tunnel test. Unlike the former blade element momentum (BEM) method, which assumes the tower drag to be constant, the proposed method demonstrates a much better agreement with the wind tunnel test, with an accuracy of up to 30%. The results show that the tower drag decreases proportionally to the rotor thrust. With regard to the sensitivity of the tower drag with respect to the rotor thrust, the following characteristics were noted:

- The drag coefficient was larger in the inboard section (50%R) than in the outboard section (80%R).
- The drag coefficient decreased as the tower was positioned closer to the rotor.
- Of the two terms influencing the deviation of the tower section drag, the effect of the decrease in wind speed was more dominant in leading to the decrease in the tower section drag.

By preparing the database of the relationship between the rotor thrust coefficient for the wind speed distribution in front of the rotor, the proposed method is expected to improve the accuracy of the load calculation in the BEM.

In future studies, the present model is planned to be extended to variable loads of wind turbine blades in succession.

Author Contributions: S.Y. organized the research, formulation, and validation. K.F. conducted the computational fluid dynamics simulations, and M.H. and A.T. conducted the wind tunnel test.

Funding: The wind tunnel test was conducted with funding from the "Advanced Practical Research and Development of Wind Power Generation/Advanced Practical Development of Wind Turbine Components/ Research on Over 10 MW Class Wind Turbines (Whole Design)" of the New Energy and Industrial Technology Development Organization (NEDO) in 2013–2014.

Acknowledgments: The authors express gratitude to Keiji Matsushima, Kimihiko Watanabe, and Yuji Ohya for their support for the wind tunnel test. The authors would also like to thank Masataka Motoyama and Omar Ibrahim for their support in the computational fluid dynamics simulation.

Conflicts of Interest: The authors declare no conflicts of interest.

Nomenclature

C_{dT}	Tower drag coefficient
C_{dT0}	Tower drag coefficient with no rotor thrust
C_P	Power coefficient
C_p	Pressure coefficient of the tower
C_T	Thrust coefficient
D	Rotor diameter
D_T	Tower diameter
f_{XT}	Tower section drag
f_{XT0}	Tower section drag with no rotor thrust
p_0	Free stream pressure
p_T	Pressure at the tower position
R	Rotor radius
r	Station radius of the blade element
T	Rotor thrust
U_0	Free stream wind speed
u_T	Wind speed at the tower position
w	Thickness of the actuator disc
T	Longitudinal (or windward) position
C_{dT}	Deviation of the tower drag coefficient from that of the isolated tower
ΔC_{dTP}	Deviation of the tower drag coefficient by the effect of the pressure gradient
ΔC_{dTV}	Deviation of the tower drag coefficient by the wind speed
Δf_{XT}	Deviation of the tower section drag from that of the isolated tower
Δf_{XTP}	Deviation of the tower section drag by the effect of the pressure gradient
Δf_{XTV}	Deviation of the tower section drag by the wind speed
θ	Blade pitch angle
λ	Tip speed ratio
μ_T	Normalized wind speed at the tower position
ξ_T	Position normalized by the tower diameter
ρ	Air density
ϕ_T	Azimuth position around the tower (0 deg in front, clockwise around the vertical axis)
ADM	Actuator disc model
ALM	Actuator line model
BEM	Blade element momentum (method)
CFD	Computational fluid dynamics

References

1. Gipe, P. *Wind Energy Comes of Age*; John Wiley and Sons: Chichester, UK, 1995.
2. *Wind Turbines—Part 1: Design Requirements*; IEC61400-1, Ed. 3; International Electrotechnical Commission: Geneva, Switzerland, 2005.
3. Germanischer Lloyd. *Guidelines for the Certifications of Wind Turbines, Edition 2010*; Germanischer Lloyd: Hamburg, Germany, 2010.
4. Burton, T.; Jenkins, N.; Srpe, D.; Bossanyi, E. *Wind Energy Handbook*, 2nd ed.; John Wiley and Sons: Chichester, UK, 2011.
5. Jonkman, J.M.; Buhl, M.L., Jr. *FAST User's Guide*; NREL/TP-500-38230; NREL: Golden CO, USA, 2005.
6. DNV GL. *Bladed, Version 4.7*; DNV GL: Bristol, UK, 2016.
7. Zhao, Q.; Sheng, C.; Afjeh, A. Computational aerodynamic analysis of offshore upwind and downwind turbines. *J. Aerodyn.* **2014**, *2014*, 860637. [CrossRef]

8. Zahle, F.; Madsen, H.A.; Sorensen, N.N. Evaluation of Tower Shadow Effects on Various Wind Turbine Concepts. Available online: http://orbit.dtu.dk/files/3550506/ris-r-1698.pdf (accessed on 19 July 2018).
9. Matiz-Chicacausa, A.; Lopez, O.D. Full downwind turbine simulations using actuator line method. *Model. Simul. Eng.* **2018**, *2018*, 2536897. [CrossRef]
10. Wang, T.; Coton, F.N. A high resolution tower shadow model for downwind wind turbines. *J. Wind Eng. Ind. Aerodyn.* **2001**, *89*, 873–892. [CrossRef]
11. Yoshida, S.; Kiyoki, S. Load equivalent tower shadow modeling for downwind turbines. *J. JSME* **2007**, *37*, 1273–1279. [CrossRef]
12. Yoshida, S.; Kiyoki, S. Load equivalent tower shadow modeling. In Proceedings of the European Wind Energy Conference 2007, Milano, Italy, 7–10 May 2007.
13. Nagao, T.; Kato, T.; Shiraishi, T.; Yoshida, S.; Sugino, J. Development of advanced wind turbine systems for remote islands. *Wind Eng.* **2006**, *28*, 729–740. [CrossRef]
14. Yoshida, S. Performance of downwind turbines in complex terrains. *Wind Eng.* **2006**, *30*, 487–501. [CrossRef]
15. Yoshida, S. *Nacelle Yaw Measurement of Downwind Turbines in Complex Terrains*; Windtech International: Groningen, The Netherlands, 2008.
16. Saeki, M.; Sano, T.; Kato, H.; Owada, M.; Yoshida, S. Concept of the HITACHI 5MW offshore downwind turbine. In Proceedings of the European Wind Energy Conference 2014, Barcelona, Spain, 10–13 March 2014.
17. Yoshida, S. Combined blade-element momentum—Lifting line model for variable loads of downwind turbine towers. *Energies* **2018**, *11*, 2521. [CrossRef]
18. Research Institute of Applied Mechanics, Kyushu University, Boundary Layer Wind Tunnel. Available online: https://www.riam.kyushu-u.ac.jp/windeng/en_equipment.html (accessed on 10 November 2018).
19. ANSYS. *ANSYS CFX*; ANSYS: Canonsburg, PA, USA, 2006.

Article

POD Analysis of Entropy Generation in a Laminar Separation Boundary Layer

Chao Jin and Hongwei Ma *

School of Energy and Power Engineering, Beihang University, Beijing 100191, China; jin_chao@buaa.edu.cn
* Correspondence: mahw@buaa.edu.cn; Tel.: +86-131-2188-5794

Received: 22 September 2018; Accepted: 26 October 2018; Published: 1 November 2018

Abstract: Separation of laminar boundary layer is a great source of loss in energy and power machinery. This paper investigates the entropy generation of the boundary layer on the flat plate with pressure gradient. The velocity of the flow field is measured by a high resolution and time related particle image velocimetry (PIV) system. A method to estimate the entropy generation of each mode extracted by proper orthogonal decomposition (POD) is introduced. The entropy generation of each POD mode caused by mean viscous, Reynolds normal stress, Reynolds sheer stress, and energy flux is analyzed. The first order mode of the mean viscous term contributes almost 100% of the total entropy generation. The first three order modes of the Reynolds sheer stress term contribute less than 10% of the total entropy generation in the fore part of the separation bubble, while it reaches to more than 95% in the rear part of the separation bubble. It indicates that the more unsteady that the flow is, the higher contribution rate of the Reynolds sheer stress term makes. The energy flux term plays an important role in the turbulent kinetic energy balance in the transition region.

Keywords: POD; entropy generation; boundary layer; laminar separation bubble

---

1. Introduction

For the energy and power machinery, such as gas turbine and other turbomachinery, the efficiency must be the most important performance parameters [1]. An effective way to improve efficiency is to reduce the loss, namely to avoid the generation of entropy [2–4]. In thermodynamics, any irreversible physical process will inevitably lead to the increase of entropy [5]. In this paper, it mainly focuses on the entropy generation in the boundary layer.

The entropy generation has been the subject of many past studies. Bejan [6], Rotta [7], and McEligot [8] studied the entropy generation in the viscous layer and analyzed the generation rates in diffident Y-plus layer. Moore [9] attempted to develop a numerical model for turbulent flow entropy generation, but Kramer-Bevan [10] verified it is not consistent near the wall due to a small temperature gradient. Adeyinka et al. [11] investigated the error of entropy generation model affected by the mesh grids in the fully-developed laminar flow.

The entropy generation is difficult to be accurately calculated [12]. Since it is difficult to predict the small fluctuating velocity and temperature in the turbulent flow through numerical and experimental [13] few experiments have had sufficient measurements to calculate the entropy generation. Thus, most studies try to estimate the entropy generation with a simplified formulation, which will be introduced in Section 3.2.

To describe and extract coherent structures in boundary layer, researchers attempt to develop new data analysis methods, such as proper orthogonal decomposition (POD), dynamic mode decomposition (DMD), and spectral POD (SPOD). POD is a method that identifies coherent structures by decomposing the flow field into orthogonal modes in space. Moreover, the dominant features in the flow is identified based on the energy rank [14]. Dynamic mode decomposition is a method that is orthogonal in time in the sense that the dynamic mode information from a given flow field is based on the Koopman analysis [15]. In contrast to POD, it is a method that gives the energy of the fluctuations at distinct frequencies meaning the modes are arranged in descending order of energy content. Spectral POD is a method that combines proper orthogonal decomposition with a spectral method to analyze and extract reduced order models of flows from time data series of velocity fields [16].

Since this paper mainly focuses on the entropy generation in the boundary layer. It is more convenient to identify the flow field based on the energy rank and care little about the time series and frequencies. Thus, the POD method is applied to the measurement to estimate the entropy generation. The POD has been widely applied for the experimental and numerical data to identify the coherent structure [17,18], such as the plate boundary layer [19], cylinder engine flow [20], and turbine rotor-stator interaction [21]. Particularly, in the case of laminar separation bubbles, POD can extract the different scale coherent structures [22,23].

While few works have been done to quantify the entropy generation of different coherent structures in the boundary layer. Calculating the entropy generation of those coherent structure helps to explore the mechanisms of entropy generation in the boundary layer. It can also identify the loss resource of the flow flied. In this paper, the entropy generation rate is analyzed by proper orthogonal decomposition (POD) applied to the PIV measurements.

2. Experimental Facility

The measurement is taken in a transparent circulating water tunnel with a 700 mm (width) $\times$ 500 mm (depth) experimental section, and the whole length of the water tunnel is 6.8 m, just as Figure 1 shows. The inlet velocity in this experiment is 0.065 m/s. The mean velocity profile in the empty experimental section was uniform except for the thin boundary layers on the walls. The turbulence of the water tunnel is less than 1%, and the Reynolds number is based on the total length of the flat plate and the inlet velocity is about 3×10^4. The flat plate is mounted horizontally in the middle of the water tunnel to ensure zero flow incidence. Total length of the flat plate is 390 mm and the geometrical structure of the leading edge is an ellipse with a ratio of 3:1 to the semi-minor axis length, just as Figure 2 shows.

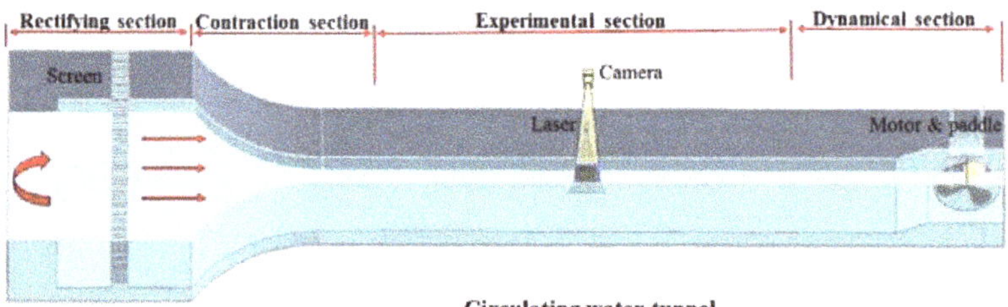

Figure 1. Water tunnel and experimental layout.

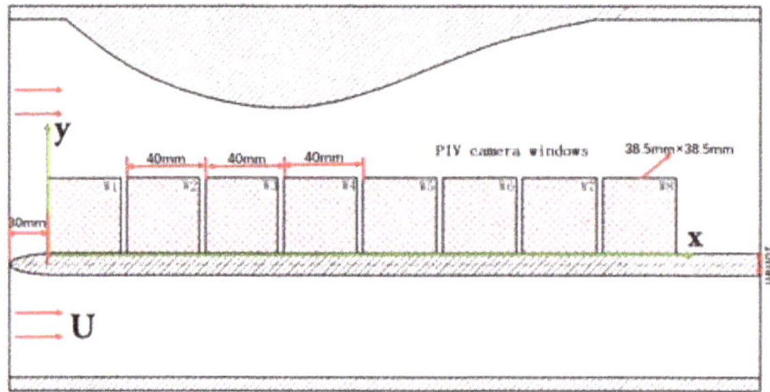

Figure 2. Placement of the PIV camera windows.

The instantaneous velocity field of eight streamwise planes (W1–8) is measured by a PIV system. The distance from the first window to the leading edge of the flat plate is 30 mm, and the distance between each windows is 40 mm. The streamwise plane is illuminated by a light sheet provided by a double cavity Nd: YAG laser, which has the maximum illumination energy of 200 mJ/pulse and a maximum repetition rate of 15 Hz. To ensure the PIV system operates in stably, the sampling rate is set as 12 Hz. One thousand pairs of signal-exposure images are continuously captured by a CCD camera with a resolution of 2072 pixels × 2072 pixels, and the reality view window is about 38.5 mm × 38.5 mm. The PIV measurement uncertainty can refer to the paper [24], in which the same PIV system is used. A more detailed investigation of the uncertainty in the turbulent boundary layer can be found in [25].

3. Dara Processing Method

3.1. POD Method

Since any turbulent flow can be viewed as a superposition of a small number of coherent structures, the equations describing these structures can be considered in a low-dimensional description of turbulence. The POD is an effective method to extract these coherent structures. The time related flow field snapshots can be decomposed into a linear basis set consisting of N basis function $\phi^{(k)}$ (POD mode) and the corresponding coefficients $\chi^{(k)}$ (POD coefficients or time coefficients). The POD mode provides the spatial information on coherent structures, and the POD coefficients retain the temporal information. In addition to those two parameters, the eigenvalues $\lambda^{(k)}$ represents the total kinetic energy that each POD mode captured. The original flow field can be reconstructed by POD mode and POD coefficients.

$$v = \sum_{n=1}^{N} \chi_n \phi_n \tag{1}$$

The instantaneous velocity of kth POD mode is:

$$v_n^{(k)} = \chi_n^{(k)} \phi^{(k)} \quad n = 1, 2 \ldots N, \tag{2}$$

The mean velocity of kth POD mode is:

$$V^{(k)} = \frac{1}{N} \sum_{n=1}^{N} \chi_n^{(k)} \phi^{(k)}, \tag{3}$$

The velocity fluctuation of kth POD mode is:

$$v'^{(k)}_n = v_n^{(k)} - V^{(k)} \quad n = 1, 2 \ldots N, \tag{4}$$

Thus, the Reynolds stresses of kth POD mode can be computed as:

$$u'^{(k)}_n v'^{(k)}_n = \left(u_n^{(k)} - U^{(k)} \right) \left(v_n^{(k)} - V^{(k)} \right) \quad n = 1, 2 \ldots N, \tag{5}$$

3.2. Estimation of Entropy Generation

Basic thermodynamics tells us that entropy generates when the process is irreversible such as viscous friction [1]. The viscous dissipation is a main frictional irreversibility in boundary layers, which includes mean viscous dissipation and turbulent dissipation. So the entropy generation can be expressed as:

$$S = \frac{\mu \Phi + \rho \varepsilon}{T}, \tag{6}$$

where the mean viscous dissipation $\mu\Phi$ is [2]:

$$\mu\Phi = 2\mu\left[\left(\frac{\partial U}{\partial x}\right)^2 + \left(\frac{\partial V}{\partial y}\right)^2\right] + \mu\left(\frac{\partial U}{\partial y} + \frac{\partial V}{\partial x}\right)^2, \tag{7}$$

The turbulent dissipation $\rho\varepsilon$ is:

$$\rho\varepsilon = 2\mu\left[\left(\frac{\partial u'}{\partial x}\right)^2 + \left(\frac{\partial v'}{\partial y}\right)^2\right] + \mu\left(\frac{\partial u'}{\partial y} + \frac{\partial v'}{\partial x}\right)^2, \tag{8}$$

Rotta [7] gives a simple estimation function of entropy generation:

$$S = \frac{\left[\mu\left(\frac{\partial U}{\partial y}\right)^2 - \rho\left(\overline{u'v'}\right)\left(\frac{\partial U}{\partial y}\right)\right]}{T}, \tag{9}$$

Since, the energy flux term plays an important role in the turbulent energy balance in the transition, so it should also take the energy flux term into account in entropy generation. Thus, the entropy generation can be calculated as following form [7,26,27]:

$$S\{\delta\} \approx \frac{\rho}{T}\left[v\int_0^\delta\left(\frac{\partial U}{\partial y}\right)^2 dy - \int_0^\delta\left(\overline{u'v'}\right)\left(\frac{\partial U}{\partial y}\right)dy - \int_0^\delta\left[\left(\overline{u'^2}\right) - \left(\overline{v'^2}\right)\right]\left(\frac{\partial U}{\partial x}\right)dy - \right.$$
$$\left.\frac{d}{dx}\int_0^\delta \frac{1}{2}U\left(\overline{q^2}\right)dy - \frac{1}{2}v'_\delta\left[(u'^2_\delta) + (v'^2_\delta) + (w'^2_\delta)\right] - \overline{v'_\delta p_\delta}\right], \tag{10}$$

where the first term on the right side of Equation (10) is the viscous term, the second term is the Reynolds shear stress term, the third term is the Reynolds normal stress term; the fourth term is the energy flux term, the fifth term is the turbulent diffusion term, and the last term is the pressure diffusion term. As the last two terms are small, comparing to the magnitudes of other terms, so it can be neglected [26]. All above equations are widely used to estimate the entropy generation especial in experimental work. Since it does not involve the pressure and temperature, only the instant velocity should be measured. It is easy to measure the instant velocity through PIV, hotline probe, and LDV.

3.3. Effect of Decomposition Region Size on POD

In order to explore the effect of decomposition region size on POD results, different decomposition region size has been investigated. The results of four different height cases are shown in Figure 3. It shows that, when the height of the decomposition region is less than 30 mm, the spatial structure of different windows is consecutive, but the symbols of the velocity in W5 and W6 are reverse at different cases. When the height reaches to 35 mm, the POD can even not extract the structure of boundary layer in window 5. This is mainly because the POD extracts a coherent structure according to the total kinetic energy of the flow field. The first-order mode captures more than 99% of the energy at W5 and W6, as shown in Figure 4; the higher-order modes naturally capture a very low energy. Therefore, the higher order modes are easily to be effect by the disturbance in main flow. In W7 and W8, the energy distribution is more uniform because of a more unsteady boundary layer, so it shows more stabilization with different decomposition region sizes. For the same reasons, the size of streamwise decomposition region has the same effect on the POD results, which is not shown in this paper. Thus, in order to extract more flow field structure and avoid the effect of main flow disturbance on the POD mode, the decomposition region size in following work is all 30 mm × 38.5 mm (only shows 20 mm × 38.5 mm).

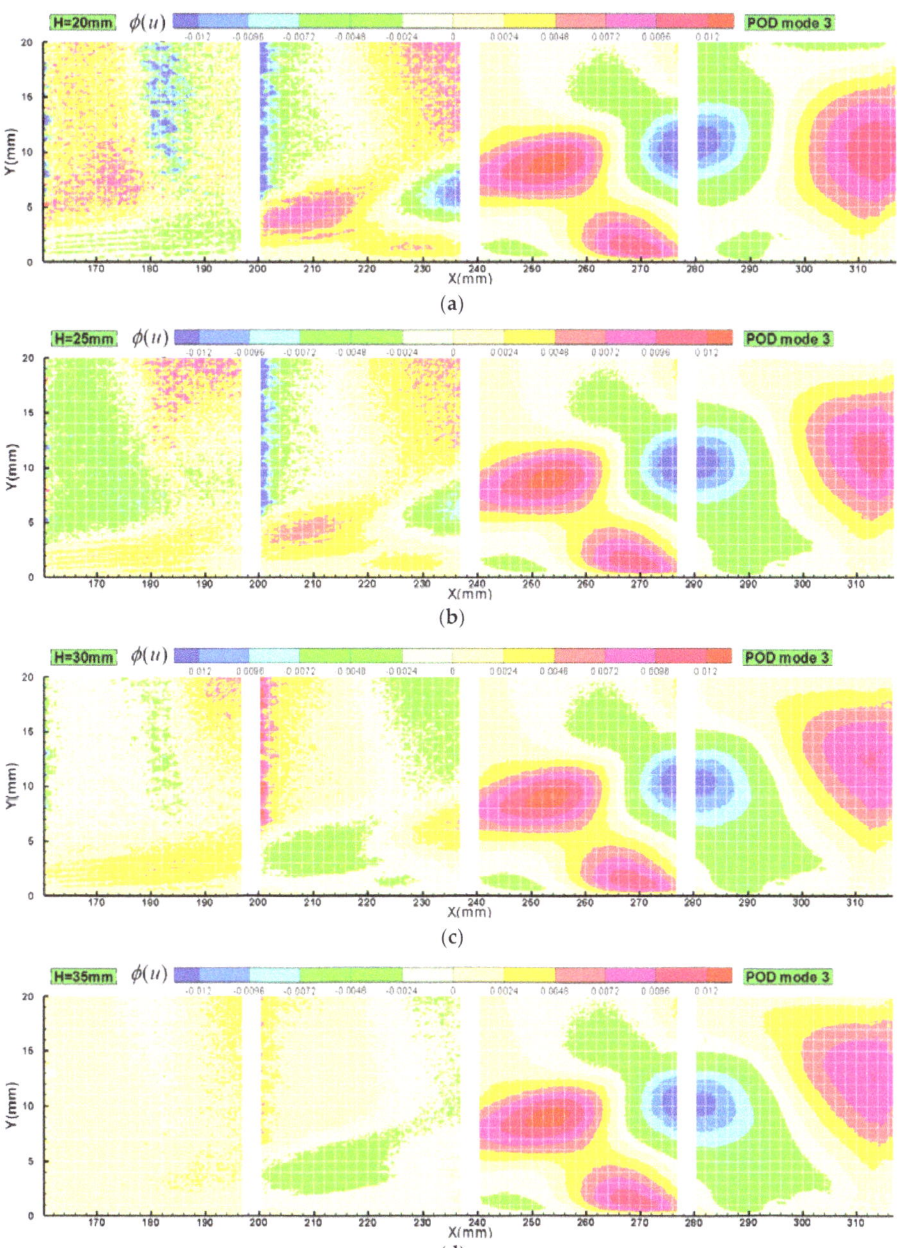

Figure 3. The third-order mode of the streamwsie velocity with different spanwise decomposition region size: (**a**) H = 20 mm; (**b**) H = 25 mm; (**c**) H = 30 mm; and (**d**) H = 35 mm.

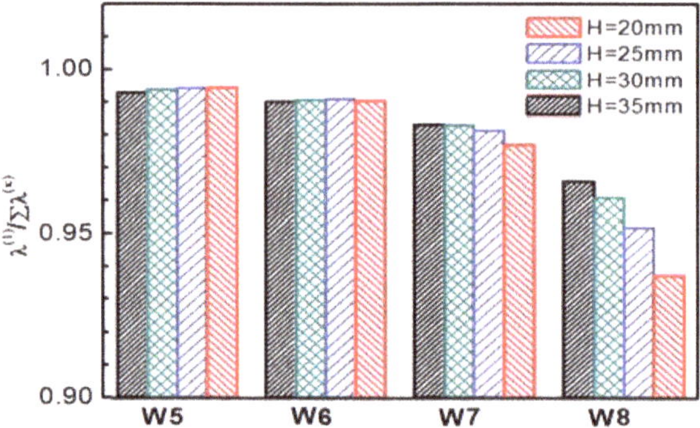

Figure 4. Contribution to total energy of the first order POD mode.

4. Results and Discussion

4.1. Time-Mean Flow Field

Figure 5 shows the time-mean flow field on the flat plate. The thickness of boundary layer decreases gradually with the velocity increase under the favorable pressure gradient, and then increases rapidly under the adverse gradient. The boundary layer separates at about $X = 200$ mm, and it reattaches again at the end of the outlet, where the adverse gradient is disappearance. It forms a time-mean separation bubble in the separated boundary layer, while for the temporary flow field, it consists of a series of vortex. In [28], some criteria can be used to assess if flow history (upstream evolution of the streamwise pressure gradient) have an impact on the development of the boundary layer. The following analyses concentrates on the aft portion of the flat plate boundary layer downstream of the transition.

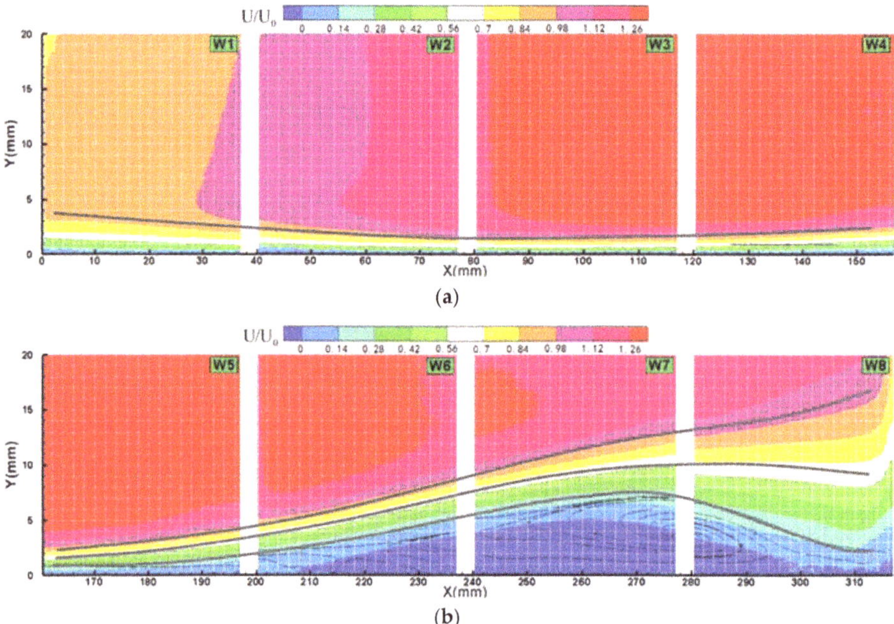

Figure 5. Time mean streamwise velocity: (**a**) the fore part of the flow field; and (**b**) the rear part of the flow field.

Figure 6 shows the normalized velocity pattern of the boundary layer along the streamwise. It is easy to judge the state of the boundary layer through the velocity pattern. The boundary layer at $X/\delta_{x=150} = 41$–46 is a typical attached boundary layer. There is no inflexion point in the velocity pattern, which indicates that the flow at those positions has not yet separated. From the position of $X/\delta_{x=150} = 47.5$, the velocity pattern has obvious inflection point. It indicates that the boundary layer has separated. The height of the inflection point also represents the boundary of the separation bubble shown as the dotted line in Figure 6.

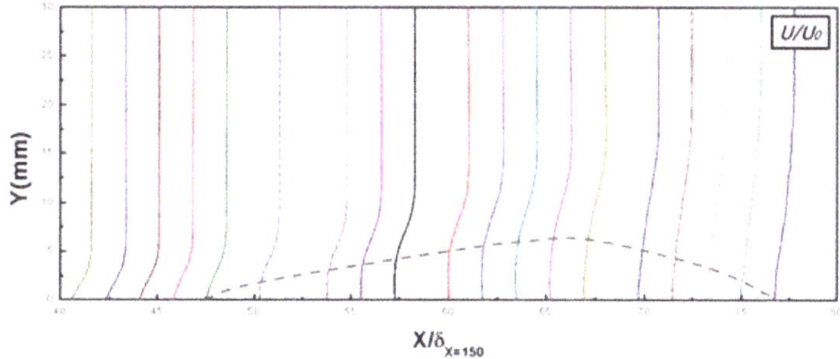

Figure 6. Normalized velocity of the boundary layer.

Figure 7 shows the variation of boundary layer thickness, displacement boundary layer thickness, momentum boundary layer thickness and the shape factor along the streamwise. What should be noted that strong pressure gradient may lead to an inconsistent boundary layer edge by the common techniques to define the boundary layer edge [29]. The shape factor maintains about 2 to 3 in W1–3, where the boundary layer keeps laminar. The boundary layer separates at the position of $X = 200$ mm, the value of the shape factor just consistent with the typical value of the separation laminar boundary. Between the laminar region and the separation region, there is a transition region. Downstream of the transition region, the shape factor increase rapidly due to the increase of the displacement boundary layer thickness. The shape factor start to decrease from the location of $X = 260$ mm, the position corresponds the maximum thickness of the separation bubble.

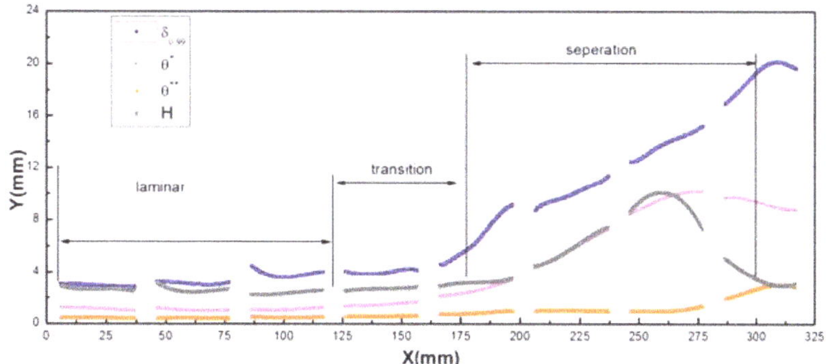

Figure 7. Streamwise variation of the boundary layer parameters.

4.2. POD Analysis of Flow Field

Figure 8 shows the different POD modes of the streamwise velocity. The flow structure of the first order mode is consistent with the time-mean flow field. Since the flow field is almost steady in W5 and W6, the flow structure of different modes is a little similar to each other, while in W7 and W8, the flow structure of different modes is much different from each other because of strong unsteady.

The second order mode is still the large-scale coherent structures that affected by the mean flow. The third–fourth-order modes represent the small coherent structures that induced by the large-scale vortex, and the coherent structure paired increases with the increase of the order number [30].

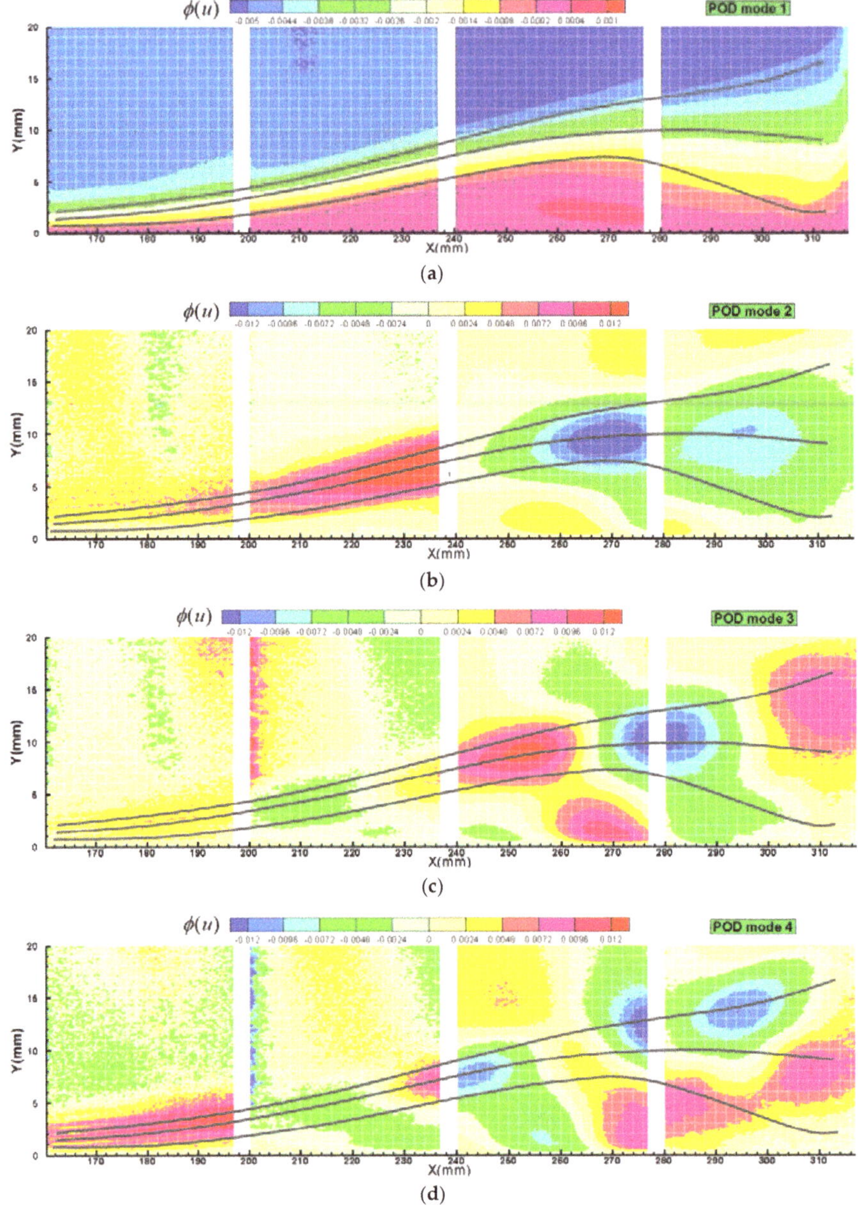

Figure 8. POD modes of streamwise velocity: (**a**) the first order mode; (**b**) the second order mode; (**c**) the third order mode; and (**d**) the forth order mode.

Figure 9 shows the power spectrum density of the POD coefficient, which represents the turbulent kinetic energy of the POD mode. There is a basic frequency (0.024 Hz) in the power spectrum of the POD coefficient, which corresponds with the frequency of the main flow turbulence. The energy mainly concentrates in a low frequency band for those four windows. The amplitude of the energy spectrum increases gradually from W5 to W8 with the increasing unsteady of the flow. The energy at high frequency decreases faster and faster from W5 to W8. The slope of the forth order mode at W8 tends to be $-3/5$, which indicates that the flow field of the forth order mode tends to be fully developed turbulent flow.

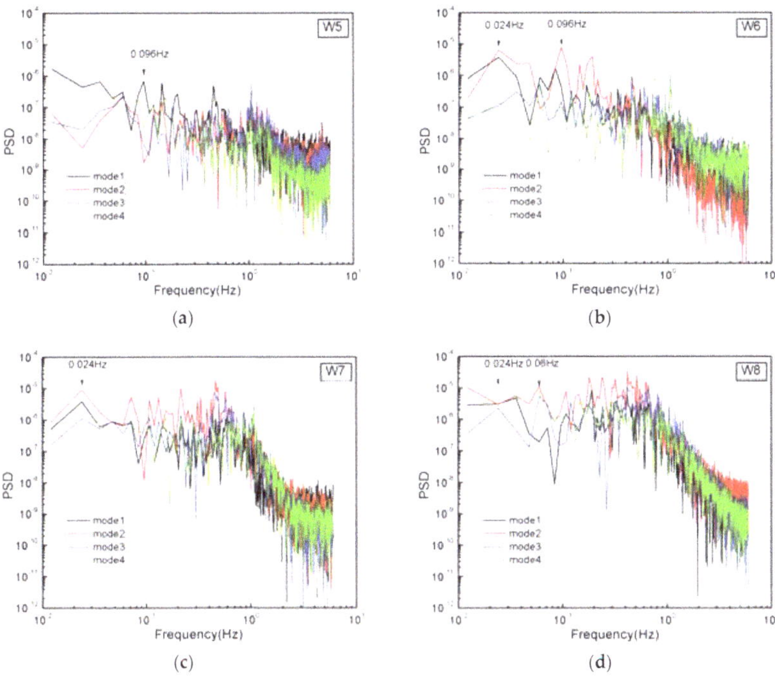

Figure 9. Power spectrum density of the POD coefficient: (**a**) W5; (**b**) W6; (**c**) W7; and (**d**) W8.

According to the Equations (2)–(5), the Reynolds stress of different POD modes are calculated and shown in Figure 10. The distribution of the first order mode Reynolds stress is basically consistent with the original flow field (not given in this paper), which also verifies the correctness of the above equations. The extracted Reynolds normal stress of mode 2 and mode 3 mainly distributes above the separation bubble, while the Reynolds shear stress distributes in the whole boundary layer.

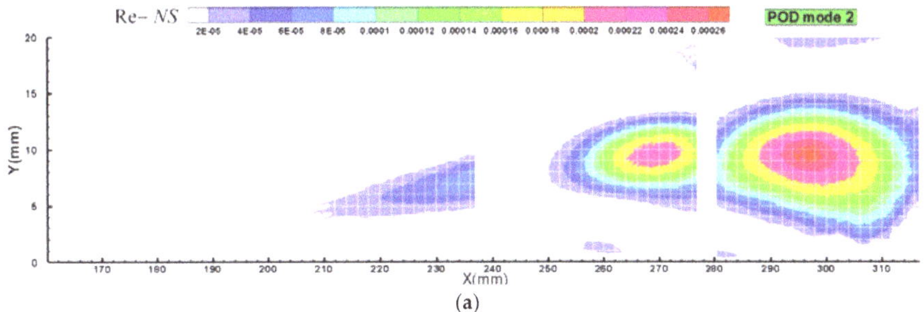

Figure 10. *Cont.*

73

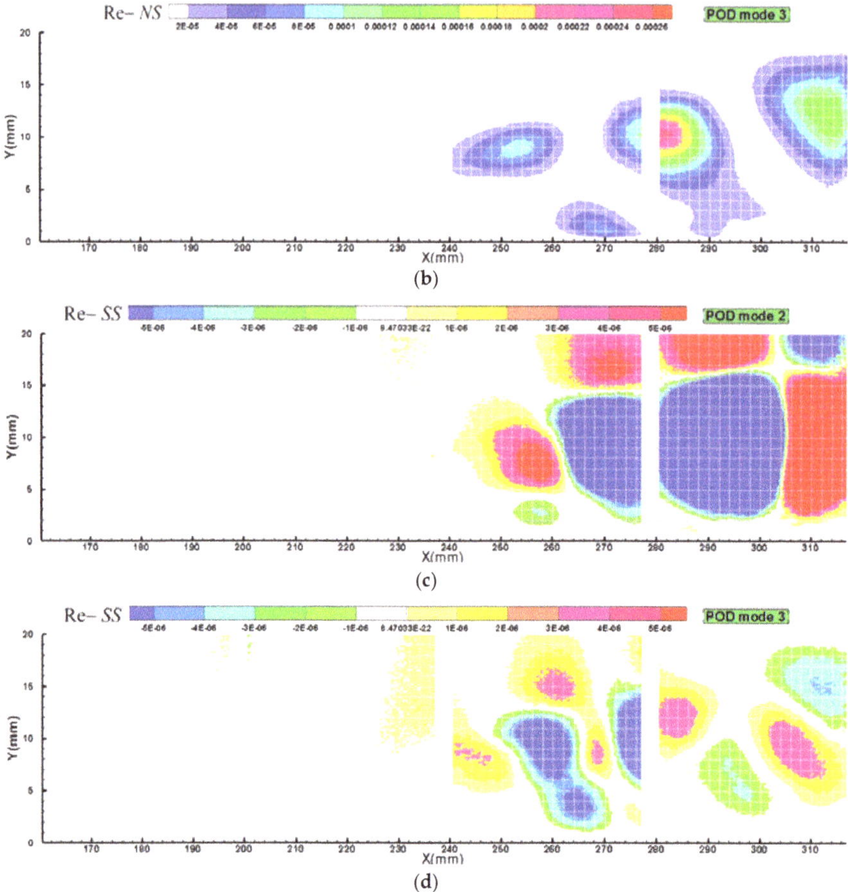

Figure 10. (**a**) The second-order mode of the Reynolds normal stress; (**b**) the third-order mode of the Reynolds normal stress; (**c**) the second-order mode of the Reynolds sheer stress; and (**d**) the third-order mode of the Reynolds sheer stress.

4.3. Entropy Generation Analysis

4.3.1. Entropy Generation of Original Flow Field

According to the Equation (10), the different entropy generation terms are calculated and shown in Figure 11. It shows that the magnitude of the mean viscous dissipation term is much smaller than other terms. The entropy of the mean viscous dissipation only generates in laminar boundary layer and the edge of the separation bubble. Since the velocity in the separation bubble is very small, the entropy generation tends to be very small in the separation bubble. Nevertheless, it does not mean that the separation bubble will not cause any loss. The separation bubble will induce a rapidly increase of the boundary layer thickness and the Reynolds stress dissipation loss in the separated boundary layer is much higher than that of the attached boundary layer, just as Figure 11 shows. The energy flux term distributes in the whole main flow and there is not a clear boundary between the positive area with the negative area.

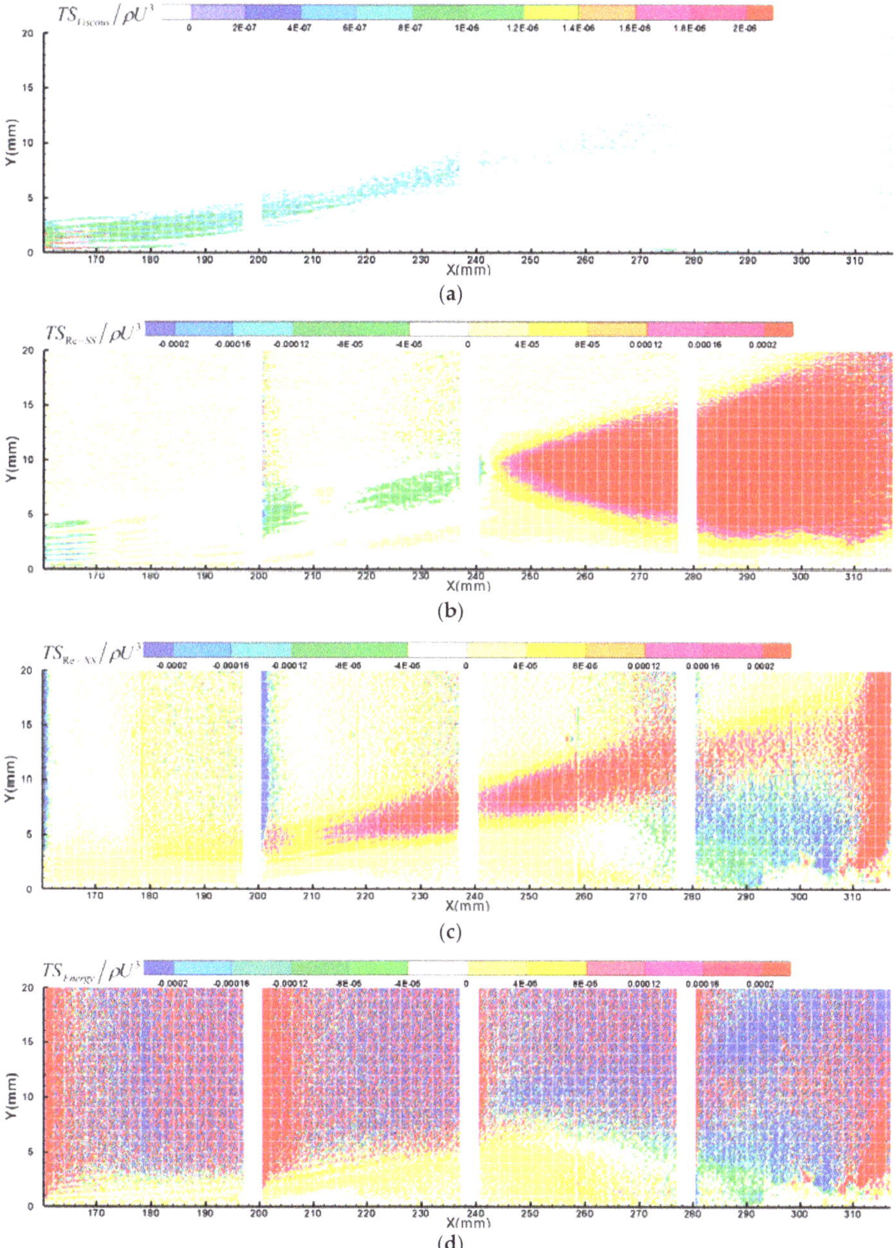

Figure 11. Distribution of different terms of entropy generation: (**a**) the mean viscous dissipation term; (**b**) the Reynolds sheer stress term; (**c**) the Reynolds normal stress term; and (**d**) the energy flux term.

Figure 12 shows the variation of the integrated entropy generation in the boundary along the streamwise. In the laminar region (W1–3), the entropy generation of the laminar boundary layer keeps at a very low level. In the transition region (W4), the amplitude of the energy flux term increases significantly. It indicates that the energy flux term plays an important role in the turbulent kinetic energy balance. The integrated energy flux shows discontinuity between windows and data sparsity, which results from the discrete distribution of energy flux term in the counter map just as the Figure 11 shows. Figure 13 shows the total integrated entropy generation in the boundary layer. It shows that the total entropy generation increases significantly when the boundary layer separated. The Reynolds shear stress term tends to be positive growth and the energy flux term is negative growth, which is consistent with the trend described in [27].

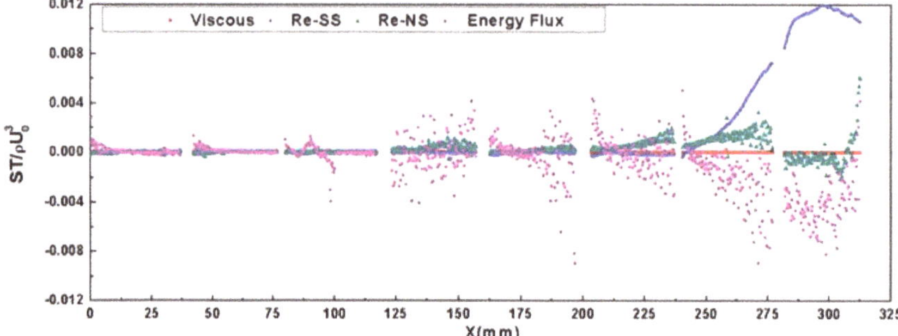

Figure 12. Streamwise variation of integral entropy generation.

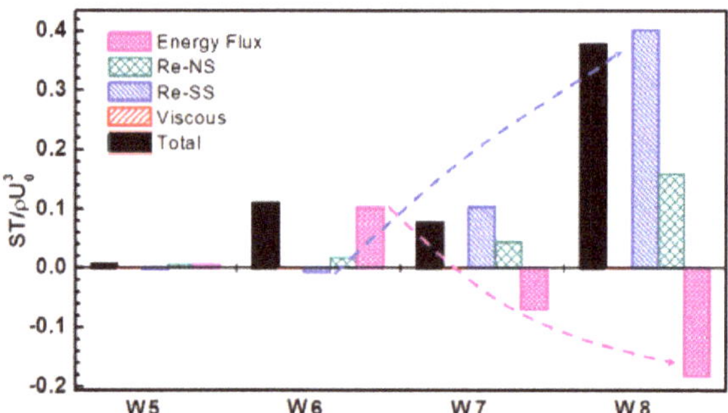

Figure 13. Total integral entropy generation of different terms.

4.3.2. Entropy Generation of POD Mode

As described above, the POD is an effective method to extract the coherent structures. The entropy generation of the coherent structures extracted by POD can be calculated through Equations (2)–(5) and (10). It significant to quantize the entropy generation of the coherent structures. Once you can identify the source of the coherent structures, then the source of the loss production can be confirmed [31].

Figures 14–16 show the entropy generation of different POD modes. The first-order mode of the mean viscous dissipation term is almost equal to that of the original flow field and the value of higher order mode tends to be zero. It also demonstrates that the first order mode represents the mean flow. The entropy generation distribution of the second and third-order mode of the Reynolds stress term and the energy flux term is similar to the distribution of the related POD mode.

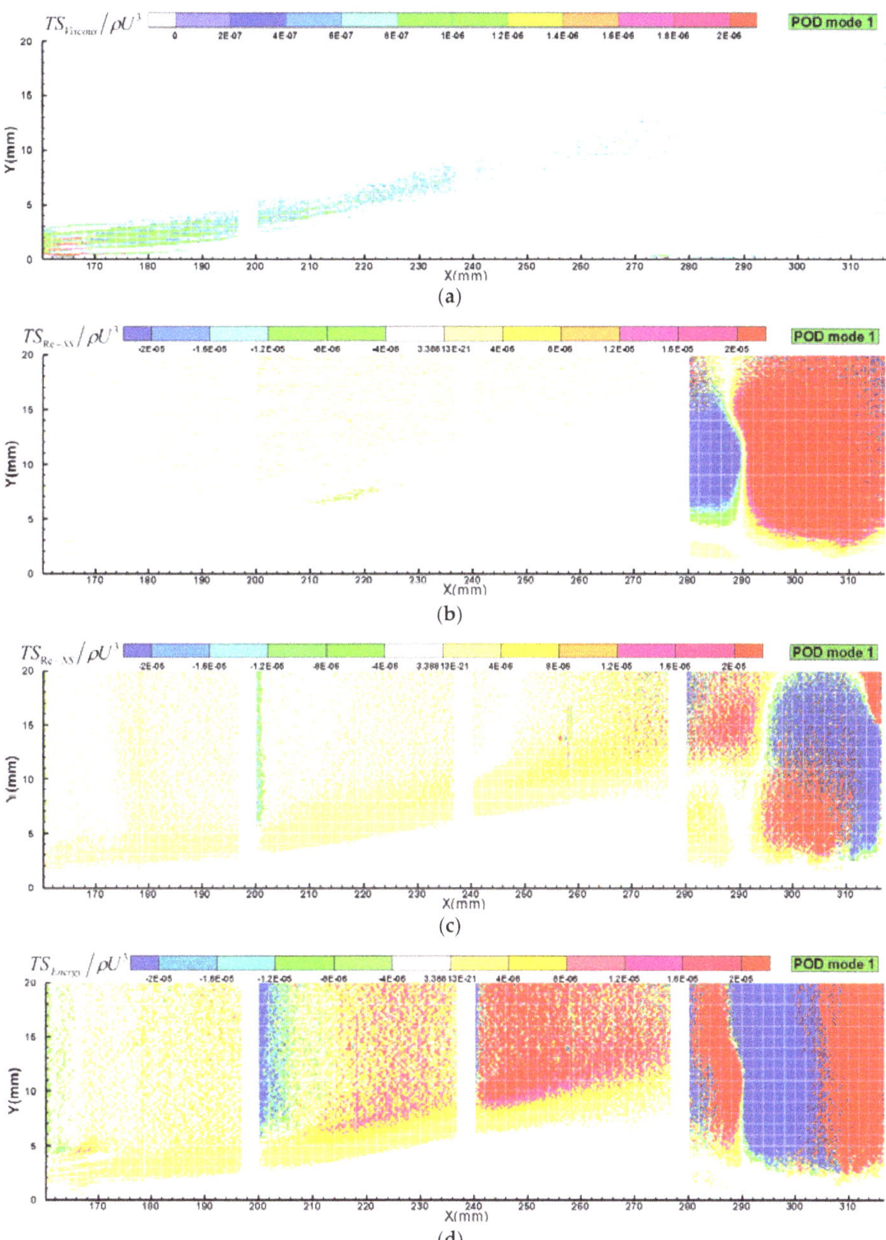

Figure 14. Entropy generation of first order mode: (**a**) the mean viscous dissipation term; (**b**) the Reynolds sheer stress term; (**c**) the Reynolds normal stress term; and (**d**) the energy flux term.

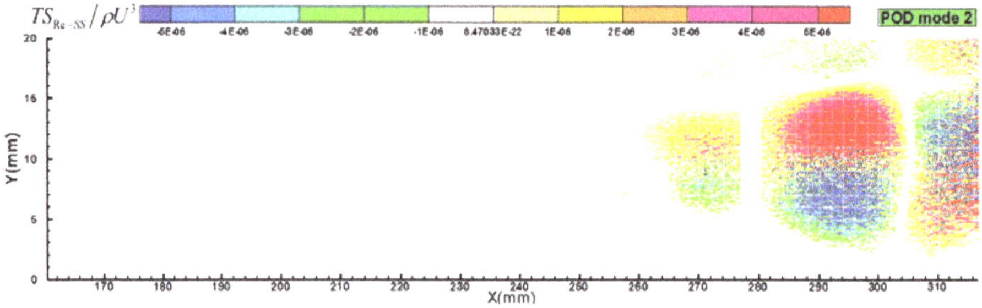

Figure 15. Entropy generation of second order mode: (**a**) the Reynolds sheer stress term; (**b**) the Reynolds normal stress term; and (**c**) the energy flux term.

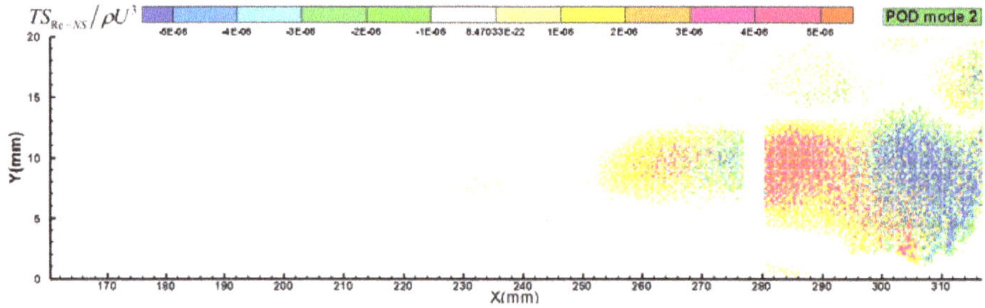

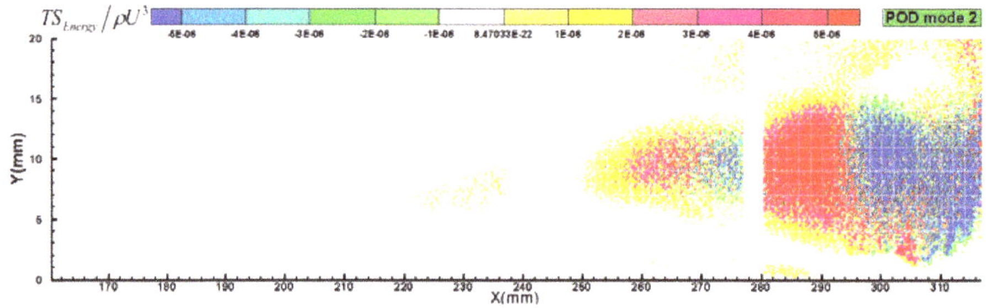

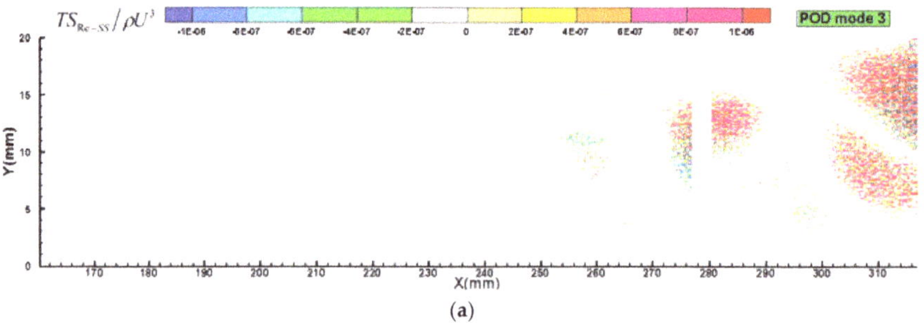

(**a**)

Figure 16. *Cont.*

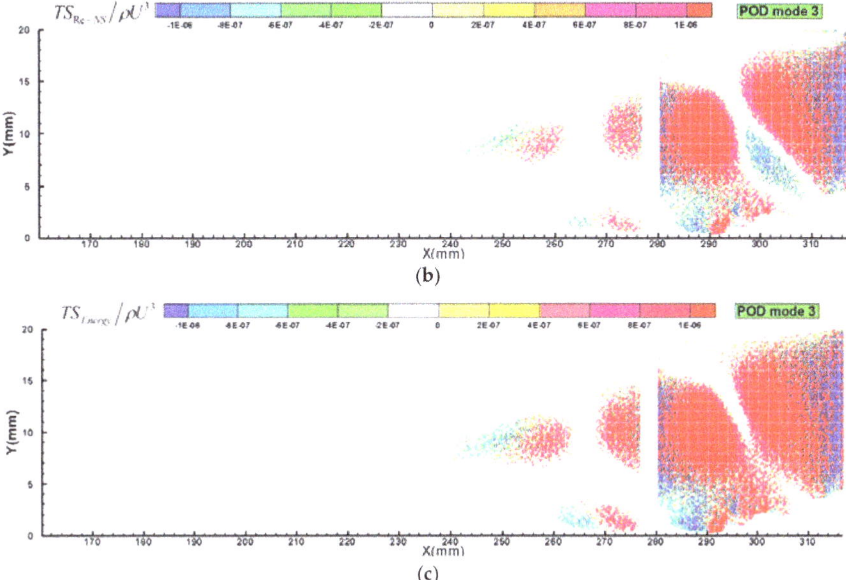

Figure 16. Entropy generation of third order mode: (**a**) the Reynolds sheer stress term; (**b**) the Reynolds normal stress term; and (**c**) the energy flux term.

Figure 17 shows the cumulative contribution to the total integral entropy generation of each POD mode in the boundary layer. The contribution rate of the first order mode of the mean viscous term reaches 100% in all of those windows. In W5, the contribution rate of the first two modes of the Reynolds stress (normal and sheer) is about 10%, the magnitude of the higher modes is almost at the same level. Thus, the contribution of each mode seems to be the same and the cumulative contribution line increases with a constant slope. The Reynolds sheer stress term in W6 has the same situation with that in W5, while the first third-order mode of the Reynolds normal stress in W6 contributes about 80% of the total entropy generation. The contribution rate of the energy flux term reaches to 90% until the cumulative order number of the POD mode is about 400 in both W5 and W6.

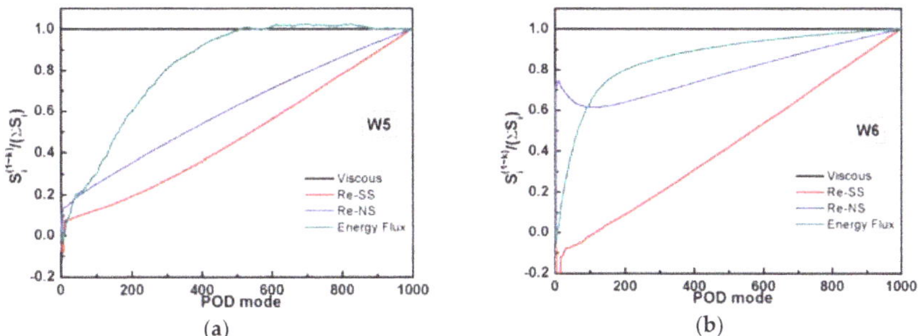

Figure 17. *Cont.*

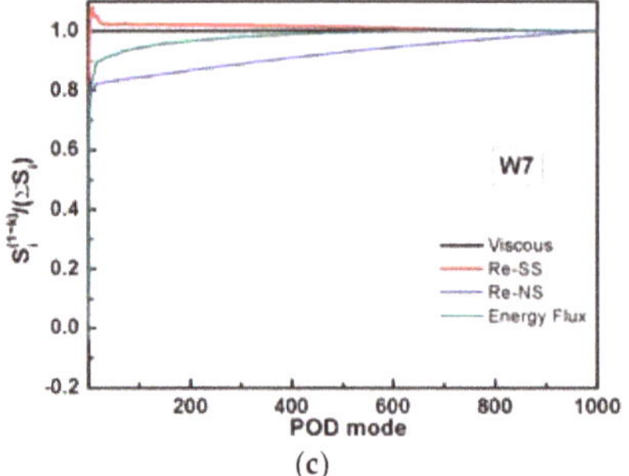
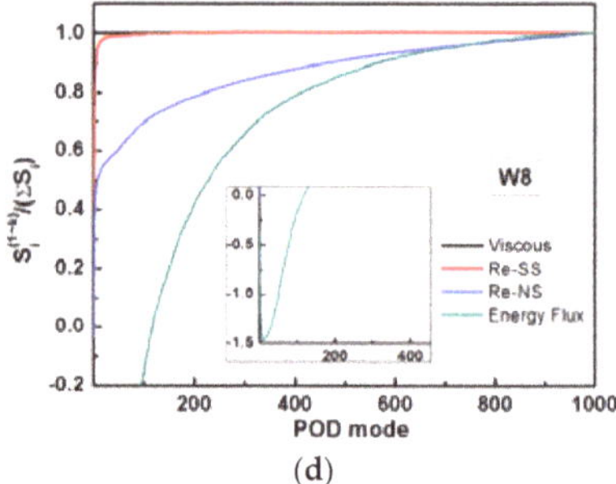

(c)   (d)

Figure 17. Cumulative contribution to the total integral entropy generation of each POD mode: (**a**) W5; (**b**) W6; (**c**) W7; and (**d**) W8.

In W7, the first third-order modes of Reynolds stress (normal and sheer) term contributes more than 80% of the total entropy generation, the higher order contributes less than 20%. In W8, the contribution of the first 30th-order modes of the Reynolds stress (normal and sheer) is about 60%. It indicates that the contribution of the Reynolds stress term of the low-order mode in the rear part of the separation bubble (W7 and W8) is much higher than that in the fore part of the separation bubble (W5). The low order mode of the energy flux term in W8 contributes the negative value, which results from the negative energy captured by the first-order mode just as shown in Figure 16.

5. Conclusions

In this paper, the boundary layer of the flat plate with a pressure gradient in the water tunnel has been studied using high-resolution particle image velocimetry (PIV). The entropy generation rate is analyzed by proper orthogonal decomposition (POD) applied to the measurements. Several conclusions can be made.

The separation bubble will dramatically increase the thickness of the boundary layer and result in a sharp increase of the loss. The loss due to the Reynolds normal stress mainly distributes above the separation bubble, and the loss due to the Reynolds shear stress distributes in the whole boundary layer. The decomposition region size has a significantly effects on the POD result of the laminar boundary layer. It should reduce the proportion of the main flow area as much as possible when carrying out the POD analysis. In the transition region, the energy flux terms plays an important role in transiting the energy from the mean flow to the turbulent flow and contributes a large ratio to the entropy generation. Estimation of the entropy generation of the POD mode helps to explore the mechanism of entropy generation and identify the source of the loss production. Additionally, it is easy to promote this method to some other complex flows, such as wake-induced separation on the suction side of blade, or the leakage flow on the blade tip from either numerical or experimental data, which is only based on the instantaneous velocity field.

Author Contributions: Conceptualization: C.J.; data curation: C.J.; investigation: C.J.; methodology: C.J.; software: C.J.; supervision: H.M.; validation: C.J.; writing—original draft: C.J.; writing—review and editing: H.M.

Funding: This study was supported by the National Natural Science Foundation of China (Grant No. 51776011).

Conflicts of Interest: The authors declare no conflict of interest.

Nomenclature

H	Shape factor
Re	Reynolds number
S	Entropy generation
T	Temperature
u	Instantaneous streamwise velocity

u'	Streamwise velocity fluctuation ($u - U$)
U	Averaged streamwise velocity
U_0	Streamwise velocity of main flow
v	Instantaneous spanwise velocity
v'	Spanmwise velocity fluctuation ($v - V$)
V	Averaged spanwise velocity
χ	POD coefficients
λ	Eigenvalue of POD
ϕ	Basis function of POD
ρ	Density
μ	Kinetic viscosity
v	Kinematic viscosity
δ	Boundary layer thickness
θ^*	Displacement boundary layer thickness
θ^{**}	Momentum boundary layer thickness

References

1. Denton, J.D. The 1993 IGTI Scholar Lecture: Loss Mechanisms in Turbomachines. *J. Turbomach.* **1993**, *115*, 621–656. [CrossRef]
2. Bejan, A.; Kestin, J. Entropy Generation through Heat and Fluid Flow. *J. Appl. Mech.* **1983**, *50*, 475. [CrossRef]
3. Naterer, G.F.; Camberos, J.A. *Entropy Based Design and Analysis of Fluids Engineering Systems*; CRC Press: Boca Raton, FL, USA, 2008.
4. Bejan, A. *Entropy Generation Minimization*; Advanced Engineering Thermodynamics; John Wiley & Sons, Inc.: Hoboken, NJ, USA, 1995.
5. Enrico, S. Calculating entropy with CFD. *Mech. Eng.* **1997**, *119*, 86–88.
6. Suder, K.L.; Obrien, J.E.; Reshotko, E. *Experimental Study of Bypass Transition in a Boundary Layer*; National Aeronautics & Space Administration Report; NASA: Washington, DC, USA, 1988.
7. Rotta, J.C. Turbulent boundary layers in incompressible flow. *Prog. Aerosp. Sci.* **1962**, *2*, 1–95. [CrossRef]
8. Mccligot, D.M.; Walsh, E.J., Laurien, E., Spalart, P.R. Entropy Generation In the Viscous Parts of Turbulent Boundary Layers. *J. Fluids Eng.* **2008**, *130*, 61205. [CrossRef]
9. Moore, J.; Moore, J.G. Entropy Production Rates from Viscous Flow Calculations: Part I—A Turbulent Boundary Layer Flow. In Proceedings of the ASME 1983 International Gas Turbine Conference and Exhibit, Phoenix, AZ, USA, 27–31 March 1983.
10. Kramer-Bevan, J.S. A Tool for Analyzing Fluid Flow Losses. Master's Thesis, University of Waterloo, Waterloo, ON, Canada, 1992.
11. Adeyinka, O.B.; Naterer, G.F. Apparent Eentropy Production Difference with Heat and Fluid Flow Irrevesibilities. *Numer. Heat Transf. Part B Fundam.* **2002**, *42*, 411–436. [CrossRef]
12. Adeyinka, O.B.; Naterer, G.F. Modeling of Entropy Production in Turbulent Flows. *J. Fluids Eng.* **2005**, *126*, 893–899. [CrossRef]
13. Hanjalić, K.; Launder, B. *Modelling Turbulence in Engineering and the Environment*; Cambridge University Press: Cambridge, UK, 2000.
14. Lumley, J.L. *Stochastic Tools in Turbulence*; Dover Publications: Mineola, NY, USA, 2007.
15. Schmid, P.; Sesterhenn, J. Dynamic mode decomposition of experimental data. In Proceedings of the 8th International Symposium on Particle Image Velocimetry, Melbourne, Australia, 25–28 August 2009.
16. Cammilleri, A.; Gueniat, F.; Carlier, J.; Pastur, L.; Mémin, E.; Lusseyran, F.; Artana, G. POD-spectral decomposition for fluid flow analysis and model reduction. *Theor. Comput. Fluid Dyn.* **2013**, *27*, 787–815. [CrossRef]
17. Zhu, J.; Huang, G.; Fu, X.; Fu, Y.; Yu, H. Use of POD Method to Elucidate the Physics of Unsteady Micro-Pulsed-Jet Flow for Boundary Layer Flow Separation Control. In Proceedings of the ASME Turbo Expo 2013: Turbine Technical Conference and Exposition, San Antonio, TX, USA, 3–7 June 2013.
18. Hammad, K.J. Coherent Structures in Turbulent Boundary Layer Flows over a Shallow Cavity. In Proceedings of the ASME 2017 International Mechanical Engineering Congress and Exposition, Tampa, FL, USA, 3–9 November 2017.

19. Anbry, N. The dynamics of coherent structures in the wall region of a turbulent boundary layer. *J. Fluid Mech.* **1988**, *192*, 115–173.
20. Chen, H.; Reuss, D.L.; Sick, V. On the use and interpretation of proper orthogonal decomposition of in-cylinder engine flows. *Meas. Sci. Technol.* **2012**, *23*, 085302. [CrossRef]
21. Cizmas, A.; Paul, G.; Palacios, A. Proper Orthogonal Decomposition of Turbine Rotor-Stator Interaction. *J. Propuls. Power* **2003**, *19*, 268–281. [CrossRef]
22. Lengani, D.; Simoni, D. Recognition of coherent structures in the boundary layer of a low-pressure-turbine blade for different free-stream turbulence intensity levels. *Int. J. Heat Fluid Flow* **2015**, *54*, 1–13. [CrossRef]
23. Lengani, D.; Simoni, D.; Ubaldi, M.; Zunino, P.; Bertini, F. Experimental Investigation on the Time–Space Evolution of a Laminar Separation Bubble by Proper Orthogonal Decomposition and Dynamic Mode Decomposition. *J. Turbomach.* **2016**, *139*, 31006. [CrossRef]
24. Tian, Y.; Ma, H.; Wang, L. An Experimental Investigation of the Effects of Grooved Tip Geometry on the Flow Field in a Turbine Cascade Passage Using Stereoscopic PIV. In Proceedings of the ASME Turbo Expo 2017: Turbomachinery Technical Conference and Exposition, Charlotte, NC, USA, 26–30 June 2017.
25. Vinuesa, R.; Schlatter, P.; Nagib, H.M. Role of data uncertainties in identifying the logarithmic region of turbulent boundary layers. *Exp. Fluids* **2014**, *55*, 1751. [CrossRef]
26. Walsh, E.J.; Mc Eligot, D.M.; Brandt, L.; Schlatter, P. Entropy Generation in a Boundary Layer Transitioning under the Influence of Freestream Turbulence. *J. Fluids Eng.* **2011**, *133*, 61203. [CrossRef]
27. Skifton, R.S.; Budwig, R.S.; Crepeau, J.C.; Xing, T. Entropy Generation for Bypass Transitional Boundary Layers. *J. Fluids Eng.* **2017**, *139*, 041203. [CrossRef]
28. Vinuesa, R.; Orlu, R.; Vila, C.S.; Ianiro, A.; Discetti, S.; Schlatter, P. Revisiting History Effects in Adverse-Pressure-Gradient Turbulent Boundary Layers. *Flow Turbul. Combust.* **2017**, *99*, 565–587. [CrossRef] [PubMed]
29. Vinuesa, R.; Bobke, A.; Örlü, R.; Schlatter, P. On determining characteristic length scales in pressure-gradient turbulent boundary layers. *Phys. Fluids* **2016**, *28*, 55101. [CrossRef]
30. Vila, C.S.; Orlu, R.; Vinuesa, R.; Schlatter, P.; Ianiro, A.; Discetti, S. Adverse-Pressure-Gradient Effects on Turbulent Boundary Layers: Statistics and Flow-Field Organization. *Flow Turbul. Combust.* **2017**, *99*, 589–612. [CrossRef] [PubMed]
31. Lengani, D.; Simoni, D.; Ubaldi, M.; Zunino, P.; Bertini, F.; Michelassi, V. Accurate Estimation of Profile Losses and Analysis of Loss Generation Mechanisms in a Turbine Cascade. *J. Turbomach.* **2017**, *139*, 121001–121007. [CrossRef]

Article

Experimental Investigation of Flow-Induced Motion and Energy Conversion of a T-Section Prism

Nan Shao, Jijian Lian, Guobin Xu, Fang Liu , Heng Deng, Quanchao Ren and Xiang Yan * 

State Key Laboratory of Hydraulic Engineering Simulation and Safety, Tianjin University, No. 92, Wei Jin Road, Nan Kai District, Tianjin 300072, China; shaonan@tju.edu.cn (N.S.); tju_luntan@126.com (J.L.); xuguob@sina.com (G.X.); fangliu@tju.edu.cn (F.L.); d_heng@yeah.net (H.D.); 2014205258@tju.edu.cn (Q.R.)
* Correspondence: xiangyan@tju.edu.cn; Tel.: +86-22-2740-1127

Received: 25 June 2018; Accepted: 2 August 2018; Published: 6 August 2018

check for updates

Abstract: Flow-induced motion (FIM) performs well in energy conversion but has been barely investigated, particularly for prisms with sharp sections. Previous studies have proven that T-section prisms that undergo galloping branches with high amplitude are beneficial to energy conversions. The FIM experimental setup designed by Tianjin University (TJU) was improved to conduct a series of FIM responses and energy conversion tests on a T-section prism. Experimental results are presented and discussed, to reveal the complete FIM responses and power generation characteristics of the T-section prism under different load resistances and section aspect ratios. The main findings are summarized as follows. (1) Hard galloping (HG), soft galloping (SG), and critical galloping (CG) can be observed by varying load resistances. When the load resistances are low, HG occurs; otherwise, SG occurs. (2) In the galloping branch, the highest amplitude and the most stable oscillation cause high-quality electrical energy production by the generator. Therefore, the galloping branch is the best branch for harvesting energy. (3) In the galloping branch, as the load resistances decrease, the active power continually increases until the prism is suppressed from galloping to a vortex-induced vibration (VIV) lower branch with a maximum active power P_{harn} of 21.23 W and a maximum η_{out} of 20.2%. (4) Different section aspect ratios (α) can significantly influence the FIM responses and energy conversions of the T-section prism. For small aspect ratios, galloping is hardly observed in the complete responses, but the power generation efficiency ($\eta_{out,0.8}$ = 27.44%) becomes larger in the galloping branch.

Keywords: flow-induced motion; sharp sections; T-section prism; load resistances; section aspect ratios; energy conversion

---

1. Introduction

The flow-induced motion (FIM) phenomenon [1] widely exists in the civil engineering field, and it can lead to the failure of oscillating structures such as solar receiver tubes [2–4], long-span bridges [5], parallel twin bridges [6], offshore risers [7], and aircraft [8]. Many early studies have devoted great efforts to the suppression [9,10] of FIM. With the study of FIM increasing significantly [11–14], researchers [15–17] are gradually paying more attention to the potential of FIM energy. Many creative structures [18–20] have been proposed for exploiting this energy, especially for the vortex-induced vibration (VIV) [21] and galloping [22,23] responses of FIM.

VIV occurs due to the alternating shedding of vortices from either side of the bluff cylinder [1]. Galloping occurs due to the forces acting on a prism as it is subjected to periodic variations in the angle of attack of the flow [24]. An isolated smooth circular cylinder can only undergo VIV, while galloping is rarely observed [25]. On the contrary, galloping is easily observed for non-circular-section prisms such as rectangular prisms, triangular prisms, and passive turbulence control (PTC) circular cylinders, etc.

There are some studies on rectangular prisms [26–29] to measure the aerodynamic forces during unsteady galloping oscillations, and some studies report that the incident angle has a significant influence on the performance of prisms. Research on triangular prisms has concluded that the stability of galloping strongly depends on the incident flow orientation and on the aspect ratio, as well as on the mass and damping [30–33]. In addition, two different types of galloping, soft galloping (SG) and hard galloping (HG), were first discovered during trials on PTC circular cylinders by Park [21]. SG is self-excited by VIV, and HG cannot be self-excited by VIV, but HG can be excited by external excitation at a high-flow velocity. Furthermore, Zhang, J. et al. [34,35] also discovered the phenomenon of SG and HG in a study on triangular prisms, and pointed out two HG types: I-type HG and II-type HG. Subsequently, Lian, J. et al. [36] found that the VIV and galloping of the triangular prism in complete FIM responses depended on the damping, stiffness, and mass of the system, as well as the section of the prism.

Research on VIV and galloping with energy conversion targets has developed rapidly in recent years. FIM power generation equipment such as VIVACE (Vortex-Induced Vibration for Aquatic Clean Energy) was successively proposed and gradually optimized. By inventing the virtual damper spring system [37,38] (V_{CK}) and the PTC [39], Bernitsas et al., in the University of Michigan, carried out many experiments to further understand FIM responses and power generation for circular cylinders at high damping [37], a high Reynolds number [21,39,40], variable stiffness [41,42], and variable mass [41], etc. It was found that a high Reynolds number, high damping, and low mass helped to improve power generation capacity and oscillation strength. In addition, non-cylindrical oscillators such as PTC cylinders performed with outstanding advantages in power generation due to the high amplitude of galloping.

At present, most researchers pay more attention to oscillators with regular cross-sections, such as cylinders, triangular prisms, square prisms, and rectangular prisms. However, less attention is paid to oscillators with irregular cross-sections, such as the T-section. Recently, FIM tests on T-section prisms were conducted in Tianjin University [43]; the published study reported that T-section prisms also presented galloping responses. Actually, the T-section prism can be regarded as a simplified triangular prism with axisymmetric structures. Until now, previous studies on the FIM of T-section prisms have not been systematic, and have only focused on oscillation responses. The complete FIM responses and the power generation characteristics of T-section prisms have not been investigated systematically. In order to better understand the complete FIM responses and their power generation characteristics, a series of tests were conducted, specifically including the following three aspects:

(1) Experimental research on FIM responses to external forces were conducted to determine the complete FIM responses of T-section prisms;

(2) FIM power generation tests with different load resistances were carried out to investigate the power generation of the T-section prism in complete FIM responses;

(3) In order to guide the optimization design of the prism, FIM responses and power generation tests at different section aspect ratios were conducted.

2. Experiment Setup

2.1. Water Channel and Calibration of Flow Velocity

All experiments were conducted in a recirculating water channel at the State Key Laboratory of Hydraulic Engineering Simulation and Safety (SKL-HESS) of Tianjin University (shown in Figure 1). The recirculating water channel consisted of a water tank, a variable frequency power pump, a 2-m wide flow channel, a bend flow channel, a contraction section, and a 1-m wide flow channel. The whole length was about 50 m. The water tank dimensions were 5 m (length) × 5 m (width) × 2 m (height). The channel was made out of transparent tempered glass and it was powered by a 90-kW variable frequency power pump with a maximum speed of 490 r/min. It could recirculate 200 cubic meters of fresh water at flow rates up to 2600 L/s. The frequency range of water pump was 0.0–50.0 Hz, controlled by a frequency conversion controller (FCC). In the 1-m wide flow channel, the velocity variation range was 0.0–1.8 m/s. The water depth was 1.34 m.

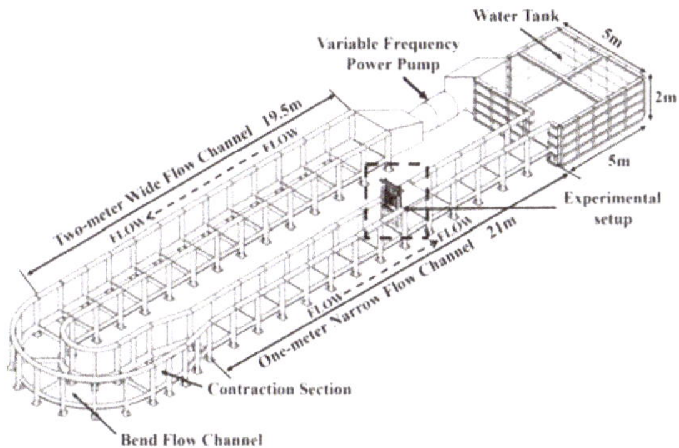

Figure 1. Recirculating water channel system.

In order to guarantee the authenticity of the results, two test instruments, a Pitot tube with a differential pressure transmitter and a propeller current meter [36], were employed to measure the flow velocity. The accuracy of the differential pressure transmitter was within ±0.1% of 6 KPa, which was the linear available range, and the resolution was within 0.01% FS (Full Scale). All of the data were recorded over a time interval of 60 s at a 40 Hz sampling rate. The probes of the Pitot tube and the propeller were placed 1 m in front of the T-section prism.

All experiments were conducted using the TrSL3 (20,000 < Re < 300,000) flow regime [13]. This study covered a range of Reynolds numbers of 45,133 $\leq$ Re $\leq$ 116,396 (0.516 m/s $\leq$ U $\leq$ 1.332 m/s). In order to describe the incoming flow, the flow velocity and the turbulence were both analyzed. We conducted three tests with different flow velocities. The velocity profile and the turbulence profile of the recirculating water channel were taken at a 15–100-cm deep area of the test section, as shown in Figure 2. In Figure 2a, the flow velocity near the bottom was smaller, due to viscous action. The average flow velocities in the vertical direction of three tests were 0.62 m/s, 0.84 m/s, and 1.22 m/s, respectively. The difference between the flow velocities in the oscillation range (40 cm–80 cm) was small. In Figure 2b, the turbulence grew as the flow velocity decreased and the depth increased. The average turbulences in the vertical direction of the three tests were 8%, 14%, and 24%, respectively. The difference in the turbulence in the oscillation range (40 cm–80 cm) was small as well. In a word, it could be ensured that the incoming flow in the oscillation range of the prism was a uniform flow.

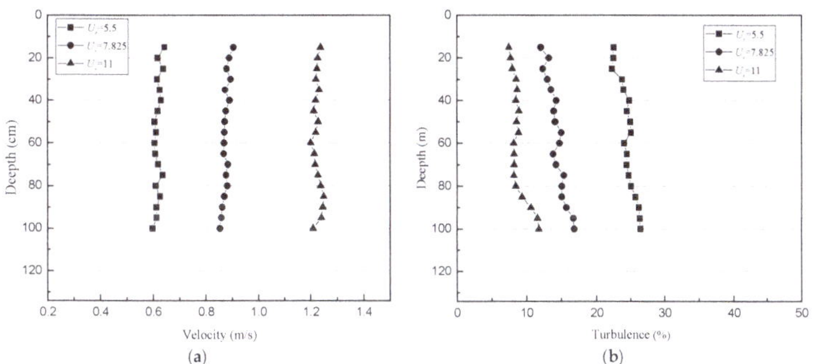

Figure 2. Incoming flow characteristics. (**a**) Velocity profile; (**b**) turbulence profile.

2.2. Test Apparatus and Energy Conversion System

2.2.1. Test Apparatus

The test apparatus consisted of two parts: the oscillation system and the transmission system. The oscillation system included the frame, linear guide ways, side struts, spring carrier structure, and springs (shown in Figure 3). The frame was made of steel and was fixed in a moving car at the top of the 1-m wide flow channel. The linear guideways were attached onto the steel frame, parallel to the side struts and perpendicular to the flow velocity direction.

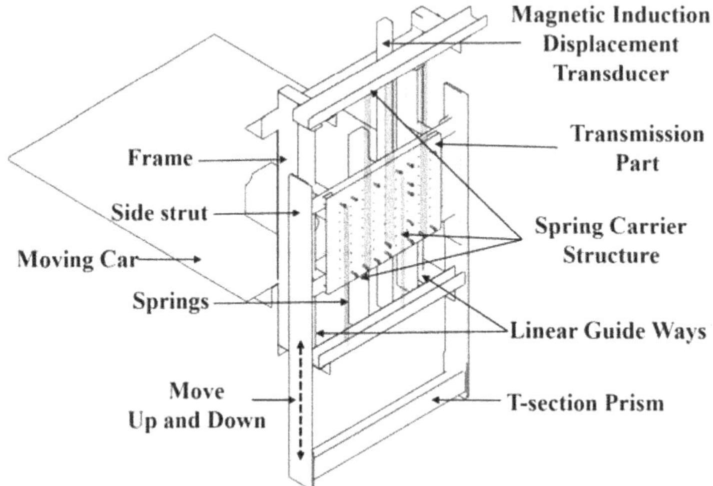

Figure 3. Test apparatus.

The transmission system included two side struts and a connective structure. The side struts, which were made of duralumin plate, were rigidly connected with the prism, which was immersed in the water. The connective structure was joined with the side struts and springs, and it was constrained to move on the linear guideways in a vertical direction by four linear bearings. Each side of the upper and the lower extensional springs was suspended vertically on the frame and the connection structure by the spring carrier structure.

2.2.2. Energy Conversion System

The rack was fixed into the transmission part that was connected to the rotor of the generator by the gears (shown in Figure 4). The linear motion of the prism was transferred to the rotational motion of the rotor. The generator was connected to the load resistances by the output wire, creating an electrical circuit. In this system, the mechanical energy of the prism was partly transformed to the electric energy of the generator and was dissipated by the load resistances.

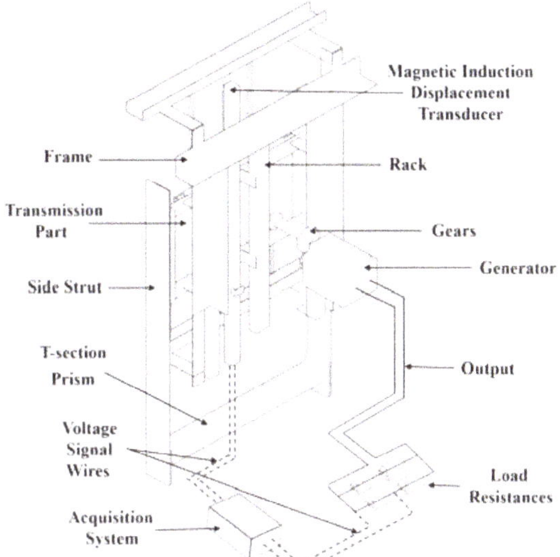

Figure 4. Energy conversion system.

2.2.3. T-Section Prisms

In the tests, the projection width of the T-section prism in the direction of the incoming flow (D) was 0.1 m, the prism length (L) was 0.9 m, the height of the prism cross-section (H) was 0.1 m, and the thickness (d) was 0.01 m. The prism was made of polymethyl methacrylate. A rectangular endplate was installed at both ends of the prism to reduce the effect of the boundary [30]. The thickness of the endplate was 0.01 m (shown in Figure 5).

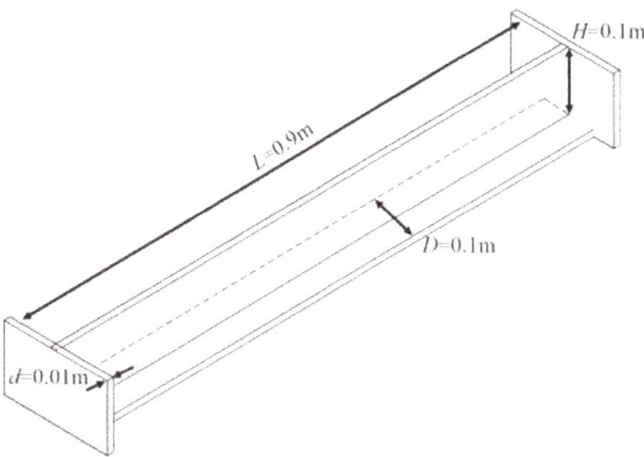

Figure 5. T-section prism size diagram.

In order to investigate the influence of the aspect ratio of the T-section prism on FIM responses and power generation characteristics, five T-section prisms with different H values (0.15 m, 0.12 m, 0.1 m, 0.09 m, and 0.08 m) were tested in the experiments. The corresponding section aspect ratios ($\alpha = H/D$) were 1.5, 1.2, 1.0, 0.9, and 0.8, respectively.

2.3. Test Methods and Sensors

2.3.1. Displacement, Frequency, and Voltage

The experimental study had two objectives. The first objective was to find the complete FIM responses of the T-section prism and to analyze the stability of each branch. The second aim was to estimate the energy conversion of the oscillation system in the complete FIM responses, and to investigate the influence of different aspect ratios on the power generation efficiency.

For oscillation characteristic tests, the main measurements were the displacements and frequencies of the prism. The displacements were tested by a magnetic induction displacement transducer with a direct current (DC) 24 V working voltage, which was supplied by external DC power. The testing range was 0–800 mm, with a sensitivity of 0.1% and an error range of ±0.05%.

For the power generation tests, the connection method for the load resistances is presented in Figure 6. The minimum of the load resistances was 1 Ω, and the maximum was 50 Ω. In this study, 10 different load resistances were applied: 4 Ω, 8 Ω, 11 Ω, 13 Ω, 16 Ω, 18 Ω, 21 Ω, 31 Ω, 41 Ω, and 51 Ω. The experimental data was collected in the form of a voltage signal (shown in Figure 7) by the data acquisition system.

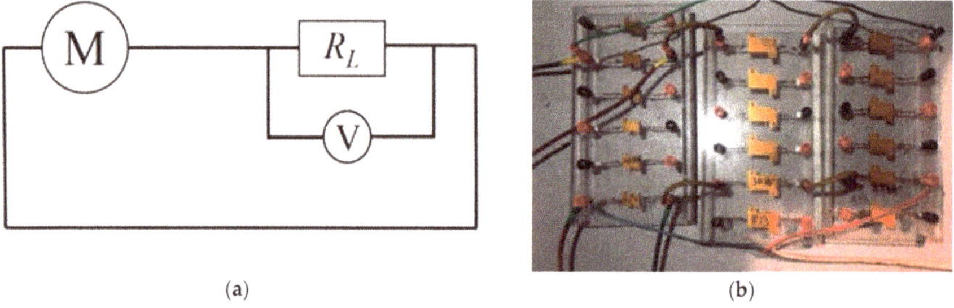

Figure 6. Voltage acquisition system: (**a**) contained circuit diagram; (**b**) connection resistances diagram.

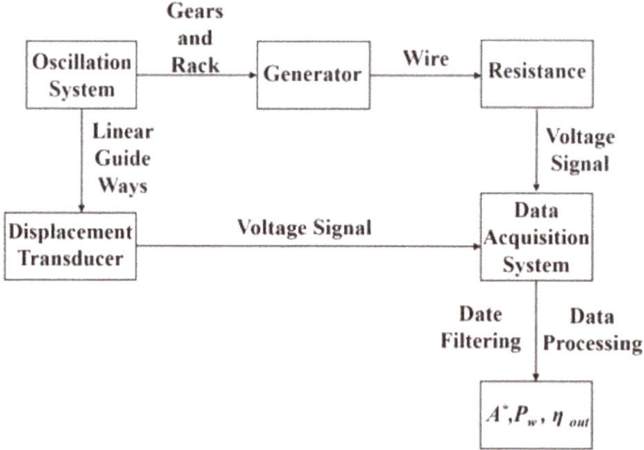

Figure 7. Test flow chart.

2.3.2. Active Power and Power Generation Efficiency

Active power (P_{harn}) and power generation efficiency (η_{out}) are the key parameters for evaluating the energy conversion capacity of the prism. In the tests, the system damping (ζ) was varied by

changing the load resistances (R_L) [34]. The output voltage (u) of the generator was measured by the data acquisition system, while P_{harn} and η_{out} of the system were calculated by the following equations:
Instantaneous power expression: Equations (1)–(3)

$$P(t) = \frac{u^2(t)}{R_L} \tag{1}$$

where $P(t)$ is the instantaneous power, $u(t)$ is the instantaneous voltage, and R_L is the load resistance. The active power is written as:

$$P_{harn} = \frac{1}{T}\int_0^T P(t)dt = \frac{1}{T}\int_0^T \frac{u^2(t)}{R_L}dt \tag{2}$$

where P_{harn} is the active power and T is a period of oscillation.
The power generation efficiency is derived as:

$$\eta_{out} = \frac{P_{harn}}{P_w} \tag{3}$$

where η_{out} is the power generation efficiency and P_w is the total power in the fluid, which is written as:

$$P_w = \frac{1}{2}\rho U^3 DL \tag{4}$$

where ρ is the water density, U is the incoming flow velocity, D is the projection width of the T-section prism in the direction of the incoming flow, and L is the prism length.

2.4. Calibration of Stiffness and Damping

A simple physical spring system designed by Xiang Yan et al. [36] was adopted to implement the stiffness of the system. First, free decay tests [35] with different spring stiffness values in the air were conducted to obtain the natural frequency (f_n) of the experimental system.

The stiffness (K) of the system varied from 800 to 1600 N/m. For each K, free decay tests were performed six times for the respective cases in air. f_n was then calculated using a simple averaging method. The test results are shown in Table 1. In these tests, the damping (C_{total}) of the experimental system was equal to the mechanical damping (C_m). The damping errors of different stiffnesses were within the allowable ranges of the tests (the average was ±5%).

Table 1. Free decay test results by varying stiffnesses.

K (N/m)	f_n (Hz)	m_{osc} (kg)	$C_{total} = C_m$ (N·s·m^{-1})	$\zeta_{total} = \zeta_m$
800	0.833	29.210	34.451	0.113
1000	0.930	29.302	34.032	0.094
1200	1.015	29.521	33.932	0.082
1400	1.087	30.046	34.903	0.085
1600	1.156	30.355	33.324	0.076

The damping ratio (ζ_{total}) of the test results could be determined by using the logarithmic decrement method, which is expressed as:

$$\zeta_{total} = \frac{1}{2\pi}\ln\left(\frac{A_i}{A_{i+1}}\right) \tag{5}$$

where A_i is the amplitude of the ith peak and ζ_{total} is the damping ratio of the experimental system.

The C_{total} of the oscillation system can be calculated as:

$$C_{total} = 2\zeta \sqrt{m_{osc} K} \tag{6}$$

where K is the system stiffness, C_{total} is the damping of the oscillation system, and m_{osc} is the system mass.

The FIM responses of the T-section prism with different ζ_{total} are described in this section. The ζ_{total} of the system is varied by changing R_L, and the parameters are shown in Table 2.

Table 2. Free decay test results by varying load resistances.

R_L (Ω)	f_n (Hz)	m_{osc} (kg)	C_{total} (N·s·m^{-1})	C_{harn} (N·s·m^{-1})	ζ_{total}
4.210	1.084	30.046	142.212	107.312	0.347
8.130	1.089	30.046	125.192	90.292	0.305
11.250	1.086	30.046	100.951	66.051	0.246
13.410	1.087	30.046	85.482	50.582	0.208
16.330	1.082	30.046	72.783	37.883	0.177
18.250	1.086	30.046	67.547	32.647	0.165
21.450	1.088	30.046	62.842	27.942	0.153
31.540	1.090	30.046	52.115	17.215	0.127
41.110	1.086	30.046	48.680	13.780	0.119
51.620	1.084	30.046	46.974	12.074	0.115
∞	1.087	30.046	34.903	0.000	0.085

3. Results and Discussion

3.1. Vibration Characteristic Analysis

3.1.1. Amplitudes and Frequencies

To explore the FIM performances of the T-section prism in water, a series of FIM tests with 10 different R_L were carried out at K = 1400 N/m (as seen in Section 2.4). In Figure 8a, A^* is the amplitude ratio ($A^* = A/D$), A denotes the average amplitude of continuous oscillation for 30 s, D is the projection width of the T-section prism in the direction of the incoming flow, U_r is the reduced velocity ($U_r = U/(D \cdot f_n)$), U is the incoming flow velocity, and f_n is the natural frequency in air. In Figure 8b, f^* is the frequency ratio ($f^* = f_{osc}/f_n$), and f_{osc} is the main frequency of oscillation obtained from the displacement time-history curves by the Fast Fourier Transformation (FFT) method. A^* and f^* vary with U_r; U and the Reynolds number Re for all R_L cases are plotted in Figures 8–11, respectively.

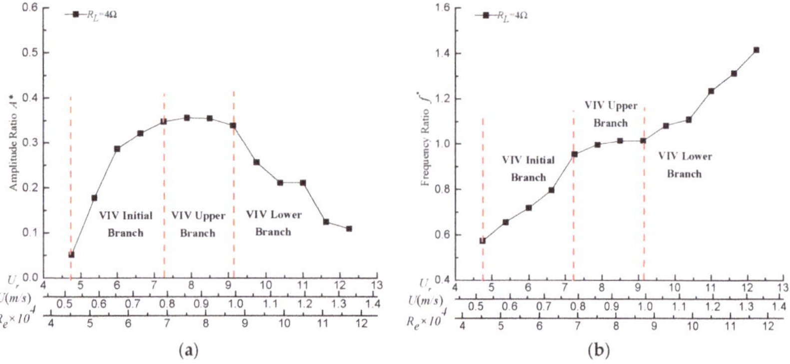

Figure 8. Vortex-induced vibration (VIV) oscillation characteristics (R_L = 4 Ω, ζ = 0.347): (**a**) response of the amplitude ratio; (**b**) response of the frequency ratio.

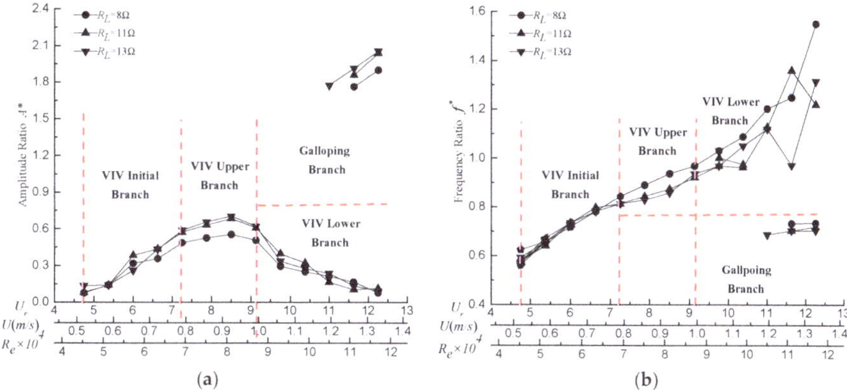

Figure 9. High galloping (HG) oscillation characteristics $8\ \Omega \leq R_L \leq 13\ \Omega$ ($0.208 \leq \zeta \leq 0.305$): (**a**) response of the amplitude ratio; (**b**) response of the frequency ratio.

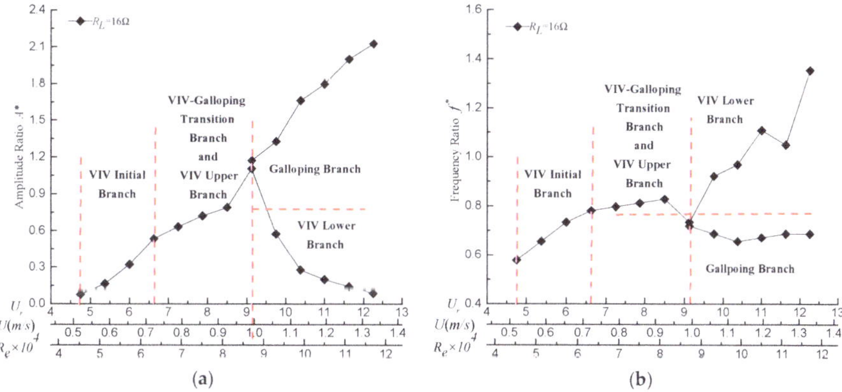

Figure 10. Critical galloping (CG) oscillation characteristics $R_L = 16\ \Omega$ ($\zeta = 0.177$): (**a**) response of the amplitude ratio; (**b**) response of the frequency ratio.

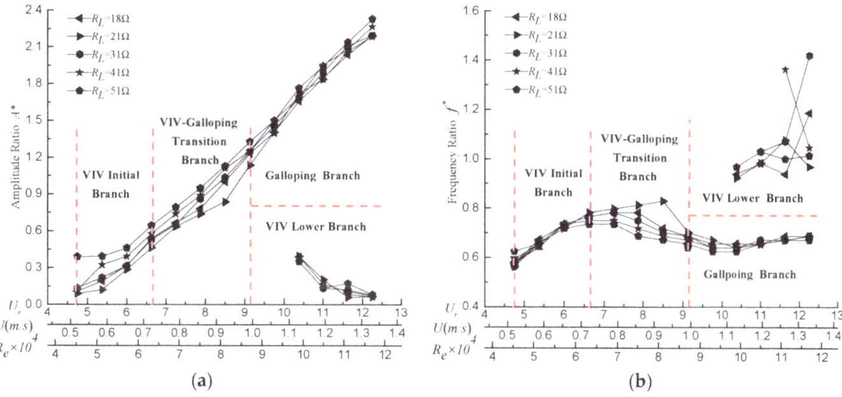

Figure 11. Soft galloping (SG) oscillation characteristics $18\ \Omega \leq R_L \leq 51\ \Omega$ ($0.115 \leq \zeta \leq 0.165$): (**a**) response of the amplitude ratio; (**b**) response of the frequency ratio.

(1) VIV: For $R_L = 4\,\Omega$ ($\zeta = 0.347$), the T-section prism was only characterized by the typical VIV in the range of $4.75 \leq U_r \leq 12.25$ (shown in Figure 8), and the prism did not present any signs of galloping. For $4.75 \leq U_r \leq 7.25$, A^* and f^* were small but increased rapidly, indicating that the prism underwent the VIV initial branch. The motion of the prism was induced by vortex shedding, but the oscillation and vortex shedding were not synchronized well. For $7.25 \leq U_r \leq 9.125$, A^* stayed at 0.35, and f^* increased sharply and was maintained at approximately 1. The oscillation entered the VIV upper branch (lock-in range), which was more stable than the VIV initial branch. For $U_r \geq 9.125$, A^* rapidly collapsed down to about 0.1, and f^* increased rapidly to a high level ($f^* \gg 1$). In this range, the oscillation and vortex shedding were not synchronized well, demonstrating that the prism underwent the VIV lower branch.

(2) HG: For $8\,\Omega \leq R_L \leq 13\,\Omega$ ($0.208 \leq \zeta \leq 0.305$), two modes of motion could be observed in the response of the prism (shown in Figure 9). If the prism oscillated freely (without any external forces acting on the prism), the prism underwent complete VIV responses as U_r increased or decreased in the range of $4.75 \leq U_r \leq 12.25$. However, when U_r decreased from 12.25 and the prism was manually pushed by a threshold initial displacement (exceeding $1 \times D$), the prism first entered the galloping branch, accompanied by a large amplitude ($A^* \approx 2.3$) and a stable frequency ($f^* = 0.7$). As U_r decreased, A^* decreased gradually and the oscillation was still in the galloping branch. When U_r reached 11–11.625, the oscillation was suddenly suppressed, and the oscillation mode transformed from galloping to VIV (lower branch), accompanied by a sudden collapse in A^* from 1.8 to 0.2, and a sudden jump in f^* from 0.7 to 1. It was concluded that the prism cannot be self-excited from VIV to galloping in any case, without any external conditions (such as a threshold initial displacement) or forces acting on the prism, no matter whether the velocity decreases or increases, demonstrating that the prism experiences HG responses in these cases.

(3) Critical galloping (CG): For $R_L = 16\,\Omega$ ($\zeta = 0.177$), the system presented both complete VIV responses and galloping while the two branches intersected (shown in Figure 10). When U_r increased from 4.75 to 12.25, the prism experienced a VIV initial branch, a VIV upper branch, and a VIV lower branch. If a threshold initial displacement was applied at $U_r = 12.25$, the oscillation was suddenly enhanced, and the oscillation mode dramatically transformed from VIV (initial branch) to galloping, accompanied by a sudden jump in A^* from 0.08 to 2.12 as well as a sudden collapse in f^* from 1.35 to 0.68. Afterwards, the oscillation entered the galloping branch. A^* gradually decreased, with the decrease of U_r, and f^* was maintained at about 0.7. No oscillation suppression phenomena occurred until the oscillation underwent the VIV upper branch where the intersection point was $U_r = 9.125$. However, at $U_r \geq 9.125$, galloping could be induced by external excitation. This condition is actually a critical oscillation mode between HG and SG, which can be referred to as "critical galloping" (CG).

(4) SG. For $18\,\Omega \leq R_L \leq 51\,\Omega$ ($0.115 \leq \zeta \leq 0.165$), two modes of motion were observed in the responses of the prism. If the prism oscillated freely (without any external forces acting on the prism), the T-section prism experienced the VIV initial branch, followed by the VIV-galloping transition branch, and ended with the galloping branch (shown in Figure 11). In the range of $4.75 \leq U_r \leq 6.625$, A^* and f^* were small but increased rapidly, indicating that the prism underwent the VIV initial branch. In the range of $6.625 \leq U_r \leq 9.125$, A^* continued to grow with a strong uptrend to 1.5, while f^* decreased to about 0.7 with a slight downward trend. For $U_r \geq 9.75$, A^* continued to increase up to 2.36, while f^* almost remained at 0.7. The oscillation underwent the fully developed galloping branch. On the other hand, at $U_r = 12.25$, external suppression was applied to the prism; thus, the oscillation presented a low $A^* \approx 0.2$ and a high $f^* \approx 1.4$, the system experienced the VIV lower branch until $U_r \leq 10.375$, and the oscillation mode returned to galloping. For $6.625 \leq U_r \leq 10.375$, the oscillation was self-excited from VIV to galloping (regardless of whether the velocity increased or decreased). If U_r continued to increase, the oscillation would be maintained within the range of galloping. For $U_r \geq 10.375$, if there was an external suppression, the oscillation would be converted to the VIV lower branch and could not be self-excited to galloping.

It can be concluded that the oscillation responses of the T-section prism were similar to those of the triangular prism and the PTC circular cylinder. The occurrence conditions of different FIM branches of the T-section prism are listed in Table 3.

Table 3. VIV and galloping occurrence conditions with $K = 1400$ N/m.

Oscillation Form	R_L	U_r	External Excitation	External Suppression	Self-Excited by VIV
VIV	4 Ω	4.75 ~12.25			√
HG	8 ~13 Ω	11 ~12.25	√ (HG)		
CG	16 Ω	9.125 ~12.25		√ (VIV Low B)	√
SG	18 ~51 Ω	10.375 ~12.25		√ (VIV Low B)	√

3.1.2. Time-History and Frequency Spectrum of Each Branch

In order to explore the oscillation characteristics of each branch, four typical time-history curves and frequency spectra are depicted in this section (plotted in Figures 12 and 13). The physical parameters of the oscillation system are listed in Table 4.

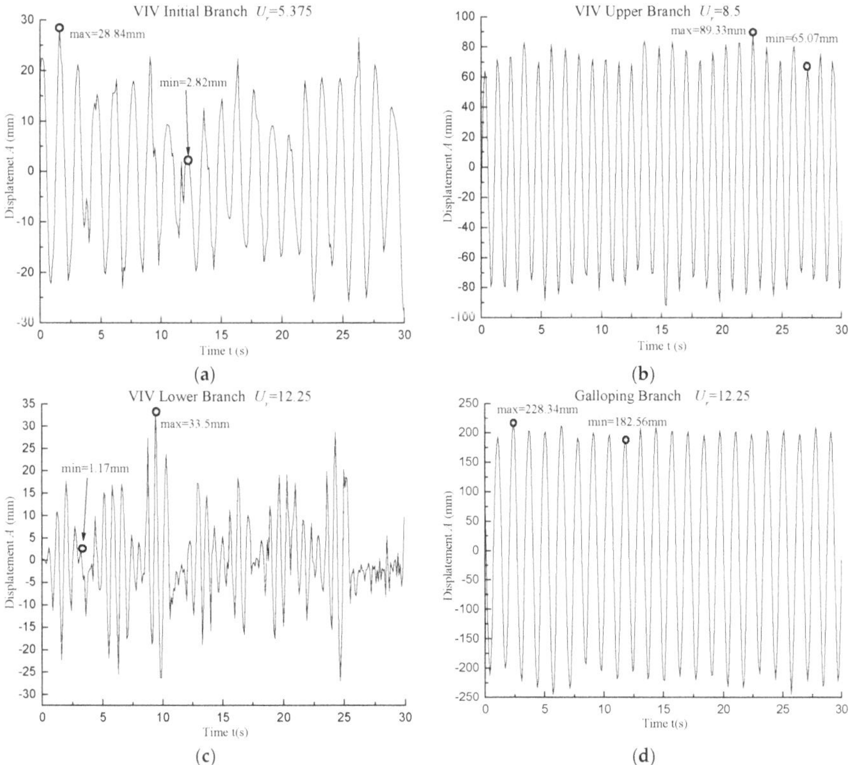

Figure 12. Typical flow-induced motion displacement time-history curves: (**a**) VIV initial branch; (**b**) VIV upper branch; (**c**) VIV lower branch; (**d**) galloping branch.

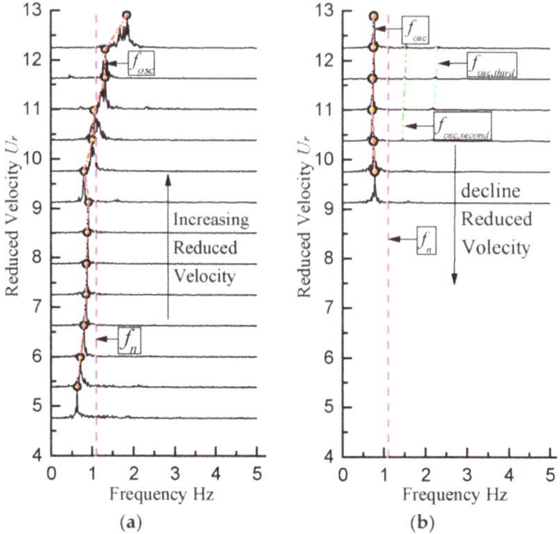

Figure 13. Typical oscillation frequency spectrum: (**a**) complete VIV branch; (**b**) galloping branch.

Table 4. Oscillation form and parameters list.

R_L (Ω)	C_{total} (N·s·m^{-1})	C_{harn} (N·s·m^{-1})	ζ	Oscillation Form
16	72.783	37.883	0.177	CG

(1) Time-History

For U_r = 5.375, the prism underwent the VIV initial branch. The time-history curve of the displacement is shown in Figure 12a. It was observed that the minimum of the positive displacement was 2.82 mm, the maximum was 28.84 mm, and the difference between the maximum and minimum was 1.25A (A = 16 mm), indicating that the amplitude was low and the oscillation was unstable.

As U_r reached 8.5, the oscillation mode entered the VIV upper branch, as shown in Figure 12b. It was observed that the minimum of the positive displacement was 65.07 mm, the maximum was 89.33 mm, and the difference between the maximum and minimum was 0.31A (A = 78 mm). Moreover, the prism underwent a more stable oscillation, although there was a certain fluctuation response amplitude.

For U_r = 12.25, it was observed that the minimum of the positive displacement was 1.17 mm, the maximum was 33.50 mm, and the difference between the maximum and minimum was 4.04A (A = 8 mm). The oscillation amplitude fluctuation was large and the performance was very unstable, as shown in Figure 12c, demonstrating that the prism underwent the VIV lower branch.

For U_r = 12.25 (galloping branch), the minimum of the positive displacement was 182.56 mm, the maximum was 228.34 mm, and the difference between the maximum and minimum was 0.21A (A = 212 mm), indicating that the oscillation amplitude was large and the fluctuation was small, as shown in Figure 12d.

Both stability and amplitude are key parameters for energy harvesting. Thus, the galloping branch is the most suitable branch for energy extraction and utilization.

(2) Frequency Spectrum

The frequency spectrum is important for describing the oscillation characteristics and mechanical energy. The oscillation frequency of the prism was extracted by using the FFT method from the displacement time-history curves (shown in Figure 13). Figure 13a plots the frequency spectrum of the complete VIV branch and Figure 13b plots the frequency spectrum of the galloping branch, while the trends of f_n and f_{osc} are marked by dashed lines.

For $4.75 \leq U_r \leq 6.625$, the prism underwent the VIV initial branch, and the frequency band was wider and exhibited double peaks in the frequency spectrum, revealing that periodicity of oscillation and mechanical energy was poor.

In the range of $6.625 \leq U_r \leq 9.125$, the prism entered the VIV upper branch, the frequency band became narrower, and the main frequency was clearly observed. The periodicity of oscillation was significantly enhanced, the prism oscillated more stably, and the main frequency was concentrated at 0.88 Hz.

For $U_r \geq 9.125$, two modes of oscillation were observed. When the prism oscillated freely (without external excitation), the prism went into the VIV lower branch, the main frequency was not obvious, the frequency band widened, and the energy was very dispersed.

On the other hand, if the prism was manually pushed by a threshold initial displacement (exceeding $1 \times D$), the prism entered the galloping branch with a narrowed frequency band and evident first dominant frequency, as shown in Figure 13b. It was noted that in the galloping branch, the oscillation was mainly caused by the instability of the lift force due to the T-section prism with sharp sections. Because of the lower energy and the higher frequency of the vortex shedding, the effect of vortex shedding on the prism was weaker, resulting in the second and third dominant frequencies being hardly observed in the frequency spectrum.

The conclusions of the time-history curves and frequency spectra are as follows: (1) different fluctuations are observed in each branch—the VIV upper branch and the galloping branch have higher amplitudes and more stable oscillations; (2) the VIV upper branch and the galloping branch have a better periodicity and frequency spectrum. In summary, the VIV upper branch and the galloping branch are more suitable for energy harvesting, but the galloping branch is better (shown in Table 5).

Table 5. Oscillation characteristics of different branches.

Branch	VIV Initial	VIV Upper	VIV Lower	Galloping
Amplitude	low	high	low	highest
Stability	bad	good	bad	best
Frequency	bad	good	bad	best

3.1.3. Summary

The specific findings are listed as follows:

(1) With an increase of damping (decrease in load resistances), the T-section prism oscillation mode gradually changes from SG ($0.115 \leq \zeta \leq 0.165$, $18\ \Omega \leq R_L \leq 51\ \Omega$) to CG ($\zeta = 0.177$, $R_L = 16\ \Omega$), and eventually to HG ($0.208 \leq \zeta \leq 0.305$, $8\ \Omega \leq R_L \leq 13\ \Omega$). The oscillation mode of SG, HG, and CG all contain VIV branches and galloping branches, the only difference being whether these can be self-excited from VIV to galloping.

(2) The analysis of the displacement time-history curves and the frequency spectra demonstrates that the stability and intensity of the VIV upper branch and the galloping branch are both better-performing, and that the maximum amplitude of the galloping branch is larger. It can be concluded that at a high velocity ($U_r \geq 10.375$, $U \geq 1.128$ m/s) the galloping branch is better for harvesting energy, and at a lower velocity ($6.625 \leq U_r \leq 9.125$, $0.720 \leq U \leq 0.992$ m/s) the VIV upper branch is better. Power generation with different load resistances will be presented in the next section.

3.2. Power Generation Analysis

3.2.1. Active Power Analysis

The variations of active power P_{harn} of the generator versus incoming flow velocity U, reduced flow U_r, and Reynolds number Re are plotted in Figure 14. P_{harn} is the average of the instantaneous power under continuous oscillation for 30 s. For $U_r \geq 6$, the oscillation system started to output electrical energy and P_{harn} was smaller, as the system experienced the complete VIV responses and the P_{harn} did not exceed 5 W. As galloping occurred, P_{harn} increased to over 20 W. The details are described as follows:

(1) For 18 Ω ≤ R_L ≤ 51 Ω, SG occurred.

As U_r increased, the prism experienced the VIV initial branch, the VIV-galloping transition branch, and the galloping branch. P_{harn} monotonically increased with U_r and the maximum P_{harn} appeared at the galloping branch ($P_{harn,max}$ = 13 W, R_L = 18 Ω, U_r = 12.25). It was noted that the higher the U_r, the lower the R_L, and the more energy was converted.

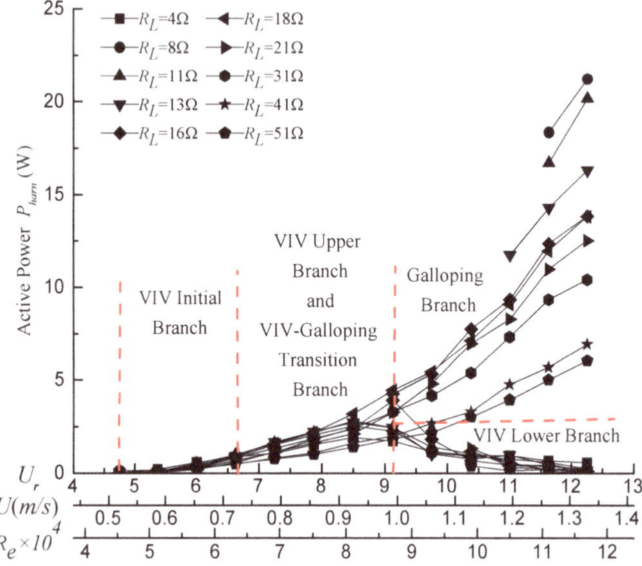

Figure 14. T-section prism power generation active power.

(2) For 8 Ω ≤ R_L ≤ 16 Ω, HG occurred, and two oscillation modes were observed.

(i) Without external excitation (such as forces, threshold initial displacement, etc.)

The prism only experienced VIV branches. P_{harn} rose with the increase of U_r at the VIV initial branch and at the VIV upper branch ($P_{harn,max}$ = 3.92 W, R_L = 16 Ω, U_r = 9.125). Afterwards, P_{harn} decreased to lower than 1 W, as U_r rose up to 12.25 in the VIV lower branch.

(ii) An external excitation was applied to the prism.

At U_r = 12.25, if the threshold initial displacement was applied to the prism, the oscillation directly jumped into the galloping branch. As U_r decreased, P_{harn} dropped rapidly. In the present tests, the maximum P_{harn} was 21.23 W (U = 1.332 m/s, Re = 116396, R_L = 8 Ω). The maximum P_{harn} was close and it was slightly lower than that of the PTC circular cylinder ($P_{harn,max}$ = 23.54 W), as reported by Lin, D [44].

(3) For R_L = 4 Ω, only VIV branches were observed.

The variation of P_{harn} versus U_r for R_L = 4 Ω was similar to those for 8 Ω ≤ R_L ≤ 16 Ω in VIV branches. Within the test flow velocity range, the prism did not experience any forms of galloping. The maximum P_{harn} was 2.5 W at U_r = 8.5 (VIV upper branch).

3.2.2. Efficiency Analysis

The variations of efficiency η_{out} versus incoming flow velocity U, reduced flow U_r, and Reynolds number Re are plotted in Figure 15. η_{out} was calculated using Equation (3) (shown in Section 2.3).

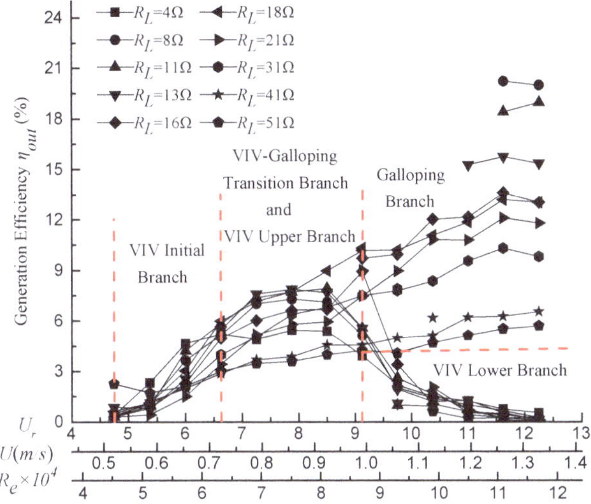

Figure 15. T-section prism power generation efficiency.

It was observed that η_{out} did not exceed 9% in VIV branches (whether in SG or HG responses). The corresponding load resistance was $R_L = 16\ \Omega$ and the corresponding velocity was $U_r = 9.125$ ($U = 0.992$ m/s, $Re = 86703$). On the contrary, the maximum η_{out} reached 20.2% in the galloping branch, which was close to the test results of the smooth (22%) and the PTC (28%) circular cylinders. The corresponding load resistance was $R_L = 8\ \Omega$, and the corresponding velocity was $U_r = 11.625$ ($U = 1.264$ m/s, $Re = 110{,}458$). It is concluded that the energy conversion efficiency in the galloping branch is much higher than that of the VIV branches for the T-section prism.

In addition, in most of the galloping branches, especially for small load resistances ($R_L < 31$) or high damping ratios ($\zeta > 0.127$), a slight declining trend was observed for U_r exceeding 11.625 in the present tests. It was indicated that $U_r = 11.625$ was the optimal velocity.

3.2.3. Stability Analysis

In order to describe the power generation stability of each oscillation branch, the instantaneous voltage time-history curves and instantaneous power time-history curves of four oscillation branches are discussed in this section (shown in Figures 16 and 17).

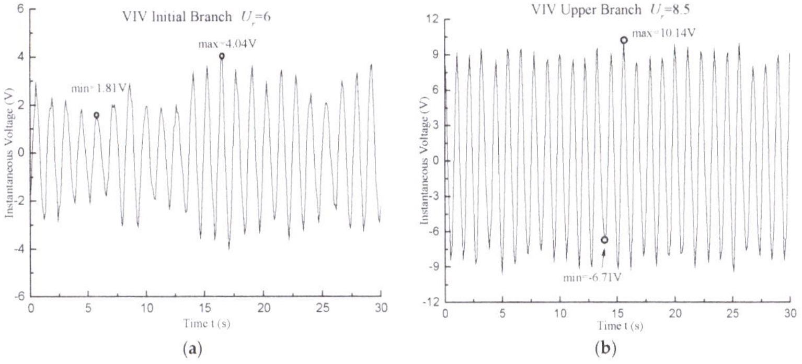

(a)

(b)

Figure 16. *Cont.*

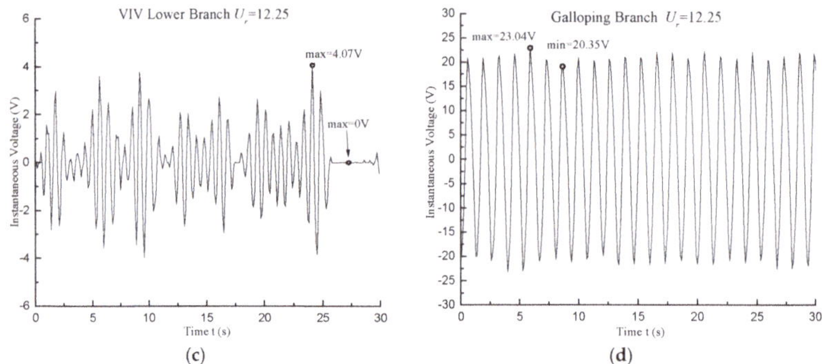

Figure 16. Instantaneous voltage time-history curves of T-section prism power generation with $R_L = 16\ \Omega$: (**a**) VIV initial branch; (**b**) VIV upper branch; (**c**) VIV lower branch; (**d**) galloping branch.

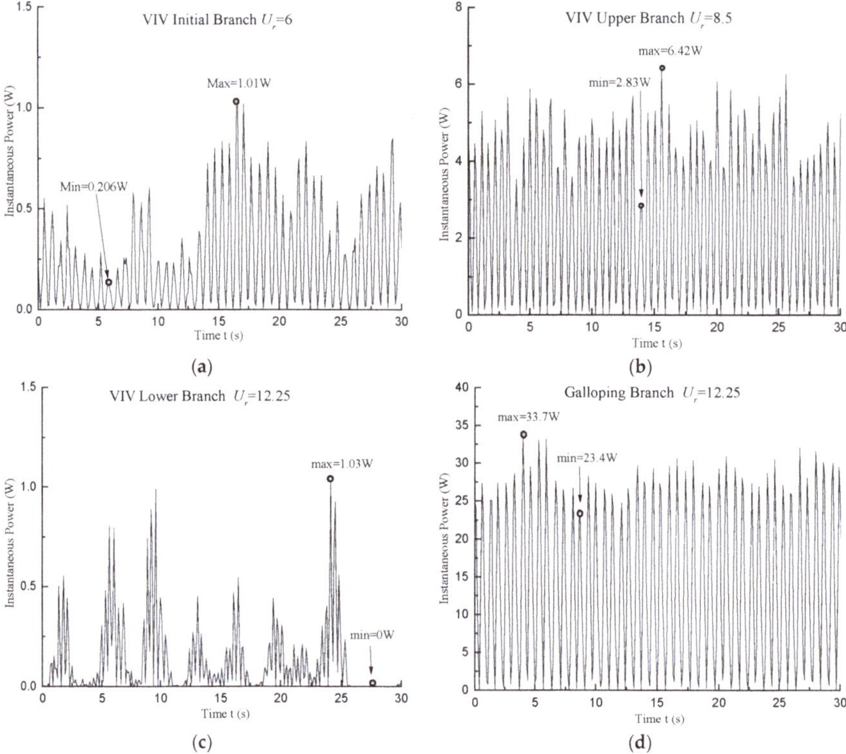

Figure 17. Power generation instantaneous power time-history curves of the T-section prism with $R_L = 16\ \Omega$: (**a**) VIV initial branch; (**b**) VIV upper branch; (**c**) VIV lower branch; (**d**) galloping branch.

Figure 16a,c show that the instantaneous voltage peaks fluctuated greatly when the prism underwent the VIV initial branch ($V_{max} = 4.04$ V, $V_{min} = 1.81$ V) and the VIV lower branch ($V_{max} = 4.07$ V, $V_{min} = 0$ V). This led directly to the instability of the instantaneous power of the VIV initial branch ($P_{max} = 1.01$ W, $P_{min} = 0.206$ W) and the VIV lower branch ($P_{max} = 1.03$ W, $P_{min} = 0$ W), as shown in Figure 17a,c. Additionally, the instantaneous voltages of the VIV initial branch and the

VIV lower branch were small, so that the powers were not as high as the instantaneous active power in the VIV upper branch (P_{max} = 6.42 W, P_{min} = 2.83 W) and in the galloping branch (P_{max} = 33.7 W, P_{min} = 23.4 W), as shown in Figure 17b,d. The peak instantaneous voltages of the VIV upper branch (V_{max} = 10.14 V, V_{min} = −6.71 V) and the galloping branch (V_{max} = 23.04 V, V_{min} = 20.35 V) had small fluctuations, as shown in Figure 16b,d. The reason for this was that the two branches had better oscillation stability. In addition, the instantaneous power of the galloping branch was much higher and the quality of the output energy was better.

In order to further discuss the stability of the harnessed power, the difference coefficient C_v was introduced as follows:

$$C_v = \frac{\sigma_p}{P_{harn,max}} \tag{7}$$

where σ_p is the square deviation of all peak powers, and $P_{harn,max}$ is the average of all peak powers. It was noted that a higher C_v corresponded to a more unstable power output; otherwise, it corresponded to a more stable power output. The power generation difference coefficient C_v was calculated based on Equation (6), and is shown in Figure 18. In different branches, the stability of the oscillation was quite different. The performances are described as follows:

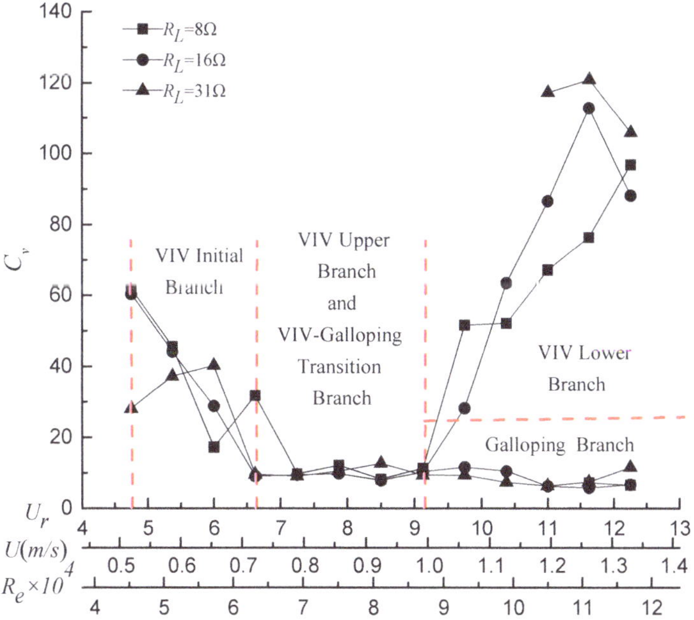

Figure 18. Power generation difference coefficient C_v value.

For $4.75 \leq U_r \leq 6.625$, the value of C_v exceeded 20 due to the unstable oscillation in the VIV initial branch, indicating that the power generation stability was poor. For $6.625 \leq U_r \leq 9.125$, as U_r increased, C_v dropped to lower than 10 and remained almost stable due to the stable oscillation of the prism. This indicated that the stability of the power output was good at the VIV upper branch. For $9.125 \leq U_r \leq 12.25$, two performances of C_v could be observed. In the VIV lower branch, C_v increased rapidly and exceeded over 100 at U_r = 11.625. This indicated that the power output was more unstable than any branches, due to the poor oscillation of the prism. On the contrary, in the galloping branch, C_v remained at approximately 10, indicating that the power output was perfectly stable due to the perfectly stable oscillation of the prism.

It is concluded that the galloping branch was the best of all FIM branches to obtain a stable power output. The results of the FIM power generation are summarized in Table 6.

Table 6. Power generation characteristics of different branches.

Branch	VIV Initial	VIV Upper	VIV Lower	Galloping
Active Power (W)	0.02–1.67	1.53–4.46	0.12–3.92	2.18–21.22
Efficiency (%)	0.38–7.6	3.45–8.9	0.11–7.9	4.07–20.2
Stability	bad	good	bad	best

3.3. Effects of Aspect Ratios on Oscillation and Energy Conversion

3.3.1. Effects of Aspect Ratios on Oscillation

Different cross-section ratios (α) can significantly influence the angle of attack and lift force, resulting in different performances by the FIM responses and the energy conversion of the T-section prism [43].

In this section, five tests for different α values were conducted. The five α values were 0.8, 0.9, 1, 1.2, and 1.5. In addition, the physical parameters of the system were summarized as: m_{osc} = 29.521 kg, K = 1200 N/m, and R_L = 16 Ω. The variation of amplitude ratio A^* and the efficiency of η_{out} versus the reduced velocity U_r, the incoming flow velocity U, and the Reynolds number Re for the five α cases are plotted in Figures 19 and 20.

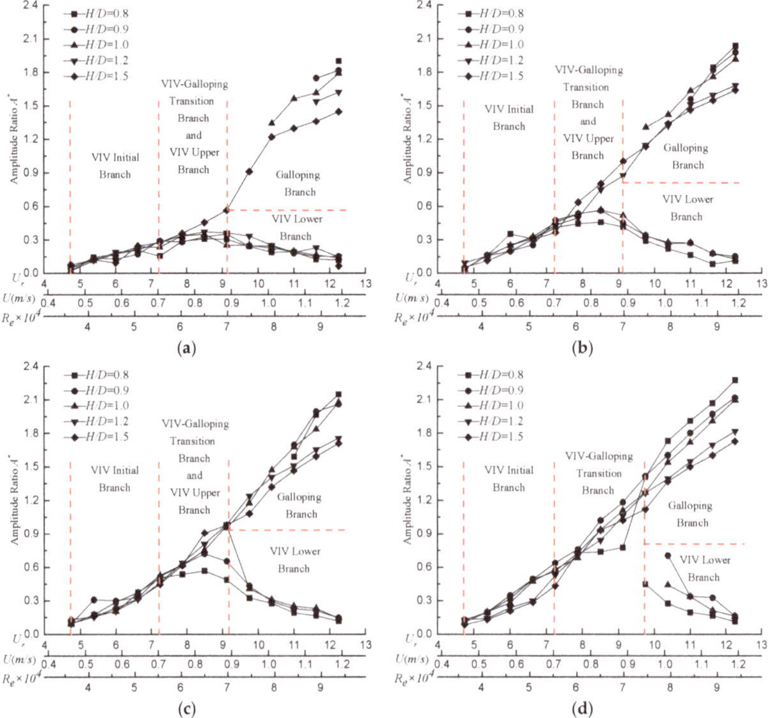

Figure 19. Oscillation characteristics with different section aspect ratios at K = 1200 N/m: (**a**) amplitude ratio (4 Ω, ζ = 0.347); (**b**) amplitude ratio (8 Ω, ζ = 0.305); (**c**) amplitude ratio (11 Ω, ζ = 0.246); (**d**) amplitude ratio (16 Ω, ζ = 0.177).

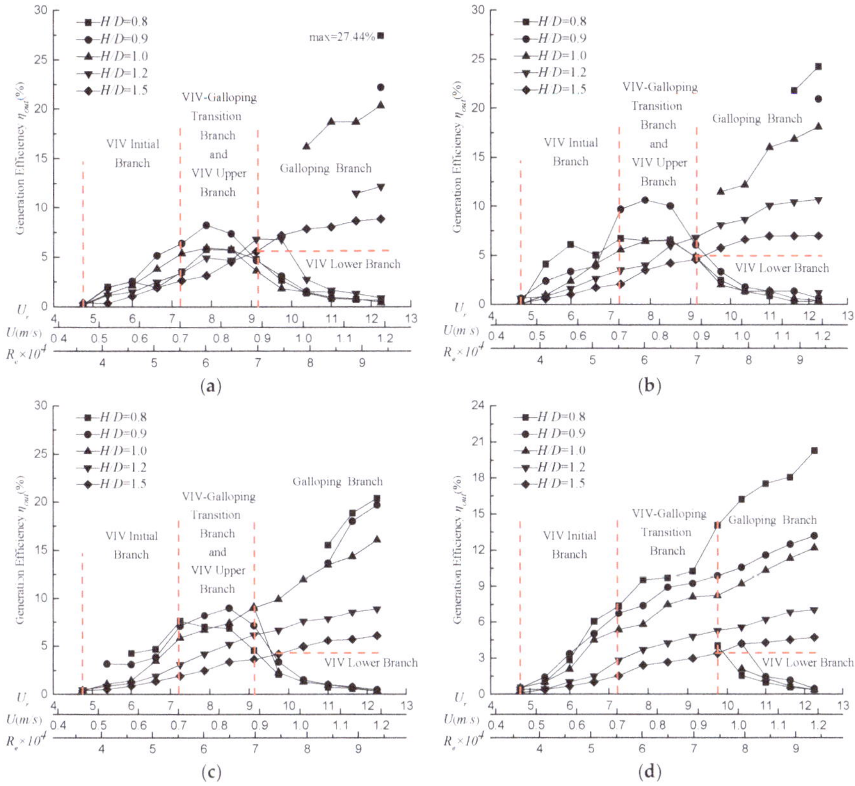

Figure 20. Power generation efficiency with different section aspect ratios at K = 1200 N/m: (**a**) amplitude ratio (4 Ω, ζ = 0.347); (**b**) amplitude ratio (8 Ω, ζ = 0.305); (**c**) amplitude ratio (11 Ω, ζ = 0.246); (**d**) amplitude ratio (16 Ω, ζ = 0.177).

From the previous results of this paper, it was found that the T-section prism could not be self-excited from VIV to galloping at high damping conditions with α = 1. In Figure 19a, for ζ = 0.347, only SG occurred at α = 1.5. For the prisms of other α values forced by initial displacement, HG occurred at a high velocity.

In Figure 19b,c, for ζ = 0.305 and ζ = 0.246 in the prisms of α = 1.5 and α = 1.2, SG was observed, while in the prisms of other α, HG was observed with external excitation. In Figure 19d, for ζ = 0.177, SG occurred under all conditions. For $1.2 \leq \alpha \leq 1.5$, the VIV lower branch could not be observed with external suppression. Based on these results, it can be concluded that a higher α value is more likely to cause SG under the same conditions. The rise of α is beneficial to the development of the oscillation mode from HG to SG. In addition, A^* decreases with the increase of α. Therefore, a reasonable α must be selected to ensure good oscillation characteristics, as well as to guarantee better electrical energy resources.

3.3.2. Effects of Aspect Ratios on Energy Conversion

Figure 20 presents the variation of the η_{out} with the five α values, which can be summarized as follows. In the present tests, T-section prisms with a higher α have a good advantage at high damping conditions in terms of oscillation, but their performances in power generation efficiency are not very prominent. On the contrary, the prisms with smaller α show a more outstanding power generation advantage (shown in Figure 20). In Figure 20a, for R_L = 4 Ω, the maximum η_{out} rises up to 27.44%, which is approximately the maximum η_{out} (28%) of the PTC circular cylinder. In Figure 20b–d, it can be concluded that regardless of both high damping and low damping conditions, a smaller α shows a better power generation efficiency in the galloping branch. The oscillation enters the VIV lower branch, resulting in a significant decrease in η_{out}.

In summary, the optimal aspect ratio should be designed based on the oscillation branch and the flow conditions, in order to ensure reasonable energy utilization. For high stable flow velocities, the smaller the α value, the stronger the electric power conversion capacity. For widely variable flow velocities, a higher α is suitable for extracting oscillation energy from a T-section prism within a certain range of the test.

4. Conclusions

(1) The HG and SG of a T-section prism with external excitation are similar to those of a triangular prism. With a decrease in load resistances, the SG of the T-section prism gradually transforms to CG, and finally HG occurs. The oscillation modes of SG, HG, and CG all contain VIV branches and a galloping branch. The only difference is whether these can be self-excited from VIV to galloping. The T-section prism exhibits good stability and high intensity in the galloping branch, which is beneficial for energy extraction and utilization.

(2) The galloping branch of the T-section prism presents an efficient and stable power output in the tests. There is an optimal power generation resistance R_L of 8 Ω, an optimum damping ratio ζ_{total} of 0.305, a maximum power generation P_{harn} of 21.23 W at U_r = 12.25, and a maximum power generation efficiency η_{out} of 20.2% at U_r = 11.625. Compared with the test results of the smooth circular cylinder and the PTC circular cylinder, η_{out} of the smooth circular cylinder is 22% [39] and the efficiency of the PTC circular cylinder is higher, at η_{out} = 28% and P_{harn} = 23.54 W [44]. In addition, the maximum η_{out} of the T-section prism with α = 0.8 is 27.44%, and it can be concluded that the T-section prism has great potential for power generation.

(3) In the present tests with aspect ratios from 0.8 to 1.5, the smaller section aspect ratio shows a stronger ability to convert electrical energy. The aspect ratio has a great influence on the energy collection for the T-section prism. Thus, it is necessary to optimize the section aspect ratio of the prism to enhance the power generation efficiency.

Author Contributions: J.L. and X.Y. designed the experiments; X.Y., N.S., Q.R., and H.D. performed the experiments; X.Y. and N.S. analyzed the data; X.Y. built and wrote the majority of the manuscript text; all authors reviewed the manuscript; F.L. and G.X. made contributions to the structure and language of the manuscript.

Funding: This research was funded by the National Key R&D Program of China grant number 2016YFC0401905.

Acknowledgments: All workers from the State Key Laboratory of Hydraulic Engineering Simulation and Safety of Tianjin University are acknowledged. The authors are also grateful for the assistance of the editor and three anonymous reviewers for their professional and pertinent comments and suggestions, which were greatly helpful for further quality improvement of this manuscript.

Abbreviations

D	Projection width of the prism in the direction of the incoming flow
H	Height of the T-section prism cross-section
L	Prism length
α	Section aspect ratio H/D
K	System stiffness
C_{total}	System damping coefficient
C_{harn}	Electromagnetic damping coefficient
m_{osc}	Original mass, prism mass, transmission mass, and one-third of the spring mass [2]
U	Incoming flow velocity
U_r	Reduced velocity $U/(f_n \cdot D)$

R_L	Load resistance
ρ	Water density
A	Average of the amplitudes under continuous oscillation for 30 s
f_{osc}	Prime frequency of oscillation
$A^* = A/D$	Amplitude ratio A/D
$f^* = f_{osc}/f_n$	Frequency ratio
f_n	Natural frequency in air
ζ	System damping ratio
C_v	Critical velocity
η_{out}	Power generation efficiency
P_{harn}	Generation active power

References

1. Blevins, R.D. *Flow-Induced Vibration*, 3rd ed.; Van Nostrand Reinhold: New York, NY, USA, 1990.
2. Barriga, J.; Ruiz-de-Gopegui, U.; Goikoetxea, J.; Coto, B.; Cachafeiro, H. Selective coatings for new concepts of parabolic trough collectors. *Energy Procedia* **2014**, *49*, 30–39. [CrossRef]
3. Cachafeiroa, H.; de Arevaloa, L.F.; Vinuesaa, R.; Lopez-Vizcaino, R.; Luna, M. Analysis of vacuum evolution inside Solar Receiver Tubes. *Energy Procedia* **2015**, *69*, 289–298. [CrossRef]
4. Cachafeiroa, H.; de Arevaloa, L.F.; Vinuesaa, R.; Goikoetxea, J.; Barriga, J. Impact of solar selective coating ageing on energy cost. *Energy Procedia* **2015**, *69*, 299–309. [CrossRef]
5. Mannini, C.; Belloli, M.; Marra, A.M.; Bayati, I.; Giappino, S.; Robustelli, F., Bartoli, G. Aeroelastic stability of two long-span arch structures: A collaborative experience in two wind tunnel facilities. *Eng. Struct.* **2016**, *119*, 252–263. [CrossRef]
6. Seo, J.-W.; Kim, H.-K.; Park, J.; Kim, K.T.; Kim, G.N. Interference effect on vortex-induced vibration in a parallel twin cable-stayed bridge. *J. Wind Eng. Ind. Aerodyn.* **2013**, *116*, 7–20. [CrossRef]
7. Assi, G.R.; Bearman, P.W.; Tognarelli, M.A. On the stability of a free-to-rotate short-tail fairing and a splitter plate as suppressors of vortex-induced vibration. *Ocean Eng.* **2014**, *92*, 234–244. [CrossRef]
8. Dowell, E.H.; Tang, D. Nonlinear aeroelasticity and unsteady aerodynamics. *AIAA J.* **2002**, *40*, 1697–1707. [CrossRef]
9. Assi, G.R.; Bearman, P.W.; Kitney, N. Low drag solutions for suppressing vortex induced vibration of circular cylinders. *J. Fluids Struct.* **2009**, *25*, 666–675. [CrossRef]
10. Park, H.; Bernitsas, M.M.; Kumar, R.A. Selective roughness in the boundary layer to suppress flow-induced motions of circular cylinder at $30{,}000 \le \text{Re} \le 120{,}000$. *J. Offshore Mech. Arct. Eng.* **2012**, *134*, 041801. [CrossRef]
11. Sarpkaya, T. Fluid forces on oscillating cylinders. *NASA STI/Recon Tech. Rep. A* **1978**, *104*, 275–290.
12. Williamson, C.H.K.; Roshko, A. Vortex formation in the wake of an oscillating cylinder. *J. Fluids Struct.* **1988**, *2*, 355–381. [CrossRef]
13. Zdravkovich, M.M. *Flow around Circular Cylinders*; Oxford University Press: New York, NY, USA, 1997; Volume 1.
14. De Meij, A.; Vinuesa, J.F.; Maupas, V.; Waddle, J.; Price, I.; Yaseen, B.; Ismail, A. Wind energy resource mapping of Palestine. *Renew. Sustain. Energy Rev.* **2016**, *56*, 551–562. [CrossRef]
15. Sousa, V.C.; de M Anicézio, M.; De Marqui, C., Jr.; Erturk, A. Enhanced aeroelastic energy harvesting by exploiting combined nonlinearities: Theory and Experiment. *Smart Mater. Struct.* **2011**, *20*, 094007. [CrossRef]
16. Zhang, M.; Zhao, G.; Wang, J. Study on Fluid-Induced Vibration Power Harvesting of Square Columns under Different Attack Angles. *Geofluids* **2017**, *2017*, 6439401. [CrossRef]
17. Zhou, S.; Wang, J. Dual serial vortex-induced energy harvesting system for enhanced energy harvesting. *AIP Adv.* **2018**, *8*, 075221. [CrossRef]
18. Bernitsas, M.M.; Raghavan, K.; Ben-Simon, Y.; Garcia, E.M.H. VIVACE (vortex induced vibration aquatic clean energy): A new concept in generation of clean and renewable energy from fluid flow. *J. Offshore Mech. Arct. Eng.* **2008**, *130*, 041101–041115. [CrossRef]
19. Chamorro, L.P.; Hill, C.; Neary, V.S.; Gunawan, B.; Arndt, R.E.A.; Sotiropoulos, F. Effects of energetic coherent motions on the power and wake of an axial flow turbine. *Phys. Fluids* **2015**, *27*, 055104. [CrossRef]

20. Chamorro, L.P.; Lee, S.J.; Olsen, D.; Milliren, C.; Marr, J.; Arndt, R.E.A.; Sotiropoulos, F. Turbulence effects on a full-scale 2.5 MW horizontal-axis wind turbine under neutrally stratified conditions. *Wind Energy* **2015**, *18*, 339–349. [CrossRef]

21. Park, H.R.; Kumar, R.A.; Bernitsas, M.M. Enhancement of fluid induced vibration of rigid circular cylinder on springs by localized surface roughness at $3 \times 10^4 \leq \mathrm{Re} \leq 1.2 \times 10^5$. *Ocean Eng.* **2013**, *72*, 403–415. [CrossRef]

22. Daqaq, M.F. Characterizing the response of galloping energy harvesters using actual wind statistics. *J. Sound Vib.* **2015**, *357*, 365–376. [CrossRef]

23. Hemon, P.; Amandolese, X.; Andrianne, T. Energy harvesting from galloping of prisms: A wind tunnel experiment. *J. Fluids Struct.* **2017**, *70*, 390–402. [CrossRef]

24. Alonso, G.; Meseguer, J. A parametric study of the galloping stability of two dimensional triangular cross-section bodies. *J. Wind Eng. Ind. Aerodyn.* **2006**, *94*, 241–253. [CrossRef]

25. Williamsona, C.H.K.; Govardhanb, R. A brief review of recent results in vortex-induced vibrations. *J. Wind Eng. Ind. Aerodyn.* **2008**, *96*, 713–735. [CrossRef]

26. Mannini, C.; Marra, A.M.; Massai, T.; Bartoli, G. Interference of vortex-induced vibration and transverse galloping for a rectangular cylinder. *J. Fluid Struct.* **2016**, *66*, 403–423. [CrossRef]

27. Liu, B.; Hamed, A.M.; Jin, Y.; Chamorro, L.P. Influence of vortical structure impingement on the oscillation and rotation of flat plates. *J. Fluid Struct.* **2017**, *70*, 417–427. [CrossRef]

28. Bokaian, A.R.; Geoola, F. Hydroelastic instabilities of square cylinders. *J. Sound Vib.* **1984**, *92*, 117–141. [CrossRef]

29. Vinuesa, R.; Prus, C.; Schlatter, P.; Nagib, H.M. Convergence of numerical simulations of turbulent wall-bounded flows and mean cross-flow structure of rectangular ducts. *Meccanica* **2016**, *51*, 3025–3042. [CrossRef]

30. Lian, J.; Yan, X.; Liu, F.; Zhang, J. Analysis on Flow Induced Motion of Cylinders with Different Cross Sections and the Potential Capacity of Energy Transference from the Flow. *Shock Vib.* **2017**, *2017*, 4356367. [CrossRef]

31. Camarri, S.; Salvetti, M.V.; Buresti, G. Large-eddy simulation of the flow around a triangular prism with moderate aspect-ratio. *J. Wind Eng. Ind. Aerodyn.* **2006**, *94*, 309–322. [CrossRef]

32. Iungo, G.V.; Buresti, G. Experimental investigation on the aerodynamic loads and wake flow features of lowaspect-ratio triangular prisms at different wind directions. *J. Fluid Struct.* **2009**, *25*, 1119–1135. [CrossRef]

33. Alonso, G.; Sanz-Lobera, A.; Meseguer, J. Hysteresis phenomena in transverse galloping of triangular cross-section bodies. *J. Fluid Mech.* **2012**, *33*, 243–251. [CrossRef]

34. Zhang, J.; Liu, F.; Lian, J.; Yan, X.; Ren, Q. Flow Induced Vibration and Energy Extraction of an Equilateral Triangle Prism at Different System Damping Ratios. *Energies* **2016**, *9*, 938. [CrossRef]

35. Zhang, J.; Xu, G.; Liu, F.; Lian, J.; Yan, X. Experimental investigation on the flow induced vibration of an equilateral triangle prism in water. *Appl. Ocean Res.* **2016**, *61*, 92–100. [CrossRef]

36. Lian, J.; Yan, X.; Liu, F.; Zhang, J.; Ren, Q.; Yan, X. Experimental Investigation on Soft Galloping and Hard Galloping of Triangular Prisms. *Appl. Sci.* **2017**, *7*, 198. [CrossRef]

37. Lee, J.H.; Bernitsas, M.M. High-damping High-Reynolds VIV tests for energy harnessing using the VIVACE converter. *Ocean Eng.* **2011**, *38*, 1697–1712. [CrossRef]

38. Hai, S.; Kim, E.S.; Bernitsas, M.P.; Bernitsas, M.M. Virtual Spring–Damping System for Flow-Induced Motion Experiments. *J. Offshore Mech. Arct. Eng.* **2015**, *137*, 061801.

39. Bernitsas, M.M.; Ben-Simon, Y.; Raghavan, K.; Garcia, E.M. The VIVACE converter: Model tests at high damping and Reynolds number around 10^5. *J. Offshore Mech. Arct. Eng.* **2009**, *131*, 011102. [CrossRef]

40. Chang, C.C.; Kumar, R.A.; Bernitsas, M.M. VIV and galloping of single circular cylinder with surface roughness at $3.0 \times 10^4 \leq \mathrm{Re} \leq 1.2 \times 10^5$. *Ocean Eng.* **2011**, *38*, 1713–1732. [CrossRef]

41. Hai, S.; Kim, E.S.; Nowakowski, G.; Mauer, E.; Bernitsas, M.M. Effect of mass-ratio, damping, and stiffness on optimal hydrokinetic energy conversion of a single, rough cylinder in flow induced motions. *Renew. Energy* **2016**, *99*, 936–959.

42. Ma, C.; Sun, H.; Nowakowski, G.; Mauer, E.; Bernitsas, M.M. Nonlinear piecewise restoring force in hydrokinetic power conversion using flow induced motions of single cylinder. *Ocean Eng.* **2016**, *128*, 1–12. [CrossRef]

43. Lian, J.; Ren, Q.; Liu, F.; Yan, X.; Zhang, J. Experimental Study of Flow-Induced Vibration Characteristics of T-shape Cross-Section Oscillator. *J. Exp. Mech.* **2017**, *32*, 216–222. (In Chinese)
44. Lin, D. *Research on Flow Induced Motion of Multiple Circular Cylinders with Passive Turbulence Control*; Chong Qing University: Chongqing, China, 2013. (In Chinese)

Article

Flow Structure and Heat Transfer of Jet Impingement on a Rib-Roughened Flat Plate

Abdulrahman H. Alenezi [1,*], **Abdulrahman Almutairi** [1], **Hamad M. Alhajeri** [1], **Abdulmajid Addali** [2] **and Abdelaziz A. A. Gamil** [3]

[1] Mechanical Power and Refrigeration Technology Department, College of Technological Studies, Shuwaikh Educational, P. O. Box 23167, Safat, Al-Asamah 13092, Kuwait; asa.almutairi@paaet.edu.kw (A.A.); hmhajeri@gmail.com (H.M.A.)

[2] Advanced Centre for Technology, Tripoli, Libya; ammaddali@gmail.com

[3] Department of Power and Propulsion, Cranfield University, Cranfield, Bedfordshire MK43 0AL, UK; a.gamil@cranfield.ac.uk

[*] Correspondence: ah.alenezi@paaet.edu.kw

Received: 5 May 2018; Accepted: 10 June 2018; Published: 13 June 2018

Abstract: The jet impingement technique is an effective method to achieve a high heat transfer rate and is widely used in industry. Enhancing the heat transfer rate even minimally will improve the performance of many engineering systems and applications. In this numerical study, the convective heat transfer process between orthogonal air jet impingement on a smooth, horizontal surface and a roughened uniformly heated flat plate is studied. The roughness element takes the form of a circular rib of square cross-section positioned at different radii around the stagnation point. At each location, the effect of the roughness element on heat transfer rate was simulated for six different heights and the optimum rib location and rib dimension determined. The average Nusselt number has been evaluated within and beyond the stagnation region to better quantify the heat transfer advantages of ribbed surfaces over smooth surfaces. The results showed both flow and heat transfer features vary significantly with rib dimension and location on the heated surface. This variation in the streamwise direction included both augmentation and decrease in heat transfer rate when compared to the baseline no-rib case. The enhancement in normalized averaged Nusselt number obtained by placing the rib at the most optimum radial location R/D = 2 was 15.6% compared to the baseline case. It was also found that the maximum average Nusselt number for each location was achieved when the rib height was close to the corresponding boundary layer thickness of the smooth surface at the same rib position.

Keywords: impingement heat transfer enhancement; orthogonal jet; turbulence; flat plate

1. Introduction

In the early 1960s, the jet impingement cooling technique was first introduced for internal cooling. It is a complex technique, but the most effective one when a high heat transfer rate is required. The complexity of impinging jet flow makes the heat transfer from/to the surface subjected to such flows difficult to resolve. However, a range of jet configurations and parameters have been investigated in terms of both heat transfer and fluid flow. The parameters which are known to influence the rate of heat transfer between the jet and the target surface includes Reynolds number, level of turbulence, jet-to-target distance, intermittency, nozzle geometry, target surface roughness, and jet temperature [1,2]. Kuraan et al. [3] performed a recent experimental study of a free water jet impinging a flat surface under the influence of jet-to-target distance of less than one. The effects of jet-to-target distance on stagnation point Nusselt number (N_u) and pressure were considered in this study under the influence of a wide range of R_e between 4000 and 8053. New correlations of stagnation point N_u and pressure were reported in this study based on the author's key findings. Zu et al. [4] presented results of a numerical study of the heat transfer behavior of a circular air jet impinging normally onto a flat plate with a nozzle to plate spacing ratios (H/D) of one to six. Their study used Fluent—a computational fluid dynamics (CFD) commercial package. Seven different turbulence models were implemented to evaluate modeling prediction capabilities by comparison with the benchmark experimental data. The Shear Stress Transport (SST k-ω) and Large Eddy Simulation (LES) models gave better accuracy in predicting both the heat transfer and fluid flow. With the high computational cost of LES, the SST k-ω turbulence model is an attractive and promising modeling option.

Heat and mass transfer as a function of surface roughness has been thoroughly considered by researchers such as Zhang, et al. [5] and Kim and Lee [6]. Basic studies using a single jet impinging on a roughened surface, with and without cross-flow, have been conducted by numerous investigators such as Beitelmal and Saad [7], Sharif and Ramirez [8], Xing and Weigand [9], Gabour and Lienhard [10], and Celik [11]. These investigations showed that a roughened surface could enhance the local N_u by up to 50% when compared to a smooth target surface, because of the turbulence induced by the surface roughness element. This enhancement due to surface roughness was investigated by many researchers employing different arrangements of jet impingement parameters [12–18]. More recent investigations [19,20] studied the effect of using novel jet impingement parameters such as solid volume fraction, Richardson number, and roughness element orientation on either local or average N_u number heat transfer rate between the roughened surface and impingement jet.

The literature to date agrees that the use of turbulence promoters (i.e., ribs) has a major impact on enhancing heat transfer rate [7–11,14,15]. However, most of these studies were conducted using turbulators with fixed dimensions and locations on the heated surface. This research employs uniform turbulence promoters with variable locations and dimensions to study the impact on the heat transfer rate between the working fluid and the heated surface.

2. Numerical Methodology

In all the cases simulated in this research paper, the jet flow had a velocity of 24.8 m/s. The air was assumed to be incompressible due to its low Mach number, and with the assumption of axisymmetric flow, the governing equations are:

$$\frac{\partial U}{\partial z} + \frac{\partial V}{\partial r} + \frac{V}{r} = 0 \tag{1}$$

$$\rho\left(U\frac{\partial U}{\partial z} + V\frac{\partial U}{\partial r}\right) = -\frac{\partial P}{\partial z} + \frac{1}{r}\frac{\partial}{\partial r}\left[r\left(\mu\frac{\partial U}{\partial r} - \rho\overline{u'v'}\right)\right] + \frac{\partial}{\partial z}\left((\mu)\frac{\partial U}{\partial z} - \rho\overline{u'u'}\right) \tag{2}$$

$$\rho\left(U\frac{\partial V}{\partial z} + V\frac{\partial V}{\partial r}\right) = -\frac{\partial P}{\partial r} + \frac{1}{r}\frac{\partial}{\partial r}\left[r\left(\mu\frac{\partial V}{\partial r} - \rho\overline{v'v'}\right)\right] + \frac{\partial}{\partial z}\left((\mu)\frac{\partial U}{\partial z} - \rho\overline{u'u'}\right) - \mu\frac{V}{r^2} \tag{3}$$

$$\rho C_p\left(U\frac{\partial T}{\partial z} + V\frac{\partial T}{\partial r}\right) = \lambda\left[\frac{\partial}{\partial r}\left(\frac{\partial T}{\partial r}\right) + \frac{1}{r}\frac{\partial T}{\partial r} + \frac{\partial}{\partial z}\left(\frac{\partial T}{\partial z}\right)\right] + \overline{\phi} - \rho C_p\left(\frac{\partial \overline{u't'}}{\partial z} + \frac{\partial \overline{v't'}}{\partial r} - \frac{\partial \overline{v't'}}{r}\right) \tag{4}$$

where $\overline{\phi}$ is the viscous dissipation heat source

$$\overline{\phi} = 2\mu\left[\left(\frac{\partial V}{\partial r}\right)^2 + \left(\frac{V}{r}\right)^2 + \left(\frac{\partial U}{\partial z}\right)^2 + \frac{1}{2}\left(\frac{\partial V}{\partial z} + \frac{\partial u}{\partial z}\right)^2\right] \tag{5}$$

Here, ρ, P, and T are the density, average pressure, and temperature, respectively. U and V are velocity components in the z and r directions, whereas u', v', and t' are the fluctuating velocity components and temperature in the z and r directions. Finally C_p is the heat capacity of air at constant pressure. All geometry and domain details are shown in Figure 1.

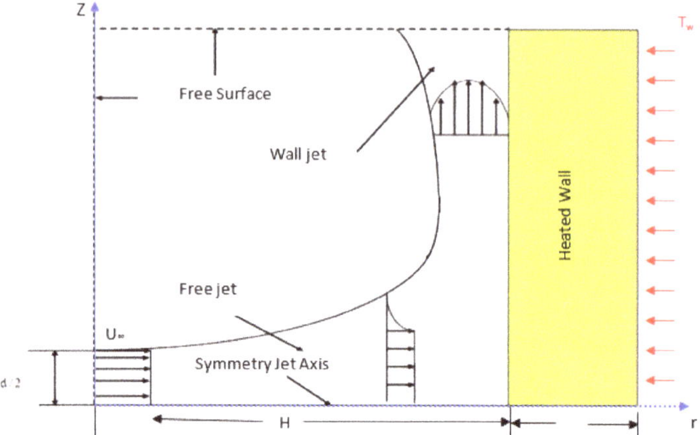

Figure 1. Geometry and computational domain.

Flow can exist in three regimes: laminar, turbulent or in the transitional phase. Laminar, or streamline flow occurs at relatively low flow velocities and can be characterized as layers of fluid flowing in parallel with no disruption between them. Turbulent flow, however, occurs at a high Reynolds numbers with the presence of random fluctuations and disruptions. The flow regime can be specified by the value of Reynolds number (R_e) which is defined as the ratio of inertial and viscous forces. Based on the bulk jet exit velocity (U) and nozzle diameter (D), Reynolds number can be defined as:

$$R_e = \frac{\rho UD}{\mu} = \frac{UD}{\nu} \tag{6}$$

where μ and ν are, respectively, the dynamic and kinematic viscosities of the fluid.

The Nusselt number, N_u, is the ratio of convective to conductive heat transfer. If its value is close to unity, that means both convection and conduction have a similar magnitude and the flow will be almost stationary or laminar. However, if N_u has a large value, it means more heat is convected than conducted, and the flow will be turbulent. N_u can be defined as:

$$N_u = \frac{\text{Convective heat transfer}}{\text{Conductive heat transfer}} = \frac{h \cdot D}{k} \tag{7}$$

where k is the thermal conductivity of the fluid, and h is the convective heat transfer coefficient which is given by:

$$h = \frac{q_w}{T_w - T_{ref}} \tag{8}$$

Usually T_{ref} is either the jet temperature (T_j) or the adiabatic wall temperature (T_{aw}), the latter can be obtained from the non-dimensional recovery factor: factor =

$$\text{Recovery factor} = \frac{T_{aw} - T_j}{U_j^2/2C_p} \tag{9}$$

The Nusselt number varies depending on the temperature chosen as the reference temperature T_j or T_{aw}. For low Reynolds numbers, the difference in the value of the Nusselt number won't be noticeable. However, for large Reynolds numbers, the reference temperature must be appropriately chosen. How to choose the reference temperature has been explained by References [21,22].

$$\overline{N_u} = \overline{h}\,\frac{D}{k} = \frac{2}{R^2}\int_0^R N_u(r)\,dr \tag{10}$$

2.1. Computational Domain, Boundary Conditions, and Grid Independence Check

A schematic representation of circular jet impingement is shown in Figure 2, which is consistent with the experimental setup described by O'Donovan and Murray [23]. An orthogonal jet impinged on an isothermal flat plate is kept at a constant temperature of 60 °C, where the nozzle exit velocity was obtained from the jet exit Reynolds number, which was first set at $R_e = 10{,}000$ and then at 20,000. Geometric dimensions were normalized relative to the nozzle hydraulic diameter ($D = 13.5$ mm), the normalized vertical distance between the flat plate (target), and the nozzle exit (jet) (H/D), and the normalized radius of the rib (R/D). The chosen value of H/D was six and the angle of impingement was 90° (orthogonal). The circular domain diameter (D_d), which included the heated impingement surface, was 40D. The assumption of using an axisymmetric model in the simulation was to save computational time and cost.

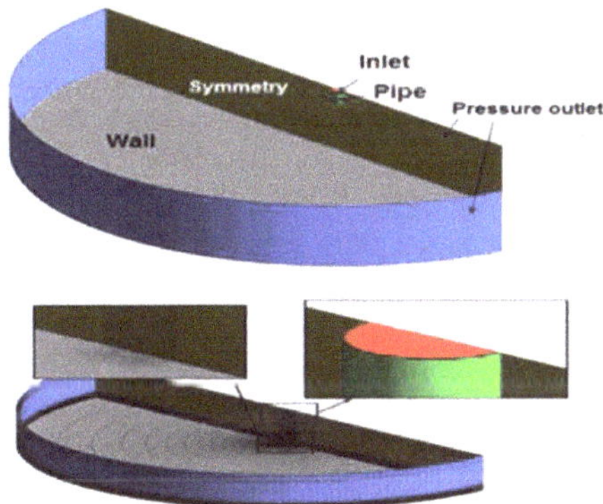

Figure 2. Jet impingement geometry, boundary conditions, and grid strategy.

Figure 3 illustrates the geometry and meshing strategy used in the case of the roughened flat plate. It is well known from the literature that the near wall regions have a large impact on the solution variables and momentum, and thus extra care was taken with near-wall meshing in this model to get accurate results. Hexahedral elements were used by block-structured grids using spatial discretization of the domain. A very fine mesh was applied in the direction normal to the heated surface as well as the roughness elements to ensure proper functionality of the turbulence model, especially low Reynolds numbers which required a dimensionless distance between the wall and the first node of less than unity. The growth rate of the cell near the heated surface where all heat transfer takes place was no greater than 1.2 in the direction normal to the heated surface. O-grid strategy was applied within and near the round nozzle to ensure high cell orthogonality. The total grid size was approximately one million for the whole computational domain.

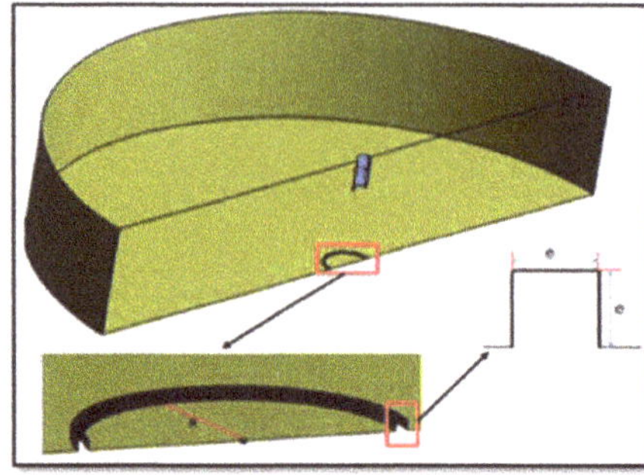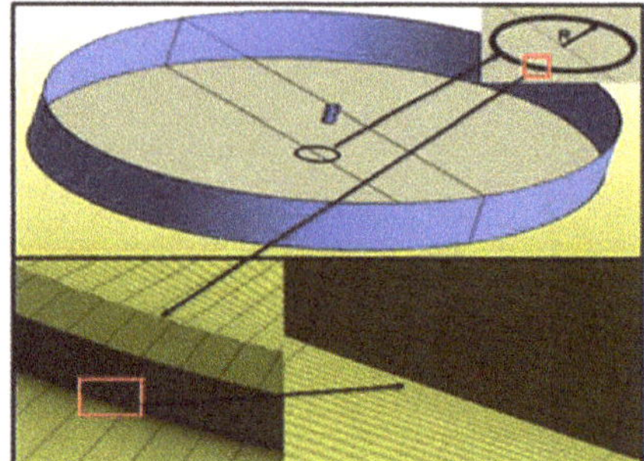

Figure 3. Geometry and meshing of circular rib of radius R, with a square cross-section of length e.

2.2. Grid Independence and Turbulence Model Validation

The mesh strategy used in this simulation was intended to resolve the wall boundary layer accurately. A fine structured mesh was adopted and then refined near the wall where pressure, velocity and turbulence gradients occur, to achieve a stable numerical solution (see Figure 3). The y^+ value was maintained at less than unity near the wall using at least ten nodes within the viscous sub-layer as recommended by Reference [24]. Here y^+ is a dimensionless quantity related to the distance between the wall and the first mesh node above it. It is important that the first cell should be fine enough to avoid positioning it in the buffer layer. The y^+ is a function of both free stream properties and velocity (U_t), $y^+ = \frac{\rho U_t y}{\mu}$.

A non-uniform finite volume mesh with collocated variable locations was used for the computations, where the mesh density was carefully scaled to investigate the sensitivity of the predicted results to the number of grid nodes. A rectangular mesh gathered toward both top and bottom walls, with a bias factor of 70, was used in the z-direction. Finer computational meshes were adopted near the heated surface and the nozzle to obtain accurate values for velocity and thermal boundary layer. The grid size modified in this study was approximately one million cells based on a systematic grid independence study. The turbulence models including Re-Normalisation Group (RNG) k-epsilon, used in this study gave good results when using a y^+ value less than five in the boundary layer region, and employing 5 to 30 nodes within this region [24].

3. Results and Discussion

3.1. Simulation Characteristics

To validate the results, comparisons were made between numerical data obtained in this research and the experimental data of O'Donovan and Murray [23]. A grid dependence study was then conducted to verify the independence of the numerical solution on mesh size. Table 1 below shows the three mesh sizes adopted in this research.

Table 1. Mesh sizes.

No.	Cell Size	y^+
Mesh 1	400,000	0.99
Mesh 2	986,000	0.64
Mesh 3	1.789.00	0.5

Figure 4 compares the simulated local N_u values for the three mesh sizes with the experimental results along the normalized radial distance. The RNG k-epsilon turbulence model was used.

The figure shows that both numerical and experimental data values were close for all grid sizes. Therefore, mesh 2 was adopted for the numerical calculations.

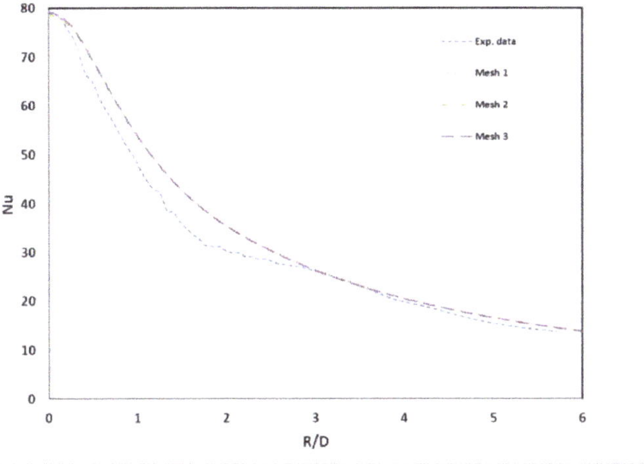

Figure 4. Nusselt number distribution for three mesh grids, jet-to-target distance $H/D = 6$, $R_e = 10,000$.

The distribution of the local Nusselt number along the pre-heated impingement surface was compared with the experimental data of O'Donovan and Murray [23] for $H/D = 6$ and $\alpha = 90°$ using different turbulence models (see Figure 5). The RNG k-epsilon turbulence model showed better overall agreement with the experimental data and succeeded in predicting the local stagnation point Nusselt number ($N_{u_{stag.}}$) with an error of only about 1.7%. However, both SST k-ω and RSM models overestimated $N_{u_{stag}}$ with errors of 18% and 21%, respectively. Moving in the radial direction, the RNG k-epsilon turbulence model gives, overall, a more accurate prediction of local N_u values. Unfortunately, none of the models discerned the slight trough in the experimental results at R/D ~2, which gave rise to the small maximum at R/D ~3. Based on this evaluation, the RNG k-epsilon turbulence model was selected to be used for this parametric study.

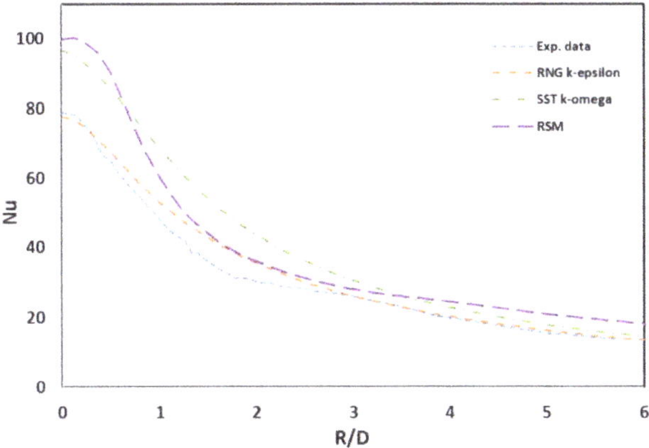

Figure 5. Turbulence model study. $H/D = 6$, $R_e = 10,000$.

3.2. Aerodynamical Results

The following sections report results for the higher flow velocity, R_e = 20,000, jet-to-target distance (H/D) = 6, and jet angle α = 90°. The study included simulating the effect of rib location and rib height on the average Nusselt number ($\overline{N_u}$). The rib was tested with four different values of radial distance: R/D = 1, 1.5, 2, and 3, where D was the jet hydraulic diameter, and for each location it was tested for six different rib heights (e) between 0.25 mm and 1.5 mm in increments of 0.25 mm to ascertain the optimum height for each location. This range of rib locations should extend from within to outside the stagnation region. Figure 6 shows the geometric details of the rib cross-section and all four rib locations (R) simulated in this research paper. These radial locations tested cover both the stagnation and wall jet regions.

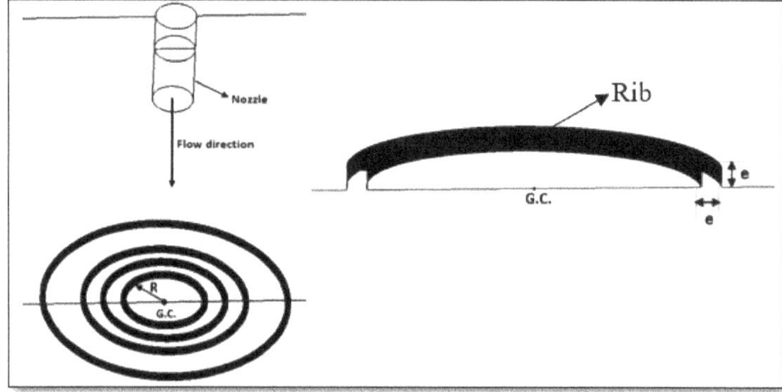

Figure 6. Geometric details of turbulence inducing rib, showing four rib positions and rib cross-section.

First, it was required to determine the effect of the new value of R_e on heat transfer for the smooth flat plate configuration. Figure 7 shows the local N_u values with the radial distance on the heated surface. Overall, the local N_u values are higher for the higher Re, with the maximum value, as expected, at the stagnation point (R/D = 0). As can be seen in the figure, a noticeable decrease in N_u values occurs as the flow travels downstream losing about 53% of its maximum value at a radial distance R/D = 2.5.

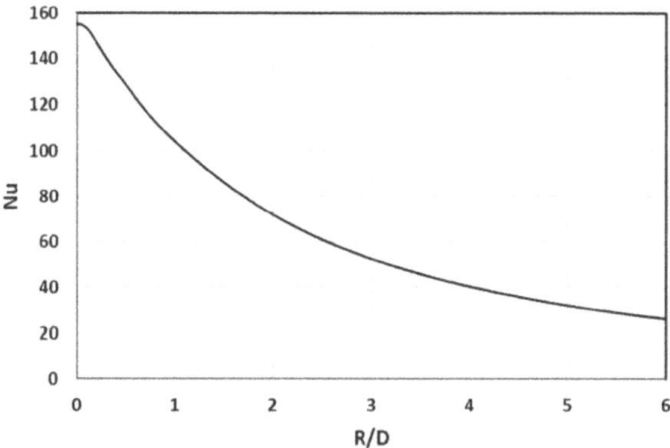

Figure 7. Local Nusselt Number (N_u) distribution for R_e = 20,000 and H/D = 6.

The value of the velocity boundary layer thickness (δ) has been estimated in the literature [25] as one-tenth of the hydraulic diameter (0.1D) for the stagnation zone. While for the wall jet region it has been estimated at 95% of that for the free stream velocity at each radial location, Table 2 demonstrates values of velocity boundary layer thickness at each tested radial location for the unobstructed flat surface.

Table 2. Velocity boundary layer thickness at different rib locations [25].

Variable	R_1	R_2	R_3	R_4
Rib Location (R)	1D	1.5D	2D	3D
δ (mm)	0.135	0.22	0.56	0.73

The rib was placed at radial distances R/D = 1 and 1.5 and positioned within or close to the stagnation region [26]. As mentioned above, the rib height (e) was varied to find the optimum rib geometry that gave maximum average N_u. Figure 8 shows the effect of rib height on the local distribution of N_u for a circular jet impinging normally on a flat plate. As shown in the figure, the local value of N_u directly behind the rib is increased because of rib induced flow separation and re-attachment. In very simple terms, it disturbs the stagnation layer which acts to insulate the surface. Depending on geometry and circumstances, the ribs may also increase heat transfer by increasing the effective area of the surface [27]. It can be seen from the figure that the higher the rib (greater e), the lower the local N_u value due to lower flow turbulence intensity as the flow travels a longer distance after passing the rib before re-attaching. The re-attachment point is where the flow hits the heated surface after passing the rib, the velocity with which the flow hits the surface at this point is known as the "arrival velocity".

For rib heights of 0.25, 0.50, and 0.75 mm, enhancement of heat transfer rate was achieved. Where e $\geq$ 1 mm, all rib heights caused a subsidiary peak in the local N_u immediately in front of the rib due to the flow recirculation that occurred before the protrusion. However, the increase in the drag resulting from the presence of ribs with e $\geq$ 1 mm caused the impinging jet to decelerate and disperse more rapidly, such that further from the stagnation point the value of N_u decreased, as shown in the figure.

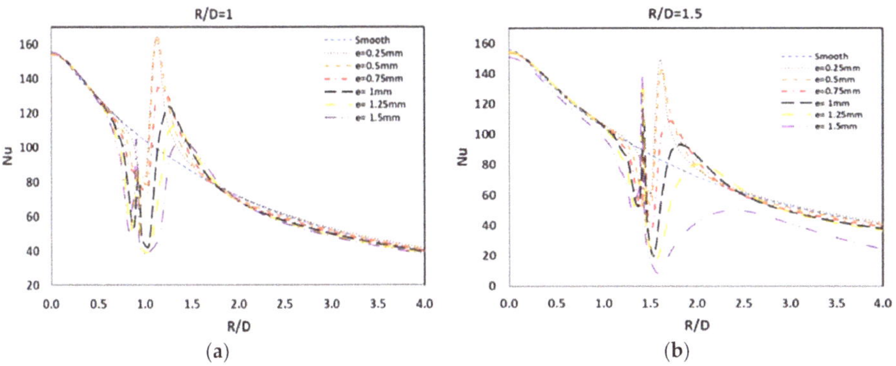

Figure 8. Local N_u distributions for six rib heights e: (**a**) R/D = 1 and (**b**) R/D = 1.5.

For the special case where R/D = 1.5 and e = 1.5 mm, the stagnation Nusselt number $N_{u_{stag}}$ starts to be noticeably affected by the rib height showing a significantly lower value than for the baseline case. This is due to the lack of re-attachment between the flow and the heated target surface, which was not the case for R/D = 1, see Figure 9.

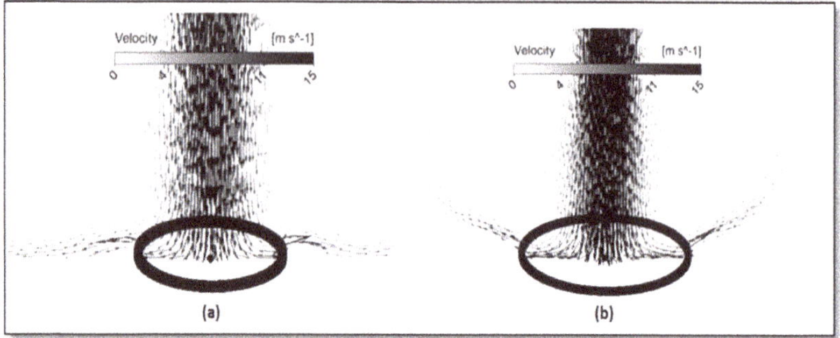

Figure 9. Velocity vectors for e = 1.5 mm. (**a**) R/D = 1 and (**b**) R/D = 1.5.

Gau, et al. [28] reported this phenomenon for a 2-D air slot jet impingement on a flat rectangular surface with straight ribs of different heights attached perpendicular to the flow. They explained that the shear layer, while separated from the surface, experiences turbulence effects which enhance momentum and mass (heat) transfer so that its re-attachment to the surface behind the rib results in an increase in heat transfer. After this impingement, the turbulent free shear layer reattaches itself to the surface, and a new boundary layer develops. Thus, placing ribs normal to the flow of the wall jet is an effective means for enhancing total heat transfer rate to/from the wall by disrupting the rather rapid decrease in convective heat transferred to/from the wall.

The formation of a boundary layer begins in the stagnation region with a thickness of no more than one-tenth of the jet hydraulic diameter [26]. The locations R/D = 2 and 3 represent the beginning of the wall jet region [27] where the flow starts to exchange momentum with the wall. The wall jet boundary layer thickness is influenced by both flow velocity gradients with respect to the no-slip wall and with respect to the stationary flow above the jet. The wall shearing layer thickness increases as the flow move downstream, while its average velocity decreases due to momentum exchange with the wall. Depending on the velocity gradients, rib location could have a different impact on the flow physics and heat transfer rate.

By comparison with the baseline case, placing the circular rib at R/D = 2 and 3, improves the local N_u significantly (see Figure 10). This enhancement is a result of the high turbulence level induced by the flow recirculation before and after the rib. As rib height increases, the distance between the heated wall and the upper edge of the rib (where flow separation occurs) also increases, causing the flow to travel a greater distance before it re-attaches to the wall (see Figure 11). The re-attachment point shifts downstream (to the right on Figure 10). Unlike the two radial locations previously discussed, the average N_u enhancement lasts till rib height, e = 1.00 mm. For e $\geq$ 1 mm, another N_u peak occurs just before the rib, due to flow recirculation in front of the rib which enhances the heat transfer rate in this region.

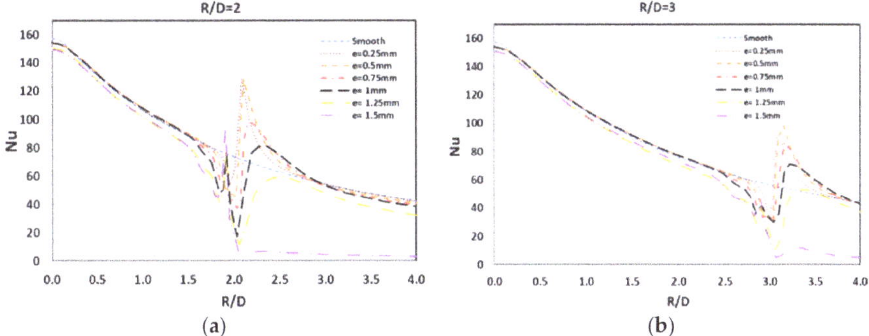

Figure 10. Local N_u distributions for six rib heights at (**a**) R/D = 2 and (**b**) R/D = 3.

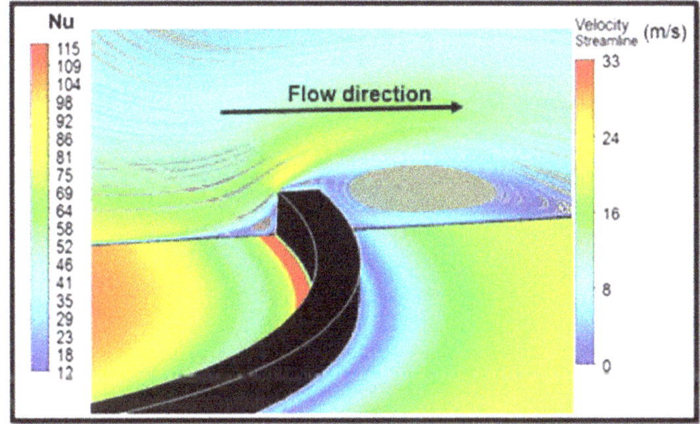

Figure 11. Velocity streamlines and local N_u contours for R/D = 2.0 and e = 1.25 mm.

Figure 12 presents velocity contours for three rib heights at R/D = 2 and 3, and shows that flow separation takes place in front of the rib resulting in a small separation region followed by a larger after-rib recirculation vortex. It also shows that the highest flow velocity is located above the top of the protrusion, while the lowest velocity is found at the bottom and around the protrusions. It can be seen that the flow with higher velocity, at R/D = 2 and 3, exists predominantly for protrusions with lower heights.

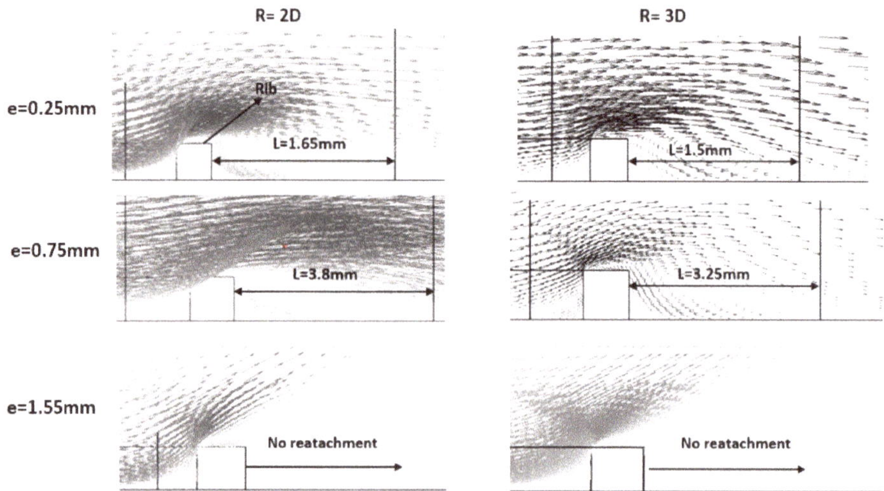

Figure 12. Velocity vector contours for three rib heights at two radial distances.

A protrusion usually exerts drag on the flow, causing a pressure build up and lower velocity, as shown in Figure 13, which shows that the higher the rib, the more the pressure build up. After the flow passes the protrusion a low-pressure region will occur behind the protrusion. Flow recirculation and higher turbulence will then occur in this low-pressure region which enhances the heat transfer rate. However, depending on the height and shape of the protrusion, this low-pressure region could be large enough to have a negative effect on heat transfer rate by preventing the main jet flow proceeding on its regular path. Figure 12 demonstrates this fact as it shows that the higher the rib, the larger the low-pressure region, and the longer distance the flow travels before it re-attaches to the heated surface.

The re-attachment length (L), which represents the length of the low-pressure region, increases as the rib height increases, this is true until above a certain height, the flow does not re-impinge on the heated surface and travels upward instead. Here, at e = 1.50 mm, the flow failed to reattach.

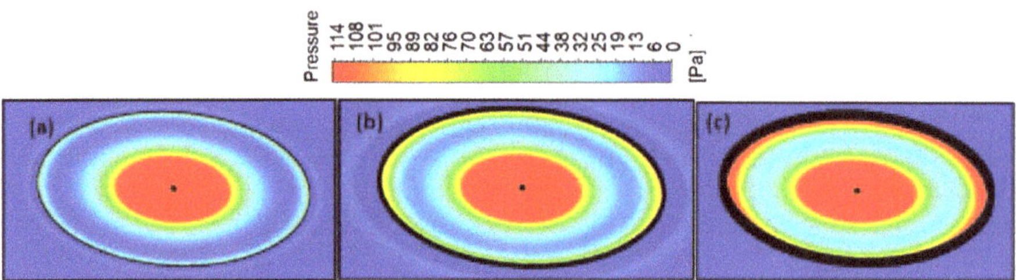

Figure 13. Pressure contours for R/D = 2, for three rib heights, (**a**) 0.25 mm; (**b**) 0.75 mm; and (**c**) 1.5 mm.

Figure 14 shows contours of local N_u distributions for three rib heights for the four radial locations the circular rib is represented by the black circle. The low-pressure region (wake), where there was flow recirculation, is shown by the blue area behind the rib. As can be seen from the figure, as the rib height increased, the extent of the low-pressure region behind the also rib increased. This process continued as the rib height increased, until at a rib height of 1.50 mm, the flow did not re-attach, but travelled upward away from the heated surface. This happened only for locations between R/D = 1.5 to 3, because R/D = 1 was within the stagnation region.

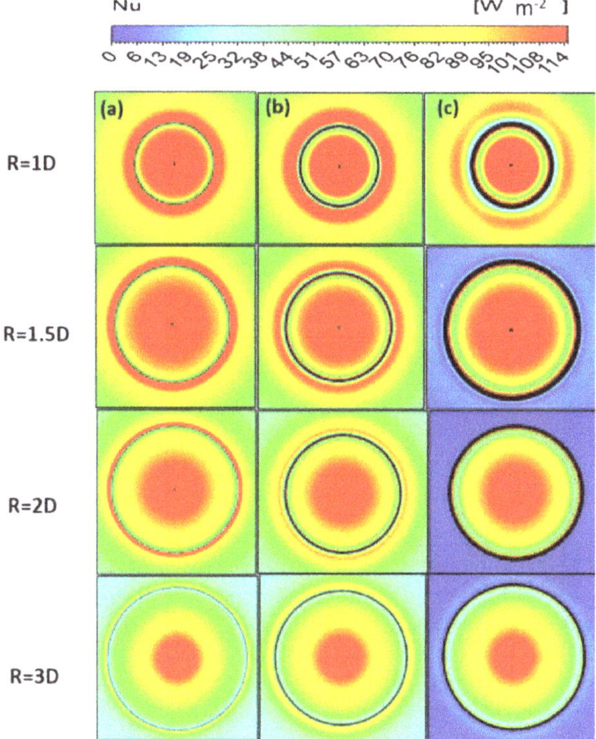

Figure 14. Local N_u contours for three selected rib heights, (**a**) 0.25 mm; (**b**) 0.75 mm; and (**c**) 1.5 mm.

3.3. Average N_u Characteristics

Figure 15 shows the effect of rib height on the average Nu of the heated surface, by presenting the average N_u ($\overline{N_u}$) normalized to the average N_u of the baseline ($\overline{N_{u_o}}$) for all rib radial locations. The average was obtained by numerically integrating the local N_u between $0 \leq R/D \leq 4$, in the downstream direction using Equation (10). It should be noted, for completeness sake, that the local heat transfer was, in fact, slightly affected at distances further downstream than $R/D \geq 4$ for rib heights $0.25 \leq e \leq 0.75$ mm, as was previously shown in Figure 8. The optimal rib height, at both $R/D = 1$ and $R/D = 1.5$, was e = 0.25 mm, which gave an increase in relative ($\overline{N_u}$) of 3.5% and 6%, respectively, compared to the baseline (no rib) case. Unlike the two locations at $R/D = 1.0$ and 1.5, discussed above, placing the rib at $R/D = 2$ enhanced heat transfer even for a rib of height, e = 1.00 mm. For $R/D = 2$, ribs in the range $0.25 \leq e \leq 1.00$ mm, the enhancements with respect to the baseline case were 10.5%, 15.6%, 12.7%, and 10.7%, respectively. Thus, there was a greater increase in heat transfer rate with the rib at this radial location compared to previous locations. For $R/D = 3$, a maximum enhancement in the average N_u of about 11.5% was achieved by introducing a rib with height 0.75 mm. Rib heights of 0.25 mm, 0.50 mm, and 1.00 mm gave enhancements of the average Nusselt number $\overline{N_u}$ of 6.2%, 8.7%, and 4.5%, respectively. A drop in heat transfer was noticeable as rib height increased above $e \geq 1.00$ mm. The use of inappropriate rib height, of e = 1.50 mm say, could lead to a loss of heat transfer of over 55% compared to the baseline case.

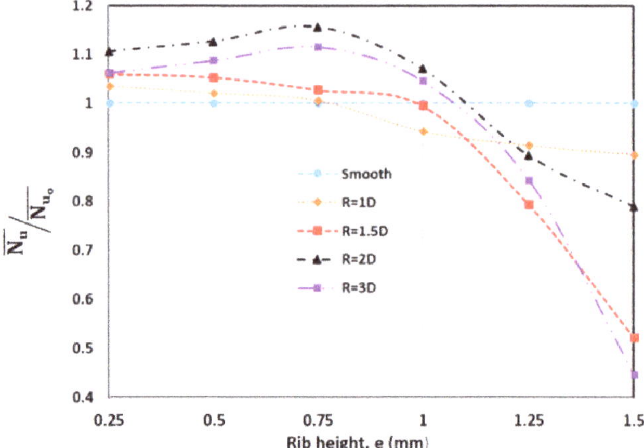

Figure 15. Effect of rib height on normalized average N_u for all rib locations, R_e = 20,000.

3.4. Comparison of Results

This research paper has presented the results of a simulation exercise on the effect of rib location and height on normalized N_u, averaged over a surface area. Figure 16 shows the effect on the local Nusselt number for ribs located at different radii from the center of the jet. Here the range of R/D was 1.0, 1.5, 2, and 3, and most of the heat transfer takes place in this region. For each rib location, the local N_u distribution determined the optimal rib height. As the radial distance of the rib from the stagnation point increased, and as R/D increased, the local maximum N_u associated with the presence of a rib decreased. This phenomenon is explained by the combined effects of the decrease in turbulent energy of the fluid as it moves away from the stagnation point and the rapid decrease in the velocity of the wall jet with radial distance from the impingement point.

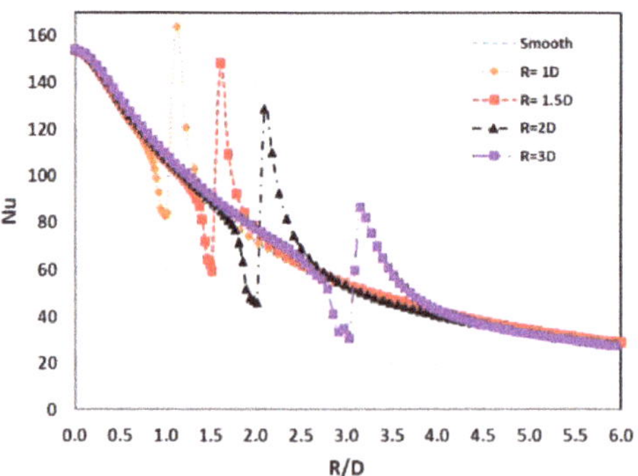

Figure 16. Effect of rib location on local N_u.

Figure 17 shows the normalized average Nusselt number for a circular rib on the target surface. The Nusselt number was normalized by dividing by the value for the averaged Nu over $0 \leq$ R/D ≤ 4 when no rib was present. The optimum rib height at each rib location was shown in this figure to determine the maximum enhancement in heat transfer between all rib locations. Clearly the most effective location for the rib was at R/D = 2 for which $\left[\dfrac{\overline{N_u}}{N_{uo}}\right]$ = 1.156. For R/D ≥ 2, the averaged

Nusselt number decreased as r increased. It is argued that maximum enhancement of the heat transfer rate was obtained when the rib was installed at a location close to the boundary between the stagnation region and the wall jet region, at R/D = 2.

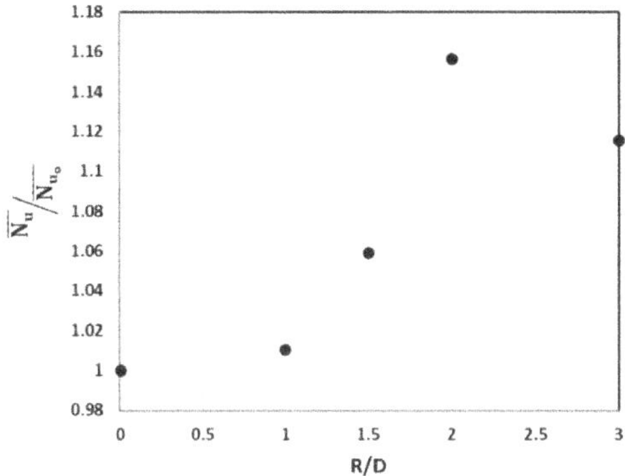

Figure 17. Effect of rib location on normalized average N_u.

Table 3 summarizes the main results reported in this research paper.

Table 3. Summary of findings.

Variable	R_1	R_2	R_3	R_4
Rib Location (R), (mm)	1D	1.5D	2D	3D
Calculated velocity boundary layer thickness (δ), (mm)	0.135	0.22	0.56	0.73
The optimum rib height (e), (mm)	0.25	0.25	0.5	0.75
Maximum heat transfer enhancement, (%)	3.5%	6.0%	15.6%	11.5%

4. Conclusions

Numerical analysis of a convective heat transfer process was conducted on the orthogonal air jet impingement on a smooth, horizontal surface, and a roughened uniformly heated flat plate. The commercial software ANSYS 17.0 was used for modelling and analyzing of both proposed cases. The continuous circular rib with the square cross-section is used as a roughness element, while the tested surface area covers the range of $0 \leq R/D \leq 4$. The circular roughness element was centered on the geometric center and tested for four different radii (R/D = 1, 1.5, 2, and 3) and six different heights (e) between 0.25 mm and 1.50 mm with an increment of 0.25 mm. In general, the rib height that matches the velocity boundary layer thickness at the rib location seems to be the most effective height for maximizing heat transfer rate. However, too high a rib gave a lower heat transfer rate than the no-rib case. It was also found that placing the rib, regardless of its height, in the stagnation region was ineffective when seeking to enhance heat transfer. The most effective rib location was at R/D = 2 which, based on the literature, is the beginning of the wall jet region. The results showed that when using the optimum rib height and location, a maximum heat transfer enhancement of 15.6% was achieved. The range of heat transfer enhancement for the rib at R/D = 2 was between 10.7% and 15.6% for rib heights e $\leq$ 1.00 mm.

Author Contributions: A.H.A. conceived the study, built the numerical models, performed the numerical solution and wrote the paper; H.M.A. helped performing the numerical solution, improved the simulations accuracy and manuscript review, A.A. (Abdulrahman Almutairi) contributed with conception of the work, data interpretation and revision of the manuscript; A.A. (Abdulmajid Addali) participated in the elaboration of the manuscript; A.A.A.G. helped analysing the numerical results. All authors read, edited and approved the manuscript.

Acknowledgments: The authors wish to express their sincere thanks to the Exergy Engineering Solutions (EES) in the state of Kuwait for its valuable support and assistance to the current work.

Conflicts of Interest: The authors declare no conflict of interest.

Nomenclature

Symbol	Description	Units
R_e	Jet Reynolds number, $\rho UD/\mu$	m^3/s
C_p	Heat capacity	$J/kg{\cdot}K$
h	Convective heat transfer coefficient	W/m^2K
$\bar{h}$	Average heat transfer coefficient	W/m^2K
D	Hydraulic diameter	m
R	Rib radial location	m
N_u	Nusselt number, hD/k	-
$N_{u_{stag}}$	Stagnation Nusselt number	-
$\overline{N_u}$	Average Nusselt number, $\bar{h}\frac{D}{k}$	-
$\overline{N_{u_o}}$	Baseline case average Nusselt number	-
q_w	Wall heat flux	W/m^2
T_{ref}	Reference temperature	$°C$
T_{aw}	Adiabatic wall temperature	$°C$
T_w	Wall temperature	$°C$
T_j	Jet temperature	$°C$
U_j	Jet velocity	m/s
y^+	Near-wall distance	m
ν	Kinematic viscosity	m^2/s
u_∞	Free stream velocity	m/s
δ	boundary layer thickness	mm
D_d	Domain diameter	m
P	Mean pressure	Pa
U, V, W	Streamwise, vertical and spanwise components of velocity	m/s
U', V', W'	Streamwise, vertical and spanwise components of fluctuating velocity	m/s
T	Mean temperature	$°C$
T'	Fluctuating temperature	$°C$

References

1. Jambunathan, K.; Lai, E.; Moss, M.A.; Button, B.L. A review of heat-transfer data for single circular jet impingement. *Int. J. Heat Fluid Flow* **1992**, *13*, 106–115. [CrossRef]

2. Almutairi, A. Computation of Conjugate Heat Transfer in Impinging Flows. Master's Thesis, University of Manchester, Manchester, UK, September 2010.

3. Kuraan, A.M.; Moldovan, S.I.; Choo, K. Heat transfer and hydrodynamics of free water jet impingement at low nozzle-to-plate spacings. *Int. J. Heat Mass Transf.* **2017**, *108*, 2211–2216. [CrossRef]

4. Zu, Y.Q.; Yan, Y.Y.; Maltson, J. Numerical Study on Stagnation Point Heat Transfer by Jet Impingement in a Confined Narrow Gap. *J. Heat Transf.* **2009**, *131*, 094504. [CrossRef]

5. Zhang, D.; Qu, H.; Lan, J.; Chen, J.; Xie, Y. Flow and heat transfer characteristics of single jet impinging on protrusioned surface. *Int. J. Heat Mass Transf.* **2013**, *58*, 18–28. [CrossRef]

6. Kim, W.S.; Lee, S.Y. Behavior of a water drop impinging on heated porous surfaces. *Exp. Therm. Fluid Sci.* **2014**, *55*, 62–70. [CrossRef]

7. Beitelmal, A.H.; Saad, M.A. Effects of surface roughness on the average heat transfer of an impinging air jet. *Int. Commun. Heat Mass Transf.* **2000**, *27*, 1–12. [CrossRef]

8. Sharif, M.A.R.; Ramirez, N.M. Surface roughness effects on the heat transfer due to turbulent round jet impingement on convex hemispherical surfaces. *Appl. Therm. Eng.* **2013**, *51*, 1026–1037. [CrossRef]
9. Xing, Y.; Weigand, B. Experimental investigation of impingement heat transfer on a flat and dimpled plate with different crossflow schemes. *Int. J. Heat Mass Transf.* **2010**, *53*, 3874–3886. [CrossRef]
10. Gabour, L.A.; Lienhard, J.H. Wall Roughness Effects on Stagnation-Point Heat Transfer Beneath an Impinging Liquid Jet. *J. Heat Transf.* **1994**, *116*, 81–87. [CrossRef]
11. Celik, N. Effects of the surface roughness on heat transfer of perpendicularly impinging co-axial jet. *Heat Mass Transf.* **2011**, *47*, 1209–1217. [CrossRef]
12. Kanokjaruvijit, K.; Martinez-Botas, R.F. Jet impingement on a dimpled surface with different crossflow schemes. *Int. J. Heat Mass Transf.* **2005**, *48*, 161–170. [CrossRef]
13. El-Gabry, L.A.; Kaminski, D.A. Experimental Investigation of Local Heat Transfer Distribution on Smooth and Roughened Surfaces under an Array of Angled Impinging Jets. *J. Turbomach.* **2005**, *127*, 532–544. [CrossRef]
14. Spring, S.; Xing, Y.; Weigand, B. An Experimental and Numerical Study of Heat Transfer from Arrays of Impinging Jets with Surface Ribs. *J. Heat Transf.* **2012**, *134*, 082201. [CrossRef]
15. Caliskan, S. Flow and heat transfer characteristics of transverse perforated ribs under impingement jets. *Int. J. Heat Mass Transf.* **2013**, *66*, 244–260. [CrossRef]
16. Wan, C.; Rao, Y.; Chen, P. Numerical predictions of jet impingement heat transfer on square pin-fin roughened plates. *Appl. Therm. Eng.* **2015**, *80*, 301–309. [CrossRef]
17. Nuntadusit, C.; Wae-hayee, M.; Bunyajitradulya, A.; Eiamsa-ard, S. Thermal visualization on surface with transverse perforated ribs. *Int. Commun. Heat Mass Transf.* **2012**, *39*, 634–639. [CrossRef]
18. Choi, E.Y.; Choi, Y.D.; Lee, W.S.; Chung, J.T.; Kwak, J.S. Heat transfer augmentation using a rib-dimple compound cooling technique. *Appl. Therm. Eng.* **2013**, *51*, 435–441. [CrossRef]
19. Hassan, M.M.; Teamah, M.A.; El-Maghlany, W.M. Numerical investigation for heat transfer enhancement using nanofluids over ribbed confined one-end closed flat-plate. *Alex. Eng. J.* **2017**, *56*, 333–343. [CrossRef]
20. Lo, Y.H.; Liu, Y.H. Heat transfer of impinging jet arrays onto half-smooth, half-rough target surfaces. *Appl. Therm. Eng.* **2018**, *128*, 79–91. [CrossRef]
21. Goldstein, R.J. Film cooling. *Adv. Heat Transf.* **1971**, *7*, 321–379.
22. Han, B.; Goldstein, R.J. Jet-Impingement Heat Transfer in Gas Turbine Systems. *Ann. N. Y. Acad. Sci.* **2001**, *934*, 147–161. [CrossRef] [PubMed]
23. O'Donovan, T.S.; Murray, D.B. Jet impingement heat transfer—Part I: Mean and root-mean-square heat transfer and velocity distributions. *Int. J. Heat Mass Transf.* **2007**, *50*, 3291–3301. [CrossRef]
24. Tu, J.; Yeoh, G.; Liu, C. *Computational Fluid Dynamics: A Practical Approach*; Elsevier: Amsterdam, The Netherlands, 2008.
25. Martin, H. Heat and mass transfer between impinging gas jets and solid surfaces. *Adv. Heat Transf.* **1977**, *13*, 1–60.
26. Katti, V.; Prabhu, S.V. Experimental study and theoretical analysis of local heat transfer distribution between smooth flat surface and impinging air jet from a circular straight pipe nozzle. *Int. J. Heat Mass Transf.* **2008**, *51*, 4480–4495. [CrossRef]
27. Zuckerman, N.; Lior, N. Jet impingement heat transfer: Physics, correlations, and numerical modeling. *Adv. Heat Transf.* **2006**, *39*, 565–631.
28. Gau, C.; Lee, I.C. Flow and impingement cooling heat transfer along triangular rib-roughened walls. *Int. J. Heat Mass Transf.* **2000**, *35*, 3009–3020. [CrossRef]

Article

Numerical Study on Thermal Hydraulic Performance of Supercritical LNG in Zigzag-Type Channel PCHEs

Zhongchao Zhao *, Yimeng Zhou, Xiaolong Ma, Xudong Chen, Shilin Li and Shan Yang

School of Energy and Power, Jiangsu University of Science and Technology, Zhenjiang 212000, China;
ymzhou@stu.just.edu.cn (Y.Z.); marlon@stu.just.edu.cn (X.M.); xudongchen@stu.just.edu.cn (X.C.);
shilinli@stu.just.edu.cn (S.L.); shanyang33@stu.just.edu.cn (S.Y.)
* Correspondence: zhongchaozhao@just.edu.cn; Tel.: +86-0511-8449-3050

Received: 18 December 2018; Accepted: 4 February 2019; Published: 11 February 2019

Abstract: In this paper, we study a promising plate-type heat exchanger, the printed circuit heat exchanger (PCHE), which has high compactness and is suitable for high-pressure conditions as a vaporizer during vaporization. The thermal hydraulic performance of supercritical produce liquefied natural gas (LNG) in the zigzag channel of PCHE is numerically investigated using the SST κ-ω turbulence model. The thermo-physical properties of supercritical LNG from 6.5 MPa to 10MPa were calculated using piecewise-polynomial approximations of the temperature. The effect of the channel bend angle, mass flux and inlet pressure on local convection heat transfer coefficient, and pressure drop are discussed. The heat transfer and pressure loss performance are evaluated using the Nusselt and Euler numbers. Nu/Eu is proposed to evaluate the comprehensive heat transfer performance of PCHE by considering the heat transfer and pressure drop characteristics to find better bend angle and operating conditions. The supercritical LNG has a better heat transfer performance when bend angle is less than $15°$ with the mass flux ranging from 207.2 kg/(m^2·s) to 621.6 kg/(m^2·s), which improves at bend angle of $10°$ and lower compared to $15°$ at mass flux above 414.4 kg/(m^2·s). The heat transfer performance is better at larger mass flux and lower operating pressures.

Keywords: printed circuit heat exchanger; supercritical LNG; zigzag type; heat transfer performance

---

1. Introduction

Natural gas (NG) is an advantageous energy source for various applications due to its clean nature and its environmental and economic advantages [1]. NG is liquefied to produce liquefied natural gas (LNG) for long-distance transportation and storage, and is regasified before terminal utilization [2]. Therefore, an efficient and reliable LNG vaporizer is a key component in a LNG vaporization system. The common LNG vaporizers, such as intermediate fluid vaporizers, open rack vaporizers (ORVs), super ORVs and submerged combustion vaporizers [3–5] do not satisfy the requirement of high efficiency and compactness in finite volume vaporization processes. Hence, the printed circuit heat exchanger (PCHE), which is a prospective plate-type heat exchanger with high compactness that can operate under high pressure and low temperature, has been investigated by many researchers [6,7].

In recently years, four PCHE channel morphologies have been studied, namely the straight, zigzag, S-shape, and airfoil shapes. Figley et al. [8] conducted numerical simulations to investigate the thermal hydraulic performance of the straight-channel PCHE using helium. The thermal effectiveness and overall heat transfer coefficient were defined and calculated to describe the overall heat transfer performance of PCHE. Kim et al. [9] predicted the thermal performance by developing a mathematical expression of geometric parameters, material properties, and flow conditions to express the effectiveness of cross, parallel, and counterflow PCHEs. Yoon et al. [10] developed a friction factor and Nusselt number relationship of laminar flow in various bend angles for a semi-circular zigzag channel PCHE. Ishizuk et al. [11] studied heat transfer and flow characteristics of zigzag-type PCHE experimentally using supercritical carbon dioxide and obtained the total heat transfer efficiency and pressure drop. Tsuzuki et al. [12] performed numerical analysis of the transition section of a zigzag-type channel PCHE, and encountered the presence of vortices and local circulation flow. Ngo et al. [13,14] conducted three-dimensional simulations of the pressure drop and Nusselt number of PCHEs with S-shaped fins in a supercritical carbon dioxide loop. Kim et al. [15] studied airfoil PCHEs using numerical analysis and indicated that the stream line was smooth and the vortices and countercurrents disappeared in the airfoil channel. Zhao et al. [16] investigated the heat transfer characteristics of an airfoil PCHE numerically using supercritical LNG and optimized the arrangement of airfoil fins.

In spite of the superior pressure drop performance of PCHEs with S-shaped and airfoil fins, compared to straight and zigzag-type PCHEs, discontinuous fins lack durability at high pressure operating conditions, which is attributed to the confined junction area of the fins in the diffuser bonding [17]. This in turn leads to an increase in manufacturing and maintenance costs [18,19]. Therefore, it makes sense to further investigate continuous channels and optimize channel shapes.

Previously, investigations were mainly concerned with the heat transfer performance and pressure drop of zigzag PCHEs, but few combined the heat transfer and pressure drop performance to study the optimization of zigzag PCHEs in terms of bend angle and operating conditions. In addition, supercritical fluids can be used to improve heat transfer performance due to their favorable properties, like high density, low viscosity, and high thermal conductivity compared to conventional fluids [20]. Most studies on the flow and heat transfer characteristics of fluids in PCHE have been conducted using helium and water [21–23], and many researchers also used supercritical water and supercritical carbon dioxide to investigate PCHEs [24,25]. However, studies on the heat transfer performance of supercritical LNG in zigzag PCHE under low temperature and high pressure conditions are rare [16].

In this study, the flow and heat transfer characteristics of supercritical LNG in zigzag-type PCHEs were numerically simulated at an operating pressure of 6.5–10 MPa. The effects of bend angle and mass flux on heat transfer performance were also investigated. The local convection heat transfer coefficient and pressure drop of supercritical LNG under different conditions are discussed. Dimensionless parameters such as the Nusselt number (Nu) and Euler number (Eu) are analyzed to assess the heat transfer and pressure loss performance, Nu/Eu of the bend angles and operating conditions are investigated by considering the performance of heat transfer and the pressure drop of supercritical LNG in PCHEs.

2. Numerical Approach

2.1. Physical Model and Boundary Conditions

In this study, the thermal hydraulic performance of supercritical LNG in zigzag PCHEs is investigated. The cross flow PCHE core model with a full length of 400 mm using supercritical LNG in the cold side and R22 in the hot side is shown in Figure 1a. The supercritical LNG and R22 flow in the semicircular channel with a diameter of 1.5 mm, the solid is composed of steel with thermal conductive coefficient of 16.27 W/(m·K). In this paper, we only study the performance of supercritical LNG in the cold channel. Considering that the cold side of the PCHE contains hundreds of channels, it is unrealistic to consider all the channels, and it is therefore necessary to simplify the cold channels' model. Supercritical LNG flows in parallel in each cold channel, so some assumptions on its flow in the cold side are made. The mass flux is the same in every channel, and there is no temperature difference and heat loss between neighboring channels. The flow of supercritical LNG is steady and uniformly distributed. Based on these assumptions, the cold channel can be simplified to a single model with a geometry of 2 mm × 1.75 mm. The cross-section of the fluid channel is semicircular with a diameter of 1.5 mm (Figure 1b). The bend angles α vary from 0° (which is a straight channel) to 45° (Figure 1c).

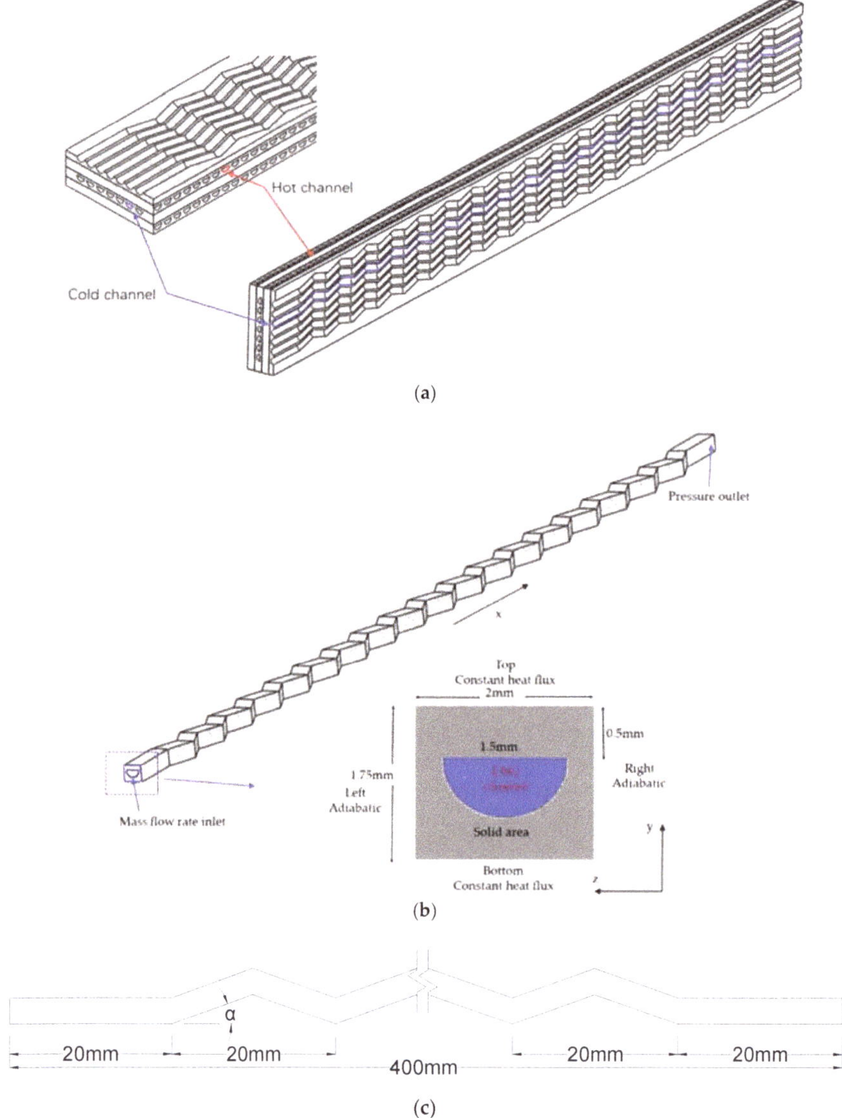

Figure 1. Schematic diagram of the physical model: (**a**) Schematic diagram of cross flow printed circuit heat exchanger (PCHE) model; (**b**) The computational single channel and boundary conditions; (**c**) Top view of zigzag channel.

The adjacent cold channels do not exhibit temperature difference and heat transfer loss; only the supercritical LNG in cold channels absorb heat from top and bottom hot channels. Three types of boundary conditions were applied in the model: fluid inlet, fluid outlet, and wall. The mass flow rate boundary condition was set at the inlet of the supercritical LNG channel whereas at the outlet the pressure-outlet boundary condition was applied (Figure 1b). The left and right walls of the single model are set to adiabatic boundary conditions, and the constant heat flux condition was used at top and bottom walls. The details of the boundary conditions are presented in Table 1.

Table 1. Boundary conditions in detail.

	Inlet		Outlet	Left/Right Walls	Top/Bottom Walls
Pressure (MPa)	Temperature (K)	Mass flux (kg/m^2·s)	Pressure outlet	Adiabatic	Constant heat flux (W/m^2)
10	121	207.2			7.5×10^4

2.2. Thermo-Physical Properties of Supercritical LNG

In this paper, the operating pressure of LNG considered varies from 6.5 MPa to 10 MPa, which is supercritical pressure. Supercritical LNG has gas-like properties, such as low viscosity, and liquid-like characteristics, like high density and high thermal conductivity. The thermo-physical properties of supercritical LNG, i.e., density, specific heat, thermal conductivity and viscosity, are affected by temperature and pressure. The properties' values were obtained from the NIST Standard Reference Database (REFPROP) [26]. For the numerical simulations, the temperature was changed from 121 K to 385 K. At such a large temperature difference, the properties of supercritical LNG change dramatically, using the average values will cause the inaccurate calculation results in ANSYS Fluent. Therefore, as shown in Figure 2, the thermal properties of supercritical LNG were approximated as piecewise-polynomial functions of temperature. The piecewise-polynomial functions of temperature at 10 MPa is shown in Table 2. The error percentages of various properties using the proposed approximation are shown in Figure 3. The errors were within ±2.5%, which indicates the fitted piecewise-polynomial function approximations are suitable.

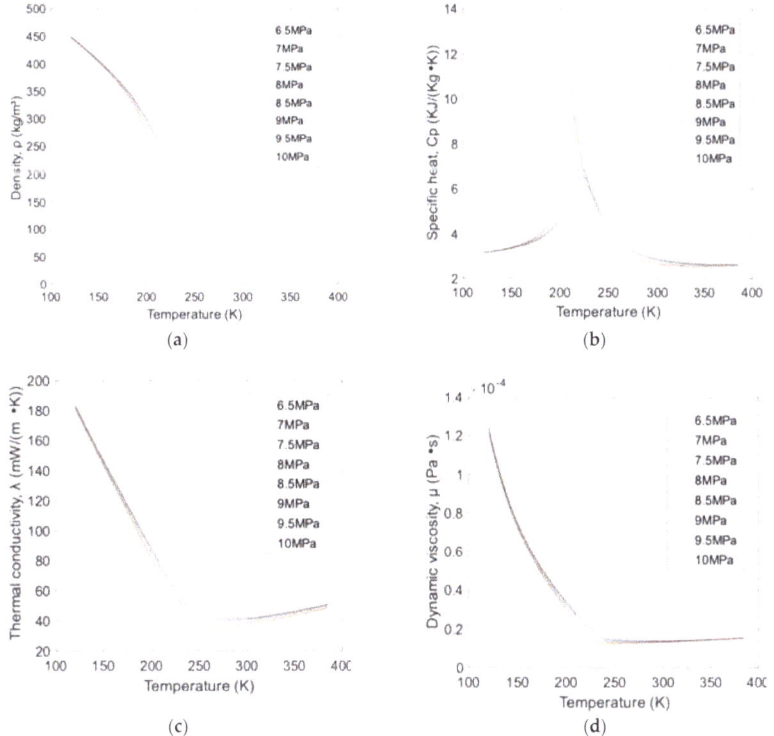

Figure 2. Thermo-physical properties of supercritical LNG at different pressures: (**a**) Density; (**b**) Specific heat; (**c**) Thermal conductivity; (**d**) Dynamic viscosity.

Table 2. Piecewise-polynomial functions at 10 MPa.

Temperature Range (K)	Density
121–223	$\rho = -1.29548 \times 10^{-6}T^4 + 7.459 \times 10^{-4}T^3 - 0.16537T^2 + 15.12763T - 2.45869$
223–271	$\rho = -5.30962 \times 10^{-4}T^3 + 0.43352T^2 - 119.03258T + 11{,}091.66885$
271–385	$\rho = 1.32923 \times 10^{-7}T^4 - 1.93454 \times 10^{-4}T^3 + 0.10675T^2 - 26.70158T + 2634.06161$
	Specific Heat
121–223	$c_p = 0.00436T^3 - 1.83425T^2 + 263.05475T - 9567.33957$
223–261	$c_p = 0.06448T^3 - 46.88971 \times 10^3 T^2 + 11{,}278.86482T - 892{,}240.75366$
261–385	$c_p = -0.00111T^3 + 1.18621T^2 - 423.72673T + 53{,}206.50774$
	Thermal Conductivity
12––235	$\lambda = 1.07039 \times 10^{-8}T^3 - 4.35253 \times 10^{-6}T^2 - 6.44568 \times 10^{-4}T + 0.30675$
235–262	$\lambda = 7.96913 \times 10^{-6}T^2 - 0.00432T + 0.62882$
262–385	$\lambda = -6.69403 \times 10^{-9}T^3 + 7.36829 \times 10^{-6}T^2 - 0.00258 \times 10^{-2}T + 0.33402$
	Viscosity
121–218	$\mu = -7.71822 \times 10^{-11}T^3 + 4.71489 \times 10^{-8}T^2 - 1.01809 \times 10^{-5}T + 8.02631 \times 10^{-4}$
218–254	$\mu = 6.9276 \times 10^{-9}T^2 - 3.52808 \times 10^{-6}T + 4.64105 \times 10^{-4}$
254–385	$\mu = -2.07823 \times 10^{-12}T^3 + 2.1834 \times 10^{-9}T^2 - 7.43585 \times 10^{-7}T + 9.67009 \times 10^{-5}$

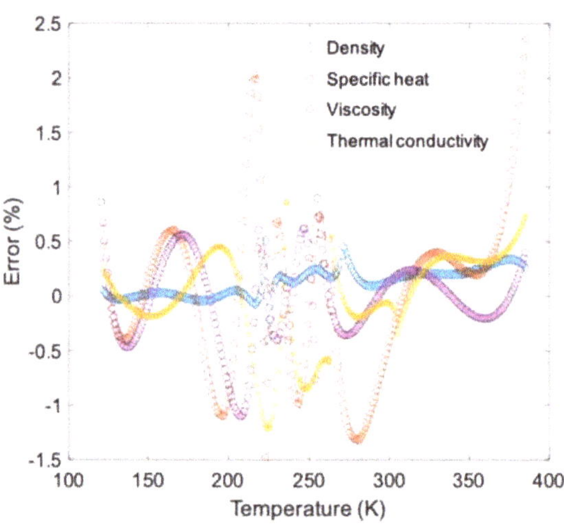

Figure 3. Error curves of linear interpolation functions.

2.3. Numerical Method and Grid Independence

The commercial software FLUENT was used to solve the 3D numerical model. Considering the inlet parameters, the flow corresponded to turbulent flow regimes. Some turbulence models have been studied and used in the literature; these include the $\kappa\text{-}\varepsilon$ standard model, the RNG $\kappa\text{-}\varepsilon$ model, the shear-stress transport (SST) $\kappa\text{-}\omega$ model and the low Reynolds number turbulence model [27,28]. In this study, the SST $\kappa\text{-}\omega$ model was used because of its more accurate results on heat transfer of supercritical fluids [29–32].

The governing equations for heat transfer were the continuity, momentum, and energy equations, respectively:

Continuity equation:

$$\frac{\partial}{\partial x_i}(\rho u_i) = 0, \tag{1}$$

where ρ is the density, and u_i is the velocity vector.

Momentum equation:

$$\frac{\partial}{\partial x_i}\left(\rho u_i u_j\right) = -\frac{\partial p}{\partial x_i} + \rho g_i + \frac{\partial}{\partial x_j}[(\mu + \mu_t)\frac{\partial u_i}{\partial x_j}], \tag{2}$$

where p is the pressure, μ and μ_t are the molecular and turbulent viscosities, respectively.

Energy equation:

$$\frac{\partial}{\partial x_i}(u_i(\rho E + p)) = \frac{\partial}{\partial x_i}\left(k_{eff}\frac{\partial T}{\partial x_i} + u_i \tau_{ij}\right), \tag{3}$$

where k_{eff} is effective conductivity, $k_{eff} = k + k_t$, and k_t is the turbulent thermal conductivity.

The transport equations are expressed as follows:

$$\frac{D(\rho\kappa)}{Dt} = \frac{\partial}{\partial x_j}\left[(\mu + \sigma_\kappa\mu_t)\frac{\partial\kappa}{\partial x_j}\right] + \tau_{ij}\frac{\partial u_i}{\partial x_j} - \beta^*\rho\omega\kappa \tag{4}$$

$$\frac{D(\rho\omega)}{Dt} = \frac{\partial}{\partial x_j}\left[(\mu + \sigma_{\omega 1}\mu_t)\frac{\partial\omega}{\partial x_j}\right] + \frac{\gamma}{\nu_t}\tau_{ij}\frac{\partial u_i}{\partial x_j} - \beta\rho\omega^2 + 2(1 - F_1)\rho\sigma_{\omega 2}\frac{1}{\omega}\frac{\partial\kappa}{\partial x_j}\frac{\partial\omega}{\partial x_j} \tag{5}$$

$$\omega = \frac{\varepsilon}{\beta^*\kappa}; \nu_t = \frac{a_1\kappa}{max(a_1\omega; \Omega F_2)} \tag{6}$$

$$F_1 = \tan h(\mathrm{arg}_1^4); F_2 = \tan h(\mathrm{arg}_2^2) \tag{7}$$

$$\mathrm{arg}_1 = min\left(max\left(\frac{\sqrt{\kappa}}{0.09\omega y}; \frac{500\nu}{y^2\omega}\right); \frac{4\rho\sigma_{\omega 2}\kappa}{CD_{\kappa\omega}y^2}\right); CD_{\kappa\omega} = max\left(2\frac{\rho\sigma_{\omega 2}}{\omega}\frac{1}{\omega}\frac{\partial\kappa}{\partial x_j}\frac{\partial\omega}{\partial x_j}, 10^{-20}\right) \tag{8}$$

$$\mathrm{arg}_2 = max\left(2\frac{\sqrt{\kappa}}{0.09\omega y}; \frac{500\nu}{y^2\omega}\right) \tag{9}$$

where ε is the turbulent kinetic energy dissipation rate, Ω is vorticity, and y is the distance from the wall. The constants and damping functions of the SST κ-ω model are shown in Table 3.

Table 3. Constants and functions used in the shear-stress transport (SST) model.

	$\sigma_{\omega 1}$	$\sigma_{\omega 2}$	κ	α_1	β^*
SST	0.5	0.865	0.41	0.31	0.09

The local convective heat transfer coefficient was calculated using Equation (10):

$$h = \frac{q}{T_w - T_b} = \frac{q}{T_w - (T_{out} + T_{in})/2}, \tag{10}$$

where q is the constant heat flux from the top and bottom walls, T_w is the wall temperature and T_b is average temperature of the inlet and the outlet.

Nu was defined as:

$$N_u = \frac{hD_h}{\lambda}; D_h = 4A/l \tag{11}$$

where D_h is the hydraulic diameter and λ is the local thermal conductivity of LNG, A is the cross-sectional area of the semicircular fluid channel and l is the circumference of the semicircular fluid channel section.

The local Fanning friction coefficient (f) was defined in terms of the pressure drop and is expressed by Equation (12):

$$f = \frac{\Delta P_f D_h}{2L\rho_b v_b^2}, \tag{12}$$

$$\Delta P_f = \Delta P - \Delta P_a = \Delta P - \left(\rho_{out} v_{out}^2 - \rho_{in} v_{in}^2 \right), \tag{13}$$

where ΔP is the total pressure and was obtained from Fluent directly, ΔP_f and ΔP_a are the frictional and accelerated pressure drops, respectively, L is the channel length, ρ_b and v_b are the bulk density and velocity of LNG, respectively.

The Reynolds number (Re) is given by Equation (14):

$$R_e = \frac{\rho_b v_d D_h}{\mu_b}, \tag{14}$$

The Euler number (Eu) is defined as Equation (15):

$$Eu = \frac{\Delta P}{\rho_b v_b^2 / 2}, \tag{15}$$

For the solution methods, the SIMPLE algorithm was applied to establish the coupling of velocity and pressure. The momentum, turbulent kinetic energy, turbulent dissipation rate and energy were discretized using the second order upwind scheme. The calculation was considered to converge when the residuals were less than 10^{-6}.

The mesh on the computational domain was generated using GAMBIT. The grid independence was verified to confirm numerical result accuracy. The mesh size of solid, fluid and boundary layer's scale in fluid affect the grid numbers. The influence of the grid numbers on the convective heat transfer coefficient is shown in Table 4. Case 4 has a larger relative error compared to the other cases. The heat transfer coefficient in Cases 1, 2, and 3 is nearly the same. The relative error of the heat transfer coefficient between Cases 1 and 7 is only 0.08%. Therefore, considering the calculation accuracy and time, the 2,988,329 grid nodes (Case 1), showing in Figure 4, was selected in the present work.

Table 4. Boundary layers study.

Case	Scale of Boundary Layer	Rows of Boundary Layer	Cells of Nodes	Heat Transfer Coefficient W/(m²·K)	Relative Error (%)
1	0.01	5	2,988,329	2678.57	3.4%
2	0.01	8	3,589,947	2680.44	3.47%
3	0.003	5	2,974,634	2678.32	3.4%
4	0.03	5	2,697,546	2669.23	3.04%
5			815,644	2590.42	0
6	0.01	5	1,962,788	2636.39	1.8%
7			4,456,851	2680.62	3.48%

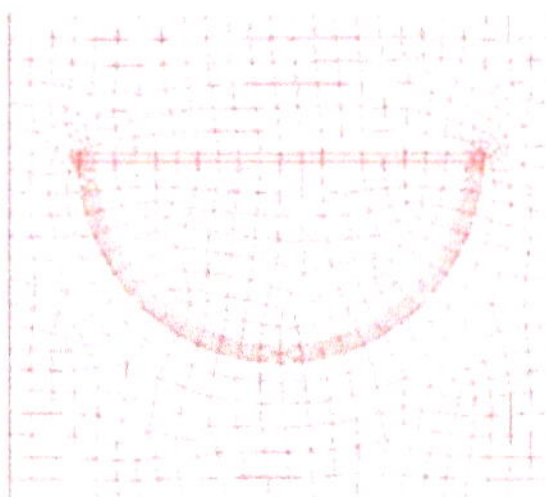

Figure 4. Cross section of computational grids.

2.4. Model Validation

To validate the accuracy and reliability of the model, the simulation results of temperature difference and pressure drop were compared to previous experimental results [30]. The experimental setup is shown in Figure 5. Since LNG is flammable and explosive, the straight-channel cross flow PCHE used supercritical nitrogen as the cold side fluid and R22 as the hot side fluid. The length of the PCHE cold channel was 520 mm, the inlet temperature was 102 K, and the operating pressure varied from 6.5 MPa to 10 MPa. In the simulation, a straight channel model with a length of 520 mm was selected, which was the same as the experimental case. The boundary conditions are shown in Figure 1b. Supercritical nitrogen was used as the working fluid to confirm the correctness the experiment. The inlet pressure changed from 6.5 MPa to 10 MPa and the inlet temperature was 102 K. The comparison of temperature difference and pressure drop between simulation and experimental results is listed in Table 5. The maximum errors of temperature difference and pressure drop are 2.1% and 10.25%, respectively. The simulation pressure drop was less than the experimental, which may be attributed to the following factors: (1) the channel was assumed to be smooth in the numerical study while the PCHE channel of the experiment was rough, (2) the header pressure drop of inlet and the outlet were neglected in the numerical study, but may have been large in the experiment, and (3) the deviation of temperature and pressure transmissions. The simulation results are in accordance with the experiment, illustrating that the simulation model and method were credible.

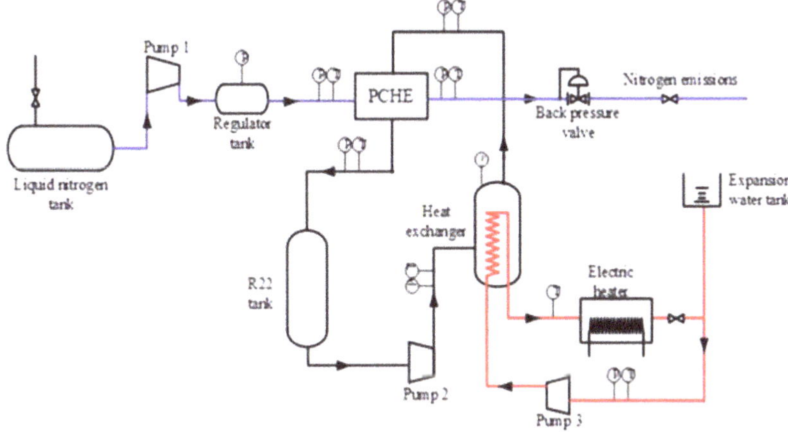

Figure 5. Schematic diagram of experimental setup.

Table 5. Relative error of simulation and experiment results.

Pressure (MPa)	Temperature Difference (K)		Relative Error (%)	Pressure Difference (Pa)		Relative Error (%)
	Experiment	Simulation		Experiment	Simulation	
6.5	178.9	175.1	2.1	16,612.35	15,167.36	10.25
7	180.3	178	1.27	15,636.09	14,521.4	9.95
7.5	182.1	180.4	0.94	14,742.24	14,071.62	7.05
8	182.6	182.3	0.16	13,847.55	13,188.32	6.62
8.5	183.5	184	0.27	13,035.22	12,504.69	6.81
9	185.7	185.5	0.11	12,156.68	11,810.66	4.70
9.5	186.1	185.8	0.16	11,342.33	10,862.6	4.42
10	186.6	186.4	0.11	10,578.6	10,189.45	3.82

3. Results and Discussion

Compared with traditional vaporizers, the heat transfer performance of PCHE is better, but the pressure drop is larger due to the small size of the channel, resulting in an increase of operating costs. A number of studies have shown that a supercritical fluid can enhance heat transfer and reduce pressure drop. The supercritical LNG vaporized by the PCHE is suitable for long distance transport and utilization. In this paper, the flow and heat transfer characteristics of supercritical LNG were studied at 6.5–10MPa, the performance of supercritical LNG at 10 MPa was discussed in detail in the following.

3.1. Effect of Bend Angles of the Zigzag Channel

The heat transfer performance of supercritical LNG is influenced significantly by the bend angle in the zigzag channel of a PCHE. The effect of the bend angles on the local convection heat transfer coefficient and bulk temperature of the LNG along the streamwise direction at a mass flux of 207.2 kg/(m^2·s) and operating pressure of 10 MPa are shown in Figure 6. As the bend angle increases, the local convection heat transfer coefficient and bulk temperature increase, which is due to the fact that with the increase of the bend angle, disturbance and turbulence will increase. Moreover, the local heat transfer coefficient increases and then decreases along the streamwise direction, and reaches a peak when the bulk temperature is near the pseudo-critical temperature at bend of 0°–15°. This is because that the thermo-physical properties of supercritical LNG vary drastically at different temperatures, and the specific heat in particular reaches an extremum near the pseudo-critical temperature (Figure 2). However, at bend angles of 25°–45°, the convection heat transfer coefficient is greater at Np = 2–4. When the LNG flows into the channel, its velocity is not large in the Np = 2–4. The flow separation is not dramatic and the velocity of vortices is not much different that of the fluid. As the flow develops, the velocity increases. The velocity difference of vortices and fluid becomes large and the vortices and flow separation increase (as shown in Figure 7b), leading to a decrease in the convection heat transfer coefficient. In addition, the heat transfer coefficient is larger at the inlet due to the entrance effect.

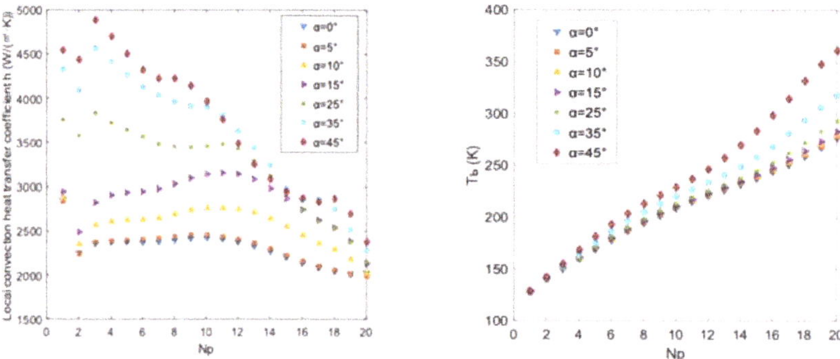

Figure 6. Local convection heat transfer coefficient and bulk temperature at different bend angles along the streamwise direction.

Figure 8 shows the pressure drop and Fanning friction coefficient (f) at different bend angles along the streamwise direction. The pressure drop increases with the bend angles and increases along the streamwise direction. The density and viscosity of the supercritical LNG decrease as the temperature increases (as shown in Figure 2a,d), resulting in a velocity increase and thus in pressure drop. Figure 7 shows the velocity vectors of different cross-sections along the channel and velocity vectors of Np = 10–12 at different bend angles. The velocity increases with the increase of bend angles, which is attributed to the increase of turbulence (Figure 7a). Flow separation and reverse flow appear in larger bend angles (Figure 7b), which increases flow resistance, resulting in an increased pressure drop. The velocity increases along the flow direction, which also increase the pressure drop along the streamwise. The Fanning friction coefficient f increases with the bend angle, which is the same trend observed in pressure drop. However, f decreases along the flow direction; this difference between pressure drop and f is due to the increase of velocity.

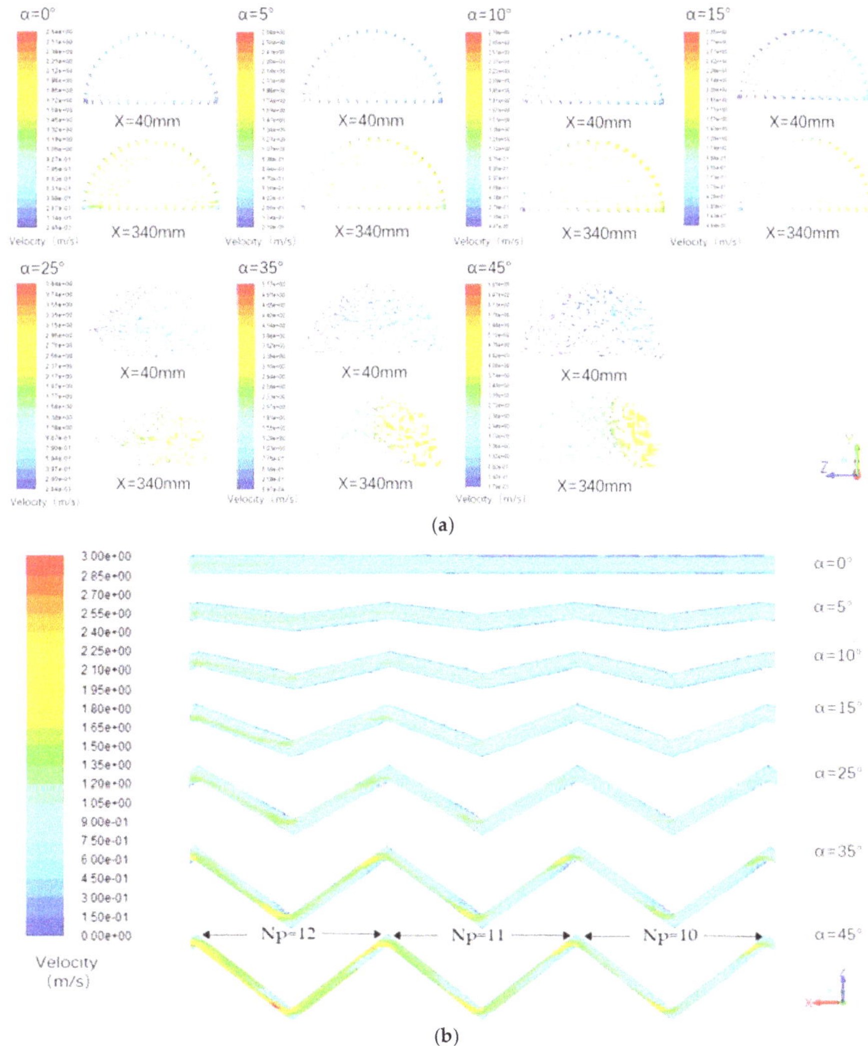

Figure 7. (**a**) Velocity vectors of different cross-sections along the channel with bend angles of 0° to 45°; (**b**) Velocity vectors of Np = 10–12 at bend angles of 0° to 45°.

Figure 9 shows the Nusselt (Nu) and Euler (Eu) numbers along the streamwise direction at different bend angles. It can be seen that Nu increases with the bend angle, and reaches its maximum value near the pseudo-critical temperature, then decreases along the streamwise direction. This is because the thermal conductivity decreases intensely as the temperature rises before pseudo-critical temperature and rises slightly when the temperature surpasses the pseudo-critical value, reaching a minimum near the pseudo-critical temperature, which leads to a maximum for Nu near the pseudo-critical temperature. The Euler number increases as the bend angle rises, which is consistent with the change in pressure drop.

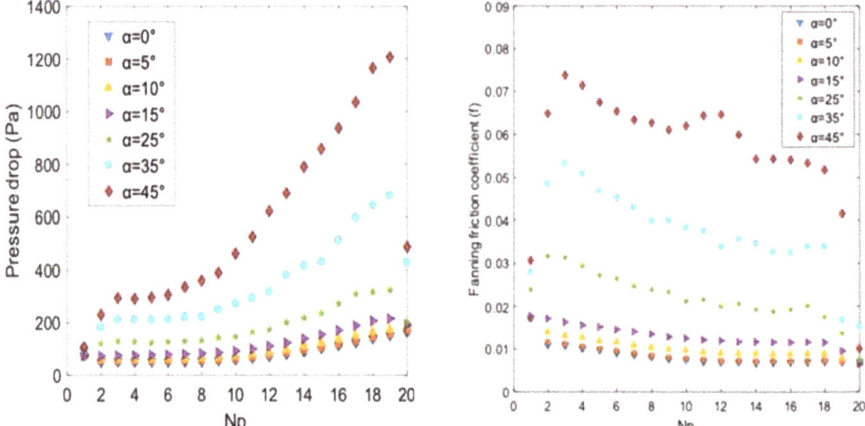

Figure 8. Pressure drop and fanning friction coefficient at different bend angles along the streamwise direction.

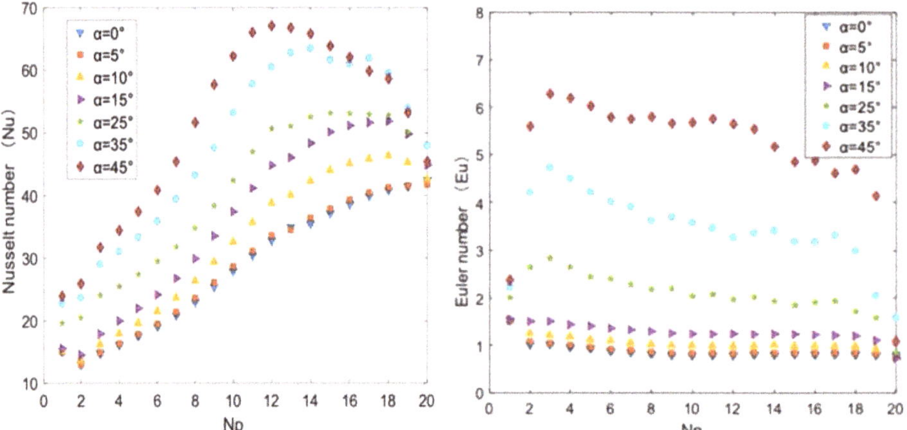

Figure 9. Local Nusselt number (Nu) and Euler (EU) numbers at different bend angles along the streamwise direction.

The objective of improving the heat exchanger's performance can be realized by increasing heat transfer performance and reducing pressure drop. It is therefore essential to comprehensively consider the heat transfer and pressure loss characteristics of supercritical LNG in PCHE. In this study, the ratio of Nusselt to Euler numbers (Nu/Eu) is proposed to evaluate the performance of supercritical LNG in the channel, where a larger ratio indicates better heat transfer performance. Figure 10 shows Nu/Eu at different bend angles. Nu/Eu reaches its peak value at a bend angle of 10°. When the bend angle exceeds 10°, the growth rate of Nu is much less than that of Eu. Nu and Eu at 0° and 5° are nearly the same, while from 5° to 10°, the increase of Nu is much greater than that of Eu.

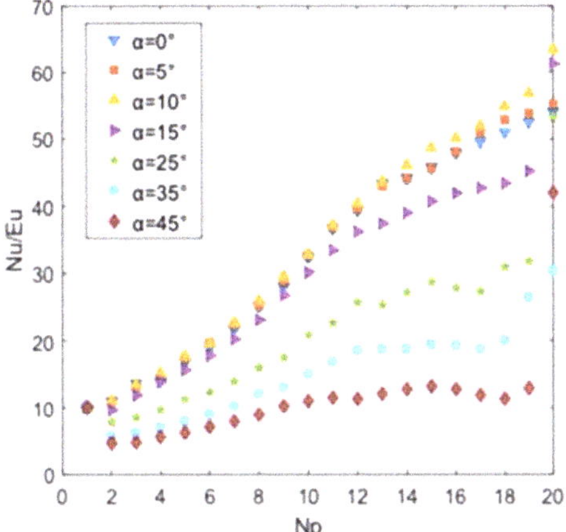

Figure 10. Nu/Eu at different bend angles along the streamwise direction.

3.2. Effect of Mass Flux

The effect of mass flux on flow and heat transfer performance of supercritical LNG was investigated at the bend angle of 10° and an operating pressure of 10 MPa. As shown in Figure 11, the local convective heat transfer coefficient and pressure drop increase significantly as the mass flux increases because of the enhancement of turbulent flow. When the mass flux is increased by 2 times, the local heat transfer coefficient increased 1.4 times, and at the same time, the pressure drop increases 3.3 times. The Nu and Eu are shown in Figure 12. The Nu increases as the mass flux is raised. However, at the last third of the channel, Nu peaks at a mass flux of 301.8 kg/(m^2·s), and then decreases at further mass flux increase. This is because as mass flux is increased, the temperature of LNG decreases, its viscosity increases and its velocity decreases. When the heat flux is kept constant, the heat absorbed by the LNG per unit volume is reduced.

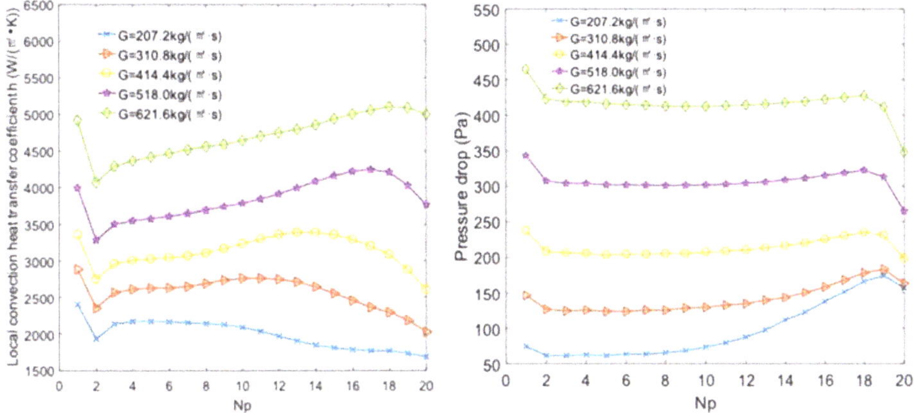

Figure 11. Effect of mass flux on local convection heat transfer coefficient and pressure drop along the streamwise direction.

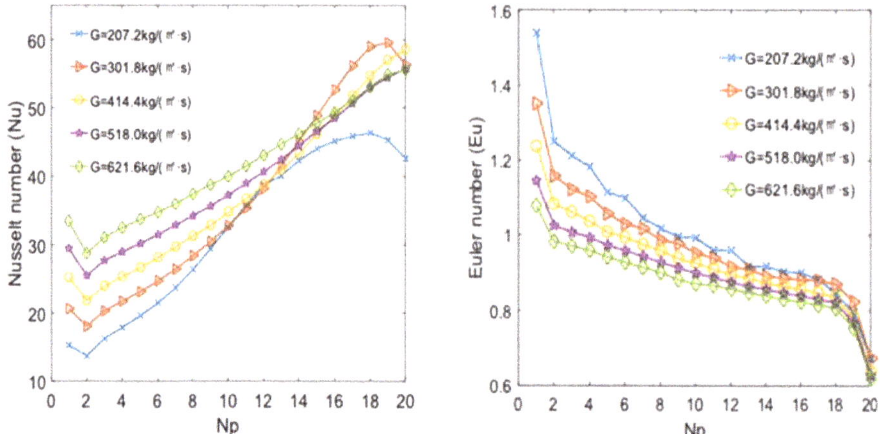

Figure 12. Effect of mass flux on Nu and Eu along the streamwise direction.

Eu increases as the mass flux decreases. The pressure drop and density increase as mass flux rises, but v_b^2 decreased. The increase rate of $\rho_b v_b^2$ is more than that of ΔP, which results in an increase of Eu as mass flux is reduced.

The influence of mass flux on Nu/Eu is shown in Figure 13. The Nu/Eu is increased as the mass flux is raised. However, at the last third of the channel, Nu/Eu peaks at a mass flux of 301.8 kg/(m²·s). The figure suggests that the heat transfer performance of the whole channel is improved as the mass flux increases, but with the development of the fluid flow, the local heat transfer performance is reduced at the last third of the channel, owing to the reduction of heat absorbed capacity by the unit volume fluid.

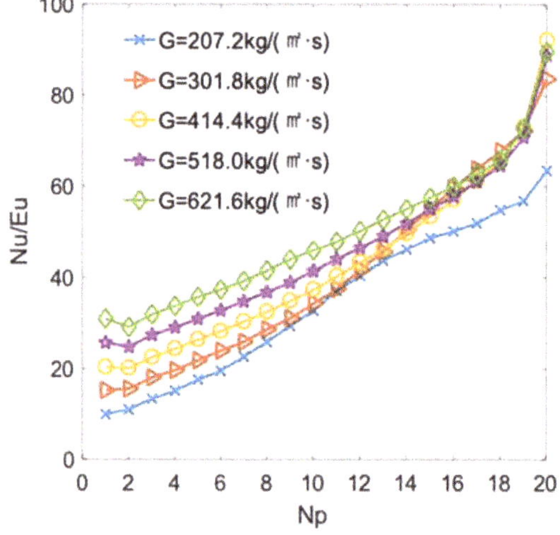

Figure 13. Effect of mass flux on Nu/Eu along the streamwise direction.

Figure 14 shows the impact of mass flux on Nu/Eu at different channel bend angles. The Nu/Eu is significantly reduced when the bend angle exceeds 15°, indicating that the increase of pressure drop is much higher than that of heat transfer performance. Consequently, the comprehensive heat transfer performance is not good when the bend angles exceeds 15°. When the mass flux varies from 207.2 kg/(m^2·s) to 621.6 kg/(m^2·s), Nu/Eu are higher when the bend angle is less than 15°. However, Nu/Eu at bend angles of 10° and lower are increased compared to 15° at mass fluxes above 414.4 kg/(m^2·s). It can be concluded that the supercritical LNG in the PCHE has better heat transfer performance when the bend angle is less than 15° with the mass flux ranging from 207.2 kg/(m^2·s) to 621.6 kg/(m^2·s), and that it improves at bend angles of 10° and lower compared to 15° at mass fluxes above 414.4 kg/(m^2·s).

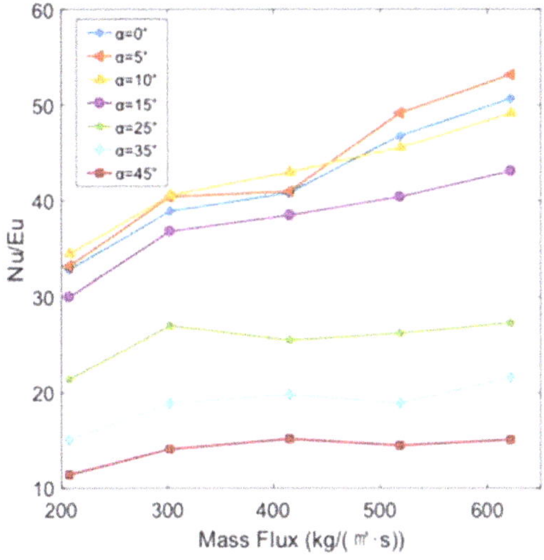

Figure 14. Effect of mass flux on Nu/Eu at different bend angles.

3.3. Effect of Inlet Pressure

At a bend angle of 10° and a mass flux of 207.2 kg/(m^2·s), the inlet pressure was varied from 6.5 MPa to 10 MP. The corresponding local heat transfer coefficient and pressure drop values are shown in Figure 15. Before the pseudo-critical temperature, the specific heat and thermal conductivity are slightly affected by the inlet pressure (as shown in Figure 2b,c), so the convection heat transfer coefficient changes slightly with the inlet pressure. After the pseudo-critical temperature, the inlet pressure effect on the local convective heat transfer coefficient is greater because of the specific heat is influenced by inlet pressure more. In addition, the reduction rate of the specific heat rises rapidly as the inlet pressure is reduced. Therefore, the local convection heat transfer coefficient decreases with the decrease of inlet pressure along the streamwise direction. This shows that the specific heat depends on inlet pressure and has a great influence on local convective heat transfer coefficient when temperature exceeds the pseudo-critical temperature.

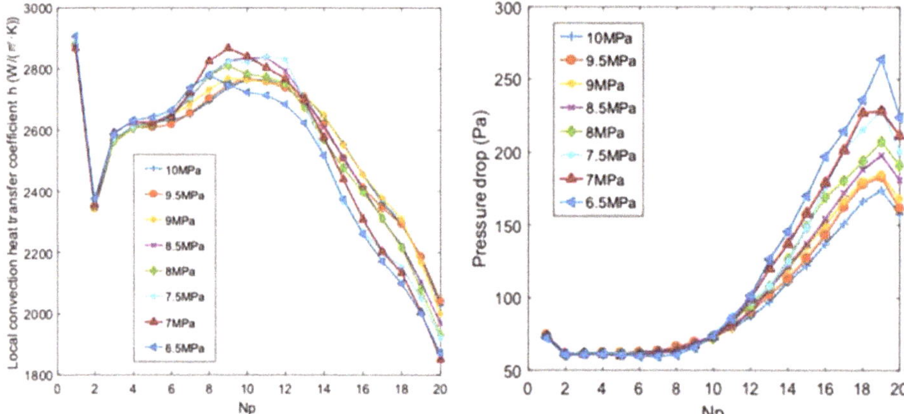

Figure 15. Effect of inlet pressure on local convection heat transfer coefficient and pressure drop along the streamwise direction.

With an increase of inlet pressure, the pressure drop decreases. As inlet pressure rises, the density and dynamic viscosity are larger (Figure 2a,d), which lowers the velocity of supercritical LNG. Hence, the pressure drop is reduced as the inlet pressure is increased. However, up until $Np = 10$, the pressure drop is uninfluenced by the changing inlet pressure. This is because the supercritical LNG has liquid-like properties, so the influence of inlet pressure on density and dynamic viscosity as well as velocity is small (as shown in Figure 2), which leads to pressure drop being only slightly effected by the inlet pressure. After $Np = 10$, the supercritical LNG has gas-like properties, so the influence of inlet pressure on density and dynamic viscosity is greater, leading to the large effect on velocity and causing the pressure drop to change more obviously with the inlet pressure. At the last portion of the channel, the pressure drop reduced, because the last pitch of the channel is straight, which reduces turbulence.

The effect of inlet pressure on Nu and Eu are shown in Figure 16, where both Nu and Eu decrease with as inlet pressure increased. Nu is inversely proportional to thermal conductivity, so as the inlet pressure increases, the thermal conductivity increased, which decreases Nu. Eu is increased as the inlet pressure is reduced, which the same behavior is observed in pressure drop, indicating that the pressure drop performance is larger when the inlet pressure is lower.

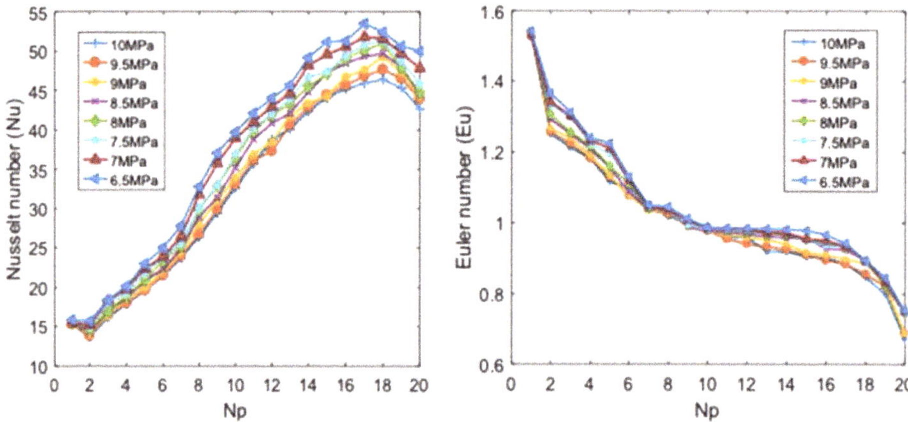

Figure 16. Effect of inlet pressure on Nu and Eu along the streamwise direction.

Although both Nu and Eu increase with a decrease in inlet pressure, Nu/Eu increases, as shown in Figure 17. This is because with the decrease of pressure, the increase rate of Nu is larger than that of Eu. Nu/Eu reaches its maximum at 6.5 MPa, indicating that supercritical LNG will have better heat transfer performance in PCHE at a lower inlet pressure.

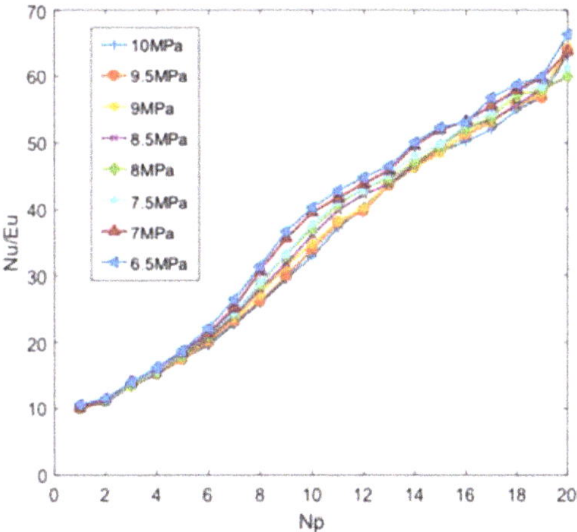

Figure 17. Effect of inlet pressure on Nu/Eu along the streamwise direction.

4. Conclusions

In this study, the flow and heat transfer characteristics of supercritical LNG in zigzag channel of PCHE are numerically investigated at different operating conditions. Some conclusions can be drawn as follows:

(1)   The local convection heat transfer coefficient rises and then falls along the streamwise direction, with the peak value appearing at the pseudo-critical temperature. The pressure drop also increases along the streamwise direction.

(2)   As the channel bend angle is increased, the local convection heat transfer coefficient and pressure drop rise, and so do the Nu and Euler numbers. The enhancement of heat transfer capability of supercritical LNG is mainly owed to increased turbulence. The increase of pressure drop is mainly due to the rising of velocity and the increase of flow resistance caused by the existence of vortices.

(3)   The local convective heat transfer coefficient and pressure drop increase significantly as the mass flux is increased due to the enhancement of turbulent flow. When the mass flux is increased by 2-fold, the local heat transfer coefficient rises by 1.4 times, and the pressure drop increases 3.3 times. The Nu increases as mass flux is increased. However, at the last third of the channel, Nu decreases as the mass flux is raised because of the decreased heat per unit volume absorbed by the LNG. This suggests that when the mass flux is raised, the heat transfer performance of the whole channel is better, but with the development of the fluid's flow, the local heat transfer performance is reduced at the last third of the channel owing to the reduction of heat-absorbed capacity by the unit volume fluid.

(4)   The improvement of heat transfer performance with bend angle depends on the mass flux. The supercritical LNG has better heat transfer performance when the bend angle is less than $15°$ when the mass flux ranges from 207.2 kg/(m^2·s) to 621.6 kg/(m^2·s), and improves at bend angles of $10°$ and lower compared to $15°$ at mass fluxes above 414.4 kg/(m^2·s).

(5)   Before the pseudo-critical temperature, the local convective heat transfer coefficient changes little with the inlet pressure, while it increases when the temperature surpasses pseudo-critical

point. The pressure drop is reduced as the inlet pressure increases. Nu and EU decrease with increasing inlet pressure, while Nu/Eu reaches a maximum at 6.5 MPa. The results show that supercritical LNG has a better heat transfer performance in zigzag channel of PCHE at lower operating pressures.

Author Contributions: Conceptualization and supervision, Z.Z.; formal analysis and data curation, Y.Z.; validation, methodology and software, X.M., X.C., S.L., S.Y. and Z.Z.; writing—original draft preparation, Y.Z. and Z.Z.

Funding: The authors gratefully acknowledge that this work was supported by Jiangsu marine and fishery science and technology innovation and extension project (HY2017-8), Zhenjiang funds for the key research and development project (GY2016002-1) and National Key R&D Program of China (2018YFC0310400).

Conflicts of Interest: The authors declare no conflict of interest.

Nomenclature

T	Temperature (K)
P	Pressure (Pa)
L	length of channel (mm)
f	Fanning factor
v	Velocity (m/s)
Re	Reynolds number
h	Convective heat transfer coefficient (W/(m^2·K))
Nu	Nusselt number
Eu	Euler number
C_p	Specific heat (KJ/(kg·K))
D_h	hydraulic diameter (m)
G	mass flux (kg/(m^2·s))
ΔP	pressure drop (Pa)
ΔP_f	pressure drop due to friction (Pa)
ΔP_a	pressure drop due to acceleration (Pa)
τ	shear stress at the wall (Pa)

Greek symbols

μ	viscosity (Pa·s)
ρ	density (kg/m^3)
λ	thermal conductivity (W/(m·K))

Subscript

w	Wall
b	Bulk mean
in	inlet
out	outlet

References

1. Pham, T.N.; Long, N.V.D.; Lee, S.; Lee, M. Enhancement of single mixed refrigerant natural gas liquefaction process through process knowledge inspired optimization and modification. *Appl. Therm. Eng.* **2017**, *110*, 1230–1239. [CrossRef]
2. Pu, L.; Qu, Z.G.; Bai, Y.H.; Qi, D.; Sun, K.; Yi, P. Thermal performance analysis of intermediate fluid vaporizer for liquefied natural gas. *Appl. Therm. Eng.* **2014**, *65*, 564–574. [CrossRef]
3. Xu, S.; Chen, X.; Fan, Z. Thermal design of intermediate fluid vaporizer for subcritical liquefied natural gas. *J. Nat. Gas Sci. Eng.* **2016**, *32*, 10–19. [CrossRef]
4. Pan, J.; Li, R.; Lv, T.; Wu, G.; Deng, Z. Thermal performance calculation and analysis of heat transfer tube in super open rack vaporizer. *Appl. Therm. Eng.* **2016**, *93*, 27–35. [CrossRef]

5. Han, C.-L.; Ren, J.-J.; Wang, Y.-Q.; Dong, W.-P.; Bi, M.-S. Experimental investigation on fluid flow and heat transfer characteristics of a submerged combustion vaporizer. *Appl. Therm. Eng.* **2017**, *113*, 529–536. [CrossRef]

6. Kim, J.H.; Baek, S.; Jeong, S.; Jung, J. Hydraulic performance of a microchannel PCHE. *Appl. Therm. Eng.* **2010**, *30*, 2157–2162. [CrossRef]

7. Chu, W.-X.; Li, X.-H.; Ma, T.; Chen, Y.-T.; Wang, Q.-W. Experimental investigation on SCO2-water heat transfer characteristics in a printed circuit heat exchanger with straight channels. *Int. J. Heat Mass Transf.* **2017**, *113*, 184–194. [CrossRef]

8. Figley, J.; Sun, X.; Mylavarapu, S.K.; Hajek, B. Numerical study on thermal hydraulic performance of a Printed Circuit Heat Exchanger. *Prog. Nucl. Energy* **2013**, *68*, 89–96. [CrossRef]

9. Kim, W.; Baik, Y.-J.; Jeon, S.; Jeon, D.; Byon, C. A mathematical correlation for predicting the thermal performance of cross, parallel, and counterflow PCHEs. *Int. J. Heat Mass Transf.* **2017**, *106*, 1294–13022. [CrossRef]

10. Yoon, S.-Y.; O'Brien, J.; Chen, M.; Sabharwall, P.; Sun, X. Development and validation of Nusselt number and friction factor correlations for laminar flow in semi-circular zigzag channel of printed circuit heat exchanger. *Appl. Therm. Eng.* **2017**, *123*, 1327–1344. [CrossRef]

11. Ishizuka, T.; Kato, Y.; Muto, Y.; Nikitin, K.; Tri Lam, N. Thermal-hydraulic characteristics of a Printed Circuit Heat Exchanger in a supercritical CO_2 loop. *Nucl. React. Therm. Hydraul.* **2006**, *30*, 109–116.

12. Tsuzuki, N.; Kato, Y.; Ishiduka, T. High performance printed circuit heat exchanger. *Appl. Therm. Eng.* **2007**, *27*, 1702–1707. [CrossRef]

13. Ngo, T.L.; Kato, Y.; Nikitin, K.; Tsuzuki, N. New printed circuit heat exchanger with S-shaped fins for hot water supplier. *Exp. Therm. Fluid Sci.* **2006**, *30*, 811–819. [CrossRef]

14. Ngo, T.L.; Kato, Y.; Nikitin, K.; Ishizuka, T. Heat transfer and pressure drop correlations of microchannel heat exchangers with S-shaped and zigzag fins for carbon dioxide cycles. *Exp. Therm. Fluid Sci.* **2007**, *32*, 560–570. [CrossRef]

15. Kim, D.E.; Kim, M.H.; Cha, J.E.; Kim, S.O. Numerical investigation on thermal hydraulic performance of new printed circuit heat exchanger model. *Nucl. Eng. Des.* **2008**, *238*, 3269–3276. [CrossRef]

16. Zhao, Z.; Zhao, K.; Jia, D.; Jiang, P.; Shen, R. Numerical Investigation on the Flow and Heat Transfer Characteristics of Supercritical Liquefied Natural Gas in an Airfoil Fin Printed Circuit Heat Exchanger. *Energies* **2017**, *10*, 1828. [CrossRef]

17. Lee, S.Y.; Park, B.G.; Chung, J.T. Numerical studies on thermal hydraulic performance of zigzag-type printed circuit heat exchanger with inserted straight channels. *Appl. Therm. Eng.* **2017**, *123*, 1434–1443. [CrossRef]

18. Yoon, S.Y.; No, H.C.; Kang, G.B. Assessment of straight, zigzag, S-shape, and airfoil PCHEs for intermediate heat exchangers of HTGRs and SFRs. *Nucl. Eng. Des.* **2014**, *270*, 334–343. [CrossRef]

19. Kim, I.H.; No, H.C. Physical model development and optimal design of PCHE for intermediate heat exchangers in HTGRs. *Nucl. Eng. Des.* **2012**, *243*, 243–250. [CrossRef]

20. Huang, D.; Wu, Z.; Sunden, B.; Li, W. A brief review on convection heat transfer of fluids at supercritical pressures in tubes and the recent progress. *Appl. Energy* **2016**, *162*, 494–505. [CrossRef]

21. Kim, I.H.; No, H.C. Thermal–hydraulic physical models for a Printed Circuit Heat Exchanger covering He, He-CO_2 mixture, and water fluids using experimental data and CFD. *Exp. Therm. Fluid Sci.* **2013**, *48*, 213–221. [CrossRef]

22. Yu, X.; Yang, X.; Wang, J. Heat Transfer and Pressure Loss of Immediate Heat Exchanger. *Inst. Nucl. New Energy Technol.* **2009**, *43*, 256–259.

23. Sung, J.; Lee, J.Y. Effect of tangled channels on the heat transfer in a printed circuit heat exchanger. *Int. J. Heat Mass Transf.* **2017**, *115*, 647–656. [CrossRef]

24. Nikitin, K.; Kato, Y.; Ngo, L. Printed circuit heat exchanger thermal-hydraulic performance in supercritical CO_2 experimental loop. *Int. J. Refrig.* **2006**, *29*, 807–814. [CrossRef]

25. Jeon, S.; Baik, Y.-J.; Byon, C.; Kim, W. Thermal performance of heterogeneous PCHE for supercritical CO_2 energy cycle. *Int. J. Heat Mass Transf.* **2016**, *102*, 867–876. [CrossRef]

26. Higashi, Y. NIST Thermodynamic and Transport Properties of Refrigerants and Refrigerant Matures (REFPROP). *Netsu Bussei* **2000**, *14*, 1575–1577.

27. Kwon, J.G.; Kim, T.H.; Park, H.J.; Cha, J.E.; Kim, M.H. Optimization of airfoil-type PCHE for the recuperate of small scale Brayton cycle by cost-based objective function. *Nucl. Eng. Des.* **2016**, *298*, 192–200. [CrossRef]

28. Han, C.L.; Ren, J.J.; Dong, W.-P.; Bi, M.-S. Numerical investigation of supercritical LNG convective heat transfer in a horizontal serpentine tube. *Cryogenics* **2016**, *78*, 1–13. [CrossRef]

29. Xu, X.; Ma, T.; Li, L.; Zeng, M.; Chen, Y.; Huang, Y.; Wang, Q. Optimization of fin arrangement and channel configuration in an airfoil fin PCHE for supercritical CO_2 cycle. *Appl. Therm. Eng.* **2014**, *70*, 867–875. [CrossRef]

30. Zhao, Z.; Zhang, X.; Zhao, K.; Jiang, P.; Chen, Y. Numerical investigation on heat transfer and flow characteristics of supercritical nitrogen in a straight channel of printed circuit heat exchanger. *Appl. Therm. Eng.* **2017**, *126*, 717–729. [CrossRef]

31. Yang, J.G.; Wu, H. Explicit Coupled Solution of Two-equation k-ω SST Turbulence Model and Its Application in Turbomachinery Flow Simulation. *Acta Aeronaut. ET Astronaut. Sinaica* **2014**, *35*, 116–124.

32. Ren, Y.; Liu, H.L.; Shu, M.H. Improvement of SST k-ω SST Turbulence Model and Numerical Simulation in Centrifugal Pump. *Trans. Chin. Soc. Agric. Mach.* **2014**, *43*, 123–128.

 *energies*

Article

A Machine Learning Approach to Correlation Development Applied to Fin-Tube Bundle Heat Exchangers

Karl Lindqvist [1,*], **Zachary T. Wilson** [2], **Erling Næss** [1] and **Nikolaos V. Sahinidis** [2]

[1] Department of Energy and Process Engineering, Norwegian University of Science and Technology, NO-7491 Trondheim, Norway; erling.nass@ntnu.no
[2] Department of Chemical Engineering, Carnegie Mellon University, Pittsburg, PA 15213, USA; ztw@andrew.cmu.edu (Z.T.W.); niksah@gmail.com (N.V.S.)
* Correspondence: karl.erik.lindqvist@ntnu.no; Tel.: +47-9682-9629

Received: 18 October 2018; Accepted: 6 December 2018; Published: 10 December 2018

Abstract: Heat exchanger designers need reliable thermal-hydraulic correlations to optimize heat exchanger designs. This work combines an adaptive sampling method with a Computational Fluid Dynamics (CFD) simulator to obtain increased accuracy and validity range of heat transfer and pressure drop predictions using a limited number of data points. Correlation efficacy was evaluated based on a steam generator case study. The sensitivity to the design parameters was analyzed in detail. The RMSE (root mean square error) of the developed correlations were reduced, through CFD sampling, from 28% to 15% for pressure drop, and from 33% to 25% heat transfer, compared to regression on experimental data only. The best reference correlations have RMSE values of 43% and 33% on pressure drop and heat transfer, respectively, on an independent validation set. Indeed, a radically different fin-tube geometry was suggested for the case study, compared to results using the Escoa correlations. The developed correlations show good to excellent agreement with trends in the CFD model. The quantitative error of predicted heat transfer and pressure drop coefficients at the case study optimum, however, was as large as those of the Escoa correlations. More data are likely needed to improve accuracy for compact heat exchanger designs further.

Keywords: numerical modeling; surrogate model; correlation; fin-tube; spiral fin-tube; CFD

---

1. Introduction

Increased energy efficiency is a key strategy to reduce anthropogenic CO_2 emissions, and often the most economical one in the industrial sector. Many energy intensive industries have already implemented measures such as heat integration and bottoming cycles up to its economic potential. An exception to this rule is the offshore oil and gas sector, where space and weight restrictions put severe limits to the amount of equipment that can be placed on each installation.

Volume and weight optimization of an offshore bottoming cycle is contingent on accurate thermal-hydraulic correlations. This is particularly true for the large and heavy waste heat recovery unit (WHRU), which typically consists of a circular fin-tube bundle. Numerical methods such as Computational Fluid Dynamics (CFD) can be used to predict the performance of such heat exchangers, but direct optimization is usually not feasible due to the large number of design variables and constraints and the computational cost (lead time) of each function evaluation.

Many thermal-hydraulic correlations for fin-tube bundles have been presented in the literature over the last half-century. Typically, correlations are algebraic expressions of non-dimensional groups with model constants fitted to experimental data by regression. The underlying data are derived from the authors' own published experimental work (e.g., [1]), from proprietary databases (e.g., [2]) or from a collection of several literature sources (e.g., [3]). In the two former cases, correlations tend to have a rather limited (or unknown) range of applicability. In the latter case, the underlying data are inherently scattered due to differences in experimental setups, data reduction methods and tube geometry details. This is particularly true for pressure drop measurements, where uncertainties are larger. Earlier work has shown that the WHRU skid weight can be reduced by scaling down the tube diameter to about 10 mm [4]. This requires new correlations with an extended validity range, to avoid extrapolation. There may also be a need to verify the accuracy of previously published work in a consistent manner.

Machine learning methods represent a contrasting approach to model building, where the model structure is less restricted. Artificial Neural Networks (ANN), Radial Basis function Neural Networks (RBNN) and Support Vector Regression (SVR) models have been used successfully to predict the thermal performance of a number of heat exchanger types. As shown in Table 1, most published studies utilize fully connected ANN models trained on experimental data. More recent publications have also considered other model setups, as well as sampling from a CFD model.

Table 1. Published work on thermal-hydraulic heat exchanger modeling using machine learning methods.

Data Source	Experimental	Correlation or CFD
Fully connected ANN	[5–10]	[11]
SVR, RBNN, Kriging	[9]	[12,13]

CFD models are increasingly being used for predictive design, even in critical applications such as nuclear reactor thermal-hydraulics [14], provided that rigorous verification and validation practices are adhered to. In the context of fin-tube bundles, CFD models can provide heat transfer and pressure drop coefficients in a consistent and time efficient manner. Comparisons with experimental data has shown good agreement, close to or within the experimental uncertainty, for a wide range of geometries [15]. CFD models are also able to provide data for extreme geometries that may not be possible to manufacture and test experimentally, but still add valuable data in the correlation development process. This includes "adversarial examples", i.e., geometries where small changes in parameters cause large changes in model output.

Given these developments, we propose to combine predictive CFD simulations with adaptive sampling and automated correlation building methodologies based on machine learning theory. We hypothesize that correlation accuracy and validity range can be increased simultaneously, with reasonable computational effort, by leveraging publicly available experimental work in conjunction with new, adaptively sampled simulation points. This paper provides the underlying simulated data points, in addition to the improved thermal-hydraulic correlations, to foster and accelerate further developments in the field. The presented methodology is applicable to a wide range of multivariate design problems where direct optimization with CFD is infeasible.

Specifically, the novelty of the investigation, as applied to fin-tube thermal-hydraulic correlation development, is the following:

- application of error estimation and adaptive sampling
- direct inclusion of predictive CFD model data in model regression
- extended validity range of geometric parameters towards the weight optimum indicated by earlier work

2. Method

The overall correlation building and verification process used in this article is shown in Figure 1. A design space relevant for waste heat recovery units was firstly defined (see Table 2). An initial database of experimental work from the open literature was fed into the model building software

ALAMO version 2018.4.3 [16]. A correlation was generated and improved through adaptive sampling of data points from the CFD simulator. The correlation was then evaluated using a separate validation set, as well as a case study where the optimal point was compared to an independent CFD simulation. The validation set consists of 30 CFD simulations selected through Latin hypercube sampling of the design space, as indicated in Figure 1. Finally, the trends in the different input variables were evaluated using CFD simulations and compared to correlations from the present and earlier work. The remainder of this section describes the components of the methodology in further detail.

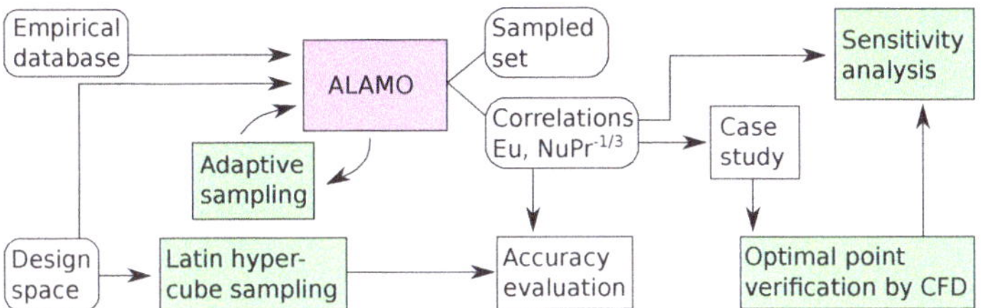

Figure 1. Process for correlation development, testing and benchmarking. Green boxes indicate where CFD simulations are employed.

Table 2. Considered design space for compact fin-tube bundles. Ancillary variables are adjusted to achieve a reasonable number of segments per fin revolution and a representative fin efficiency. Geometric parameters are shown in Figure 2.

Design Variables	Min	Max
$u_{F_{min}}$ [m/s@500 °C]	2.53	30.4
d_o [mm]	9.65	50.8
h_f [mm]	1.4	25.4
h_s/h_f [-]	0.0	1.0
$\hat{s}_f$ [mm]	0.49	4.9
c_f [mm]	0.39	8.0
Ancillary and Derived Variables		
Re [-]	310	19,000
t_f [mm]	0.3	0.75
w_s [mm]	2.0	4.0
β [1] [deg]	30.0	30.0

[1] The tube bundle layout angle is defined as $\beta = \tan^{-1}\left(\frac{P_t}{2P_l}\right)$.

2.1. Initial Database

A database of published experimental work has previously been established at the Norwegian University of Science and Technology [17]. The database contains data for 248 different fin-tube bundles from 21 publications, including both plain and serrated fin geometries. Several criteria were used to single out and prepare relevant data points for this study:

- Data points outside the ranges defined in Table 2 were omitted. The upper limit on $U_{F_{min}}$ was relatively restrictive since kinematic viscosity for air is about three times higher at 500 °C compared to usual test conditions ($\sim$100 °C). Hence, many experimental data points were excluded, but the resulting Reynolds number range (cf. Table 2). was considered representative of the possible operating conditions of a WHRU
- Geometries with only heat transfer or only pressure drop data were removed. A power law function was fitted to the heat transfer data of each remaining geometry and interpolated to the Reynolds numbers at which the (adiabatic) pressure drop was measured. This is necessary because the chosen model building method requires both outputs to be defined at each data point.

- A tube bank array of 30° was considered in this work, as it is the most compact arrangement. Tube banks with array angles in the range 25°–35° were corrected using Equation (1), derived from the Escoa correlation [18], to obtain data corresponding to $\beta = 30°$. The maximum applied corrections were 5% for the heat transfer data and 9% for the pressure drop data, respectively. Other tube bank data were discarded.
- The number of streamwise tube rows was not considered as a parameter in this work. Data point duplicates were removed such that only data for the largest number of tube rows were retained.

$$\text{NuPr}_{30°} = \text{NuPr} \cdot \frac{1 + e^{-1/(2\tan\beta)}}{1 + e^{-1/(2\tan 30°)}},$$

$$\text{Eu}_{30°} = \text{Eu} \cdot \frac{1.1 + 1.8e^{-1/\tan\beta}}{1.1 + 1.8e^{-1/\tan 30°}}$$

(1)

After this procedure, the remaining database contained 108 experimental data points.

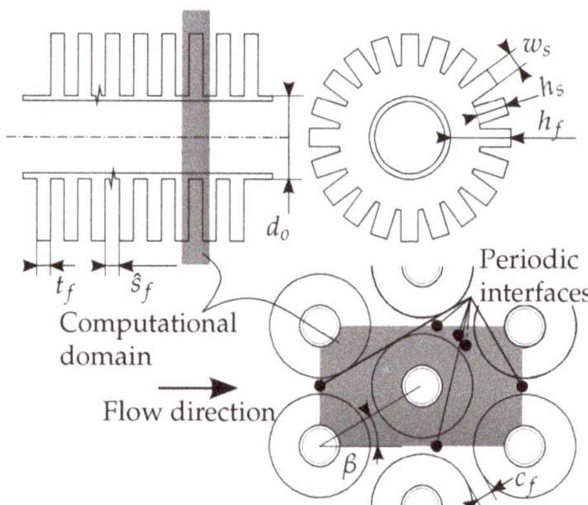

Figure 2. Geometric parameters of the fin-tube bundle and CFD computational domain.

2.2. Correlation Development (ALAMO)

ALAMO is a learning software that identifies simple, accurate surrogate models (correlations) using a minimal set of sample points from black box emulators such as experiments, simulations, and legacy code. ALAMO initially builds a low-complexity surrogate model using a best subset technique that leverages a mixed-integer programming formulation to consider a large number of potential functional forms. The model is subsequently tested, exploited, and improved through the use of derivative-free optimization solvers that adaptively sample new simulation or experimental points. For more information about ALAMO, see Cozad et al. [16,19] and Wilson and Sahinidis [20].

The functional form of a regression model was assumed to be unknown to ALAMO. Instead, several simple basis functions were proposed, e.g., x, x^2, $1/x$, $\log(x)$, and a constant. Once a set of potential basis functions was collected, ALAMO attempted to construct the lowest complexity function that accurately models the initial training data. This model was obtained by minimizing the Bayesian information criterion, $BIC = SSR_p/\hat{\sigma} + p\log(n)$, where p is the number of included basis functions, SSR_p is the sum of squared residuals for the selected model, and $\hat{\sigma}$ is an estimation for the variance of the residuals, which is obtained using the least squares solution of the full basis set or can be specified a priori. The model fitness metric (BIC) was rigorously minimized using a combination of enumeration, heuristics, and eventual global optimization using the BARON solver [21].

Combinations of linear terms and fractions of the input variables were considered for this particular application, as well as selected powers of these functions with an exponent smaller than unity. Binomial terms, and powers of these, were also considered in modeling the Nusselt number.

Once a model was identified, it was improved systematically in ALAMO using an adaptive sampling technique that added new simulation or experimental points to the training set. New sample points were selected to maximize model inconsistency in the original design space, as defined by box constraints on x, using derivative-free optimization methods [22]. It was observed that a higher correlation accuracy was obtained if the outputs and the velocity-related input were log-transformed. It is well known that the velocity dependence is essentially logarithmic and therefore easier to model in log space. As an additional benefit, correlation terms become multiplicative rather than additive. This facilitates interpretation of the correlation as consisting of one velocity-dependent term multiplied with a number of "geometry correction" terms.

2.3. Numerical Model

Numerical simulations in this work followed the methodology described in a previous article [15], where thorough validation with experimental data are given. The main characteristics of the numerical model are as follows:

- Fully periodic computational domain (Figure 2) was discretized primarily with hexahedral cells. A graded boundary layer grid was used in the wall normal direction in the space between the fins ($y^+ < 1$).
- Density and thermophysical properties were considered constant, properties for air were used for the external fluid and the fin thermal conductivity was set corresponding to carbon steel ($48.5\,\mathrm{W\,m^{-1}\,K^{-1}}$).
- The steady-state Reynolds Averaged Navier–Stokes (RANS) equations were solved together with the energy equation and the Spalart–Allmaras turbulence model equation [23] using the open source CFD toolbox OpenFOAM v4.1.
- The Spalart–Allmaras turbulence model was selected due to its simplicity, robustness and suitability for simulating boundary layers under adverse pressure gradient conditions. It also yields similar results as other eddy viscosity turbulence models when applied to finned tube bundles [24]. Model constants were kept at their default value, including the turbulent Prandtl number.
- Second order upwind discretization was used for all convective terms.
- The conjugate heat transfer between the fin and the external fluid was modeled explicitly, resolving the temperature field in the fin. The tube wall thermal resistance was neglected—a uniform temperature was applied at the fin root and on the tube surface. The fluid bulk temperature was specified at the leftmost periodic boundary, avoiding source terms in the energy equation.
- Fin efficiency was evaluated by solving the energy equation a second time, subsequent to RANS model convergence, assuming a frozen flow field and a uniform temperature boundary condition on one fin-tube row. The resulting heat flux was used to compute the fin efficiency in the first simulation having finite thermal conductivity in the fin.
- The computed heat flux, bulk temperature and total pressure drop were normalized into Nusselt and Euler numbers according to standard practice (see, e.g., [17]).

2.4. Accuracy Evaluation

The accuracies of the correlations were evaluated based on the coefficient of determination (R^2) and root mean square error (RMSE) values on an independent dataset sampled using CFD ($N = 30$). Some samples struggled to reach iterative convergence during CFD simulation, in which case the flow velocity was reduced. The RMSE is expressed in terms of the deviation from the observed values, viz.

$$\mathrm{RMSE} = \sqrt{\frac{1}{N}\sum_N \left(\frac{y_i - f_i}{y_i}\right)^2} \tag{2}$$

for the predicted values f_i (from correlations) and the observed values y_i (from CFD simulations). Note that the RMSE is equal to the standard deviation for an unbiased estimator.

2.5. Case Study and Verification of Optimal Point

A case study was defined to test the developed correlation on a realistic optimization task and compare the computed optimal point to that obtained using a reference correlation. The predicted heat transfer and pressure drop coefficients were also compared to those predicted by CFD simulations. A boiler section of a generic offshore heat recovery steam generator was considered, with the steam and exhaust gas specifications and constraints given in Table 3. The tube wall temperature was considered constant, and heat conduction through the tube wall, as well as steam/water pressure drop, was neglected. This leads to a computationally cheap heat balance, which can be evaluated using the ϵ-NTU method as a function of the exhaust mass flow, cold end temperature difference, and gas side heat transfer coefficient. The objective function is the fin-tube bundle weight, where the tube wall thickness is calculated as

$$t_w = 1.1 \frac{p_s(d_o/2)}{0.85\sigma_y + 0.4p_s} \tag{3}$$

where p_s is the steam pressure and σ_y is the yield stress of carbon steel at $500\,^\circ$C. The fins were assumed to be made from carbon steel ($\lambda = 48.5\,\mathrm{W\,m^{-1}\,K^{-1}}$, $\rho = 7850\,\mathrm{kg\,m^{-3}}$). A theoretical fin efficiency according to Hashizume et al. [25] was used. The ideal gas law was used to evaluate the exhaust gas density at the average bulk temperature, and all other physical properties were considered constant.

Table 3. Boiler section case study: Definitions and constraints.

Exhaust mass flow [kg/s]	75.0
Exhaust pressure drop [Pa]	1500
Steam pressure [Pa]	25×10^5
Steam/water temperature [$^\circ$C]	224
Cold end temperature difference [$^\circ$C]	20
Transferred heat [W]	$\geq 15 \times 10^6$
Narrow gap flow velocity [m/s]	≤ 30
Frontal cross-section	square

The number of tube rows in the transverse and longitudinal directions were modeled as real numbers to obtain a smooth objective function. N_r was calculated directly from the pressure drop constraint, whereas N_t was a free variable. Remaining free variables and bounds were equal to the design variables and bounds in Table 2, with the fin thickness lower bound set to 0.5 mm and w_s adjusted to give 20 segments per revolution for serrated fins. A tube bundle layout angle of 30° was used throughout.

The optimization problem was solved in MATLAB R2017a using the built-in function *fmincon* that implements the SQP algorithm. The optimization was repeated 100 times for a given case, using random starting points within the design space, to ensure that a global, feasible optimum point was reached.

3. Results and Discussion

3.1. Correlation Development

The accuracy evaluation (Table 4) confirmed a relatively good fit between the developed correlations and the independently sampled validation set. The coefficient of determination is high for the Euler number, but less impressive for the Nusselt number. The RMSE is acceptable for the Euler number, but quite large for the Nusselt number, particularly when considering that the 95% confidence interval is much wider than the RMSE.

Table 4. Accuracy on validation dataset, computed from Latin hypercube sample of design space (30 CFD simulations).

Model	Eu		NuPr$^{-1/3}$	
	R^2 [-]	RMSE [%]	R^2 [-]	RMSE [%]
This work				
database only	0.75	28	0.70	33
after sampling	0.94	15	0.76	25
Holfeld [17]	−0.08	58	0.75	35
Escoa [18]	−0.05	43	0.42	34
PFR [26]	0.79	72	0.59	33

The difficulty in modeling the Nusselt number may be due to complex changes in the flow field, whereby the flow bypasses the aperture between the fins and flow outside the fin diameter for certain geometric configurations (typically low fin apertures and large fin tip clearances). This phenomenon has been more thoroughly discussed in [27], but ultimately necessitates a different modeling approach than the current one due to the large nonlinearity involved.

The predictive accuracies for the reference correlations (Holfeld [17], Escoa [18] and PFR [26]) are, in general, poor due to the severe extrapolation induced by the design space definition. As shown in Figure 3, the general scatter for the reference correlations on all data in this work (empirical database, adaptively sampled simulations and simulations used for validation) is large, particularly for low Reynolds numbers. The PFR correlation for Eu has a relatively high coefficient of determination compared to the Escoa and Holfeld correlations, indicating that the hydraulic diameter (which is the unique feature of the PFR Eu correlation) may be a more robust length scale than the tube diameter. The high RMSE associated with the PFR Eu correlation is due to a positive bias (overprediction) that would be simple to rectify with a constant correction factor.

The detailed functional form of the developed correlations is further discussed in Section 3.3. Algebraic expressions of the correlations are provided in the Appendix A.

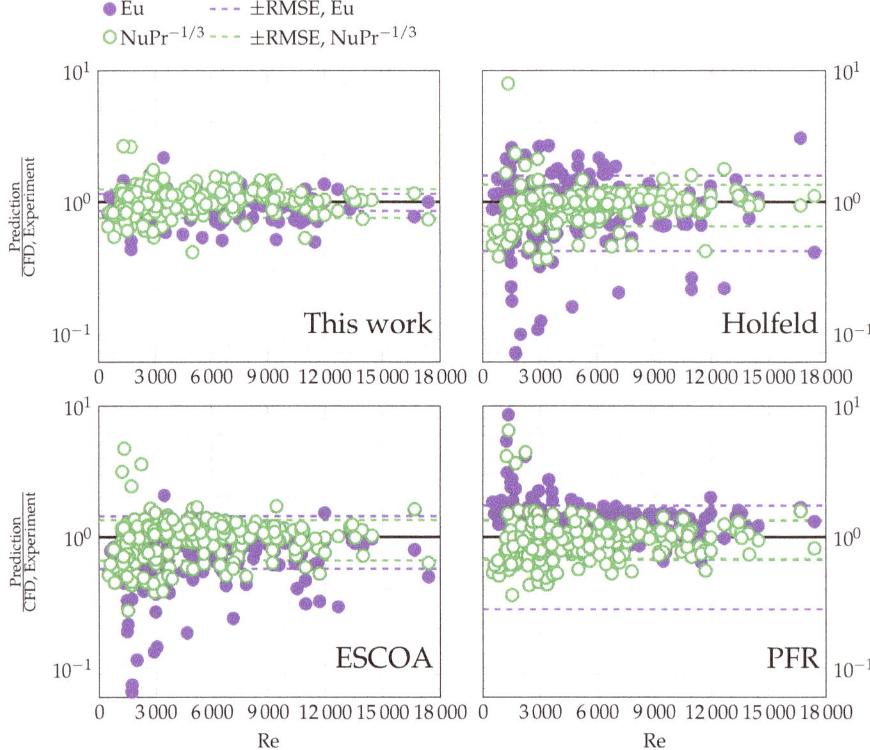

Figure 3. Predictive accuracy on regression and validation data (208 data points): correlations developed in this work (Supplementary Materials) and correlations from Holfeld [17], Escoa [18] and PFR [26].

3.2. Case Study

Results from the case study consist of four optimized geometries, two for each correlation with and without an explicit tube diameter constraint, respectively. Geometries and thermal-hydraulic results are given in Table 5 and 3D representations of the two geometries without diameter constraints are shown in Figure 4.

Clearly, two different strategies towards fulfilling the heat duty under a pressure drop constraint emerge for the two different correlations. The optimal geometry for the correlations developed in this work is extremely compact (dense) with a few number of tube rows and a high surface area per tube, but a large frontal area. The optimal geometry using the Escoa correlations, on the other hand, aim to reduce the Euler number such that a larger number of tube rows can be afforded under the pressure drop constraint. This results in a smaller frontal area and a more "open" geometry with less heat transfer area per tube.

Note that the tube diameter reaches its lower bound, and the flow velocity its upper bound, for both correlations. This result is as expected, since the Nusselt number ($=\alpha_o d_o/\lambda_{air}$) scales with approximately $Re^{0.7}$ and therefore $\alpha_o \propto Re^{0.7}/d_o \propto d_o^{-0.3}$ for a constant velocity. The heat transfer area scales with the perimeter of the tube (disregarding the fins for a moment), leading to $\alpha_o A_t \propto d_o^{0.7}$. The Euler number (i.e., the pressure drop), on the other hand, scales with $d_o^{-0.3}$ at best. This mean that the heat transfer coefficient decreases at the same rate as the pressure drop. A hypothetical doubling of the tube diameter changes the transferred heat by a factor of $2^{0.7} = 1.62$ and the pressure drop by a factor of $2^{-0.3} = 0.812$. If the number of tube rows are increased accordingly (to utilize the available pressure drop), the transferred heat can be increased to $1.62/0.812 = 2.0$ and the transferred heat per unit surface area is the same as for the smaller tubes. The *weight* of the larger tubes, however, is higher than the smaller tubes due to a larger wall thickness and larger internal fluid volume. The only caveat to the argument for a small tube diameter is that restrictions on the steam side pressure drop, steam side heat transfer coefficient, boiling stability and heat exchanger frontal area are not considered here. Moreover, large diameter tubes are less susceptible to flow induced vibrations due to a higher bending stiffness.

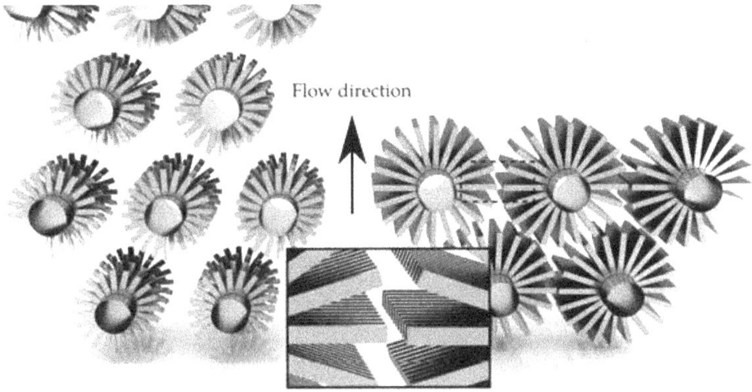

Figure 4. Optimized boiler geometry using the Escoa correlations (**left**) and correlations developed in this work (**right**), without explicit diameter constraints

Table 5. Case study: Geometry optima for correlations from this work and Escoa, without (left column) and with explicit diameter constraint of 25.4 mm (right column). Eu and $\mathrm{NuPr}^{-1/3}$ data are computed at the closest geometry possible to simulate with CFD

Correlations: Geometry	This Work		Escoa	
Re [-]	3600	9400	3600	9400
d_o [mm]	9.65	25.4	9.65	25.4
h_f [mm]	10.9	12.7	6.28	15.8
h_s/h_f [mm]	1.0	1.0	0.96	1.0
t_f [mm]	0.50	0.50	0.50	0.50
$\hat{s}_f$ [mm]	0.49	0.49	4.90	4.90
w_s [mm]				
c_f [mm]	0.39	0.39	8.00	8.00
N_r [-]	2.2	1.2	21.8	17.1
N_t [-]	140	102	105	52
Normalized objective function	1.00	1.49	0.87	2.33
Eu [-]				
This work	5.64	10.8		
Escoa			0.594	0.76
CFD	3.81	6.28	0.52	0.75
Deviation [%]	+48	+72	+13	−2.2
$\mathrm{NuPr}^{-1/3}$ [-]				
This work	66.6	125.0		
Escoa			56.8	96.5
CFD	67.7	136	38.9	74.4
Deviation [%]	−1.7	−8.0	−46	+30

Optimization with a fix tube diameter show that similar geometries are obtained as when the tube diameter is free, only with larger frontal area, fewer tube rows and a higher total weight.

The correlation accuracy at the optima, quantified by the deviations from the CFD simulation results, showed mixed performance. The pressure drop was grossly overpredicted by the correlation developed in this work, whereas the heat transfer coefficient was grossly overpredicted by the Escoa correlation. The normalized objective function should therefore be interpreted with care. The optimized geometry using correlations from this work turn out to be conservative (would satisfy heat duty and pressure drop constraints), whereas the optimized Escoa geometries would transfer too little heat.

However, the *trends* in the prediction of the design variables may be just as important as the accuracy at the optimum, since these trends determine the *location* of the optimum to begin with. This is the topic of the sensitivity analysis.

3.3. Sensitivity Analysis

The trends in Eu and Nu with respect to the flow velocity and the five geometric variables are shown, for perturbations around the midpoint of the design space, in Figures 5 and 6. These figures also include independent CFD simulations (not included in the training data) at the optimum and at perturbations around the optimum (with remaining variables held constant). Corresponding figures for perturbations around the optimal point in the case study are Figures 7 and 8.

The Euler number exhibits a high sensitivity to the flow velocity and the fin aperture compared to the other four variables (Figures 5 and 7). The velocity dependence can be explained as a transition from friction dominated drag to a mix of friction and form drag as velocity increases. At the design space midpoint, both of these pressure losses seem to be of equal importance, since the sensitivity to increases in wake size (d_o) and increases in friction area (h_f) is about the same. At the optimum point, a higher sensitivity to the wake size relative to friction area can be noticed (Figure 7, top middle and top right panels) due to the higher flow velocity.

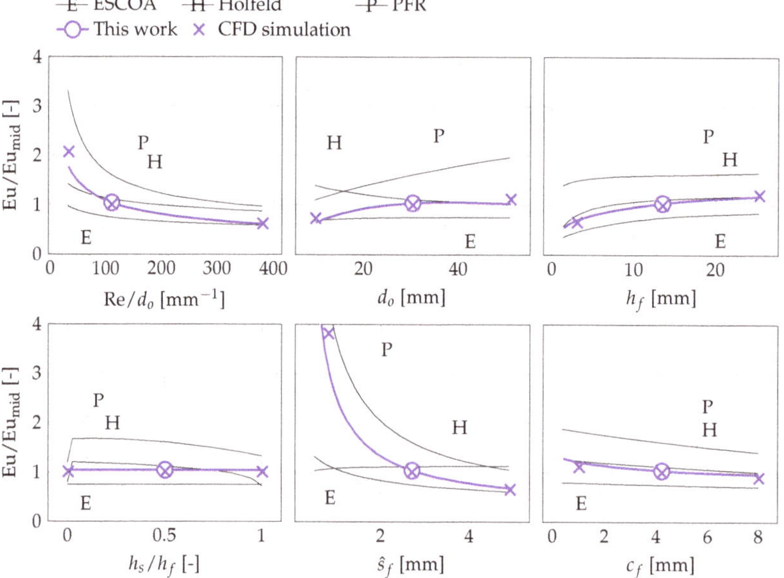

Figure 5. Trends in correlations and CFD simulator around the midpoint in the design space (marked by a circle). CFD simulations were independently sampled (not part of the dataset used for model development). Each parameter was varied independently, with remaining parameters held constant.

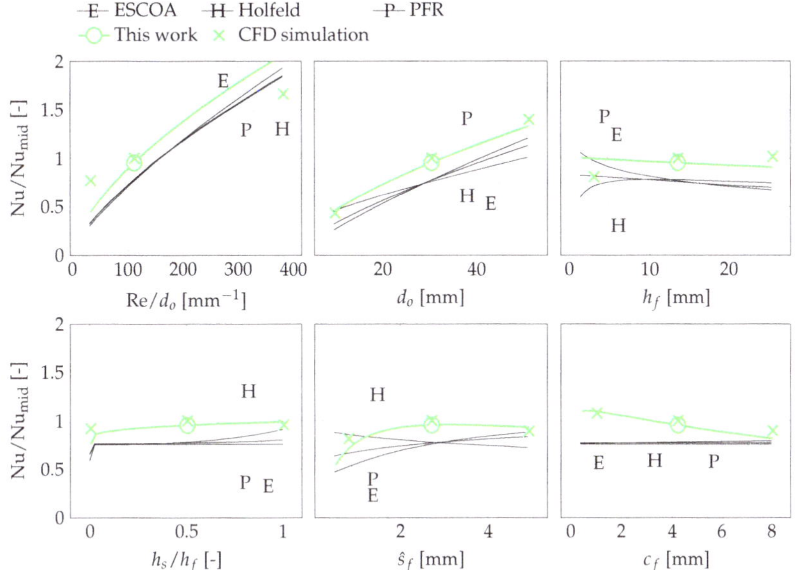

Figure 6. Trends in correlations and CFD simulator around the midpoint in the design space (marked by a circle). CFD simulations were independently sampled (not part of the dataset used for model development). Each parameter was varied independently, with remaining parameters held constant.

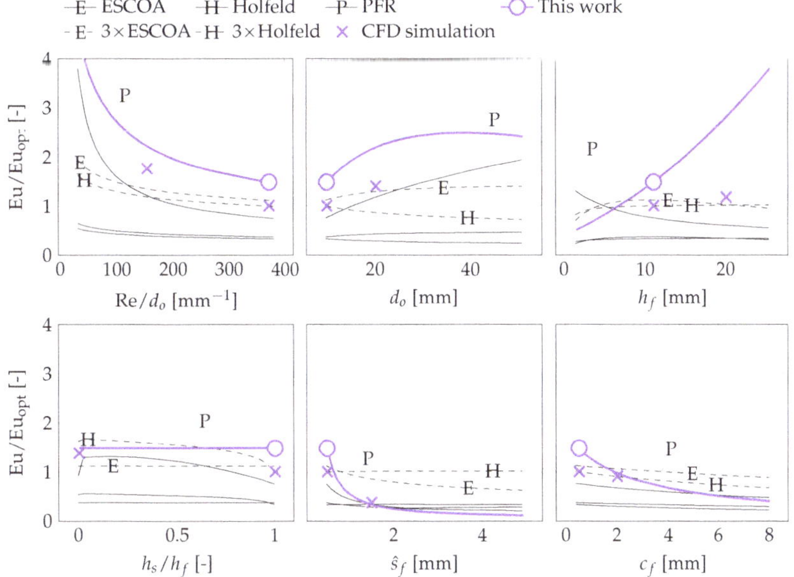

Figure 7. Trends in correlations around the case study optimum found using correlations developed in this work (marked by a circle). CFD simulations were independently sampled (not part of the dataset used for model development). Each parameter was varied independently, with remaining parameters held constant at the optimum.

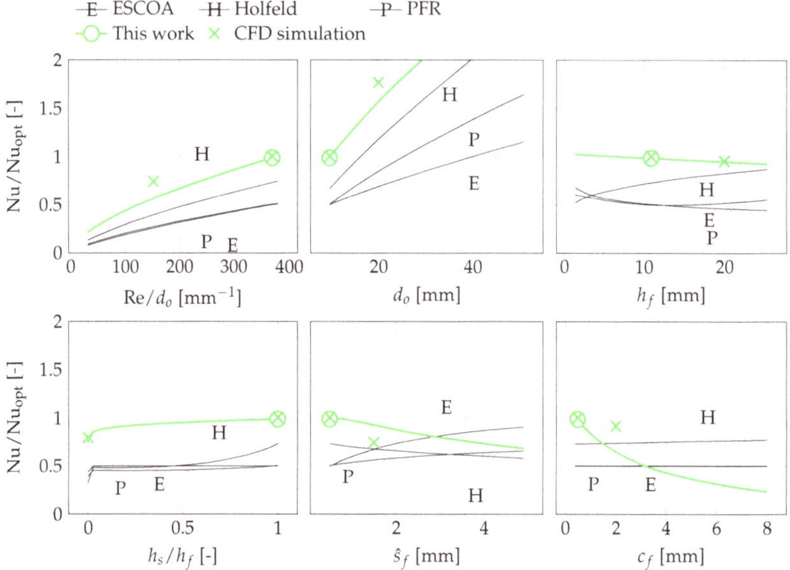

Figure 8. Trends in correlations around the case study optimum found using correlations developed in this work (marked by a circle). CFD simulations were independently sampled (not part of the dataset used for model development). Each parameter was varied independently, with remaining parameters held constant at the optimum.

The sensitivity to the fin aperture at constant mean flow velocity should be interpreted in light of the small fin pitches used in this work. Not only does a smaller fin aperture mean more friction area per unit tube length, but also corresponds to a higher maximum flow velocity between the fins required to maintain the same mean velocity. In other words, the blockage caused by the boundary layers on the fins is significant compared to the available cross-sectional (flow) area for the considered range of fin apertures. The lack of a positive correlation between the Euler number and the degree of serration (h_s/h_f) is unexpected, given that serrations break up the fin boundary layer and hence decrease the average boundary layer thickness. On the other hand, the correlation between h_s/h_f and the Nusselt number is also small and, therefore, consistent with the Euler number results. These observations point towards a conclusion that the thermal-hydraulic benefit of serrated fins lie primarily in the increased fin efficiency.

The Nusselt number is primarily a function of the flow velocity and the tube diameter (Figures 6 and 8). Considering that Nu $= \alpha_o d_o / \lambda_{\text{air}}$, i.e., linear in d_o, the heat transfer coefficient is approximately constant with the tube diameter as well as with the other geometric parameters. This can be expected in this case, since the heat transfer resistance consists of external boundary layers.

A trend change in $\hat{s}_f$ can be noted when comparing sensitivities at the midpoint and the optimal point (lower middle panels, Figures 6 and 8): The Nusselt number is relatively constant at the midpoint, but shows a clear negative trend at the optimal point. A possible explanation involves the already mentioned bypass effect. At the midpoint, a decrease in $\hat{s}_f$ increases the pressure drop (Figure 5), but also redistributes the flow to the passage outside the fin perimeter. The increased pressure drop does not translate to an increased heat transfer coefficient (Figure 6). A small fin tip gap on the other hand, such as at the optimal point, suppresses the bypass effect and forces the flow to pass through the fin aperture. Hence, a decrease in $\hat{s}_f$ does translate into both increased pressure drop and heat transfer coefficient. The effect is further discussed in [27].

The reference correlations exhibit the correct variable *trend* in most cases. The PFR correlations captures the trend around the optimal point to a surprising degree, given its simplicity. The quantitative accuracy of the reference correlations, however, is not satisfactory at the optimal point. The Holfeld correlations, which are the most recently developed, grossly underpredict the pressure drop and show incorrect trends for Eu as a function of d_o and Nu as a function of $\hat{s}_f$.

The correlations developed in this work generally agree well with the CFD simulations. Trends in h_f and c_f are exaggerated at the case study optimum, most likely due to limited amount of data in the design space, but generally provides improved prediction accuracy.

4. Conclusions

Machine learning methods, including adaptive sampling using a CFD simulator, has been used to improve the accuracy and validity range of thermal-hydraulic correlations for fin-tube bundles, in terms of both the coefficient of determination (R^2) and the root mean square error. The applicability to geometry optimization was verified through a case study and the accuracy of the modeling of trends in the design space was confirmed by comparison with CFD simulations. The following specific conclusions can be drawn:

- The choice of correlation is decisive for the outcome of tube bundle weight optimization, at least for the boiler section considered in the case study. The developed correlations suggest a radically different design compared to the Escoa correlations.
- The trends of the developed correlations generally match well with data from the CFD model. The sensitivity to the design variables close to the optimal point for the case study is, however, exaggerated for some variables.
- The PFR correlation for the Euler number is the most robust reference correlation with regards to the trends in the design variables, indicating that the hydraulic diameter can be an appropriate length scale for pressure drop modeling.
- The Nusselt number is relatively insensitive to all design parameters other than the flow velocity and the tube diameter (i.e., the Reynolds number) around the studied design points.
- In general, the Nusselt number appears more difficult to model accurately, compared to the Euler number. A possible explanation, given the preceding bullet point, is that particular geometries cause complex flow redistribution that only a highly nonlinear model can represent.
- Quantitative accuracy on the case study is good for the developed heat transfer correlation, but disappointing for the pressure drop correlation. The accuracy of the Escoa correlations is also poor at the case study optimum. More data are most likely needed in the range of compact designs with low tube diameter, if further accuracy improvements are to be achieved.

The machine learning approach appears to be a viable method to extend the validity range of thermal-hydraulic correlations, with relatively moderate resource usage. As ever, the dataset size limits the model nonlinearity that can be used without overfitting to the training data. Further increase in accuracy will, most likely, require significantly larger datasets created by a combination of structured sampling methods (e.g., Latin hypercube) and adaptive sampling methods.

A limitation of the current study is that the correlation accuracy is restricted to the accuracy of the numerical model. The numerical model is successfully validated over a large range of geometric parameters for which experimental data exist [15]; Future directions of this work should therefore include experimental investigations of previously untested geometries indicated by the correlations, such as at the case study optimum indicated in this work.

Supplementary Materials: The following are available online at http://www.mdpi.com/1996-1073/11/12/3450/s1, Table S1: Underlying CFD simulation data: regression data points, Table S2: Underlying CFD simulation data: validation data points.

Author Contributions: Conceptualization, K.L., E.N. and N.V.S.; Methodology, Z.W., N.V.S. and K.L.; Software, Z.W. and N.V.S.; Formal analysis, K.L. and Z.W.; Investigation, K.L. and Z.W.; Data curation, K.L. and E.N.; Writing—original draft preparation, K.L. and Z.W.; Writing—review and editing, N.V.S. and E.N.; Visualization, K.L.; and Supervision, N.V.S. and E.N.

Funding: K.L. and E.N. acknowledge the partners: Neptune Energy Norge AS, Alfa Laval, Equinor, Marine Aluminium, NTNU, SINTEF and the Research Council of Norway, strategic Norwegian research program PETROMAKS2 (#233947) for their support KL further acknowledges the travel support from Hans Mustad og Robert og Ella Wenzins legat ved Norges teknisk- naturvitenskapelige universitet that enabled the visit at Carnegie Mellon University.

Conflicts of Interest: The authors declare no conflict of interest. The funding sponsors has approved the decision to publish the results, but had no role in the design of the study; in the collection, analyses, or interpretation of data or in the writing of the manuscript.

Abbreviations

The following abbreviations are used in this manuscript:

ALAMO	Automated Learning of Algebraic Models for Optimization
ANN	Artificial Neural Network
BIC	Bayesian Information Criterion
CFD	Computational Fluid Dynamics
RANS	Reynolds Average Navier–Stokes
RBNN	Radial Basis function Neural Network
RMSE	Root Mean Square Error
SVR	Support Vector Regression
WHRU	Waste Heat Recovery Unit

Nomenclature

Roman symbols

A_f	fin heat transfer area [m^2]
A_t	tube heat transfer area [m^2]
c_f	fin tip-to-tip clearance [m]
d_o	outer tube diameter [m]
h_f	total fin height [m]
h_s	segmented height [m]
N_r	number of streamwise tube rows [-]
N_t	number of transverse tube rows [-]
p	total pressure [Pa]
P_t	transverse tube pitch [m]
P_l	longitudinal tube pitch [m]
s_f	fin pitch [m]
$\hat{s}_f$	fin aperture ($=s_f - t_f$) [m]
t_f	fin thickness [m]
t_w	tube wall thickness [m]
$u_{F_{min}}$	mean velocity in minimum free flow area [m s^{-1}]
w_s	segment width [m]

Greek symbols

α_o	outer heat transfer coefficient [W m^{-2} K^{-1}]
β	tube bundle layout angle [°]
η_f	fin efficiency [-]
λ	thermal conductivity [W m^{-1} K^{-1}]
ν	kinematic viscosity [m^2 s^{-1}]
$\tilde{\nu}$	modified turbulent viscosity [m^2 s^{-1}]
ρ	density [kg m^{-3}]
σ_y	yield stress [-]

Dimensionless numbers

$\mathrm{Re} = u_{F_{min}} d_o / \nu$	Reynolds number
$\mathrm{Eu} = \Delta p / \left(N_r \frac{1}{2} \rho u_{F_{min}}^2 \right)$	Euler number
$\mathrm{Nu} = \alpha_o d_o / \lambda$	Nusselt number
$\mathrm{Pr} = \nu \rho c_p / \lambda$	Prandtl number

Appendix A

Algebraic expressions for the regression models developed in this work are given in Equations (A1) and (A2). All dimensions (d_o, h_f, etc.) *must* be in millimeters due to the dimensional nature of some of the regression constants.

$$\mathrm{Eu} = \left(\frac{\mathrm{Re}}{d_o}\right)^{-0.420} 0.990^{d_o} 0.971^{h_f} 1.04^{c_f}$$

$$\times\, 0.00246^{\left(\frac{\log_{10}(\mathrm{Re}/d_o)}{d_o}\right)^{0.2}} 137^{\left(\frac{\log_{10}(\mathrm{Re}/d_o)}{h_f}\right)^{0.2}} 12.5^{\left(\frac{h_f}{\hat{s}_f}\right)^{0.2}}$$

$$\times\, 0.778^{\left(\frac{d_o}{h_f}\right)^{0.6}} 1.27^{\left(\frac{h_f}{\hat{s}_f}\right)^{0.6}} 0.685^{\left(\frac{c_f}{\hat{s}_f}\right)^{0.6}} \tag{A1}$$

$$\mathrm{NuPr}^{-1/3} = \mathrm{Re}^{0.637} 0.996^{h_f} 0.511^{(c_f/\hat{s}_f)^{0.5}}$$

$$\times\, 1.26^{\left(\log_{10}(\mathrm{Re}/d_o)*\frac{h_s}{h_f}\right)^{0.2}} 0.262^{\left(\log_{10}(\mathrm{Re}/d_o)*\hat{s}_f\right)^{0.2}}$$

$$\times\, 2.14^{\left(\log_{10}(\mathrm{Re}/d_o)*c_f\right)^{0.2}} \tag{A2}$$

References

1. Ma, Y.; Yuan, Y.; Liu, Y.; Hu, X.; Huang, Y. Experimental investigation of heat transfer and pressure drop in serrated finned tube banks with staggered layouts. *Appl. Therm. Eng.* **2012**, *37*, 314–323. [CrossRef]
2. Weierman, C. Correlations ease the selection of finned tubes. *Oil Gas J.* **1976**, *74*, 10–94.
3. Nir, A. Heat transfer and friction factor correlations for crossflow over staggered finned tube banks. *Heat Transf. Eng.* **1991**, *12*, 43–58. [CrossRef]
4. Skaugen, G.; Walnum, H.T.; Hagen, B.A.L.; Clos, D.P.; Mazzetti, M.; Nekså, P. Design and optimization of waste heat recovery unit using carbon dioxide as cooling fluid. In Proceedings of the ASME 2014 Power Conference, Baltimore, MD, USA, 28–31 July 2014; pp. 1–10.
5. Pacheco-Vega, A.; Sen, M.; Yang, K.T.; McClain, R.L. Neural network analysis of fin-tube refrigerating heat exchanger with limited experimental data. *Int. J. Heat Mass Transf.* **2001**, *44*, 763–770. [CrossRef]
6. Islamoglu, Y. A new approach for the prediction of the heat transfer rate of the wire-on-tube type heat exchanger - Use of an artificial neural network model. *Appl. Therm. Eng.* **2003**, *23*, 243–249. [CrossRef]
7. Islamoglu, Y.; Kurt, A. Heat transfer analysis using ANNs with experimental data for air flowing in corrugated channels. *Int. J. Heat Mass Transf.* **2004**, *47*, 1361–1365. [CrossRef]
8. Xie, G.N.; Wang, Q.W.; Zeng, M.; Luo, L.Q. Heat transfer analysis for shell-and-tube heat exchangers with experimental data by artificial neural networks approach. *Appl. Therm. Eng.* **2007**, *27*, 1096–1104. [CrossRef]
9. Peng, H.; Ling, X. Predicting thermal-hydraulic performances in compact heat exchangers by support vector regression. *Int. J. Heat Mass Transf.* **2015**, *84*, 203–213. [CrossRef]
10. Hassan, M.; Javad, S.; Amir, Z. Evaluating different types of artificial neural network structures for performance prediction of compact heat exchanger. *Neural Comput. Appl.* **2017**, *28*, 3953–3965. [CrossRef]
11. Thibault, J.; Grandjean, B.P.A. A neural network methodology for heat transfer data analysis. *Int. J. Heat Mass Transf.* **1991**, *34*, 2063–2070. [CrossRef]
12. Koo, G.W.; Lee, S.M.; Kim, K.Y. Shape optimization of inlet part of a printed circuit heat exchanger using surrogate modeling. *Appl. Therm. Eng.* **2014**, *72*, 90–96. [CrossRef]
13. Shi, H.N.; Ma, T.; Chu, W.X.; Wang, Q.W. Optimization of inlet part of a microchannel ceramic heat exchanger using surrogate model coupled with genetic algorithm. *Energy Convers. Manag.* **2017**, *149*, 988–996. [CrossRef]
14. Bauer, R.C. A framework for implementing predictive-CFD capability in a design-by-simulation environment. *Nuclear Technol.* **2017**, *200*, 177–188. [CrossRef]
15. Lindqvist, K.; Næss, E. A validated CFD model of plain and serrated fin-tube bundles. *Appl. Therm. Eng.* **2018**, *143*, 72–79. [CrossRef]

16. Cozad, A.; Sahinidis, N.V.; Miller, D.C. Learning surrogate models for simulation-based optimization. *AIChE J.* **2014**, *60*, 2211–2227. [CrossRef]

17. Holfeld, A. Experimental Investigation of Heat Transfer and Pressure Drop in Compact Waste Heat Recovery Units. Ph.D. Thesis, Norwegian University of Science and Technology, Trondheim, Norway, 2016.

18. ESCOA Fintube Corp. ESCOA Engineering Manual (Electronic Version). Technical Report. 2002. Available online: http://www.fintubetech.com/escoa/ (accessed on 1 September 2005).

19. Cozad, A.; Sahinidis, N.V.; Miller, D.C. A combined first-principles and data-driven approach to model building. *Comput. Chem. Eng.* **2015**, *73*, 116–127. [CrossRef]

20. Wilson, Z.T.; Sahinidis, N.V. The ALAMO approach to machine learning. *Comput. Chem. Eng.* **2017**, *106*, 785–795. [CrossRef]

21. Kılınç, M.; Sahinidis, N.V. Exploiting integrality in the global optimization of mixed-integer nonlinear programming problems in BARON. *Optim. Methods Soft.* **2019**, *33*, 540–562. [CrossRef]

22. Rios, L.M.; Sahinidis, N.V. Derivative-free optimization: A review of algorithms and comparison of software implementations. *J. Glob. Optim.* **2013**, *56*, 1247–1293. [CrossRef]

23. Spalart, P.; Allmaras, S. A one-equation turbulence model for aerodynamic flows. *La Reserche Aérospatiale* **1994**, *1*, 5–21.

24. Nemati, H.; Moghimi, M. Numerical Study of Flow Over Annular-Finned Tube Heat Exchangers by Different Turbulent Models. *CFD Lett.* **2014**, *6*, 101–112.

25. Hashizume, K.; Morikawa, R.; Koyama, T.; Matsue, T. Fin Efficiency of Serrated Fins. *Heat Transf. Eng.* **2002**, *23*, 6–14. [CrossRef]

26. PFR Engineering Systems Inc. *Heat Transfer and Pressure Drop Characteristics of Dry Tower Extended Surfaces*; Technical Report, PFR Report BNWL-PFR-7-100; PFR Engineering Systems Inc.: Sante Fe Springs, CA, USA, 1976.

27. Lindqvist, K.; Næss, E. On correction factors in thermal-hydraulic correlations for compact fin-tube bundles. *Heat Mass Transf.* **2018**, submitted.

 *energies*

Article

A Numerical Study on the Light-Weight Design of PTC Heater for an Electric Vehicle Heating System

Hyun Sung Kang [1], Seungkyu Sim [2] and Yoon Hyuk Shin [1,*]

[1] Green Car Power System R&D Division, Korea Automotive Technology Institute, 303 Pungse-Ro, Pungse-Myeon, Cheonan-Si 330-912, Chungnam, Korea; hskang@katech.re.kr

[2] Donga High Tech Company, Dongtangiheung-Road, Dongtan-Myeon, Hwaseong-Si 445-813, Gyunggi-Do, Korea; shimsk76@dongah.biz

*   Correspondence: yhshin@katech.re.kr; Tel.: +82-41-559-3284; Fax: +82-41-559-3235

Received: 4 May 2018; Accepted: 15 May 2018; Published: 16 May 2018

Abstract: As the market for electric vehicles grows at a remarkable rate, various models of electric vehicles are currently in development, in parallel to the commercialization of components for diverse types of power supply. Cabin heating and heat management components are essential to electric vehicles. Any design for such components must consider the requirements for heating capacity and power density, which need to reflect both the power source and weight reduction demand of any electric vehicle. In particular, design developments in electric heaters have predominantly focused on experimental values because of structural characteristics of the heater and the variability of heat sources, requiring considerable cost and duration. To meet the ever-changing demands of the market, an improved design process for more efficient models is essential. To improve the efficacy of the design process for electric heaters, this study conducted a Computational Fluid Dynamics (CFD) analysis of an electric heater with specific dimensions by changing design parameters and operating conditions of key components. The CFD analysis modeled heat characteristics through the application of user-defined functions (UDFs) to reflect temperature properties of Positive Temperature Coefficient (PTC) elements, which heat an electric heater. Three analysis models, which included fin as well as PTC elements and applied different spaces between the heat rods, were compared in terms of heating performance. In addition, the heat performance and heat output density of each analysis model was analyzed according to the variation of air flow at the inlet of the radiation section of an electric heater. Model B was selected, and a prototype was fabricated based on the model. The performance of the prototype was evaluated, and the correlation between the analysis results and the experimental ones was identified. The error rate between performance change rates was approximately 4%, which indicated that the reliability between the design model and the prototype was attained. Consequently, the design range of effective performance and the guideline for lightweight design could be presented based on the simulation of electric heaters for various electric vehicles. The fabrication of prototypes and minimum comparison demonstrated opportunities to reduce both development cost and duration.

Keywords: performance characteristics; Positive Temperature Coefficient (PTC) elements; heat transfer; thermal performance; Computational Fluid Dynamics (CFD) simulation; air heater

---

1. Introduction

Issues such as energy shortages and environmental pollution are currently being addressed across all industries. The automobile industry accounts for over 10% of both global energy consumption and greenhouse gas emissions. Such a contribution cannot be neglected [1]. According to the World

Business Council for Sustainable Development, the number of passenger cars in the world will reach about 2 billion by 2050 [2].

The replacement of the existing combustion engine vehicles with electric vehicles (EVs) is one of the solutions to the above problem [3–5]. However, one obstacle to the spread of electric vehicles is the reduction of electric vehicle mileage by almost 50% due to the operation of cabin heating system [6]. Accordingly, cabin heating systems must be improved through the development of enhanced heating capacity and a reduction in energy consumption [7].

Heating systems in conventional combustion engines use waste heat from the engines. On the other hand, EVs have no engine and require a separate heating system. To improve the heating system of an EV, some studies have proposed heat pump systems for EVs, which reduce energy consumption [8,9]. Cho and Lee utilized the energy discharged from electrical components such as the motor, battery, or inverter to develop a heat pump that was suitable for the heating condition of EV [10]. Shin verified the performance of an electric heater using high-voltage Positive Temperature Coefficient (PTC) elements to improve the energy efficiency of EVs [11].

With regard to the above heating systems for EVs, the performance of the heat pump heating system showed considerable deterioration during prolonged low temperature conditions, as in winter. Studies which used waste heat from electrical components such as the motor, battery, or inverter found that the heat capacity was far below the cabin heating capacity and thus was largely useless for heating. A promising and realistic alternative is to improve the performance of the electric heater, which is a key heating component. In this regard, it is necessary to determine an optimal heating capacity, designing an electric heater according to the performance requirements of the system in which the heater is used.

An electric heater offers a simple structure as well as good compatibility and a fast response time due to PTC elements. In addition, because PTC elements drastically increase the resistance at or above a set temperature to maintain it, electric heaters include both temperature control and safety functions of its own, unlike other heating systems. However, one significant drawback to this feature is its inability to predict the power consumption of a PTC element based electric heater according to external environment and boundary conditions. Moreover, even if an electric heater with optimal heating capacity is designed in accordance to various specifications (weight, volumes, etc.) of the system where the heater is to be used, the reliable (accurate) performance of real products is difficult to attain.

The performance of an air-heating electric heater is significantly affected by the structural designs of heat rods (Figure 1) and fins. Heat transfer is caused by the dispersion of wake flow which occurs as a result of periodic friction in the boundary layer between air and fin. Accordingly, if the finite area of fin increases, heat transfer is improved; however, this also increases friction and drag, which facilitates a drop in pressure [12]. As mentioned above, the heat transfer mechanisms are under development not only in heaters but throughout various fields, such as nanoscales [13,14]. Any successful design must satisfy the requirements for heat transfer and pressure drop as well as the weight reduction of components, which have great effect on the fuel economy of EV. An analytical approach can effectively consider these factors.

A three-dimensional Computational Fluid Dynamics (CFD) simulation can easily reflect physical conditions without the need for an expensive tester or fabrication of a prototype; accordingly, various design options can be effectively tested at a low cost. Lalot and Florent used a CFD simulation to examine the non-uniformity of flow in an electric heater and demonstrated its impact on the non-uniformity of heat exchange [15]. Zhang and Li applied a CFD method to ensure uniform heat distribution according to fluid flow inside a heat exchanger. They could easily predict physical phenomena caused by the inlet shape of a heat exchanger [16].

This study conducted a comparative analysis of power density (heating performance/weight) according to the configuration of key components by using a 3D heat flow analysis model. The analysis examined the design of 6 kW electric heaters which are conventionally used in the cabin of EVs. The heat flow simulation model for an electric heater formulated PTC characteristic curves to model

PTC heating. In this way, the number of heat rods inserted into the radiation fin and the heating performance could be simulated. Additionally, to verify the reliability of the simulation, a prototype of the electric heater was fabricated based on the analysis model which tested performance. Both the heating performance and the output density characteristic of the electric heater were analyzed according to the radiation fin and heat rod. On the basis of this analysis, a guideline on weight reduction design, which satisfied reference performance factors, was proposed.

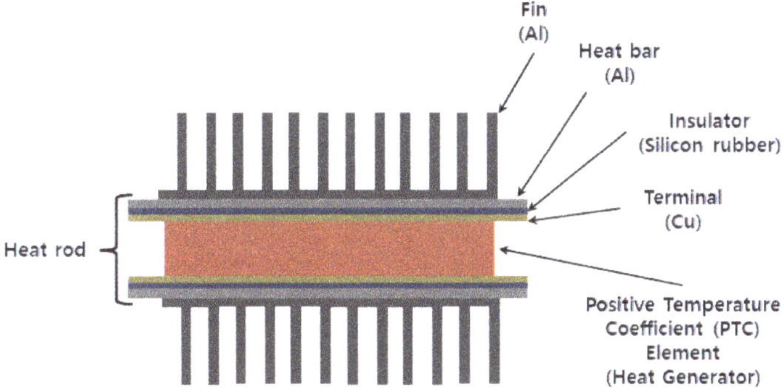

Figure 1. Schematic diagram of heat rod and fin for electric heater. PTC: Positive Temperature Coefficient.

2. CFD Model Details

2.1. Physical Model

Figure 2 shows a geometric model of an electric heater including the plate fins and heat rods which were adopted in this study [17]. Different spaces between heat rods were applied by changing the number of heat rods used in the electric heater. The radiation performance of each model was compared. Seven PTC heating elements were inserted into each heat rod and the elements were connected in parallel. The heat core, which consisted of heat rods and fins, was divided into two zones. In each heat core, heat rods were inserted at regular intervals into 112 fin layers.

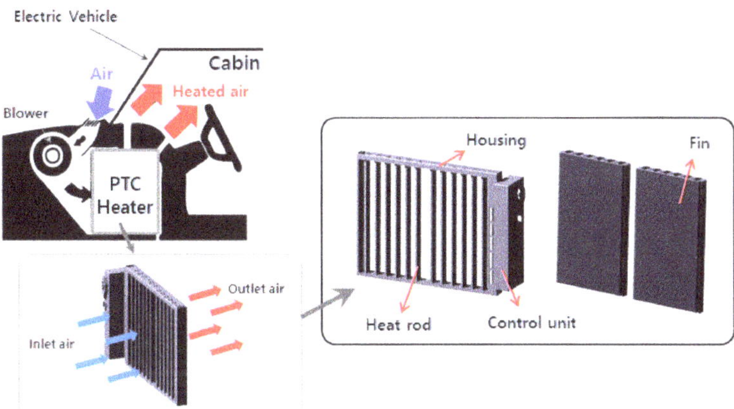

Figure 2. Schematic diagram of PTC heater in an electric vehicle (geometry of housing, heat rod, and fin array).

Comprising a repeating shape of fins and heat rods, the electric heater model required a considerable number of meshes, which demanded a significant period of time for the convergence process of analysis; it also decreased the reliability of results because of its complex structure. To reduce the number of iterative calculations and required analysis time, an analysis domain was set, as shown in Figure 3. The analysis domain indicated half of the fin pitch and heat rod pitch. The open ratio O_r of this domain was calculated by the ratio of the inlet area of the domain to that of the heater model. F_w and F_p indicate the width and height of the analysis domain, respectively. To improve the reliability of the analysis, the flow path of the inlet was reduced while that of the outlet was set at over five times the hydraulic diameter, which could prevent reverse flow. As shown in Figure 3, the analysis domain included air, heat bar, PTC elements, fin, and insulator. Table 1 presents the common geometric design values of each model. Table 2 shows the properties of each part of the domain.

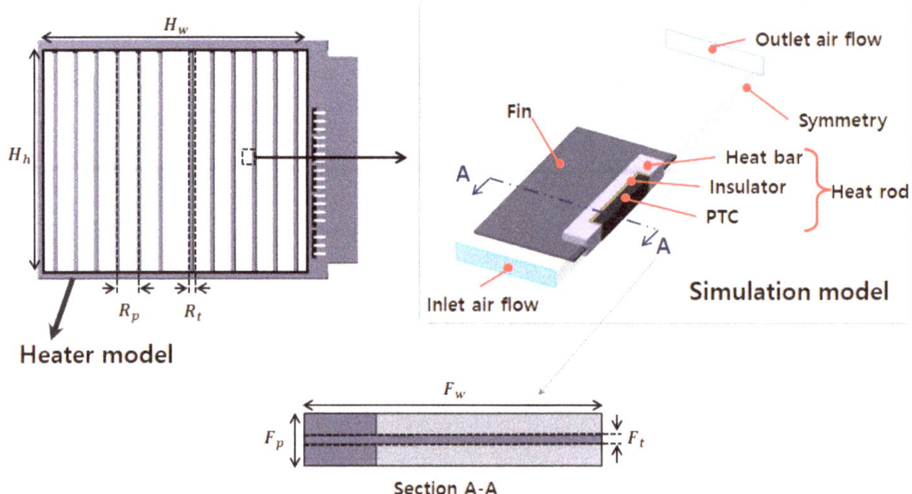

Figure 3. Schematic of simulation model with boundary conditions.

Table 1. Geometrical parameter for heater models.

Parameters	Values
Full size Heater width, H_w (mm)	265.5
Full size Heater height, H_h (mm)	215
Heat rod pitch, R_p (mm)	21
Heat rod thickness, R_t (mm)	5.1
Fin pitch and Inlet face height, F_p (mm)	1.8
Fin thickness, F_t (mm)	0.3

Table 2. Material properties used in the Computational Fluid Dynamics (CFD).

Material Properties	Air (Fluid)	Aluminum (Fin, Heat Rod)	Silicon Series (Insulator)
Density (kg/m^3)	Incompressible-ideal-gas	2719	2329
Specific Heat (J\kg·K)	1006.43	871	720
Thermal conductivity (w\m·K)	0.0242	202.4	0.25
Viscosity (kg/m·s)	1.7894×10^{-5}	-	-

A three-dimensional simulation was implemented through the commercial program, Fluent. The grid was built and the meshes were improved by applying the proximity and curvature method.

An inflation layer was applied near the fin surface to create the necessary amount of meshes to study whether the heat transfer according to air flow on the fin surface was more effective. In addition, on the basis of the initial analysis, the adaption function was applied to points with high temperature variation to generate meshes with higher density. Finally, approximately $1 \times 10^6 \sim 2 \times 10^6$ meshes were made.

2.2. Analysis Model

Navier-Stokes and SIMPLE algorithm were used to solve three-dimensional energy equations for steady-state fluid mechanics and the heat transfer zone (1)–(3).

Continuity equation:

$$\nabla \cdot \rho u = 0 \tag{1}$$

Momentum equation:

$$\rho(u \cdot \nabla u) = -\nabla p + \mu \nabla^2 u \tag{2}$$

Energy equation in steady state condition:

$$\rho C_p(u \cdot \nabla)T = k \nabla^2 T \tag{3}$$

Among turbulence models provided by Ansys Fluent, the Transition SST (Shear Stress Transport) model was used as the analysis model. This model was the combination of the *k*-omega model for analyzing wall flow and the *k*-epsilon, which was effective in predicting flow behaviors occurring far from the wall. To implement the Transition SST model, meshes need to satisfy the requirement of $Y^+ \leq 1$. The *y* value, which was the distance from a wall to the first mesh-formation layer, was calculated by using the Equation (4). The prism mesh was applied to the layer.

$$Y^+ = \frac{\rho U_\tau y}{\mu} = \frac{U_\tau y}{v} \tag{4}$$

2.3. Model Parameters and Boundary Condition

For the boundary condition for the inlet of the heater, the mass flow rate of 300 kg\h, which is conventionally used to test a heater, was adopted; the temperature condition was 0 °C, which acted as the reference temperature of winter. However, this boundary condition corresponded to the area of $H_w \times H_h$. Accordingly, it had to be adjusted to the simulation domain. Table 3 presents mass flow rates and open area ratios of each simulation model. The open area ratio was the ratio of inlet area of a model to the radiation section area of the electric heater.

Open area ratio (O_r):

$$O_r = \frac{F_w \times F_p}{H_w \times H_h} \tag{5}$$

The length of fin (that is, the distance between heat rods) F_w was set as a model parameter to predict the heater performance according to the number of heat rods in the same area of heater. Figure 3 illustrates Model B. The fin lengths of Models A and C were set to the half and twice of that of Model B, respectively. Table 3 provides parameters of each model.

Table 3. Specification of each simulation model.

Parameter	Model A	Model B	Model C
Inlet face width, F_w (mm)	5.25	10.5	21
Open ratio, O_r	1.655×10^{-4}	3.311×10^{-4}	6.622×10^{-4}
Mass flow rate, M_d (kg/s)	1.38×10^{-5}	2.76×10^{-5}	5.52×10^{-5}
Hydraulic diameter, D_h (mm)	2.68	3.07	3.31

User define functions (UDFs) were used to apply real radiation characteristics of PTC to the simulation. On the basis of the experimental data, curve fitting was performed to obtain the resistance-temperature curve of PTC elements, as shown in Figure 4. Two sections were distinguished to effectively formulate the drastic change of resistance according to temperature. For the section of the heating temperatures ranging from 25 °C to 140 °C, Equation (6) was applied. Equation (7) was used for the heating temperature over 145 °C. To calculate the equation of PTC elements characteristic curve as power consumption, Equation (10) was used, which applied Ohm's law (9) to Joule's law (8). Power consumption of PTC elements was assumed to be converted to thermal energy without loss. The power consumption was obtained by using the applied voltage of 330 V and the resistance values functionalized by temperature variation, as shown in Equations (6) and (7). This was set to the boundary condition of the calorific value of PTC elements.

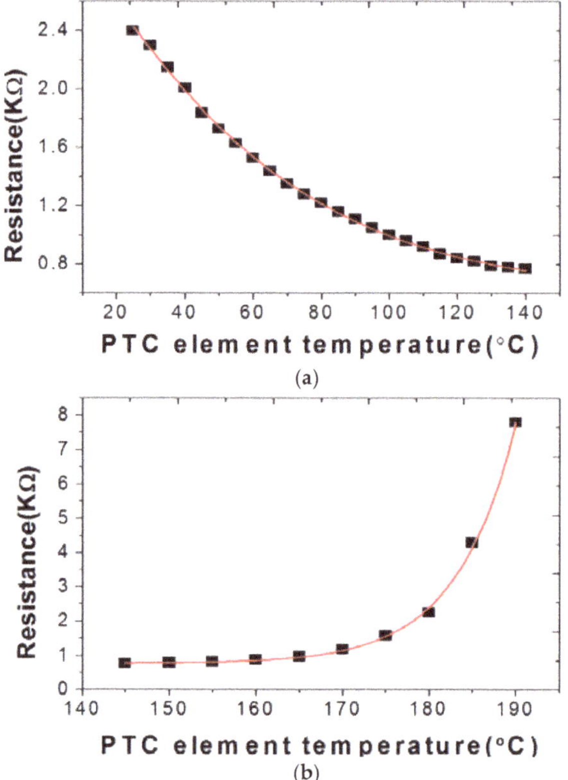

Figure 4. Resistance–Temperature curve of PTC, (**a**) exponential fitting form 25 °C to 140 °C, (**b**) cubic fitting after 145 °C.

Resistance–Temperature equation

$$(25\,°\text{C} < T \le 140\,°\text{C}):$$
$$R_1 = 3.39653 - 0.04493T_c + \left(2.66467 \times 10^{-4}\right)T_c^2 - \left(5.72879 \times 10^{-7}\right)T_c^3 \tag{6}$$

$$(140\,°\text{C} < T \le 145\,°\text{C}):$$
$$R_2 = 0.77$$

$$(145\,^{\circ}\mathrm{C} < T):$$
$$R_3 = 0.77297 + \left(6.39769 \times 10^{-12}\right)e^{0.14594 * T_c} \tag{7}$$

Heat generation

$$P = V \times I \tag{8}$$

$$V = I \times R \tag{9}$$

$$P_p = \frac{V_p^2}{R_p}\left(V_p = 330,\ Constant\right) \tag{10}$$

3. Simulation Result and Analysis

3.1. Analysis of Each Model According to Test Condition

To design a lightweight electric heater for EVs, which could satisfy the target performance (6 kW), this study conducted a simulation of heating performance by applying different heat rod distances to the heater radiation section (heat core). Figure 5 illustrates distributions of air, fin, and PTC temperature for three models.

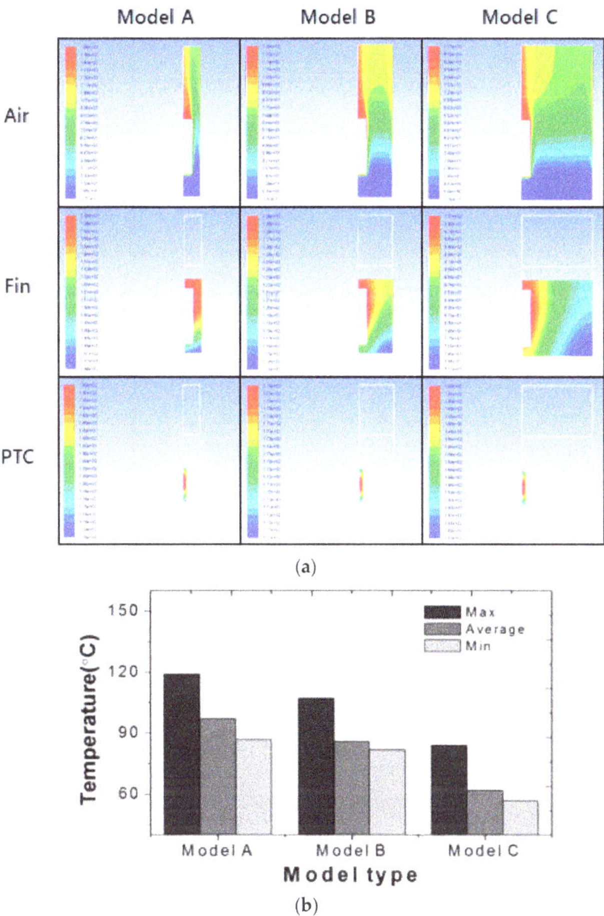

(a)

(b)

Figure 5. *Cont.*

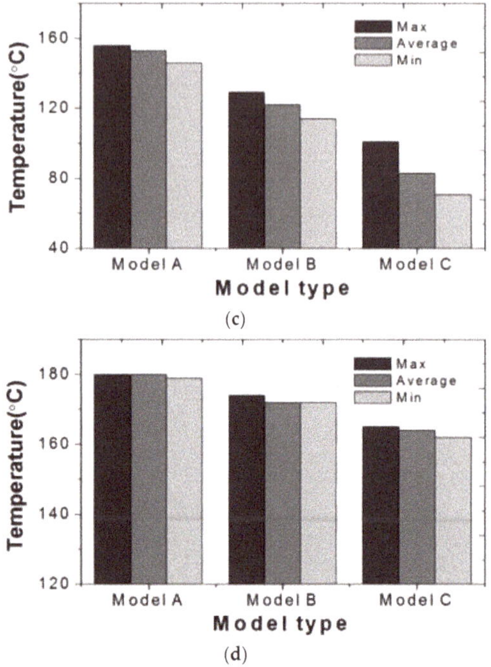

(c)

(d)

Figure 5. Total temperature of heater, (**a**) Temperature contour for different heat rod interval, (**b**) temperature on air outlet, (**c**) temperature in fin, (**d**) temperature in PTC.

Figure 6 shows temperature distributions from the center of PTC elements in the width direction of fins, an essential component of heaters. Models A and B had variations of 0.8% and 6.2%, respectively, both of which were lower than 24.2% of Model C.

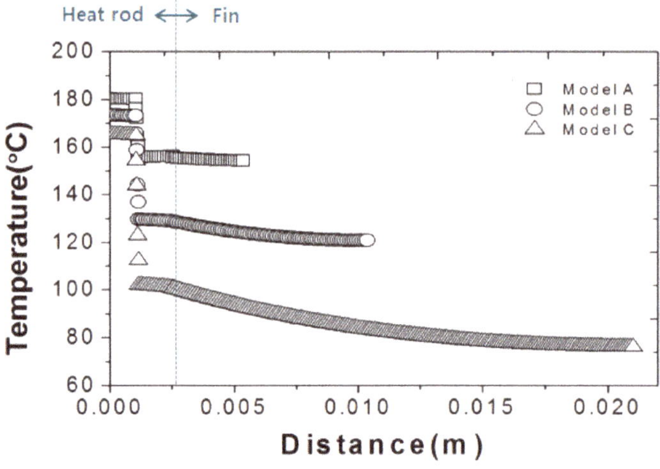

Figure 6. Temperature distribution of fin (Vertical direction).

In the case of a drop in pressure, as shown in Table 4, Model A had the highest value of 43 Pa, as it included the largest number of heat rods in the same area. Model A showed the lowest calorific value of 1.34 W in the analysis domain. However, this value was converted to 7203.84 W by the unit of the whole heater model, which was the highest value. The comparison of heating capacity in Figure 7a revealed that Models A and B had similar levels. As shown in Figure 7c, Model B had the highest heating power density, when heating capacities were compared according to the unit weight of heater per each model. Also, Model C showed a similar value but it did not satisfy the target performance (6 kW) of electric heaters.

Table 4. Analysis results of simulation. Positive Temperature Coefficient.

	Description	Model A	Model B	Model C
Simulation model	Pressure drop (Pa)	43	24	19
	PTC elements temperature (°C)	180	172	164
	PTC resistance (kΩ)	2.412	1.283	0.932
	Heating capacity (w)	1.34	2.39	3.45
Heater model	Number of PTC (ea)	168	84	42
	Power consumption of a PTC element (w)	42.88	76.48	110.4
	Weight (kg)	1.659	1.081	0.792
	Heating capacity (w)	7203.84	6424.32	4636.8

When an electric heater is designed, heating capacity must be secured that is suitable for its required size and performance; the use of main component materials can achieve a lightweight and cost-effective design. On the basis of these considerations, Model B was selected as the most appropriate option as a result of the simulation.

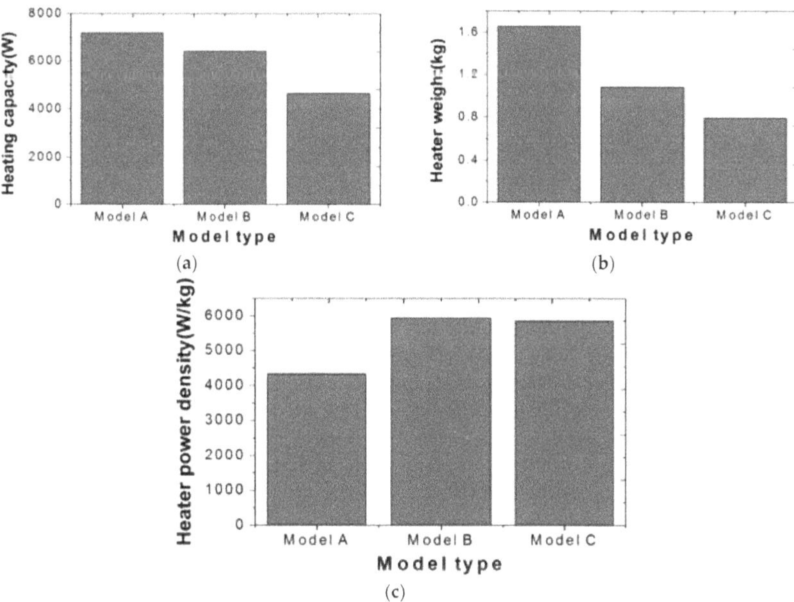

Figure 7. Result of CFD analysis for heater for each model, (**a**) heating capacity, (**b**) weight, (**c**) heater power density.

3.2. Analysis of Each Model According to Flow Rate Variation

The above result was obtained under the ideal condition that the outlet air temperature and mass flow rate were constant according to locations of heater radiation section. For this reason, the heating performance attained was beyond expectation.

However, existing studies have already demonstrated that the entire radiation section of an electric heater does not have a uniform temperature distribution; temperature variation may be as much as 10 °C~20 °C in areas [18]. Such non-uniformity of temperature distribution points to the non-uniformity of air density, which transfers heat, in relationship to the location of a heater radiation section. This also implies that the non-uniformity of mass flow rate depends on the location of the heater.

The analysis domain of this study was an ideal position, which was a local area of a real heater, as shown in Figure 8. If the results of the analysis can be converted to data for an entire heater, the ideal heater could be designed and fabricated. However, as is clear from Figure 8, the real heater included an edge position that exhibited relatively inferior performance. Accordingly, temperature and mass flow rate had different values, depending on the distance from the aforementioned edge position. Analysis results for the heating performance, which were derived from the ideal position, were more positive than the experimental results reflecting flow and heat loss, which were applicable to the entire heater.

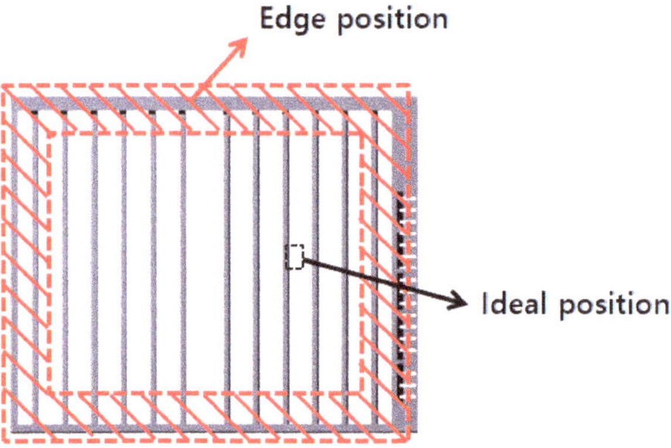

Figure 8. Schematic diagram of individual position in PTC heater.

To verify such a gap with experimental performance, the mass flow rates of each model were modified by applying 150 kg/h, 225 kg/h, 300 kg/h, 375 kg/h, and 450 kg/h to the basic flow rate, respectively, as shown in Table 5; the heating performance was also verified. The distribution of heating performance was obtained reflecting different flow rates, as shown in Figure 9. Figure 9a shows the heating capacities of the entire heater, which were calculated by multiplying the heating capacity of the simulation domain. In Models A and B, the heating capacity is proportional to the mass flow rate. Figure 9b illustrates the density of heat output, which is an important factor for securing sufficient heating performance and lightweight design for an entire heater. The above results show that the heating capacity of all three models increases when the mass flow rate increases. In addition, it was confirmed that the heating power density of Model B was the highest when the mass flow rate was 300 kg/h or higher. On the basis of this consideration, an effective number of heat rods and the distance between radiation fins could be determined to achieve the target performance and lightweight design of an electric heater with specific dimensions.

Table 5. Air mass flow conversion for heating performance according to each model.

Inlet Air Mass Flow Rate of Heater (kg/h)	Inlet Air Mass Flow Rate of Each Simulation Model (kg/s)		
	Modle A	Modle B	Modle C
150	6.90×10^{-6}	1.38×10^{-5}	2.76×10^{-5}
225	1.04×10^{-5}	2.07×10^{-5}	4.14×10^{-5}
300	1.38×10^{-5}	2.76×10^{-5}	5.52×10^{-5}
375	1.73×10^{-5}	3.45×10^{-5}	6.90×10^{-5}
450	2.07×10^{-5}	4.14×10^{-5}	8.28×10^{-5}

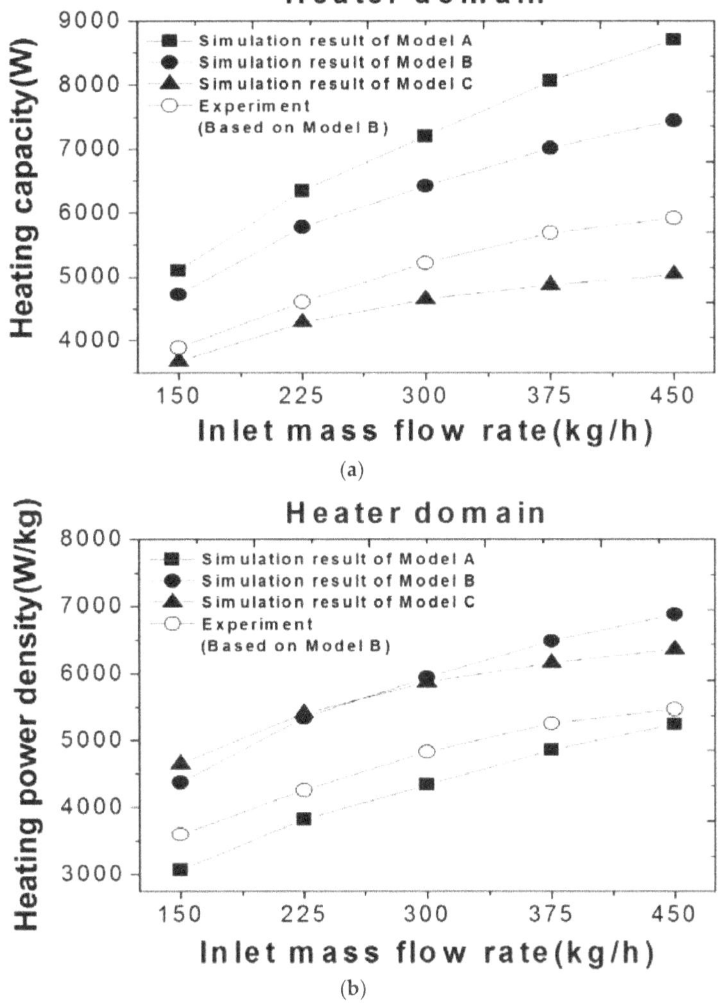

Figure 9. Variation of inlet air mass flow rate, (**a**) heater heating capacity, (**b**) heater power density.

4. Experimental Results for the Electric Heater

To measure and compare the heating performance, efficiency, and pressure values of the prototype electric heater fabricated in this study, a wind tunnel and environmental chamber system (Figure 10) were constructed. Each operational condition was applied and compared. The inlet/outlet air temperature of the electric heater in the winder tunnel was measured using 25 T-type thermocouples with ± 0.1 °C error rate. The pressure drop of the radiation section was measured by using a pressure gauge. The data loggers of Ganter and Yokogawa were used to collect temperature and pressure values. To prevent the influence of the radiation section, the temperature sensors were installed in a 4×6 arrangement, with equal spacing at a distance over 3 cm from the outlet surface of the radiation section. Accordingly, temperature distributions of heat cores could be compared. A 12 V power supply was used to operate the controller of the radiation section. The electric heater ran based on the duty control using a can analyzer. The inlet flow rate was set by considering air density (0.99~1.28 kg/m^3) according to temperature variation. The test conditions of Table 6 were applied.

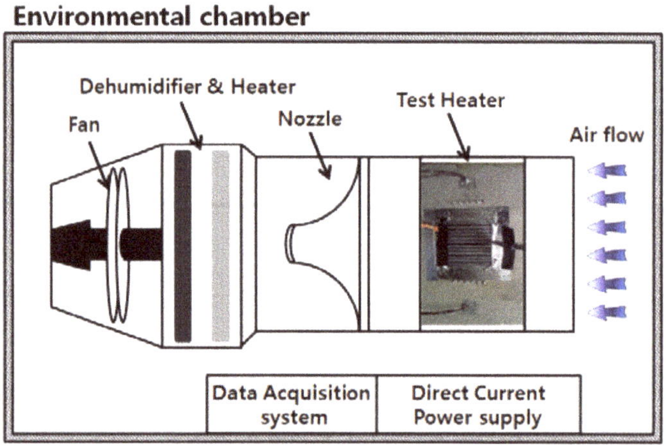

Figure 10. Schematic diagram of the PTC heater experimental apparatus.

Table 6. PTC heater experimental conditions.

	Parameter	Value
Inlet air	Temperature (°C)	0 ± 2
	Mass flow (kg/h)	300 ± 10
Power	Input Voltage (V)	330 ± 5

As shown in Figure 11, the results of the prototype electric heater test show a difference in heating performance according to the heat dissipation location of the heater. This effect was likely caused by the difference in contact of thermal resistance of each part (heat rod and fin). Therefore, to determine qualitative characteristics, we compared the rate of change in heating performance by changing the inlet air mass flow rate between the experiment and the analysis results. Figure 9 illustrates the comparison of heating performance results according to flow rate between analysis and experiment. The variations of heating performance according to flow rates at the level of the entire heater ranged from 3% to 22% in the analysis and experiment, respectively. When the variations were compared for each flow rate range in both the analysis and experiment, the maximum error rate was 4%, which indicated that the error rate of performance variation between the analysis model and the prototype was generally reliable.

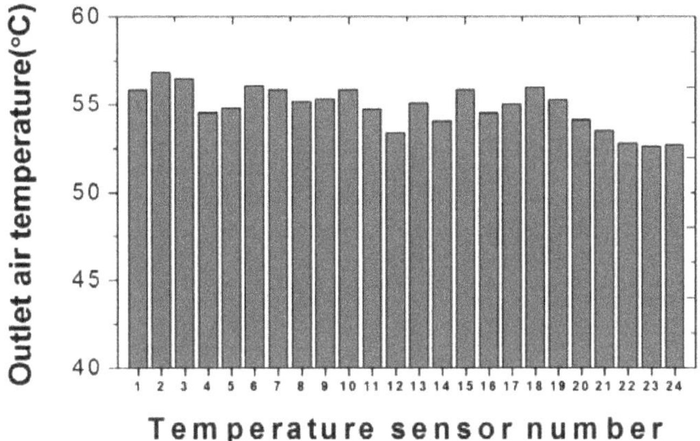

Figure 11. Experiment result of outlet air temperature distribution according to the heat dissipation location of the heater (inlet mass flow rate 300 kg/h).

5. Conclusions

To ensure an effective design of an electric heater for an EV heating system, this study conducted a simulation of heating performance by applying temperature characteristics of PTC elements. A comparative analysis applied different design parameters of main components and operational conditions to an electric heater with specific dimensions. In this way, candidate models were compared in terms of heating performance and heat output density, which was an indicator of lightweight design. A prototype heater was fabricated based on the selected optimal model, and its performance was evaluated. The conclusions of this study can be summarized as follows.

(1) To obtain an optimal design of an electric heater using PTC elements, a three-dimensional heat transfer analysis was performed by applying simple models that reflected the radiation characteristics of PTC elements and the structural characteristics of the heater. The performance of each model was compared according to different configurations of heat rods and fins, which are dominant components in terms of heat transfer performance.

(2) Each model was analyzed to realize heating capacities over 6 kW, which was the target performance at the level of the entire heater. As the indicator of lightweight design, the heat output densities of Models A, B, and C were 4.34, 5.94, and 5.87 kW/kg. In addition, when the inlet air flow rate varied, the performances (Delta Q, the performance variation in the same flow rate rage) of Models A, B, and C were 3.6, 2.71, and 1.36 kW, respectively. On the basis of these results, Model B was selected as the optimal design option that could achieve a high heat output density as well as proportional and stable performance variation.

(3) A prototype electric heater was fabricated through the application of Model B. Under the reference conditions, the prototype was evaluated to have a heating capacity of approximately 5.23 kW. The correlation between the simulation results and the heating performance results of the prototype according to inlet flow rate showed that the error rate between performance variations was about 4%. This indicated that sufficient reliability between the prototype and the design model had been secured.

In the development of various electric heaters for EVs, a simulation reflecting radiation elements and heater characteristics can provide a guideline to an effective performance design range and lightweight design. Moreover, cost will be reduced by minimizing the comparative fabrication of prototype heaters.

Author Contributions: Hyun Sung Kang designed the research, Hyun Sung Kang, Yoon Hyuk Shin discussed the results and contributed to writing the paper. Thanks to Seungkyu Sim from Donga High Tech Company for helping us with the experiment.

Acknowledgments: This work was supported by the Energy Efficiency & Resources of the Korea Institute of Energy Technology Evaluation and Planning (KETEP), a grant funded by the Korean Government Ministry of Knowledge Economy (20152000000370), and supported by the New Product Development Program with Conditional Purchase (S2422112) funded by the Ministry of Small and Medium-sized Enterprises (SMEs) and Startups.

Conflicts of Interest: The authors declare no conflict of interest.

Nomenclature

F	Fin
H	Heater
R	Heat rod
T	Temperature
y	Distance from wall (mm)
O_r	Simulation domain open ratio
U_τ	Velocity in shear stress direction (m/s)
M_d	Mass flow rate (kg/s)
D_h	Inlet hydraulic diameter (mm)
R_1	$25 < T_c \leq 140$, Resistance of PTC (kΩ)
R_2	$140 < T_c \leq 145$, Resistance of PTC (kΩ)
R_3	$145 < T_c$, Resistance of PTC (kΩ)
P	Power (W)
V	Voltage (V)
I	Current (A)

Greek symbols

∇	Vector operator
ρ	Density (kg/m^3)
u	Velocity (m/s)
μ	Coefficient of dynamic viscosity (kg/m·s)
v	Coefficient of kinematic viscosity (m^2/s)

Subscripts

h	height (mm)
p	pitch (mm)
w	width (mm)
t	thickness (mm)
c	PTC

References

1. Solomon, S.; Qin, D.; Manning, M.; Chen, Z.; Marquis, M.; Averyt, K.B.; Tignor, M.; Miller, H.L. *Contribution of Working Group I to the Fourth Assessment Report of the Intergovernmental Panel on Climate Change*; Cambridge University Press: Cambridge, UK; New York, NY, USA, 2007.

2. Stigson, B. *Mobility 2030: Meeting the Challenges to Sustainability*; World Business Council for Sustainable Development: Geneva, Switzerland, 2004.

3. Kambly, K.R.; Bradley, T.H. Estimating the HVAC energy consumption of plug-in electric vehicles. *J. Power Sources* **2014**, *259*, 117–124. [CrossRef]

4. Panchal, S.; Dincer, I.; Agelin-Chaab, M.; Fraser, R.; Fowler, M. Experimental and theoretical investigations of heat generation rates for a water cooled LiFePO4 battery. *Int. J. Heat Mass Transf.* **2016**, *101*, 1093–1102. [CrossRef]

5. Panchal, S.; Dincer, I.; Agelin-Chaab, M.; Fraser, R.; Fowler, M. Thermal modeling and validation of temperature distributions in a prismatic lithium-ion battery at different discharge rates and varying boundary conditions. *Appl. Therm. Eng.* **2016**, *96*, 190–199. [CrossRef]

6.   Kim, K.; Lee, W.S.; Kim, Y.Y. Investigation of Electric Vehicle Performance Affected by Cabin Heating. *J. Korea Acad. Ind. Coop. Soc.* **2013**, *14*, 4679–4684. [CrossRef]

7.   Yokoyama, A.; Osaka, T.; Imanishi, Y.; Sekiya, S. Thermal management system for electric vehicles. *SAE Int. J. Mater. Manuf.* **2011**, *4*, 1277–1285. [CrossRef]

8.   Torregrosa, B.; Payá, J.; Corberán, J.M. Modeling of mobile air conditioning systems for electric vehicles. In Proceedings of the 4th European Workshop—Mobile Air Conditioning and Vehicle Thermal Systems, Torino, Italy, 1–2 December 2011.

9.   Lee, D.Y.; Cho, C.W.; Won, J.P.; Park, Y.C.; Lee, M.Y. Performance characteristics of mobile heat pump for a large passenger electric vehicle. *Appl. Therm. Eng.* **2013**, *50*, 660–669. [CrossRef]

10.   Cho, C.W.; Lee, H.S.; Won, J.P.; Lee, M.Y. Measurement and evaluation of heating performance of heat pump systems using wasted heat from electric devices for an electric bus. *Energies* **2012**, *5*, 658–669. [CrossRef]

11.   Shin, Y.H.; Ahn, S.K.; Kim, S.C. Performance Characteristics of PTC Elements for an Electric Vehicle Heating System. *Energies* **2016**, *9*, 813. [CrossRef]

12.   Wang, C.C.; Chang, Y.J.; Hsieh, Y.C.; Lin, Y.T. Sensible heat and friction characteristics of plate fin-and-tube heat exchangers having plane fins. *Int. J. Refrig.* **1996**, *19*, 223–230. [CrossRef]

13.   Vo, T.Q.; Kim, B. Transport phenomena of water in molecular fluidic channels. *Sci. Rep.* **2016**, *6*, 33881. [CrossRef] [PubMed]

14.   Vo, T.Q.; Kim, B. Interface thermal resistance between liquid water and various metallic surfaces. *Int. J. Precis. Eng. Manuf.* **2015**, *16*, 1341–1346. [CrossRef]

15.   Lalot, S.; Florent, P.; Lang, S.K.; Bergles, A.E. Flow maldistribution in heat exchangers. *Appl. Therm. Eng.* **1999**, *19*, 847–863. [CrossRef]

16.   Zhang, Z.; Li, Y. CFD simulation on inlet configuration of plate-fin heat exchangers. *Cryogenics* **2003**, *43*, 673–678. [CrossRef]

17.   Park, M.H.; Kim, S.C. Heating Performance Characteristics of High-Voltage PTC Heater for an Electric Vehicle. *Energies* **2017**, *10*, 1494. [CrossRef]

18.   Shin, Y.H.; Sim, S.; Kim, S.C. Performance Characteristics of a Modularized and Integrated PTC Heating System for an Electric Vehicle. *Energies* **2016**, *9*, 18. [CrossRef]

 *energies*

Article

Unsteady Simulation of a Full-Scale CANDU-6 Moderator with OpenFOAM

Hyoung Tae Kim [1], Se-Myong Chang [2],* and Young Woo Son [2]

[1] Thermal Hydraulic and Severe Accident Research Division, Korea Atomic Energy Research Institute, 989-111 Daedeok-daero, Yuseong-gu, Daejeon 34057, Korea; kht@kaeri.re.kr

[2] School of Mechanical Convergence Systems Engineering, Kunsan National University, 558 Daehak-ro, Gunsan, Jeonbuk 54150, Korea; ywson@kunsan.ac.kr

* Correspondence: smchang@kunsan.ac.kr; Tel.: +82-63-469-4724

Received: 21 November 2018; Accepted: 16 January 2019; Published: 21 January 2019

Abstract: Three-dimensional moderator flow in the calandria tank of CANDU-6 pressurized heavy water reactor (PHWR) is computed with Open Field Operation and Manipulation (OpenFOAM), an open-source computational fluid dynamics (CFD) code. In this study, numerical analysis is performed on the real geometry model including 380 fuel rods in the calandria tank with the heat-source distribution to remove uncertainty of the previous analysis models simplified by the porous media approach. Realizable k-ε turbulence model is applied, and the buoyancy due to temperature variation is considered by Boussinesq approximation for the incompressible single-phase Navier-Stokes equations. The calculation results show that the flow is highly unsteady in the moderator. The computational flow visualization shows a circulation of flow driven by buoyancy and asymmetric oscillation at the pseudo-steady state. There is no region where the local temperature rises continuously due to slow circulating flow and its convection heat transfer.

Keywords: CANDU-6; PHWR; moderator; turbulence; OpenFOAM

---

1. Introduction

CANadian Deuterium Uranium (CANDU) reactors have been introduced in Korea since the late 1980s, and four units of CANDU-6 reactors were constructed in the Wolsong areas [1]. The horizontal fuel channels in a CANDU-6 reactor (a pressurized heavy water reactor, PHWR) are submerged in the heavy water (D_2O) pool which is contained by a cylindrical tank called calandria. One of the important design features of the CANDU-6 reactor is the use of moderator as a heat sink during some postulated accidents such as a large-break Loss of Coolant Accident (LOCA). Therefore, it is one of the major concerns in the CANDU safety analyses to estimate the local subcooling margin of the moderator inside the calandria tank.

Previous experimental studies [2] showed that the film boiling on the outside surface of fuel channels would be unlikely to occur if the local moderator subcooling is sufficient. Therefore, an accurate prediction of the moderator temperature distribution in the calandria tank is needed to confirm the channel integrity [3]. To predict the local temperature of the calandria tank, numerous experimental and numerical studies have been performed so far. Huget et al. [4,5] conducted two-dimensional moderator circulation tests at the STERN Laboratories Inc. (STERN Lab.), and they validated a specific code, MODerator TURbulent Circulation (MODTURC) [4] and its advanced version, Co-Located Advance Solution (MODTURC_CLAS) [5] against the experimental results [6] of the velocity and the temperature distributions.

Temperature distribution in the moderator is highly affected by flow patterns and circulation characteristics which themselves are generated as a result of interactions between the inertia forces (produced by inlet jets) and buoyancy forces (resulting from heat addition) in the calandria (see Figure 1). Given the differences in the moderator heat load, flow rate and inlet nozzle distribution and design, each CANDU reactor has a different flow pattern and temperature distribution during normal operation. Therefore, Korea Atomic Energy Research Institute (KAERI) installed a 1/4 scaled Moderator Circulation Test (MCT) facility [7] that is representative of CANDU-6 reactors with 380 fuel channels. These test results [8] showed that the moderator circulation flow has a mixed flow patterns with combination of inertial forces and buoyancy forces under the CANDU-6 operation conditions. Furthermore, the flow oscillation and unsteady flow behavior were observed, which were not reported in the previous studies [4,6].

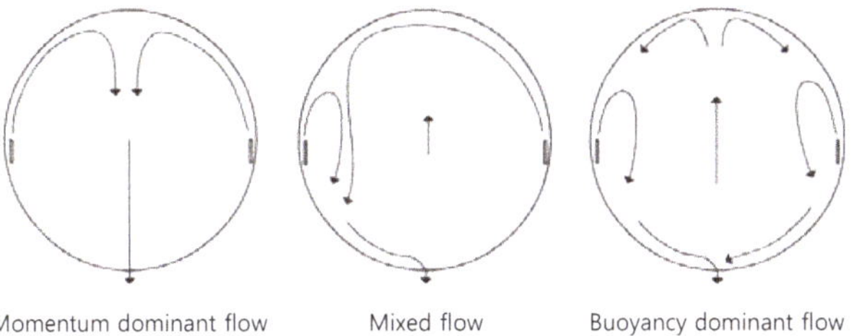

Figure 1. Flow pattern inside a calandria by balance of buoyancy and momentum forces.

There have been numerous computational efforts to estimate the thermal hydraulics in the calandria tank using CFD codes. Hadaller et al. [9] obtained a tube bank pressure drop model for tube bundle region of the calandria tank and implemented it into the MODTURC_CLAS code. Yoon et al. [10] used a commercial code, CFX to develop a CFD model with a porous media approach for the core region. However, it is known that porous media modeling provides only average values of flow velocities and temperatures in the moderator and do not give any information about 3-D local flow variables near tube solid walls, which are necessary to implement accurate heat transfer calculations. Recently, porous media modeling in the tube bank region of core using economic computing resources are replaced by the full geometric model of calandria tubes requiring high computing resources. Sarchami et al. [11] used another FLUENT code to model all the calandria tubes as they are without any approximation for the core region. They could show the nature of moderator temperature fluctuations by dynamic flow behavior with completion between the upward moving buoyancy driven flows and the downward moving momentum driven flows. Teyssedou et al. [12] conducted FLUENT code simulation of moderator flow around calandria tubes of CANDU-6 and showed that the standard $k\text{-}e$ model is appropriate for turbulence model to perform this kind of simulation. Application of FLUENT and CFX code is successfully performed for the reduced-scale CFD models for various thermal hydraulics problems in nuclear engineering also by the authors [13,14].

In this study, Open Field Operation and Manipulation (OpenFOAM) [15], an open-source CFD solver, is used to simulate the three-dimensional flows improving the computational efficiency by parallel computing which does need no proprietary license. The feasibility on the computation of 3-D flow has been tested and validated by the comparison with other codes by the authors [16], but the models are just focused on the pressure drop in a straight channel. In this paper, the full capacity of OpenFOAM CFD is tested for a turbulent unsteady flow as observed in the 1/4 scale of test [8] to resolve the 3-D structure of circulation flow in the moderator system of a real-scale CANDU-6 reactor.

We have studied the suitable grid levels and the validation of pressure drops with the comparison with various commercial codes such as ANSYS-CFX and COMSOL (COMputer SOLution) Multiphysics as well as experimental data using OpenFOAM [16]. However, the full simulation of CANDU-6 is not yet attempted because of its high complexity in three dimensions. From the dimensional analysis, the complex scale effects between prototype CANDU-6 and model MCT should be considered [7]. Therefore, the full-scale simulation is expected to show the overall flow physics with proper predictions of subcooling margin in this research. In the system codes, the difference of temperature or pressure between inlet and outlet is given as a lumped input parameter. For example, a system analysis code for PHWR, CATHENA [17] can consider a pipe network for tube bundles of fuel channels, but a three-dimensional numerical model is made for the present study to understand sophisticated flow physics such as turbulence diffusion and mixing, convective heat transfer, buoyancy forces, etc. in a moderator pool.

2. Simulation Method

2.1. Open Source Code

OpenFOAM has been developed by Henry Weller and Hrvoje Jasak in Imperial College. The source code has been opened to the public since 2004. This code is operated on the Linux-based O/S such as Ubuntu, so the copyright is absolutely free for every CFD program developer. This code is originated from the object-oriented programming (OOP) concept based on C++ program language. Solvers and libraries are defined as C++ classes. With the post processor ParaView, the graphical visualization becomes possible with a command paraFoam [15]. In this study, OpenFOAM version 2.3.1 (The OpenCFD Ltd., London, UK) is used. The numerical calculations are conducted with two OpenFOAM standard solver, "buoyantBoussinesqSimpleFoam" and "buoyantBoussinesqPimpleFoam".

The governing equations of the solvers are incompressible continuity equation, the Navier-Stokes equations and energy equation for the heat transfer where the buoyant force is related in the source term in the momentum equation with Boussinesq approximation. The realizable k-ε turbulence model is also applied for the low Reynolds number turbulent flow in the moderator. Computation is performed with two stages to save the settling time for the pseudo-steady state: "buoyantBoussinesqSimpleFoam" is for steady flow using Semi-Implicit Method for Pressure-Linked Equations (SIMPLE) algorithm, while "buoyantBoussinesqPimpleFoam" is for unsteady flow using PISO, Pressure Implicit with Splitting of Operator, and SIMPLE (PIMPLE) algorithm because the latter one is known to be better for the time-accurate computation [15].

2.2. Governing Equations and Discretization

The hydraulic governing equations based on the single-phase incompressible flow are written in the vector form:

$$\nabla \cdot \mathbf{V} = 0 \tag{1}$$

$$\rho \left\{ \frac{\partial \mathbf{V}}{\partial t} + (\mathbf{V} \cdot \nabla) \mathbf{V} \right\} = -\nabla p + \rho \mathbf{g} + (\mu + \mu_t) \nabla^2 \mathbf{V} + \mathbf{f}_V \tag{2}$$

where $\mathbf{V}$, ρ, p are velocity vector, density, and pressure while the constants μ and $\mathbf{f}_V$ are dynamic viscosity and body force per unit volume. Equation (1) is the continuity equation for incompressible flow, and the Navier-Stokes momentum equation, Equation (2) is decoupled from energy equation in the source term, or buoyancy force of Boussinesq approximation, $\mathbf{f}_V \approx -\rho \mathbf{g} \beta (T - T_0)$, where β is the thermal expansion in the unit of $1/K$, and $T - T_0$ is the difference of temperature from the reference condition.

A realizable k-ε model, which is better for rotational flow, is used for the simulation of turbulent flow. This model includes two additional equations in a tensor form:

$$\rho \left\{ \frac{\partial k}{\partial t} + (\mathbf{V} \cdot \nabla) k \right\} = \frac{\partial}{\partial x_j} \left\{ \left(\mu + \frac{\mu_t}{\sigma_k} \right) \frac{\partial k}{\partial x_j} \right\} + P_k + P_b - \rho \varepsilon - Y_M + S_k \tag{3}$$

$$\rho \left\{ \frac{\partial \varepsilon}{\partial t} + (\mathbf{V} \cdot \nabla) \varepsilon \right\} = \frac{\partial}{\partial x_j} \left\{ \left(\mu + \frac{\mu_t}{\sigma_\varepsilon} \right) \frac{\partial \varepsilon}{\partial x_j} \right\} + \rho \, \Phi S \varepsilon - \frac{\rho C_{2\varepsilon} \varepsilon^2}{k + \sqrt{\nu \varepsilon}} + C_{1\varepsilon} \frac{\varepsilon}{k} C_{3\varepsilon} P_b + S\varepsilon \tag{4}$$

where $S = \sqrt{2S_{ij}S_{ij}}$ is the modulus of mean rate-of-strain tensor, and $S_{ij} = \frac{1}{2}\left(\frac{\partial u_j}{\partial x_i} + \frac{\partial u_i}{\partial x_j}\right)$. In Equations (3) and (4), the turbulent eddy viscosity is defined as:

$$\mu_t = \rho C_\mu \frac{k^2}{\varepsilon} \tag{5}$$

where Equation (5) is substitute to Equation (2) for the consideration of turbulence. C_μ is not a constant like that of standard k-ε model but a function of S, ω_k (angular velocity in the reference frame), ε_{ijk} (dissipation tensor), and $\overline{\Omega}_{ij}$ (mean rate-of-rotation tensor), and Φ in Equation (4) is specified as

$$\Phi = \max\left[0.43, \frac{kS}{kS + 5\varepsilon}\right] = \begin{cases} 0.43 & if\ kS/(kS + 5\varepsilon) \leq 0.43 \\ kS/(kS + 5\varepsilon) & else \end{cases} \tag{6}$$

Other coefficients in Equations (3) and (4) are listed as $C_{1\varepsilon} = 1.44$, $C_{2\varepsilon} = 1.9$, $C_{3\varepsilon} = -0.03$, $\sigma_k = 1.0$, $\sigma_\varepsilon = 1.2$. The energy equation to get the temperature field for the computation of f_V in Equation (2) is

$$\rho C_p \left\{\frac{\partial T}{\partial t} + (\mathbf{V} \cdot \nabla)T\right\} = \left(\lambda + \frac{\mu_t}{\Pr_t}C_p\right)\nabla^2 T + Q_s \tag{7}$$

where T is temperature; C_p is heat capacity; λ is thermal conductivity; $\Pr_t$ is turbulent Prandtl number, assumed as a constant of 0.85 for all the range of fluid, and Q_S is volumetric heat source, which should be specified in next section, Equation (8).

The convection terms of the governing equations are discretized with second order upwind scheme and diffusion terms are calculated with second order centered difference scheme. Turbulence equations and heat transfer equation were discretized with first order upwind scheme.

In the computation using OpenFOAM, SIMPLE algorithm, a kind of finite volume method (FVM) is applied for the iteration until the steady state for Equations (1) and (2). In this method, the pressure gradient term in Equation (2) is isolated, and sub-iterations should be performed between predictor and corrector [14]. The PIMPLE method is used for unsteady time marching, which is specified as no under-relaxation and multiple corrector steps in the calculation of momentum. PIMPLE is far accurate in time and applied to the unsteady computation instead of SIMPLE.

2.3. Boundary and Initial Conditions

The essential boundary conditions in this problem are listed as follows:

- Velocities: no-slip conditions at walls, and the mass flow rate is specified on the inlet, fixed to 127.4 kg/s per each inlet nozzle, or 1019 kg/s in total for the present problem. The inlet turbulent intensity is fixed as 5%, which can make the additional uncertainty for the turbulent flow linked with the full system;
- Pressure: zero pressure gradient conditions at walls and inlet, which should be valid under the assumption that the thickness of boundary layer is very thin. The outlet pressure is fixed by the moderator system;
- Temperature: the inlet temperature is fixed to 47.3 °C.

Total thermal power exerted to the whole system is 100 MW, which should be processed as the source term, Q_S in Equation (7) where the factor 1.089 (of course, the volume blockage of tubes is considered). The equivalent temperature, or the energy dived by density and heat capacity, should be considered in the energy equation of OpenFOAM where the temperature should be specified instead of power. The power distribution is defined as $Q_s(r,z) = Q_s f_r(r) f_z(z)$, and the shape functions are, in the dimensionless form [18,19],

$$f_r(r) = 0.94588 - 0.01989r + 0.0995r^2 - 0.03888r^3 - 0.00256r^4\ (0.0 \leq r[1/m] \leq 3.8)$$
$$f_z(z) = 1.0 - 0.1111z^2\ (-3.0 \leq z[1/m] \leq 3.0) \tag{8}$$

where Equation (8) is obtained from group distributions of fuel bundles measured from the plant data in a Wolsong PHWR [10], and the correlation is regressed with a fourth-order least-square curve.

The initial temperature of the whole computational domain is 47.3 °C, and the flow is assumed stationary in the beginning of computation. Actually, the CANDU-6 moderator is liked with the system network, but we did an independent simulation for the moderator only. The properties of the fluid (D_2O) for simulations are summarized in Table 1.

Table 1. Material properties of the heavy water.

Definition (Symbol)	Symbol	Value	Unit
Density	ρ	1085	kg/m^3
Thermal expansion	β	5×10^{-4}	K^{-1}
Dynamic viscosity	μ	5.5×10^{-4}	$kg/(m{\cdot}s)$
Heat Capacity	C_p	4207	$J/(kg{\cdot}K)$
Thermal conductivity	λ	0.659	$W/(m{\cdot}K)$

2.4. Grid Generation

The prototype of CANDU-6 is such as Figure 2. The 380 circular rods called calandria tubes are allocated symmetry from the central line of tank; the inlet holes are four along each side part, i.e., eight in total with feeding nozzles consisting of four radial diffusers; and there are two outlet exits at the bottom. This prototype has an asymmetric shape for the cross section along the longitudinal direction because the outlet vent hole is tilted from the vertical midline.

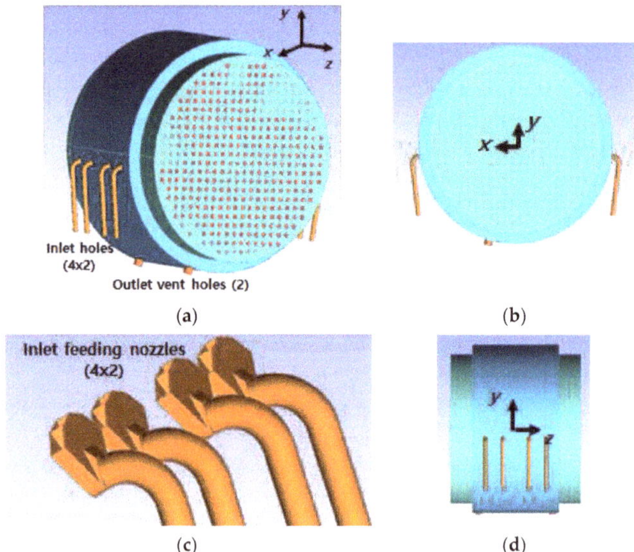

Figure 2. 3-D modeling of the prototype: (**a**) 3-D shape; (**b**) axial view; (**c**) feeding nozzles; (**d**) side view.

Figure 3 shows the three-dimensional unstructured grids at the view of lateral and longitudinal direction. The total grids are 6,740,446 consisting of 5,112,270 for the hexahedral, 13,112 for pyramids, and 1,615,064 for the tetrahedral. They are concentrated at the wall boundary with 15 stretched layers to increase the accuracy in turbulent boundary layers. The computation is done with a message passing interface (MPI) parallel machine where 24 processors are used. Each computational result is stored at multiple folders to assemble them in the post processor.

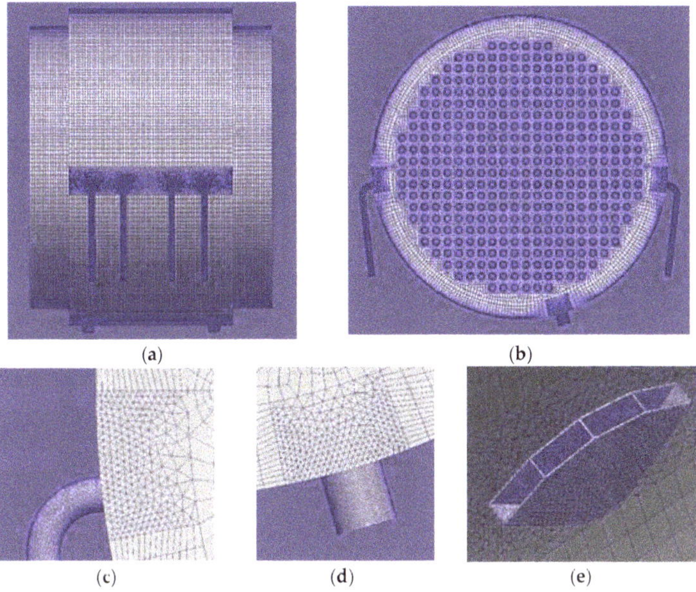

(a)

(b)

(c)   **(d)**   **(e)**

Figure 3. Grids of the prototype: (**a**) side longitudinal; (**b**) planform sectional; (**c**) inlet plumbing; (**d**) outlet exit; and (**d**) feeding nozzle.

3. Result and Discussion

The numerical method is verified and validated in the previous research by the authors [14,16]. The pressure drop with comparison of STERN laboratory experiment shows an error within 16.3% from the experimental data [16] (see Figure 4). The pressure drop is measured for isolated four-row bank of aligned cylinders of 33.02 mm diameter and 71.4 mm spacing. The pressure sensors are in the distance of sixteen blocks of cylinders along the central axis. Three sets of experiments are used for this comparison, specified with the Reynolds number based on the tube diameter, $Re_d = \rho V d / \mu$. Among various codes such as ANSYS-CFX and COMSOL, the open source code OpenFOAM displayed similar or better level of coincidence for all kinds of turbulence models, and k-ε model was the best result. The modeling of two-dimensional heat flow can predict the temperature with a maximum local error of 3.5 °C, which can be a reduced model of CANDU-6 moderator [14].

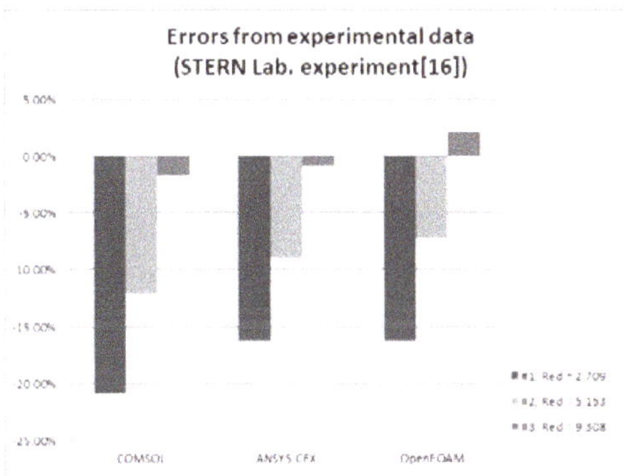

Figure 4. Comparison of numerical results from various codes for the experiment of STERN laboratory.

3.1. Quasi-Steady State

The solution is not converged to a steady state with segregated solvers such as SIMPLE or PIMPLE, but instead it fluctuates with oscillation [2]. In the earlier stage, steady solution is obtained with SIMPLE algorithm. After the computation is stabilized, in the later stage, the solution is time-marched to get the unsteady one. Figure 5 is the temperature at two outlets and the origin of (a) quasi-steady, or the center of calandria, before 12,000 time steps and (b) unsteady procedure to 850 s of physical time. The temperature of two outlets are slightly different from each other because of the asymmetry from flow instability. At the center, the time-averaged temperature is about 85 °C.

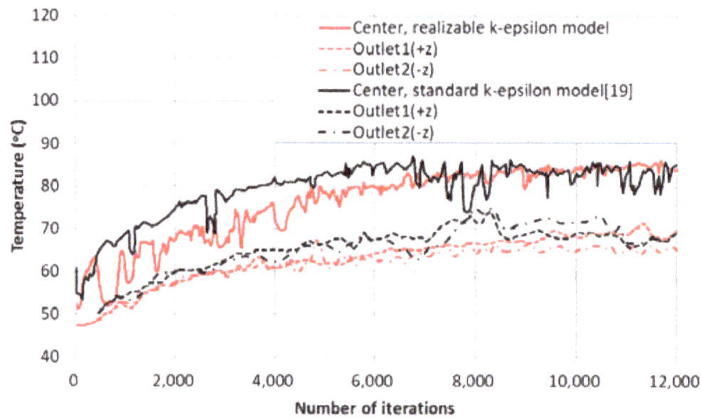

(**a**) Earlier stage, or steady solution.

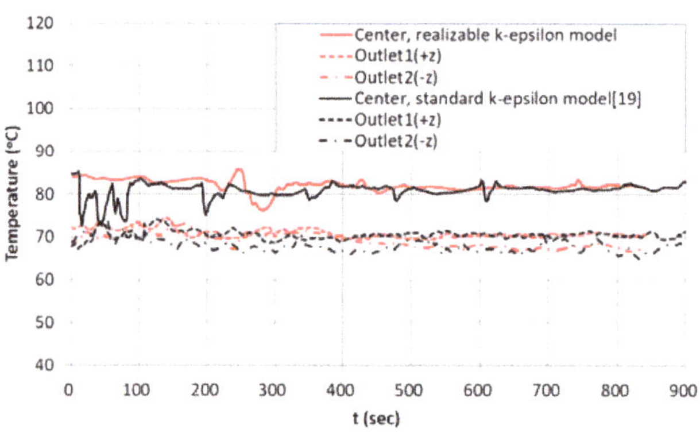

(**b**) Later stage, or unsteady solution.

Figure 5. Temperature at the center and two outlets.

3.2. Turbulent Model and Scale

In Figure 5, the present computational results are compared with those from the standard k-ε model [19]. With the realizable k-ε turbulent model, the convergence is slower than that with standard one in Figure 5a but shows overall better stable temperature in the pseudo-steady stage in Figure 5b. The inlet turbulence intensity is fixed to 5% [16].

To show how the grid system in Figure 4 can capture turbulent physics in a proper scale, the normalized wall distance, y^+ is plotted in Figure 6, which is ranged widely. At the outer wall, the maximum y^+ exceeds 100 where the fast waterjet sweeps injected from the nozzles. The y^+ is distributed from 10 to 60 at the cylinder walls. However, with the use of the wall function, the value of $y^+ < 80$ should be enough in the most of computational domain of the present problem since the value less than 30 can often make the turbulent wake flow unstable even though that at the tube wall boundary must be maintained near unity [18].

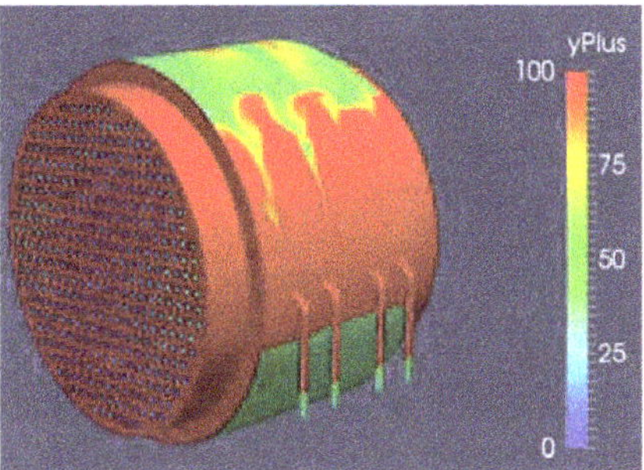

Figure 6. Distribution of y^+ at 840 s.

3.3. Velocity and Temperature Fields in the Unsteady Solution

After the quasi-steady state after 12,000 iterations, the time is reset to zero, and the fields of velocity and temperature are visualized in Figures 7–10 from 615 to 840 s.

Figures 7 and 8 are plotted at the sectional plane $z = 0$ (x-y plane), and the change of velocity and temperature are observed in the series of figures, respectively. In Figure 7, the cooling water from nozzles, initially to the upper direction or the positive y-axis, in both sides meets at a stagnation point, denoted with S in the upper right-hand side, the same tilt direction of vent hole. Please note that it is not symmetric. The flow field seems to be periodic for 225 s time difference. However, the temperature field in Figure 8 presents much more turbulent diffusion, so it becomes very difficult to find the obvious regularity. The period is not resolved from the figures, but the similar flow patterns are repeated with time passing: the cooling waterjet falls from the stagnation point, soaked into gaps of cylinders until the outflow at the vent hole. The heated water maintains balance of temperature at the upper region of tube bundles because of the buoyancy in the momentum transfer. The flow velocity is very slow less than 1 m/s in most of the domain, and no local region is found for the rapid increase of temperature thanks to the mixing of diffusive turbulent flow.

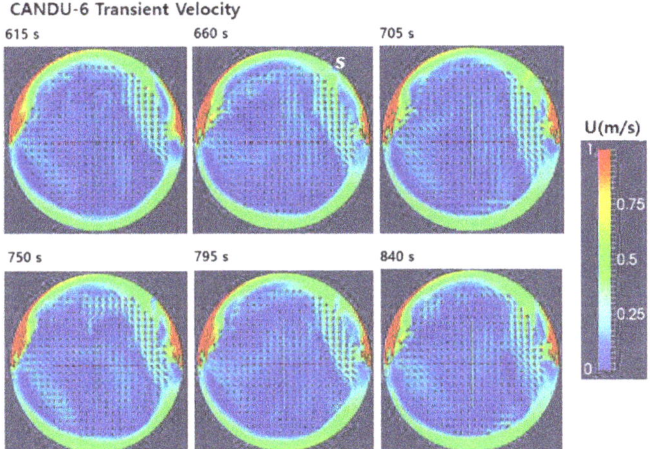

Figure 7. Velocity distribution at $z = 0$; 615–840 s.

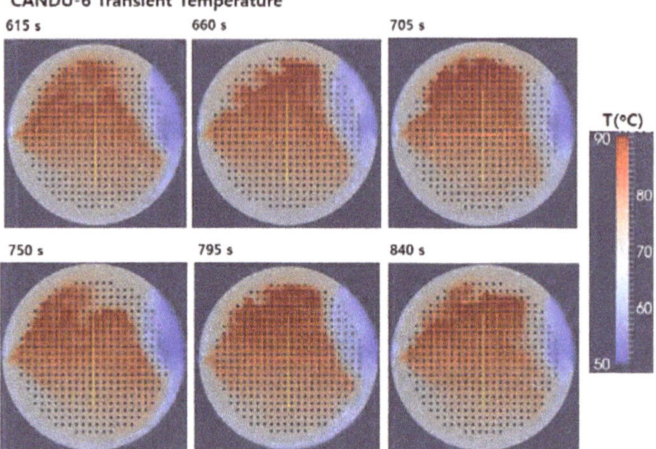

Figure 8. Temperature distribution at $z = 0$; 615–840 s.

Figures 9 and 10 are plotted at the sectional plane $x = 0$ (longitudinal), and the flow is not simple, too. In Figure 9, the flow velocity is so slow, but the marks of calandria tubes are dimly visible like stripes as they decelerate the circulation flow from the no-slip boundary condition. The maximum temperature stays about 89 °C in Figure 10, and cannot be found the region of successive increase of temperature.

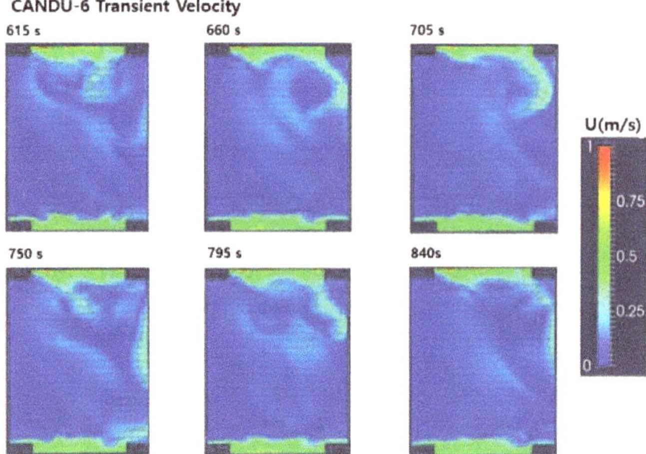

Figure 9. Velocity distribution at $x = 0$; 615–840 s.

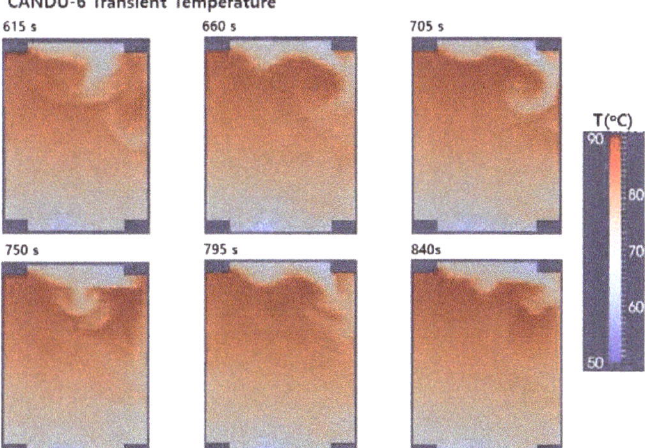

Figure 10. Temperature distribution at $x = 0$; 615–840 s.

The velocity distribution in Figures 7 and 9 show obviously that the flow circulation penetrates the interval of circular rods decelerating the flow with a pressure drop. The diffused flow makes the temperature increase at the upper central region in Figure 8 because flow resistance takes the worse cooling efficient. In the temperature field view of Figure 8, the largest turbulent eddies can be discerned at the interface of different temperature at the upper half plane such as the mushroom shape. They merge and separate continuously, developing a highly complex turbulent structure, so the high temperature difference of about 20 °C is dramatically visualized in both Figures 8 and 10.

In Figure 11, the mean inlet velocity at the nozzle is approximately 2 m/s, and speed at the central section is slower than the side one where a nozzle consists of four sections because the expansion ratio is greater. This fact compensates for the inlet jet flow to maintain a uniform flow along the curve of outer wall, approximately.

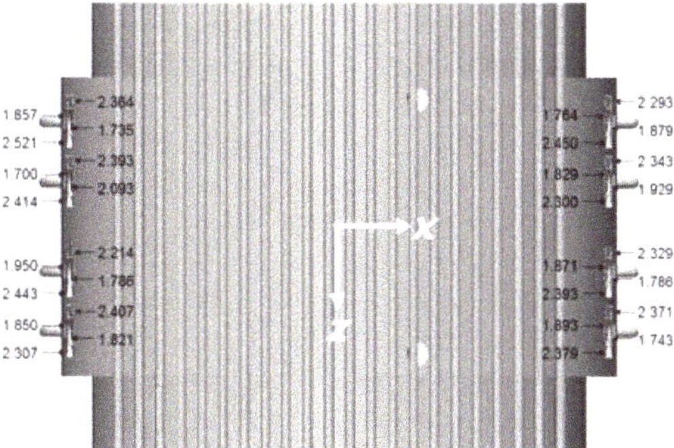

Figure 11. Mean velocities at the inlet nozzle, final time of the simulation (top view, unit: m/s).

3.4. Validation of Numerical Data

In Figure 12, the vertical axis at the center is plotted on the temperature for the last one of Figures 8 and 10 at 840 s. As we had no measures data for the prototype CANDU-6, the temperature distribution is normalized with the reduced-scale model test [14], and compared with other methods of computation as well as a set of coarse but experimental data: the numerical data from ANSYS-CFX, and

MODTURC-CLAS [5,6], etc. Although position and temperature are normalized, the circulating flow derived from buoyancy force reaches the equilibrium of maximum temperature at $0.25 < y/D < 0.3$.

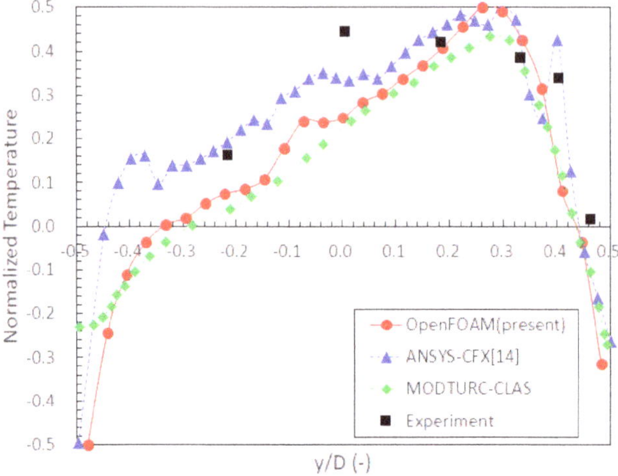

Figure 12. Validation of numerical data with other codes and experiment for the normalized temperature distribution along the vertical axis.

4. Conclusions

A prototype of CANDU-6 reactor is numerically analyzed around a three-dimensional moderator flow in calandria tank with OpenFOAM, an open-source CFD code. The three-dimensional shape including 380 rods in the calandria tank is precisely modeled without porous approximation to avoid parasite errors. The buoyancy term in the incompressible Navier-Stokes equation is considered with Boussinesq approximation of the temperature variation. Turbulence effect is reflected to energy equation as well as momentum equation with the realizable k-ε model.

The computational result shows that there should be no steady solution about the circulation flow, and therefore the unsteady simulation is achieved after getting a quasi-steady with oscillation of flow properties. The flow field is not converged to a steady solution. Instead, it oscillates in the regime of quasi-steady state. After 12,000 iterations from initial condition to the quasi-steady state, the unsteady simulation within 840 s shows no evidence of exact periodic oscillation for physical properties. The observation for 225 s, an approximate period of flow pattern, presents a complex structure of turbulent mixing despite uncertainties originated from the high intensity of turbulence. There are no regions where the temperature rises more than 90 degrees Celsius due to very slow transferring flow. Most of computational region marks the velocity less than 1 m/s. As the inlet nozzle flow going down from the stagnation point, it is highly diffused with the pressure drop due to the calandria tubes. Turbulent eddies were found in the temperature field, continuously developing to merge or separate at the interface of hot and cool fluid. The dimensionless wall distance of the first grid from wall, y^+ was checked as less than 80 in the most of computational domain but should be reduced with finer grids free of wall functions, especially for the outer wall of calandria tank.

Overall, this research presents that the use of open-source software is also very feasible for the application of analysis on the moderator system of PHWR such as CANDU-6. Compared with other commercial codes, the equivalent computation could be obtained from cheaper price and free copyright. However, the use of CFD alone provides a limited perspective. In practice, the CFD boundary condition should be supported by system analysis for possible transient phenomena.

Author Contributions: Conceptualization, H.T.K. and S.-M.C.; Methodology, S.-M.C.; Software, S.-M.C.; Validation, H.T.K., S.-M.C. and Y.W.S.; Formal Analysis, Y.W.S.; Investigation, H.T.K.; Resources, H.T.K.; Data Curation, H.T.K. and S.-M.C.; Writing—Original Draft Preparation, H.T.K.; Writing—Review & Editing, S.-M.C.; Visualization, S.-M.C.; Supervision, S.-M.C.; Project Administration, S.-M.C.; Funding Acquisition, H.T.K.

Funding: This work was supported by the National Research Foundation of Korea (NRF) grant funded by the Korea government (Ministry of Science, ICT, and Future Planning) (No. NRF-2016R1D1A3-B01015543), and the Human Resources Development Program (Grant No. 20174010201350) of the Korea Institute of Energy Technology Evaluation and Planning (KETEP) grants funded by the Korea government (Ministry of Trade, Industry, and Energy).

Conflicts of Interest: The authors declare no conflict of interest. The funders had no role in the design of the study; in the collection, analyses, or interpretation of data; in the writing of the manuscript, and in the decision to publish the results.

References

1. Wolsong Units. *2/3/4 Final Safety Analysis Report*; Korea Electric Power Corporation: Naju, Korea, 1995; Chapter 15.
2. Gillespie, G.E. An Experimental Investigation of Heat Transfer from a Reactor Fuel Channel: To Surrounding Water. In Proceedings of the 2nd Annual Conf. Canadian Nuclear Society, Ottawa, ON, Canada, 10 June 1981.
3. Fan, H.Z.; Aboud, R.; Neal, P.; Nitheanandan, T. Enhancement of the Moderator Subcooling Margin using Glass-peened Calandria Tubes in CANDU Reactors. In Proceedings of the 30th Annual Conference of the Canadian Nuclear Society, Calgary, AB, Canada, 31 May–3 June 2009.
4. Huget, R.G.; Szymanski, J.; Midvidy, W. Status of Physical and Numerical Modelling of CANDU Moderator Circulation. In Proceedings of the 10th Annual Conference of the Canadian Nuclear Society, Ottawa, ON, Canada, 4–7 June 1989.
5. Huget, R.G.; Szymanski, J.; Galpin, P.F.; Midvidy, W.I. MODTURC-CLAS: An Efficient Code for Analyses of Moderator Circulation in CANDU Reactors. In Proceedings of the 3rd International Conference on Simulation Methods in Nuclear Engineering, Montreal, QC, Canada, 18–20 April 1990.
6. Khartabil, H.F.; Inch, W.W.; Szymanski, J.; Novog, D.; Tavasoli, V.; Mackinnon, J. Three-dimensional moderator circulation experimental program for validation of CFD code MODTURC_CLAS. In Proceedings of the 21th CNS Nuclear Simulation Symposium, Ottawa, ON, Canada, 24–26 September 2000.
7. Kim, H.T.; Rhee, B.W. Scaled-down moderator circulation test facility at Korea Atomic Energy Research Institute. *Sci. Technol. Nucl. Install.* **2016**, *2016*, 5903602. [CrossRef]
8. Im, S.; Kim, H.T.; Rhee, B.W.; Sung, H.J. PIV measurement of the flow patterns in a CANDU-6 model. *Ann. Nucl. Eng.* **2016**, *98*, 1–11. [CrossRef]
9. Hadaller, G.I.; Fortman, R.A.; Szymanski, J.; Midvidy, W.I.; Train, D.J. Frictional Pressure Drop for Staggered and In Line Tube Bank with Large Pitch to Diameter Ratio. In Proceedings of the 17th Annual Conference of the Canadian Nuclear Society, Fredericton, NB, Canada, 9–12 June 1996.
10. Yoon, C. Development of a CFD Model for the CANDU-6 Moderator Analysis Using a Coupled Solver. *Ann. Nucl. Eng.* **2007**, *35*, 1041–1049. [CrossRef]
11. Sarchami, A.; Ashgriz, N.; Kwee, M. Three Dimensional Numerical Simulation of a Full Scale CANDU Reactor Moderator to Study Temperature Fluctuations. *Int. J. Eng. Phys. Sci.* **2012**, *6*, 275–281. [CrossRef]
12. Teyssedou, A.; Necciari, R.; Reggio, M.; Zadeh, F.M.; Étienne, S. Moderator Flow Simulation around Calandria Tubes of CANDU-6 Nuclear Reactors. *Eng. Appl. Comput. Fluid Mech.* **2014**, *8*, 178–192. [CrossRef]
13. Gim, G.H.; Chang, S.M.; Lee, S.; Jang, G. Fluid-Structure Interaction in a U-Tube with Surface Roughness and Pressure Drop. *Nucl. Eng. Technol.* **2014**, *46*, 633–640. [CrossRef]
14. Kim, H.T.; Chang, S.M. Computational Fluid Dynamics Analysis of the Canadian Deuterium Uranium Moderator Tests at the Stern Laboratories Inc. *Nucl. Eng. Technol.* **2015**, *47*, 284–292. [CrossRef]
15. OpenFOAM User Guide; CFD Direct Ltd.: 2019. Available online: https://openfoam.com/documentation/user-guide/ (accessed on 20 January 2019).
16. Kim, H.; Chang, S.M.; Shin, J.H.; Kim, Y.G. The Feasibility of Multidimensional CFD Applied to Calandria System in the Moderator of CANDU-6 PHWR Using Commercial and Open-Source Codes. *Sci. Technol. Nucl. Install.* **2016**, *2016*, 3194839. [CrossRef]

17. Hanna, B.N. CATHENA: A thermalhydraulic code for CANDU analysis. *Nucl. Eng. Des.* **1998**, *180*, 113–131. [CrossRef]
18. Seo, Y.S.; Chang, S.M.; Yeom, G.S. *CFD Analysis on the Validation Experiment with MCT 1/4 Model*; KAERI Report; Mirae Engineering Co.: Jeonju, Jeonbuk, Korea, 2015.
19. Kim, H.T.; Chang, S.M. OpenFOAM Analysis of CANDU-6 Moderator Flow. In Proceedings of the Transactions of the Korean Nuclear Society Autumn Meeting, Gyeongju, Korea, 29–30 October 2015.

Article

Visualization Study on Thermo-Hydrodynamic Behaviors of a Flat Two-Phase Thermosyphon

Chao Wang [1], Feng Yao [1,2,*], Juan Shi [3], Liangyu Wu [1,*] and Mengchen Zhang [3]

[1] School of Hydraulic, Energy and Power Engineering, Yangzhou University, Yangzhou 225127, China; cwang@microflows.net

[2] Jiangsu Key Laboratory of Micro and Nano Heat Fluid Flow Technology and Energy Application, School of Environmental Science and Engineering, Suzhou University of Science and Technology, Suzhou 215009, China

[3] Key Laboratory of Energy Thermal Conversion and Control of Ministry of Education, School of Energy and Environment, Southeast University, Nanjing, Jiangsu 210096, China; 158065072@seu.edu.cn (J.S.); 220130421@seu.edu.cn (M.Z.)

* Correspondence: yaofeng@mail.usts.edu.cn (F.Y.); lywu@yzu.edu.cn (L.W.); Tel.: +86-514-8797-1315 (F.Y.)

Received: 20 July 2018; Accepted: 30 August 2018; Published: 31 August 2018

Abstract: The coupled effect of boiling and condensation inside a flat two-phase thermosyphon has a non-negligible influence on the two-phase fluid flow behavior and heat transfer process. Therefore, a flat two-phase thermosyphon with transparent wall was manufactured. Based on this device, a visualization experiment system was developed to study the vapor–liquid two-phase behaviors and thermal performance of the flat two-phase thermosyphon. A cross-shaped wick using copper mesh was embedded into the cavity of two-phase thermosyphon to improve the heat transfer performance. The effects of heat flux density, working medium, and wick structure on the thermal performance are examined and analyzed. The results indicated that a strong liquid disturbance is caused by the bubble motions, leading to the enhancement of both convective boiling and condensation heat transfer. More bubbles are generated as the heat flux increases; therefore, the disturbance of bubble motion on liquid pool and condensation film becomes stronger, resulting in better thermal performance of the flat two-phase thermosyphon. The addition of the wick inside the cavity effectively reduces the temperature oscillation of the evaporator wall. In addition, the wick structure provides backflow paths for the condensate owing to the effect of capillary force and enhances the vapor–liquid phase change heat transfer, resulting in the improvement of thermal performance for the flat two-phase thermosyphon.

Keywords: thermosyphon; phase change; two-phase flow; visualization

1. Introduction

Highly efficient cooling technologies have always been the subject of both scientific and engineering investigation in high-powered electronics circuits. Aiming to achieve highly efficient heat dissipation of high-heat-flux electronics, several advanced cooling technologies, including boiling cooling [1–3], liquid cooling [4,5], functional surface [6,7], microchannels heat sink [8–10], heat pipes [11–13], microfluidic engineering [14,15], metal foam [16], etc., have been introduced and applied in every field. Among these advanced cooling technologies, the heat pipes (such as grooved heat pipes [17], pulsation heat pipes [18], thermosyphons, etc.) are most widely used for the heat dissipation of microelectronic devices under high heat flux density due to high heat transfer capacity, good temperature uniformity, and no power consumption [19]. Among those heat pipes, the flat two-phase thermosyphon possesses higher heat transfer limit and superior temperature uniformity over other heat pipes, due to the operation principle of gravity-driven vapor–liquid flow and boiling phase change heat transfer [20,21]. The flat two-phase thermosyphon is considered a highly efficient heat spreader and shows good application prospects in solving uniform heat dissipation under high-heat-flux electronics [22–26]. Therefore, study on the flat two-phase thermosyphon's thermal properties is of significance and has become a topic of growing technological interest during the past decades.

It should be noted that the flat two-phase thermosyphon is a confined structure, where the vapor–liquid two-phase flow self-circulates between the evaporator and condensation surface. The confinement effect in turn affects the vapor–liquid flow dynamics of the flat heat pipe. The previous investigation indicated that the bubbles generated by the evaporation surface survive on the liquid surface for a while and continue to grow, merge, and burst when the liquid level inside the cavity is low [27]. In addition, the rapid rise of bubbles can promote the liquid convection heat transfer on the evaporator surface [28,29]. Liu et al [30] add a hollow foam metal block into the cavity of a flat two-phase thermosyphon; they observed that the dense bubbles generated in the foam metal fly to the condenser surface and break, and the liquid is caused to splash to the condenser surface by the intense movement of the bubbles under the optimal filling rate. As a result, the convective boiling and film condensation heat transfer are enhanced. In summary, the coupled effect of boiling and condensation inside the flat two-phase thermosyphon has a non-negligible influence on the two-phase fluid flow behavior and heat transfer process.

Although several experimental attempts have been devoted to investigating the thermal performance of flat two-phase thermosyphons, special attention is paid to thermal performance indices, equivalent thermal conductivity, and the maximum heat transfer capacity [31,32]. For example, Lips et al. [33] found that the thermal resistance of the flat two-phase thermosyphon reduces due to the liquid retention induced by the small cavity height. Ju et al. [34] developed a flat two-phase thermosyphon incorporating hybrid wicks, which is composed of a spreading layer with low thermal resistance and a supply structure for the working medium; then, a low thermal resistance and high heat flux limit over large heating areas can be achieved. In addition, the vapor–liquid two-phase flow behaviors are complex due to the interaction between the boiling and condensation. The thermo-hydrodynamic characteristics of the two-phase thermosyphon are still waiting to be explored by visualization experiment. Still, there are few studies have been conducted to visually study the internal vapor–liquid two-phase flow behaviors and thermal performance.

Apart from experimental investigation, limited theoretical attempts on vapor–liquid two-phase flow and phase change heat transfer in the two-phase thermosyphon have been conducted. Wu et al. investigated the co-existing boiling and condensation heat transfer in a confined space using lattice Boltzmann simulation [35]. It is indicated that the interaction of condensation and boiling in a two-phase thermosyphon is embodied in the condensate droplet contact the growing bubble that contributes to bubble detachment or motion.

The heat is mainly transferred by the process of boiling and condensation inside the flat two-phase thermosiphon. However, these two processes coexist in the confined cavity and interact with each other, resulting complex vapor–liquid two-phase flow behaviors. Compared with the study of boiling or condensation in a confined space separately, the research of the coupled effect of boiling and condensation is more helpful for understanding of the thermo-hydrodynamic behaviors inside a confined space. Therefore, an experimental system is developed to visually study the vapor–liquid two-phase flow and phase change heat transfer of the thermosyphon, especially to elucidate the coupled boiling and condensation heat transfer. Through the experiment, the effects of heat flux density, working medium, and wick structure on the vapor–liquid two-phase flow behavior and thermal performance are examined and investigated.

2. Description of Experiment

2.1. Experimental Apparatus

The heat transfer performance test system is shown in Figure 1. As shown, two electric heating rods (6 mm in diameter and 50 mm in length) are embedded inside the bottom side of the copper column (20 mm in diameter and 105 mm in length). The top side of the copper column is closely contacted with the evaporator surface of the flat two-phase thermosyphon. The heating power of the electric heating rods is provided by the DC power supply. The working condition adjustment is realized by adjusting the voltage regulator and the power meter. The electric heating power ranges from 20 W to 90 W in the experiment. The constant-temperature water bath provides cooling water with a constant temperature (± 0.1 °C of temperature fluctuation) for the condenser section of the flat two-phase thermosyphon. The flow meter is used to control the flow rate of the circulating water. In the experiment, the temperature of the cooling water was set to room temperature of 25 °C, and the circulating water flow rate was set to 80 mL/min.

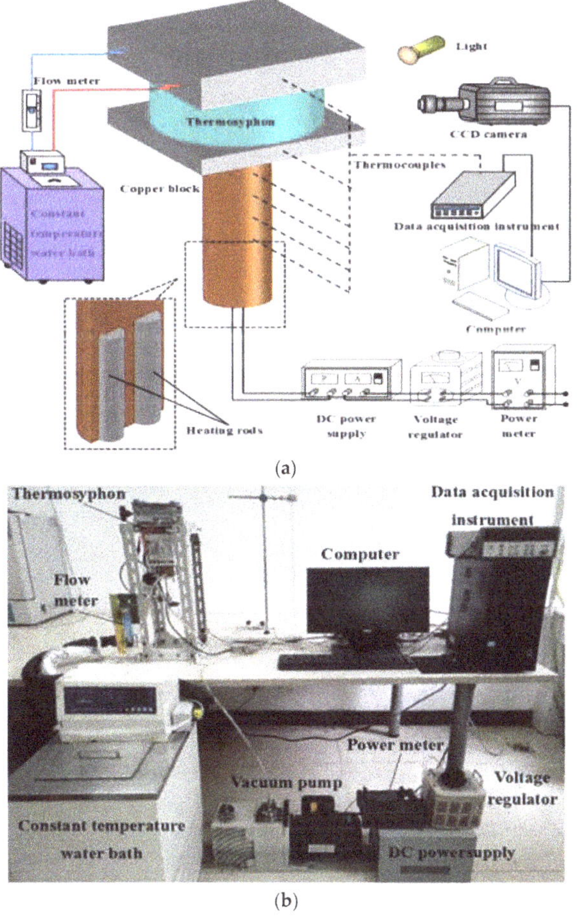

Figure 1. Schematic of experimental setup: (**a**) system diagram; (**b**) experimental rig.

As shown in Figure 2, the flat two-phase thermosyphon is assembled using an evaporator plate, a transparent quartz glass tube, a condenser plate, and a cooling water tank sealing plate. The glass tube is of 50 mm in outer diameter, 44 mm in inner diameter, and 3 mm in thickness. Both the evaporator and condenser plate are made of square brass plates with 54 mm in side length. The thicknesses of the evaporator and condenser plate are 5 mm and 10 mm, respectively. The evaporator and condenser surfaces are a smooth plane with a milled annular groove. Both ends of the glass tube are tightly embedded in the groove on the plate, so a closed cavity is formed. The cooling water tank is set at the back of the condenser plate and sealed by a sealing plate. The working medium is charged into the cavity through a charging hole drilled in the condenser plate. In the experiment, unless specified, deionized water was selected as the working liquid.

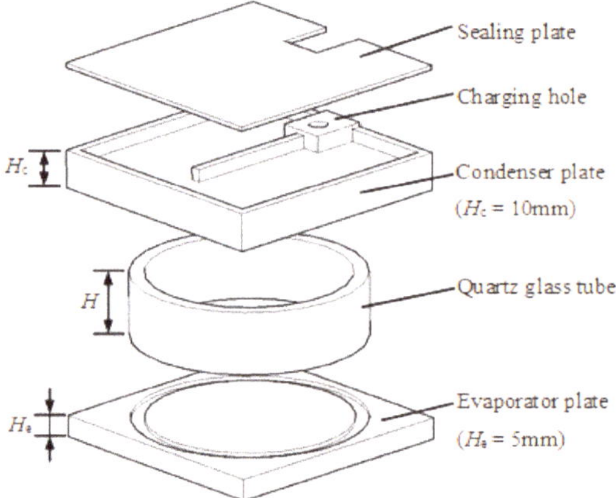

Figure 2. Schematic of the flat two-phase thermosyphon.

The CCD camera and light are used to monitor the vapor–liquid two-phase flow behavior in the cavity, and the thermocouples and data acquisition instrument are used to test the temperature changes during the experiment. Both the images and the tested temperature are recorded on the computer. In order to monitor the working conditions of the testing system, thermocouples are arranged in all parts of the system. Four thermocouple holes with a diameter of 0.5 mm and a depth of 10 mm are arranged in the upper side of the copper column, with respective distances of 5 mm, 20 mm, 35 mm, and 50 mm from the upper surface of the copper column. During operation, the heat load of the copper column imposed on the heat pipe is obtained according to the axial temperature distribution measured by the thermocouples in the copper column. The thermocouples are also arranged at the inlet and outlet of the cooling water tank for measuring the temperature change of the circulating water before and after the condenser section. Through the sensible heat gain of circulation fluid, the heat load input into the heat pipe is also obtained. The experiment test indicated that the difference of the heat load obtained by above two approaches is within 3%. Furthermore, six thermocouple holes have been arranged in the evaporator and condenser plate with the locations shown in Figure 3, so that the temperature changes of the evaporator and condenser surfaces of the flat two-phase thermosyphon can be monitored.

Through the performance test on the above experimental setup, it was indicated that there is an obvious temperature oscillation during the operation of the flat two-phase thermosyphon owing to the intermittent boiling. The presence of temperature oscillation is unfavorable for the application of flat two-phase thermosyphons. In order to eliminate the temperature oscillation, a cross-shaped wick formed by folding 200-mesh copper screen was installed into the cavity of the flat two-phase thermosyphon, as shown in Figure 4. The wick height is 15 mm, which is the same as the cavity height.

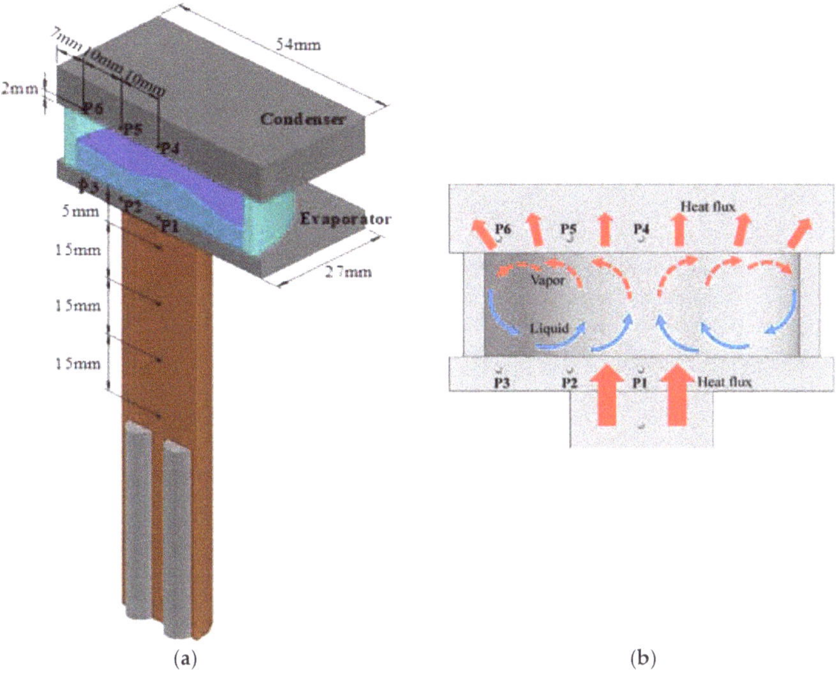

Figure 3. Schematic of temperature measuring points: (**a**) thermocouple positions; (**b**) diagram of temperature measuring in the two-phase thermosyphon.

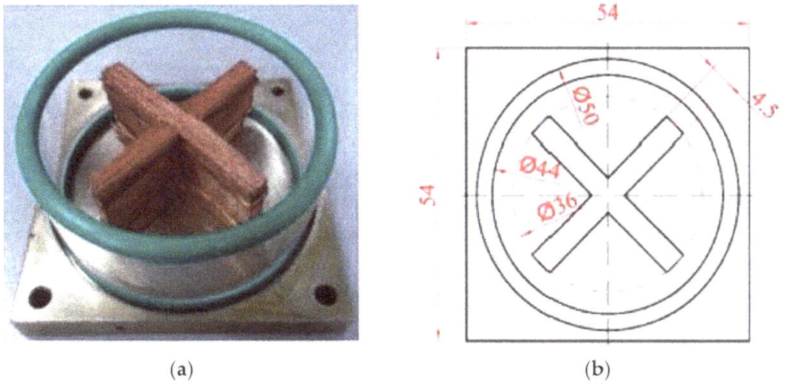

Figure 4. Structure of the cross-shaped wick: (**a**) picture of cross-shaped wick; (**b**) dimensions of cross-shaped wick.

2.2. Data Reduction

The total thermal resistance of the flat two-phase thermosyphon characterizes the temperature difference required for transporting unit heat power, which is an important parameter for measuring the heat transfer performance of the flat two-phase thermosyphon. In this paper, the thermal resistance is adopted to quantitatively evaluate the heat transfer performance of the flat two-phase thermosyphon, which is defined as

$$R = \Delta \overline{T}/Q \qquad (1)$$

where $\Delta \overline{T}$ is the average temperature difference between the evaporator section and the condenser section, as follows:

$$\Delta \overline{T} = \overline{T}_e - \overline{T}_c$$

$$\overline{T}_e = \frac{T_1 + T_2 + T_3}{3}, \ \overline{T}_c = \frac{T_4 + T_5 + T_6}{3}$$

where R is the total thermal resistance, and $\overline{T}_e$ and $\overline{T}_c$ represent the average temperatures of the evaporator section and condenser section, respectively. The subscripts 1–6 represent the temperature measurement points of the evaporator and condenser sections (P1–P6, see Figure 3). Q is the heat load. This definition characterizes the resistance of heat transfer from the evaporator surface to the condenser surface, and smaller thermal resistance means better heat transfer performance of the flat two-phase thermosyphon. The heat load can be obtained as $Q = \rho G C_p(T_{\text{out}} - T_{\text{in}})$, where ρ is density, G is the volume flow rate, C_p is specific heat, and T_{out} and T_{in} are the outlet temperature and inlet temperature, respectively, of the cooling water. The uncertainties of $\Delta \overline{T}$, Q, and R are 0.8%, 2.62%, and 2.71% according to the data error analysis.

2.3. Experiment Procedures

Before each experiment, the evaporation surface and the condensation surface were polished with metallographic abrasive paper. Then, in order to remove copper rust and dust, the surfaces were cleaned using deionized water, absolute ethanol, acetone, and deionized water in sequence. After ensuring that the evaporation plate, the condensation plate, and the quartz glass tube were clean, the two-phase thermosyphon was assembled. Then, the two-phase thermosyphon was subjected to vacuuming, liquid filling, and leak detection.

In order to ensure good repeatability of the experimental results, the experiment steps are as follows:

(1) Connect each unit of the experiment system, and then perform circuit and connection checks on every unit to ensure that the experimental system is correct.
(2) Open constant-temperature water bath, and adjust the temperature and flow rate to the preset values. Turn on the high-speed CCD camera to debug the shooting effect.
(3) Turn on the data acquisition instrument, and observe the temperature change until the temperature at each measuring point reaches a steady state.
(4) Maintain the operation of every other piece of equipment, reset the data acquisition instrument, and record the temperature data of the temperature measurement point.
(5) Observe and record the gas–liquid two-phase behavior in the confined space during the start-up and quasi-steady operation in real time. Collect the temperature data. When the change of temperature is less than 0.5 °C over 20 min, the heat balance can be considered to be reached, and the experiment for the current working condition can be finished.
(6) After each working condition is completed, adjust the input voltage to zero, and keep the constant-temperature water bath cooling water circulating until the temperature falls back to the initial temperature, then perform the experiment on the next working condition.

3. Results and Discussion

When there is a heat load imposed on the flat two-phase thermosyphon, the gravity-driven vapor–liquid two-phase self-circulation flow is generated in the cavity, accompanied by sensible heat transport and latent heat transport (evaporation/boiling and condensation). During this vapor–liquid phase change process, the evaporation, boiling, and condensation phase change, vapor–liquid two-phase flow, and phase interface fluctuations are very complex and coupled with each other in the flat two-phase thermosyphon. This paper began an observation of the coupled boiling and condensation phase change inside the confined cavity, and now the effects of heat flux, working medium, and wick structure on two-phase flow behaviors and thermal performance of the flat two-phase thermosyphon are examined and analyzed.

3.1. Coupled Boiling and Condensation Phase Change

The nucleate boiling occurs on the evaporator section as soon as a certain degree of superheat is achieved when the liquid continues to be heated in a two-phase thermosyphon. Figure 5 shows the typical process of nucleate boiling, namely, the bubbles generate on the evaporator surface, grow, and then detach from the surface, accompanying intense gas–liquid two-phase flow and phase change heat transfer in the cavity. As shown in the figure, bubble nucleation is first generated on the heated surface when the energy accumulates to a certain degree. Then, the bubble starts to grow according to the evaporation of the liquid microlayer at the bottom of the bubble and the heat transferring from the superheated liquid around the bubble. When the bubble grows to a certain size on the evaporator surface, it will depart from the surface by the combined effect of surface tension, buoyancy, and inertial force. After that, the bubble continues to grow to a diameter equal to the liquid level. When the bubble rises, the liquid film will be lifted directly against the condenser surface, coursing radial motion of the liquid on the condenser surface by the scour of the gas flow. The rising movement of bubbles causes strong disturbance of the liquid pool, which is beneficial for enhancing convective boiling phase change heat transfer. In addition, the condensation heat transfer can also be increased due to the scour phenomenon caused by the bubble motion on the condenser surface.

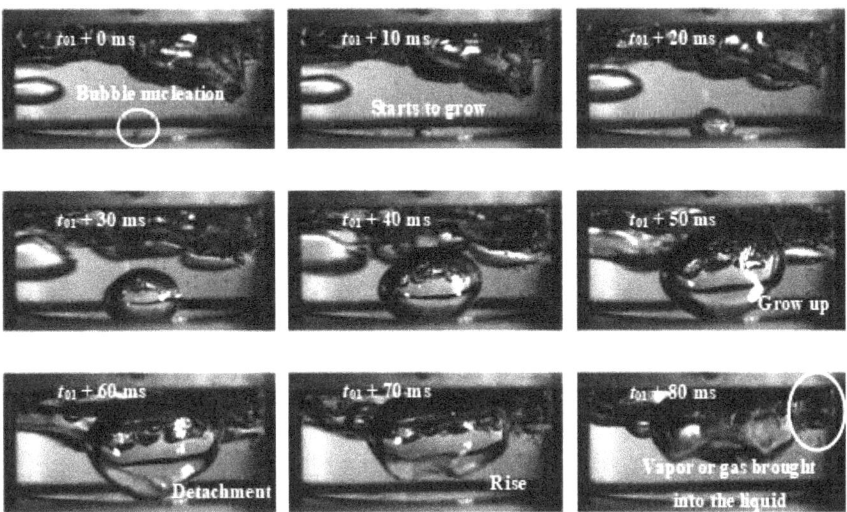

Figure 5. The evolution of gas–liquid two-phase flow pattern in cavity (φ = 80%, q = 17.7 W/cm^2, H = 20 mm).

It is conceivable that during the operation of the flat two-phase thermosyphon, the boiling and condensation phase change heat transfer processes coexist and occur simultaneously inside the cavity, accompanied by complex gas–liquid two-phase flow behavior, such as bubble generation, growth, movement, coalescence, rupture, condensation, and fluctuation of the gas–liquid interface, etc. Therefore, these complex thermo-hydrodynamic behaviors play an important role in the operation and heat transfer of the flat two-phase thermosiphon.

3.2. Effect of Heat Flux

The heat flux directly determines the intensity of vapor–liquid phase change inside a confined cavity. Figure 6 shows the variation of temperature on the center point of the evaporator surface over time under different heat inputs. Figure 7 plots the total thermal resistance as a function of heat flux, and the corresponding vapor–liquid two-phase flow behaviors affected by heat flux are illustrated in Figure 7. It can be seen from the figures that, with the increase of heat flux, the thermal resistance of the flat two-phase thermosyphon shows a decreasing trend (see Figure 7), and the boiling heat transfer intensifies as indicated by increasing bubble number (see Figure 8).

When the heat flux increases, the temperature on the center point of the evaporator surface becomes higher when it reaches steady state, as shown in Figure 6. The superheat degree of the evaporator surface temperature increases, so more nucleation sites are activated, leading to generation of more bubbles. In this case, the liquid disturbance is intensified and, hence, the convective heat transfer between the liquid and solid surface is improved. As a result, the boiling heat transfer is enhanced. In addition, the condensate backflow into the liquid pool is promoted by the bubble rupture as well as the liquid flushing on the condenser surface. Induced by the stronger disturbance of the bubble by a larger heat load, the thickness of the condensate film is thinned, so the condensation heat transfer is enhanced. Therefore, as the heat load increases, the thermal resistance of the flat two-phase thermosyphon gradually becomes smaller.

It also can be seen from Figure 7 that the thermal resistance of the two-phase thermosyphon charged with water is smaller than that charged with ethanol as the working medium. This can be explained by the fact that the viscous resistance of ethanol flowing through the cavity is greater because of its higher viscosity, which causes weaker liquid disturbance and condensate backflow. Furthermore, the heat transport capacity of the ethanol is smaller than that of water due to its lower latent heat of vaporization.

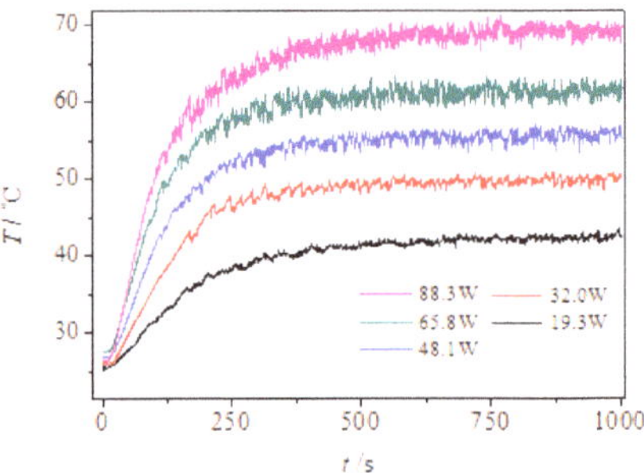

Figure 6. Effect of heat input on the temperature of the center point on evaporator surface.

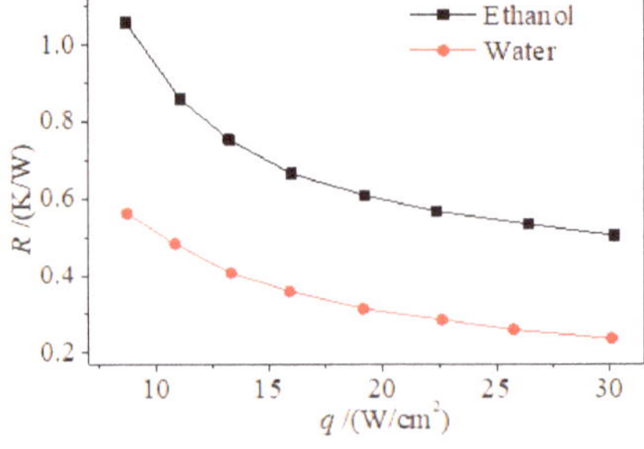

Figure 7. Effect of heat input on thermal resistance ($\varphi = 65\%$).

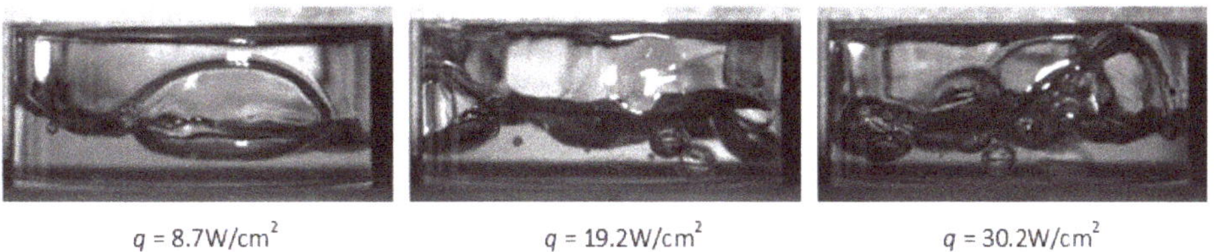

Figure 8. Effect of heat flux on two-phase flow behavior (working medium: water).

3.3. Effect of Wick Presence

The experimental results indicated that the temperature of the evaporator wall oscillates with time when the flat two-phase thermosyphon operates. As shown in Figure 9a, there is large temperature oscillation during the quasi-steady state process, and the temperature oscillation magnitude is about 5 °C when the heat flux is 10.9 W/cm². Note that the temperature mainly arises from the intermittent boiling when the heat flux imposed on the heat pipe is not sufficiently large. Considering the unfavorable temperature oscillation, the cross-shaped wick is introduced in this paper to eliminate the temperature oscillation. As expected, the temperature oscillation is obviously reduced once the cross-shaped wick using copper mesh is embedded into the cavity of the two-phase thermosyphon, as shown in Figure 9b. Under the case with the wick, the temperature oscillation magnitude is only about 0.15 °C, indicating the feasibility of the wick structure to enhance the operation stability of the flat two-phase thermosyphon.

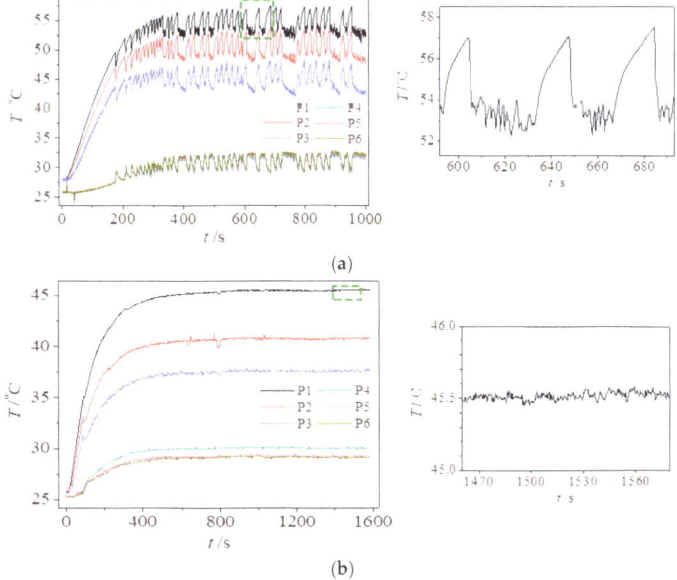

Figure 9. Effect of wick presence on evaporator wall temperature evolution: (**a**) no wick (φ = 50%, q = 10.9 W/cm², H = 15 mm); (**b**) cross-shaped wick (φ = 50%, q = 9.9 W/cm², H = 15 mm).

Apart from the reduction of temperature oscillation, the presence of the wick structure can also improve the thermal performance of the flat two-phase thermosyphon. Figure 10 describes the thermal resistance of the flat two-phase thermosyphon with a cross-shaped wick as a function of heat flux.

In order to highlight the influence of the wick structure, the thermal resistance of the flat two-phase thermosyphon without a wick is also included in the figure. It can be seen from the figure that the thermal performance of the flat two-phase thermosyphon is better than that with no wick structure. This is due to the reason that the heat is more easily transferred from the evaporator heat spot to the upper and side liquid owing to the high thermal conduction of the metal mesh wick. As shown in Figure 11, the liquid disturbance is increased as the detachment frequency and number of generated bubbles on the evaporator surface are significantly increased, so the convective boiling heat transfer of the liquid pool is improved. Besides, since the wick top contacts with the condenser surface, a shortcut of condensate backflow is provided by the wick, i.e., the condensate backflow can be promoted by the capillary force provided by the wick.

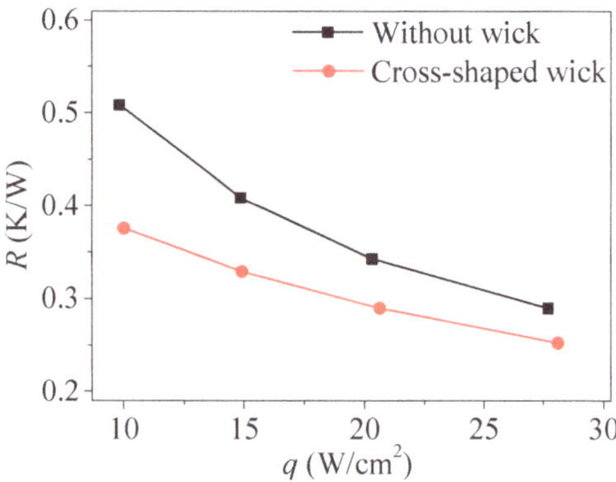

Figure 10. Effect of wick presence on thermal resistance (H = 15 mm, φ = 50%).

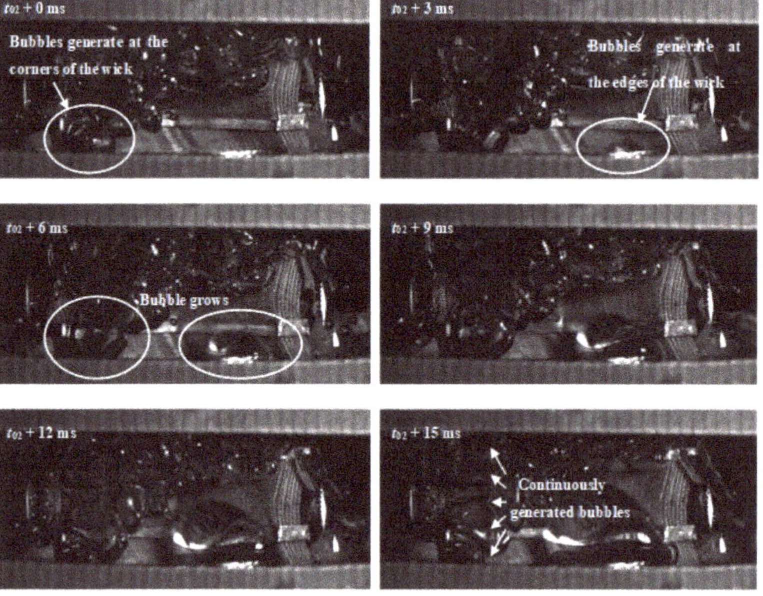

Figure 11. Evolution of vapor–liquid two-phase flow inside the wick thermosyphon (H = 15 mm, φ = 50%, q = 20 W/cm^2).

4. Conclusions

A flat two-phase thermosyphon with transparent wall was manufactured and assembled in this paper, and a visualization experiment system was developed for the visual experimental study of the boiling and condensation phase change heat transfer process inside the cavity. Considering the unfavorable temperature oscillation, a cross-shaped wick was introduced to eliminate the temperature oscillation. The effects of heat flux and wick structure on the two-phase flow behaviors and thermal performance were analyzed. The following main conclusions were obtained:

(1) Contributed by the evaporation of the liquid microlayer and the heat transferring from the superheated liquid around, bubbles grow to a certain size on the evaporator surface and then depart from the surface by the combined effect of surface tension, buoyancy, and inertial force. As a result, a strong liquid disturbance caused by the bubble motions such as growth, rise, coalescence, and rupture can be observed during the experiment, which is beneficial for enhancing the phase change heat transfer processes including boiling and condensation.

(2) More bubbles are generated as more nucleation sites are activated on the heated surface when the heat flux increases; therefore, the liquid disturbance is intensified and the thickness of the condensate film is thinned, leading to higher convective heat transfer and condensation heat transfer. In addition, the thermal resistance of the two-phase thermosyphon charged with water is smaller than that charged with ethanol because of lower viscosity and higher latent heat of vaporization.

(3) When a wick is embedded into the cavity, more bubbles are generated and the detachment frequency is significantly increased, causing stronger liquid disturbance. Therefore, the heat is more easily transferred from the evaporator heat spot to the liquid in the cavity, leading to a decrease in the degree of superheat of the liquid. As a result, the temperature oscillation is obviously reduced.

(4) As a shortcut path is constructed when a wick is used in the cavity, the backflow efficiency can be improved, resulting in the improvement of thermal performance for the flat two-phase thermosyphon.

Author Contributions: F.Y. provided guidance and supervision; C.W. implemented the main research, discussed the results, and wrote the paper; M.C.Z. collected the data; L.Y.W. and J.S. checked and revised the paper; all authors read and approved the final manuscript.

Funding: This research was funded by National Natural Science Foundation of China (grant number 51706194 and 51606037), Natural Science Foundation of Jiangsu Province (grant number BK20160687) and Natural Science Foundation of Yangzhou City (grant number YZ2017103).

References

1. Liang, G.T.; Mudawar, I. Review of pool boiling enhancement with additives and nanofluids. *Int. J. Heat Mass Transf.* **2018**, *124*, 423–453. [CrossRef]
2. Al-Zareer, M.; Dincer, I.; Rosen, M.A. Novel thermal management system using boiling cooling for high-powered lithium-ion battery packs for hybrid electric vehicles. *J. Power Source* **2017**, *363*, 291–303. [CrossRef]
3. Zhang, C.; Chen, Y.; Wu, R.; Shi, M. Flow boiling in constructal tree-shaped minichannel network. *Int. J. Heat Mass Transf.* **2011**, *54*, 202–209. [CrossRef]
4. Amalfi, R.L.; Salamon, T.; Lamaison, N.; Marcinichen, J.B.; Thome, J.R. Two-phase liquid cooling system for electronics—Part 3: Ultracompact liquid-cooled condenser. In Proceedings of the 16th Intersociety Conference on Thermal and Thermomechanical Phenomena in Electronic Systems (ITherm), Orlando, FL, USA, 30 May–2 June 2017; pp. 687–695.

5. Ling, Z.; Cao, J.; Zhang, W.; Zhang, Z.; Fang, X.; Gao, X. Compact liquid cooling strategy with phase change materials for li-ion batteries optimized using response surface methodology. *Appl. Energy* **2018**, *228*, 777–788. [CrossRef]

6. Zhang, C.; Deng, Z.; Chen, Y. Temperature jump at rough gas–solid interface in couette flow with a rough surface described by cantor fractal. *Int. J. Heat Mass Transf.* **2014**, *70*, 322–329. [CrossRef]

7. Zhang, C.; Chen, Y.; Deng, Z.; Shi, M. Role of rough surface topography on gas slip flow in microchannels. *Phys. Rev. E* **2012**, *86*. [CrossRef] [PubMed]

8. Chen, Y.; Zhang, C.; Shi, M.; Yang, Y. Thermal and hydrodynamic characteristics of constructal tree-shaped minichannel heat sink. *AIChE J.* **2009**, *56*, 2018–2029. [CrossRef]

9. Najim, M.; Feddaoui, M.B.; Nait Alla, A.; Charef, A.; Kabeel, A.E. New cooling approach using successive evaporation and condensation of a liquid film inside a vertical mini-channel. *Int. J. Heat Mass Transf.* **2018**, *122*, 895–912. [CrossRef]

10. Chen, Y.; Shen, C.; Shi, M.; Peterson, G.P. Visualization study of flow condensation in hydrophobic microchannels. *AIChE J.* **2014**, *60*, 1182–1192. [CrossRef]

11. Chen, X.; Ye, H.; Fan, X.; Ren, T.; Zhang, G. A review of small heat pipes for electronics. *Appl. Therm. Eng.* **2016**, *96*, 1–17. [CrossRef]

12. Qu, J.; Li, X.J.; Cui, Y.Y.; Wang, Q. Design and experimental study on a hybrid flexible oscillating heat pipe. *Int. J. Heat Mass Transf.* **2017**, *107*, 640–645. [CrossRef]

13. Naphon, P.; Wongwises, S.; Wiriyasart, S. Application of two-phase vapor chamber technique for hard disk drive cooling of pcs. *Int. Commun. Heat Mass Transf.* **2013**, *40*, 32–35. [CrossRef]

14. Chen, Y.; Gao, W.; Zhang, C.; Zhao, Y. Three-dimensional splitting microfluidics. *Lab Chip* **2016**, *16*, 1332–1339. [CrossRef] [PubMed]

15. Chen, Y.P.; Deng, Z.L. Hydrodynamics of a droplet passing through a microfluidic t-junction. *J. Fluid Mech.* **2017**, *819*, 401–434. [CrossRef]

16. Deng, Z.; Liu, X.; Zhang, C.; Huang, Y.; Chen, Y. Melting behaviors of pcm in porous metal foam characterized by fractal geometry. *Int. J. Heat Mass Transf.* **2017**, *113*, 1031–1042. [CrossRef]

17. Chen, Y.; Zhang, C.; Shi, M.; Wu, J.; Peterson, G.P. Study on flow and heat transfer characteristics of heat pipe with axial "Ω"-shaped microgrooves. *Int. J. Heat Mass Transf.* **2009**, *52*, 636–643. [CrossRef]

18. Deng, Z.; Zheng, Y.; Liu, X.; Zhu, B.; Chen, Y. Experimental study on thermal performance of an anti-gravity pulsating heat pipe and its application on heat recovery utilization. *Appl. Therm. Eng.* **2017**, *125*, 1368–1378. [CrossRef]

19. Qu, J.; Wu, H.Y.; Cheng, P.; Wang, Q.; Sun, Q. Recent advances in mems-based micro heat pipes. *Int. J. Heat Mass Transf.* **2017**, *110*, 294–313. [CrossRef]

20. Lv, L.; Li, J. Managing high heat flux up to 500w/cm^2 through an ultra-thin flat heat pipe with superhydrophilic wick. *Appl. Therm. Eng.* **2017**, *122*, 593–600. [CrossRef]

21. Zhang, M.; Liu, Z.; Ma, G. The experimental investigation on thermal performance of a flat two-phase thermosyphon. *Int. J. Therm. Sci.* **2008**, *47*, 1195–1203. [CrossRef]

22. Chang, S.W.; Yu, K.C. Thermal performance of reciprocating two-phase thermosyphon with nozzle. *Int. J. Therm. Sci.* **2018**, *129*, 14–28. [CrossRef]

23. Fertahi, S.E.D.; Bouhal, T.; Agrouaz, Y.; Kousksou, T.; El Rhafiki, T.; Zeraouli, Y. Performance optimization of a two-phase closed thermosyphon through cfd numerical simulations. *Appl. Therm. Eng.* **2018**, *128*, 551–563. [CrossRef]

24. Narcy, M.; Lips, S.; Sartre, V. Experimental investigation of a confined flat two-phase thermosyphon for electronics cooling. *Exp. Therm. Fluid. Sci.* **2018**, *96*, 516–529. [CrossRef]

25. Wang, X.Y.; Wang, Y.F.; Chen, H.J.; Zhu, Y.Z. A combined cfd/visualization investigation of heat transfer behaviors during geyser boiling in two-phase closed thermosyphon. *Int. J. Heat Mass Transf.* **2018**, *121*, 703–714. [CrossRef]

26. Fadhl, B. Modelling of the Thermal Behaviour of a Two-Phase Closed Thermosyphon. Ph.D. Thesis, Brunel University London, London, UK, March 2016.

27. Zhang, G.; Liu, Z.; Wang, C. An experimental study of boiling and condensation co-existing phase change heat transfer in small confined space. *Int. J. Heat Mass Transf.* **2013**, *64*, 1082–1090. [CrossRef]

28. Chen, Y.; Yu, F.; Zhang, C.; Liu, X. Experimental study on thermo-hydrodynamic behaviors in miniaturized two-phase thermosyphons. *Int. J. Heat Mass Transf.* **2016**, *100*, 550–558. [CrossRef]

29. Zhang, G.; Liu, Z.; Wang, C. A visualization study of the influences of liquid levels on boiling and condensation co-existing phase change heat transfer phenomenon in small confined spaces. *Int. J. Heat Mass Transf.* **2014**, *73*, 415–423. [CrossRef]

30. Liu, Z.L.; Zheng, F.W.; Li, Y.X. Enhancing boiling and condensation co-existing heat transfer in a small and closed space by copper foam inserts. *Int. J. Heat Mass Transf.* **2017**, *108*, 961–971. [CrossRef]

31. Peng, H.; Li, J.; Ling, X. Study on heat transfer performance of an aluminum flat plate heat pipe with fins in vapor chamber. *Energy Convers. Manag.* **2013**, *74*, 44–50. [CrossRef]

32. Hassan, H.; Harmand, S. An experimental and numerical study on the effects of the flat heat pipe wick structure on its thermal performance. *Heat Transf. Eng.* **2015**, *36*, 278–289. [CrossRef]

33. Lips, S.; Lefevre, F.; Bonjour, J. Combined effects of the filling ratio and the vapour space thickness on the performance of a flat plate heat pipe. *Int. J. Heat Mass Transf.* **2010**, *53*, 694–702. [CrossRef]

34. Ju, Y.S.; Kaviany, M.; Nam, Y.; Sharratt, S.; Hwang, G.S.; Catton, I.; Fleming, E.; Dussinger, P. Planar vapor chamber with hybrid evaporator wicks for the thermal management of high-heat-flux and high-power optoelectronic devices. *Int. J. Heat Mass Transf.* **2013**, *60*, 163–169. [CrossRef]

35. Wu, S.; Yu, C.; Yu, F.; Chen, Y. Lattice boltzmann simulation of co-existing boiling and condensation phase changes in a confined micro-space. *Int. J. Heat Mass Transf.* **2018**, *126*, 773–782. [CrossRef]

 *energies*

Article

Gas–Liquid Two-Phase Upward Flow through a Vertical Pipe: Influence of Pressure Drop on the Measurement of Fluid Flow Rate

Tarek A. Ganat [1,*] and **Meftah Hrairi** [2]

1   Department of Petroleum Engineering, Universiti Teknologi PETRONAS,
    Seri Iskandar, Perak 32610, Malaysia
2   Department of Mechanical Engineering, International Islamic University Malaysia,
    P.O. Box 10, Kuala Lumpur 50728, Malaysia; meftah@iium.edu.my
*   Correspondence: tarekarbi.ganat@utp.edu.my; Tel.: +60-53687111

Received: 23 June 2018; Accepted: 31 August 2018; Published: 27 October 2018

Abstract: The accurate estimation of pressure drop during multiphase fluid flow in vertical pipes has been widely recognized as a critical problem in oil wells completion design. The flow of fluids through the vertical tubing strings causes great losses of energy through friction, where the value of this loss depends on fluid flow viscosity and the size of the conduit. A number of friction factor correlations, which have acceptably accurate results in large diameter pipes, are significantly in error when applied to smaller diameter pipes. Normally, the pressure loss occurs due to friction between the fluid flow and the pipe walls. The estimation of the pressure gradients during the multiphase flow of fluids is very complex due to the variation of many fluid parameters along the vertical pipe. Other complications relate to the numerous flow regimes and the variabilities of the fluid interfaces involved. Accordingly, knowledge about pressure drops and friction factors is required to determine the fluid flow rate of the oil wells. This paper describes the influences of the pressure drop on the measurement of the fluid flow by estimating the friction factor using different empirical friction correlations. Field experimental work was performed at the well site to predict the fluid flow rate of 48 electrical submersible pump (ESP) oil wells, using the newly developed mathematical model. Using Darcy and Colebrook friction factor correlations, the results show high average relative errors, exceeding ±18.0%, in predicted liquid flow rate (oil and water). In gas rate, more than 77% of the data exceeded ±10.0% relative error to the predicted gas rate. For the Blasius correlation, the results showed the predicted liquid flow rate was in agreement with measured values, where the average relative error was less than ±18.0%, and for the gas rate, 68% of the data showed more than ±10% relative error.

Keywords: pressure loss; pressure drop; friction factor; multiphase flow; flow rate; flow regime

1. Introduction

In the oil and gas industry, multiphase flow in vertical pipes often occurs. The flow of fluids through the vertical pipe string causes a loss of energy through friction losses, where the value of this loss depends on the fluid flow viscosity and the size of the conduit. Often, the friction loss is an important part of the oil well completion design [1]. The pressure drop occurs as a result of the changes in potential and kinetic energy of the fluid due to the friction on the pipe walls [2]. Generally, the total pressure drop in the vertical conduit is basically related to four main components: frictional, hydrostatic, acceleration, and pressure drop. Among these four components, calculation of the pressure drop is the most complex component and has received extensive attention by researchers [3,4]. Many researchers have attempted to determine the two-phase frictional pressure drop over the whole range of flow patterns through a vertical pipe. A substantial number of experiments have been carried out to determine fluid flow friction losses in both Newtonian [1,2,5–11] and non-Newtonian systems [5,12]. A large number of experimental works was made in short tubes. Consequently, a lot of engineering problems come up when efforts are made to extend these experimental results to real oil field conditions where a longer pipe is used. In those experiments, the data shows only a limited number of variables, and as a result, imprecisions are introduced when the friction correlations are applied outside the limitations of the experimental data. As a consequence of the limited amount of data available for these experiments, the effects of some significant variables were ignored in the early studies [13–17]. The accuracy of the pressure drop prediction in flowing wells has a significant influence on the fluid flow measurement. There are many particular solutions, but they are valid only for some specific conditions. This is due to the complexity of two-phase flow analysis. In some conditions, the gas travels at a much higher velocity than the liquid. Accordingly, the flowing density of the gas–liquid mixture is higher than the corresponding density. Moreover, the liquid's velocity inside the pipe wall can be different over a short distance and can cause a variable friction loss. The difference in velocity and flow regime of the two phases strongly affect pressure drop computations [13], meaning that slippage is a consequence of the difference between the combined velocities of the two phases, which is caused by the physical properties of the fluids involved. For single-phase flow, the frictional pressure losses do not normally increase with a decrease in the tubing size or an increase in well production flow rate. This refers to the existence of a gas phase, which tends to slip by the liquid phase without essentially contributing to its lift. Many researchers have tried to show a relationship between the slippage losses and the friction losses [15–18]. A method for the estimation of gas–liquid flow rates in the vertical pipe has been proposed [19]. The method was used to calibrate a differential pressure sensor to predict the flow rates of both phases in air–water flow. The estimations were in good agreement with real flow rate measurements. A study by Daev and Kairakbaev [20] proposed a new model of the liquid flow through pipes that incorporated flow straighteners. The prediction of the flow rate of liquid was studied and the parameters affecting the process of measuring the flow rate of liquid were considered. An experimental study of the two-phase flow regime and frictional pressure drop inside the pipe was done by Cai et al. [21]. The flow patterns were defined and recorded by a high-speed camera. A new empirical correlation was proposed based on the experimental results to predict the liquid multiplier factor of the test channel. A two-phase flow measurement applying a resistive void fraction meter combined to a venturi, or orifice plate, was suggested by Oliveira et al. [22]. This method was applied to determine the fluid mass flow rates using an air–water experimental apparatus. The results showed that the flow path has no important effect on the meters in relation to the frictional pressure drop in the experimental process range. The outcomes of the experimental work displayed a mean slip ratio of less than 1.1, when slug and bubbly flow patterns were lower than 70%.

This research work aims to evaluate the influence of a pressure drop on the measurement of the fluid flow rate in ESP oil wells. A new mathematical model was developed to determine the fluid flow rate of the oil wells through the prediction of multiphase flow parameter variations inside a vertical pipe based on local temperature and pressure changes with depth and applying multiphase flow physics equations and empirical correlations. The objective of this study was to obtain data from well tests conducted in a long vertical pipe and utilize this data to evaluate the effects of slippage and friction factor, in different flow regimes, on the calculation accuracies of the fluid flow rate of the oil wells. The approach measured the liquid hold-up along the conduit and used different friction correlations such as Blasius, Darcy, and Colebrook friction factor correlations to compare the predicted fluid flow rate with the measured fluid flow rate for each oil well. Generally, the results show that any errors in pressure drop calculation will generate inaccuracies in the prediction of fluid flow rate.

2. Experimental Arrangements and Measurement Procedure

The experiments conducted in the present study were carried out for two-phase flow through a vertical pipe of 48 oil wells using ESP pumps. A schematic of the experimental system is shown in Figure 1. The flow measurement starts at the surface wellhead, and then down to the bubble point pressure location depth in the well. Wellhead flowing pressure was measured at normal production conditions before and after the wing valve shut-in, leaving the ESP pump running, to measure the build-up of pressure at the wellhead. The total shut-in time period of the wellhead valve was then recorded. The first free gas bubbles started liberating from the bubble point location depth inside the tubing string. This occurred in the production flowing well before and after the wellhead wing valve shut-in, and the changes of flow patterns inside the pipe were reallocated once again, due to variations of temperature and pressure along the conduit. As a consequence, the liberated gas was dissolved in the oil phase, and the location depth of the bubble point pressure relocated to another position after the wellhead wing valve shut-in. The column of liquid that replaced the liberated gas column space, during the shut-in time period, was the difference between the first and second bubble point location depths. Figure 2 shows the bubble point location depths before and after the wellhead wing valve shut-in. A conceptual basis of physics for prediction of fluid flow rate in the conduit was employed along with multiphase empirical correlations to compute the variations of fluid flow parameters inside the tubing.

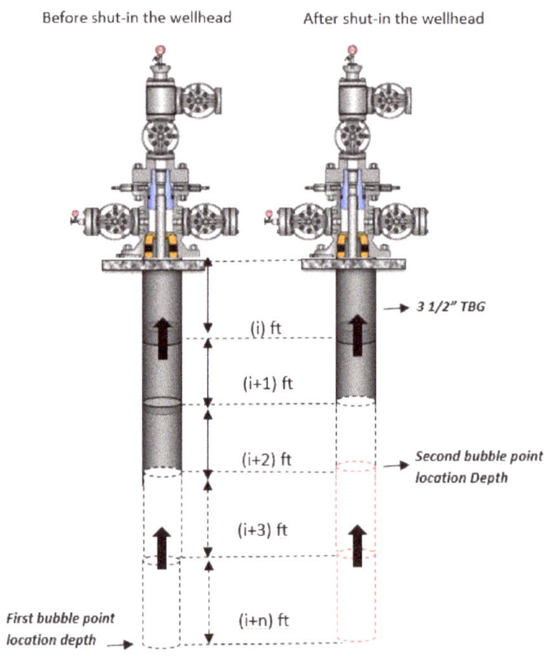

Figure 1. Schematic of the flow measurement stages in a vertical pipe before and after the well head wing valve shut-in.

Several assumptions were made to conduct the calculations such as: assumed one-dimensional flow in the conduit, assumed uniform cross-sectional area of the pipe, the phase's properties varied with depth, the frictional factor varied along the conduit, and the effect of the liquid compressibility was neglected.

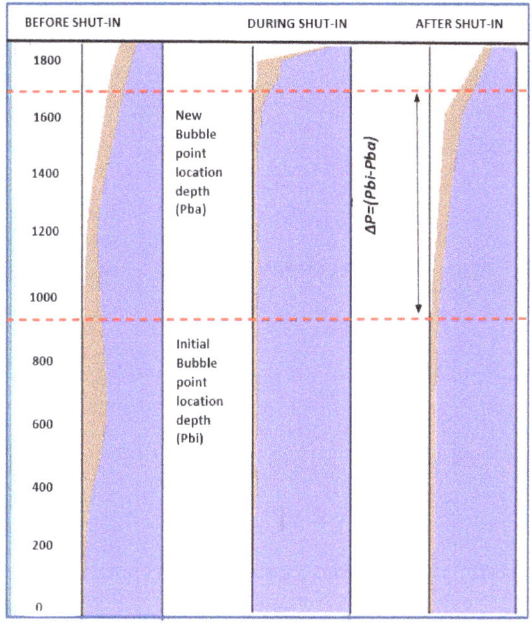

Figure 2. Bubble point location depths before and after closing the well head wing valve.

2.1. Required Input Data

The input data required were the well parameters data and physical properties data of the fluid, as seen in Table 1. To carry out this study, 48 ESP oil wells were selected where the wells were producing from four different reservoirs using same production pipe diameter. Also, these reservoirs had almost the same reservoir fluid properties: the bubble point pressure ranged from 924 psi to 1124 psi, the American Petroleum Institute (API) oil gravity ranged from 36 to 37 @ 60 °F, the oil viscosity ranged from 0.784 cP to 1.0119 cP, and the reservoir temperature ranged from 157 °F to 186 °F. Furthermore, Figure 3 classifies the input data required.

Table 1. Well and physical properties of the fluid.

Well Name	WHPb	WHPa	WHT	GOR	WC	Total Shut-in Time
	(PSIA)	(PSIA)	(F)	(SCE/STB)	(%)	(min)
A33	140	200	98	360.92	93	1.55
A125	100	200	95	360.92	91.62	0.83
A64	180	250	107	360.92	81.52	1.06
A29	250	270	107	360.92	84.88	0.80
A23	210	260	127.7	360.92	82.11	0.24
A135	210	250	100	360.92	59.88	0.56
A126	250	300	98	360.92	66.91	0.20
A12	175	270	107	360.92	82.9	0.84
A108	260	300	97	360.92	81.31	0.28
5J5	150	300	95	360.92	4	3.36
5J2	100	170	101	360.92	52	1.39
5J4	250	300	101.6	360.92	58.95	0.27
5J7	250	300	98	360.92	30	0.47
E89	150	190	140	300	79	0.27
E210	80	120	110	300	83	0.33

Table 1. *Cont.*

Well Name	WHPb (PSIA)	WHPa (PSIA)	WHT (F)	GOR (SCE/STB)	WC (%)	Total Shut-in Time (min)
E211	80	120	129	300	74	0.96
E286	70	100	146	300	90	0.25
E192	80	110	146	300	83	0.24
E327	80	110	115.5	300	77	0.35
E325	70	110	124.6	300	83	0.44
E197	90	110	146.1	300	82	0.11
E208	95	110	146.8	300	81	0.07
E226	80	110	138.4	300	91	0.25
E284	80	120	124.5	300	76	0.33
E258	65	90	142	300	86	0.23
E326	60	100	113	300	82	0.48
E227	100	150	142	300	84	0.36
4E_3	130	300	146	300	87	1.18
B56	120	170	120	384	42	2.3
B70	160	230	120	384	29.9	2.9
B121	100	160	120	364	67.29	0.5
B119	100	160	120	364	76.18	1.1
B50	180	250	110	364	68.65	0.55
B88	100	160	110	364	63.59	2.1
B14	250	270	110	364	76.28	0.15
B151	180	230	110	364	55.78	0.66
B164	100	170	120	364	26.3	2.1
B51	240	310	120	364	59.05	0.44
Q89	100	150	120	364	0	3.7
Q21	80	150	120	364	79.22	2.1
Q53	80	150	120	364	71.41	2.3
Q14	75	130	120	364	74.83	1.4
Q100	80	130	110	364	78.27	0.55
Q12	80	150	110	364	80.33	0.58
Q85	100	150	110	364	18.18	2.5
Q82	100	150	110	364	75.3	0.5
Q78	80	150	120	364	37.27	2.5
Q76	80	150	120	364	80.5	1.3

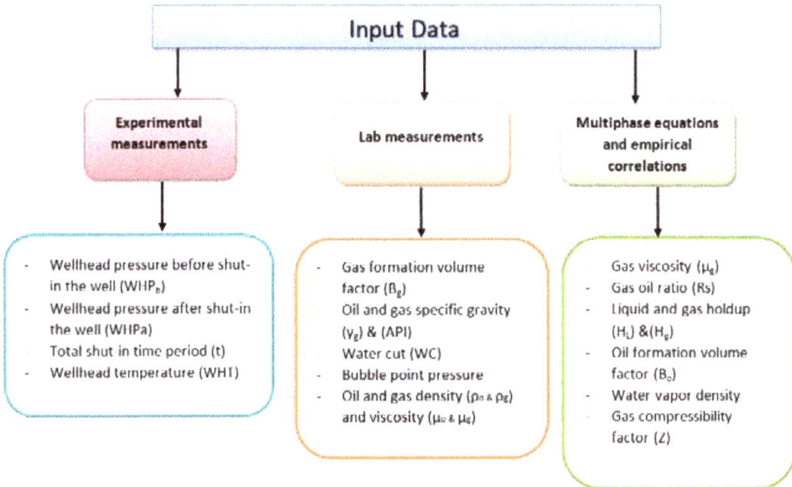

Figure 3. Input data required.

2.2. Computational Algorithm

Figure 4 shows the algorithm steps to evaluate the mathematical model. The algorithm classified all the main stages and sub-steps in the model. In this process, the calculations were performed to obtain the bubble point pressure location depth before and after the wellhead wing valve shut-in. The fluid flowing pressure gradient could be calculated anywhere inside the pipe. All the variables needed to be identified to correctly evaluate the physics interactions between all the fluid parameters using the suitable multi-physics equations and empirical correlations.

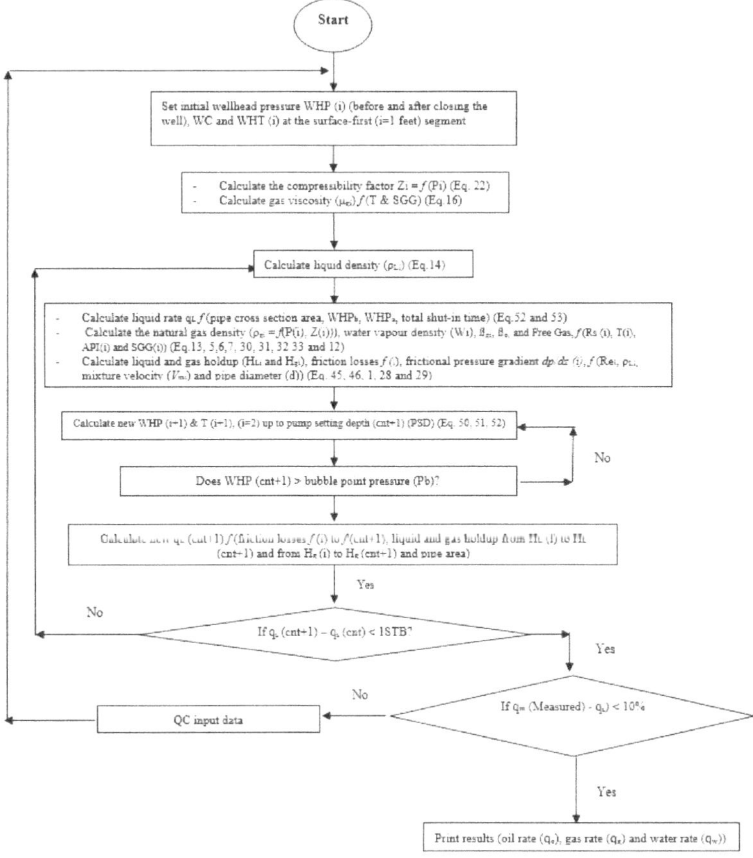

Figure 4. Flowchart of the new mathematical model algorithm.

The calculation starts at the surface wellhead and then down to the location depth of the bubble point pressure as a function of temperature and pressure variations with depth. To consider the fact that flow regimes vary depending on the in situ flow rates of gas/liquid, the model calculates, at each foot along the vertical pipe, the variations of supercritical velocities, viscosities, and densities for both phases (liquid and gas). The in situ flow rate can also be calculated by the mathematical model at any flow regime at any depth. As shown in Figures 5 and 6, the calculation iteration can stop at any depth (i, ..., i + n) using all the equations (from Equation (1) to Equation (56)), where there is a different flow regime along the vertical pipe.

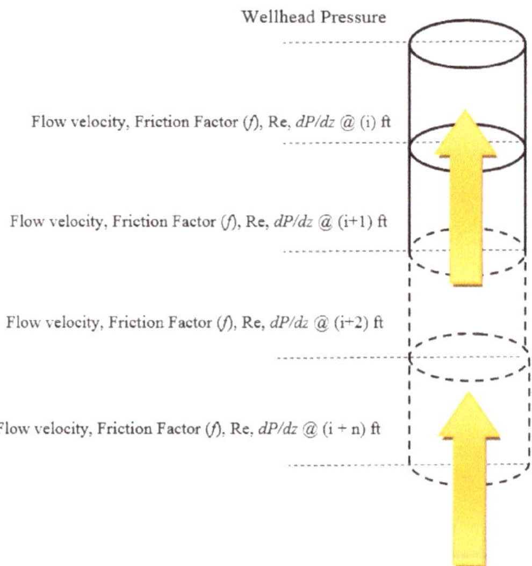

Figure 5. Stages of computational methodology.

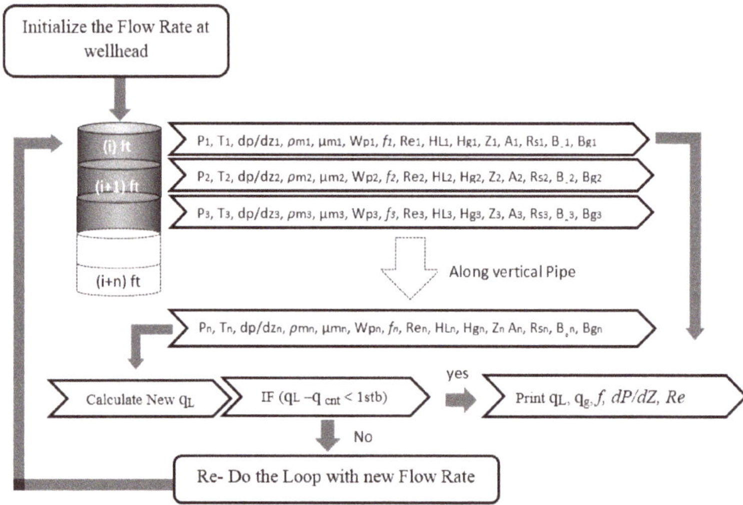

Figure 6. Flow diagram of new computational method procedure.

The following are the physics equations and the correlations applied to determine each independent variable at every single foot.

Total pressure losses expressed as

$$\Delta P_{Total} = \Delta P_{HH} + \Delta P_{Frictional}$$ (1)

Hydrostatic head is expressed as

$$\Delta P_{HH} = \frac{\rho_m g \Delta Z}{144 g_c}$$ (2)

Darcy–Weisbach equation [23] was used to calculate the frictional pressure loss

$$\Delta P = f \frac{L}{D} \frac{\rho V^2}{2} \tag{3}$$

Reynolds number is given by

$$Re = \frac{2.2 \times 10^{-2} m_t}{D \mu_L{}^{H_L} \mu_g{}^{(1-H_L)}} \tag{4}$$

Three different friction factor correlations were applied to evaluate the impact of the friction on the computation of the fluid flow rate. The first correlation is the Blasius empirical correlation for turbulent flow [24].

$$f = 0.316 \left(Re\right)^{-0.25} \tag{5}$$

The second friction correlation applied is Darcy correlation [23]

$$f = \frac{64}{Re} \tag{6}$$

The third friction factor correlation applied is from Colebrook [25]

$$\frac{1}{\sqrt{f}} = 2log_{10}\left(\frac{\varepsilon/D_h}{3.7} + \frac{2.51}{R_e\sqrt{f}}\right) \tag{7}$$

for $Re <\approx 2300$ and $Re >\approx 4000$.

The gas density is expressed as

$$\rho_g = \frac{m_g}{V_R} = \frac{M_g P}{ZRT} \tag{8}$$

The density at wellbore condition, is given by

$$\rho_g = \frac{\rho_{gs}}{B_g} \tag{9}$$

The oil density is expressed as

$$\rho_o = \frac{62.428\gamma_o + 0.014\gamma_g R_s}{B_o} \tag{10}$$

where

$$\gamma_o = \rho_o/\rho_w \tag{11}$$

$$\gamma_g = \frac{\rho_g}{\rho_{air}} = \frac{\rho_g}{0.077} \tag{12}$$

$$\rho_g = 0.077\gamma_g \tag{13}$$

Liquid density is given by

$$\rho_L = \rho_W WC + \rho_o (1 - WC) \tag{14}$$

Mixture density is expressed as

$$\rho_m = \rho_L H_L + \rho_g (1 - H_L) \tag{15}$$

The gas viscosity is determined by the following equation [26]:

$$\mu_g = K_1 exp\left(X\rho^Y\right) \tag{16}$$

where

$$\rho = \frac{pM_g}{zRT} = 0.0015\frac{pM_g}{zT} \tag{17}$$

$$K_1 = \frac{\left(0.001 + 2 \times 10^{-6} M_g\right) T^{1.5}}{\left(209 + 19M_g + T\right)} \tag{18}$$

$$X = 3.5 + \frac{986}{T} + 0.01M_g \tag{19}$$

$$Y = 2.4 - 0.2X \tag{20}$$

Mixture viscosity is given by

$$\mu_m = \mu_L^{H_L} + \mu_g^{(1-H_L)} \tag{21}$$

Beggs and Brill equation [27] was applied to estimate the gas compressibility factor (Z)

$$Z = A + \frac{(1-A)}{e^B} + CPr^D \tag{22}$$

Using Standing and Katz equations [28] to obtain the pseudo critical temperature and pressure of the gas mixture

$$Pr = 688.634 - 21.983\gamma_g - 13.886\gamma_g^2 \tag{23}$$

$$Tr = 158.01 + 342.12\gamma_g - 16.04\gamma_g^2 \tag{24}$$

and

$$A = 1.39(Tr - 0.92)^{0.5} - 0.36Tr - 0.101 \tag{25}$$

$$B = (0.62 - 0.23Tr)Pr + \left(\frac{0.066}{Tr - 0.86} - 0.037\right)Pr^2 + \frac{0.32}{10^{9(Tr-1)}}Pr^2 \tag{26}$$

$$C = 0.132 - 0.32log(Tr) \tag{27}$$

$$D = 10^{(0.302 - 0.49Tr + 0.182Tr^2)} \tag{28}$$

Superficial gas velocity is expressed as

$$V_{sg} = \frac{4q_g B_g}{\pi D^2} \tag{29}$$

Superficial liquid velocity is expressed as

$$V_{sL} = \frac{4q_L}{\pi D^2} \tag{30}$$

The water vapor density using the Sloan correlation [29] is expressed as

$$W = exp\left(c_1 + \frac{c_2}{T} + c_3 ln(p) + \frac{c_4}{T^2} + \frac{c_5 ln(P)}{T} + c_6(ln(P))^2\right) \tag{31}$$

where the values of constants c_1 to c_6 are shown in Table 2.

Table 2. Constants c_1 to c_6.

Constants	Value
c_1	28.911
c_2	−9668.146
c_3	−1.663
c_4	−130,823.5
c_5	205.323
c_6	0.0385

The gas formation volume factor is expressed as

$$B_g = \frac{P_{sc}ZT}{T_{sc}P} = 0.028\frac{ZT}{P}$$

(32)

Using the Vasquez and Beggs equation [30] to obtain the oil formation volume factor

$$B_{ob} = 1 + C_1 R_{sb} + C_2(T - 60)\left(\frac{\gamma_{API}}{\gamma_g}\right) + C_3 R_{sb}(T - 60)\left(\frac{\gamma_{API}}{\gamma_g}\right)$$

(33)

and oil gas ratio

$$R_{sb} = \frac{\gamma_g Pb^{C_2}}{C_1} 10^{\left(\frac{C_3 \gamma_{API}}{T + 459.67}\right)}$$

(34)

where the coefficients C_1, C_2 and C_3 are given by

Coefficient	°API ≤ 30	°API ≥ 30
C_1	27.64	56.060
C_2	1.0937	1.187
C_3	11.172	10.393

To make sure that the obtained liquid and gas hold-up is accurate, some popular correlations, used by the industry and are included in almost every commercial software package, were considered to predict the liquid and gas hold-up inside each well. The correlations considered in this study are the ones developed by Hagedorn and Brown [31], Duns and Ros [32], Orkiszewski [33], and Aziz et al. [34]. The statistical results for the various prediction methods when applied to all 25 well tests are shown in Figure 7 and Table 3. These results indicate that the Hagedorn and Brown correlation seems to predict liquid and gas hold up better than the other correlations selected in this study. However, the overall results show minor differences between the different correlations. This is because each correlation was developed based on certain assumption and for a particular range of data.

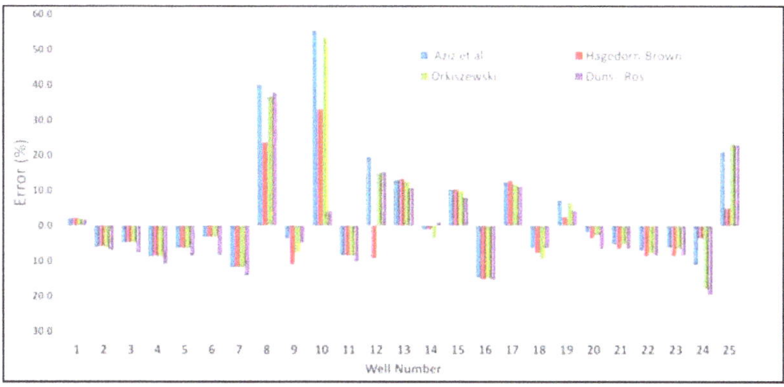

Figure 7. Hold-up prediction accuracy using some popular correlations.

Table 3. Statistical results for the various prediction correlations.

Prediction Method	Average Error (%)	Standard Deviation (%)
Duns and Ros	−1.06	13.06
Hagedorn and Brown	−0.86	11.57
Orkiszewski	1.8	16.52
Aziz et al.	2.9	16.66

Using the Hagedorn-Brown empirical correlation [31] to obtain liquid and gas hold-up (H_L and H_g)

$$N_{Lv} = 1.938 v_{sL} \frac{\sqrt[4]{\rho}}{\sigma} \tag{35}$$

$$N_{gv} = 1.938 v_{sg} \sqrt[4]{\frac{\rho_L}{\sigma}} \tag{36}$$

$$N_d = 120.872 D \frac{\sqrt{\rho_L}}{\sigma} \tag{37}$$

$$N_L = 0.157 \mu_L \sqrt[4]{\frac{1}{\rho_L \sigma^3}} \tag{38}$$

$$Y = -2.699 + 0.158 X_1 - 0.551 X_1{}^2 + 0.548 X_1{}^3 - 0.122 X_1{}^4 \tag{39}$$

where

$$X_1 = log(N_L + 3) \tag{40}$$

$$CN_L = 10^Y \tag{41}$$

$$\frac{H_L}{\psi} = -0.103 + 0.618(log X_2 + 6) - 0.633(log X_2 + 6)^2 + 0.296(log X_2 + 6)^3 - 0.04(log X_2 + 6)^4 \tag{42}$$

where

$$X_2 = \frac{N_{VL} P^{0.1} CN_L}{N_{Vg}{}^{0.575} P a^{0.1} N_D} \tag{43}$$

$$\psi = 0.912 - 4.822 X_3 + 1232.25 X_3^2 - 22253.6 X_3^3 + 116174.3 X_3^4 \tag{44}$$

where

$$X_3 = \frac{N_{Vg} N_L{}^{0.38}}{N_D{}^{2.14}} \tag{45}$$

The liquid hold-up is

$$H_L = \psi \left(\frac{H_L}{\psi} \right) \tag{46}$$

and

$$H_g = (1 - H_L) \tag{47}$$

The liquid flow rate is expressed as

$$q_L = \frac{\Delta H \cdot A}{t} \tag{48}$$

the cross section of the conduit is given by

$$A = \frac{\pi r^2}{4} \tag{49}$$

$$\Delta H = H_2 - H_1 \tag{50}$$

as

$$\Delta P = \Delta H \cdot \rho_L \tag{51}$$

then

$$\Delta H = \frac{\Delta P}{\rho_L} \tag{52}$$

and

$$\Delta P = WHP_a - WHP_b \tag{53}$$

then

$$q_L = \Delta P \frac{A}{\rho_L} t \tag{54}$$

The flow rates for gas, oil, and water are expressed as

$$q_o = q_L (1 - WC) \tag{55}$$

$$q_g = q_o \, Rs \tag{56}$$

$$q_w = q_o \, WC \tag{57}$$

3. Results and Discussion

The experiments were run on 48 ESP oil wells from four different reservoirs. For each friction factor correlation, the measured oil flow rate values for each oil well were compared against the predicted flow rate values. It should be noted that as the points near the dotted straight line drawn at 45° (i.e., $y = x$) in the graph, the more accurate the prediction was. The results show that the pressure drop value was the significant parameter that had the main influence on the fluid flow rate computation. Indeed, any errors in pressure drop values would lead to high uncertainty errors of fluid flow rate prediction. For this reason, the properties of independent variables needed to be considered. Likewise, the interactions between each phase needed to be taken into account along with mixture properties and in situ volume fractions of oil and gas inside the conduit. Each multiphase flow correlation found the friction factor differently. Typically, each friction correlation made its own assumptions and modifications to make them useable to multiphase conditions. The prediction of frictional pressure drop in two-phase flow was usually complicated due to pressure and temperature variations along the flow path. When estimating the friction factor, there were a number of methods for calculating the Reynolds number depending on how much of the two-phase flow mixture was defined. Therefore, the oil and water were considered as a single liquid phase while the gas was considered as a separate phase.

By using the Blasius friction factor correlation, the differences between the predicted flow rate and the measured flow rate were very small. R-squared (R^2) explained exactly how the data points were fitted close to the regression line ($y = x$). Figures 8–10 displayed the regression model for oil, water, and gas flow rate measurements. It can be seen that the plots show that most data points lie on or close to the unit slope line (e.g., best fit line), indicating that the predicted and actual values were in excellent agreement and illustrated an accurate flow rate prediction for oil, water, and gas with good correlating coefficients of 0.994, 0.993, and 0.966, respectively. This means that 99.4%, 99.3%, and 96.6% of the variance in the oil, water, and gas data, respectively, was explained by the line and 0.6%, 0.7%, and 3.4% of the variance was due to unexplained effects. The figures show that the predicted wells flow rates fell within the accepted uncertainty when compared with the measured flow rates.

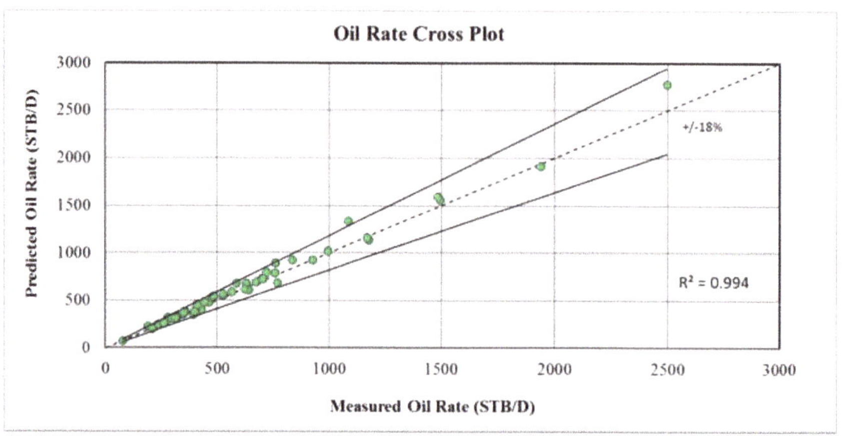

Figure 8. Predicted vs measured oil rate using the Blasius correlation.

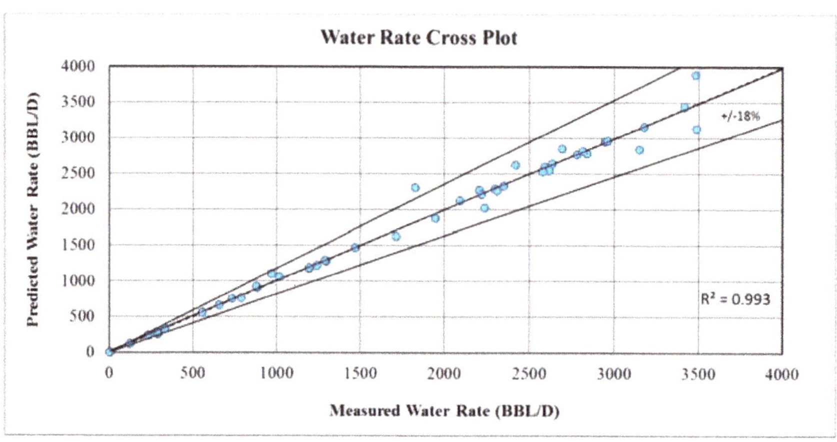

Figure 9. Predicted vs measured water rate using the Blasius correlation.

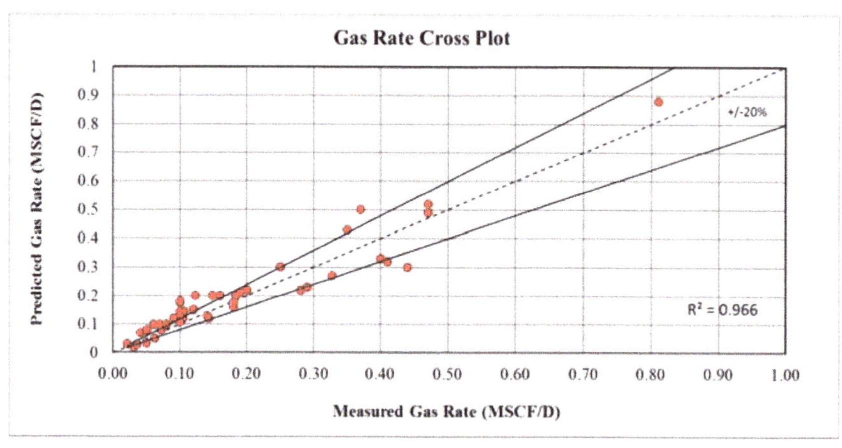

Figure 10. Predicted vs measured gas rate using the Blasius correlation.

By using the Darcy friction factor correlation, the differences between the predicted fluid flow rates with the measured flow rates were larger than those of the Blasius correlation. Figures 11–13 displayed the regression model for oil, water, and gas flow rate measurements. From these figures, one can easily recognize that the data plotted is under-estimated for oil and water flow rates and scattered around the best fit line for gas flow rates. This discrepancy was more evident for high flow rates where the correlation coefficients for oil, water, and gas flow rates accounted for 90.6%, 86.6%, and 78.7% of the variance, respectively. The figures show that the predicted well flow rates did not fall within the accepted uncertainty when compared with the measured flow rates.

By using the Colebrook friction factor correlation, the differences between the predicted fluid flow rates with the measured flow rates were slightly better than the Darcy correlation performance, but still less than the Blasius correlation performance. Figures 14–16 displays the data fitting for oil, water, and gas flow rate measurements. Similar to the performance of the Darcy correlation, one can easily recognize that the data plotted is under-estimated for oil and water flow rates and scattered around the best fit line for gas flow rates. This discrepancy was more evident for high flow rates where the correlation coefficients for oil, water, and gas flow rates accounted for 93.0%, 87.1%, and 80.8% of the variance, respectively. The figures showed that the predicted wells flow rates did not fall within the accepted uncertainty when compared with the measured flow rates.

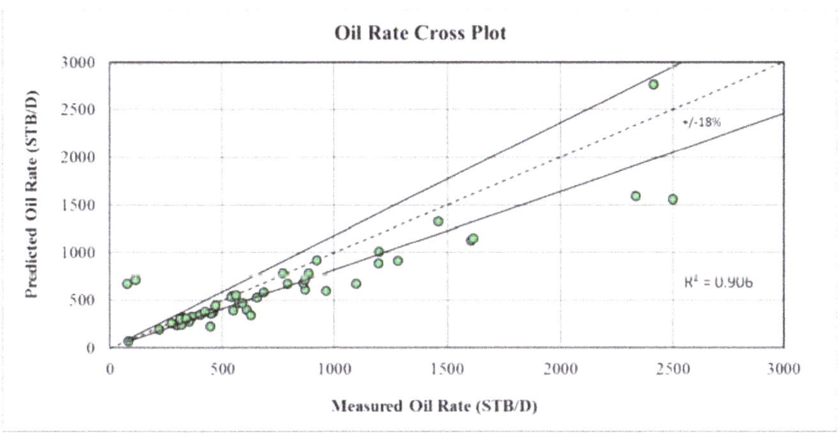

Figure 11. Predicted vs measured oil rate using the Darcy correlation.

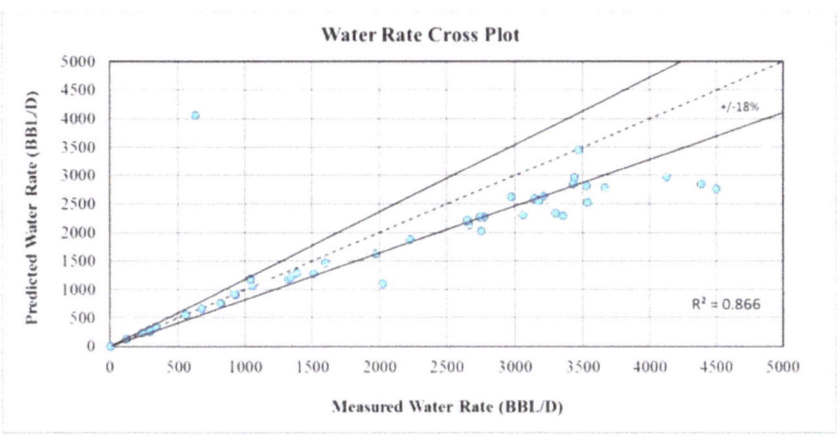

Figure 12. Predicted vs measured water rate using the Darcy correlation.

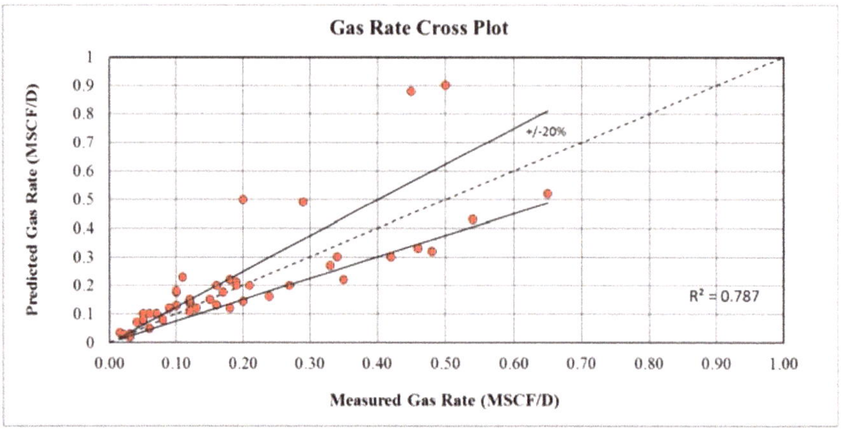

Figure 13. Predicted vs measured gas rate using the Darcy correlation.

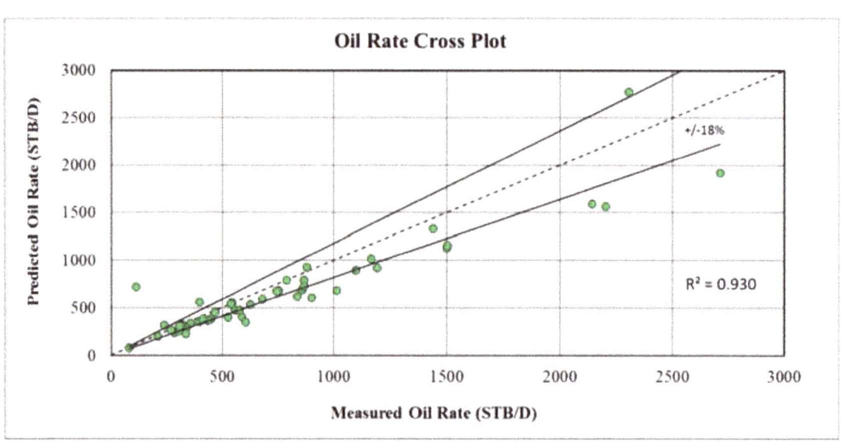

Figure 14. Predicted vs measured oil rate using the Colebrook correlation.

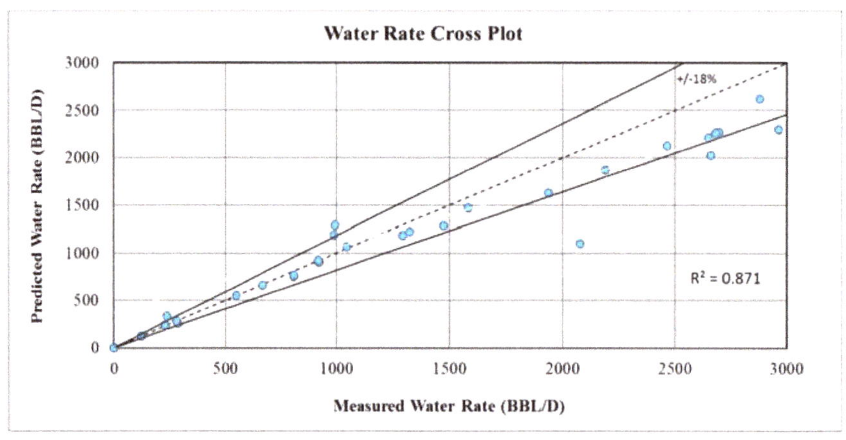

Figure 15. Predicted vs measured water rate using the Colebrook correlation.

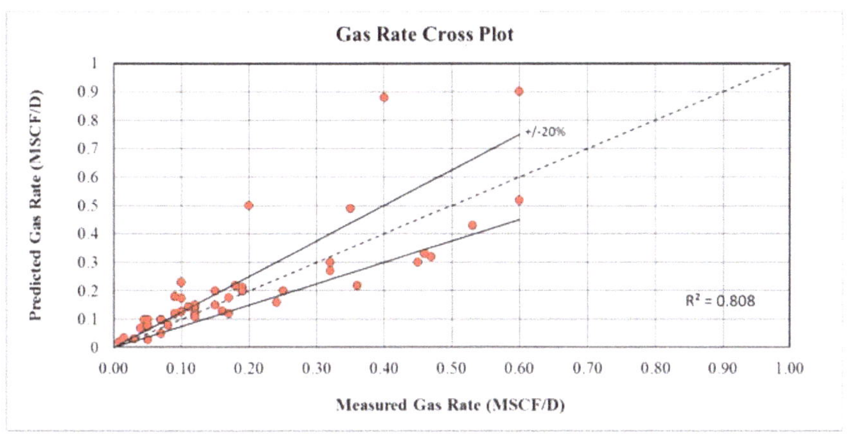

Figure 16. Predicted vs measured gas rate using the Colebrook correlation.

In general, the validation results of the predicted fluid flow rates were satisfactory when using the Blasius correlation rather than the Darcy or Colebrook correlations, where 96% and 98% of the predicted fluid flow rates were in good agreement with the real measured oil and water flow rate, respectively. Furthermore, the relative errors were less than ±18%, which were still within the reasonable uncertainty, as shown in Figures 17 and 18. For the predicted and measured gas rates, 68% of the wells showed about ±10% relative errors, as shown in Figure 19. By using the Darcy correlation, 63% and 70% of the wells were not in good agreement with the predicted and measured oil and water rate, respectively, with more than ±18% for relative errors, as shown in Figures 20 and 21. For predicted and measured gas rates, 79% of the wells showed more than ±10% relative errors, as shown in Figure 22. By using the Colebrook correlation, 67% and 75% of the wells were not in good agreement with the predicted and measured oil and water rate, respectively, with more than ±18% relative errors, as shown in Figures 23 and 24. For predicted and measured gas rates, 77% of the wells showed more than ±10% relative errors, as shown in Figure 25.

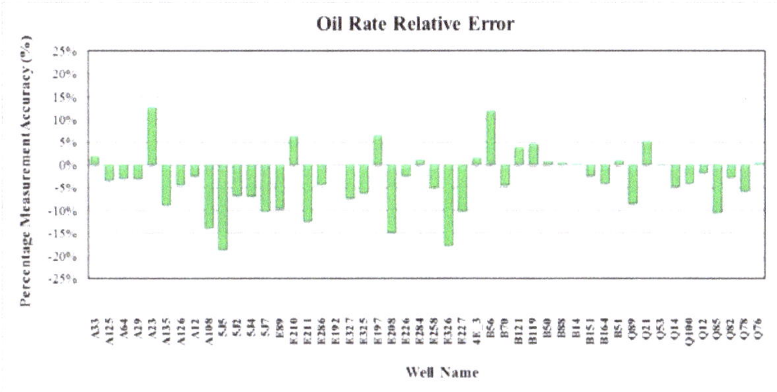

Figure 17. Oil rate measurement accuracy using the Blasius correlation.

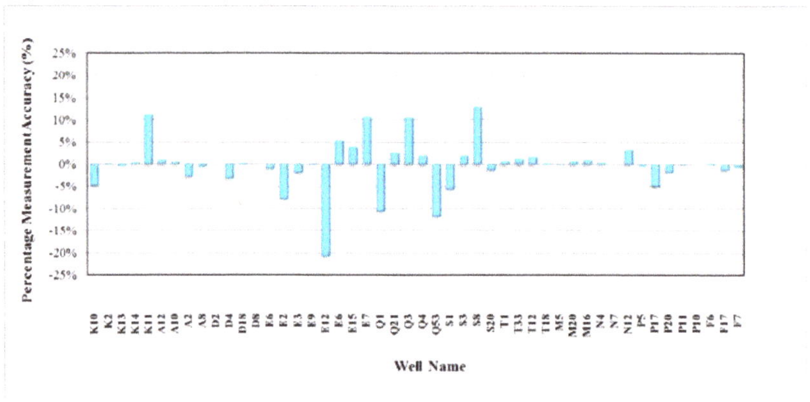

Figure 18. Water rate measurement accuracy using the Blasius correlation.

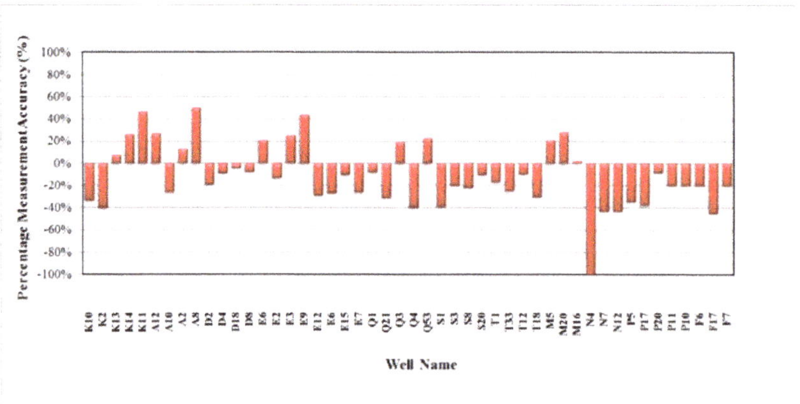

Figure 19. Gas rate measurement accuracy using the Blasius correlation.

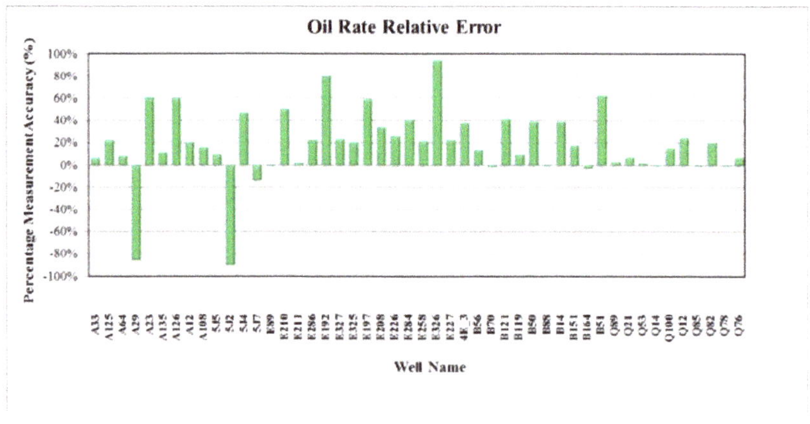

Figure 20. Oil rate measurement accuracy using the Darcy correlation.

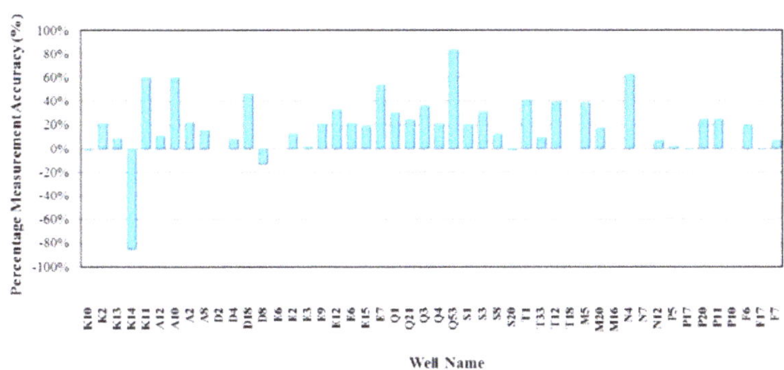

Figure 21. Water rate measurement accuracy using the Darcy correlation.

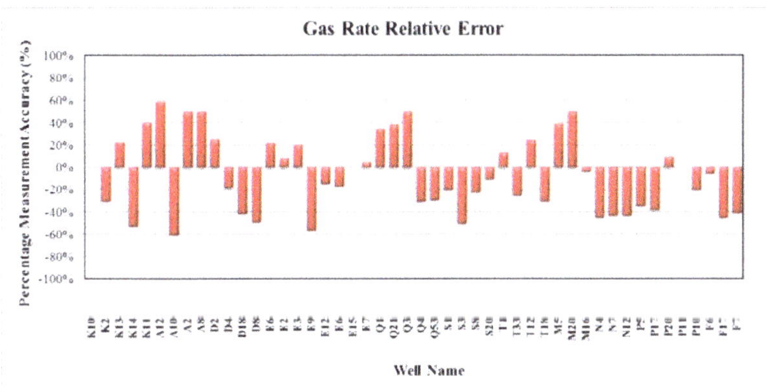

Figure 22. Gas rate measurement accuracy using the Darcy correlation.

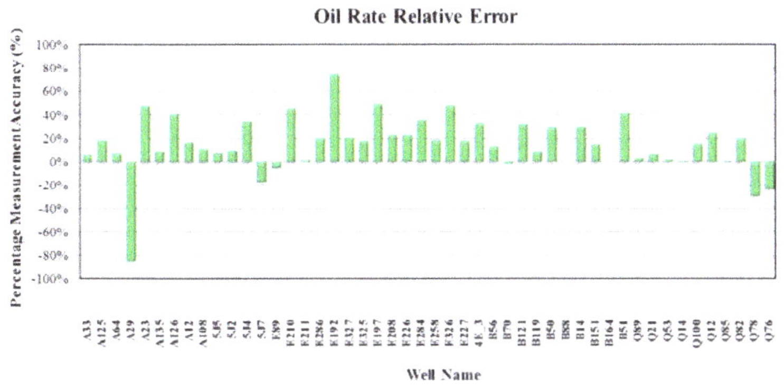

Figure 23. Oil rate measurement accuracy using the Colebrook correlation.

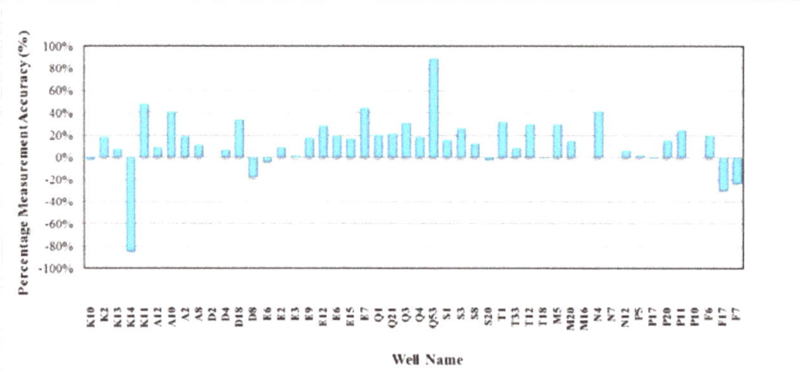

Figure 24. Water rate measurement accuracy using the Colebrook correlation.

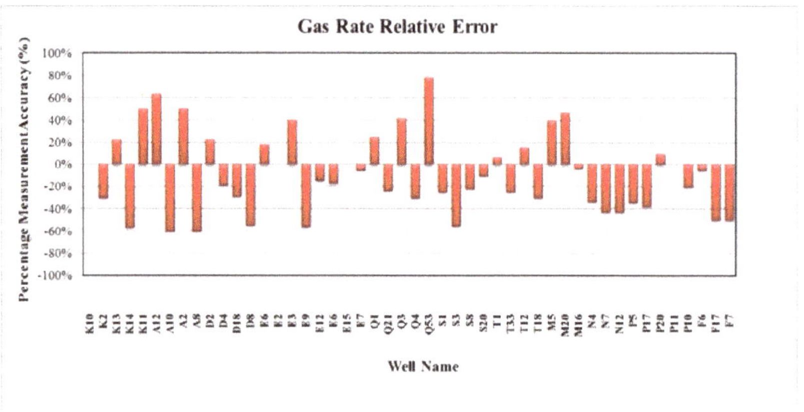

Figure 25. Gas rate measurement accuracy using the Colebrook correlation.

The results showed high relative errors in gas rate prediction which can happen due to the oil separator meters being insufficiently accurate. Also, these errors may occur due to fixed orifice plate meters used to measure the gas flow rate despite the fact that orifice plates are not appropriate to measure low gas rates. Besides, wear and corrosion can increase the orifice size and cause excessive loss.

4. Summary and Conclusions

The prediction of the fluid flow rate of oil wells using the new mathematical model has been made and validated with experimentally measured fluid flow rate data. To evaluate the influence of the frictional pressure drop value on the measurement of fluid flow rate of oil wells, Blasius, Darcy, and Colebrook friction correlations were applied. Using the Blasius correlation, the analysis showed that the predicted fluid flow rate values were in accord with the measured values, while by using the Darcy and Colebrook friction correlations, the results were not in good agreement with the measured values. This discrepancy was due to the fact that each friction correlation found the friction factor differently. To determine the friction factor, many expressions were used to compute the Reynolds number. Essentially, each empirical correlation states its own assumptions and modifications to defend the variable components in order to be applicable to multiphase conditions. The two-phase flow significantly complicated the pressure drop calculations, where any errors in determining the frictional pressure drop values would generate some inaccuracies in predicting the fluid flow rate of the oil wells. Consequently, mixture properties and the interactions between the existing phase's properties must be considered. Therefore, the gas and liquid volume fractions throughout the conduit needed to be determined. Overall, the performance of the new mathematical model indicated that the selection of the appropriate friction factor correlation would lead to predicting the gas and liquid flow rate within the acceptable accuracy. However, the friction loss dominated only with very high flow rates. For relatively small flow rates, the hydrostatic pressure played the key role in the overall pressure drop in the vertical tubing. Thus, different multiphase flow models, either empirical or mechanistic model, used in the computation would output different predictions. That being said, the Blasius equation may be superior to other models coupled with the Hagedorn-Brown empirical correlation, as it has been shown in this work. Indeed, a very reasonable average relative error of 4.6% was observed between the predicted and measured flow rates. However, it may not be as good as it is when coupled with other mechanistic models that may further reduce this error. Further research is needed to further validate the developed model by accounting for other sophisticated multiphase models.

Author Contributions: T.A.G. derived the mathematical model, carried out the experimental validation and drafted the manuscript. M.H. contributed to data presentation, analysis of results and revision preparation.

Funding: The authors would like to acknowledge the Petroleum Engineering Department and Centre of Research in Enhanced Oil Recovery (COEOR) at Universiti Teknologi PETRONAS for the funding (YUTP-0153AA-E70) and the technical assistance.

Acknowledgments: Special thanks are due to production engineering staff of Waha Oil Company for their support. The authors also gratefully acknowledge assistance received from Lynn Mason for editing this manuscript.

Conflicts of Interest: The authors declare no conflict of interest.

Nomenclature

A	cross-sectional area, (sq ft)
API	American Petroleum Institute
B_o	oil formation volume factor, (bbl/stb)
B_{ob}	oil formation volume (at bubble point pressure), (bbl/STB)
B_g	gas formation volume factor, (cf/scf)
Cnt	count
dp/dz	pressure gradient, (psi/ft)
d	inside diameter, (ft)
ESP	electrical submersible pump
f	friction factor, (unitless)
g	Gravity, (ft/s^2)
H_L	liquid hold-up
H_G	gas hold-up
$H1$	bubble point pressure (at location depth before shut-in the well head valve), (ft)
$H2$	bubble point pressure (at location depth after shut-in the well head valve), (ft)
mt	mass flow rate, (lb/day)
N_{gv}	gas velocity number, (unitless)
N_{Lv}	liquid velocity number, (unitless)
N_d	pipe diameter number, (unitless)
N_{CL}	coefficient number of viscosity correction, (unitless)
N_L	liquid viscosity number, (unitless)
q_o	oil rate, (stb/day)
q_g	gas rate, (stb/day)
q_w	water rate, (stb/day)
q_L	liquid rate, (stb/day)

q_m	measured flow rate, (stb/day)
Q_C	quality check
P	pressure, (psia)
P_r	pseudo-critical pressure (for gas mixture), (psia)
P_b	bubble point pressure, (psia)
P_{sc}	pressure at standard conditions (P = 14.7 atm, T = 60 °F), (psia)
PSD	pump setting depth
SGG	gas specific gravity
STB	stock tank barrel (for liquid)
r_w	wellbore radius, (ft)
R_s	gas-oil ratio, (scf/stb)
R_{sb}	gas oil ratio at bubble point pressure, (cf/scf)
R_e	Reynolds number, (unitless)
T	temperature, (°F)
t	total shut-in time, (min)
T_r	pseudo-critical temperature (for gas mixture), (psia)
T_{sc}	temperature at standard condition, (°R)
T_r	reservoir fluid temperature, (°F)
VR	gas volume at reservoir conditions, (ft^3)
V_{sc}	gas volume at standard condition, (ft^3)
V_{SL}	superficial velocity for liquid, (ft/sec)
V_{Sg}	superficial velocity for gas, (ft/sec)
V_m	total mixture velocity, (ft/sec)
WHP_a	well head pressure (after shut-in the well), (psia)
WHP_b	well head pressure (before shut-in the well), (psia)
WC	water cut (unitless)
WHT	well head temperature, (°F)
W	water vapor density, (unitless)
Z	gas compressibility factor (unitless)
Greek Symbols	
ΔP	pressure drop, (psia)
H_L/ψ	hold-up correlation factor
γ_o	oil gravity
γ_w	water gravity
γ_g	gas gravity
σ	surface tension, (dyne/m)
ΔH	differences between bubble point pressure location depths (before and after shut-in the well head valve), (ft)
ρ_o	oil density, (lbm/ft^3)
ρ_g	gas density, (lbm/ft^3)
ρ_w	water density, (lbm/ft^3)
ρ_L	liquid density, (lb/ft^3)
ρ_m	mixture density, (lbm/ft^3)
μ_o	oil viscosity, cP
μ_g	gas viscosity, cP
μ_L	liquid viscosity, cP
Subscripts	
g_{sc}	gas (at standard condition)
h	hydrostatic
L	liquid
m	mixture (liquid and gas)
o	oil
sc	standard condition
w	water

References

1. Saeb, M.B.; Philip, D.M.; David, C.C.; Ali, J. Modeling friction factor in pipeline flow using a GMDH-type neural network. *Cogent Eng.* **2015**, *2*. [CrossRef]
2. Shannak, B.A. Frictional pressure drop of gas liquid two-phase flow in pipes. *Nuclear Eng. Des.* **2008**, *238*, 3277–3284. [CrossRef]
3. Jiang, J.Z.; Zhang, W.M.; Wang, Z.M. Research progress in pressure-drop theories of gas-liquid two-phase pipe flow. In Proceedings of the China International Oil & Gas Pi (CIPC 2011), Langfang, China, 5 September 2011.
4. Xu, Y.; Fang, X.D.; Su, X.H.; Zhou, Z.R.; Chen, W.W. Evaluation of frictional pressure drop correlations for two-phase flow in pipes. *Nuclear Eng. Des.* **2012**, *253*, 86–97. [CrossRef]
5. Mittal, G.S.; Zhang, J. Friction factor prediction for Newtonian and non-Newtonian fluids in pipe flows using neural networks. *Int. J. Food Eng.* **2007**, *3*, 1–18. [CrossRef]
6. Fadare, D.A.; Ofidhe, U.I. Artificial neural network model for prediction of friction factor in pipe flow. *J. Appl. Sci.* **2009**, *5*, 662–670.
7. Bilgil, A.; Altun, H. Investigation of flow resistance in smooth open channels using artificial neural networks. *Flow Meas. Instrum.* **2008**, *19*, 404–408. [CrossRef]
8. Özger, M.; Yildirim, G. Determining turbulent flow friction coefficient using adaptive neuro-fuzzy computing techniques. *Adv. Eng. Softw.* **2009**, *40*, 281–287. [CrossRef]
9. Shayya, W.H.; Sablani, S.S.; Campo, A. Explicit calculation of the friction factor for non-Newtonian fluids using artificial neural networks. *Dev. Chem. Eng. Miner. Process.* **2005**, *13*, 5–20. [CrossRef]
10. Yuhong, Z.; Wenxin, H. Application of artificial neural network to predict the friction factor of open channel flow. *Commun. Nonlinear Sci. Numer. Simul.* **2009**, *14*, 2373–2378. [CrossRef]
11. Yazdi, M.; Bardi, A. Estimation of friction factor in pipe flow using artificial neural networks. *Can. J. Autom. Control Intell. Syst.* **2011**, *2*, 52–56.
12. Sablani, S.S.; Shayya, W.H. Neural network based non-iterative calculation of the friction factor for power law fluids. *J. Food Eng.* **2003**, *57*, 327–335. [CrossRef]
13. Griffith, P. Two-Phase Flow in Pipes. In *Special Summer Program*; Massachusetts Institute of Technology: Cambridge, MA, USA, 1962.
14. Raxendell, P.B. The Calculation of Pressure Gradients in High-Rate Flowing Wells. *J. Pet. Technol.* **1961**, *13*, 1023. [CrossRef]
15. Fancher, G.H.; Brown, K.E. Prediction of Pressure Gradients for Multiphase Flow in Tubing. *Soc. Pet. Eng. J.* **1963**, *3*, 59. [CrossRef]
16. Hagedorn, A.R.; Brown, K.E. The Effect of Liquid Viscosity in Vertical Two-Phase Flow. *J. Pet. Technol.* **1964**, *16*, 203. [CrossRef]
17. Poettmann, F.H.; Carpenter, P.G. The Multiphase Flow of Gas, Oil and Water through Vertical Flow Strings with Application to the Design of Gas-Lift Installations. In Proceedings of the Drilling and Production Practice, New York, NY, USA, 1 January 1952; Volume 52, p. 257.
18. Tek, M.R. Multiphase Flow of Water, Oil and Natural Gas Through Vertical Flow Strings. *J. Pet. Technol.* **1961**, *13*, 1029. [CrossRef]
19. Shaban, H.; Tavoularis, S. Identification of flow regime in vertical upward air–water pipe flow using differential pressure signals and elastic maps. *Int. J. Multiph. Flow* **2014**, *61*, 62–72. [CrossRef]
20. Daev, Z.A.; Kairakbaev, A.K. Measurement of the Flow Rate of Liquids and Gases by Means of Variable Pressure Drop Flow Meters with Flow Straighteners. *Meas. Tech.* **2017**, *59*, 1170–1174. [CrossRef]
21. Cai, B.; Guo, D.X.; Jing, F.W. Study on gas-liquid two-phase flow patterns and pressure drop in a helical channel with complex section. In Proceedings of the 23rd International Compressor Engineering Conference, West Lafayette, IN, USA, 11–14 July 2016.
22. Oliveira, J.L.G.; Passos, J.C.; Verschaeren, R.; Van der Geld, C. Mass flow rate measurements in gas-liquid flows by means of a venturi or orifice plate coupled to a void fraction sensor. *Exp. Therm. Fluid Sci.* **2009**, *33*, 253–260. [CrossRef]
23. Brown, G.O. The history of the Darcy-Weisbach equation for pipe flow resistance. *Environment.* Available online: https://ascelibrary.org/doi/abs/10.1061/40650 (accessed on 17 September 2018).
24. Blasius, H. *Das Ähnlichkeitsgesetz bei Reibungsvorgängen in Flüssigkeiten, Mitteilung 131 über Forschungsarbeiten auf dem Gebiete des Ingenieurwesens*; Springer: Berlin, Germany, 1913.

25. Colebrook, C.F. Turbulent flow in pipes, with particular reference to the transition region between the smooth and rough pipe laws. *J. Inst. Civ. Eng.* **1939**, *11*, 133–156. [CrossRef]

26. Lee, A.L.; Gonzalez, M.H.; Eakin, B.E. The Viscosity of Natural Gases. *J. Pet. Technol.* **1966**, *18*, 997–1000. [CrossRef]

27. Beggs, D.H.; Brill, J.P. A study of two-phase flow in inclined pipes. *J. Pet. Technol.* **1973**, *25*, 607–617. [CrossRef]

28. Standing, M.B.; Katz, D.L. Density of natural gases. *Trans. AIME* **1942**, *146*, 140–149. [CrossRef]

29. Sloan, E.; Khoury, F.; Kobayashi, R. Measurement and interpretation of the water content of a methane-propane mixture in the gaseous state in equilibrium with hydrate. *Ind. Eng. Chem. Fundam.* **1982**, *21*, 391–395.

30. Vazquez, M.; Beggs, H.D. Correlations for Fluid Physical Property Prediction. *J. Pet. Technol.* **1980**, *32*, 968–970. [CrossRef]

31. Hagedorn, A.R.; Brown, K.E. Experimental Study of Pressure Gradients Occurring during Continuous Two-Phase Flow in Small Diameter Vertical Conduit. *J. Pet. Technol.* **1965**, *17*, 475–484. [CrossRef]

32. Duns, H., Jr.; Ros, N. Vertical Flow of Gas and Liquid Mixtures in Wells. In Proceedings of the 6th World Petroleum Congress, Frankfurt am Main, Germany, 19–26 June 1963; p. 451.

33. Orkiszewski, J. Predicting Two-Phase Pressure Drops in Vertical Pipe. *J. Pet. Technol.* **1967**, *19*, 829–838. [CrossRef]

34. Aziz, K.; Govier, G.; Fogarasi, M. Pressure Drop in Wells Producing Oil and Gas. *J. Can. Pet. Technol.* **1972**, *11*, 38. [CrossRef]

Article

Investigation on the Handling Ability of Centrifugal Pumps under Air–Water Two-Phase Inflow: Model and Experimental Validation

Qiaorui Si [1,*], Gérard Bois [2], Qifeng Jiang [3], Wenting He [1], Asad Ali [1] and Shouqi Yuan [1]

[1] National Research Center of Pumps, Jiangsu University, Zhenjiang 212013, China; 741070483@qq.com (W.H.); 3360719084@qq.com (A.A.); shouqiy@ujs.edu.cn (S.Y.)
[2] LMFL, FRE CNRS 3723, ENSAM, 8 boulevard Louis XIV, 59046 Lille CEDEX, France; Gerard.BOIS@ENSAM.EU
[3] Key Laboratory of Fluid and Power Machinery, Xi Hua University, Chengdu 610039, China; qifeng.jiang@mail.xhu.edu.cn
* Correspondence: siqiaorui@ujs.edu.cn; Tel.: +86-136-5529-3881

Received: 10 October 2018; Accepted: 2 November 2018; Published: 6 November 2018

Abstract: The paper presents experimental and numerical investigations performed on a single stage, single-suction, horizontal-orientated centrifugal pump in air–water two-phase non-condensable flow conditions. Experimental measurements are performed in a centrifugal pump using pressure sensor devices in order to measure the wall static pressures at the inlet and outlet pump sections for different flow rates and rotational speeds combined with several air void fraction (*a*) values. Two different approaches are used in order to predict the pump performance degradations and perform comparisons with experiments for two-phase flow conditions: a one-dimensional two-phase bubbly flow model, and a full "Three-Dimensional Unsteady Reynolds Average Navier–Stokes" (3D-URANS) simulation using a modified *k-epsilon* turbulence model combined with the Euler–Euler inhomogeneous two-phase flow description. The overall and local flow features are presented and analyzed. Limitations concerning both approaches are pointed out according to some flow physical assumptions and measurement accuracies. Some additional suggestions are proposed in order to improve two-phase flow pump suction capabilities.

Keywords: two-phase flow; pump performance; computational fluid dynamics; centrifugal pump; flow behavior

---

1. Introduction

Pumps play an important role in many energy engineering applications. Their main advantages are their high operational efficiency and reliability. In case of inlet two-phase flow conditions such as irrigation, refrigeration, petroleum, nuclear power industries, sewage treatment, etc., pump performance always fails under air–water two-phase inflow compared to single-phase operations. Especially, the need for the safe operation and design of a nuclear power reactor is of paramount importance to assure the safety evaluation of hypothetical loss-of-coolant accidents (LOCA) in light water reactors. Therefore, the investigation of pump behavior during blowdown is very important.

The degree of head degradation depends on the geometrical, physical, and thermal conditions. Compared with an axial flow pump that is able to suck high inlet gas volume fraction values, centrifugal pumps are less powerful when handling two-phase flow conditions (Gülich [1]). When the inlet gas void fraction increases, the pump head can be totally lost, and this often results in huge system instabilities and pump failure. Such a situation should be controlled using adequate prediction methods. Some improvements on pump ability to suck two-phase flow can also be achieved by a better understanding of local flow features. Several models have been developed in order to predict the two-phase flow performance of both the axial and centrifugal types of pumps. A first correlation was proposed in 1978 by Mikielewicz et al. [2], based on the concept of apparent loss coefficient. Among several theoretical approaches, the one-dimensional two-phase flow models proposed by Murakami et al. [3] and Minemura et al. [4] are still considered sufficiently accurate for engineering purposes for the prediction of pump surge or shut-off conditions.

A non-exhaustive list of important published works is proposed in the paper of Si et al. [5]. Related proposed models can be considered to be valid for low values of inlet void fraction (maximum 6–7% in volume). They cannot accurately predict the surge operating conditions that correspond to a rapid decrease in the pump performance, just because they only give mean flow characteristics along one mean streamline with no local information inside the impeller whole passage. Just before such severe conditions, several investigations have also detected the presence of stationary bubbles at impeller entrance channels for high gas fractions, being responsible for the starting point of the strong performance degradation of the pump (Sekoguchi et al. [6], Estevam et al. [7] and Barrios [8]). Numerical simulations, using the so-called "Three-Dimensional Unsteady Reynolds Average Navier–Stokes" (3D-URANS) approach, have also been applied to describe local phenomena more precisely in such flow conditions. They generally show significant deviations between predicted and experimental overall results when inlet void fraction values are higher than 6%. Recently, Müller et al. [9] have performed numerical calculations using a monodispersed phase distribution model, the results of which are compared with the experimental results obtained by Suryawijaya and Kosyna [10]. The comparison of impeller blade static pressure distributions showed that some improvements should be done for a high inlet gas flow rate in order to get a better fit between the numerical and experimental results. This means that there is still a need for experimental results on low specific speed pump geometries in order to obtain more data for numerical assessments.

The present paper presents a compilation of new experimental and numerical analysis performed by the same research team, starting from three years ago by Si et al. [5,11]. In the first part, the measured overall performances of the pump (head and global efficiency versus flow coefficients) are presented for three different rotational speeds and different inlet void fraction values up to pump breakdown. The results are then compared with the one-dimensional two-phase flow model developed by Minemura et al. [4]. In the second part, a comparison is proposed between the experimental results and 3D-URANS simulations using inhomogeneous flow model assumption. Based on the local flow feature analysis obtained from 3D-URANS simulations, radial impeller design modification is proposed to improve pump ability with respect to multiphase flows.

2. Pump Geometry and Experimental Test Rig

A single stage, low specific speed ($\Omega s = 0.65$) centrifugal commercial pump is chosen for two-phase flow investigations. The impeller casing is combined with a vaneless spiral volute. The design parameters of the pump, given by the manufacturer, are as follows: design volume flow rate: $Q_d = 50.6 \text{ m}^3/\text{h}$, design head: $H_d = 20.2 \text{ m}$, design speed $N_d = 2910 \text{ r/min}$, pump horizontal inlet tube diameter: $D = 65 \text{ mm}$. The impeller inlet and outlet diameters are equal to 79 mm and 140 mm, respectively, and the impeller outlet width is equal to 15.5 mm. The six blades' inlet geometry corresponds to a three-dimensional twisted shape. The volute's law evolution obeys the Archimedes' spiral shape. The test rig is shown in Figure 1. As shown in this open loop, the air injection system is able to provide 0.2 mm diameter bubbles using a specific compressor device (more details can be found in Si et al. [5,11]). The air flow rate is measured by a microelectromechanical flow sensors system that could supply the volume air flow rate value on standard conditions (25 °C, 101.326 kPa). A gas–liquid flow mixture is sucked by the pump, goes through the regulating valve, and finally goes into the downstream tank. This open tank allows air bubbles to be separated from the water up to a second tank (upstream tank). The water flow rate was measured by an electromagnetic flowmeter set between the upstream tank and the air supply device. Measurements were performed using the followed procedure: a constant inlet void fraction was set by changing the throttle vane position and the corresponding water flow rate was consequently obtained. Measurement uncertainties are evaluated as a 1.2% error on the pump head and a 2.4% error regarding pump global efficiency. Figure 2 shows a general view of the sensor locations close to the pumping environment. Pump head and global efficiency values were obtained following ISO 9906:2012. It has to be noticed that only the air volume flow rate was measured at this step, as the bubble diameter distribution, including bubble number per unit volume at pump inlet, was not available.

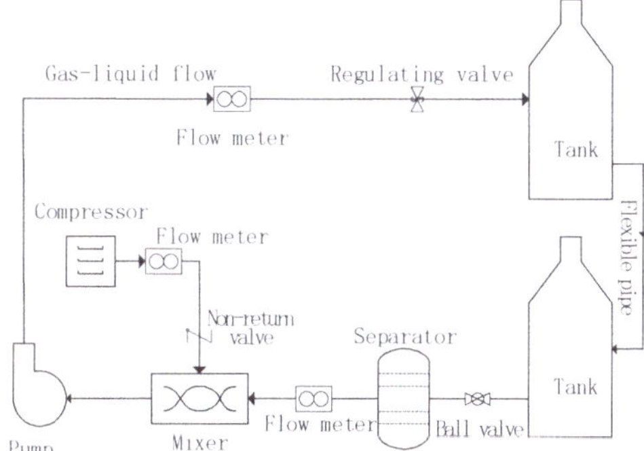

Figure 1. Test rig.

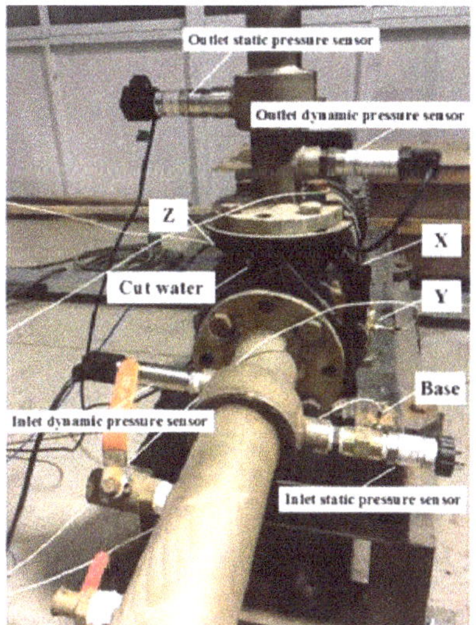

Figure 2. Sensor locations.

3. Numerical Model and Setups

3.1. Model and Mesh

Software Pro/E 5.0 was used for three-dimensional modeling and assembly. The computational domain included five separated domains: inlet pipe, ring, impeller, volute, and disc chambers, as illustrated in Figure 3. The hexahedron structured mesh was used to easily control the density of each node and properly adjust all of the nodes. In order to improve the accuracy of the simulation, the boundary layer grid was added on the volute shell wall after the local encryption of the volute tongue. The resulting pump model that consisted of 2,775,915 elements was chosen for rotating and stationary domains in total after grid independence analysis (see the paper of Si et al. [11]), the result of which is shown in Figure 4.

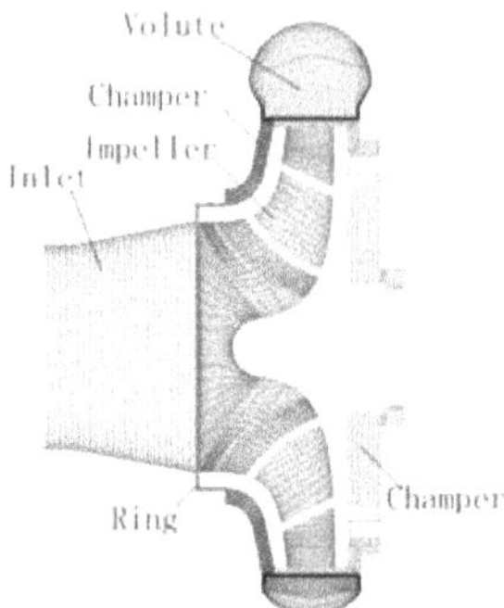

Figure 3. Three-dimensional (3D) numerical domain (partial) with local meshing.

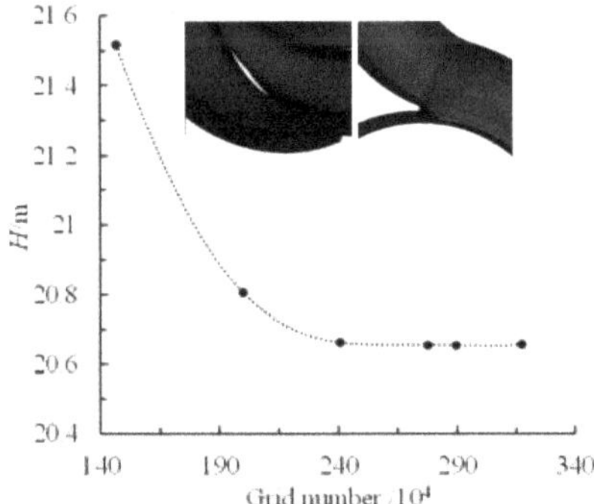

Figure 4. Grid independence results.

3.2. The Eulerian–Eulerian Inhomogeneous Two-Phase Flow Model

There are two kinds of multiphase flow models: homogeneous and inhomogeneous ones. The former assumes that the two-phase velocities are the same, not considering any velocity slip between the two phases. The latter considers not only the velocity slip, but also the interphase mass and momentum transfer terms. In the inhomogeneous model, each phase has its own fluid field and passes through the phase transfer unit. That is to say, each phase has its own velocity and temperature field, if any. This paper adopts the inhomogeneous model regardless of the temperature field, for which the liquid phase is continuous and the gas phase is discrete. The particle model assumes that the gas–liquid two-phase flow pattern corresponds to a bubbly flow, meeting the principle of the mass and momentum conservation:

$$\frac{\partial}{\partial t}(\alpha_k \rho_k) + \nabla \cdot (\alpha_k \rho_k \boldsymbol{w}_k) = 0 \tag{1}$$

$$\frac{\partial}{\partial t}(\alpha_k \rho_k \boldsymbol{w}_k) + \nabla \cdot (\alpha_k \rho_k \boldsymbol{w}_k \otimes \boldsymbol{w}_k) = -\alpha_k \nabla p_k + \nabla \cdot (\alpha_k \mu_k (\nabla \boldsymbol{w}_k + (\nabla \boldsymbol{w}_k)^{\mathrm{T}})) + \boldsymbol{M}_k + \boldsymbol{f}_k \tag{2}$$

where k—phase (l-liquid, g-gas); ρ_k—density of the k phase, kg/m³; p_k—pressure of k phase, Pa; α_k—void fraction of k phase; μ_k—dynamic viscosity of k phase, Pa·s; $\boldsymbol{w}_k$—relative velocity of the k phase fluid, m/s; $\boldsymbol{M}_k$—due to interphase drag force; and $\boldsymbol{f}_k$—added mass force related to the contribution of the impeller rotation.

This also satisfied:

$$\alpha_g = 1 - \alpha_l = \frac{Q_g}{Q_g + Q_l} \tag{3}$$

where α_g—gas void fraction; α_l—liquid void fraction; Q_g—volume flow rate of the gas, m^3/h; and Q_l—volume flow rate of the liquid, m^3/h.

For this two-phase flow approach, the liquid phase is considered the continuous phase using a modified turbulence model that is based on the renormalization group (RNG) k-ε model [12]. Meanwhile, the gas phase is considered as the discrete phase using the zero-equation theoretical model, which means the action between the two phases only considers the so-called interfacial drag coefficient through the following relations [13]:

$$M_l = -M_g = \frac{3}{4} c_D \frac{\rho_l}{d_B} \alpha_g (w_g - w_l) |w_g - w_l| \tag{4}$$

with:

$$c_D = \begin{cases} \frac{24}{R_e}(1 + 0.15Re^{0.687}) & (R_e \leq 1000) \\ 0.44 & (R_e > 1000) \end{cases} \tag{5}$$

$$\text{And } Re = \rho_l \frac{|w_g - w_l|}{\mu_l} d_B \tag{6}$$

where d_B—diameter of the bubble, c_D—resistance coefficient.

3.3. The Modified Turbulence Model

As known, the turbulence model plays a significant role in the prediction of pumping two-phase flow. Since two-phase flow in centrifugal pumps is also an unsteady multiphase compressible flow concerning a liquid phase and a bubble phase, a corrected method was applied to solve that physical phenomenon. This method was proved effective on the prediction of cavitation flow in centrifugal pumps by Wang et al. [14]. The liquid density is defined as:

$$\rho_l = \rho_{ref} \sqrt[n]{\frac{p + C}{p_{ref} + C}} \tag{7}$$

where P_{ref} stands for the reference pressure at different void fraction, and ρ_{ref} is set as 998.2 kg/m^3. As for constant values of C and n, they are set to 300 MPa and seven for water, respectively, according to Dular and Coutier-Delgosha [15].

Moreover, the standard k-e model is deemed to overestimate the eddy viscosity in the mixture region. It cannot effectively resolve the detachment of the flow separation from the solid surface, and excessively attenuates the two-phase flow instability. Therefore, a modified RNG k-e model, as proposed by Coutier-Delgosha et al. [16], was employed in this work. It can successfully reduce the eddy viscosity by defining the turbulent viscosity, and based on RNG k-ε turbulence model, Johansen's filter-based model is used to correct the turbulent viscosity μ_t:

$$\mu_t = f(\rho_m) C_\mu \frac{k^2}{\varepsilon} \tag{8}$$

where $f(\rho_m)$ is the expression of the density correction which can be determined as:

$$f(\rho_m) = \begin{cases} 998.2 \ kg/m^3 & (n = 1) \\ \rho_g + \frac{(\rho_m - \rho_g)^n}{(\rho_l - \rho_g)^{n-1}} & (n > 1) \end{cases} \tag{9}$$

where $f(\rho_m)$ changed with the density of the mixed flow following the rules, as shown in Figure 5. Coutier-Delgosha et al. [16] validated that n should be equal to 10 after comparing the experimental and numerical results.

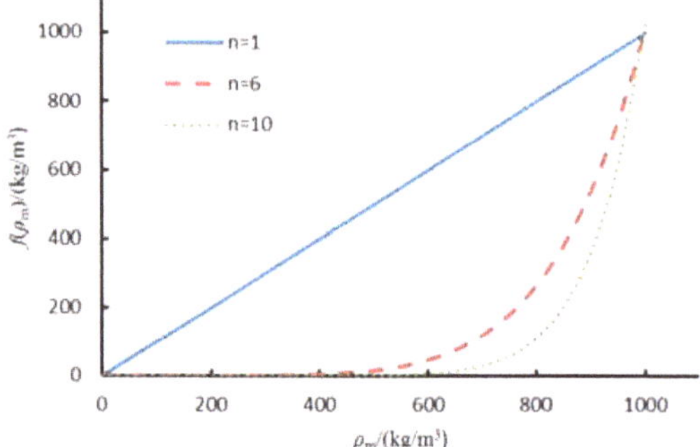

Figure 5. The law of the $f(\rho_m)$ with the density in the mixed phase.

Finally, the modified RNG k-ε turbulence model becomes as follows:

$$\frac{\partial(\rho_m k)}{\partial t} + \frac{\partial(\rho_m k \overline{u_j})}{\partial x_j} = \frac{\partial}{\partial x_j}\left[\left(\mu_m + \frac{\mu_t'}{\sigma_k}\right)\frac{\partial k}{\partial x_j}\right] + P_t - \rho_m \varepsilon \tag{10}$$

$$\frac{\partial(\rho_m \varepsilon)}{\partial t} + \frac{\partial(\rho_m \varepsilon u_i)}{\partial x_i} = \frac{\partial}{\partial x_j}\left[\left(\mu_m + \frac{\mu_t'}{\sigma_\varepsilon}\right)\frac{\partial \varepsilon}{\partial x_j}\right] + C_{\varepsilon 1}\frac{\varepsilon}{k} - C_{\varepsilon 2}\rho_m\frac{\varepsilon^2}{k} - R \tag{11}$$

Solutions of the simulations were processed in Ansys CFX 14.1 by adding the modified coefficients programs using CFX Expression Language (CEL).

3.4. Boundary Conditions

Considering the requirements of the test, the inlet boundary conditions are set according to the pressure measured in the experiment. A certain amount of gas is imported into the inlet, and the outlet boundary conditions are set as the mass flow rate. The liquid boundary adopts the non-slip solid wall condition, and the gas boundary adopts the free-slip solid wall condition. The smooth wall condition is used for the near-wall function. The initial inlet bubble diameter is set as 0.2 mm, according to the experimental setup device. The time step was set as 1.718×10^{-4} s, corresponding to a 3° impeller rotation for each step. A total time set was 0.206 s for simulated results storage in order to further process a data reduction of the unsteady flow field, corresponding to 10 impeller revolutions for which stable results were obtained.

4. Overall Pump Performances

In previous experimental work [5], as well as for pure water, pump head and global efficiency curves were found to be independent of rotational speed inside the range of 1800–2910 r/min. This result agrees with the similarity laws that have been applied for incompressible flow for impeller Reynolds number values (based on impeller diameter and angular speed) that are higher than the critical one that is usually adopted for turbomachinery applications. The rotational speed of 2910 r/min corresponds to the rated one. This result was achieved by considering that the maximum experimental error is 6% for the head at partial flow rates, and 2% around the maximum efficiency point (experimental error combines both measurement device accuracy and local unsteadiness due to two-phase flow stability problems).

Figure 6 shows the experimental pump performance head coefficient ψ_{tp} (defined in Equation (12)) versus flow coefficient φ for different inlet void fraction values for the rated rotational speed.

The corresponding theoretical head curve coefficient obtained with different rotational speeds is shown in Figure 7.

$$\psi_{tp} = \frac{gH}{u_2^2}\left(\frac{\rho_l}{\rho_g \times \alpha + \rho_l \times (1-\alpha)}\right) \tag{12}$$

$$\psi_{tp,th} = \psi_{tp}/\eta \tag{13}$$

As already pointed out by several previous researchers, pump performance starts to be significantly lower when the void fraction reaches 3% or more. The value of the head performance degradation also depends on the flow coefficient. For a given loop characteristic curve, both the head and volume flow rate decrease when a increases. A decrease in 20% in the head coefficient compared with single phase shut-off conditions is achieved for all of the water flow rates that are below nominal conditions when a reach 7%. A maximum a value, which is close to 10%, can be achieved without pump surge for a water flow coefficient around 0.065–0.077. These values correspond to flow rate values between 34–40 m^3/h, which are close to the pump-measured best efficiency (reached at $\varphi = 0.084$) for single-flow operations. Pump performance sharply drops when a is more than 8% for all of the flow rates.

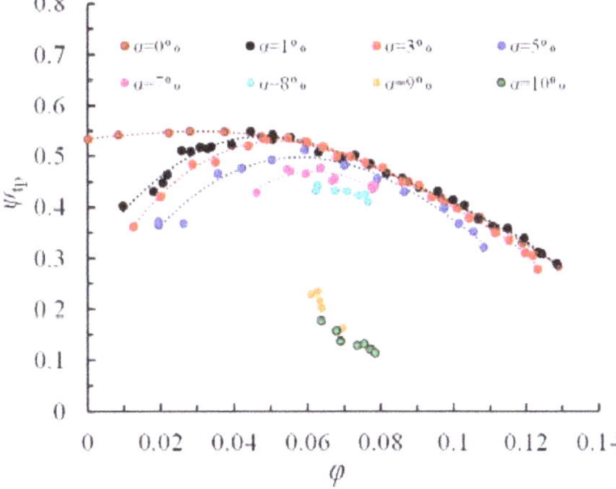

Figure 6. Head coefficient at $n = 2910$ r/min.

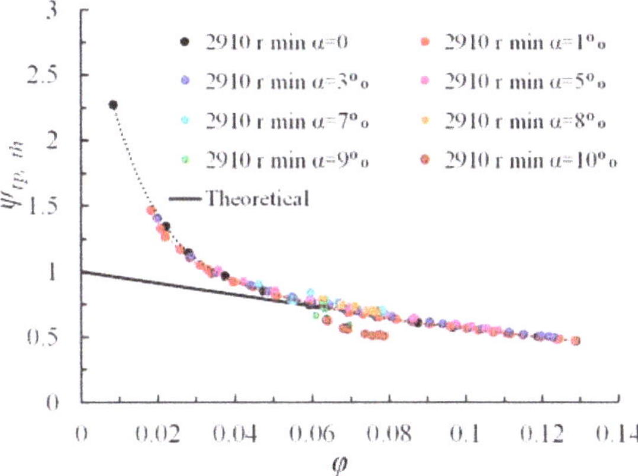

Figure 7. Theoretical head coefficient.

The concept of apparent losses is confirmed by the theoretical head coefficient curve always being the same whatever the inlet void fraction is (up to 8%), as shown in Figure 7. This is the reason why all of the following performance modifications due to two-phase flow conditions are defined by a head ratio ψ^* between the actual head coefficient ψ divided by the head coefficient ψ_0, which is related to the measurement results that are obtained only with water.

The starting point of the severe pump degradation rate is related to a specific flow coefficient corresponding to the change of the negative constant slope of the theoretical head coefficient, as shown in Figure 7. Below a flow coefficient approximately equal to 0.04, the theoretical head curve does not follow a straight line anymore, which corresponds to the hypothesis of a no swirl flow condition at the inlet and a constant outlet relative flow angle at the impeller exit plane. According to this point, an analysis on the influence of rotational speed is proposed in the following section for three different flow coefficients higher than $\varphi = 0.038$ in order to avoid strong swirling recirculation at the pump inlet.

5. Influence of Rotational Speed

Two additional rotational speeds have been selected (respectively: 2300 r/min and 1800 r/min) in order to get pump performance modifications under two-phase flow conditions. Pump experimental degradation performance ratios (ψ/ψ_0) are shown on Figure 8a–c for three decreasing flow coefficient values (solid black lines represent results from one-dimensional models, which are discussed later). The more the rotational speed decreases, the more important the degradation ratio. This is more pronounced when the flow coefficient also decreases. Below $\varphi = 0.058$ and for the lowest rotational speed, the degradation ratio is close to 50% when $a = 8\%$ and more. The combination of low flow rates and high inlet void fraction is probably one of the possible causes for the high level of pump degradation, as already observed by Schäfer et al. [17] in another pump's geometry. One also has to consider two additional important parameters: the increase of local losses due to the bubble drag coefficient, and the Froude number effects. When the flow coefficient decreases (by reducing the flow rate for a given rotational speed and/or reducing the rotational speed and flow rate), the inlet pipe mean velocity decreases accordingly. In the present test of the pump inlet pipe, the limiting value of the inlet velocity, for which a transition from bubbly flow to plug or slug flow occurs, is equal to 2 m/s according to Baker's chart for horizontal pipes [18]. This means that such a transitional flow pattern may happen not only inside the impeller blade and at its blade passages, but also inside the incoming pump pipe, with non-uniform and non-homogeneous inlet flow conditions. According to this analysis, it is not the rotational speed that is the effective parameter, but rather the mean velocity level at the pump inlet. In this respect, the experimental results obtained from the nominal rotational speed are considered to be in fair agreement with the bubbly flow regime assumption (up to a certain value of a around 7–8%), except for below a flow coefficient value $\varphi = 0.045$, which is evaluated as a limiting value. For 2300 r/min, the limiting value of φ is 0.06, and for 1800 r/min's set of results, only the ones corresponding to $\varphi = 0.07$ can be considered to fulfill the hypothesis of the bubbly flow regime. These experimental results will be compared with the one-dimensional approach proposed by Minemura and URANS calculations in a specific section of the present paper.

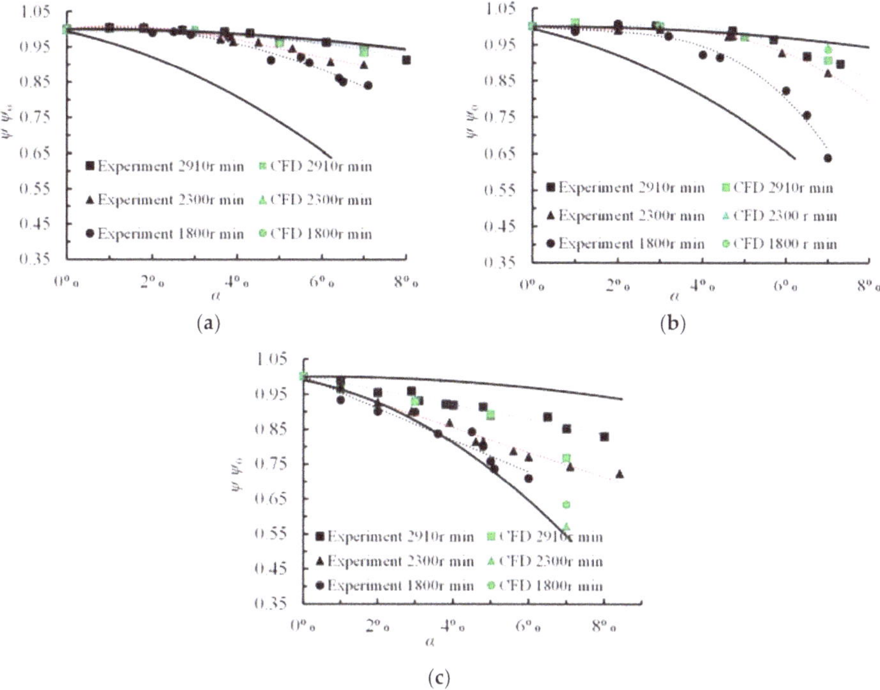

Figure 8. Pump degradation level for three flow coefficients. Comparison between experiments, one-dimensional (1D) model and "Three-Dimensional Unsteady Reynolds Average Navier–Stokes" (3D-URANS) calculations. Solid black lines represent the theoretical evolution from Minemura's approach using two different bubble drag coefficients (upper line for bubbly flow, lower line for churn flow): (**a**) $\varphi = 0.077$; (**b**) $\varphi = 0.058$; (**c**) $\varphi = 0.038$.

6. Numerical Results

6.1. Pump Head Deterioration Ratio ψ^*: Comparisons between Experiments and One-Dimensional Two-Phase Flow Model

Results deduced from Minemura's model [4] are first compared with experimental ones. They correspond to the solid lines plotted in Figure 8a–c. Up to $a = 6\%$, they all give the same performance deterioration ratio as the model used by Tomiyama et al. [19] (see upper solid black curve) and Hench [20] (see lower solid black curve) for all of the flow coefficients at nominal speed (2910 r/min), and the two highest ones for 2300 r/min. The drag coefficient in Equation (2) is also used in the 3D-URANS numerical simulation. Finally, for the smallest rotational speed and flow coefficient values ($\varphi = 0.038$), the evolution of ψ^* looks quite similar to the lower solid black curve (Figure 8c) corresponding to the churn flow regime used for the drag coefficient proposed by Zuber and Hench [21], the value of which depends on the bubble diameter and local void fraction. Since all of the experimental results regarding the pump deterioration ratio are located between these two curves, thus, it is believed that the complete flow structure should be considered not only inside the impeller, but also at the pump inlet section. Possible combinations of bubble and churn flow may arise due to the internal three-dimensional flow structure and void fraction inside the impeller [22]. In addition, it may arise when the flow rate and/or rotational speed are too low, including inlet recirculation. This was obviously not taken into consideration in previous studies when using a one-dimensional model. This is the reason why the URANS approach results are presented in the next section.

6.2. 3D-URANS Overall Performance Results

6.2.1. Comparison between the Initial and Modified Turbulence Model (See Section 3.3)

For pure water (Figure 9) and different inlet void fraction conditions (Figure 10a–d), comparisons of overall pump performance between experiments and 3D-URANS results are presented. For numerical and experimental comparisons on pump overall performance, the final mesh of 2,775,915 elements can be considered quite good (Figure 9) for a wide range of flow coefficients. A small shift in the flow coefficient can be seen probably due to some flow leakage during the experiments.

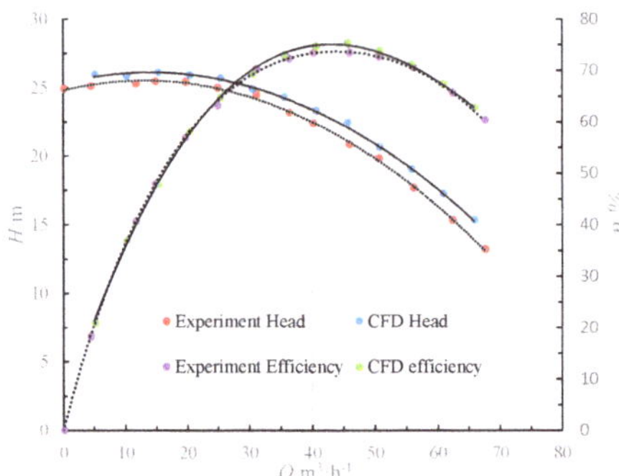

Figure 9. Comparison between experiment and numerical head and efficiency ($a = 0$).

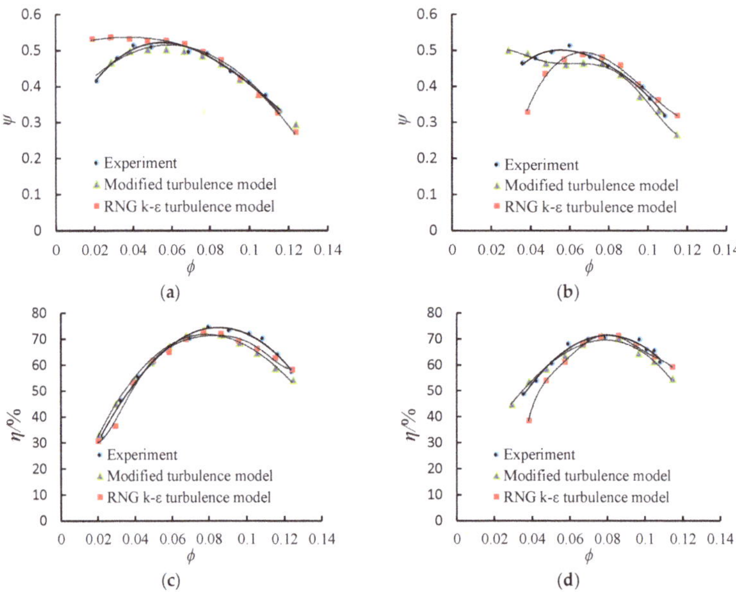

Figure 10. Comparison between experiment and numerical head coefficient and efficiency for four different *a* values: (**a**) Head coefficients when $a = 3\%$; (**b**) Head coefficients when $a = 5\%$; (**c**) Efficiency when $a = 3\%$; (**d**) Efficiency when $a = 5\%$.

When using the modified turbulence model, the numerical results show a better overall performance of the pump than the initial model, especially for low flow coefficients, as shown in Figure 10a,b with quite lower scattering between the experiments and the 3D-URANS calculation. Consequently, the efficiency curves are quite similar to the experimental curves (Figure 10c,d).

6.2.2. 3D-URANS Results Comparison for Different Inlet Void Fraction and Flow Coefficients

These results are also plotted in Figure 8a–c, using green symbols for comparisons with experimental ones. Compared with the one-dimensional models already presented in the previous section, the calculation results give slightly better results when the values of *a* increase up to 7% at the rated rotational speed. The effects of rotational speed are not well predicted, which are probably for the same reasons that have been highlighted previously.

Some additional effects must be taken into consideration when the flow coefficient is more reduced, such as for example pressure level effects on bubble diameter, which is not discussed in the present work. Concerning this particular point, one has shown that the main loss coming from the two-phase flow character appears in the first 20% of the blade passage, where the pressure increase is generally low. There is probably an effect when the rotational speed is decreasing, but it has not been evaluated in the present analysis.

6.3. Local Impeller Passage Flow Structures

6.3.1. Flow at Inlet Impeller Section

A particular focus on inlet flow conditions is shown in Figure 11 for low flow rates in order to illustrate the effects of *a*. Strong reverse inlet pre-rotation flow with high tangential absolute velocities can be seen for pure water. When *a* = 7%, the flow pattern strongly changed at the pump inlet tube section; recirculation does not exist anymore, or is quite small. This is because air is going toward the pipe external radius due to the lower air velocities and the axial adverse pressure gradient along the streamline paths. Consequently, the reverse flow mainly takes place inside the first part of the impeller channel. This may explain what has been experimentally observed when a rapid breakdown of the pump occurs.

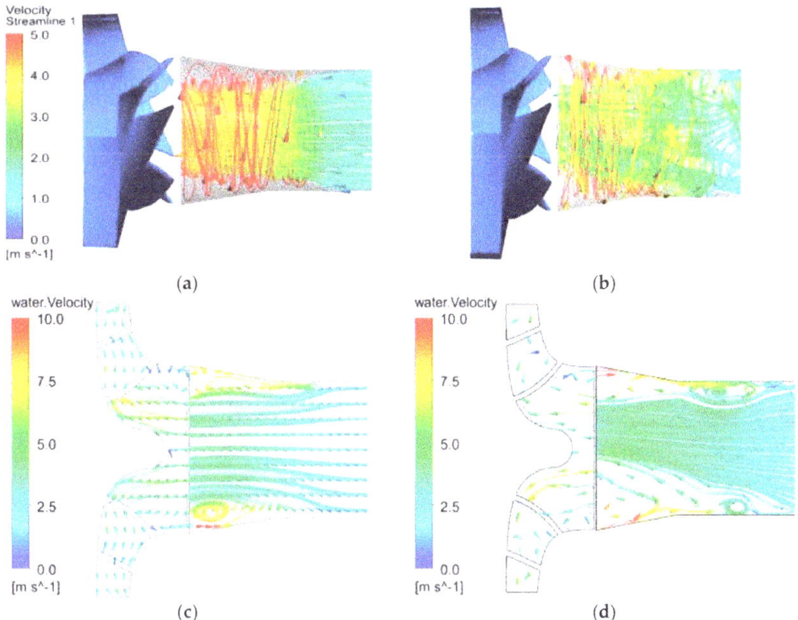

Figure 11. Local two-phase flow pattern inside the impeller passage (not including impeller wall regions), nominal rotational speed, φ = 0.038: (**a**) *a* = 0% (**b**) *a* = 7%; (**c**) *a* = 0%; (**d**) *a* = 7%.

The following pictures in Figure 12 describe for a constant value of *a* = 7%, the evolution of the surfaces of the constant local void fraction α = 15% for two different water flow rates. Although interpretation difficulties arise in such complicated flow patterns, one can detect the air–water mixture locations at the upstream inlet bend of the impeller hub, and the shroud sections inside the regions indicated by red color areas. The more the flow rate decreases, the more it spreads and tends to block the inlet shroud region. This is the reason why the local void fraction appears partly on the suction side of the blade's leading edge, and not close to the shroud, but rather in the second part of the suction side toward the mid-height and the hub. When the flow rate decreases, because of a higher inlet incidence angle, the void fraction extends to the entire suction-side blade height. At the impeller inlet section, more flow blockage can be seen; this may explain why no water inlet recirculation occurs when the inlet void fraction is important, as already discussed earlier. Further downstream, this void fraction value extends toward the mid-impeller channel (green color circles). For $Q = 30$ m^3/h, a separation zone can be seen starting from the pressure side of the hub region, which migrates toward the mid-impeller passage as well. As pointed by Gülich [1], such a complicated flow pattern results from the dominant body forces and inertia for the liquid behavior (due to the density ratio between water and gas), while gas distribution is mainly determined by the pressure field and secondary flows. In the next section, the void fraction on the pressure and suction sides of the impeller is going to be analyzed more specifically.

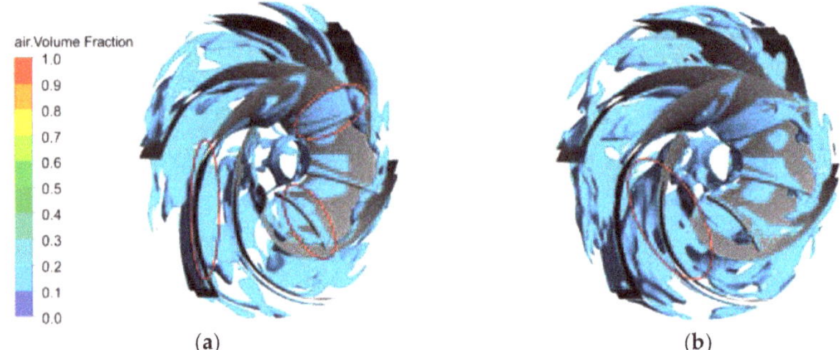

Figure 12. Constant void fraction iso-lines inside the impeller passage when $a = 7\%$: (**a**) $\varphi = 0.077$; (**b**) $\varphi = 0.058$.

6.3.2. Flow inside the Impeller Section

Void fraction distributions for decreasing flow rate values on the impeller blade and at the outlet radius of the impeller are shown in Figure 13a–d. Looking at the leading-edge regions of the impeller, a high local void fraction (between 20–30%) appears on the suction side mainly close to the shroud. High α values (up to 100%, corresponding to full air) appear on the pressure side of the shroud corner for $Q = 30$ m^3/h, just after the inlet throat section of the impeller passage. The corresponding drag force, acting on a bubble, combined with the adverse pressure gradient inside the impeller passage, forces the gas bubble to move toward the pressure side where the local liquid relative velocity tends to be lower.

For a flow rate of 40 m^3/h, the local void fraction values remain relatively small at both sides of the impeller leading edge. This is perhaps the reason why the experimental results show that the pump is still working at a high inlet a (up to 10%; see Figure 2) without gas locking. Finally, for 20 m^3/h, a quite high local void fraction can be seen on the suction side of the leading edge due to incidence angle effects (see again Figure 11b for the meridional plane).

Further downstream, and for all of the flow rates, high void fractions are located near the pressure side, close to the second throat. This has been already mentioned by Mirakami and Minemura [3]. This is attributed to the increase in the velocity difference between the water and the gas when the void fraction increases. The corresponding drag force, acting on a bubble, combined with the adverse pressure gradient inside the impeller passage, forces the gas bubble to move toward the pressure side where the local liquid relative velocity tends to be lower. As an example, this is clearly obtained by the CFD results given in Figure 14 for the mid-section between the hub and the shroud, for $Q = 30$ m^3/h.

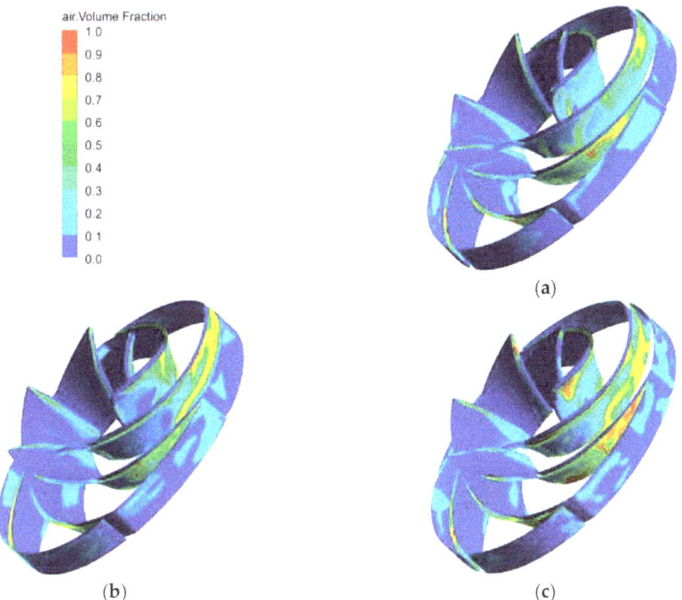

Figure 13. Void fraction pattern at the impeller outlet section, $a = 7\%$: (**a**) $\varphi = 0.077$; (**b**) $\varphi = 0.058$; (**c**) $\varphi = 0.038$.

By looking at the additional maps for the specific local void distribution (Figure 15), we can have a better insight on the pressure and suction sides of the blade. For instance, at a given flow rate ($Q = 30$ m^3/h), it can be observed that on the pressure sides, the pattern of air bubbles differs between $a = 5$–7%. For $a = 5\%$, high values of local void fraction are reached and spread along the blade height just after the leading edge. Then, due to the Coriolis force, the liquid is mainly present on this pressure side, and finally, due to the radial pressure gradient and slip velocity, air accumulates again close to the trailing edge. When a increases to 7%, air locations are shifted downstream in the hub region, and the liquid (and low local void fractions) area takes a larger surface and pushes air closer to the outlet blade section. As already mentioned, this last case exhibits strong 3D effects with flow separations and reverse flow regions.

One can observe that the more the flow rate decreases, the more of high void fractions are shifted toward the pressure side due to the decrease of the mean relative velocities. For $Q = 20$ m^3/h, all of the passages are filled with air. All of the blade passages do not have the same pattern due to the presence of the volute tongue.

At the impeller outlet plane, the void fraction does not exceed 30%, instead of reaching the lowest flow rate, for which the values may reach up to 60%. Just downstream of the impeller, due to strong mixing, the void fraction strongly decreases (not shown here). Close to nominal conditions, the void fraction values obtained from CFD are quite similar to the values predicted by the model developed by Minemura et al. [3]. On the pressure side, there is always a thin layer that develops with high void fraction values.

The next set of figures gives the pattern of the local void fraction in the mid-section plane between the hub and the shroud, for decreasing flow rates and a given a value of 7%. This corresponds to the maps given in Figure 16.

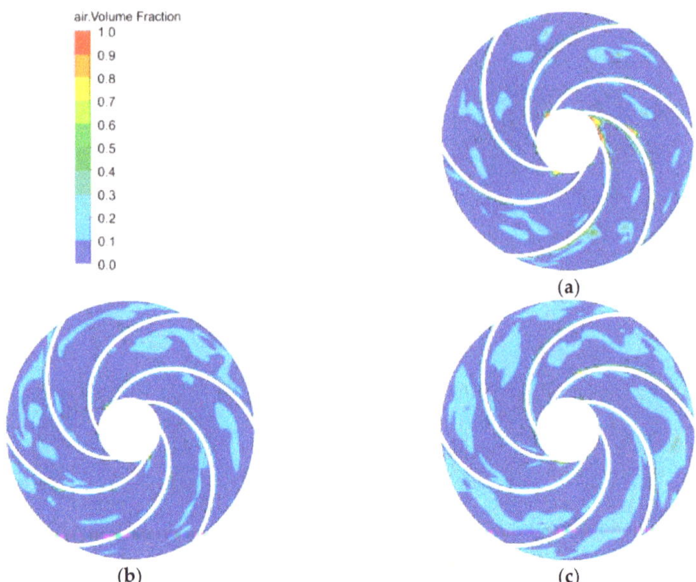

Figure 14. Void fraction pattern in the blade-to-blade mid-section plane, $Q = 30$ m^3/h, $\varphi = 0.058$: (**a**) $a = 3\%$; (**b**) $a = 5\%$; (**c**) $a = 7\%$.

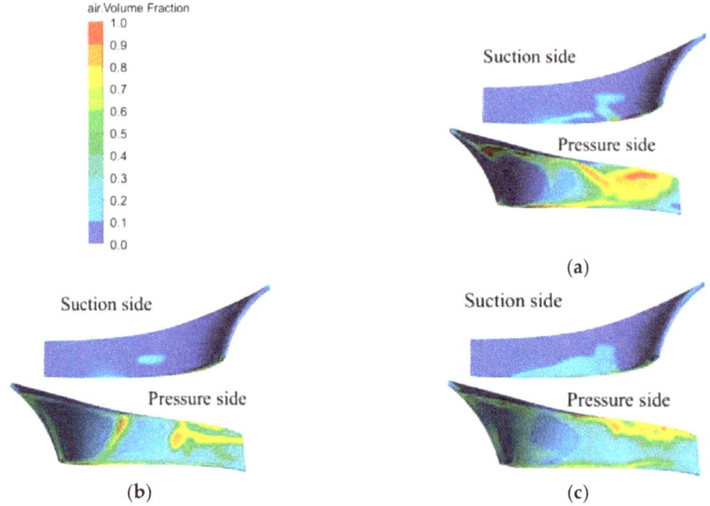

Figure 15. Void fraction pattern on the blade surface, $\varphi = 0.058$: (**a**) $a = 3\%$; (**b**) $a = 5\%$; (**c**) $a = 7\%$.

All of the flow patterns that have been found and described inside the impeller channels are the consequences of what happens at the impeller inlet section just before and after the blade leading-edge area. This means that one has to find the local design modifications forcing high two-phase flow values to be mixed with the main flow using additional local body forces. For this reason, it is expected that the blade clearance between the hub and the shroud can be an efficient way to create a local vortex for a better mixing, avoiding a local accumulation of the high void fraction values on the meridional plane inlet sections and in the blade-to-blade plane further downstream of the inlet passage throat. This will be carried out in a future work.

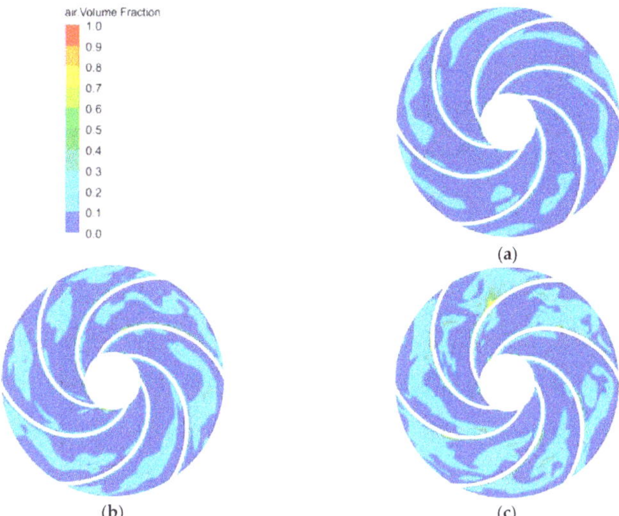

Figure 16. Void fraction pattern in the blade-to-blade mid-section plane for different flow rates, a = 7%: (**a**) φ = 0.077; (**b**) φ = 0.058; (**c**) φ = 0.038.

7. Conclusions

Experimental and numerical analysis has been performed in a centrifugal pump for different flow coefficients and inlet void fractions up to 10%. The effect of rotational speed has also been investigated. The main conclusions are the following:

1. Pump performance degradation is more pronounced for low flow rates compared to high flow rates when the inlet void fraction increases. The starting point of a severe pump degradation rate is related to a specific flow coefficient, whose value corresponds to the change of the slope of the theoretical head curve. When increasing the inlet void fraction, the degradation slope curves increase (with a negative sign) with the decreasing flow coefficient. The more the rotational speed decreases, the more the experimental pump performance is affected at a given inlet void fraction value.
2. Existing one-dimensional models can be considered quite good tools for the first step of a two-phase flow analysis. They give good global indications based on the mean values along one streamline, but attention should be taken when using non-dimensional flow coefficients. The chosen particle fluid model with interface transfer terms looks quite suitable for evaluating pump performance degradation up to an a value of 7%.
3. For a higher a value, CFD can't always correctly predict the sudden breakdown of the pump performance as obtained by measurements. This is probably due to the flow regime inside the impeller, which does not correspond to a bubbly one anymore. However, for the lowest flow rate, the performance breakdown is well predicted with the modified turbulence model.
4. The difference between experimental and numerical results exist not because of rotational speed, but because of its consequence on the local velocity values that decrease according to the flow coefficient, and more specifically at the pump inlet tube and at the pump inlet section. Numerical simulations must take churn flow characteristics into account in order to get better results. Both bubbly and churn flow conditions may be present for inlet flow conditions depending on the experimental setup, rotational speed, and the pump flow coefficient, even at nominal conditions.
5. The numerical simulation gives interesting local flow information that would be taken into consideration for new design approaches at the pump inlet section for an improved two-phase flow suction capability of centrifugal pumps.

Author Contributions: Data curation, W.H.; Formal analysis, Q.S.; Funding acquisition, S.Y.; Investigation, Q.S. and B.G.; Writing—original draft, Q.S.; Writing—review & editing, Q.J. and A.A.

Funding: National Natural Foundation of China: 51609164, 51779107; China Postdoctoral Science Foundation Funded Project: 2015M581735, 2016T90422; Senior Talent Foundation of Jiangsu University: 15JDG048.

Acknowledgments: The authors gratefully acknowledge the financial support by National Natural Foundation of China (51609164, 51779107), China Postdoctoral Science Foundation Funded Project (2015M581735, 2016T90422) and Senior Talent Foundation of Jiangsu University (15JDG048).

Conflicts of Interest: The authors declare no conflict of interest.

Nomenclature

a	inlet void fraction $a = Q_{air,inlet} / (Q_{air,inlet} + Q_{water,inlet})$
b	impeller blade width
C	constant pressure value when modified liquid density
D	diameter
H	pump head
n	exponent
N	rotational speed
p	local pressure
P	shaft power
Q	volume water flow rate
R	radius
$R_{e,imp}$	Impeller Reynolds number $R_{e,imp} = u_2 \cdot R_2 / v$
t	time
th	theoretical
tp	related to two-phase condition
u	circular velocity
v	water cinematic viscosity
z	impeller blade

Greek symbols

η	global efficiency of the pump $\eta = \rho g Q H / P$
ρ	density
ρ_m	density of fluid mixture $\rho_m = \rho_l (1 - \alpha) + \alpha \rho_g$
φ	flow coefficient $\varphi = Q/(2\pi \cdot R_2 \cdot b_2 \cdot u_2)$
ψ	head coefficient $\psi = gH/(u_2)^2$
Ωs	specific speed $\Omega s = \omega \cdot \dfrac{Q^{0.5}}{(gH)^{0.75}}$
ω	angular velocity

Subscripts

α	local void fraction
B	bubble
d	design condition
g	gas
l	liquid
m	mixture fluid
ref	reference
0	related to $a = 0$
1	Impeller pump inlet
2	Impeller pump outlet

References

1. Gülich, J.F. *Centrifugal Pumps*; Springer: Berlin/Heidelberg, Germany; New York, NY, USA, 2010; ISBN 978-3-642-12823-3.
2. Mikielewicz, J.; Wilson, D.G.; Chan, T.C.; Goldfinch, A.L. A method for correlating the characteristics of centrifugal pumps in two-phase flow. *J. Fluids Eng.* **1978**, *100*, 395–409. [CrossRef]

3. Murakami, M.; Minemura, K. Effects of entrained air on the performance of a centrifugal pump First report-Performance and Flow Conditions. *Bull. JSME* **1974**, *17*, 1047–1055. [CrossRef]

4. Minemura, K.; Uchiyama, T.; Shoda, S.; Egashira, K. Prediction of air-water two-phase flow performance of a centrifugal pump based on one-dimensional two-fluid model. *J. Fluids Eng.* **1998**, *120*, 237–334. [CrossRef]

5. Si, Q.R.; Cui, Q.L.; Zhang, K.Y.; Yuan, J.P.; Bois, G. Investigation on centrifugal pump performance degradation under air-water inlet two-phase flow conditions. *La Houille Blanche* **2018**, *3*, 41–48. [CrossRef]

6. Secoguchi, N.; Takada, S.; Kanemori, Y. Study of air-water two-phase centrifugal pump by means of electric resistivity probe technique for void fraction measurement-First report-measurement of void fraction distribution in a radial flow impeller. *Bull. JSME* **1984**, *27*, 931–939. [CrossRef]

7. Estevam, V.; França, F.A.; Alhanati, F.J. Mapping the performance of centrifugal pumps under two-phase conditions. In Proceedings of the 17th International Congress of Mechanical Engineering, Sao Paulo, Brazil, 10–14 November 2003.

8. Barrios, L.; Prado, M.G. Experimental visualization of two-phase flow inside an electrical submersible pump stage. *J. Energy Resour. Technol.* **2011**, *133*. [CrossRef]

9. Müller, T.; Limbach, P.; Skoda, R. Numerical 3D RANS simulation of gas-liquid flow in a centrifugal pump with an Euler-Euler two-phase model and a dispersed phase distribution. In Proceedings of the 11th European Conference on Turbomachinery, Fuild Dynamics and Thermodynamics, Madrid, Spain, 23–25 March 2015.

10. Suryawijaya, P.; Kosyna, G. Unsteady measurement of static pressure on the impeller blade surfaces and optical observation on centrifugal pumps under varying liquid/gas two-phase flow condition. *J. Comput. Appl. Mech.* **2001**, *6759*, D9–D18.

11. Si, Q.R.; Bois, G.; Zhang, K.Y.; Yuan, J.P. Air-water two-phase flow experimental and numerical analysis in a centrifugal pump. In Proceedings of the 12th European Conference on Turbomachinery, Fluid Dynamics and Thermodynamics, Stockholm, Sweden, 3–7 April 2017.

12. Yakhot, V.; Orszag, S.; Thangam, S.; Gatski, T.; Speziale, C. Development of turbulence models for shear flows by a double expansion technique. *Phys. Fluids A Fluid Dyn.* **1992**, *4*, 1510–1520. [CrossRef]

13. Clift, R.; Grace, J.; Weber, M. *Bubbles, Drops and Particles*; Academic Press: New York, NY, USA, 1978.

14. Wang, J.; Wang, Y.; Liu, H.; Huang, H.; Jiang, L. An improved turbulence model for predicting unsteady cavitating flows in centrifugal pump. *Int. J. Numer. Methods Heat Fluid Flow* **2015**, *25*, 1198–1213. [CrossRef]

15. Dular, M.; Coutier-Delgosha, O. Numerical modelling of cavitation erosion. *Int. J. Numer. Methods Fluids* **2009**, *61*, 1388–1410. [CrossRef]

16. Coutier-Delgosha, O.; Fortes-Patella, R.; Reboud, J.L. Evaluation of the turbulence model influence on the numerical simulations of unsteady cavitation. *ASME J. Fluids Eng.* **2003**, *125*, 38–45. [CrossRef]

17. Schäfer, T.; Bieberle, A.; Neumann, M.; Hampel, U. Application of gamma-ray computed tomography for the analysis of gas holdup distributions in centrifugal pumps. *Flow Meas. Instrum.* **2015**, *46*, 262–267. [CrossRef]

18. Baker, O. Simultaneous Flow of Oil and Gas. *Oil Gas J.* **1954**, *53*, 185–190.

19. Tomiyama, A.; Kataoka, I.; Sakaguchi, T. Drag coefficients of bubbles. 1st report. drag coefficients of a single bubble in a stagnant liquid. *Nihon Kikai Gakkai Ronbunshu B Hen/Trans. Japan Soc. Mech. Eng. Part B* **1995**, *61*, 2357–2364.

20. Hench, J.E.; Johnston, J.P. Two-dimensional diffuser performance with subsonic, two-phase, air-water flow. *ASME J. Basic Eng.* **1972**, *107*, 105–121. [CrossRef]

21. Zuber, N.; Hench, J.E. *Steady State and Transient Void Fraction of Bubbling Systems and Their Operating Limits*; General Electric: Schenectady, NY, USA, 1962.

22. Zhu, J.J.; Zhang, H.Q. A review of experiments and modeling of gas-liquid flow in electrical submersible pumps. *Energies* **2018**, *11*, 180. [CrossRef]

 *energies*

Article

Resonant Pulsing Frequency Effect for Much Smaller Bubble Formation with Fluidic Oscillation

Pratik Devang Desai [1,2] , **Michael John Hines [2]**, **Yassir Riaz [1]** and **William B. Zimmerman [1,*]**

[1]   Department of Chemical and Biological Engineering, University of Sheffield, Mappin Street, Sheffield S1 3JD, UK; pratik@perlemax.com (P.D.D.); y.riaz@shef.ac.uk (Y.R.)

[2]   Perlemax Ltd., Kroto Innovation Centre, 318 Broad Ln, Sheffield S3 7HQ, UK; michael@perlemax.com

*   Correspondence: w.zimmerman@sheffield.ac.uk; Tel.: +44-(0)-114-222-7517

Received: 21 July 2018; Accepted: 27 September 2018; Published: 9 October 2018

Abstract: Microbubbles have several applications in gas-liquid contacting operations. Conventional production of microbubbles is energetically unfavourable since surface energy required to generate the bubbles is inversely proportional to the size of the bubble generated. Fluidic oscillators have demonstrated a size decrease for a system with high throughput and low energetics but the achievable bubble size is limited due to coalescence. The hypothesis of this paper is that this limitation can be overcome by modifying bubble formation dynamics mediated by oscillatory flow. Frequency and amplitude are two easily controlled factors in oscillatory flow. The bubble can be formed at the displacement phase of the frequency cycle if the amplitude is sufficient to detach the bubble. If the frequency is too low, the conventional steady flow detachment mechanism occurs instead; if the frequency is too high, the bubbles coalesce. Our hypothesis proposes the existence of a resonant mode or 'sweet-spot' condition, via frequency modulation and increase in amplitude, to reduce coalescence and produce smallest bubble size with no additional energy input. This condition is identified for an exemplar system showing relative size changes, and a bubble size reduction from 650 μm for steady flow, to 120 μm for oscillatory-flow, and 60 μm for resonant condition (volume average) and 250 μm for steady-flow, 15 μm for oscillatory-flow, 7 μm for the resonant condition. A 10-fold reduction in bubble size with minimal increase in associated energetics results in a substantial reduction in energy requirements for all processes involving gas-liquid operations. The reduction in the energetic footprint of this method has widespread ramifications in all gas-liquid contacting operations including but not limited to wastewater aeration, desalination, flotation separation operations, and other operations.

Keywords: microbubbles; fluidics; flow oscillation; oscillators; energetics

1. Introduction

Gas-liquid contacting operations are arguably among the most important processing operations. The oil we produce, the air, the food, the drinks (fizzy drinks, beer, and fermented beverages), chemical dyeing processes, mixing operations, wastewater aeration (WWA). and remediation, and several operations require good gas-liquid contacting [1–4]. One way to achieve such contacting is by increasing the surface area of the system corresponding to its volume. This results in an increase in surface area with respect to volume, and for a bubble, results in a slower rise velocity and a substantial increase in contact time. The major problem with smaller bubble generation is the energy required due to the large surface energy involved in generating these smaller bubbles.

Two examples where the bubble size in terms of number and volume contribution matters are dissolved/dispersed air flotation (DAF) and wastewater aeration (WWA). Both these operations are important, mandatory for any wastewater remediation, and highly energy intensive. DAF uses 70–100 μm size bubbles, generated by involving high pressure nozzles (14–15 bar (g)) [5], in order to separate systems out for remediation or further processing. Variants of this method include froth flotation such as for mineral processing—copper ore processing, effluent removal from gas processing/petrochemical plants, paper mills, and drinking water plants. The second is WWA which accounts for nearly 0.25–0.4% of the UK's total energy consumption [6]. This is due to the barriers to implementing high energy microbubbles and the inability to produce these in an energy efficient manner.

If smaller bubbles could be generated with less energy, it could lead to a fundamental change in gas-liquid contacting equipment design. Of particular importance to the interplay between fluid flow and heat transfer are applications of microbubbles in phase change separations, particularly vaporisation and distillation [7–10]. Experiments and models [7,10] demonstrate the layer height of liquid is extremely important, with fixed microbubble size, for determining the degree of non-equilibrium separation and the overall rate of vaporization. The highest vaporization and the greatest enrichment for 100 μm diameter microbubbles occur with a contact time of ~1 ms. Since bubble rise rate at terminal velocity is well established to depend on bubble size, hitting the optimum vaporisation rates and enrichment (joint heat and mass transfer at the microbubble interface) strongly depend on tuning the microbubble size. The purpose of this paper is to explore how microbubble size depends on the fluid dynamics of fluidic oscillator induced microbubble generation.

There are several methods to generate microbubbles such as ablative technology, ultrasound, microfluidic devices, nozzles, and other techniques, but each faces problems either due to scalability or energy input [11–17]. Microbubbles have been divided into several size classes and depend on the application [18]. In this paper, microbubbles are gas-liquid interfaces (bubbles) ranging from 1 μm to 1000 μm in size. Several applications have been identified for them including microalgal separation [19], wastewater clean-up [20,21], theranostics [22–24], algal growth [25–28], oil emulsion separation [29] and for heat and mass transfer applications (due to the vastly increased surface area to volume ratios) [30–32]. A major advantage is gained if a different bubble generation regime can be formulated such that it does not specifically depend on the conventional form of detachment.

The fluidic oscillator is a fluidic device that creates hybrid synthetic jets which help engender microbubbles in an economical fashion via pulsatile flow through the aerator. The adherence of the jet to the wall, due to the Coandă effect, and its subsequent detachment to the other leg due to a switch over created by a geometric cusp [33–35] generates the pulsatile flow. The actual mechanism of bubble formation via fluidic oscillation resulting in the back flow into the membrane due to the net positive displacement is responsible for the smaller bubble size and has been demonstrated by Tesař [35]. Zimmerman et al. [9,30,35] have shown the efficacy of the fluidic oscillator with respect to bubble size reduction via aeration.

Although introduction of the negative feedback can be achieved using both the Warren and the Spyropoulos configuration for the fluidic oscillator [36], for the experiments herein, to facilitate the use of discrete frequencies, a Spyropoulos type feedback loop has been used. This type of negative feedback has several advantages. Firstly, frictional losses are kept to a minimum and secondly, the frequency of oscillation can be easily controlled by changing the negative feedback configuration i.e., with the use of different feedback loop lengths and volumes.

The Spyropoulos loop has been adequately described in Tesař et al. [37], which introduces a negative feedback to the system and in physical form is a single loop connecting the control terminals of the fluidic oscillator shown in Figure 1 as X_1 and X_2. S (supply nozzle), X_1 and X_2 (control terminals) and Y_1 and Y_2 (outlets). The incoming jet enters via the supply nozzle, is amplified at the throat via a constriction of appropriate size. Control terminals aid in switching the flow due to a pressure differential formed in order to switch on outputs Y_1 and Y_2 at relevant frequencies. Y_1 and Y_2 can be connected to microporous diffusers placed in a liquid stream which result in bubble formation when oscillatory gas exits Y_1 and Y_2 and into the microporous spargers. Typical widths for X_1 and X_2 are 2–4 mm. Fluidic oscillation is a nozzle free bubble generation method which also allows generation in the laminar flow mode. This is contrasted with conventional nozzles used for microbubbles in dissolved air flotation microbubble generation or droplet generation [38].

An added advantage of using a single feedback loop over the two required for the Warren type oscillator is that there is a reduction in the degrees of freedom, which results in a simpler system to control and quantitatively understand it.

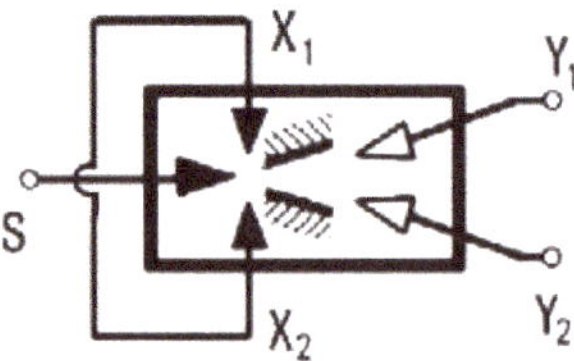

Figure 1. The Spyropoulos loop configuration for the fluidic oscillator, reproduced with permission [37]. S is the supply nozzle where gas enters, X_1 and X_2 are the control terminals where gas switches, Y_1 and Y_2 are the outlets where microporous spargers can be connected with oscillatory gas output. These spargers are placed in liquid and generate bubbles.

The usual method for bubble generation using fluidic oscillation is using two outlets connected to a set of bubble generating microporous membranes (spargers/aerators) placed under liquid of interest and gas entering via supply nozzle S. The entire flow can be utilised for the generation for bubbles and this results in an unvented condition for the fluidic oscillator.

Venting this jet, increases the momentum of the pulse post the fluidic oscillator as the oscillator is a flow amplifier. The flow switches from each leg, and if the momentum of the jet and therefore the pulse strength associated with the jet can be increased whilst maintaining the appropriate flow into the aerator.

Bubble formation conventionally requires the bubble to detach when the forces acting on it are balanced. According to Zimmerman et al., the proposed mechanism for bubble formation mediated by the fluidic oscillator typically takes place in the pulse cycle of the frequency switch. Therefore, technically this should lead to a bubble formed at every pulse and the throughput determined by the frequency of the oscillator. However, this does not take place as seen in the previous papers [9,30,35]. Each pulse of air is based on the frequency of oscillation. This means that higher the frequency, shorter the oscillatory pulse, and therefore in theory should lead to a smaller bubble. This has not been observed and led to one of the hypotheses proposed in this paper. The first part of the hypothesis is that the amplitude of the pulse must be high enough for detachment to occur. Post the bubble detachment, the bubbles may coalesce if the frequency is too high as they are close to each other. The frequency and amplitude of the fluidic oscillator are the two control parameters capable of producing the bubbles at the required frequencies. To increase the amplitude of the flow, higher flow rates will be used and then vented such that the actual flow into the aerator is as desired but the amplitude of the flow has increased. Different feedback conditions will be used in order to see the variations of the amplitude. Higher feedback conditions will have a higher amplitude whilst lower feedback conditions will introduce a higher friction loss and therefore lower resultant amplitude of the flow.

The second part of the hypothesis is that there is a resonant condition for the system which depends on a specific frequency which determines the bubble size. Even when the amplitude condition is met and bubbles are generated at each pulse, just increasing the frequency will not result in a smaller bubble size. At a specific frequency, there will be the presence of the resonant mode condition, where the bubbles are detached, but not too quickly so as to coalesce, and not too slowly so as to resemble a conventional steady flow bubble formation. The paper aims to explore this new regime of bubble formation and check if the hypothesis is supported with experimental evidence.

The flow has to be vented in order to generate higher amplitude of the jet and only partially diverted into the aerators used for bubble generation. This helps in two ways—it increases the effect of the oscillatory flow (by virtue of an increase in amplitude by increasing momentum of the wave) in order to observe the difference between discrete oscillatory flow conditions (frequencies) for the lab scale and minimises the bubble coalescence due to the increased momentum imparted to the newly engendered microbubble. This fluidic oscillator mediated oscillation has an amplitude and frequency dependence on the inlet flow rate to the oscillator provided that all other conditions are maintained. This proposed addition to the experiment enhances the fluidic oscillator mediated microbubble size reduction.

Several industrial applications can afford to have gas wastage for a significantly larger bubble throughput concomitant with reduced bubble size. Since air, in particular, is not that expensive with respect to the decreased energetics, this is justified by the increase in reaction surface area and the lack of maintenance requirements with the no-moving part fluidic oscillation.

Figure 2 shows the vented schematic of the fluidic oscillator with V_1 and V_2 acting as the vents to the system. Additional venting ensures that the appropriately controlled flow can pass through the aerator whilst maintaining the appropriate flow into oscillator in order for it to actuate the oscillation. Additionally, this increases the momentum of the jet and the amplitude of the oscillatory flow.

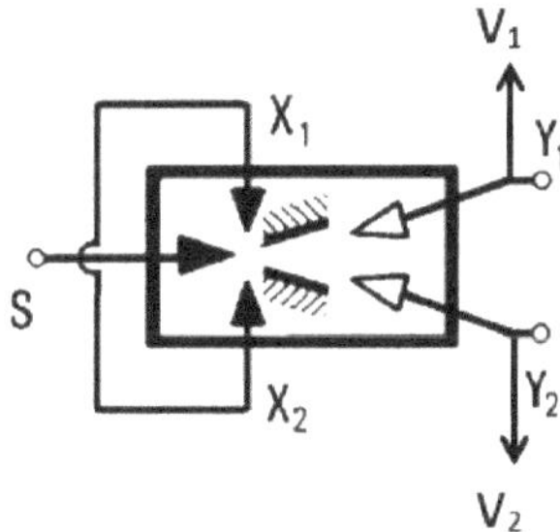

Figure 2. Vented Fluidic Oscillator (adapted from [17]) S is the supply nozzle where gas enters, X_1 and X_2 are the control terminals where gas switches , Y_1 and Y_2 are the outlets where microporous spargers can be connected with oscillatory gas output. These spargers are placed in liquid and generate bubbles, V_1 and V_2 (vents).

Conventional bubble formation depends on a host of other factors such as surface energy of the bubble engendering surface and the liquid, liquid and gas viscosity, momentum of the gas, height of liquid, pressure exerted by the system and acting upon it, and size of the bubble-engendering orifice. Fluidic devices such as the Tesař-Zimmerman fluidic oscillator generate a net positive hybrid synthetic oscillatory jet that results in a specific reduction in bubble size compared to conventional or steady flow as discussed previously.

2. Methods and Materials

2.1. Fluidic Oscillator

The fluidic oscillator is oscillated at 86 L per min (lpm) and 92 lpm (corrected for pressure and temperature) with most of the flow being vented and a frequency sweep is performed in order to support the hypothesis posited earlier. As discussed previously, vents have been introduced in order to increase the momentum of jet pulse and amplitude of oscillation whilst controlling appropriate flow into the aerator (MBD 75, Point Four Systems, Coquitlam, BC, Canada). Rotameters have been utilised to act as metered valves. The frequency of oscillation and the amplitude of the oscillation are measured using an Impress G-1000 pressure transducer (Impress Sensors and Systems, Ltd., Berkshire, UK) controlled and recorded using characterisation software developed in LabView (National Instruments, Austin, TX, USA). The pressure drop across the fluidic oscillator is 100 mbar. The total pressure drop depends on the combined pressure drop across the fluidic oscillator and aerator.

2.2. The Aerator

The proprietary Point Four Systems MBD 75 aerator produces a cloud of fine bubbles approximately 500 μm in size under steady flow. MBD 75 has ultrafine ceramic pores and a flat surface, thereby retarding bubble coalescence as compared to other types of material such as sintered glass or steel membranes. Ceramic, being inert, hydrophilic and robust, is a preferred surface for bubble generation in water.

2.3. System Set up

The system has been set up according to the schematic shown in Figure 3.

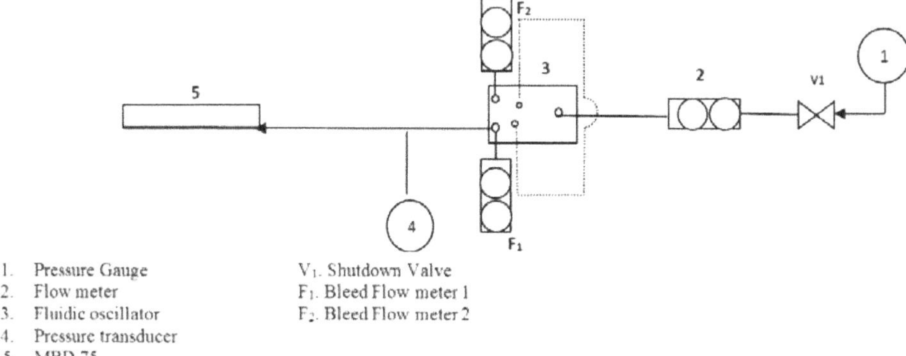

1. Pressure Gauge
2. Flow meter
3. Fluidic oscillator
4. Pressure transducer
5. MBD 75

V₁. Shutdown Valve
F₁. Bleed Flow meter 1
F₂. Bleed Flow meter 2

Figure 3. Schematic of the setup.

2.4. Pneumatic Set Up

Figure 3 shows the system schematic described here. The pressure regulator (Norgren, Littleton, CO, USA) controls the systemic pressure which is set at 2 bar(g)—the required pressure for the particular aerator to bubble. A different aerator would require lesser pressure. The flow controller (FTI Instruments, Sussex, UK) has been corrected for the pressure being used in the system. The air enters the system from the compressor via the pressure regulator and the flow controller regulates the global flow entering the fluidic oscillator. The fluidic oscillator is connected to two vent rotameters (F_1 and F_2) to act as metered valves and these are set up in order to vent appropriately and send the appropriate amount of flow into the aerator. The aerator is placed in a tank wherein the bubble size is measured using acoustic bubble spectrometry.

The frequency and amplitude of pulse from fluidic oscillator is measured simultaneously using a combination of pressure transducers and Fast Fourier Transform (FFT) code developed in LabView (*cf.* frequency measurement and FFT).

The aerator is kept in the centre with the hydrophones set around it as shown in Figure 4 with the set operational conditions. Bubble sizing is performed continuously and the distilled deionized (DDI) water in the tank is replaced after each reading.

The frequency is changed whilst all other conditions are kept constant and different feedback configurations are used coupled with an additional flow rate. This is an exemplar system and we are just demonstrating the ability for the system to change with different systems and the aim of this paper is to show that relative changes are possible for the system without any additional energy input. Several configurations of the Spyropoulos type fluidic oscillator being used are also able to engender the same frequency which helps observe the effect of the frequency.

2.5. Bubble Sizing Using Acoustic Bubble Spectrometry

The hydrophones are placed over at a height 5 cm above the diffuser and equidistant at 15 cm. These data were repeated 7 times and performed at 2 different flow rates—86 lpm and 92 lpm (standard conditions—293.15 K and 101,325 Pa). Twenty two frequencies were used in the study performing a frequency sweep and 3 different configurations of feedback for fluidic oscillator. Bubble sizing was performed using Acoustic Bubble Spectrometry commercially available from Dynaflow Inc.™ (Jessup, MD, USA). This has been found to be an effective method for visualising cloud bubble dynamics. The Acoustic Bubble Spectrometer (ABS) with 4 pairs of hydrophones—50, 150, 250, and 500 kHz were used in this study, with the capability to collate a size distribution from 3 μm to 600 μm in size (radius). The ABS is then set up along with an octaphonic set up and Figure 4 schematically represents the flow diagramme of the bubble visualisation set up.

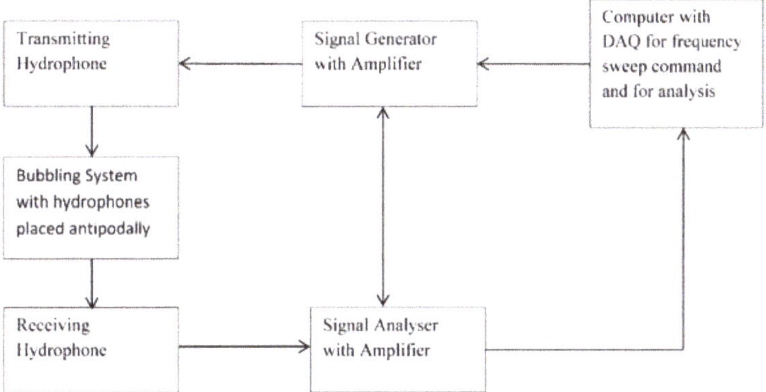

Figure 4. Information flow diagram of the Acoustic Bubble Spectrometer (ABS) and the hydrophone set up.

Two equiresponsive hydrophones are placed antipodal to each other with the aerator bubbling in the centre i.e., bubble cloud in the centre. The data acquisition module and computer (controlled using software) send out a frequency sweep via a signal generator with amplifier into the transmitting hydrophone. The signal passes through the bubble cloud (calibrated under no bubble condition) and into the receiving hydrophone which is demodulated using an amplifier too. This is used in conjunction with the no bubble condition in order to generate a bubble size distribution.

Acoustic bubble sizing relies on the principle of bubble resonance upon frequency insonation and the resonant bubble approximation. Upon insonation by a specific frequency, a bubble starts to oscillate and this frequency is specific to the bubble by a sixth power to the radius, i.e., [39]:

$$f \propto r^6 \tag{1}$$

Hydrophone pairs are used, with one being a transmitter and the other acting as the receiver, and each hydrophone pair has a specific resonant frequency at which it works best and a range of frequencies that it can operate reasonably. When a hydrophone insonates a bubble cloud with a specific frequency, the bubble corresponding to that size resonates and therefore oscillates, thereby attenuating the signal due to the pressure change caused by the oscillating bubble as compared to a clear/bubble-free solution. This attenuation is then measured by the receiving hydrophone and the signal is inverted in order to garner a bubble size distribution. A frequency sweep is performed at equally spaced frequencies between 5 kHz to 950 kHz, from which the bubble size distribution is compiled. Chahine et al. [40–46] describe the methodology underpinning ABS and the algorithm for transformation of the raw data into a bubble size distribution.

2.6. Frequency Measurement and Fast Fourier Transform

Impress G-1000 pressure transducers were used in this experiment. Fast Fourier Transforms are simple algorithms designed to convert a signal from one domain (time or space) and convert it to the frequency domain and vice versa. Figure 5 shows an exemplar waveform with the FFT. This provides a quick and easy way to determine the frequency of the oscillatory flow from a fluidic oscillator. LabView is used to acquire the signal and process it. The oscillatory pulse from a fluidic oscillator is composed of the amplitude and frequency of oscillation.

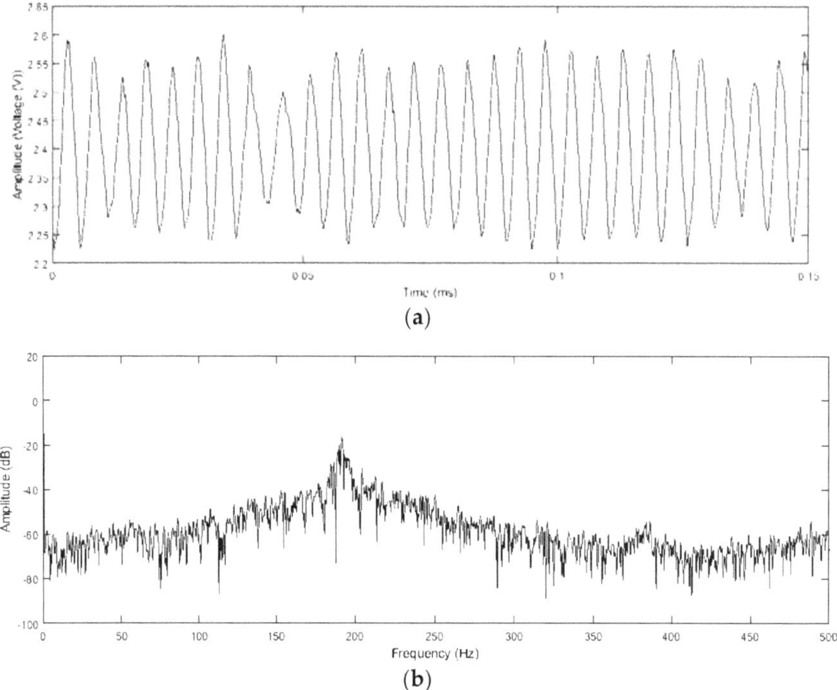

Figure 5. An exemple of a raw waveform from a fluidic oscillator (**a**) and its translation in frequency domain post FFT (**b**). Aliasing effects of FFT are mitigated by having a high acquisition rate (512 kHz with 512 k samples acquisition window).

2.7. Bubble Sizing Analyses

Two factors are readily computable for the bubble size analysis obtained from the ABS- average bubble size in terms of number of bubbles and average bubble size in terms of void fraction contribution (volume contribution) of the bubble. Depending on the application, bubble sizing is usually reported using either of these two factors. Since volume contributions (due to their association with increased mass transfer/heat transfer for microbubbles) are more relevant for a majority of industrial processes, this paper discusses bubble sizes in terms of volume contributions.

Table 1 provides an exemplar for a case wherein there are three classes of bubbles—Class A—with a size of 1 μm and 600 in number, Class B with size of 100 μm and 200 in number and Class C of size 500 μm and 200 in number. This would lead to a total of 1000 bubbles. The surface area of the bubbles is calculated and so is the volume. The bubble size can be computed by either using average bubble size in terms of numbers i.e. weighted bubble size divided by total numbers, or in terms of average bubble size in terms of volume contribution i.e., weighted bubble volume divided by total bubble volume:

$$N_{av} = \sum_{i=1}^{n} \frac{n_i x_i}{n} \tag{2}$$

$$N_{vc} = \sum_{i=1}^{n} \frac{n_i V_i}{nV} \tag{3}$$

with:

$$V_i = \frac{4}{3}\pi x_i^3 \tag{4}$$

where average bubble size in terms of number (2) volume (3) is shown, n is the total number of bubbles and n_i is the bubble contribution for number (2) volume (3) of each bubble of size x_i represented by V_i (4).

Table 1. Exemplar.

S.No.	Bubble	Size	Number	Volume of Individual Bubbles	Total Volume Contribution	Surface Area	Total Surface Area	Surface Area/Volume
1	A	1	600	5.24×10^{-1}	3.14×10^2	3.14	1.88×10^3	6.00
2	B	100	200	5.24×10^5	1.05×10^8	3.14×10^4	6.28×10^6	6.00×10^{-2}
3	C	500	200	6.54×10^7	1.31×10^{10}	7.85×10^5	1.57×10^8	1.20×10^{-2}
			1000		1.32×10^{10}			
N_{Av}	120.6 µm	N_{VC}		200 µm				

This also means that 1 million 1 µm bubbles would be required to occupy the same volume as a single 100 µm bubble. This brings about a massive disparity in bubble size in terms of volume contribution. However, the volume contribution would be a useful tool for estimating any transport phenomena exercise over number contribution. Generally speaking, size distributions collated from membranes are narrow and the difference in the two averages is lower. A large difference in bubble sizes is observed for a highly dispersed distribution and it is beneficial to the system to have a narrower size distribution. This exemplar demonstrates how these two values need not be the same and their dispersity results in the width of the bubble size distribution. Work by Allen [47] and Merkus [48] explain the nuances associated with particle sizing and statistical calculations performed for them in detail.

2.8. Results and Discussion

In order to prove the hypotheses, two experimental modalities were tested. Bubble sizes were measured at various frequencies and under different conditions. Bubbles were sized at 22 frequencies, for three feedback conditions (to test amplitude variations) and two higher flowrates (vented, so flow through the fluidic oscillator could be 86 lpm and 92 lpm). Bubble size for steady flow at the same conditions resulted in a bubble size of approximately 350 µm and 450 µm, respectively. This confirmed previous work performed in literature [19,30,49] that fluidic oscillation resulted in a significant decrease in bubble size when compared to conventional methods of microbubble generation.

An approximate 60% reduction in bubble size than the average bubble size estimated from oscillatory flow at other frequencies was observed for all configurations, supporting the proposed hypothesis. The variations between the amplitudes provided additional information on this new bubble formation dynamic under the resonant mode regime.

With the configuration that induced the highest negative feedback, there is a suggestion that two dips observed and this is probably due to the higher feedback introduced for oscillatory control thereby changing the effect of the system.

Figure 6 shows the length of feedback loop compared to the average bubble size at two amplitudes. This shows that although the conditions have been slightly changed, the resonant condition observed is at the same frequency as can be seen in Figure 7. The frequency of the fluidic oscillator can be changed by changing the feedback loop length. This changes the amount of feedback introduced into the system. Figure 6 shows that although the dip is at the same frequency, the actual feedback loop is different for both cases for different conditions of feedback loop lengths. This causes the slight shift in the frequency observed for the dip.

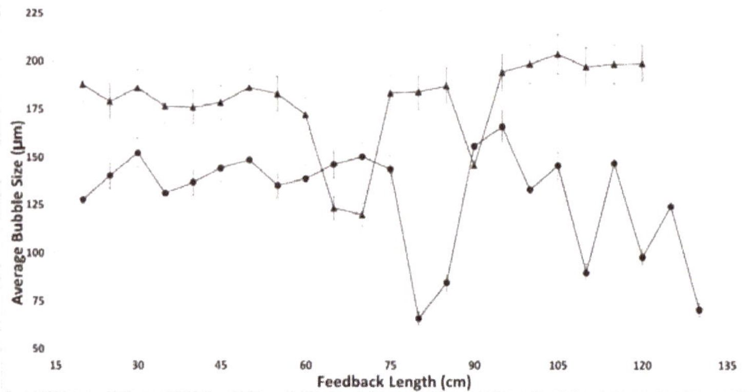

Figure 6. Feedback length vs. average bubble size (—●—92 lpm OD4 mm, —▲—0 86 lpm OD 4 mm).

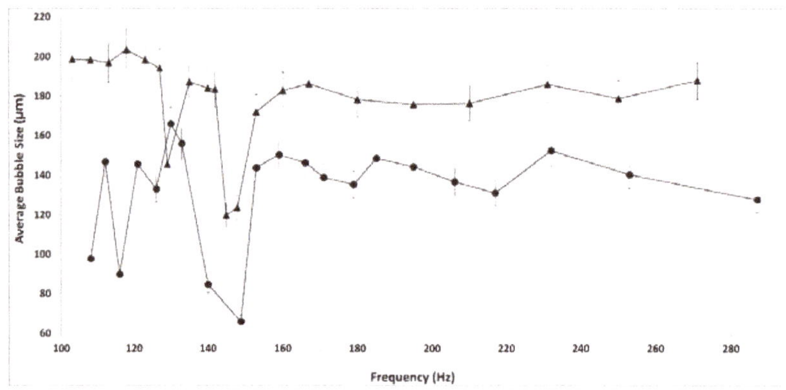

Figure 7. Frequency vs. average bubble size at different flow rates (—●—92 lpm OD4 mm, —▲—86 lpm OD 4 mm).

Figure 7 shows the average bubble size at different flowrates when the frequency is modulated for the smallest feedback configuration. At these two conditions, it is observed that the primary dip occurs circa 150 Hz. The dip is large and observable and there is a suggestion of another dip prior to that. This is shifted slightly for different amplitudes. This is what is seen herein. The frequency remains the same for different configurations of feedback, resulting in a change in bubble size even for different conditions.

Varying the lengths of the tube were used to introduce a change in the feedback. Figures 6 and 7 show that even though the feedback loop lengths were different, the frequency change observed due to the change in flow rate with respect to feedback loop length resulted in a minimum bubbles size at the same frequency. Figures 6 and 7 are plots from a reading taken for single tubing, lowest feedback configuration, at two different inlet global flow rates at 22 different frequencies. The bubble size distribution showed the dip in bubble size and the resonant mode—'sweet spot' was observed here.

Figure 6 shows the presence of the 'sweet spot' at different lengths of the feedback loop. Figure 7 shows that although the feedback loop lengths were different for the different dips in bubble size, the actual frequency remained the same and was approximately 150 Hz.

Figure 8 shows the resonant mode for the medium feedback condition. The dip is quite close to each other for this scenario. The frequency for the sweet spot is similar to the low feedback condition and is at 150 Hz. There is a significant dip observed due to potential matching of the flow rate with the frequency of the system and bubble formation.

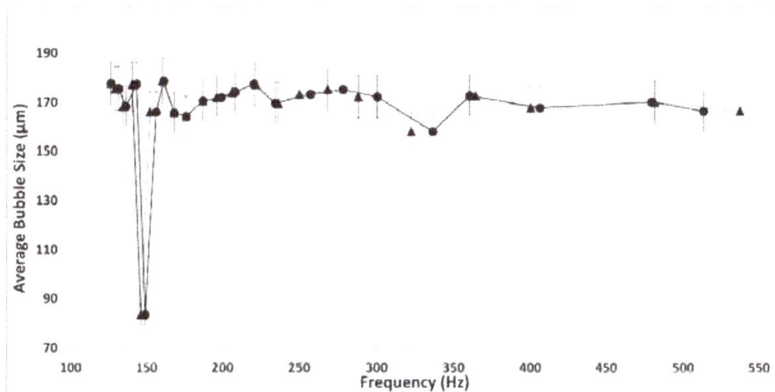

Figure 8. Frequency for bubble size—resonant mode observed—medium feedback condition—(●— 92 lpm OD 6 mm, ▲—86 lpm OD 6 mm).

Figure 9 shows the bubble size versus frequency at the higher feedback configuration. Resonant condition dips are observed for all three conditions as seen in Figures 7–10.

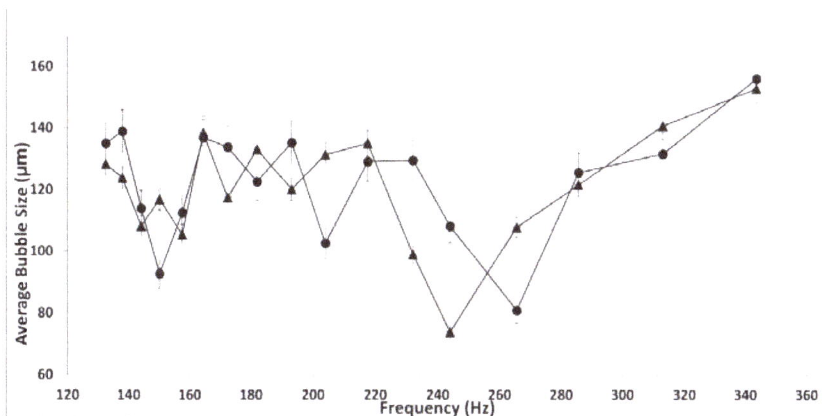

Figure 9. Bubble Size vs. frequency for two flow rates (higher feedback configuration) (●—92 lpm OD 10mm, ▲—86 lpm OD 10 mm).

Under higher feedback condition, i.e., 10 mm OD, the frequency is higher for the dip aside from the initial dip. This is probably due to the higher amplitude observed for the higher feedback condition which results in lesser coalescence for the system. The dip observed is still consistent with the average bubble size formed at other conditions of the resonant modes.

The presence of the resonant condition or 'sweet spot' was observed for the higher feedback condition. However, the magnitude of the decrease in bubble was not as significant as for the lower feedback condition (i.e., feedback loop length is constant but volume is lower) when considered in a relative manner. However, due to the greater amplitude, there is a higher frequency and several smaller bubbles being formed which results in the dampening of the 'sweet spot'. There is another 'sweet spot' formed at 250 Hz.

These figures also show a skew observed in terms of frequency and flow rate, with the skew being less for the larger feedback condition than for the lower feedback configuration indicating that even the shift observed in the frequency is due to the combination of the length of the feedback tube and the flow rate and this skews the bubble size and the resonant condition.

These three conditions (amplitude variations) consistently show the resonant mode condition. The different vent flows also show the resonant mode condition and the resonant condition changes based on the variations introduced. The flow rate is an additional variable influencing the frequency [49] can be adjusted to achieve the same average bubble size. This is because the change in flow rate has a much larger impact on the frequency of the smaller feedback configuration than that of the larger one due to the difference in the frictional losses for both conditions. The dip in bubble size is observed at different feedback conditions when the global flow rate is different. However, it is the same frequency at which bubble size reduction is observed, demonstrating that the change in global flow results in no significant shift in the frequency 'sweet spot', ~150 Hz. Of note is that not only is a similar trend being observed at these different flow rates but there is also a clear indication of the dip in bubble size at the specific flow rate and frequency that seems to be characteristic of this newly discovered 'sweet spot'. It is interesting to note that the extent of the dip in bubble size varies with the flow rate and therefore on the amplitude of fluidic oscillator. The 'sweet spot' depends on the fluidic circuit i.e., aerator, liquid, fluidic oscillator, and gas aside from incoming flow rate. Amplitude is one of the major causes for a bubble size reduction which results in higher frequencies and bubble throughput.

Figure 10 shows the amplitudes and the differences seen for the different resonant conditions and shows the resonant condition for these configurations. It is seen that there is a greater decrease in bubble size for the higher feedback condition (i.e., higher amplitude), as also observed for higher fluidic oscillator incoming flow rate. This happens at a higher frequency because of the change in the 'sweet spot' condition. This means that the amplitude is higher for the same condition. This results in the lowest feedback condition having the 'largest' bubble size for its sweet spot condition. There is a smaller difference between OD 4 mm and OD 6 mm condition due to smaller differences in the feedback introduced as compared to the OD 10 mm. This increases the condition substantially resulting in an increase in the effect observed.

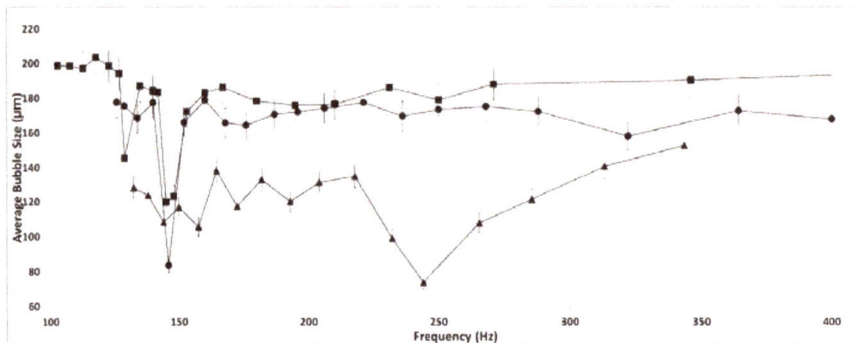

Figure 10. Comparison between amplitudes for the resonant conditions or 'sweet spot' (■—OD 4 mm, ●—OD 6 mm , ▲—OD 10 mm).

2.9. Mechanism for the Resonant Mode Condition or 'Sweet Spot'

Bubble size is directly proportional to the rise velocity and inversely proportional to the bubble wake and liquid present in the system. There are two ways that can be used to increase the amplitude to test the hypothesis we have proposed—increase it by imposing a higher feedback (i.e., 10 mm OD volume) or by increasing the flow through the fluidic oscillator (86/92 lpm). Both conditions have been used in this study and this provides several conditions to observe for fluidic oscillation mediated bubble formation. The change between the OD 4 mm and OD 6 mm is not as significant as between the two and OD 10 mm. This is due to the large feedback introduced in the system in order to cause an improved performance regime in terms of momentum of the jet as well as amplitude of the oscillatory pulse.

This results in different systems being imposed whilst keeping the rest of the system as constant as possible. The reason why different aerators were not used was because it would be difficult to compare two aerators (they can have several different properties—wettability, porosity, thickness, mesoporosity, polydispersity of orifice sizes, material of fabrication, and pressure drop across the membrane.

3. Dimensionless Analysis

The dynamics of bubbles generated in an oscillatory system can be defined by dimensionless quantities such as the Weber Number—*We*, Stokes Number—*Sk*, Strouhal Number—*Sh* and Reynolds Number—*Re*.

We is defined as the ratio of the inertia of fluid to its surface tension and determines the curvature of the bubble which means smaller the bubble greater is the curvature and higher is the surface tension and higher the *We*. *Sk* relates the bubble size to the rise velocity of the system whereas the *Sh* is used for oscillatory systems and for the fluidic oscillator, helps determine the frequency of bubble generation and the characteristics of the fluidic oscillator, especially the oscillatory flow and bistability.

Conjunctions occur when either of these values is unbalanced, leading to a bigger bubble size being observed. Greater unbalance leads to coalescence in the system. Tesař et al. [50] describes the case of the smallness of microbubble being limited due to coalescence:

$$Sh_t = \frac{fL}{V} \tag{5}$$

$$Re = \frac{\rho V D_h}{\mu} = \frac{Vb}{v} = v\frac{\dot{M}}{hv} \tag{6}$$

$$Sk = \frac{fb^2L}{V} \tag{7}$$

$$We = \frac{w^2 D_b}{2v\sigma} \tag{8}$$

$$We = \frac{f^2 D^3{}_b}{2v\sigma}n \tag{9}$$

f = oscillation frequency (Hz), L = length of feedback loop (m), V = supply nozzle bulk exit velocity (m/s), b = Constriction width (m), ρ = density (kg/m^3), η = viscosity (Pa.s), D_h = Hydraulic diameter (m), v = specific fluid volume (m^3), w = bubble rise velocity (m/s), D_b = bubble diameter (m), and σ = surface tension (N/m).

Sanada et al., [51] discussed the coalescence observed in rising bubbles and the interactions between two rising bubbles. This interaction depends on the rise velocity, (therefore size) and the rate of formation of the bubble (oscillation velocity in case of fluidic oscillator). Therefore this means that coalescence is directly related to the *We*, (bubble formation and size), *Sk* (bubble rise and size), *Sh* (bubble formation via oscillation and oscillatory flow) and *Re* (determining the momentum carried by the bubble due to the pulse). This, coupled with the resultant bubble wake and zeta potential (if any due to presence of ions/ surfactant layers on the system), defines the ultimate bubble size. This does not account for the fact that the surface of the membrane generating the bubble (involves the We and $We_{Oscillatory}$ since surface tension is involved) and the orifice size (dependent on Sh since it is the constriction that determines the bulk exit velocity of the orifice).

A conclusion to be drawn from this is that if there is the appropriate balance in the system, (with respect to rise velocity, frequency of bubble generation, bubble size and orifice diameter), analogous to a resonant mode of the system where these conditions are balanced just about correctly, then bubble conjunctions would be avoided leading to a significant size reduction in the bubbles. This is a substantial reduction in size by optimising the parameters of an existing system without any modifications or retrofitting. The amplitude of the pulse is also important as it increases the Sk and Sh which results in lesser conjunctions and coalescence.

- *Force Balance:*

Balancing the forces on the bubble being formed as seen in Figure 11.

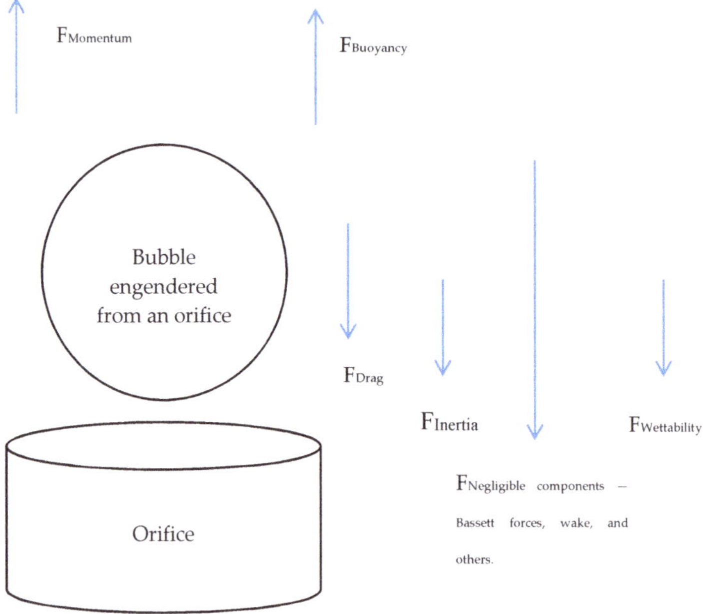

Figure 11. Bubble forces resolved (adapted from [52]).

Figure 11 shows the forces on a bubble detaching and being formed. The buoyancy force and momentum force act upwards, the cosine component of the wettability (anchoring force), drag force, surface tension act downwards. For an orifice of diameter D, density of liquid—ρ_L, density of gas—ρ_g and volume of bubble former—V_b, the following equations are obtained for the forces acting upwards:

- *Bouyancy Force*

$$F_{Bouyancy} = V_b \left(\rho_L - \rho_g\right)g \tag{10}$$

$$V_b = \frac{1}{6}\pi d_b^3 \tag{11}$$

resulting in:

$$F_{Bouyancy} = \frac{1}{6}\pi d_b^3 \left(\rho_L - \rho_g\right)g \tag{12}$$

- *Momentum Force*

$$F_{Momentum} = \frac{1}{4} \pi D_o^2 u_o^2 \rho_g \tag{13}$$

taking downward forces into account.

- *Wettability Force*

$$F_{Wettability} = \pi D_o \sigma \cos\theta \tag{14}$$

where θ is the wetting angle / contact angle made by the engendering bubble and the surface and σ is the surface tension force / anchoring force/ wetting force and D_o is the diameter of the orifice.

- *Drag Forces*

$$F_{Drag\ Force} = \frac{1}{2} C_d \rho_L \frac{\pi}{4} d_b^2 u_b^2 \tag{15}$$

with the rise velocity denoted by u_b.

- *Bubble Inertial Force*

$$F_{IF} = \frac{d\left(u_b V_b \rho_g\right)}{dt} = \frac{\rho_g Q^2 V_b^{\frac{-2}{3}}}{12\ \pi\left(\frac{3}{4\pi}\right)^{\frac{2}{3}}} \tag{16}$$

where V_b is the bubble volume, ρ_g is the gas density, ρ_l is the liquid density, g is the acceleration due to gravity, D_b is the bubble diameter, D is the orifice diameter, u_b is the rise velocity of the bubble centre , C_d is the drag coefficient and Q is the volumetric gas flow rate.

Our hypothesis comes from the fact that the bubble formation is most dependent on frequency of the system for oscillatory flow and the amplitude associated with it. If these two are appropriate, such that the bubbles are formed at regular intervals, fast enough to have substantial size reduction (reduction of pulse length and increase in throughput) and not too fast as to coalesce, with the increase in amplitude such that the bubble will cut off instantaneously and not coalesce. Balancing these forces together would result in pinch off. Compensation must be provided for the bubble rise and pinch off due to the oscillatory flow. The force balance turns out to be complicated due to the oscillatory waves and the hybrid synthetic jet engendered by the oscillator, resulting in a highly non-linear system.

- *Prediction of Bubble Size at Resonant Frequency (Volume-based Bubble Size)*

It is seen in Tesař et al. [53] that the bubble rise is dominated by the coalescence and each individual coalescing bubble leads to an increased probability for another staged coalescence which leads to largeness in bubble size as compared to the orifice. Once two bubbles merge together, due to the increase in size and the change in the surface energy as well as the energy associated with the ascent, it is easier for the other bubble to catch up with it. This is better explained by the concept of bubble wake. Each bubble creates a wake (region of lower pressure) upon being created and this allows other bubbles to catch up, coalesce and result in larger bubbles.

Therefore, the smaller the bubble that is created will result in a smaller wake. However, it is easier for the small bubble to be affected by a wake of another bubble (especially a larger one—which is why a smaller bubble generated after a larger one usually results in coalescence.

The active diffusing area of the aerator is 0.15×0.03 m^2. The bubble flux recorded by the acoustic bubble spectrometry is based on the capture rate or acquisition rate that is set for the system at 200 ms. The 200 ms ensures that the flow in the system is captured at a fixed rate and this results in a delimiting information due to the resultant lower acquisition observed. The capture results in determining the flow to be 0.5 lpm. Using these results, a size of 74 μm approximately is observed. This is thus the minimum size achieved for this. Taking this value and placing it in the equations (10–16) results in the bubble formation force, and taking a frequency of 150 Hz as the bubbling frequency, and the bubble flux from the ABS (120,000) it results in a pulse requirement of 0.007 bar to detach the bubble. This has been achieved by the oscillator as observed in Figure 5 and therefore explains the sweet spot possibility for these conditions of resonance.

Figure 5 shows that the amplitude of the pulse is approximately 0.2 V, which is equivalent to 0.02 bar. This is roughly twice the required pulse strength for the bubble formation and this is why the bubble is detached at the frequency. Judging by that, the frequency band is slightly wide due to the lack of coalescence at this point.

The amplitude of the pulse reduces as the frequency increases for the same flow resulting in intermittent pinch off (since the amplitude is important for imparting sufficient force for bubble detachment). Lower frequency has a larger amplitude but the rate of generation is not fast enough. Additionally as can be seen in Tesař ([34,50]), the coalescence perpetuates the higher rise velocity of the larger bubble. This results in larger bubbles for the non-resonant conditions at lower frequencies. Therefore, the resonant condition or 'sweet spot' is possibly the only condition where the amplitude, frequency, and size are balanced so as not to have bubble coalescence.

There is greater amplitude for the higher feedback condition which results in the 'sweet spot' occurs at a higher frequency (250 Hz). This is what results in the larger increase in the number of bubbles throughput.

- *Prediction of Bubble Size at Resonant Frequency (Number of Bubbles-Based Bubble Size)*

This paper has placed a lot of emphasis on bubble size suitable for transport phenomena but for finding out what the actual bubble size is, when compared to those that have been reported previously, such as Hanotu et al. [19], the size distribution based on number of bubbles calculation—N_{Av}, provides a better idea for those applications concerned with size of the bubbles (flotation/DAF for example) in Figure 12.

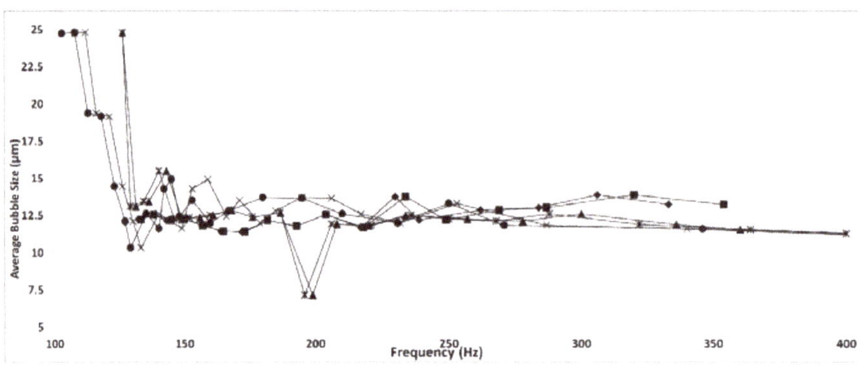

Figure 12. This is an example of the sweet spot for the average bubble size when one considers number of bubbles as opposed to the volume fraction of bubbles. The sweet spot changes slightly but there is still a dip observed at the higher frequencies (◆—OD 10 mm (92 lpm), ▲—OD 6 mm (92 lpm), ✳—OD 4 mm (92 lpm), ■—OD 10 mm (86 lpm), ✳—OD 6 mm (86 lpm), ◆—OD 4 mm (86 lpm)).

This brings down the average bubble size reduction producing about 7 μm size bubbles in terms of average size distribution which makes it suitable for generating the bubbles for flotation studies. The 'sweet spot' changes to 200 Hz. This is because the size distribution changes due to the change in flow dynamics. The flow results in this size change as the effect of the number of bubbles cannot be discounted just as the volume average bubble size depends on the dispersity of the size distribution.

An example is provided herewith: The outlet flow of the diffuser measured at 0.1 lpm (1.67×10^{-6} m^3/s) for which, when corrected for the 200 ms acquisition rate of the ABS is $Q = 3.33 \times 10^{-7}$ m^3/200 ms. N = No of Bubbles— Approximating total number of bubbles as 120,000 as an average measured from the readings, assuming that the distribution is narrow (for the purposes of a general understanding).

Assuming that since bubble pinch-off force (by using Equations (10)–(16)) results in 0.007 bar, and the pulse of the oscillator is an average of 0.02 bar, it exceeds the force required for bubble pinch-off (*cf.* Figure 5). As seen from the FFT of the pulse for the oscillator, the pulse strength is at 0.02 bar per pulse which ensures that there is sufficient momentum and amplitude to generate the bubbles and ensure that the bubble detaches at each oscillatory pulse, so it is an accurate assumption providing no coalescence takes place. This is why the sweet spot is likely at 200 Hz as for the same momentum the force of the bubble pinch-off seems to be at the appropriate level. Figure 5 is shown herein for reference. This has been done in order to reduce the bubble size by matching the bubble formation characteristics to achieve the lowest possible bubble size and by increasing the amplitude of the oscillation in order to impart momentum to the jet and therefore the bubble so that it has a higher rise velocity. This reduces the bubble coalescence when the appropriate conditions are met. Too slow, and the bubble does not detach quickly enough and therefore coalesces, too fast, and the bubble cannot detach due to reduced amplitude and pinch-off. At the right frequency and amplitude, what we term—resonant condition or 'sweet spot', one can detach the bubble at the smallest possible size for that system and it is significantly smaller than what was originally possible via conventional steady flow. This is achieved for the frequency at 200 Hz. Since f = 200 Hz means that it is per second, for 200 ms, the equivalent frequency to be considered would be 40 Hz.

Using the A = diffusing area (0.15×0.03 m^2), and assuming that all the flow forms a bubble (there are no leaks and the bubble sizes are small enough to form bubbles rather than slugs:

$$V_{av} = Average\ Volume\ of\ bubble\ formed = \frac{Q}{ANf_{eq}}$$

which results in D_B i.e., (2r) = 6.8 μm for 200 Hz.

This is in close agreement to the size obtained in the system. Depending on changes introduced to the system, these values will change. Whilst these values will change depending on the system, the work supports the presence of the resonant bubbling condition that can occur in any oscillatory flow mediated bubble generating system.

Figure 13 shows the calculations for a sweep for bubble formation via frequencies juxtaposed with the size for the bubbles that would be formed. There is an extension but this is likely due to the fact that at lower frequencies, steady bubble formation dominates as it is results in faster bubble pinch-off.

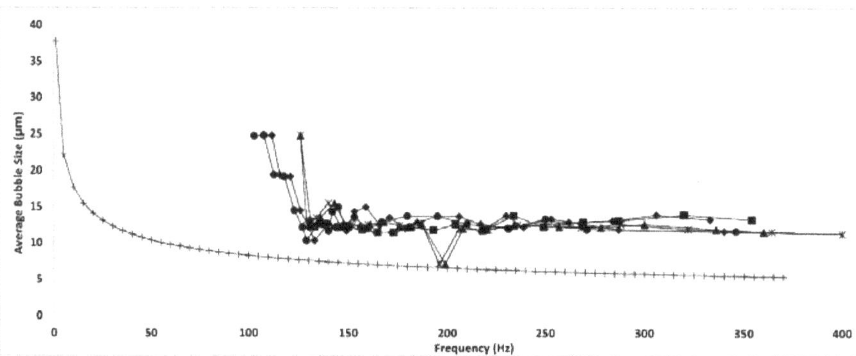

Figure 13. Calculated value as compared to the average bubble size garnered experimentally. Deviation from the sweet spot is less than 0.0001%. ─┼─—Calculated ─◆─—OD 10 mm (92 lpm), ─▲─—OD 6 mm (92 lpm), ─✳─—OD 4 mm (92 lpm), ─■─—OD 10 mm (86 lpm), ─✳─—OD 6 mm (86 lpm), ─●─—OD 4 mm (86 lpm).

As discussed earlier, 0.25–0.4% of the U.K.'s energy use is utilised for WWA operations. This is not discussed earlier These energy requirements could be substantially reduced by bubble size reduction. The 10 fold size reduction would provide a significant increase in the transport phenomena associated with the bubbles. The cost of adding the fluidic oscillator is based on the pressure drop of the oscillator, which for an industrial plant would typically be about 400 mbar for a large scale system. The cost of frequency modulation in terms of energy is negligible. This is also applicable for several other applications and remediation steps including aeration of waste streams, oxidation of volatiles, advanced oxidation processes, and other treatment techniques. Several other processes exist where gas-liquid operations are involved and bubble sizes are required to be small.

4. Conclusions

This paper reports several scientific results. One of the major ones is the exploration of a new regime for bubble generation mediated by oscillatory flow which reduces coalescence and results in the smallest bubble formation for a given system (number averaged or volume averaged) at specific conditions of detachment with no additional energy.

The ability to generate substantially smaller bubbles, useful for several applications, by simply engineering the conditions for bubble formation under oscillatory flow such that the frequency and amplitude are maintained is a shift in understanding for bubble formation dynamics. Whilst oscillatory flow results in bubble size reduction over the gamut of frequencies, at a particular frequency, specific to a particular set up (but valid for other set ups as seen with different configurations), a substantial reduction in bubble size (≈60%) is observed. It has been hypothesised that tuning of all the different parameters controlling bubble formation will result in a smaller overall bubbles size, presumably by the combined effect of more efficient pinch off and reduced coalescence. This has resulted in a large size reduction from a typical bubble size distribution produced in a system. Conventional bubble generation would have produced bubbles of sizes: 650 µm (volume averaged) and 250 µm (number averaged), whereas the fluidic oscillator would result in an average size of 120 µm (volume averaged) and 14 µm (number averaged), and the sweet spot would result in 60 µm (volume averaged) and 7 µm (number averaged).

The addition of the fluidic oscillator reduces the bubble size to a certain extent but coalescence prevents the size being smaller than a certain range. This adds an additional pressure drop across the system by 20–150 mbar in the system (system dependent) which is insignificant compared to the actual savings in terms of surface energy. A further reduction of bubble size is achieved, without any additional cost to the system, which then brings about a step change in any gas liquid operation and overcomes any of the limitations that had been posed earlier for using microbubbles for mass transfer in gas-liquid operations. A 60% reduction in bubble size from what was already a reduced bubble size, resulting in a 10-fold reduction in bubble size with suitable tuning and adding a pressure drop of 150 mbar maximum, is paradigm shifting and has the potential to change the processing industry.

An example can be seen in DAF, which has been discussed at the start. In order to achieve a 70–100 μm bubble size (number averaged), 12–14 bar of pressure is used to generate the flux and throughput required. We can achieve a 10-fold reduction for this size (7 μm) with a tuned frequency, fluidic oscillator, and aerator totalling 2.15 bar(g) total pressure which means that the energetics involved in order to get 10 times larger size bubbles will result in substantial reduction. A simplistic approach will be to relate energy with pressure as a proxy and the energetics for a specific process can be taken into account. The fact the frequency modulation for a bistable diverter valve (the fluidic oscillator) does not require any additional energy input, but results in a 60% reduction in bubble size (increase in associated transport phenomena) [9,11,29], and is a 10-fold reduction in bubble size (100–fold increase in interfacial area) by using this tuned oscillator as opposed to a conventional steady flow system. The substantial energy saving and increase in transfer efficiencies for any gas-liquid operation involved results in widespread ramifications. This is not limited to only bistable diverters but all oscillatory flow systems as it is actually dependent on oscillatory flow and the control of frequency and amplitude as opposed to using a specific system.

These wider implications are not only limited to DAF, but for all operations that can be thought of as gas-liquid contacting or processing including aeration, gas-liquid contacting operations which constitute a large proportion of industrial processes, biochemical reactors, fermenters, remediation processes, digesters, incubators, bioreactors, and others.

Author Contributions: W.B.Z. and P.D.D. jointly proposed the central hypothesis. P.D.D. augmented the hypothesis with the inclusion of higher amplitude by artificial venting, created the experimental design and setup, and conducted the first three experiments and the design for the analysis, hence is the principal author. Y.R. provided the individual bias removal, substantial experimental work via repeats (5 in total), equation input, and collated the data for analysis. M.J.H. and W.B.Z. provided organisational and logistical support and compositional critique.

Funding: This research was funded by EPSRC grant number [EP/K001329/(1)] and part of the IUK Energy Catalyst, IB Catalyst programmes as well as the FP7 R3 water programme.

Acknowledgments: This work was carried out as part of the "4CU" programme grant, aimed at sustainable conversion of carbon dioxide into fuels, led by The University of Sheffield and carried out in collaboration with The University of Manchester, Queens University Belfast and University College London. The authors acknowledge gratefully the Engineering and Physical Sciences Research Council (EPSRC) for supporting this work financially (Grant no. EP/K001329/1). M.J.H. and P.D.D. would like to thank R3Water FP7 grant. W.B.Z. and P.D.D. would like to thank IUK IBCatalyst and Energy Catalyst award. P.D.D. would like to thank Professor Ray Allen for discussions, Audrey YY Tan for bias studies and GK for inspiration.

Conflicts of Interest: The authors declare no conflict of interest.

References

1. Bird, R.B.; Stewart, W.E.; Lightfoot, E.N. *Transport Phenomena*; Wiley: Hoboken, NJ, USA, 2007.
2. Incropera, F.P.; Lavine, A.S.; Bergman, T.L.; DeWitt, D.P. *Fundamentals of Heat and Mass Transfer*; Wiley: Hoboken, NJ, USA, 2007.
3. Green, D.; Perry, R. *Perry's Chemical Engineers' Handbook*, 8th ed.; McGraw-Hill Education: New York, NY, USA, 2007.
4. Treybal, R.E. *Mass-Transfer Operations*; McGraw-Hill Education: New York, NY, USA, 1980.
5. Tesař, V. What can be done with microbubbles generated by a fluidic oscillator? (survey). *EPJ Web Conf.* **2017**, *143*. [CrossRef]
6. Metcalf, E.I.; Tchobanoglous, G.; Burton, F.; Stensel, H.D. *Wastewater Engineering: Treatment and Reuse*; McGraw-Hill Education: New York, NY, USA, 2002.

7.	Abdulrazzaq, N.; Al-Sabbagh, B.; Rees, J.M.; Zimmerman, W.B. Purification of Bioethanol Using Microbubbles Generated by Fluidic Oscillation: A Dynamical Evaporation Model. *Ind. Eng. Chem. Res.* **2016**, *55*, 12909–12918. [CrossRef]

8.	Abdulrazzaq, N.; Al-Sabbagh, B.; Rees, J.M.; Zimmerman, W.B. Separation of zeotropic mixtures using air microbubbles generated by fluidic oscillation. *AIChE J.* **2015**, *62*, 1192–1199. [CrossRef]

9.	Al-Yaqoobi, A.; Zimmerman, W.B. Microbubble distillation studies of a binary mixture. In *USES—University of Sheffield Engineering Symposium 2014*; The University of Sheffield: Sheffield, UK, 2014.

10.	Zimmerman, W.B.; Al-Mashhadani, M.K.H.; Bandulasena, H.C.H. Evaporation dynamics of microbubbles. *Chem. Eng. Sci.* **2013**, *101*, 865–877. [CrossRef]

11.	Fiabane, J.; Prentice, P.; Pancholi, K. High Yielding Microbubble Production Method. *BioMed Res. Int.* **2016**, 3572827. [CrossRef] [PubMed]

12.	Kantarci, N.; Borak, F.; Ulgen, K.O. Bubble column reactors. *Process Biochem.* **2005**, *40*, 2263–2283. [CrossRef]

13.	Tesař, V. What can be done with microbubbles generated by a fluidic oscillator? (survey). *EPJ Web Conf.* **2017**, *143*. [CrossRef]

14.	Makuta, T.; Takemura, F.; Hihara, E.; Matsumoto, Y.; Shoji, M. Generation of micro gas bubbles of uniform diameter in an ultrasonic field. *J. Fluid Mech.* **2006**, *548*, 113–131. [CrossRef]

15.	Shirota, M.; Sanada, T.; Sato, A.; Watanabe, M. Formation of a submillimeter bubble from an orifice using pulsed acoustic pressure waves in gas phase. *Phys. Fluids* **2008**, *20*. [CrossRef]

16.	Stride, E.; Edirisinghe, M. Novel microbubble preparation technologies. *Soft Matter* **2008**, *4*, 2350–2359. [CrossRef]

17.	Zimmerman, W.B.; Tesař, V.; Bandulasena, H.C.H. Towards energy efficient nanobubble generation with fluidic oscillation. *Curr. Opin. Colloid Interface Sci.* **2011**, *16*, 350–356. [CrossRef]

18.	Parmar, R.; Majumder, S.K. Microbubble generation and microbubble-aided transport process intensification—A state-of-the-art report. *Chem. Eng. Process. Process Intensif.* **2013**, *64*, 79–97. [CrossRef]

19.	Hanotu, J.; Bandulasena, H.C.; Zimmerman, W.B. Microflotation performance for algal separation. *Biotechnol. Bioeng.* **2012**, *109*, 1663–1673. [CrossRef] [PubMed]

20.	Agarwal, A.; Ng, W.J.; Liu, Y. Principle and applications of microbubble and nanobubble technology for water treatment. *Chemosphere* **2011**, *84*, 1175–1180. [CrossRef] [PubMed]

21.	Rehman, F.; Medley, G.; Bandalusena, H.C.H.; Zimmerman, W.B. Fluidic oscillator-mediated microbubble generation to provide cost effective mass transfer and mixing efficiency to the wastewater treatment plants. *Environ. Res.* **2015**, *137*, 32–39. [CrossRef] [PubMed]

22.	Mulvana, H.; Eckersley, R.J.; Tang, M.X.; Pankhurst, Q.; Stride, E. Theoretical and experimental characterisation of magnetic microbubbles. *Ultrasound Med. Biol.* **2012**, *38*, 864–875. [CrossRef] [PubMed]

23.	Lukianova-Hleb, E.Y.; Hanna, E.Y.; Hafner, J.H.; Lapotko, D.O. Tunable plasmonic nanobubbles for cell theranostics. *Nanotechnolohy* **2010**, *21*. [CrossRef] [PubMed]

24.	Cai, X. Applications of Magnetic Microbubbles for Theranostics. *Theranostics* **2012**, *2*, 103–112. [CrossRef] [PubMed]

25.	Zimmerman, W.B.; Hewakandamby, B.N.; Tesař, V.; Bandulasena, H.C.H.; Omotowa, O.A. Design of an airlift loop bioreactor and pilot scales studies with fluidic oscillator induced microbubbles for growth of a microalgae Dunaliella salina. *Appl. Energy* **2011**, *88*, 3357–3369. [CrossRef]

26.	Zimmerman, W.B.; Hewakandamby, B.N.; Tesař, V.; Bandulasena, H.C.H.; Omotowa, O.A. On the design and simulation of an airlift loop bioreactor with microbubble generation by fluidic oscillation. *Int. Sugar J.* **2010**, *112*, 90–103. [CrossRef]

27.	Ying, K.; Gilmour, D.J.; Shi, Y.; Zimmerman, W.B. Growth Enhancement of Dunaliella salina by Microbubble Induced Airlift Loop Bioreactor (ALB)—The Relation Between Mass Transfer and Growth Rate. *J. Biomater. Nanobiotechnol.* **2013**, *4*, 1–9. [CrossRef]

28.	Ying, K.; AlMashhadani, K.H.; Hanotu, J.O.; Gilmour, D.J.; Zimmerman, W.B. Enhanced Mass Transfer in Microbubble Driven Airlift Bioreactor for Microalgal Culture. *Engineering* **2013**, *5*, 735–743. [CrossRef]

29.	Hanotu, J.; Bandulasena, H.C.H.; Chiu, T.Y.; Zimmerman, W.B. Oil emulsion separation with fluidic oscillator generated microbubbles. *Int. J. Multiph. Flow* **2013**, *56*, 119–125. [CrossRef]

30.	Zimmerman, W.B.; Tesař, V.; Butler, S.; Bandulesena, H.C.H. Microbubble Generation. *Recent Pat. Eng.* **2008**, *2*. [CrossRef]

31. Zimmerman, W.B.; Hewakandamby, B.N.; Tesař, V.; Bandulasena, H.C.H.; Omotowa, O.A. On the design and simulation of an airlift loop bioreactor with microbubble generation by fluidic oscillation. *Food Bioprod. Process.* **2009**, *87*, 215–227. [CrossRef]

32. Tesař, V. *Pressure-Driven Microfluidics*; Artech House: Norwood, UK, 2007.

33. Tesař, V.; Bandalusena, H. Bistable diverter valve in microfluidics. *Exp. Fluids* **2011**, *50*, 1225–1233. [CrossRef]

34. Tesař, V. Mechanisms of fluidic microbubble generation part Ii: Suppressing the conjunctions. *Chem. Eng. Sci.* **2014**. [CrossRef]

35. Zimmerman, W.B.; Tesař, V.; Bandulasena, H.C.H. *Efficiency of an Aerator Driven by Fluidic Oscillation. Part I: Laboratory Bench Scale Studies*; The University of Sheffield: Sheffield, UK, 2010.

36. Warren, R.W. Negative Feedback Oscillator. U.S. Patent 3,158,166, August 1962.

37. Tesař, V. Microbubble generation by fluidics. part I: Development of the oscillator. In Proceedings of the Fluid Dynamics 2012, Prague, Czech Republic, 24–26 October 2012.

38. Kooij, S.; Sijs, R.; Denn, M.M.; Villermaux, E.; Bonn, D. What determines the drop size in sprays? *Phys. Review X* **2018**, *8*. [CrossRef]

39. Leighton, T.G. *The Acoustic Bubble*; Academic Press: Cambridge, MA, USA, 1994.

40. Chahine, G.L. Numerical Simulation of Bubble Flow Interactions. In *Cavitation: Turbo-machinery and Medical Applications-WIMRC FORUM 2008*; Warwick University: Coventry, UK, 2008.

41. Chahine, G.L.; Duraiswami, R.; Frederick, G. *Detection Of Air Bubbles In Hp Ink Using Dynaflow's Acoustic Bubble Spectrometer (Abs) Technology*; Hewlett Packard: Palo Alto, CA, USA, 1998.

42. Chahine, G.L.; Gumerov, N.A. An inverse method for the acoustic detection, localization and determination of the shape evolution of a bubble. *Inverse Probl.* **2000**, *16*, 1–20.

43. Duraiswami, R.; Prabhukumar, S.; Chahine, G.L. Bubble counting using an inverse acoustic scattering method. *J. Acoust. Soc. Am.* **1998**, *104*, 2699–2717. [CrossRef]

44. Chahine, G.L.; Kalamuck, M.K.; Cheng, J.Y.; Frederick, G.S. Validation of Bubble Distribution Measurements of the ABS with High Speed Video Photography. In Proceedings of the CAV 2001: Fourth International Symposium on Cavitation, Pasadena, CA, USA, 20–23 June 2001.

45. Tanguay, M.; Chahine, G.L. Acoustic Measurements of Bubbles in Biological Tissue. In *Cavitation: Turbo-machinery and Medical Applications*; Warwick University: London, UK, 2008.

46. Wu, X.-J.; Chahine, G.L. Development of an acoustic instrument for bubble size distribution measurement. *J. Hydrodyn. Ser. B* **2010**, *22*, 330–336. [CrossRef]

47. Allen, T. *Particle Size Measurement: Volume 1: Powder Sampling and Particle Size Measurement*; Springer: New York, NY, USA, 1996.

48. Merkus, H.G. *Particle Size Measurements: Fundamentals, Practice, Quality*; Springer: New York, NY, USA, 2009.

49. Tesař, V. Microbubble smallness limited by conjunctions. *Chem. Eng. J.* **2013**, *231*, 526–536. [CrossRef]

50. Sanada, T.; Sato, A.; Shirota, M.; Watanabe, M. Motion and coalescence of a pair of bubbles rising side by side. *Chem. Eng. Sci.* **2009**, *64*, 2659–2671. [CrossRef]

51. Pinczewski, W.V. The formation and growth of bubbles at a submerged orifice. *Chem. Eng. Sci.* **1981**, *36*, 405–411. [CrossRef]

52. Tesař, V. Shape Oscillation of Microbubbles. *Chem. Eng. J.* **2013**, *235*, 368–378. [CrossRef]

53. Tesař, V.; Hung, C.-H.; Zimmerman, W.B. No-moving-part hybrid-synthetic jet actuator. *Sens. Actuators A Phys.* **2006**, *125*, 159–169. [CrossRef]

Article

Bubble Size and Bubble Concentration of a Microbubble Pump with Respect to Operating Conditions

Seok-Yun Jeon [1,2], Joon-Yong Yoon [1] and Choon-Man Jang [2,*]

[1] Department of Mechanical Engineering, Hanyang University, 55 Hanyangdeahak-ro, Sangnok-gu, Ansan, Gyeonggi-do 15588, Korea; lo21c@hanyang.ac.kr (S.-Y.J.); joyoon@hanyang.ac.kr (J.-Y.Y.)

[2] Department of Land, Water and Environment Research, Korea Institute of Civil Engineering and Building Technology, 283, Goyangdae-ro, Ilsanseo-gu, Goyang-si, Gyeonggi-do 10223, Korea

* Correspondence: jangcm@kict.re.kr; Tel.: +82-31-910-0494

Received: 8 June 2018; Accepted: 12 July 2018; Published: 17 July 2018

Abstract: The present paper describes some aspects of the bubble size and concentration of a microbubble pump with respect to flow and pressure conditions. The microbubble pump used in the present study has an open channel impeller of a regenerative pump, which generates micro-sized bubbles with the rotation of the impeller. The bubble characteristics are analyzed by measuring the bubble size and concentration using the experimental apparatus consisting of open-loop facilities; a regenerative pump, a particle counter, electronic flow meters, pressure sensors, flow control valves, a torque meter, and reservoir tanks. To control the intake, and the air flowrate upstream of the pump, a high precision flow control valve is introduced. The bubble characteristics have been analyzed by controlling the intake air flowrate and the pressure difference of the pump while the rotational frequency of the pump impeller was kept constant. All measurement data was stored on the computer through the NI (National Instrument) interface system. The bubble size and concentration are mainly affected by three operating parameters: the intake air flowrate, the pressure difference, and the water flowrate supplied to the pump. It is noted that the operating conditions that can most effectively generate microbubbles in the range of 20 to 30 micrometers are at the pressure of 5 bar and at the air flowrate ratio of 4.0 percent for the present pump. Throughout the experimental measurements, it was found that the pump efficiency changed by less than 1.2 percent, depending on the intake air supply. The performance characteristics of microbubble generation obtained by experimental measurements are analyzed and discussed in detail.

Keywords: microbubble pump; bubble generation; pump efficiency; bubble size; concentration; particle counter

---

1. Introduction

A microbubble defined as a bubble having the size of 50 micrometers or less and having several unique characteristics resulting from their fine size [1]. It has the characteristics of a large surface area due to its very small diameter, a slow rising velocity inside the liquid due to the low buoyancy force, self-pressurized dissolution, a superior mass transfer efficiency, and a high dissolution rate. Due to these characteristics, microbubbles are broadly applied in various fields; sewage treatment, water purification technology, ozone water disinfection, mineral processing, natural ecology restoration, bio-pharmaceutical production, fine chemical reactions, and so on [2,3]. With further applications of microbubble technology in human life and industry, it has been getting more attention in recent years. In addition to the field of application using the inherent characteristics of the microbubbles above, they have the potential to be applied in applications such as non-detergent cleaning, skin care, sterilization cleaning and anion effect generation.

The formation and growth mechanism of microbubbles is closely related to cavitation which is one of the hydrodynamic phenomena. Although it is able to be used as a method for generating microbubbles in research area, it is not used in large-scale bubble generation applications due to high cost and technical reasons. Therefore, the following methods are generally used for bubble generation.

Typical bubble generation methods include pressurized dissolution type, gas–water circulation type, and Venturi-type. For the pressurized dissolution type bubble generator, microbubbles are generated as the pressure decreases due to injecting pressurized air into a water tank. The method uses a supersaturated condition to dissolve gas under high pressure of 3–4 bars and the gas becomes very unstable and escapes from the water [4]. As a result, microbubbles are generated in a very short time. Kim et al. [5] investigated the effect of swirling chamber and breaker disk on bubble generation by pressurized dissolution method. For the gas–water circulation type bubble generator, Takahashi [6] proved that gas mixed in a vortex flow changes continuously into microbubbles as it is broken down by the vortex. Ikeura et al. [7] found that ozone microbubbles generation by pressurized dissolution type is more efficient than gas–water circulation type to remove the residual pesticide in vegetables. The Venturi-type bubble generator is similar to the pressurized dissolution type in the bubble generation method and has been widely used due to its simple structure. Yoshida et al. [8] found that microbubbles were generated through the collision of a gas with the pressure wall formed by a shock wave. Yin et al. [9] conducted an experimental study on the bubble size distribution in the turbulence flow for the Venturi-type bubble generator and proved that the averaged bubble diameter has a dependence on the Reynolds number. Uesawa et al. [10] researched the mechanism of the bubble breakup phenomenon in a Venturi tube and obtained experimental results that the bubbles expanded once into a divergence region of the Venturi tube and then contracted rapidly and broke up into a great number of fine bubbles. Gordiychuk et al. [11] investigated the effects of the flowrates of air and water on the bubble size for a Venturi tube connected bubble generator. They measured bubble size by picturing the flow and using a post-processing algorithm. Zhao et al. [12] visualized the details of the transportation of individual bubbles in a Venturi tube by using a digital image analysis (DIA) method. An extremely rapid deceleration and intense breakup process was observed at the entrance of the diverging section of the bubble generator.

The bubble generator types described above are based on the assumption that the growth and release of a sphere bubble is formed from nozzles or orifices, and this process consists of bubble expansion stage and collapse stage [13]. This complex process of the bubble formation necessarily involves a variety of additional techniques for precisely controlling the bubble size and amount of production. On the other hand, a microbubble pump generates microbubbles by forming a vortex flow using centrifugal force in the radial direction and reaction force in the circumferential direction by rotation of impeller.

The microbubble pump with a regenerative pump type impeller (hereinafter, the term 'microbubble pump' is used.) has the advantages of continuous micro-sized bubble generation and a being compact single unit. In addition, the microbubble pump is able to be applied to any small-scale application requiring microbubbles, and is easy to maintain, due to its simple structure. The microbubble pump adopted in the present study has a disk-type impeller with many vanes on its periphery rotating in an annular flow path as that is usual in conventional regenerative pumps. Therefore, it follows inherent characteristics of a regenerative pump, such as having high pressure at a low flow capacity [14]. Despite the relatively low mechanical efficiency compared to the other types of pumps having an equivalent impeller diameter, the regenerative pump has advantages, such as versatile capacity and scalability, low production cost, and robust reliability. Until now, researchers on regenerative turbomachinery have been interested in improving efficiency and performance. Karlsen-Davies and Aggidis [15] reviewed overall aspects of regenerative pumps and their basic principle of operation, areas of application, performance challenges, and the improvement of geometrical parameters. Quail et al. [16] carried out parametric design through one- and three-dimensional numerical techniques and Jeon et al. [17] tried to enhance the pump performance using an optimized design. Shimizu et al. [18] experimentally analyzed the flow patterns and the bubble size of methane–water two-phase flows.

The microbubble pump has both functions of a regenerative pump and a microbubble generation in a pressurized condition. The regenerative pump having multi impeller blades generates high suction pressure with the rotation of the impeller, and thus sucks gas through tube installed upstream of the pump. The inhaled gas mixes with water inside the pump and dissolves. It is noted that the microbubble pump is able to remove water-soluble gases effectively in chemical plant environments without complicated processes [19]. Also, it has the potential to replace the bubble generator applied to conventional dissolved air flotation (DAF) systems and has an advantage in terms of operating cost and energy saving [20].

Despite the various advantages described above, research on the microbubble pump has been rare until now. This is especially true when considering the bubble size and the concentration generated from the microbubble pump with respect to operating conditions which are important in evaluating the pump performance.

In the present study, the bubble size and the concentration of a microbubble pump are investigated by experimental measurements. To measure the bubble characteristics with respect to flowrate and pressure conditions, an open-looped measuring facility has been designed. A particle counter having optical sensors and an infrared laser and a tube-linked peristaltic pump are used for analyzing the microbubbles generated from the pump. Bubble characteristics are also measured by controlling the intake air flowrates and the pressure rise of the pump, while the rotational frequency of the pump impeller was kept constant. All measured data is stored on the computer through the NI interface system. The characteristics of generated microbubbles are obtained by experimental measurements and are analyzed and discussed in detail.

2. Experimental Setup

2.1. Microbubble Pump

The microbubble pump introduced in the present study has an open channel impeller, as shown in Figure 1. It has a duct inlet for water and a port for air supply, and a duct outlet for discharging water with microbubbles. The pump is composed of a regenerative impeller with multiple blades, a casing, and a dissolution tank for discharging the pressurized and mixed bubble water. Jang et al. [21,22] conducted a numerical analysis of a regenerative blower and observed that the fluid passes through each impeller blade with recirculation flow. In the same manner as the microbubble pump, the regenerative impeller induces higher a pressure at the inlet than the outlet and the local recirculation flow at each blade passage produces microbubbles by mixing and crushing the water and air supplied.

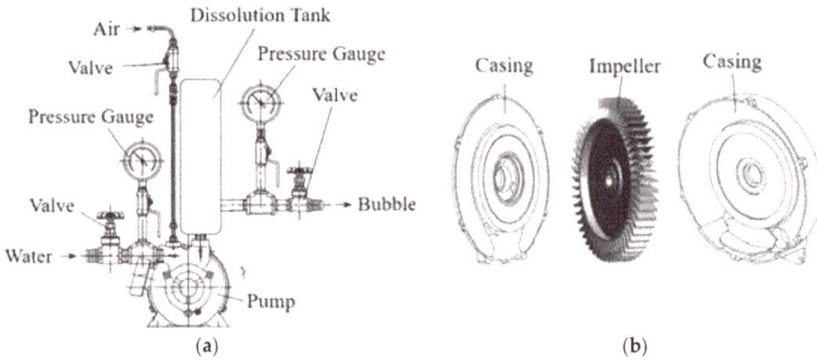

Figure 1. Schematic view of test microbubble pump: (**a**) overall view; (**b**) pump.

The pump is made of stainless steel with capacity of 1.1 kW at rated operating conditions. A mechanical seal is installed between the pump casing and the drive shaft to prevent water leakage at high-pressurized operating conditions. The detailed specifications of the test pump are shown in Table 1. The flow and pressure coefficients of the pump from performance test at rated operating condition are 0.39 and 2.04, respectively. The outer diameter of the impeller is 72.6 mm and the number of blades is 48.

Table 1. Specifications of test microbubble pump.

Parameters	Value
Flow coefficient	0.39
Pressure coefficient	2.04
Rotational speed, rpm	3550
Diameter of impeller, mm	72.6
Number of impeller blades, ea	48

The flow coefficient (ϕ) and pressure coefficient (ψ) of the pump are defined as follows.

$$\phi = \frac{Q}{AU_t} \tag{1}$$

$$\psi = \frac{2\Delta P}{\rho_w U_t{}^2} \tag{2}$$

where Q is the volumetric flowrate of water, ΔP is the pressure rise between the inlet and outlet of the microbubble pump, U_t is the rotational velocity of the pump impeller at the blade tip, ρ_w is the density of water, and A is the cross-sectional area of the side channel of the microbubble pump.

2.2. Experimetal Apparatus and Method

Figure 2 shows the schematic diagram of the experimental apparatus of the microbubble pump. Using the experimental apparatus of the pump, the generation feature of the microbubbles and the bubble sizes are measured and analyzed with respect to the pump's operating conditions. A water tank having the capacity of 1 ton is installed upstream of the pump to stabilize the water supply to the microbubble pump. The experimental apparatus has an open-loop water supply system, so the large water tank is necessary to keep the water supply continuous. Upper and lower reservoirs made of acrylic are installed upstream and downstream of the test pump for the checks of the water level and the visual monitoring of the microbubbles. The water level of the upper reservoir is kept constant by the optimal operation of the pump installed between the water tank and the upper reservoir. Mixed water containing microbubbles flows into the lower reservoir through the pipeline at the bottom of it, and the vertical height from the discharge port to the free surface is 0.42 m.

Two electronic flow meters are installed upstream of the microbubble pump to measure the flowrate of the supplied water and the air sucked to the microbubble pump, respectively. As for measuring the water flowrate supplied to the pump, the electronic flow meter having a range between 7 and 150 LPM (liter per minute) has been used (VVX25, Sika, Kaufungen, Germany). Considering the small air flowrate to the pump, a thermal flowrate sensor with high sensitivity and high response characteristics is introduced. The air flowrate meter has a measuring range between 0 and 2 LPM (CMS0002, Azbil, Tokyo, Japan). Two electronic pressure sensors are installed, one upstream and one downstream of the test pump to measure the pressure difference of the pump. A pressure sensor (model: PSAN-(L)V01C(P)A, Autonics, Busan, Korea) having a range between -101.3 and 0 kPa is installed upstream of the pump. On the other hand, a pressure sensor up to 1000 kPa has been used for measuring high positive pressure (model: PSAN-(L)1C(P)A, Autonics, Busan, Korea). Two valves are installed to control the suction air flowrate and the pressure of the pump.

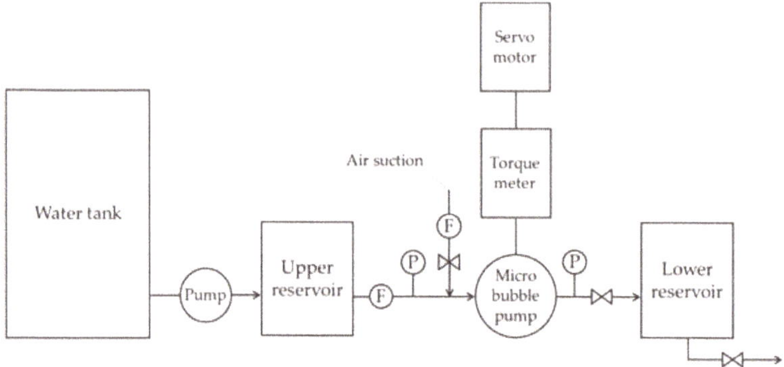

Figure 2. Schematic diagram of a microbubble pump experimental apparatus. (F: electronic flow meter, P: electronic pressure sensor).

The pump driving system is introduced to control the rotational speed of the pump and to measure the pump driving torque. A servo motor (FMA-CN10-AB00, Higen, Changwon, Korea) of 1.0 kW is installed to drive the pump at a constant rotational speed. A torque meter with a tachometer (SS-100, Ono Sokki, Yokohama, Japan) is selected for measuring the rotational speed and the torque of the pump.

A microbubble counting system is used to measure the size and the concentration of the microbubbles that are generated by the microbubble pump in respect to time. As shown in Figure 3, a microbubble counting system consists of a peristaltic pump, a particle counter, and a data processing PC. A particle counter (PC3400, Chemtrac, Norcoss, GA, USA) having optical sensors and processing electronics, including an infrared laser, is introduced. The measuring range of the bubble size is between 2 and 125 μm while the pump flowrate of the microbubble into the particle counter is kept at 75 mL/min. A constant flow pump (JWS-600, JenieWell, Seoul, Korea) is used to send a constant amount of the microbubbles to the counting system through a tube-linked peristaltic pump (JWS-600, JenieWell, Seoul, Korea).

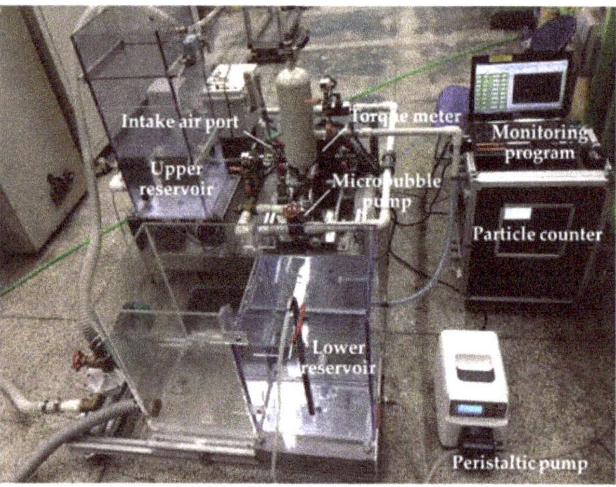

Figure 3. Picture of a microbubble pump experimental apparatus.

Microbubbles have very sensitive characteristics, depending on the pump pressure and suction air flowrate. In order to measure the performance characteristics of microbubble generation, the measuring ranges of variables, such as pressure and air flowrate, are determined through preliminary experiments. The experimental conditions are shown in Table 2. In the table, the air flowrate ratio, Ra, represents the ratio of the air flowrate and the water flowrate.

Table 2. Experimental conditions.

Variables	Value
Air Flowrate Ratio (*Ra*), %	0.8–4.8
Pump pressure, bar	4–6
Rotational speed of impeller, RPM	3550

3. Result and Discussion

3.1. Performance Curve of Test Microbubble Pump

Prior to measuring the microbubbles generated, the pump performance is obtained by an open loop pump test facility, as shown in Figure 4. It is noted that the pump performance is determined by the water supply only, in the absence of air suction. Mechanical efficiency (η) of the pump is derived from following Equation (3).

$$\eta = \frac{Q\Delta P}{T\omega} \tag{3}$$

where T and ω are the torque acting on the rotational axis and the angular velocity of the microbubble pump impeller, respectively.

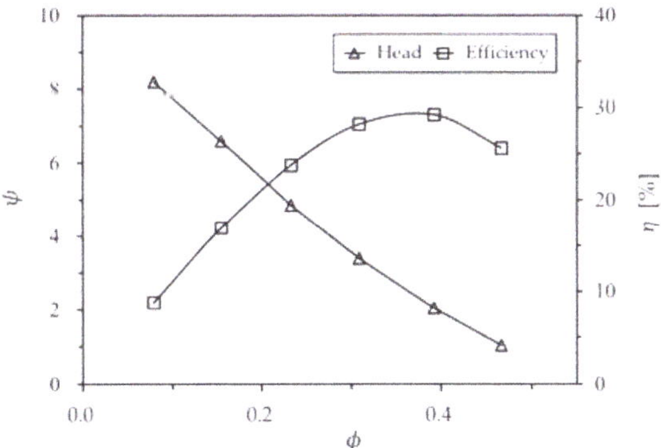

Figure 4. Performance curve of test microbubble pump.

3.2. Bubble Size and Distribution

Since the diameter of microbubbles are usually considered to be 50 micrometers or less, the present study also analyzes the bubble size range. Figures 5–7 show bubble concentration (*c*) with respect to the bubble diameter (*d*) at the three different pump pressure conditions, while the air flowrate ratio (*Ra*) also being changed. The particle counter used for the measurement of the bubble size and concentration automatically converts the total counts of particles to the volume of them and displays the concentration of particles in ppb. That is, the bubble concentration in the unit of ppb means the relative volume ratio of air and water with a scale of one billionth. It is noted that the bubble concentration is determined at the fixed rotational speed of impeller of 3550 rpm. As shown in Figure 6, the bubble concentration is higher when the pump pressure is at 5 bar, than the other two pressures in Figures 5 and 7. A higher bubble concentration is distributed between the 20 and 30 micrometer bubble diameters at a pump pressure of 5 bar, as shown in Figure 6. When considering the microbubble generation by mixing the water and suction air at the blade passage with the rotation of the impeller, it is understood that optimal pump pressure is important to generate a higher concentration of microbubbles. On the other hand, the amount of suction air is also an important factor for bubble generation.

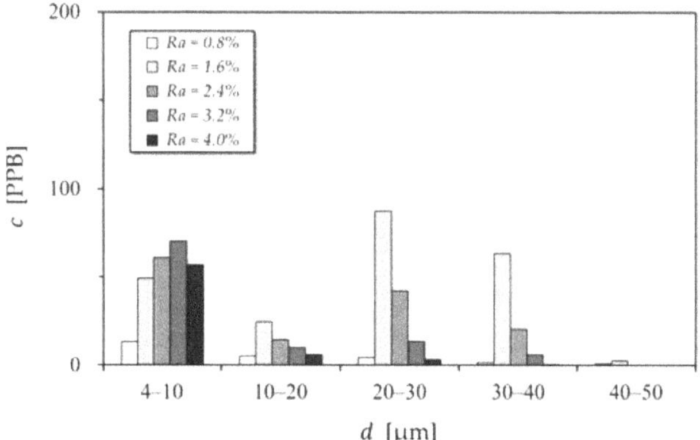

Figure 5. Bubble concentrations at the pump pressure of 4 bar.

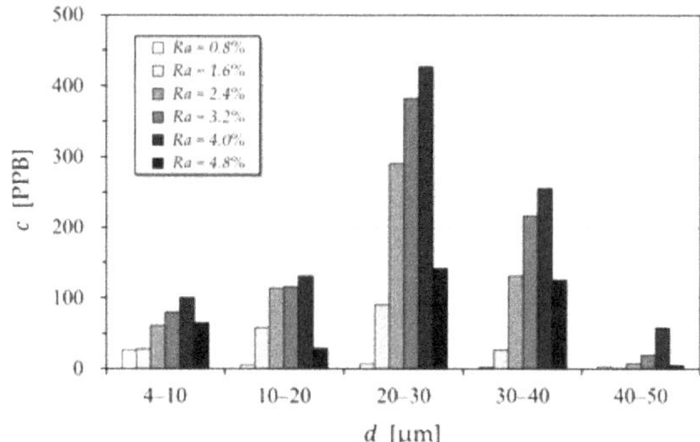

Figure 6. Bubble concentrations at the pump pressure of 5 bar.

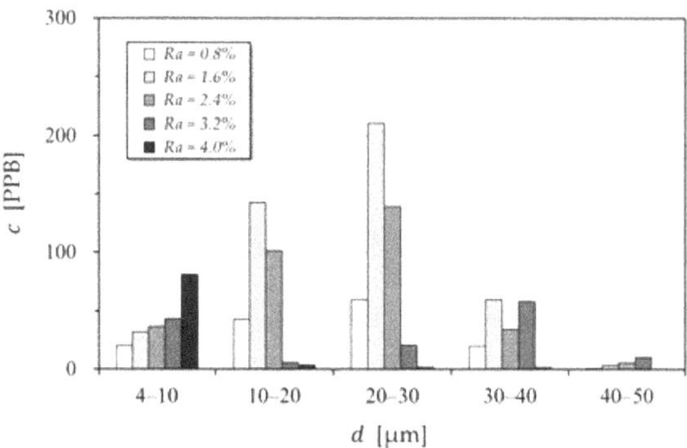

Figure 7. Bubble concentrations at the pump pressure of 6 bar.

Figure 8 shows the bubble concentration with respect to the air flowrate ratios at the pump pressure of 5 bar, which is the optimal operation condition for microbubble generation. As shown in the figure, the microbubble concentration at the range of 20–30 micrometers is the most prevalent when compared to other ranges. It is noted that the most efficient operational condition for microbubble generation is located near the 4.0 percent air flowrate ratio. The bubble concentration increases with the increase of the intake air flowrate ratios from 0.8 to 4.0, and then rapidly decreases at 4.8 percent. Thus, the optimal intake air amount for the maximum bubble concentration can be determined.

Figure 9 shows microbubbles from the maximum generation condition with the pump pressure of 5 bar and the air flowrate ratio of 4.0 percent, and the minimum condition with the pump pressure of 4 bar and the air flowrate ratio of 3.2 percent. Unlike to the picture of Figure 9a, the microbubbles have a gray milky color under the best operational condition in Figure 9b.

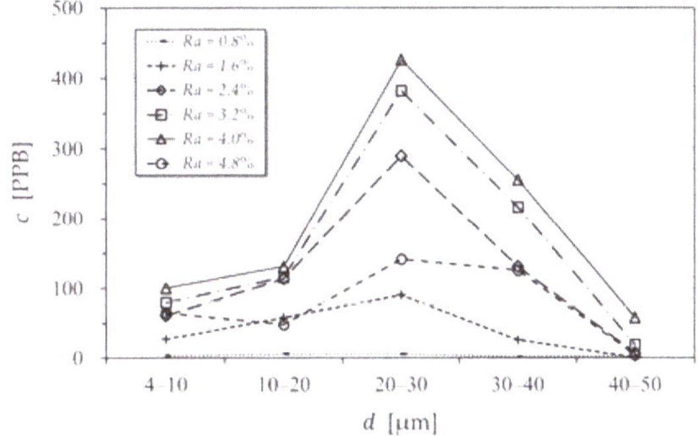

Figure 8. Bubble concentration with respect to the airflow ratios at the pump pressure of 5 bar.

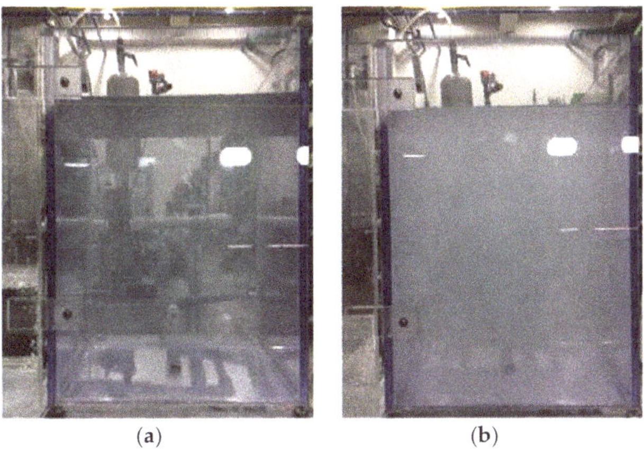

Figure 9. Picture of microbubbles: (**a**) pump pressure of 4 bar and air flowrate ratio of 3.2 percent; (**b**) pump pressure of 5 bar and air flowrate ratio of 4.0 percent.

3.3. Effect of Intake Air Flowrate on Pump Performance

In order to understand the effect of the suction air flowrate on the pump performance, the pump performance is compared at the operational conditions of the pump, with the pressure of 5 bar and the air flowrate ratio of 4.0 percent, as shown in Figure 10. This operation condition corresponds to the maximum generation of microbubbles. As shown in the figure, pressure and efficiency have a similar tendency for both air suction conditions, while the maximum difference of the pump efficiency is less than 1.2 percent. Therefore, the performance difference of the microbubble pump can be neglected depending on the air flowrate ratio.

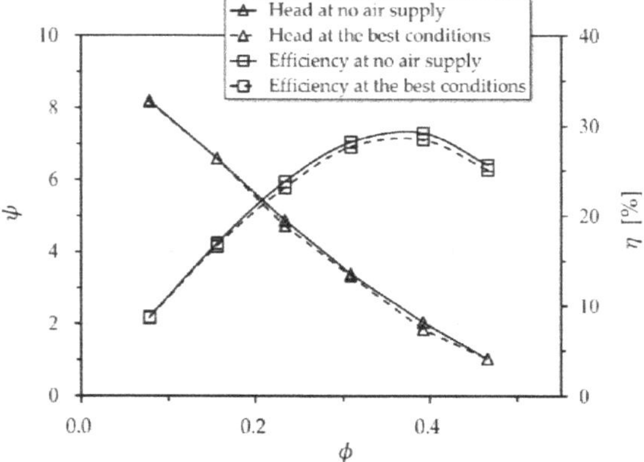

Figure 10. Effect of the suction air flowrate on the pump performance.

3.4. Relationship between Microbubble Generation and Dissolved Oxygen

For measuring the dissolved oxygen concentration during microbubble generation, a dissolved oxygen meter with a range between 0 and 90 mg/L has been used (Orion Star A223, Thermo Scientific, Waltham, MA, USA). Its resolution is 0.01 or 0.1 mg/L and has a relative accuracy is ±0.2 mg/L.

Figure 11 shows the dissolved oxygen and microbubble concentration with respect to time at the pump pressure of 5 bar and the air flowrate of 4.0 percent, which is the maximum microbubble generation condition. The dissolved oxygen concentration is measured during microbubble generation. As shown in the Figure, dissolved oxygen rapidly increases up to 180 s after pump operation, then increases more gradually, subsequently decreasing rapidly after stopping pump operations at 420 s. It can be seen that the tendency of the oxygen dissolved amount changing in resects to time is the same as the generated amount of microbubbles. From these results, it can be said that the time variation of the oxygen concentration and the concentration of microbubbles measured in the present study have a close relationship from generation to extinction.

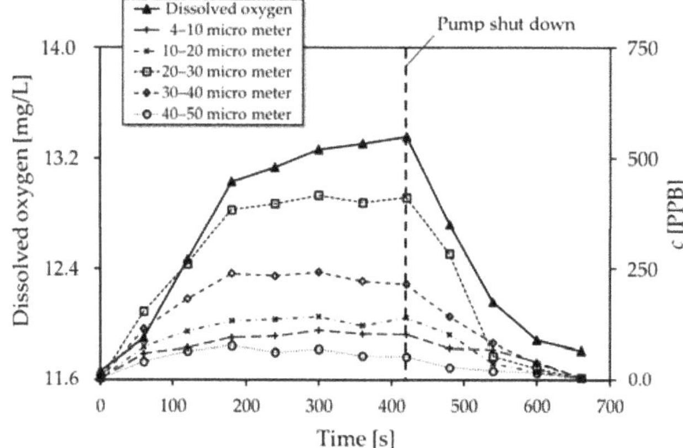

Figure 11. Dissolved oxygen and microbubble concentration in respect to time at the pump pressure of 5 bar and the air flowrate ratio of 4.0 percent.

3.5. Correlation between Microbubble Diameter and Terminal Rise Velocity

The terminal rise velocity estimated at the lower reservoir where the microbubbles generated from the pump are stored as shown in Figure 3 is compared to theoretical Equation for spherical fluid body derived by Hadamard and Rybczynski as follows [23,24].

$$U_{t(H-R)} = \frac{2\Delta\rho g r^2}{9\mu} \frac{\mu + \mu'}{2\mu + 3\mu'} \tag{4}$$

where $U_{t(H-R)}$, $\Delta\rho$, g, r, μ and μ' are the terminal rise velocity, the difference in density between air and water, gravitational acceleration, the radius of the microbubble, viscosity of water and internal viscosity of air, respectively. In case of negligible internal viscosity like air, Equation (4) and diameter of microbubble become following Equations:

$$U_{t(H-R)} = \frac{\Delta\rho g r^2}{3\mu} \tag{5}$$

$$d = \left[\frac{12\mu U_{t(H-R)}}{\Delta\rho g}\right]^{1/2} \tag{6}$$

where d is the diameter of microbubble.

As shown in Equation (6), the bubble diameter can be predicted from the terminal velocity. In the methodology to determine it, Parkinson et al. have introduced visualization and image analysis by attaching a high-speed camera [25]. In this study, it was estimated from the time required for the concentration change of microbubbles at the conditions under which the microbubble was generated at the maximum for pump pressure of 4–6 bar.

Figure 12 shows the terminal velocity with respect to the microbubble diameter. In the figure, Hadamard and Rybczynski indicate the theoretical relation between the microbubble diameter and the terminal velocity. The microbubble diameters for three different pump pressures shows a similar tendency to the theoretical formula of Hadamard and Rybczynski. In particular, the microbubble diameter of about 20 μm determined at the pump pressure of 5 bar shows a relatively good agreement to the value acquired from the theoretical formula. It is noted that the highest concentration of microbubbles measured by the particle counter is at the pump pressure of 5 bar. Throughout the theoretical comparison of the microbubble size based on the terminal velocity, it can be said that the experimental study on the characteristics of microbubbles generated from the pump with respect to operating conditions has been carried out properly.

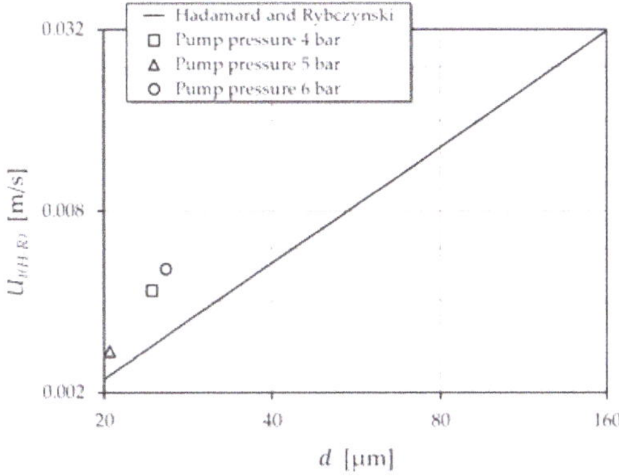

Figure 12. Terminal rise velocity with respect to microbubble diameter.

4. Conclusions

Bubble size and the concentration of a microbubble pump have been investigated by controlling intake air flowrates and pump pressure of 4, 5 and 6 bar, while the rotational frequency of the pump impeller was kept constant. The results are summarized as follows:

(1) Microbubble pumps with a regenerative impeller induce higher pressure differences from the inlet to the outlet and local recirculation flow at each blade passage. Considering the microbubble generation by mixing the water and suction air at the blade passage with the rotation of the impeller, optimal pump pressure is important to generate a higher concentration of microbubbles. It is noted that the amount of air suction is also an important factor for optimal bubble generation.

(2) Among the three different pump pressures, the bubble concentration of the test microbubble pump is highest when the pump pressure is 5 bar. The highest bubble concentration is distributed between the 20 and 30 micrometer bubble diameters, while the air flowrate ratio (*Ra*) is 4.0 percent.

(3) In the condition of higher bubble concentrations, it can be seen that pressure and efficiency of the microbubble pump have a similar tendency, regardless of the air supply. In the present pump, the maximum difference of the pump efficiency is less than 1.2 percent. So, the performance differences of the microbubble pump can be neglected, depending on the air flowrate ratio.

(4) Throughout the measurement of the dissolved oxygen concentration during microbubble generation, the tendency of the amount of oxygen dissolved to change with respect to time is the same as the generated amount of microbubbles. It is noted that the time variation of the oxygen concentration and the concentration of microbubbles have a close relationship from generation to extinction.

(5) Using a theoretical comparison of the terminal rise velocity with respect to the bubble size derived by Hadamard and Rybczynski, the microbubble diameters obtained from present measurements show a similar tendency to the theoretical formula.

Author Contributions: C.-M.J. designed the experiments; J.-Y.Y. advised throughout the research; S.-Y.J. performed the experiments, analyzed the data, and wrote the paper.

Funding: This work is supported by a grant (13IFIP-B06700801) from Industrial Facilities & Infrastructure Research Program funded by the Ministry of Land, Infrastructure and Transport of Korea government.

Conflicts of Interest: The authors declare no conflicts of interest.

Nomenclature

c	concentration of microbubbles
d	diameter of microbubble
g	gravitational acceleration
r	radius of microbubble
A	cross-sectional area of side channel of microbubble pump
ΔP	pressure rise between inlet and outlet of microbubble pump
Q	volumetric flowrate of water
Ra	air flowrate ratio
T	torque acting on rotational axis of microbubble pump impeller
U_t	rotational velocity of microbubble pump impeller
$U_{t(H-R)}$	terminal rise velocity

Greek Letters

η	machinical efficiency of microbubble pump
μ	viscosity of water
$\mu\prime$	internal viscosity of air
$\Delta\rho$	difference in density between air and water
ρ_w	density of water
ϕ	flow coefficient
ψ	pressure coefficient
ω	angular velocity of microbubble pump impeller

References

1. Khuntia, S.; Majumder, S.K.; Ghosh, P. Microbubble-aided water and wastewater purification: A review. *Rev. Chem. Eng.* **2012**, *28*, 191–221. [CrossRef]
2. Takahashi, M. Effect of shrinking microbubble on gas hydrate formation. *J. Phys. Chem. B* **2003**, *107*, 2171–2173. [CrossRef]
3. Takahashi, M. Potential of Microbubbles in Aqueous Solutions: Electrical Properties of the Gas-Water Interface. *J. Phys. Chem. B* **2005**, *109*, 21858–21864. [CrossRef] [PubMed]
4. Agarwal, A.; Ng, W.J.; Liu, Y. Principle and application of microbubble and nanobubble technology for water treatment. *Chemosphere* **2011**, *84*, 1175–1180. [CrossRef] [PubMed]
5. Kim, H.S.; Lim, J.Y.; Park, S.Y.; Kim, H.J. Effects on swirling chamber and breaker d3isk in pressurized-dissolution type micro-bubble generator. *KSCE J. Civ. Eng.* **2017**, *21*, 1102–1106. [CrossRef]
6. Takahashi, M. Base and technological application of micro-bubble and nanobubble. *Mater. Integration* **2009**, *22*, 2–19.
7. Ikeura, H.; Kobayashi, F.; Tamaki, M. Removal of residual pesticide, fenitrothion, in vegetables by using ozone microbubbles generated by different methods. *J. Food Eng.* **2011**, *103*, 345–349. [CrossRef]
8. Yoshida, A.; Takahashi, O.; Ishii, Y.; Sekimoto, Y.; Kurata, Y. Water purification using the adsorption characteristics of microbubbles. *Jpn. J. Appl. Phys.* **2008**, *47*, 6574–6577. [CrossRef]
9. Yin, J.; Li, J.; Li, H.; Liu, W.; Wang, D. Experimental study on the bubble generation characteristics for an venturi type bubble generator. *Int. J. Heat Mass Transf.* **2015**, *91*, 218–224. [CrossRef]
10. Uesawa, S.; Kaneko, A.; Nomura, Y.; Abe, Y. Study on bubble Beakup behavior in a venturi tube. *Multiphane Sci. Technol.* **2012**, *24*, 257–277. [CrossRef]
11. Gordiychuk, A.; Svanera, M.; Benini, S.; Poesio, P. Size distribution and Sauter mean diameter of micro bubbles for a Venturi type bubble generator. *Exp. Therm. Fluid Sci.* **2016**, *70*, 51–60. [CrossRef]
12. Zhao, L.; Mo, Z.; Sun, L.; Xie, G.; Liu, H.; Du, M.; Tang, J. A visualized study of the motion of individual bubbles in a venture-type bubble generator. *Prog. Nuclear Energy* **2017**, *97*, 74–89. [CrossRef]
13. Rodríguez-Rodríguez, J.; Sevilla, A.; Martínez-Bazán, C.; Gordillo, J.M. Generation of Microbubbles with Applications to Industry and Medicine. *Ann. Rev. Fluid Mech.* **2015**, *47*, 405–429. [CrossRef]
14. Nesbitt, B. *Handbook of Pumps and Pumping*; Elsevier: London, UK, 2005.
15. Karlsen-Davies, N.D.; Aggidis, G.A. Regenerative liquid ring pumps review and advances on design and performance. *Appl. Energy* **2016**, *164*, 815–825. [CrossRef]
16. Quail, F.J.; Scanlon, T.; Baumgartner, A. Design study of a regenerative pump using one-dimensional three-dimensional numerical techniques. *Eur. J. Mech. B/Fluids* **2012**, *31*, 181–187. [CrossRef]
17. Jeon, S.Y.; Kim, C.K.; Lee, S.M.; Yoon, J.Y.; Jang, C.M. Performance enhancement of a pump impeller using optimal design method. *J. Them. Sci.* **2017**, *26*, 119–124. [CrossRef]
18. Shimizu, T.; Yamamoto, Y.; Tenma, N. Experimental analysis of two-phase flows and turbine pump performance. *Int. J. Offshore Polar Eng.* **2016**, *26*, 371–377. [CrossRef]
19. Ahmadun, F.-R.; Pendashedh, A.; Abdullah, L.C.; Biak, D.R.A.; Madaeni, S.S.; Abidin, Z.Z. Review of technologies for oil and gas produced water treatment. *J. Hazard. Mater.* **2009**, *170*, 530–551.
20. Lundh, M.; Jonsson, L.; Dahlquist, J. The influence of contact zone configuration on the flow structure in a dissolved air flotation pilot plant. *Water Res.* **2002**, *36*, 1585–1595. [CrossRef]
21. Jang, C.M.; Han, G.Y. Enhancement of Performance by Blade Optimization in Two-Stage Ring Blower. *J. Them. Sci.* **2010**, *19*, 383–389. [CrossRef]
22. Jang, C.M.; Lee, J.S. Shape Optimization of a Regenerative Blower Used for Building Fuel Cell System. *Open J. Fluid Dyn.* **2012**, *2*, 208–214. [CrossRef]
23. Hadamard, J.S. Hadamard, Mouvement permanent lent d'une sphère liquide et visqueuse dans un liquide visqueux. *Acad. C. R. Sci.* **1911**, *152*, 1735–1752.

24. Rybczynski, W. Über die fortschreitende Bewegung einer flüssigen Kugel in einem zähen medium. *Bull. Acad. Sci. Cracovie Ser. A* **1911**, *1911*, 40–46.
25. Parkinson, L.; Sedev, R.; Fornasiero, D.; Ralston, J. The terminal rise velocity of 10–100 μm diameter bubbles in water. *J. Colloid Interface Sci.* **2008**, *322*, 168–172. [CrossRef] [PubMed]

Article

Spherical Shaped ($Ag - Fe_3O_4/H_2O$) Hybrid Nanofluid Flow Squeezed between Two Riga Plates with Nonlinear Thermal Radiation and Chemical Reaction Effects

Naveed Ahmed [1] , Fitnat Saba [1] , Umar Khan [2] , Ilyas Khan [3,*] , Tawfeeq Abdullah Alkanhal [4] , Imran Faisal [5] and Syed Tauseef Mohyud-Din [1]

[1] Department of Mathematics, Faculty of Sciences, HITEC University, Taxila Cantt 47080, Pakistan; naveed.ahmed@hitecuni.edu.pk (N.A.); fitnat_saba89@gmail.com (F.S.); syedtauseefs@hitecuni.edu.pk (S.T.M.-D.)

[2] Department of Mathematics and Statistics, Hazara University, Mansehra 21300, Pakistan; umar_jadoon@hu.edu.pk

[3] Faculty of Mathematics and Statistics, Ton Duc Thang University, Ho Chi Minh City 736464, Vietnam

[4] Department of Mechatronics and System Engineering, College of Engineering, Majmaah University, Majmaah 11952, Kingdom of Saudi Arabia; t.alkanhal@mu.edu.sa

[5] Department of Mathematics, Taibah University, Universities Road, P.O. Box 344 Medina, Kingdom of Saudi Arabia; mfaisal@taibahu.edu.sa

* Correspondence: ilyaskhan@tdt.edu.vn; Tel.: +92-332-890-2728

Received: 17 November 2018; Accepted: 20 December 2018; Published: 27 December 2018

Abstract: The main concern is to explore an electro-magneto hydrodynamic (EMHD) squeezing flow of ($Ag - Fe_3O_4/H_2O$) hybrid nanofluid between stretchable parallel Riga plates. The benefits of the use of hybrid nanofluids, and the parameters associated to it, have been analyzed mathematically. This particular problem has a lot of importance in several branches of engineering and industry. Heat and mass transfer along with nonlinear thermal radiation and chemical reaction effects have also been incorporated while carrying out the study. An appropriate selection of dimensionless variables have enabled us to develop a mathematical model for the present flow situation. The resulting mathematical method have been solved by a numerical scheme named as the method of moment. The accuracy of the scheme has been ensured by comparing the present result to some already existing results of the same problem, but for a limited case. To back our results further we have also obtained the solution by anther recipe known as the Runge-Kutta-Fehlberg method combined with the shooting technique. The error analysis in a tabulated form have also been presented to validate the acquired results. Furthermore, with the graphical assistance, the variation in the behavior of the velocity, temperature and concentration profile have been inspected under the action of various ingrained parameters. The expressions for skin friction coefficient, local Nusselt number and local Sherwood number, in case of ($Ag - Fe_3O_4/H_2O$) hybrid nanofluid, have been derived and the influence of various parameters have also been discussed.

Keywords: ($Ag - Fe_3O_4/H_2O$) hybrid nanofluid; nonlinear thermal radiation; heat transfer; chemical reaction; mass transfer; method of moment; numerical results

---

1. Introduction

An unprecedented and staggering development in the field of microfluidics, microelectronics, optical devices, chemical synthesis, transportation, high power engines and microsystems, including mechanical and electrical components, transforms the underpinnings of human life. These expansions further demand efficient cooling techniques, in order to manage the thermal performance, reliability and long-term operational devices. The primitive thermal management techniques (like cooling through liquids) seem to be deficient, in order to meet the challenges of thermal efficiency. Later on, this issue has been resolved by dispersing nano-meter sized structures, within the host fluid, which certainly influences its thermo-mechanical properties. In this regard, Choi [1,2] was considered as the pioneer, who gave this concept and calls it 'Nanofluid'. Many researchers have proposed various theoretical models for thermal conductivity, by following his footsteps. Maxwell [3] worked on a model for the thermal conductivity which is suitable only for the spherical shaped nanoparticles. Further studies in this area lead us to a variety of models, containing the impact of, particle–particle interactions (i.e., Bruggeman model, 1935) [4], particles shapes (i.e., Hamilton and crosser model 1962) [5] and particles distribution (i.e., Suzuki et al. 1969) [6]. Furthermore, researchers have found a number of articles in the literature that covers the different aspects of the nanofluid. Some of them can be found in the references [7–12].

In recent past years, a new class of nanofluids, entitled "Hybrid nanofluid", have come into existence that bears high thermal conductivity as compared to that of mono nanofluid. They have brought a revolution in various heat transfer applications like nuclear system cooling, generator cooling, electronic cooling, automobile radiators, coolant in machining, lubrication, welding, solar heating, thermal storage, heating and cooling in buildings, biomedical, drug reduction, refrigeration, and defense etc. In the case of a regular nanofluid, the critical issue is either they possess a good thermal conductive network or display a better rheological properties. The nanocomposites (single handedly) do not possess all the possible features which are required for a certain application. Therefore, by an appropriate selection of two or more nanoparticles, hybrid nanofluid can lead us to a homogeneous mixture, which possesses all physicochemical properties of various substances that can hardly be found in an individual substance [13,14].

The distinctive features of hybrid nanofluid have gained the attention of worldwide researchers and therefore a number of research articles have been published over the past few years. By employing a new material design concept, Niihara [15] discussed that the mechanical and thermal properties of the host fluid can be greatly enhanced, by the inclusion of nanocomposites.. Jana et al. [16] examined the thermal efficiency of the host fluid, by incorporating single and hybrid nanoparticles. Suresh et al. [17] takes into account a two-step method in order to synthesize water-based $(Al_2O_3 - Cu)$ hybrid nanofluid. Their experimental results reveal an improvement in the viscosity and thermal properties of the prepared hybrid nanofluid. In their next study [18], the effects of $(Al_2O_3 - Cu)$ hybrid nanofluid on the rate of heat transfer have been investigated. Momin [19], in 2013, conducted an experiment to study the impact of mixed convection on the laminar flow of hybrid nanofluid inside an inclined tube. By employing a numerical scheme, Devi and Devi [20] investigated the influence of magneto hydrodynamic flow of H_2O based $(Cu - Al_2O_3)$ hybrid nanofluid, over a porous dilating surface. With the aid of entropy generation, the magneto hydrodynamic flow of water based $(Cu - Al_2O_3)$ hybrid nanofluid, inside a permeable channel, has been discussed by Das et al. [21]. Chamkha et al. [22], numerically analyzed, the time dependent conjugate natural convection of water based hybrid nanofluid, within a semicircular cavity The Blasius flow of hybrid nanofluid with water, taken as a base fluid over a convectively heated surface, has been examined by Olatundun and Makinde [23]. Besides, in [24], Hayat and Nadeem incorporated the silver (Ag) and copper oxide (CuO) as nanoparticles within the water, to enhance the rate of heat transfer, over the linearly stretching surface.

These days, researchers have been attracted, to analyze the squeezing flows in various geometries. Due to their significance, they have been involved in many practical and industrial situations, like biomechanics, food processing, and chemical and mechanical engineering. They have also been utilized, in order to examine the formation of lubrication, polymer processing, automotive engines, bearings, injection, gear, appliances etc. These flow phenomena have been observed in different hydro dynamical machines and devices, where the normal velocities are enforced by the moving walls of the channel. Stefan [25] was the pioneer behind this concept. Later on, Shahmohamadi et al. [26]

employed an analytical technique, to examine the time-dependent axisymmetric flow of a squeezed nature. Recently, the effects of squeezing flow on nanofluid, confined between parallel plates, have been investigated by M. Sheikholeslami et al. [27]. They also utilized the Adomian's decomposition method to find the solution of the respective flow model. Khan et al. [28] have taken into account, the viscous dissipation effects along with slip condition, to analyze the two-dimensional squeezing flow of copper-water based nanofluid. For solution methodology, they have employed a variation of the parameters method. In 2017, the squeezing effects on the magneto hydrodynamic flow of Casson fluid (inside a channel) have been thoroughly inspected by Ahmed et al. [29]. They have modelled the respective flow problem and then solved it both numerically (Runge-Kutta scheme of fourth order) and analytically (Variation of parameters method).

Gallites and Lilausis [30] came up with the idea of an electromagnetic actuator device, in order to set up the crossed magnetic and electric fields, that appropriately provoked the wall's parallel Lorentz forces. The purpose of that device was to control the flow characteristics, which usually have a span wise arrangement of alternating and invariable magnets that specifically mounted a plane surface. The device, sometimes indicated as Riga plate [31], provided an aid to reduce the pressure drag, as well as the friction of submarines, that can be achieved by reducing the turbulence production and a boundary layer separation. A number of research articles have been published, in order to explore the distinctive features of the laminar flow of a fluid due to Riga plate. By assuming the least electrical conductivity effects, Pantokratoras and Magyari [32] investigated the flow behavior along with free convection. In 2011, Pantokratoras [33] reported the performance of Blasius flow, enforced by the Riga plate. He also encounterd the Sakiadis flow in his study. Later on, Magyari and Pantokratoras [34] took into account the Blasius flow of the liquid, which at the same time is electrically conducting, induced by Riga surface. The electro magneto hydrodynamic flow of nanofluid, induced by Riga plate along with the slip consequences, have been examined by Ayub et al. [35]. In 2017, Hayat et al. [36], discussed the squeezing flow of a fluid between two parallel Riga plates, together with convective heat transfer. The thermal radiative effects accompanied by chemical reaction, were also a part of their study. Moreover, Hayat et al. [37] investigated the electro magneto squeezing flow of carbon nanotube's suspended nanofluid between two parallel rotatory Riga plates along with viscous dissipation effects. They have considered the melting heat transfer condition, which basically revealed that the heat conducting process to the solid surface, involved the combine effects of both sensible and melting heat, which significantly enhances the temperature of the solid surface to its melting temperature.

The thermal radiation is a significant mode of heat transfer [38,39], which seems to be dominant, in order to transfer the net amount of heat, even in the existence of free or forced convection. The transfer of heat via radiation have been significantly found in many engineering and industrial applications, including airplanes, space vehicles, satellites, and atomic-force plant. In this context, many researchers have comprehensively discussed the radiative heat transfer phenomena. Some of the most relevant have been found in [40–44].

The literature survey revealed the fact that no single step has been taken in order to analyze the salient features of $(Ag - Fe_3O_4/H_2O)$ hybrid nanofluid, between two parallel Riga plates. This article encounters the influential behavior of the viscid flow of $(Ag - Fe_3O_4/H_2O)$ hybrid nanofluid between two parallel Riga plates, where the lower plate experiences a stretching velocity, while the upper plate enforces a squeezing flow. The transfer of heat and mass along with nonlinear thermal radiative and chemical reaction effects would also be a part of this study. By employing the suitable similarity transforms, a mathematical model for the present flow situation have been accomplished. Method of moment along with Runge-Kutta-Fehlberg method have been considered to find the solution of the model. Tables have been provided which presents the validity of the acquired results. Furthermore, the graphical aid has been provided, to demonstrate the influence of various ingrained entities, on the velocity and temperature along with concentration profiles. The expressions related to the coefficient of skin friction, local Nusselt number and local Sherwood number have also been developed and discussed with the help of graphs.

2. Formulation of the Governing Equations

Two parallel Riga plates have been under consideration, among which an electro-magneto hydrodynamic (EMHD) flow of $(Ag - Fe_3O_4/H_2O)$ hybrid nanofluid has been flowing. The flow is also time dependent and incompressible. Cartesian coordinates have been chosen in such away, that the $\check{x}-$axis coincides with the horizontal direction, whereas the $\check{\mathscr{Y}}-$axis is placed normal to it. The lower plate positioned at $\check{\mathscr{Y}} = 0$, experiences a stretching velocity $\mathscr{U}_w(\check{x}) = a\check{x}/(1-\lambda\check{t})$. Besides, the upper Riga plate, owing the place at $\check{\mathscr{Y}} = \ell(\check{t}) = \frac{-\lambda}{2}\left(a/v_f(1-\lambda\check{t})\right)^{-0.5}$. It is further assumed that the flow of $(Ag - Fe_3O_4/H_2O)$ hybrid nanofluid is a squeezing flow, having the velocity $\check{v}_\ell = d\ell/d\check{t}$. Moreover, the nonlinear thermal radiation and chemical reaction effects are also considered. Figure 1 displays the configuration of the flow model.

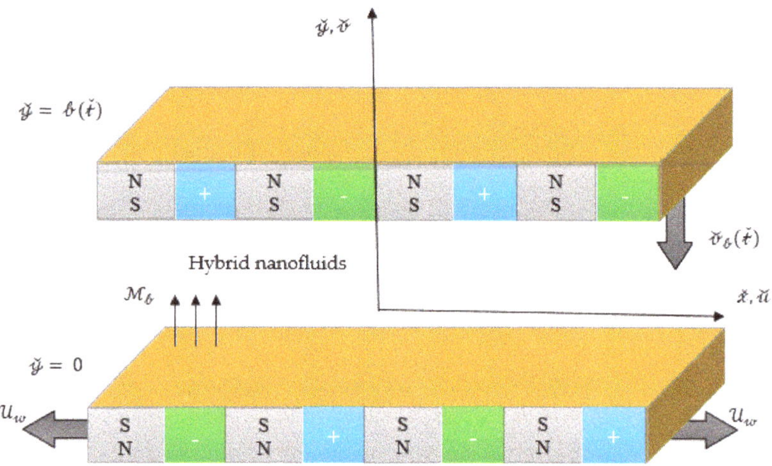

Figure 1. Physical model of the present flow situation.

The Navier-Stokes equations, suitable for the present flow situation, are given as [36]:

$$\frac{\partial \check{v}}{\partial \check{\mathscr{y}}} + \frac{\partial \check{u}}{\partial \check{x}} = 0, \tag{1}$$

$$\frac{\partial \check{p}}{\partial \check{x}} + \check{\rho}_{hnf}\left(\frac{\partial \check{u}}{\partial \check{t}} + \frac{\partial \check{u}}{\partial \check{x}}\check{u} + \frac{\partial \check{u}}{\partial \check{\mathscr{y}}}\check{v}\right) = \mu_{hnf}\left(\frac{\partial^2 \check{u}}{\partial \check{\mathscr{y}}^2} + \frac{\partial^2 \check{u}}{\partial \check{x}^2}\right) + \frac{Exp\left(-\pi\check{\mathscr{y}}/\ell\right)}{8(\pi \mathcal{I}_0 \mathcal{M}_0)^{-1}}, \tag{2}$$

$$\frac{\partial \check{p}}{\partial \check{\mathscr{y}}} + \check{\rho}_{hnf}\left(\frac{\partial \check{v}}{\partial \check{t}} + \frac{\partial \check{v}}{\partial \check{x}}\check{u} + \frac{\partial \check{v}}{\partial \check{\mathscr{y}}}\check{v}\right) = \mu_{hnf}\left(\frac{\partial^2 \check{v}}{\partial \check{\mathscr{y}}^2} + \frac{\partial^2 \check{v}}{\partial \check{x}^2}\right), \tag{3}$$

$$\frac{\partial \check{T}}{\partial \check{t}} + \frac{\partial \check{T}}{\partial \check{x}}\check{u} + \frac{\partial \check{T}}{\partial \check{\mathscr{y}}}\check{v} = \frac{k_{hnf}}{(\rho C_p)_{hnf}}\left(\frac{\partial^2 \check{T}}{\partial \check{\mathscr{y}}^2} + \frac{\partial^2 \check{T}}{\partial \check{x}^2}\right) - \frac{1}{(\rho C_p)_{hnf}}\left(\frac{\partial \check{q}_r}{\partial \check{\mathscr{y}}} + \frac{\partial \check{q}_r}{\partial \check{x}}\right), \tag{4}$$

$$\frac{\partial \check{\mathfrak{C}}}{\partial \check{t}} + \frac{\partial \check{\mathfrak{C}}}{\partial \check{x}}\check{u} + \frac{\partial \check{\mathfrak{C}}}{\partial \check{\mathscr{y}}}\check{v} = \mathfrak{D}_{hnf}\left(\frac{\partial^2 \check{\mathfrak{C}}}{\partial \check{\mathscr{y}}^2} + \frac{\partial^2 \check{\mathfrak{C}}}{\partial \check{x}^2}\right) - c_1\left(\check{\mathfrak{C}} - \check{\mathfrak{C}}_\ell\right), \tag{5}$$

where, $\check{u}$, signifies the horizontal component of velocity, while the vertical one is symbolized by $\check{v}$. The dimensional pressure, temperature and concentration, are respectively shown by $\check{p}$, $\check{T}$ and $\check{\mathfrak{C}}$. Furthermore, ℓ denotes the width between magnets and electrodes. $\mathcal{M}_0(Tesla)$ represents the magnetization of the permanent magnets, while, $\jmath_0(m^{-2}A)$ is the applied current density in the electrodes. The first order coefficient for a chemical reaction, is presented by c_1. In addition, $\check{q}_r$ symbolizes the rate of heat flux. The expression for the thermal radiative term has been successively proposed by Rosseland [38], which is given as:

$$\check{q}_r = -\frac{16\check{\sigma}\breve{T}^3}{3\Bbbk}\frac{\partial\breve{T}}{\partial\breve{y}},\tag{6}$$

where, the coefficient for mean absorption is given by $\Bbbk$, while $\check{\sigma}$ stands for Stefan-Boltzmann constant. Therefore, after incorporating the Equation (6) in Equation (5), the energy equation can be generalized as follows:

$$\frac{\partial\breve{T}}{\partial\check{t}} + \frac{\partial\breve{T}}{\partial\check{x}}\check{u} + \frac{\partial\breve{T}}{\partial\breve{y}}\check{v} = \frac{k_{hnf}}{(\check{\rho}C_p)_{hnf}}\left(\frac{\partial^2\breve{T}}{\partial\breve{y}^2} + \frac{\partial^2\breve{T}}{\partial\check{x}^2}\right) + \frac{1}{(\check{\rho}C_p)_{hnf}}\frac{16\check{\sigma}}{3\Bbbk}\frac{\partial}{\partial\breve{y}}\left(\breve{T}^3\frac{\partial\breve{T}}{\partial\breve{y}}\right).\tag{7}$$

The auxiliary conditions, specifying the current flow situation, are given as:

$$\check{u} = \mathcal{U}_w(\check{x}) = a\check{x}/(1-\lambda\check{t}), \check{v} = 0, \left(\breve{T}-\breve{T}_0\right) = 0, \left(\breve{\mathbb{C}}-\breve{\mathbb{C}}_0\right) = 0 \quad \text{at} \quad \breve{y} = 0,\tag{8}$$

$$\check{v} = \frac{d\theta}{d\check{t}} = \frac{\lambda}{2}\left(\frac{a(1-\lambda\check{t})}{v_f}\right)^{-0.5}, \check{u} = 0, \left(\breve{T}-\breve{T}_\theta\right) = 0, \left(\breve{\mathbb{C}}-\breve{\mathbb{C}}_\theta\right) = 0 \quad \text{at} \quad \breve{y} = \theta(\check{t}),\tag{9}$$

where, $\breve{T}_0$ and $\breve{T}_\theta$ simultaneously, indicates the temperatures of the plates situated at $\breve{y} = 0$ and $\breve{y} = \theta(\check{t})$. The concentration of nanoparticles at the bottom plate is denoted by $\breve{\mathbb{C}}_0$, while, $\breve{\mathbb{C}}_\theta$ is the nanoparticles concentration at the top wall. Moreover, the rate, with which the lower surface is being stretched is a, while, λ represents the constant characteristics parameter.

In the aforementioned equations, $v_f - \mu_f/\check{\rho}_f$ denotes the effective kinematic viscosity. Furthermore, μ_{hnf} and μ_f simultaneously represents the effective dynamic viscosities of the hybrid nanofluid and mono nanofluid that significantly influence the flow behavior of the host fluid. Brinkman [45], in 1952, proposed a model for the effective dynamic viscosity $\left(\mu_{nf}\right)$ of a mono nanofluid which is given below:

$$\mu_{nf} = \frac{\mu_f}{(1-\varphi)^{5/2}},\tag{10}$$

where, φ denotes the nanoparticle volume fraction. Thus, in the case of hybrid nanofluid, the effective dynamic viscosity $\left(\mu_{hnf}\right)$ is defined as [22]:

$$\mu_{hnf} = \frac{\mu_f}{(1-\varphi_h)^{5/2}},\tag{11}$$

where, $\varphi_h = \varphi_1 + \varphi_2$ (in case of hybrid nanofluid) is a net volume fraction of distinct nanoparticles.

The effective density $\left(\check{\rho}_{nf}\right)$ presented by Pak and Cho [46] and the heat capacity $(\check{\rho}C_p)_{nf}$ [47] of mono nanofluid, can be respectively given by:

$$\check{\rho}_{nf} = \check{\rho}_f + \varphi\left(\check{\rho}_p - \check{\rho}_f\right),\tag{12}$$

$$(\check{\rho}C_p)_{nf} = (\check{\rho}C_p)_f + \varphi\left((\check{\rho}C_p)_p - (\check{\rho}C_p)_f\right).\tag{13}$$

By following the rules of mixture principle, the effective density $\left(\check{\rho}_{hnf}\right)$ [22,48] and heat capacity $\left(\check{\rho}C_p\right)_{hnf}$ [22,48] of hybrid nanofluid, can be estimated via Equations (14) and (15).

$$\check{\rho}_{hnf} = \varphi_1\check{\rho}_{p1} + \varphi_2\check{\rho}_{p2} + (1 - \varphi_h)\check{\rho}_f, \tag{14}$$

$$\left(\check{\rho}C_p\right)_{hnf} = \varphi_1\left(\check{\rho}C_p\right)_{p1} + \varphi_2\left(\check{\rho}C_p\right)_{p2} + (1 - \varphi_h)\left(\check{\rho}C_p\right)_f. \tag{15}$$

The thermal conductivity $\left(\hbar_{nf}\right)$ is the fundamental property, defining the heat transfer characteristics of the mono nanofluid. Maxwell suggested a correlation [3], for the mono nanofluid, by considering the spherical shaped nanoparticles, whose mathematical expression is given by:

$$\hbar_{nf} = \hbar_f \frac{\hbar_p(1 + 2\varphi) + 2\hbar_f(1 - \varphi)}{\hbar_p(1 - \varphi) + \hbar_f(2 + \varphi)}. \tag{16}$$

In the case of hybrid nanofluid, the thermal conductivity ratio can be accomplished by modifying the Maxwell correlation [22] as:

$$\frac{\hbar_{hnf}}{\hbar_f} = \frac{\frac{\varphi_1\hbar_{p1} + \varphi_2\hbar_{p2}}{\varphi_h} + 2\hbar_f + 2\left(\varphi_1\hbar_{p1} + \varphi_2\hbar_{p2}\right) - 2\varphi_h\hbar_f}{\frac{\varphi_1\hbar_{p1} + \varphi_2\hbar_{p2}}{\varphi_h} + 2\hbar_f - \left(\varphi_1\hbar_{p1} + \varphi_2\hbar_{p2}\right) + \varphi_h\hbar_f}. \tag{17}$$

In 1935, another correlation, for spherical nanoparticles, has been introduced by Bruggeman [4], which usually considers the impact of nano clusters on the thermal conductivity. By mixture principle, this model can be extended for the estimation of thermal conductivity ratio of the hybrid nanofluid and is given by:

$$\frac{\hbar_{hnf}}{\hbar_f} = \frac{1}{4}\left[(3\varphi_h - 1)\left(\frac{\frac{\varphi_1\hbar_{p1} + \varphi_2\hbar_{p2}}{\varphi_h}}{\hbar_f}\right) + (2 - 3\varphi_h) + (\Delta)^{1/2}\right], \tag{18}$$

where,

$$\Delta = \left[\begin{array}{l} (3\varphi_h - 1)^2\left(\frac{\frac{\varphi_1\hbar_{p1} + \varphi_2\hbar_{p2}}{\varphi_h}}{\hbar_f}\right)^2 + (2 - 3\varphi_h)^2 + \\[2ex] 2\left(2 + 9\varphi_h - 9\varphi_h^2\right)\left(\frac{\frac{\varphi_1\hbar_{p1} + \varphi_2\hbar_{p2}}{\varphi_h}}{\hbar_f}\right) \end{array}\right]. \tag{19}$$

The molecular diffusivity [22,49–51], of the species concentration, for mono nanofluid and hybrid nanofluid are simultaneously defined as:

$$\mathfrak{D}_{nf} = (1 - \varphi)\mathfrak{D}_f, \tag{20}$$

$$\mathfrak{D}_{hnf} = (1 - \varphi_h)\mathfrak{D}_f. \tag{21}$$

In all the above expressions, φ_1 and φ_2 simultaneously, represents the volume concentration of magnetite (Fe_3O_4) and silver (Ag) nanoparticles in hybrid nanofluids. The viscosity, density and specific heat of host fluid are respectively denoted by μ_f, $\check{\rho}_f$ and $\left(C_p\right)_f$. At constant pressure, $\left(C_p\right)_{p1}$ and $\left(C_p\right)_{p2}$ respectively, denotes the specific heat of magnetite and silver nanoparticles. The densities, of magnetite and silver nanoparticles, are specified by $\check{\rho}_{p1}$ and $\check{\rho}_{p2}$ respectively. $\hbar_f$ and $\mathfrak{D}_f$ represents the thermal conductivity and mass diffusivity of the water (H_2O). The thermal conductivities of magnetite and silver nanocomposites, are respectively symbolized by $\hbar_{p1}$ and $\hbar_{p2}$.

The prescribed form of similarity transforms, which deals with the process of conversion of Equations (1)–(3) and (7) into a nonlinear set of ordinary differential equations (ODE), are given as:

$$\Psi = \left(\frac{(1-\lambda \check{t})}{a v_f}\right)^{-0.5} \check{x}\check{F}(\chi), \quad \chi = \check{\mathcal{Y}}(\theta(\check{t}))^{-1}, \quad \check{u} = \frac{\partial}{\partial \check{\mathcal{Y}}}(\Psi) = \mathcal{U}_w \check{F}'(\chi),$$

$$\hat{v} = -\frac{\partial}{\partial \check{x}}(\Psi) = -\left(\frac{(1-\lambda \check{t})}{a v_f}\right)^{-0.5}\check{F}(\chi), \quad \check{\mathscr{T}}(\chi) = \frac{\check{T}-\check{T}_\theta}{\check{T}_0-\check{T}_\theta}, \quad \check{\mathscr{C}}(\chi) = \frac{\check{\mathfrak{C}}-\check{\mathfrak{C}}_\theta}{\check{\mathfrak{C}}_0-\check{\mathfrak{C}}_\theta}. \tag{22}$$

where, the superscript $\prime$ stands for $d/d\chi$. Thus, by opting Brinkman (11) and Bruggeman (18) models, the dimensionless mode of a system of nonlinear ordinary differential equations, for $(Ag - Fe_3O_4/H_2O)$ hybrid nanofluid, along with radiation and chemical reaction parameters has been accomplished that can be written as:

$$\check{F}^{iv} + \check{\Upsilon}_1\left[\check{F}\check{F}'' - \check{F}\check{F}''' - \frac{\gamma}{2}\left(3\check{F}'' + \chi\check{F}'''\right)\right] - (1-\varphi_h)^{5/2}\mathcal{M}_\theta \mathscr{P} e^{-\mathscr{P}\chi} = 0, \tag{23}$$

$$\left(\left(\frac{\mathcal{k}_{hnf}}{\mathcal{k}_f} + Rd\left((1-\check{\mathscr{T}}) + \check{\mathscr{T}}\theta_w\right)^3\right)\check{\mathscr{T}}\prime\right)' + Pr\check{\Upsilon}_2\left(\check{F} - \frac{\gamma}{2}\chi\right)\check{\mathscr{T}}' = 0. \tag{24}$$

$$\check{\Theta}'' + \frac{Sc}{(1-\varphi_h)}\left(\check{F} - \frac{\gamma}{2}\chi\right)\check{\Theta}' - \frac{Sc}{(1-\varphi_h)}c_{\mathscr{R}}\check{\Theta} = 0. \tag{25}$$

where, $\check{F}$, $\check{\mathscr{T}}$ and $\check{\Theta}$, all are the dependent functions of dimensionless variable χ. Furthermore, the dimensionless auxiliary conditions, supporting the present flow situation, are therefore suggested as:

$$\check{F}(0) = 0, \quad \check{F}(1) = \frac{\gamma}{2}, \quad \check{F}\prime(0) - 1 = 0, \quad \check{F}\prime(1) = 0, \tag{26}$$

$$\check{\mathscr{T}}(0) - 1 = 0, \quad \check{\mathscr{T}}(1) = 0, \tag{27}$$

$$\check{\Theta}(0) - 1 = 0, \quad \check{\Theta}(1) = 0. \tag{28}$$

In the above-mentioned system of Equations (23) and (25), $\gamma = \lambda/a$ represents a dimensionless squeeze number, while, $\mathcal{M}_\theta = \pi \check{j}_0 \mathcal{M}_0 \check{x}/8\check{\rho}_f \mathcal{U}_w^2$ is the modified Hartman number and $\mathscr{P} = \pi\theta(\check{t})/\ell$ is the dimensionless parameter. Moreover, the radiation parameter is denoted by $Rd = 16\check{\sigma}\check{T}_\theta^3/3\check{\mathcal{k}}\,\mathcal{k}_f$. Prandtl number is symbolized by $Pr = \left(\mathcal{k}_f/(\check{\rho}C_p)_f v_f\right)^{-1}$. Besides, $Sc = v_f/\mathscr{D}_f$ signifies, the Schmidt number. The chemical reaction is indicated by $c_{\mathscr{R}} = c_1(1 - \lambda\check{t})/a$.

Moreover, the constants $\check{\Upsilon}_1$ and $\check{\Upsilon}_2$, embroiled in the governing dimensionless model, can be mathematically stated as:

$$\left.\begin{aligned}
\check{\Upsilon}_1 &= \frac{v_f}{v_{hnf}} = \frac{\left(1-\varphi_h+\varphi_1\frac{\check{\rho}_{p1}}{\check{\rho}_f}+\varphi_2\frac{\check{\rho}_{p2}}{\check{\rho}_f}\right)}{(1-\varphi_h)^{-5/2}}, \\
\check{\Upsilon}_2 &= \frac{(\check{\rho}C_p)_{hnf}}{(\check{\rho}C_p)_f} = 1 - \varphi_h + \varphi_1\frac{(\check{\rho}C_p)_{p1}}{(\check{\rho}C_p)_f} + \varphi_2\frac{(\check{\rho}C_p)_{p2}}{(\check{\rho}C_p)_f}.
\end{aligned}\right\} \tag{29}$$

The coefficient of skin friction, local heat transferal rate (i.e., local Nusselt number) and local Sherwood number, for the present flow situation, opt the following dimensionless expressions:

$$\hat{\mathscr{C}}_{f\check{x}} = \frac{\tau_w}{\check{\rho}_{hnf}\mathcal{U}_w^2}, \quad Nu_{\check{x}} = \frac{\theta \mathcal{k}_f^{-1}}{\left(\check{T}_0 - \check{T}_\theta\right)}(\check{q}_w + \check{q}_r) \quad \text{and} \quad Sh_{\check{x}} = \frac{\theta \mathscr{D}_f^{-1}}{\left(\check{\mathfrak{C}}_0 - \check{\mathfrak{C}}_\theta\right)}\check{q}_m, \tag{30}$$

where, τ_w indicates the shear stress, while, the heat and mass fluxes, at both of the walls, are simultaneously signifies by $\check{q}_w$ and $\check{q}_m$. They are respectively defined as:

$$\tau_w = \mu_{hnf}\left(\frac{\partial \check{u}}{\partial \check{y}}\right)_{\check{y}\,=\,\{\begin{smallmatrix}0\\\theta(\check{t})\end{smallmatrix}} \quad , \quad \check{q}_w = -k_{hnf}\left(\frac{\partial \check{T}}{\partial \check{y}}\right)_{\check{y}\,=\,\{\begin{smallmatrix}0\\\theta(\check{t})\end{smallmatrix}} \quad , \quad \check{q}_m = -\mathfrak{D}_{hnf}\left(\frac{\partial \check{\mathfrak{C}}}{\partial \check{y}}\right)_{\check{y}\,=\,\{\begin{smallmatrix}0\\\theta(\check{t})\end{smallmatrix}} \tag{31}$$

Subsequently, by incorporating Equations (6) and (31) into Equation (30), we finally achieved the dimensionless forms of skin friction, the Nusselt number, and the Sherwood number, both at the top and bottom walls, which can be expressed as:

$$Re_{\check{x}}^{0.5}\hat{\mathscr{C}}_{lower} = \frac{1}{\check{\Upsilon}_1}\check{F}''(0), \qquad\qquad Re_{\check{x}}^{0.5}\hat{\mathscr{C}}_{upper} = \frac{1}{\check{\Upsilon}_1}\check{F}''(1), \tag{32}$$

$$\begin{aligned}(1 - \lambda\check{t})^{0.5}Re_{\check{x}}^{-0.5}Nu_{lower} &= -\left(\frac{k_{hnf}}{k_f} + Rd(\theta_w)^3\right)\check{\mathscr{T}}'(0),\\(1 - \lambda\check{t})^{0.5}Re_{\check{x}}^{-0.5}Nu_{upper} &= -\left(\frac{k_{hnf}}{k_f} + Rd\right)\check{\mathscr{T}}'(1),\end{aligned} \tag{33}$$

and
$$Re_{\check{x}}^{-0.5}Sh_{lower} = -(1 - \varphi_h)\check{\Theta}'(0), \qquad\qquad Re_{\check{x}}^{-0.5}Sh_{upper} = -(1 - \varphi_h)\check{\Theta}'(1), \tag{34}$$

where, $Re_{\check{x}} = \check{x}\mathscr{U}_w/v_f$ denotes the local Reynolds number.

3. Solution Procedure

Method of moments (MM), one of the sub-class of the method of weighted residual (MWR), has been considered, in order to tackle the system of differential equations coupled with boundary conditions. From an accuracy point of view, a comparison has also been made, between the results achieved by Method of moments (MM) and Runge-Kutta-Fehlberg method (RKF). For this purpose, a mathematical software Maple 16 has been used.

Method of Moments

Let $\mathscr{D}$, an arbitrary differential operator, acting upon $\check{F}(\chi)$ generate a function $g(\chi)$, which is given as:

$$\mathscr{D}\left(\check{F}(\chi)\right) = g(\chi). \tag{35}$$

In order to approximate the solution of the above-mentioned problem, a trial solution has been defined, which is in the form of a linear combination of base function. These basis functions, also hold the property of linearly independence. Mathematically, it can be expressed as:

$$\check{F}(\chi) \cong \tilde{F}(\chi) = \psi_0 + \sum_{i=1}^{n} c_i\psi_i, \tag{36}$$

where, the essential boundary conditions are usually incorporated in ψ_0. By substituting back Equation (36) in Equation (35), one can acquired an exact solution, in the form of the trial solution that satisfies the given problem (35), which is an extremely rare situation. More often, it does not satisfies the given problem and therefore, left an expression that represents the error or the residual as under:

$$\tilde{\mathscr{R}}(\chi) = \mathscr{D}\left(\check{F}(\chi)\right) - g(\chi) \neq 0. \tag{37}$$

The proper selection of weights enabled us to construct weighted residual error. The values of unknown constants c_i's have been accomplished after the minimization procedure, that is:

$$\int_{\chi} \tilde{\mathscr{R}}(\chi)\,\mathscr{W}_i(\chi)d\chi = 0, \qquad i = 1, 2, \ldots, n. \tag{38}$$

The above equation generate a system of algebraic equations, whose solution finally lead us to determine the unknown constants c_i's, and thus, a numerical solution has been obtained after plugging them back into the trial solution.

It is pertinent to mention that the weight functions involved in the method of moments (MM), are defined as:

$$\mathscr{W}_i(\chi) = \frac{\partial}{\partial c_j} c_j \chi^j, \qquad j = 0, 1, \ldots, n-1. \tag{39}$$

For the present flow problem, the system of the trial solution, under consideration, are defined as:

$$\check{F}(\chi) = \frac{4}{5}\chi^3 - \frac{17}{10}\chi^2 + \chi + \sum_{i=1}^{5} c_i \chi (\chi - 1)^i, \tag{40}$$

$$\check{\mathscr{T}}(\chi) = 1 - \chi + \sum_{i=1}^{5} d_i \chi (\chi - 1)^i, \tag{41}$$

$$\check{\Theta}(\chi) = 1 - \chi + \sum_{i=1}^{5} e_i \chi (\chi - 1)^i. \tag{42}$$

By following the procedure as suggested above, the numerical solution has been achieved by substituting the above set of trial solutions into the governing dimensionless system of equations, which are nonlinear in nature. Thus, by assigning some specific values to the parameters, the approximate solution for the velocity and temperature along with concentration profiles are as under:

$$\begin{aligned}
\check{F}(\chi) \cong \tilde{F}(\chi) = {}& 0.0387970586596651373\chi^6 - 0.135831460629211753\chi^5 + \\
& 0.150915945544620123\chi^4 + 0.750688356622255526\chi^3 - \\
& 1.704678566551171612\chi^2 + 1.00010866635438700\chi,
\end{aligned} \tag{43}$$

$$\begin{aligned}
\check{\mathscr{T}}(\chi) \cong \tilde{\mathscr{T}}(\chi) = {}& -0.0866389688520037421\chi^6 + 0.500946012785362327\chi^5 - \\
& 1.06954770759623097\chi^4 + 0.911080929453323640\chi^3 - \\
& 0.0448789189043188319\chi^2 - 1.21096134688613266\chi + 1.0,
\end{aligned} \tag{44}$$

$$\begin{aligned}
\check{\Theta}(\chi) \cong \tilde{\Theta}(\chi) = {}& -0.00281815275323297988\chi^6 + \\
& 0.0301378227343711141\chi^5 - 0.0834665005070162475\chi^4 + \\
& 0.0635487269722848497\chi^3 + 0.0509906671544004958\chi^2 - \\
& 1.05839256360080736\chi + 1.0.
\end{aligned} \tag{45}$$

The above solutions are obtained for certain values of parameters, which are given as:

$$\gamma = c_{\mathscr{R}} = Rd = 0.2, \quad Sc = 0.5, \quad \theta_w = 1.1, \quad \mathscr{M}_{\beta} = 1.5, \quad \mathscr{P} = 10, \quad \varphi_1 = \varphi_2 = 0.01. \tag{46}$$

Table 1 displays some important thermal and physical properties of carrier fluid (H_2O) [52] and the nanoparticles. These values play a key role in order to obtain the above solutions.

Table 1. Thermo-mechanical properties of H_2O, Fe_3O_4 and Ag nanoparticles [51–53].

	$H_2O(f)$	$Fe_3O_4\ (\varphi_1)$	$Ag\ (\varphi_2)$
$\check{\rho}\ (kg\ m^{-3})$	997.1	5180	10,500
$C_p\ (J\ kg^{-1}K^{-1})$	4179	670	235
$\check{k}\ (Wm^{-1}K^{-1})$	0.613	9.7	429
Pr	6.2	–	–

The subsequent Tables 2–4 respectively, provide a comparison between the results obtained via MM and RKF, for velocity, temperature, and concentration profiles. The values, as suggested above, remains the same for ingrained parameters. From these tables, one can clearly visualize the validity of the acquired results. Furthermore, for the tabulated values, the significant digit is set to 4.

Table 2. Comparison of the results obtained for $\breve{F}(\chi)$ for $(Ag - Fe_3O_4/H_2O)$ hybrid nanofluid with $(\varphi_1 = 0.01)$.

χ	NM	MM	Abs Error
0.0	0	0	0
0.1	0.08361971055	0.0837285414	0.0001.088308511
0.2	0.1377601897	0.1380405799	0.0002803901712
0.3	0.1674522669	0.1678007463	0.0003484793851
0.4	0.1776981193	0.1779703975	0.0002722782048
0.5	0.1734032258	0.1735144536	0.000111227772
0.6	0.1593783977	0.1593361686	0.00000422290531
0.7	0.1403592885	0.1402398353	0.0001194532992
0.8	0.121027384	0.1209214235	0.0001059605161
0.9	0.1060297995	0.1059871531	0.0000042646395
1.0	0.1	0.1	0.0000000000000

Table 3. Comparison of the results obtained for $\breve{\mathcal{J}}(\chi)$ for $(Ag - Fe_3O_4/H_2O)$ hybrid nanofluid with $(\varphi_1 = 0.01)$.

χ	NM	MM	Abs Error
0.0	1	1	0.0000000000000
0.1	0.8793088955	0.8792641251	0.0000044770436
0.2	0.7618050723	0.7617447028	0.0000060369490
0.3	0.6498328728	0.6497624809	0.0000070391950
0.4	0.5442201803	0.5441384063	0.0000081773904
0.5	0.4447153684	0.4446388103	0.0000076558102
0.6	0.3504084469	0.3503582131	0.0000050233744
0.7	0.2600611457	0.2600397494	0.0000213962680
0.8	0.1723422761	0.1723332129	0.0000090632458
0.9	0.08599850137	0.08599072095	0.0000007780414
1.0	0	0	0.0000000000000

Table 4. Comparison of the results obtained for $\breve{\Theta}(\chi)$ for $(Ag - Fe_3O_4/H_2O)$ hybrid nanofluid with $(\varphi_1 = 0.01)$.

χ	NM	MM	Abs err
0.0	1	1	0.0000000000000
0.1	0.8947297917	0.8947261509	0.0000003640704
0.2	0.7907519656	0.7907454211	0.0000006544490
0.3	0.688190526	0.6881823084	0.0000008217550
0.4	0.5870371838	0.5870289256	0.0000008258260
0.5	0.4871828347	0.4871760929	0.0000006741763
0.6	0.3884468482	0.3884424019	0.0000004446265
0.7	0.2906036897	0.29060125	0.0000002439726
0.8	0.1934071528	0.1934058456	0.0000001307163
0.9	0.09661287932	0.09661218516	0.0000000694153
1.0	0	0.0000000000000	0.0000000000000

The reliability of the obtained results have been further checked by reproducing the results for Skin friction coefficient, which were previously presented by Hayat et al. [36]. The results were obtained for the regular fluid ($\varphi_1 = \varphi_2 = 0$). Table 5 has been prepared to check the validity of the obtained results. It has been observed that the results obtained via Method of Moments are in good agreement with the previously existing results. Moreover, Method of Moments offers less computational complexity as compared to Homotopy analysis method. From the table, it has also been detected that the skin friction coefficient displays a decline with the increasing squeezing parameter (γ). However, a reversed behavior has been observed for increasing values of Modified Hartmann number $\mathcal{M}_b$.

Table 5. Comparison of the results obtained for Skin friction coefficient with ($\varphi_1 = \varphi_2 = 0$).

$\mathcal{M}_b$	γ	NM [36]	HAM [36]	MM
1.5	0.5	0.467511	0.467511	0.467511
1.0	-	0.452395	0.452395	0.452395
0.0	-	0.422159	0.422159	0.422159
1.5	0.3	1.08543	1.08543	1.08543
-	0.1	1.69635	1.69634	1.69634

4. Results and Discussions

The goal is to graphically elucidate the influential behavior of velocity, temperature and concentration profiles, due to the various ingrained entities. A pictorial view, from Figures 2–20, has been presented for the above-mentioned purpose. Figures 2–4 displays the performance of velocity profile, under the action of the squeezing parameter, Modified Hartmann number and solid volume fraction. The variations in velocity component $\check{F}'(\chi)$, due to squeezing parameter γ, have been depicted in Figure 2a. For $\gamma > 0$, i.e., when the upper plate moves in the downward direction, the fluid nearby the upper wall experiences a force, which in turn enhances the fluid velocity in that region. As γ increases sufficiently, the velocity component $\check{F}'(\chi)$ also increases and gradually depreciates the reversal behavior of the flow. The velocity component $\check{F}(\chi)$ also experiences an increment in the region, adjacent to the upper wall, which is mainly due to the squeezing behavior of the upper plate and this phenomena has been clearly observed through Figure 2b. Figure 3 demonstrates the impact of Modified Hartmann number $\mathcal{M}_b$ on the axial and normal components of the velocity distribution. Since the magnetic field experiences an exponential decline, therefore velocity component $\check{F}'(\chi)$ seems to be increased in the lower region of the channel. The fact behind is that the application of magnetic field generates the Lorentz forces, which in turn opposes the fluid flow. But in the present situation, the magnetic field decreases, so the Lorentz forces decreases and consequently, an increment in velocity has been perceived in the region close to the lower Riga plate. Besides, in the upper half, the velocity displays an opposite behavior as compared to the lower half of the channel, which may be due to the downward squeezing motion of the upper plate. Figure 3b exhibits an increment in the normal component of velocity $\check{F}(\chi)$ with the increasing Modified Hartmann number, which is primarily be due to the decreasing effects of Lorentz force. It can be detected from Figure 4a that the axial velocity decreases in the lower region with the increasing nanoparticles concentration, while an opposite behavior has been perceived in the upper portion of the channel. The reason behind is that the nanoparticle's concentration resists the fluid to move and therefore decreases the fluid velocity. Figure 4b depicts a decline in the normal component of the velocity $\check{F}(\chi)$ with increasing nanoparticle's concentration, which opposes the fluid motion. Moreover, the inset pictures reveal the fact that the velocity for the (Fe_3O_4/H_2O) nanofluid mostly attains the higher values as compared to the ($Ag - Fe_3O_4/H_2O$) hybrid nanofluid.

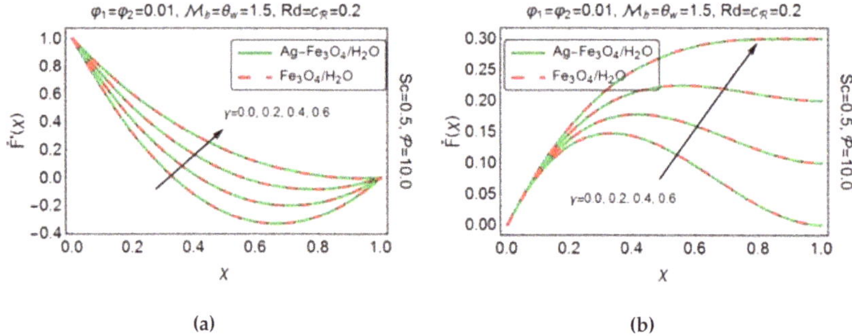

(a)

(b)

Figure 2. Impact of particular values of γ on (**a**) $\breve{F}'(\chi)$ and (**b**) $\breve{F}(\chi)$.

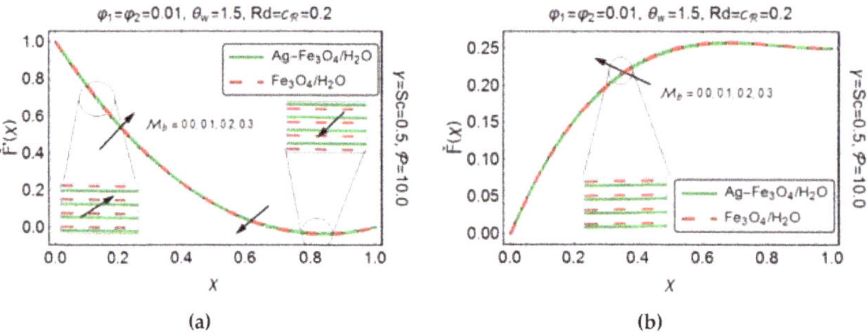

(a)

(b)

Figure 3. Impact of particular values of $\mathcal{M}_b$ on (**a**) $\breve{F}'(\chi)$ and (**b**) $\breve{F}(\chi)$.

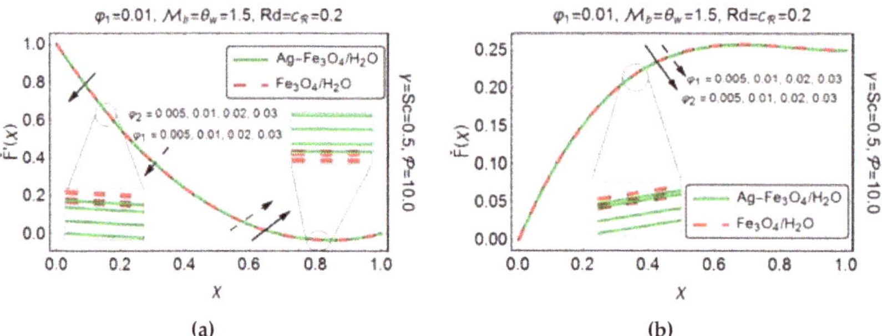

(a)

(b)

Figure 4. Impact of particular values of φ_1 and φ_2 on (**a**) $\breve{F}'(\chi)$ and (**b**) $\breve{F}(\chi)$.

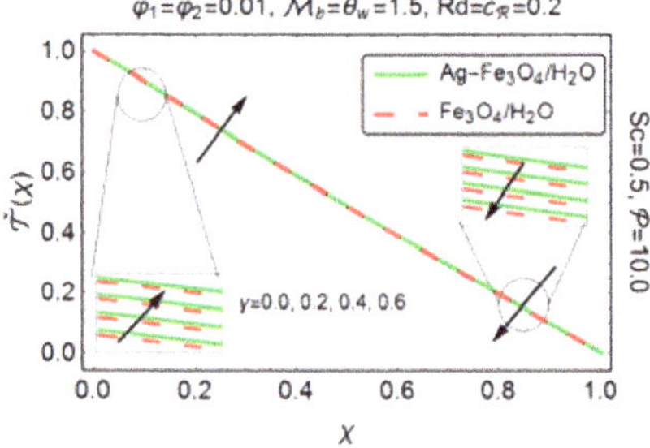

Figure 5. Impact of particular values of γ on $\breve{\mathcal{T}}(\chi)$.

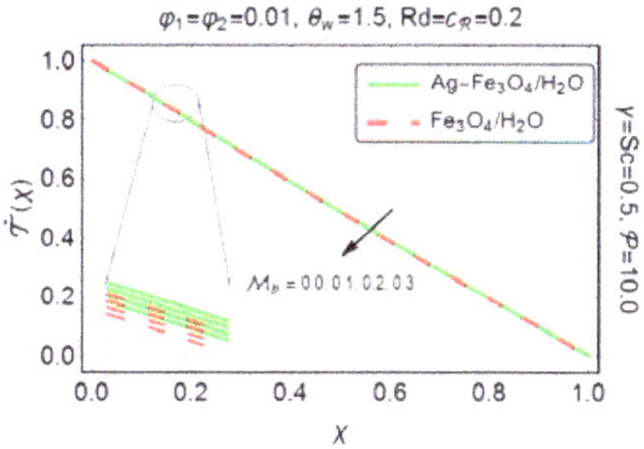

Figure 6. Impact of particular values of $\mathcal{M}_\beta$ on $\check{\mathcal{T}}(\chi)$.

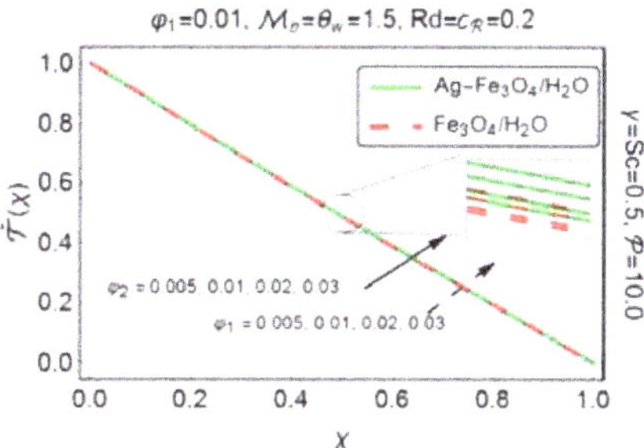

Figure 7. Impact of particular values of φ_1 and φ_2 on $\check{\mathcal{T}}(\chi)$.

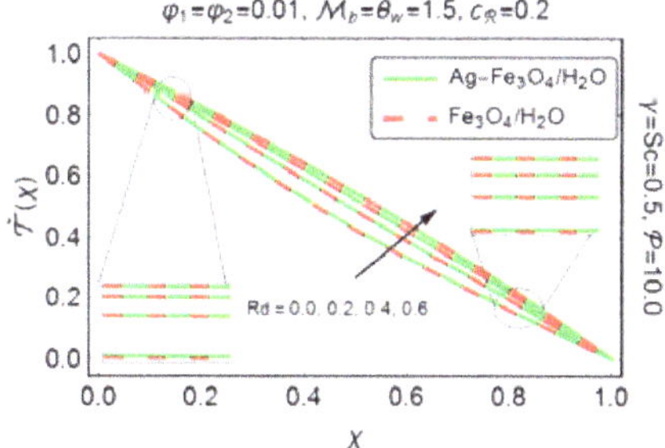

Figure 8. Impact of particular values of Rd on $\check{\mathcal{T}}(\chi)$.

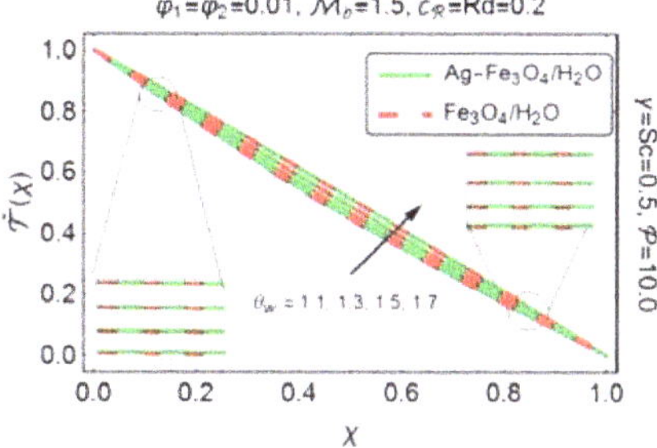

Figure 9. Impact of particular values of θ_w on $\check{\mathcal{T}}(\chi)$.

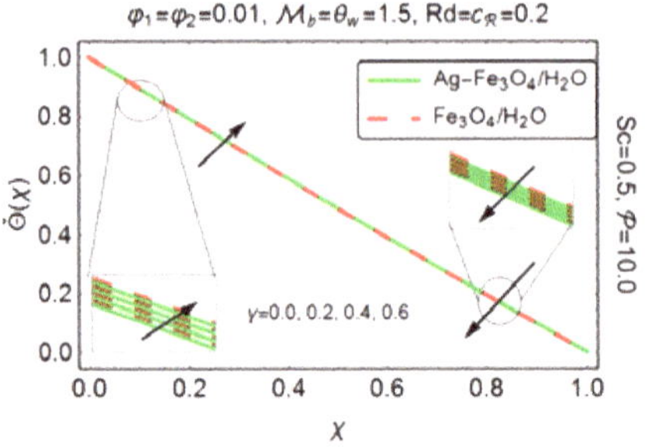

Figure 10. Impact of particular values of γ on $\breve{\Theta}(\chi)$.

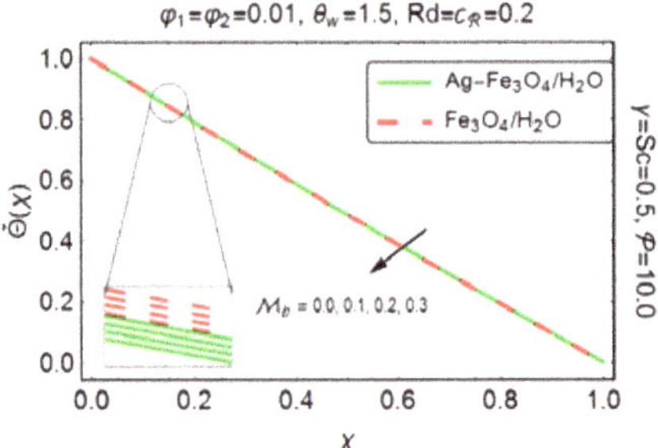

Figure 11. Impact of particular values of $\mathscr{M}_\beta$ on $\breve{\Theta}(\chi)$.

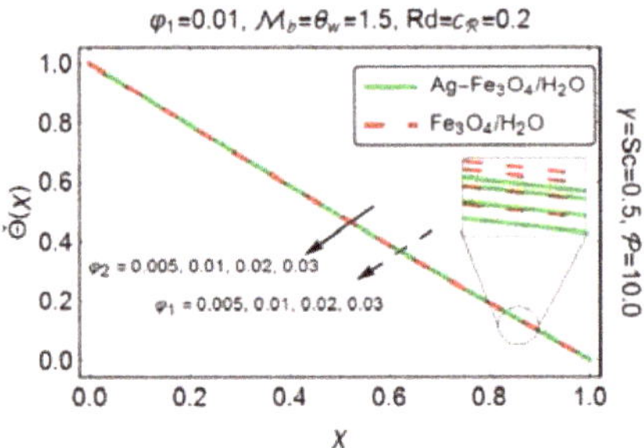

Figure 12. Impact of particular values of φ_1 and φ_2 on $\breve{\Theta}(\chi)$.

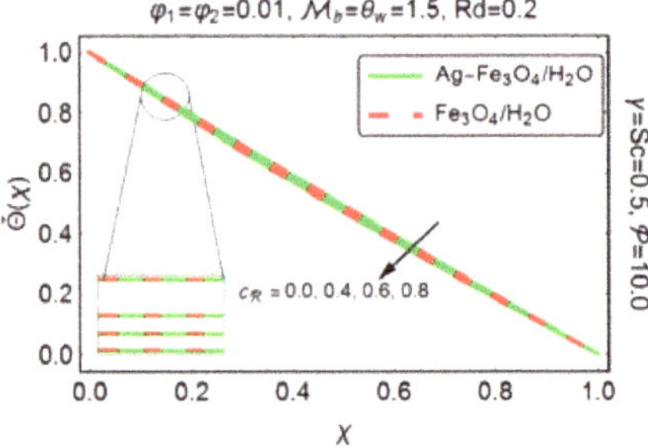

Figure 13. Impact of particular values of $c_{\mathscr{R}}$ on $\breve{\Theta}(\chi)$.

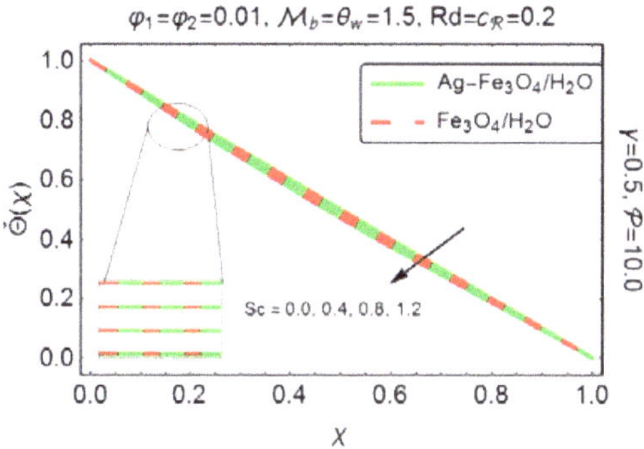

Figure 14. Impact of particular values of Sc on $\check{\Theta}(\chi)$.

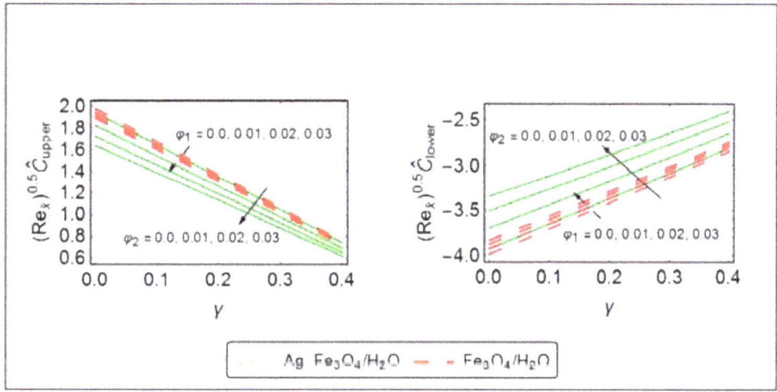

Figure 15. Coefficient of skin friction drag for particular values of φ_1 and φ_2.

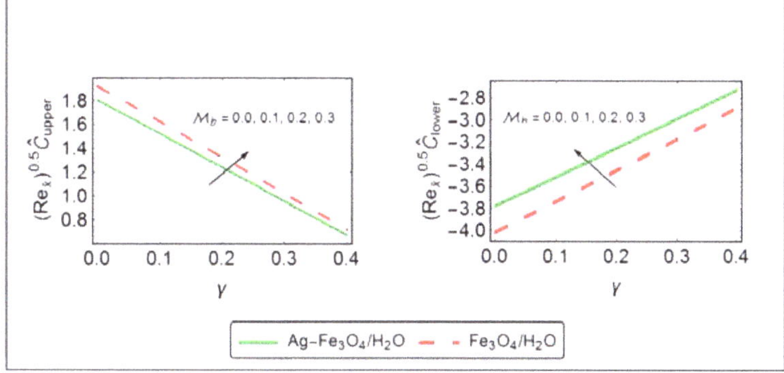

Figure 16. Coefficient of skin friction drag for particular values of $\mathcal{M}_b$.

$$\varphi_1 = 0.01, \ \mathcal{M}_b = \theta_w = 1.5, \ Rd = c_{\mathcal{R}} = 0.2, \ Sc = 0.5, \ \mathcal{P} = 10.0$$

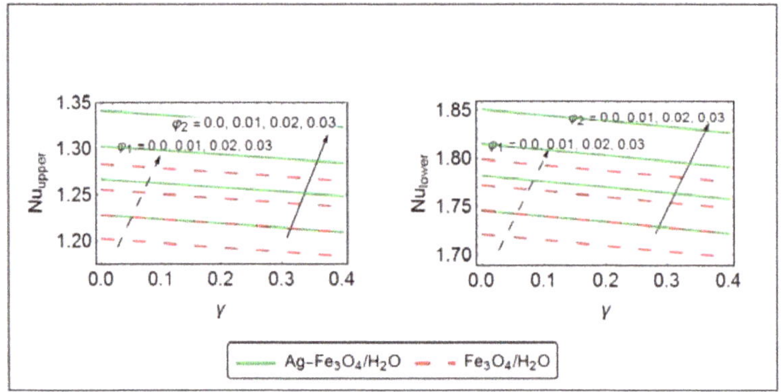

Figure 17. Local Nusselt number for particular values of φ_1 and φ_2.

$$\varphi_1 = \varphi_2 = 0.01, \ \gamma = Sc = 0.5, \ \mathcal{M}_b = 1.5, \ c_{\mathcal{R}} = 0.2, \ \mathcal{P} = 10.0$$

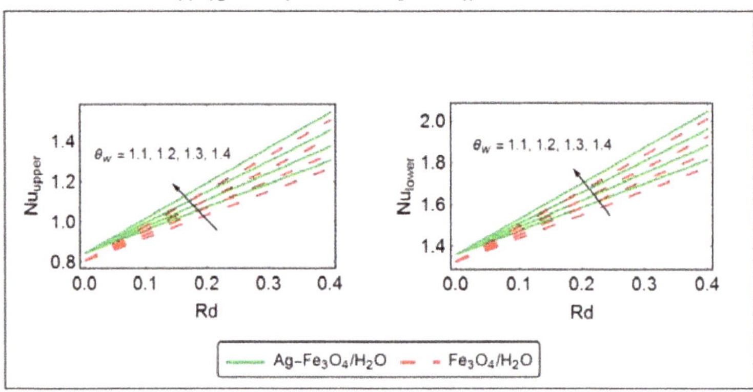

Figure 18. Local Nusselt number for particular values of θ_w.

$$\varphi_1 = 0.01, \ \mathcal{M}_b = \theta_w = 1.5, \ Rd = c_{\mathcal{R}} = 0.2, \ Sc = 0.5, \ \mathcal{P} = 10.0$$

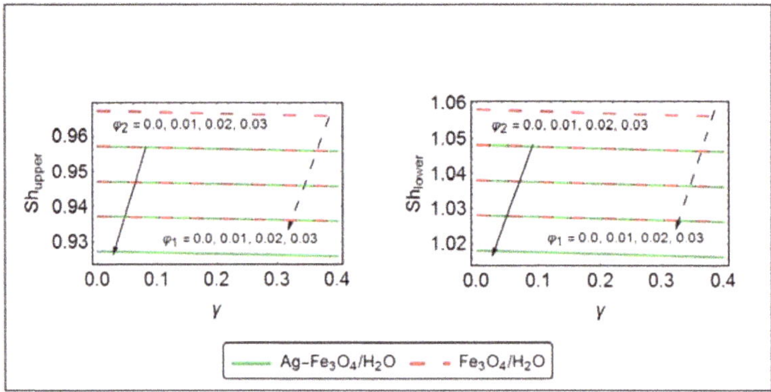

Figure 19. Sherwood number for particular values of φ_1 and φ_2.

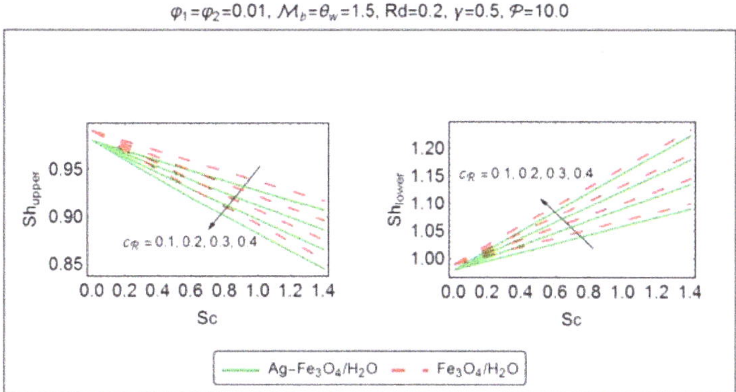

Figure 20. Sherwood number for particular values of $c_{\mathcal{R}}$.

The upcoming figures give a pictorial description of the variations in temperature distribution, for various embedded parameters. Figure 5 displays the impact of squeezing parameter γ on temperature profile. When the upper plate squeezed down, i.e., $\gamma > 0$, it exerts a force on the nearby fluid and enhances its velocity, but since the temperature seems to be dominant at the lower wall therefore the fluid in the adjacent region experiences the higher temperature values as compared to the region nearby the channel's upper wall. To demonstrate the impact of Modified Hartmann number $\mathcal{M}_\beta$ on temperature distribution, Figure 6 has been plotted. It has been found that temperature reveals lower values, as $\mathcal{M}_\beta$ increases. As the impact of Lorentz force on velocity profile produce a friction on the flow, which mainly be responsible to produce more heat energy. In the present flow situation, since the magnetic field exponentially decreases, so the Lorentz force decreases which in turn generates less friction force and consequently, decreases the heat energy and therefore decreases the fluid's temperature as well as the thermal boundary layer thickness. From Figure 7, one can clearly observe an increment in temperature profile with increasing nanoparticles concentration. The fact behind is that, the inclusion of nanoparticles with different volume fractions augments the thermal properties of the host fluid and therefore increases its temperature. It has also been observed that the temperature of $(Ag - Fe_3O_4/H_2O)$ hybrid nanofluid shows its supremacy over the (Fe_3O_4/H_2O) nanofluid, which definitely be due to the rising values of the thermal conductivity for $(Ag - Fe_3O_4/H_2O)$ hybrid nanofluid.

Figure 8 has been sketched, to highlight the temperature behavior under the influence of radiation parameter Rd. An upsurge has been encountered in temperature, for increasing Rd. The fact behind is that, the increasing Rd corresponds to the decrement in mean absorption coefficient, which in turn raises the fluid temperature. The temperature also depicts a rising behavior with increasing θ_w (see Figure 9). The increasing θ_w implies that the temperature differences between the lower and upper walls significantly rises and subsequently, an increment in temperature has been recorded.

The next set of figures provide us an aid, to visualize the deviations, in concentration profile, caused by various embedded parameters. Figure 10 demonstrates the influence of squeezing parameter γ on the concentration profile. When the upper plate moves vertically downward, i.e., $\gamma > 0$, it suppresses the adjacent fluid layers and enhances its velocity, but since the concentration shows its supremacy at the lower wall, therefore the concentration profile shows its dominancy in the region, close to the lower wall, as compared to the region adjacent to the upper wall. To demonstrate the impact of Modified Hartmann number $\mathcal{M}_\beta$ on concentration profile, Figure 11 has been painted. As explained earlier that the Lorentz forces, in present flow situation, experience a decline, which as a result generate less friction force and therefore, decrease the concentration profile along with concentration boundary layer thickness. From Figure 12, one can clearly detect a decline in concentration profile, as nanoparticle fraction increases. Moreover, it has been noticed that the concentration profile for $(Ag - Fe_3O_4/H_2O)$ hybrid nanofluid possesses lower values as compared to the (Fe_3O_4/H_2O) nanofluid.

Figure 13 portrays the influence of chemical reaction parameter $c_{\mathscr{R}}$ on concentration profile. A clear decline has been perceived in the concentration of species with the growing values of chemical reaction parameter $c_{\mathscr{R}}$. Since the chemical reaction, in the present flow analysis, is due to the consumption of the chemicals, therefore, the concentration profile experiences a decline with the increasing values of $c_{\mathscr{R}}$. The variations in concentration profile, under the action of Schmidt number Sc, has been presented in Figure 14. It has been observed that the increasing values of Schmidt number Sc causes a decline in the concentration of the species. Since the Schmidt number is the ratio of momentum diffusivity to mass diffusivity. Therefore, the increment in Schmidt number consequently implies a decline in mass diffusivity, which in turn decreases the concentration profile.

Figures 15 and 16 display the impact of various ingrained parameters on the skin friction coefficient, both at the upper and lower Riga surfaces. It has been detected from Figure 15 that increasing the nanoparticles concentration certainly enhances the coefficient of skin friction drag at the lower Riga plate. However, at the upper plate, an opposite behavior has been clearly visible. As far as squeezing parameter γ is concerned, the skin friction coefficient exhibits an increasing behavior, in the region adjacent to the lower plate. However, a decline has been perceived at the upper wall. From Figure 16, one can clearly observes an increment in skin friction coefficient, with the increasing $\mathscr{M}_\beta$, both at the upper and lower Riga plates. Moreover, the skin friction coefficient for $(Ag - Fe_3O_4/H_2O)$ hybrid nanofluid possesses higher values, at the bottom of the channel, as compared to the (Fe_3O_4/H_2O) nanofluid.

Figures 17 and 18 have been plotted, to assess the consequences of various embedded entities on the local rate of heat transfer i.e., Nusselt number. From Figure 17, one can clearly detect an increment in heat transfer, with increasing nanoparticle volume fraction, at both of the plates. Since the nanoparticle's inclusion, in the base fluid, is responsible for rising its temperature, therefore an augmentation in the heat transfer rate is quite obvious. By varying the squeezing number γ horizontally, the local Nusselt number at the upper as well as on the lower plates, indicate a decreasing behavior. Figure 18 depicts the variations in heat transfer rate, with growing values of radiation parameter Rd and temperature difference parameter θ_w. Since both the parameters (Rd and θ_w) significantly amplifies the temperature of the fluid, therefore, they play a key role in enhancing the local Nusselt number, both at the upper and lower Riga plates. Besides, it has been observed from both the figures that the $(Ag - Fe_3O_4/H_2O)$ hybrid nanofluid shows its supremacy in transferring the heat, both at the upper and lower Riga plates.

Figures 19 and 20 depict the variations in the rate of mass transfer, i.e., Sherwood number under the action of various involved parameters. Figure 19 reveals a decline in the Sherwood number with increasing nanoparticle concentration, both at the upper and lower Riga plates. Since the increasing nanoparticle's volume fraction certainly opposes the fluid motion, therefore a decrement in Sherwood number is quite obvious. The rate, with which mass flows, also shows a decreasing behavior when the squeezing parameter γ increases horizontally. From Figure 20, one can observes a clear enhancement in the rate of mass flow in the region nearby the lower plate, when the chemical reaction parameter $c_{\mathscr{R}}$ increases curve wise and Schmidt number Sc varies along the horizontal axis. On the other hand, a reverse behavior has been perceived at the upper plate. Moreover, (Fe_3O_4/H_2O) nanofluid remains dominant in transferring the mass, both at the upper and lower Riga plates.

Hybrid nanofluids, being advanced version of nanofluids, considerably influences the thermo-mechanical properties of the working fluid, particularly the thermal conductivity. For the said purpose, Tables 6 and 7 have been designed, to see the deviations in thermo-mechanical properties of the $(Ag - Fe_3O_4/H_2O)$ hybrid nanofluid and (Fe_3O_4/H_2O) nanofluid. It has been detected that the density of $(Ag - Fe_3O_4/H_2O)$ hybrid nanofluid depicts an increment, as compared to (Fe_3O_4/H_2O) nanofluid. While, the specific heat clearly experiences a decline with the increasing nanoparticles fraction. As far as thermal conductivity is concerned, $(Ag - Fe_3O_4/H_2O)$ hybrid nanofluid shows a dominant behavior against the (Fe_3O_4/H_2O) nanofluid. Besides, the Bruggeman model (18), for thermal conductivity shows its proficiency over the Maxwell's model (17). The reason is that, the Bruggeman model is more focused on the maximum interactions between randomly dispersed particles. It usually involves the spherical shaped particles, with no limitation on the particles concentration. On the other hand, the Maxwell's model depends on the nanoparticle's volume fraction and the thermal conductivity of the base fluid and the spherical shaped particles.

Table 6. Variation in thermo-physical properties of $(Ag - Fe_3O_4/H_2O)$ hybrid nanofluid with $(\varphi_1 = 0.01)$.

φ_2	$\check{\rho}$	$\check{\rho}C_p \times 10^6$	$\mathcal{k}_{hnf}$ (14)	$\mathcal{k}_{hnf}$ (13)
0.00	1038.929	4.159918091	0.628640157	0.628422974
0.02	1228.987	4.125930473	0.67317743	0.6695051201
0.04	1419.045	4.091942855	0.72042306	0.7092486827
0.06	1609.103	4.057955237	0.77474778	0.7506805172
0.08	1799.161	4.023967619	0.83787931	0.7939214931
0.10	1989.219	3.989980001	0.91213707	0.8390950667

Table 7. Variation in Thermo-physical properties of (Fe_3O_4/H_2O) nanofluid with $(\varphi_2 = 0)$.

φ_1	$\check{\rho}$	$\check{\rho}C_p \times 10^6$	$\mathcal{k}_{nf}$ (14)	$\mathcal{k}_{nf}$ (13)
0.00	997.1	4.1668809	0.613	0.613
0.02	1080.758	4.152955282	0.644998753	0.6441068291
0.04	1164.416	4.139029664	0.680047429	0.6762841143
0.06	1248.074	4.125104046	0.718527614	0.7095880769
0.08	1331.732	4.111178428	0.760870799	0.7440789486
0.10	1415.39	4.09725281	0.80756218	0.779821329

5. Conclusions

This article discloses the salient features of nonlinear thermal radiation, in the squeezing flow of $(Ag - Fe_3O_4/H_2O)$ hybrid nanofluid, between two Riga plates along with a chemical reaction. Method of moment has been employed for the solution point of view. The obtained results are then compared with the numerical results (obtained via Runge-Kutta-Fehlberg algorithm). Both the methods depict an excellent agreement between the results.

Further investigations are as follows:

- Velocity profile seems to be an increasing function of both squeezing parameter γ and modified Hartmann number $\mathcal{M}_\beta$.
- A decrement in the velocity behavior has been perceived, with increasing nanoparticle concentration.
- The velocity profile for $(Ag - Fe_3O_4/H_2O)$ hybrid nanofluid mostly remains on the lower side.
- The amplification in temperature has been recorded for increasing squeezed number γ and nanoparticle concentration, while a reversed behavior has been noticed for increasing modified Hartman number $\mathcal{M}_\beta$.
- The temperature behaves in an increasing manner with the rising Rd and θ_w. Besides, the temperature profile possesses a dominant behavior for $(Ag - Fe_3O_4/H_2O)$ hybrid nanofluid.
- The concentration profile demonstrates a decreasing behavior, with increasing modified Hartman number $\mathcal{M}_\beta$ and nanoparticle volume fraction.
- The increment in chemical reaction parameter $c_\mathcal{R}$ and Schmidt number Sc depicts a clear decline in the concentration profile.
- Skin friction coefficient for $(Ag - Fe_3O_4/H_2O)$ hybrid nanofluid displays an increasing behavior, in the region adjacent to the lower Riga plate, against the varying squeezing parameter γ, modified Hartman number $\mathcal{M}_\beta$ and the nanoparticle concentration.

- The local heat transfer rate, for $(Ag - Fe_3O_4/H_2O)$ hybrid nanofluid, shows its proficiency for varying nanoparticle concentration, radiation parameter Rd and temperature difference parameter θ_w and this phenomena has been detected at both the plates.
- The augmentation of Schmidt number Sc and chemical reaction parameter $c_{\mathscr{R}}$ enhances the Sherwood number, at the lower plate, while a reversed phenomenon has been observed at the upper plate.

Author Contributions: All the authors equally contributed to the paper. Final draft has been read and approved by all the authors.

Funding: This research received no external funding.

Acknowledgments: We are thankful to the anonymous reviewers for their valuable comments which really improved the quality of presented work.

Conflicts of Interest: The authors declare no conflict of interest.

Nomenclature

ℓ	Width between magnets and electrodes
$\mathscr{M}_0$	Magnetization of the permanent magnets, *Tesla*
$\dot{J}_0$	Applied current density in the electrodes, $m^{-2}A$
c_1	First order chemical reaction coefficient
$\mathfrak{D}$	Molecular diffusivity
$\mathscr{k}$	Thermal conductivity, W/mK
$\breve{p}$	Pressure
C_p	Specific heat at constant pressure, $J/kg.K$
$\breve{u}$	Axial velocity component, m/s
$\breve{v}$	Normal velocity component, m/s
$\breve{k}$	Coefficient for mean absorption
$\mathscr{M}_\beta$	Modified Hartman number
a	Rate of stretching
$c_{\mathscr{R}}$	Chemical reaction parameter
Pr	Prandtl number
Rd	Radiation parameter
Nu	Nusselt number
Sc	Schmidt number
Re	Reynolds number
Sh	Sherwood number
Ag	Silver nanoparticles
H_2O	Water
Fe_3O_4	Magnetite nanoparticles
EMHD	Electro-magneto hydrodynamic

Greek Symbols

φ	Solid volume fraction
μ	Dynamic viscosity, $N.s/m^2$
$\breve{\rho}$	Density, kg/m^3
v	Kinematic viscosity, m^2/s
χ	Similarity variable
$\breve{\sigma}$	Stefan-Boltzmann constant
λ	Constant characteristics parameter
γ	Dimensionless squeeze number
θ_w	Temperature difference parameter
$\breve{\rho}C_p$	Heat capacitance

Subscripts

hnf	Hybrid Nanofluid
nf	Nanofluid
f	Base fluid
p1	Solid nanoparticles of Fe_3O_4
p2	Solid nanoparticles of Ag

References

1. Choi, S.U.S. Enhancing thermal conductivity of fluids with nanoparticles. In *Developments and Applications of Non-Newtonian Flows*; Siginer, D.A., Wang, H.P., Eds.; ASME: New York, NY, USA, 1995; Volume 231, pp. 99–105.
2. Choi, S.U.S.; Zhang, Z.G.; Yu, W.; Lockwood, F.E.; Grulke, E.A. Anomalous thermal conductivity enhancement in nanotube suspensions. *Appl. Phys. Lett.* **2001**, *79*, 2252–2254. [CrossRef]
3. Maxwell, J.C. *A Treatise on Electricity and Magnetism*, 3rd ed.; Clarendon; Oxford University Press: Oxford, UK, 1904.
4. Bruggeman, D.A.G. Berechnzcrrg verschCedcner physikalducher Eonstanten von heterogenew Yuhstan.xen 1. *Ann. Phys.* **1935**, *416*, 636–664. [CrossRef]
5. Hamilton, R.L.; Crosser, O.K. Thermal conductivity of heterogeneous two-component systems. *Ind. Eng. Chem. Fundam.* **1962**, *1*, 187–191. [CrossRef]
6. Suzuki, A.; Ho, N.F.H.; Higuchi, W.I. Predictions of the particle size distribution changes in emulsions and suspensions by digital computation. *J. Colloid Interface Sci.* **1969**, *29*, 552–564. [CrossRef]
7. Buongiorno, J. Convective transport in nanofluids. *J. Heat Transf.* **2006**, *128*, 240–250. [CrossRef]
8. Xue, Q.Z. Model for thermal conductivity of carbon nanotube-based composites. *Phys. B Condens. Matter.* **2005**, *368*, 302–307. [CrossRef]
9. Iijima, S. Helical microtubules of graphitic carbon. *Nature* **1991**, *354*, 56–58. [CrossRef]
10. Timofeeva, E.V.; Gavrilov, A.N.; McCloskey, J.M.; Tolmachev, Y.V.; Sprunt, S.; Lopatina, L.M.; Selinger, J.V. Thermal conductivity and particle agglomeration in alumina nanofluids: Experiment and theory. *Phys. Rev. E* **2007**, *76*, 061203. [CrossRef]
11. Masoumi, N.; Sohrabi, N.; Behzadmehr, A. A new model for calculating the effective viscosity of nanofluids. *J. Phys. D Appl. Phys.* **2009**, *42*, 055501. [CrossRef]
12. Thurgood, P.; Baratchi, S.; Szydzik, C.; Mitchell, A.; Khoshmanesh, K. Porous PDMS structures for the storage and release of aqueous solutions into fluidic environments. *Lab Chip* **2017**, *17*, 2517–2527. [CrossRef]
13. Sarkar, J.; Ghosh, P.; Adil, A. A review on hybrid nanofluids: Recent research, development and applications. *Renew. Sustain. Energy Rev.* **2015**, *43*, 164–177. [CrossRef]
14. Ranga Babu, J.A.; Kumar, K.K.; Srinivasa Rao, S. State-of-art review on hybrid nanofluids. *Renew. Sustain. Energy Rev.* **2017**, *77*, 551–565. [CrossRef]
15. Niihara, K. New Design Concept of Structural Ceramics. *J. Ceram. Soc. Jpn.* **1991**, *99*, 974–982. [CrossRef]
16. Jana, S.; Salehi-Khojin, A.; Zhong, W.H. Enhancement of fluid thermal conductivity by the addition of single and hybrid nano-additives. *Thermochim. Acta* **2007**, *462*, 45–55. [CrossRef]
17. Suresh, S.; Venkitaraj, K.P.; Selvakumar, P.; Chandrasekar, M. Synthesis of Al_2O_3-Cu/water hybrid nanofluids using two step method and its thermo physical properties. *Colloids Surf. A Physicochem. Eng. Asp.* **2011**, *388*, 41–48. [CrossRef]
18. Suresh, S.; Venkitaraj, K.P.; Selvakumar, P.; Chandrasekar, M. Effect of Al_2O_3–Cu/water hybrid nanofluid in heat transfer. *Exp. Therm. Fluid Sci.* **2012**, *38*, 54–60. [CrossRef]
19. Momin, G.G. Experimental investigation of mixed convection with water-Al_2O_3 & hybrid nanofluid in inclined tube for laminar flow. *Int. J. Sci. Technol. Res.* **2013**, *2*, 195–202.
20. Devi, S.P.A.; Devi, S.S.U. Numerical investigation of hydromagnetic hybrid Cu-Al_2O_3/water nanofluid flow over a permeable stretching sheet with suction. *Int. J. Nonlinear Sci. Numer.* **2016**, *17*, 249–257. [CrossRef]
21. Das, S.; Jana, R.N.; Makinde, O.D. MHD Flow of Cu-Al_2O_3/Water Hybrid Nanofluid in Porous Channel: Analysis of Entropy Generation. *Defect Diffus. Forum* **2017**, *377*, 42–61. [CrossRef]

22. Chamkha, A.J.; Miroshnichenko, I.V.; Sheremet, M.A. Numerical analysis of unsteady conjugate natural convection of hybrid water-based nanofluid in a semi-circular cavity. *J. Therm. Sci. Eng. Appl.* **2017**, *9*, 1–9. [CrossRef]

23. Olatundun, A.T.; Makinde, O.D. Analysis of Blasius flow of hybrid nanofluids over a convectively heated surface. *Defect Diffus. Forum* **2017**, *377*, 29–41. [CrossRef]

24. Hayat, T.; Nadeem, S. Heat transfer enhancement with Ag–CuO/water hybrid nanofluid. *Results Phys.* **2017**, *7*, 2317–2324. [CrossRef]

25. Stefan, M.J. Versuch Uber die Scheinbare Adhasion, sitzungsber. *Abt. II, Osterr. Akad. Wiss., MathNaturwiss.kl.* **1874**, *69*, 713–721.

26. Shahmohamadi, H.; Rashidi, M.M.; Dinarvand, S. Analytic approximate solutions for unsteady two-dimensional and axisymmetric squeezing flows between parallel plates. *Math. Probl. Eng.* **2008**, *2008*, 1–13. [CrossRef]

27. Sheikholeslami, M.; Ganji, D.D.; Ashorynejad, H.R. Investigation of squeezing unsteady nanofluid flow using ADM. *Powder Technol.* **2013**, *239*, 259–265. [CrossRef]

28. Khan, U.; Ahmed, N.; Asadullah, M.; Mohyud-Din, S.T. Effects of viscous dissipation and slip velocity on two-dimensional and axisymmetric squeezing flow of Cu-water and Cu-kerosene nanofluids. *Propul. Power Res.* **2015**, *4*, 40–49. [CrossRef]

29. Ahmed, N.; Khan, U.; Khan, S.I.; Bano, S.; Mohyud-Din, S.T. Effects on magnetic field in squeezing flow of a Casson fluid between parallel plates. *J. King Saud Univ. Sci.* **2017**, *29*, 119–125. [CrossRef]

30. Gailitis, A.; Lielausis, O. On a possibility to reduce the hydrodynamic resistance of a plate in an electrolyte. *Appl. Magnetohydrodyn.* **1961**, *12*, 143–146.

31. Avilov, V.V. *Electric and Magnetic Fields for the Riga Plate*; Technical Report; FRZ: Rossendorf, Germany, 1998.

32. Pantokratoras, A.; Magyari, E. EMHD free-convection boundary-layer flow from a Riga-plate. *J. Eng. Math.* **2009**, *64*, 303–315. [CrossRef]

33. Pantokratoras, A. The Blasius and Sakiadis flow along a Riga-plate. *Prog. Comput. Fluid Dyn. Int. J.* **2011**, *11*, 329–333. [CrossRef]

34. Magyari, E.; Pantokratoras, A. Aiding and opposing mixed convection flows over the Riga-plate. *Commun. Nonlinear Sci. Numer. Simul.* **2011**, *16*, 3158–3167. [CrossRef]

35. Ayub, M.; Abbas, T.; Bhatti, M.M. Inspiration of slip effects on electromagnetohydrodynamics (EMHD) nanofluid flow through a horizontal Riga plate. *Eur. Phys. J. Plus* **2016**, *131*, 193. [CrossRef]

36. Hayat, T.; Khan, M.; Imtiaz, M.; Alsaedi, A. Squeezing flow past a Riga plate with chemical reaction and convective conditions. *J. Mol. Liq.* **2017**, *225*, 569–576. [CrossRef]

37. Hayat, T.; Khan, M.; Khan, M.I.; Alsaedi, A.; Ayub, M. Electromagneto squeezing rotational flow of Carbon (C)-Water (H2O) kerosene oil nanofluid past a Riga plate: A numerical study. *PLoS ONE* **2017**, *12*, e0180976. [CrossRef] [PubMed]

38. Rosseland, S. *Astrophysik und Atom-Theoretische Grundlagen*; Springer-Verlag: Berlin, Germany, 1931.

39. Magyari, E.; Pantokratoras, A. Note on the effect of thermal radiation in the linearized Rosseland approximation on the heat transfer characteristics of various boundary layer flows. *Int. J. Heat Mass Transf.* **2011**, *38*, 554–556. [CrossRef]

40. Rashidi, M.M.; Mohimanian pour, S.A.; Abbasbandy, S. Analytic approximate solutions for heat transfer of a micropolar fluid through a porous medium with radiation. *Commun. Nonlinear Sci. Numer. Simul.* **2011**, *16*, 1874–1889. [CrossRef]

41. Noor, N.F.M.; Abbasbandy, S.; Hashim, I. Heat and mass transfer of thermophoretic MHD flow over an inclined radiate isothermal permeable surface in the presence of heat source/sink. *Int. J. Heat Mass Transf.* **2012**, *55*, 2122–2128. [CrossRef]

42. Mohyud-Din, S.T.; Khan, S.I. Nonlinear radiation effects on squeezing flow of a Casson fluid between parallel disks. *Aerosp. Sci. Technol.* **2016**, *48*, 186–192. [CrossRef]

43. Khan, U.; Ahmed, N.; Mohyud-Din, S.T.; Bin-Mohsin, B. Nonlinear radiation effects on MHD flow of nanofluid over a nonlinearly stretching/shrinking wedge. *Neural Comput. Appl.* **2017**, *28*, 2041–2050. [CrossRef]

44. Saba, F.; Ahmed, N.; Hussain, S.; Khan, U.; Mohyud-Din, S.T.; Darus, M. Thermal Analysis of Nanofluid Flow over a Curved Stretching Surface Suspended by Carbon Nanotubes with Internal Heat Generation. *Appl. Sci.* **2018**, *8*, 395. [CrossRef]

45. Brinkman, H.C. The viscosity of concentrated suspensions and solutions. *J. Chem. Phys.* **1952**, *20*, 571. [CrossRef]
46. Pak, B.C.; Cho, Y.I. Hydrodynamic and heat transfer study of dispersed fluids with submicron metallic oxide particles. *Exp. Heat Transf.* **1998**, *11*, 151–170. [CrossRef]
47. Xuan, Y.; Roetzel, W. Conceptions for heat transfer correlation of nanofluids. *Int. J. Heat Mass Transf.* **2000**, *43*, 3701–3707. [CrossRef]
48. Ho, C.J.; Huang, J.B.; Tsai, P.S.; Yang, Y.M. Preparation and properties of hybrid water-based suspension of Al2O3 nanoparticles and MEPCM particles as functional forced convection fluid. *Int. J. Heat Mass Transf.* **2010**, *37*, 490–494. [CrossRef]
49. Mamut, E. Characterization of heat and mass transfer properties of nanofluids. *Rom. J. Phys.* **2006**, *51*, 5–12.
50. Singh, P.; Kumar, M. Mass transfer in MHD flow of alumina water nanofluid over a flat plate under slip conditions. *Alexandria Eng. J.* **2015**, *54*, 383–387. [CrossRef]
51. Reddy, N.; Murugesan, K. Numerical Investigations on the Advantage of Nanofluids under DDMC in a Lid-Driven Cavity. *Heat Tran. Asian Res.* **2017**, *46*, 1065–1086. [CrossRef]
52. Bergman, T.L.; Lavine, A.S.; Incropera, F.P.; Dewitt, D.P. *Fundamentals of Heat and Mass Transfer*, 4th ed.; John Wiley & Sons: New York, NY, USA, 2002.
53. Sheikholeslami, M.; Ganji, D.D. Free convection of Fe_3O_4-water nanofluid under the influence of an external magnetic source. *J. Mol. Liq.* **2017**, *229*, 530–540. [CrossRef]

Article

Numerical Study of the Magnetic Field Effect on Ferromagnetic Fluid Flow and Heat Transfer in a Square Porous Cavity

Mohamed F. El-Amin [1,2,*], **Usama Khaled [3,4]** and **Abderrahmane Beroual [5]**

[1] College of Engineering, Effat University, 21478 Jeddah, Saudi Arabia
[2] Department of Mathematics, Faculty of Science, Aswan University, Aswan 81528, Egypt
[3] Department of Electrical Engineering, College of Engineering, King Saud University, P.O. Box. 800, 11421 Riyadh, Saudi Arabia; ukhaled@ksu.edu.sa
[4] Department of Electrical Engineering, Faculty of Energy Engineering, Aswan University, Aswan 81528, Egypt
[5] Ecole Centrale de Lyon, University of Lyon, AMPERE CNRS UMR 5005, 36 Avenue Guy de Collongue, 69134 Ecully, France; Abderrahmane.Beroual@ec-lyon.fr
* Correspondence: mohamed.elamin.kaust@gamil.com; Tel.: +966-54-256-7083

Received: 12 August 2018; Accepted: 19 November 2018; Published: 21 November 2018

Abstract: A numerical study of ferromagnetic-fluid flow and heat transfer in a square porous cavity under the effect of a magnetic field is presented. The water-magnetic particle suspension is treated as a miscible mixture and, thus, the magnetization, density and viscosity of the ferrofluid are obtained. The governing partial-differential equations were solved numerically using the cell-centered finite-difference method for the spatial discretization, while the multiscale time-splitting implicit method was developed to treat the temporal discretization. The Courant–Friedrichs–Lewy stability condition (CFL < 1) was used to make the scheme adaptive by dividing time steps as needed. Two cases corresponding to Dirichlet and Neumann boundary conditions were considered. The efficiency of the developed algorithm as well as some physical results such as temperature, concentration, and pressure; and the local Nusselt and Sherwood numbers at the cavity walls are presented and discussed. It was noticed that the particle concentration and local heat/mass transfer rate are related to the magnetic field strength, and both pressure and velocity increase as the strength of the magnetic was increased.

Keywords: magnetic field; ferrofluid; porous cavity; heat transfer; mass transfer; numerical modeling

---

1. Introduction

Ferromagnetic fluids are smart fluids [1,2], composed of magnetized nanoparticles suspended in a liquid-based medium, such as oil and water. Ferromagnetic fluids have been used in different engineering and environmental applications, such as enhanced oil recovery (EOR). The idea of using ferrofluids, such as iron oxide, Fe_2O_3, and zinc oxide, ZnO, under an external magnetic field is to control the ferrofluids' movement in porous media [3–5]. Heat transfer in porous media involves a wide range of applications such as heat exchangers, oil/gas recovery, geophysical systems, nuclear waste disposal, chemical reactors, thermal insulation of buildings, and drying processes. Chegenizadeh et al. [6] provided a classification of the most popular nanoparticles under optimal operational conditions. El-Amin et al. [7,8] conducted some studies on modeling nanoparticle transport in porous media. Suleimanov et al. [9] and Ryoo et al. [10] conducted an experimental investigation of using nanoparticles in enhanced oil recovery application.

Flows in cavities have many technological applications, such as heat exchangers, cooling systems for electronic equipment, and environmental flows. Sheremet and Pop [11] investigated the steady laminar mixed convection inside a lid-driven square cavity filled with a water-based nanofluid. Chalambaz et al. [12] studied the heat and mass transfer in a square porous cavity with differential temperature and concentration at the sidewalls, and they found that the heat transfer of the mixture and the mass transfer of the other phase can be maximized for specific values of the Lewis number of one phase. Carvalho and de Lemos [13] presented the problem of laminar free convection within a square porous cavity filled with a saturated fluid. Javed et al. [14] presented a numerical study for free convection through a square enclosure filled with a ferrofluid-saturated porous medium under a uniform magnetic field.

Numerical simulation is an important tool enabling engineers and scientists to predict the transport phenomena, test, and optimize an appropriate intervention strategy for heat transfer [15]. The time-stepping method has been presented for the problems of fluid dynamics by using implicit-type time-marching procedures to resolve transients [16]. Martinez [17] solved the shallow water equations by using the time-splitting technique. The time viscosity-splitting method was used for the Boussinesq problem by Zhang and Qian [18]. A time-splitting Fourier spectral method has been developed for approximating singular solutions of the Gross–Pitaevskii equation [19]. El-Amin et al. [20–23] used a multiscale time-splitting strategy to manage different time-step sizes for different physics. In this research, a multiscale adaptive time-splitting scheme is introduced to simulate the problem of magnetic-field effects on ferrofluids and heat transfer with a single-phase flow in a porous cavity considering variation of the nanoparticle concentration.

2. Problem Definition

The problem of a single-phase flow with ferromagnetic fluid flow and heat transfer in a square porous cavity under the effect of an external magnetic field is considered in this study. The ferromagnetic fluids gain properties of both liquid and magnetized solid particles. If the ferromagnetic fluid is subjected to an external magnetic field, it flows toward the magnetic field, and the flow resistance increases. In the absence of the magnet, nearby ferrofluids act as normal liquid. Figure 1 shows the schematic diagram of the 2D square porous cavity of Cases 1 and 2. In Case 1, the cavity walls are kept at a low temperature/concentration, T_0, C_0, except the inlet at the center of the left wall is kept at a higher temperature/concentration T_h, C_h (Dirichlet boundary condition). Initially, the cavity was filled with a pure water, then the ferrofluid suspension injection begins. The flow boundary condition of the inlet is defined by the existing velocity, $u_{x,in}$. Above and below inlet, the no-flow boundary condition, $\frac{\partial p}{\partial x} = 0$, is assumed, as circulation may exist close to the inlet. However, far away from the inlet region, the circulation is expected to vanish, and a constant pressure boundary condition, p_0, is assumed. In Case 2, the walls are assumed adiabatic ($\frac{\partial C}{\partial x} = 0$, $\frac{\partial C}{\partial y} = 0$), ($\frac{\partial T}{\partial x} = 0$, $\frac{\partial T}{\partial y} = 0$) as indicated in Figure 1b. The constant pressure, p_0, is assumed to represent the flow boundary condition. The inlet is kept at high temperature, T_h, high concentration, C_h, velocity $u_{x,in}$. Above and below inlet are kept at low temperature T_c, concentration, C_c, and no-flow ($\frac{\partial p}{\partial x} = 0$) boundary condition.

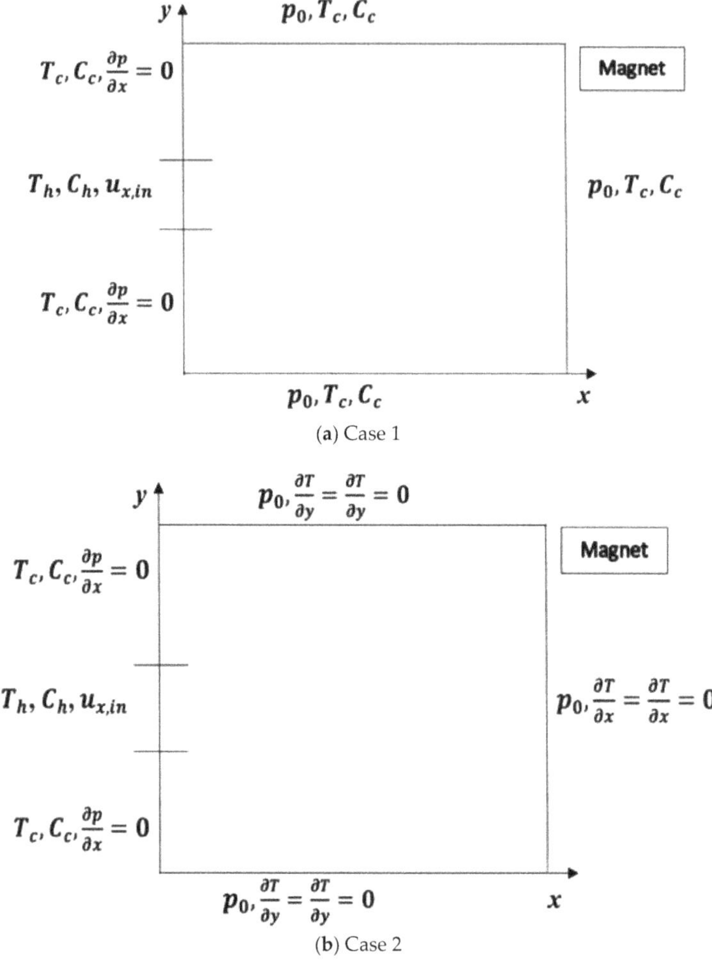

Figure 1. Schematic diagram of the 2D square porous cavity—(**a**) Case 1: cavity walls are kept at a low temperature/concentration except the inlet at center of the left wall is kept at a higher temperature/concentration (Dirichlet boundary condition) and inlet velocity (Neumann boundary condition); (**b**) Case 2: the cavity walls are adiabatic and isothermal with constant pressure (Dirichlet boundary condition), except that the inlet at center of the left wall is kept at a high temperature/concentration and the upper/lower of left wall is kept at a low temperature/concentration with no-flow Neumann boundary condition. The z-axis of the local magnet coordinate system is set perpendicular to the poles, the x-y plane.

3. Mathematical Modeling

The external permanent magnetic field produces attractive forces on the fluid. The external magnetic force with a variable concentration, and hence density, acts as a body force per unit volume, F_{mag} [4,24]:

$$F_{mag} = -\nabla \left[\mu_0 \int_0^H \left(\frac{\partial M}{\partial V} \right)_{H,T} dH \right] + \mu_0 \, M \nabla H \tag{1}$$

where μ_0 is the magnetic permeability. The specific volume, V, is defined by:

$$\partial V = \partial \left(\frac{1}{\rho} \right) = -\frac{1}{\rho^2} \partial \rho$$

Therefore, Equation (1) can be rewritten as:

$$F_{mag} = \mu_0 \rho^2 \nabla \left[\int_0^H \left(\frac{\partial M}{\partial \rho} \right)_{H,T} dH \right] + \mu_0 \, M \nabla H \tag{2}$$

Such that:

$$\left(\frac{\partial M}{\partial \rho} \right)_{H,T} = \left(\frac{\partial M}{\partial a_1} \right)_{b_1,H,T} \frac{\partial a_1}{\partial \rho} + \left(\frac{\partial M}{\partial b_1} \right)_{a_1,H,T} \frac{\partial b_1}{\partial \rho}$$

The quantities $\frac{\partial a_1}{\partial \rho}$ and $\frac{\partial b_1}{\partial \rho}$ may be approximated around the points $(a_{1,0}, b_{1,0}) = (0,0)$, which represent the non-magnetized case, as:

$$\frac{\partial a_1}{\partial \rho} \approx \frac{a_1 - a_{1,0}}{\rho - \rho_0} = \frac{a_1}{\rho - \rho_0}$$

$$\frac{\partial b_1}{\partial \rho} \approx \frac{b_1 - b_{1,0}}{\rho - \rho_0} = \frac{b_1}{\rho - \rho_0}$$

The magnetization M is a function of the magnetic field strength H:

$$M = a_1 \, tan^{-1} (b_1 H) \tag{3}$$

The two parameters a_1 and b_1 depend on the type of the ferromagnetic material and controlled both the initial susceptibility and the saturation magnetization. Therefore, Equation (1) can be written as:

$$F_{mag} = \frac{\mu_0 \rho^2}{\rho - \rho_0} M \nabla H + \frac{\mu_0 \rho^2}{\rho - \rho_0} \frac{a_1 b_1 H}{1 + b_1^2 H^2} \nabla H + \mu_0 \, M \nabla H \tag{4}$$

The resulting magnetic induction (B) is a sum of the magnetic field strength (H) and the magnetization (M), i.e., they are related by the following linear relationship:

$$B = \mu_0 \, (H + M) \tag{5}$$

Equation (5) is expressed in a scalar formulation, such that B, H, and M are scalars representing vector magnitudes. In this study, there is no free electric charge flow, so the Maxwell equations are reduced to the second Gauss law:

$$\nabla \cdot B = 0 \tag{6}$$

which means that the magnetic field induction is divergence-free. Ampere's law is reduced to the form:

$$\nabla \times H = 0 \tag{7}$$

which means that the magnetic field is vorticity-free.

Assuming that the medium is free of ferromagnetic materials and a permanent magnet is the source of the magnetic field, we can use the simple Equations (8) and (9), to calculate the components of H directly [4]. This assumption decouples the calculation of the external magnetic field from the ferrofluid distribution. At a given location the gradient of H can be calculated by simple first-order differencing in each of the coordinate directions. The z-axis of the local magnet coordinate system is set perpendicular to the poles, and the magnet is located at the upper right corner as indicated in Figure 1.

The flow simulation is in the x-y plane of the coordinate system (Equations (8) and (9) represent H_x and H_y in the x-axis and y-axis, respectively). The 2D magnetic field strength is given by [4]:

$$
\begin{aligned}
H_x &= \frac{Br}{4\pi\mu_0}\left(\ln\left[\frac{y+b+\sqrt{(y+b)^2+(x-a)^2}}{y-b+\sqrt{(y-b)^2+(x-a)^2}} \ \frac{y-b+\sqrt{(y-b)^2+(x+a)^2}}{y+b+\sqrt{(y+b)^2+(x+a)^2}} \right] \right. \\
&\quad \left. - \ln\left[\frac{y+b+\sqrt{(y+b)^2+(x-a)^2+L_p^2}}{y-b+\sqrt{(y-b)^2+(x-a)^2+L_p^2}} \ \frac{y-b+\sqrt{(y-b)^2+(x+a)^2+L_p^2}}{y+b+\sqrt{(y+b)^2+(x+a)^2+L_p^2}} \right] \right)
\end{aligned}
\tag{8}
$$

$$
\begin{aligned}
H_y &= \frac{Br}{4\pi\mu_0}\left(\ln\left[\frac{x+a+\sqrt{(y-b)^2+(x+a)^2}}{x-a+\sqrt{(y-b)^2+(x-a)^2}} \ \frac{x-a+\sqrt{(y+b)^2+(x-a)^2}}{x+a+\sqrt{(y+b)^2+(x+a)^2}} \right] \right. \\
&\quad \left. - \ln\left[\frac{x+a+\sqrt{(y-b)^2+(x+a)^2+L_p^2}}{x-a+\sqrt{(y-b)^2+(x-a)^2+L_p^2}} \ \frac{x-a+\sqrt{(y+b)^2+(x-a)^2+L_p^2}}{x+a+\sqrt{(y+b)^2+(x+a)^2+L_p^2}} \right] \right)
\end{aligned}
\tag{9}
$$

where L_p is the distance between the poles of the magnet, and B_r is the residual magnetization. If the ferromagnetic materials are heated to temperatures above the critical point (Curie temperature T_C), they become paramagnetic, whereas, below T_C, spontaneous magnetization occurs. The Curie temperature T_C for ferrofluids is high (e.g., for Fe_2O_3 is 948 K). The temperature dependence of the spontaneous magnetization at low temperatures, which is the case in this study, is given by Bloch's law [25]:

$$
M(T) = M_0\left(1 - \left(\frac{T}{T_C} \right)^{\frac{3}{2}} \right)
\tag{10}
$$

where $M_0 = M(T = 0)$ is the spontaneous magnetization at absolute zero. In fact, there are several other mechanisms of temperature influence on the resulting magnetization of the magnetic fluid. One of the most significant models is the Langevin model which takes into consideration the role of the thermal fluctuations of the magnetic moments of the particles. The Langevin model is considered one the simplest cases, which can be written as:

$$
M(T) = M_0\phi\left(\text{cth}\,\xi - \frac{1}{\xi} \right)
\tag{11}
$$

such that:

$$
\xi = \frac{\mu_0\rho H}{kT}
\tag{12}
$$

The thermal fluctuations of the magnetic moments are related to the Brownian diffusion by the second fluctuation-dissipation theorem, which will be ignored in this study for simplicity.

The magnetization increases linearly with the ferrofluid mass fraction, C [4]:

$$
M\left(C \right) = M(C = 1)\,C
\tag{13}
$$

Combining Equations (10) and (13), one may write:

$$
M(T,C) = M(T = 0, C = 1)\left(1 - \left(\frac{T}{T_C} \right)^{\frac{3}{2}} \right)C
\tag{14}
$$

It is assumed that the volumes of pure water and ferrofluid are additive; therefore, the mixture density is defined as [4,24,26]:

$$
\frac{1}{\rho} = \frac{1-C}{\rho_w} + \frac{C}{\rho_f}
\tag{15}
$$

where ρ_w is the density of the pure water component, and ρ_f is the density of ferromagnetic-particle component. The viscosity of the water–magnetic-particle mixture is calculated by the following linear relationship [4]:

$$\mu = \mu_w(1 + 1.35C) \tag{16}$$

where μ_w is the viscosity of the pure water. At a Reynolds number greater than 10, the inertial effect becomes significant. The inertial term, which is known as the Forchheimer term, can account for the nonlinear behavior of the pressure difference versus flow. Momentum conservation in porous media is represented by the extended non-Darcy's law [27]:

$$\mathbf{u} + \frac{F\sqrt{K}}{\nu}|\mathbf{u}|\mathbf{u} = -\frac{K}{\mu}\left(\nabla p - F_{mag} + g\rho_0(1 - \beta(T - T_r) - \beta^*(C - C_0))\nabla z\right) \tag{17}$$

where F is the inertia parameter, and ν is kinematic viscosity. In the above equation, the second term on the left-hand side represents the inertial term in Darcy's equation. The magnetic field effect on the fluid flow acts as a body force in the extended Darcy's law. Although this is an unsteady problem, however, the momentum (Darcy's law) has no time derivative term which is experimentally based. Actual fluid velocity varies throughout the pore space, due to the connectivity and geometric complexity of that space. This variable velocity can be characterized by its mean or average value. The convective flows are caused by the temperature difference in terms of the Boussinesq approximation, in which the density of the body force term is calculated by:

$$\rho = \rho_0(1 - \beta(T - T_0) - \beta^*(C - C_0)), \ \beta = -\frac{1}{\rho_0}\frac{\partial\rho}{\partial T}, \ \beta^* = -\frac{1}{\rho_0}\frac{\partial\rho}{\partial C}$$

The mass conservation equation is represented by:

$$\nabla\cdot\mathbf{u} = 0 \tag{18}$$

The energy conservation equation in porous media can be represented as [25]:

$$\frac{\partial\left[(1 - \varphi)\rho_s c_{p,s} + \varphi\rho c_p\right]T}{\partial t} + \rho\mathbf{u}\cdot\nabla T = \nabla\cdot[(1 - \varphi)h_s + \varphi h]\nabla T \tag{19}$$

The concentration (solute transport) equation may be represented as [26]:

$$\varphi\frac{\partial C}{\partial t} + \nabla\cdot(\mathbf{u}C - D\nabla C) = 0 \tag{20}$$

In the above equations, C is the concentration, C_0 is the initial concentration, c_p is the heat capacity, D is the diffusion-dispersion tensor, g is the gravitation acceleration, h is the thermal conductivity of the fluid, h_s is the thermal conductivity of the solid, K is the permeability, T is the temperature, T_0 is the initial temperature, p is the fluid pressure, $\mathbf{u}$ is the fluid velocity vector, β is the thermal expansion coefficient, β^* is the solute expansion coefficient, φ represents the porosity of the porous media, ρ is the density of the fluid mixture, ρ_s is the density of the solid phase, and μ is the fluid viscosity. Substituting Equation (16) into Equation (18), one may obtain:

$$\nabla\cdot\mathbf{u} = -\nabla\cdot\frac{K}{\mu\left(1 + \frac{F\sqrt{K}}{\nu}|\mathbf{u}|\right)}\left(\nabla p - F_{mag} + g\rho_0(1 - \beta(T - T_r) - \beta^*(C - C_0))\right) = 0 \tag{21}$$

Calculating the heat transfer rate at boundaries/surfaces within fluids is of great interest in industrial and technological applications. The Nusselt number (Nu), a dimensionless number, is the ratio of convective to conductive heat transfer across the boundary. Unlike in other studies in the literature, the local Nusselt number was calculated on the wall instead of the average Nusselt number, because it gives the direct heat transfer rate between the wall and the adjacent fluid. The local Nusselt number measures the competition between convection and conduction heat flows and is defined by:

$$Nu_x = \frac{q_x L}{(T_{in} - T_0)k_T} = -\frac{L}{(T_{in} - T_0)} \frac{\partial T}{\partial x}\bigg|_{x=0} \tag{22}$$

$$Nu_y = \frac{q_y L}{(T_{in} - T_0)k_T} = -\frac{L}{(T_{in} - T_0)} \frac{\partial T}{\partial y}\bigg|_{y=0} \tag{23}$$

where L is the characteristic length; q_x, q_y are the local surface heat fluxes and defined as:

$$q_x = -k_T \frac{\partial T}{\partial x}\bigg|_{x=0} \tag{24}$$

$$q_y = -k_T \frac{\partial T}{\partial y}\bigg|_{y=0} \tag{25}$$

where k_T is the thermal conductivity of the fluid. Similarly, the local surface mass fluxes, j_x, j_y maybe defined as:

$$j_x = -D \frac{\partial C}{\partial x}\bigg|_{x=0} \tag{26}$$

$$j_y = -D \frac{\partial C}{\partial y}\bigg|_{y=0} \tag{27}$$

Therefore, the local Sherwood number which represents the ratio of the convective mass transfer to the rate of diffusive mass transport and given by:

$$Sh_x = \frac{j_x L}{(C_w - C_0)D} = -\frac{L}{(C_w - C_0)} \frac{\partial C}{\partial x}\bigg|_{x=0} \tag{28}$$

$$Sh_y = \frac{j_y L}{(C_w - C_0)D} = -\frac{L}{(C_w - C_0)} \frac{\partial C}{\partial y}\bigg|_{y=0} \tag{29}$$

Initially, the cavity is filled with pure water and, then, the ferrofluid suspension injection begins. In the following, two cases based on the types of boundary condition are described:

Case 1. All the walls of the square cavity are kept cold at temperature T_c and low concentration C_c except that the center of the left wall is kept at a high temperature T_h and a high concentration C_h, as shown in Figure 1a. The boundary conditions are described mathematically as follows:

at $x = 0, 0 \leq y \leq L, T = T_c, C = C_c, \frac{\partial p}{\partial x} = 0$,
at inlet $T_{in} = T_h, C_{in} = C_h, u = u_{x,in}$;
at $0 \leq x \leq L, y = 0, T = T_c, C = C_c, p = p_0$;
at $x = 1, 0 \leq y \leq L, T = T_c, C = C_c, p = p_0$;
at $0 \leq x \leq L, y = L, T = T_c, C = C_c, p = p_0$;

where *in* refers to the injection location (inlet), and the velocity is taken to be zero on all cells (no-flow boundary conditions, i.e., impermeable wall) except for the central cells. The fluid exits from constant pressure boundary, p_0, which is a kind of exit boundary, as it is smaller than the entire pressure.

Case 2. All the walls of the square cavity are adiabatic and isothermal with constant Dirichlet pressure, except that the center of the left wall is kept hot at temperature T_h and high concentration C_h. The upper and lower left walls are kept at a cold temperature and a low concentration with no-flow Neumann boundary condition (impermeable wall), as shown in Figure 1b. The boundary conditions are described mathematically as follows.

at $x = 0, 0 \leq y \leq L, T = T_c, C = C_c, \frac{\partial p}{\partial x} = 0,$

at inlet $T_{in} = T_h, C_{in} = C_h, u = u_{x,in}.$

at $0 \leq x \leq L, y = 0, \frac{\partial T}{\partial y} = 0, \frac{\partial C}{\partial y} = 0, p = p_0.$

at $x = L, 0 \leq y \leq L, \frac{\partial T}{\partial x} = 0, \frac{\partial C}{\partial x} = 0, p = p_0.$

at $0 \leq x \leq L, y = L, \frac{\partial T}{\partial y} = 0, \frac{\partial C}{\partial y} = 0, p = p_0.$

4. Numerical Method

The governing equations were discretized numerically using the CCFD method for the spatial discretization. The CCFD method is considered one of the most commonly used finite-difference methods for flow and transport in porous media, because it is locally conservative. In the CCFD method, one uses velocity on edges of the cell, and pressure on the center of the cell (cell-centered). The permeability is calculated using the harmonic averaging on the center as well.

The backward Euler time discretization was used for the time derivative of equations of temperature and concentrations, with a time-splitting implicit technique [17,20]. The outer time discretization, which had a relatively large time step, was used for pressure. The subscript $k + 1$ represents the current time step, while the subscript k represents the previous time step. The temperature and concentration had the same level of time discretization. The following procedures have been considered,

- The total time interval, $[0, T]$, is divided into a number of N_p time steps, namely, $0 = t_0 < t_1 < \ldots < t_{N_p=T}$, with a time-step of length $\Delta t^k = t^{k+1} - t^k$.

- For temperature and concentration, each interval, $\left(t^k, t^{k+1} \right)$, was divided into a number of $N_{p,C}$ subintervals, i.e., $\left(t^k, t^{k+1} \right) = \cup_{l=0}^{N_{p,c}-1} \left(t^{k,m}, t^{k,m+1} \right).$

- The Courant–Friedrichs–Lewy stability condition (CFL < 1) is used to accomplish the time step-size adaptation.

- The time-step size of the pressure was taken to be larger than the time-step sizes of temperature and concentration.

The algorithm can be stated as:

- Calculating the pressure implicitly by coupling the continuity and momentum equations.
- Compute the velocity explicitly.
- Solve energy and concentration equations implicitly.
- Update porosity, permeability, and density.

Now, let us express the pressure equation as:

$$-\nabla \cdot \frac{K}{\mu \left(1 + \frac{F\sqrt{K}}{\nu}|\mathbf{u}^k| \right)} \left(\nabla \cdot p^{k+1} - F_{mag}^k + g\rho_0 \left(1 - \beta \left(T^k - T_r \right) - \beta^* \left(C^k - C_0 \right) \right) \nabla z \right) = q^{k+1} \tag{30}$$

The energy equation is computed implicitly as:

$$\frac{\left[(1-\varphi)\rho_s c_{p,s} + \varphi\rho\left(T^k, C^k \right) c_p \right] T^{k+1,m+1} - \left[(1-\varphi)\rho_s c_{p,s} + \varphi\rho\left(T^k \right) c_p \right] T^{k+1,m}}{\Delta t} + \rho \mathbf{u}^{k+1} \cdot \nabla T^{k+1,m+1} = \nabla \cdot \left[(1 - \varphi)h_s + \varphi h \right] \nabla T^{k+1,m+1} \tag{31}$$

Similarly, the concentration equation is computed implicitly as:

$$\varphi \frac{C^{k+1,m+1} - C^{k+1,m}}{\Delta t^m} + \nabla \cdot \left[\mathbf{u}^{k+1} C^{k+1,m+1} - D\nabla C^{k+1,m+1} \right] = 0 \tag{32}$$

In order to control the time-step size, the CFL condition (CFL < 1) is used. The CFL condition of the temperature and concentration equations is expressed as:

$$\text{CFL}_x = \frac{u_x \Delta t^{k,m}}{x}, \ \text{CFL}_y = \frac{u_y \Delta t^{k,m}}{\Delta y} \tag{33}$$

The numerical stability procedures can be summarized as follows:

- The pressure time-step, $\Delta t^{k,0} = \Delta t^k$ is taken as an initial time step for both temperature and concentration equations.
- After that, the conditions $\text{CFL}_x > 1$ and $\text{CFL}_y > 1$ are examined. If one of them is satisfied, the temperature/concentration time-step is divided into smaller steps.
- Therefore, the conditions CFL_x and CFL_y are recalculated based on the new time steps and so on, until satisfying the conditions $\text{CFL}_x < 1$ and $\text{CFL}_y < 1$.

5. Results and Discussion

To obtain physical insights, the system of the governing partial differential equations was solved numerically using the above numerical method. Two different cases were considered based on different boundary conditions, as in Figure 1a,b. In Case 1, all boundaries were of a Dirichlet type except that the west boundary was a no-flow boundary condition, which was a Neumann type. The physical and computational parameters are presented in Table 1. A value for the outer time loop k was selected, and the step size Δt^k was determined; then, the time-step size Δt^m was calculated based on the CFL conditions. In our calculation the velocity inlet was taken as 0.4756×10^{-8} m/s and the corresponding Reynolds number is 0.00143, which is suitable for flow in porous media.

Table 1. Physical parameter values.

Parameter	Description	Value	Units
a_1	Constant	10^4–10^5	m A^{-1}
b_1	Constant	10^{-6}–10^{-5}	m A^{-1}
a, b	Half of width/height of the magnet	0.02	m
Br	Residual magnetization	0:0.2	Tesla
C_0	Initial concentration	0	-
C_{in}	Inlet concentration	1	-
c_p	Heat capacity	800	J/Kg K
D	Diffusion coefficient	5×10^{-3}	m^2/S
h_s	Thermal conductivity of the solid	0.718	W/(m/K)
h	Thermal conductivity of the ferrofluid	0.6	W/(m/K)
g	Gravity acceleration	9.81	m/s^2
K	Permeability	100	md
L_p	Distance between poles	2.4	m
L_{in}	Inlet width	0.3	m
L	Cavity side length	5	m
md	Millidarcy	9.86923×10^{-16}	m^2
p_0	Initial pressure	1×10^6	N/m^2
T_0	Initial temperature	300	K
T_{in}	Inlet temperature	360	K
$u_{x,in}$	Inlet velocity	0.4756×10^{-8}	m/s
β	Thermal expansion coefficient	0.005	K^{-1}
β^*	Solute expansion coefficient	0.001	K^{-1}
φ	Porosity	0.3	-
μ	Water viscosity	0.001	Pa.s
μ_0	Magnetic permeability	1×10^{-7}	N·A^{-2}
ρ_s	Solid media density	2500	kg/m^3
ρ_w	Pure water density	1000	kg/m^3
ρ_f	Particles density	8933	kg/m^3

Some error estimates are listed in Table 2 to examine the accuracy of the proposed scheme. The errors were calculated for temperature and concentration at various values of the time-step iteration, k. The results in Table 2 indicate that the error decreases as the number of time steps increases. In addition, the temperature error is very small and approaches zero; however, the concentration error needs bigger time step numbers to reach the desired approximation. Moreover, the calculated time step size (Δt^m) is plotted against its number m in Figure 2. Moreover, concentration profiles for different spatial grid size n_x and n_y of Case 1 are plotted in Figure 3 to show the grid independence test. The spatial discretization, $n_x \times n_y$, was chosen, respectively, as 50×50, 100×100, and 150×150. It is clear from Figure 3 that the differences between the three curves are very small, therefore, we choose the course grid, 50×50, for the computations in this study.

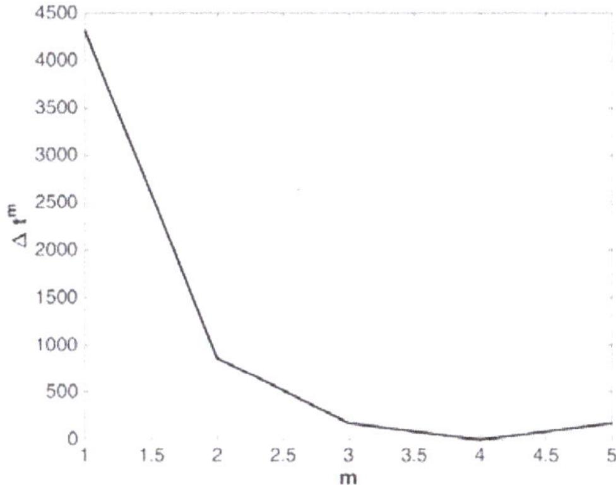

Figure 2. Variation of the adaptive time step size (Δt^m) against its number m of Case 1.

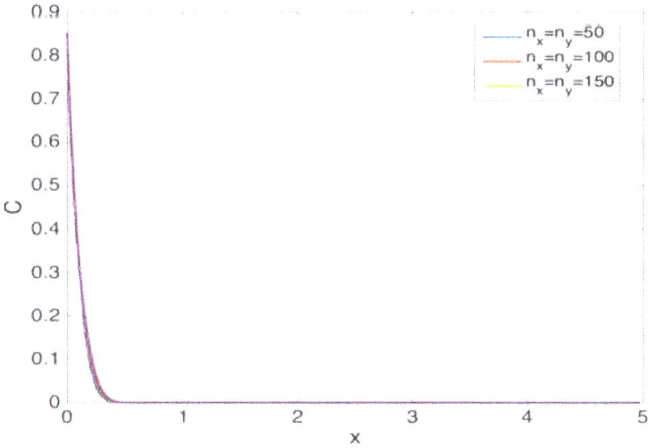

Figure 3. Concentration profiles for different spatial grid size n_x and n_y of Case 1.

Table 2. Error estimates for various values of k, Case 2.

k	$\|T^{k+1,l+1} - T^{k+1,l}\|$	$C^{k+1,m+1} - C^{k+1,m}$
50	5.0317×10^{-11}	0.0198
100	6.9070×10^{-11}	0.0099
300	3.0481×10^{-12}	0.0033

5.1. Results for Case 1

Figure 4a,b illustrates the temperature difference, $\Delta T = T - T_0$, contours of Case 1 without the magnetic-field effect (Br = 0.0) and with the magnetic-field effect (Br = 1.2), respectively. It seems that the magnetic field slightly enhances the temperature. The differences between of the two cases are similar, and it seems to be hard to notice the small differences (around 1 K); however, the differences are shown clearly in Figure 5, which shows the temperature profiles against the x-axis at $y = 1$ m, for various values of Br of Case 1. In Figure 6, the concentration profiles are plotted against the x-axis at $y = 1$ m, for various values of Br of Case 1. It is clear from Figure 6 that as the parameter Br increases, the ferrofluid concentration increases.

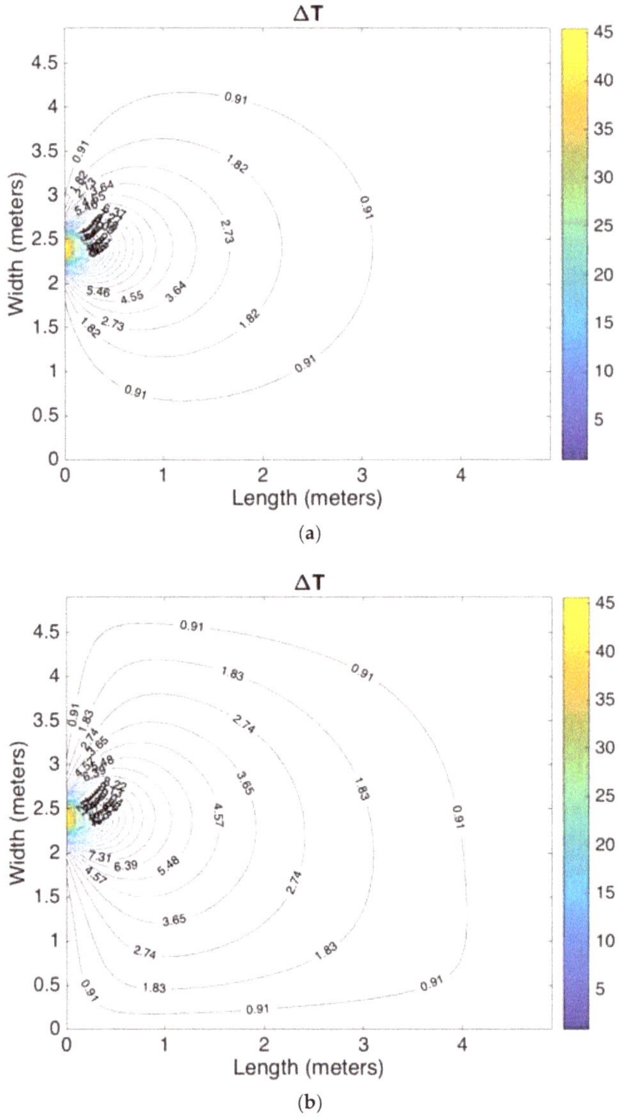

Figure 4. Temperature difference, $\Delta T = T - T_0$, contours of Case 1 with (**a**) Br = 0.0 and (**b**) Br = 1.2.

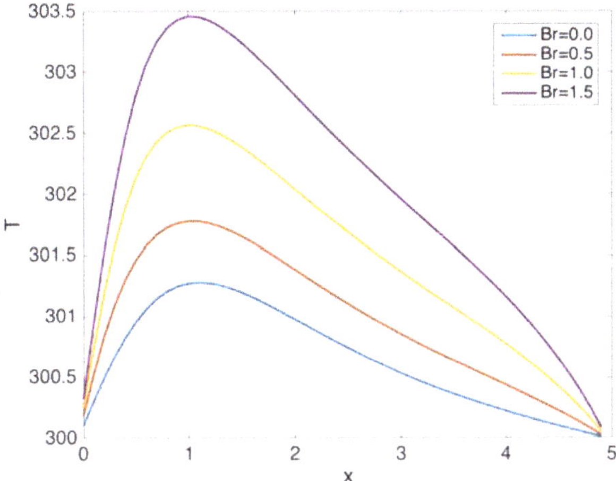

Figure 5. Temperature profiles against x-axis at $y = 1$ m, for various values of Br of Case 1.

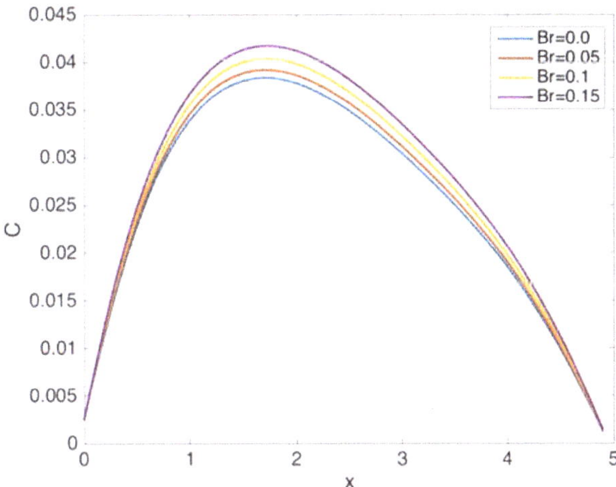

Figure 6. Concentration profiles against the x-axis at $y = 1$ m, for various values of Br of Case 1.

The pressure and velocity magnitude ($u_w = \sqrt{u_x^2 + u_y^2}$) profiles are plotted in Figure 7a,b, respectively, against the x-axis at $y = 1$ m, for various values of Br of Case 1. These figures show that, as Br increases, pressure and velocity increase. The maximum differences are located in the inlet vicinity and gradually decrease far away from it.

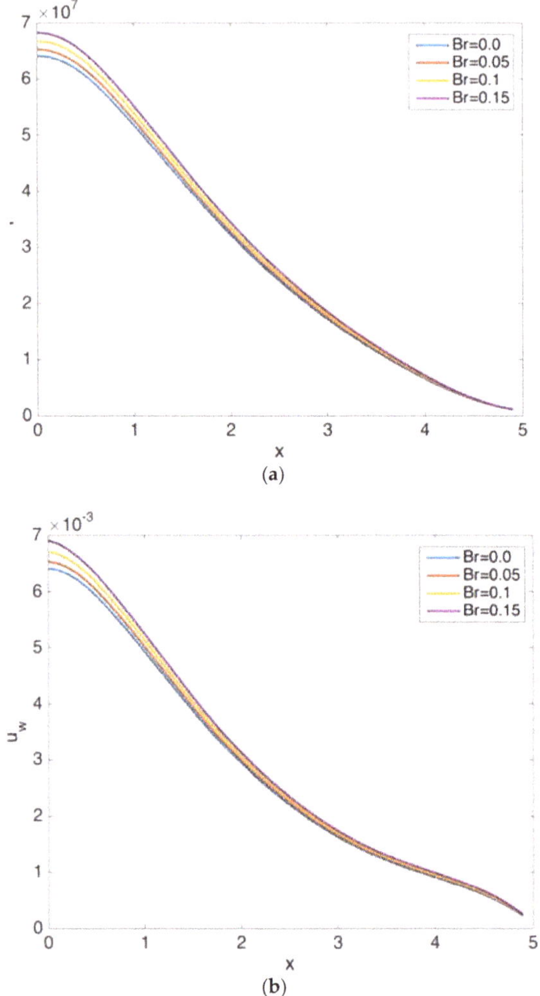

Figure 7. (**a**) Pressure and (**b**) average velocity profiles against x-axis at $y = 1$ m, for various values of Br of Case 1.

In Figure 8, the local Nusselt numbers are plotted along the left wall, right wall, upper wall, and lower wall of the porous cavity, for various values of the parameter Br of Case 1. The magnetic field has no effect on the left wall because the advection effect is much larger than the magnetic effect. The variation in magnetic-field strength has a small effect on the other walls; however, the differences are clear. The higher local Nusselt number is located in the center of the left wall, which indicates that the heat transfer rate is high in that region. Moreover, the heat transfer rate on the right wall has high values around the center of the wall and decreases gradually in the edge closure. The negative sign indicates that the heat transfer direction is from the fluid to the solid wall. The behavior of the heat transfer rate on the upper wall is similar to that on the right wall, but the maximum heat transfer rate is shifted to the bottom edge. Finally, the local Nusselt number on the lower wall has maximum values close to the left portion and then decreases gradually near the edges.

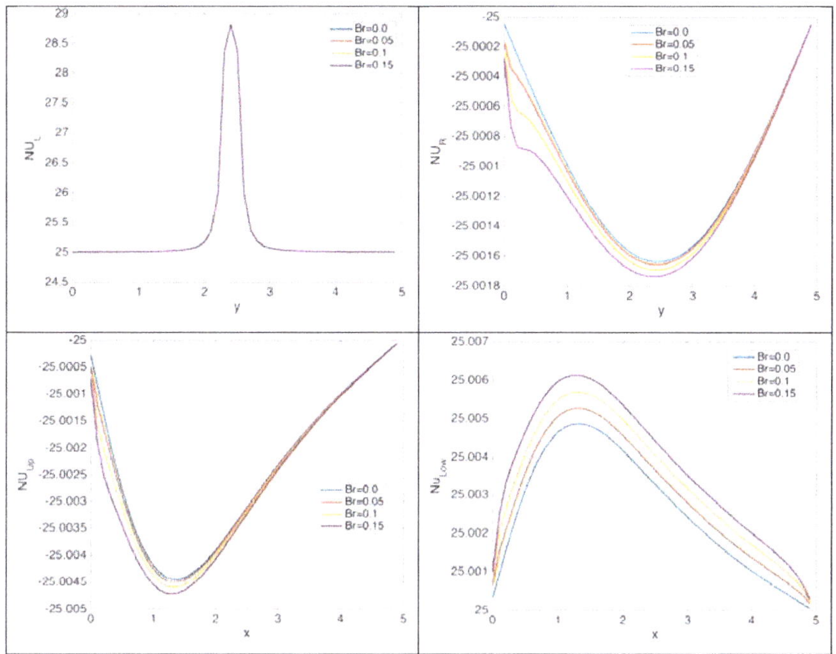

Figure 8. Local Nusselt number distributions along the left wall, right wall, upper wall, and lower wall for various values of the parameter Br of Case 1.

Figure 9 illustrates the local Sherwood number profiles along the left-, right-, upper-, and lower-walls. It can be seen from this figure that the local Sherwood number profiles along the left and right walls are symmetric around the center of y-axis and they decrease gradually far away from the center toward the upper to the lower corners. The mass transfer rate on the left-wall is high close to the inlet region and decreases from bottom to top. The negative sign indicates that mass transfer direction is from the fluid to the solid wall. It also may be noticed that the mass transfer rate on the right-wall is very small. The local Sherwood number on the upper- and left-walls has a similar behavior and same order of magnitude with different signs. The local mass transfer rate has maximum values at $x \approx 0.1$.

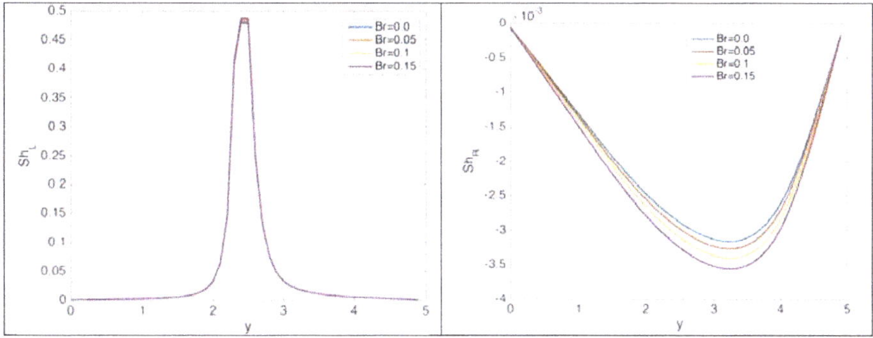

Figure 9. *Cont.*

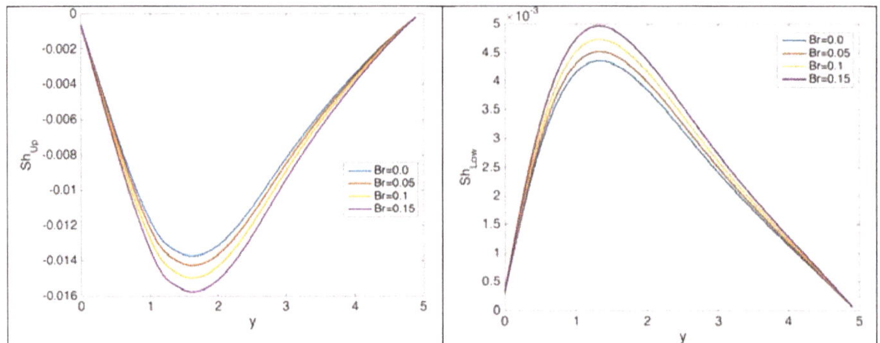

Figure 9. Local Sherwood number profiles along the **left-**, **right-**, **upper-**, and **lower-**walls for various values of the parameter Br; Case 1.

In Figure 10, the transient temperature and concentration profiles are plotted against the x-axis at $y = 4$, at different times of the ferrofluid injections. It is interesting to observe from Figure 10a that the temperature reaches the steady-state after around 2 days of injection, while the concentration reaches its steady-state after 2000 days of injection as indicated in Figure 10b.

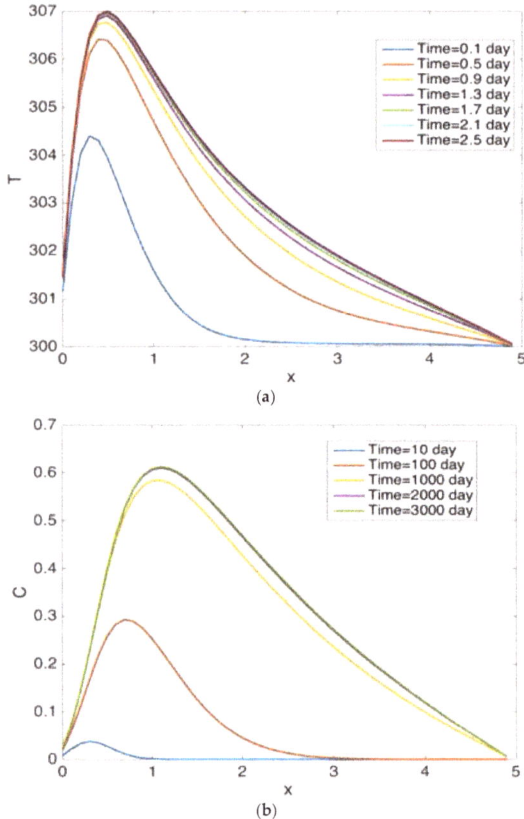

Figure 10. Transient (**a**) temperature, and (**b**) concentration profiles against x-axis at $y = 4$, of Case 1.

5.2. Results for Case 2

In Figure 11, the temperature profiles are plotted against the x-axis, for various values of Br of Case 2. This figure shows that the existence of the external magnetic field increases the temperature. The lowest temperature values are in the vicinity of the left wall, while the opposite happens at $y = 2.5$ m. The magnetic field has a small effect at the centerline as the advection dominates, whereas it has a clear effect far away from the high advection region (e.g., $y = 1$ m). As the parameter Br increases, the temperature increases, which may be interpreted as the magnetic field enhancing the kinetic energy, which, in turn, enhances the heat energy.

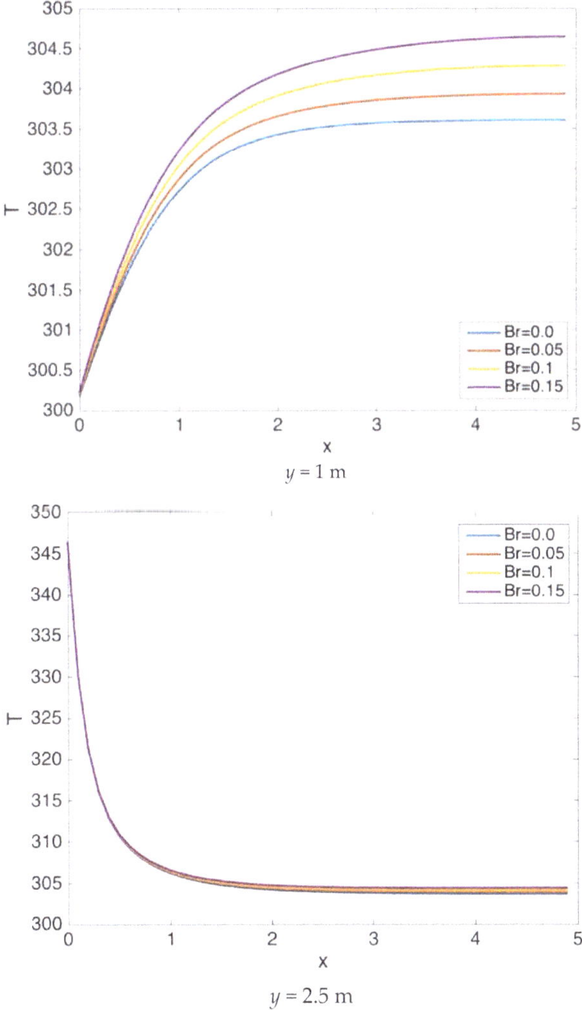

Figure 11. Temperature profiles against x-axis at $y = 1$ m, 2.5 m for various values of Br of Case 2.

The concentration profiles are plotted in Figure 12 against the x-axis at $y = 4$ m, 2.5 m, for various values of Br of Case 2. At the centerline ($y = 2.5$ m), the concentration has maximum values close to the inlet and decreases gradually far away from the inlet, whereas then opposite is true at $y = 4$ m. The concentration has minimum values close to the left wall and increases gradually far away from it.

As the magnetic field pulls the particles toward it based on its strength, the parameter Br increases the concentration of the ferrofluid.

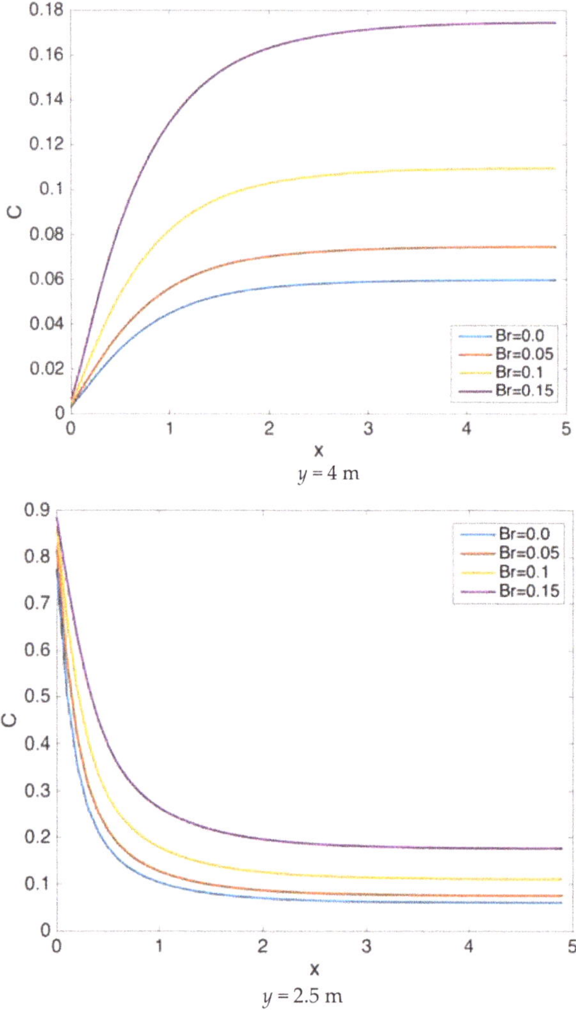

Figure 12. Concentration profiles against x-axis at $y = 4$ m, 2.5 m, for various values of Br of Case 2.

The temperature distribution of Case 2 is shown in Figure 13a. It can be seen from this figure that the temperature is high around the inlet location. Then it decreases gradually far away from the inlet because the other walls are adiabatic with a constant pressure. This may be interpreted as the heat transfer mode changes from the mixed convection region close to inlet to the purely conduction region far away. Figure 13b shows the concentration distribution of Case 2. The behavior of the concentration is similar to the temperature distribution which can be interpreted similarly but with respect to mass transfer and solute convection.

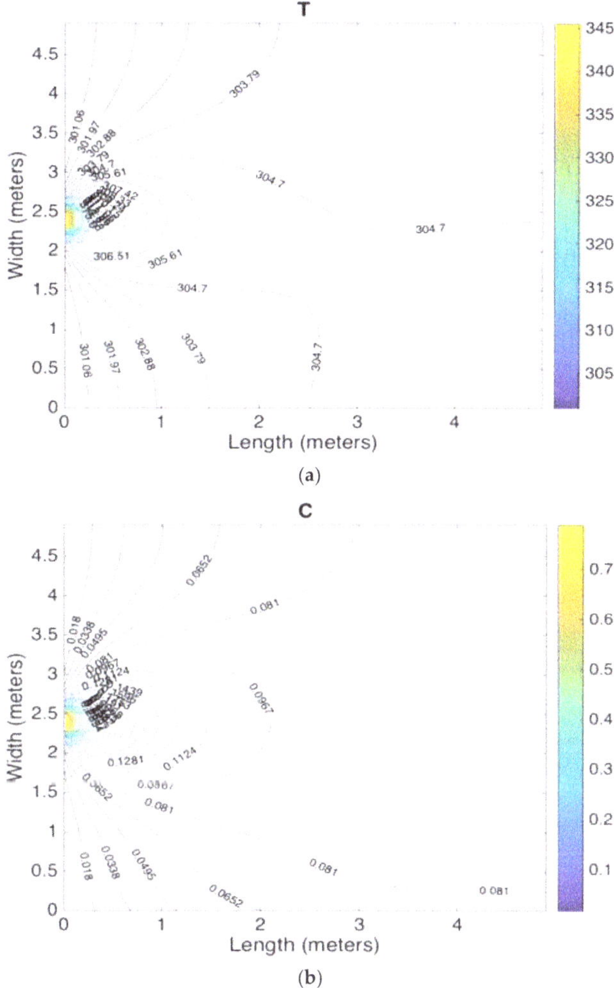

Figure 13. Contours of (a) temperature, and (b) concentration, of Case 2.

6. Conclusions

In this study, the effect of a magnetic field on ferromagnetic fluid flow and heat transfer in a square porous cavity was investigated. A mathematical model which consists of a set of governing equations and set of axillary algebraic equations, along with initial and boundary conditions, was developed. Two different cases of boundary conditions were considered. The governing equations were solved numerically using the multiscale time-splitting implicit method. An adaptive algorithm was built based on calculation of the Courant–Friedrichs–Lewy stability condition number. Error estimates for temperature and concentration at various values of iteration number were presented. Also, the grid independence of spatial discretization was examined. Numerical experiments were carried out to simulate some physical scenarios. It was found that the magnetic field increases the temperature and enhances the heat transfer rate. Also, the distribution of the particle concentration is related to the magnetic field strength. The pressure and velocity increase as the magnetic field strength increases. Close to the inlet the magnetic field had no effect on the local heat transfer rate because the advection force is much larger than the magnetic force. Moreover, the magnetic field strength had a small effect on the local heat transfer rate at the other cavity walls. The highest local Nusselt number occurs in the center of the left wall. Similar conclusions can be drawn for the mass transfer rate behavior under the effect of the magnetic field.

Author Contributions: Conceptualization, M.F.E., U.K. and A.B.; methodology, M.F.E. and U.K.; validation, M.F.E. and A.B.; analysis, M.F.E.; writing—original draft preparation, M.F.E.; writing—review and editing, M.F.E.; project administration, U.K.; funding acquisition, U.K.

Funding: This research was funded by the International Scientific Partnership Program (ISPP) at King Saud University, grant number ISPP#47.

Acknowledgments: The authors extend their appreciation to the International Scientific Partnership Program (ISPP) at King Saud University for funding this research work through ISPP#47. The authors also thank the Deanship of Scientific Research and RSSU at King Saud University for their technical support.

Conflicts of Interest: The authors declare no conflict of interest.

Nomenclature

A	[m]	half of width of the magnet,
a_1, b_1	[A m^{-1}]	constants depend on the type of the ferromagnetic material,
b	[m]	half of height of the magnet,
B_r	[Tesla]	residual magnetization,
C	[m m^{-3}]	ferrofluid concentration,
C_0	[m m^{-3}]	ferrofluid initial concentration,
c_p	[J Kg^{-1} K^{-1}]	heat capacity,
D	[m^2 s^{-1}]	diffusion coefficient,
F_{mag}	[N]	external magnetic force,
g	[m s^{-2}]	gravitation acceleration,
h	[J K^{-1}m^{-1}s^{-1}]	thermal conductivity,
H	[A m^{-1}]	magnetic field strength,
K	[m^2]	permeability of the porous medium,
L	[m]	distance between the poles of the magnet,
Nu	[-]	Nusselt number,
T	[K]	temperature,
T_r	[K]	reference temperature,
p	[Pa]	fluid pressure,
q	[kg s^{-1}]	external mass flow rate,
Q_c	[m^{3}s^{-1}]	rate of change of particle volume of a source/sink term,
Q_t	[m^{3}s^{-1}]	heat source term,
u	[m s^{-1}]	fluid velocity vector,
x, y	[m]	Cartesian coordinates,
β	[K^{-1}]	thermal expansion coefficient,
β^*		solutal expansion coefficient,
Δt^k	[s]	time step for the loop k,
Δt^l	[s]	time step for the loop l,
Δt^m	[s]	time step for the loop m,
φ	[-]	porosity of porous media,
ρ	[Kg m^{-3}]	density of fluid mixture,
ρ_w	[Kg m^{-3}]	density of pure water component,
ρ_f	[Kg m^{-3}]	density of ferro-particles component,
ρ_s	[Kg m^{-3}]	density of solid phase,
μ	[m s^{-2}]	viscosity of water-magnetic-particles mixture,
μ_w	[m s^{-2}]	viscosity of water,
μ_0	[N A^{-2}]	magnetic permeability.

References

1. Scherer, C.; Figueiredo Neto, A.M. Ferrofluids: Properties and applications. *Braz. J. Phys.* **2005**, *35*, 718–727. [CrossRef]
2. Borglin, S.E.; Moridis, G.J.; Oldenburg, C.M. Experimental studies of the flow of ferrofluid in porous media. *Transp. Porous Media* **2000**, *41*, 61–80. [CrossRef]
3. Huh, C.; Nizamidin, N.; Pope, G.A.; Milner, T.E.; Wang, B. Hydrophobic Paramagnetic Nanoparticles as Intelligent Crude Oil Tracers. U.S. Patent Application No. 14/765,426, 15 January 2014.
4. Oldenburg, C.M.; Borglin, S.E.; Moridis, G.J. Numerical simulation of ferrofluid flow for subsurface environmental engineering applications. *Transp. Porous Media* **2000**, *38*, 319–344. [CrossRef]
5. Zahn, M. Magnetic fluid and nanoparticle applications to nanotechnology. *J. Nanopart. Res.* **2001**, *3*, 73–78. [CrossRef]
6. Chegenizadeh, N.; Saeedi, A.; Quan, X. Application of nanotechnology for enhancing oil recovery—A review. *Petroleum* **2016**, *2*, 324–333.
7. El-Amin, M.F.; Salama, A.; Sun, S. Numerical and dimensional analysis of nanoparticles transport with two-phase flow in porous media. *J. Pet. Sci. Eng.* **2015**, *128*, 53–64. [CrossRef]
8. El-Amin, M.F.; Salama, A.; Sun, S. Modeling and simulation of nanoparticles transport in a two-phase flow in porous media. In Proceedings of the SPE International Oilfield Nanotechnology Conference, Noordwijk, The Netherlands, 12–14 June 2012; SPE-154972-MS.
9. Suleimanov, B.A.; Ismailov, F.S.; Veliyev, E.F. Nanofluid for enhanced oil recovery. *J. Pet. Sci. Eng.* **2011**, *78*, 431–437. [CrossRef]
10. Ryoo, S.; Rahmani, A.; Yoon, K.Y.; Prodanovi, M.; Kots-mar, C.; Milner, T.E.; Johnston, K.P.; Bryant, S.L.; Huh, C. Theoretical and experimental investigation of the motion of multiphase fluids containing paramagnetic nanoparticles in porous media. *J. Pet. Sci. Eng.* **2012**, *81*, 129–144. [CrossRef]
11. Sheremet, M.A.; Pop, I. Mixed convection in a lid-driven square cavity filled by a nanofluid: Buongiorno's mathematical model. *App. Math. Comput.* **2015**, *266*, 792–808. [CrossRef]
12. Ghalambaz, M.M.; Sheremet, F.; Mikhail, A.; Pop, I. Triple-diffusive natural convection in a square porous cavity. *Transp. Porous Media* **2016**, *111*, 59–79. [CrossRef]
13. Carvalho, P.H.S.; de Lemos, M.J.S. Passive laminar heat transfer across porous cavities using thermal non-equilibrium model. *Numer. Heat Transf.* **2014**, *66*, 1173–1194. [CrossRef]
14. Javed, T.; Mehmood, Z.; Abbas, Z. Natural convection in square cavity filled with ferrofluid saturated porous medium in the presence of uniform magnetic field. *Phys. B Condens. Matter* **2017**, *506*, 122–132. [CrossRef]
15. Kleinstreuer, C.; Xu, Z. Mathematical modeling and computer simulations of nanofluid flow with applications to cooling and lubrication. *Fluids* **2016**, *1*, 16. [CrossRef]
16. Bhallamudi, S.M.; Panday, S.; Huyakorn, P.S. Sub-timing in fluid flow and transport simulations. *Adv. Water Res.* **2003**, *26*, 477–489.
17. Martinez, V. A numerical technique for applying time splitting methods in shallow water equations. *Comput. Fluids* **2018**, *169*, 285–295. [CrossRef]
18. Zhang, T.; Qian, Y. The time viscosity-splitting method for the Boussinesq problem. *J. Math. Anal. Appl.* **2017**, *445*, 186–211. [CrossRef]
19. Caliari, M.; Zuccher, S. Reliability of the time-splitting Fourier method for singular-solutions in quantum fluids. *Comput. Phys. Commun.* **2018**, *222*, 46–58. [CrossRef]
20. El-Amin, M.F.; Kou, J.; Sun, S. Discrete-fracture-model of multi-scale time-splitting two-phase flow including nanoparticles transport in fractured porous media. *J. Comput. Appl. Math.* **2018**, *333*, 327–349. [CrossRef]
21. El-Amin, M.F.; Kou, J.; Salama, A.; Sun, S. An iterative implicit scheme for nanoparticles transport with two-phase flow in porous media. *Procedia Comput. Sci.* **2016**, *80*, 1344–1353. [CrossRef]
22. El-Amin, M.F.; Kou, J.; Sun, S.; Salama, A. Adaptive time-splitting scheme for two-phase flow in heterogeneous porous media. *Adv. Geo-Energy Res.* **2017**, *1*, 182–189. [CrossRef]
23. El-Amin, M.F.; Kou, J.; Salama, A.; Sun, S. Multiscale adaptive time-splitting technique for nonisothermal two-phase flow and nanoparticles transport in heterogeneous porous media. In Proceedings of the SPE Reservoir Characterisation and Simulation Conference and Exhibition, Abu Dhabi, UAE, 8–10 May 2017; SPE-186047-MS.
24. Rosensweig, R.E. *Ferrohydrodynamics*; Cambridge University Press: Cambridge, UK, 1985.

25. Ashcroft, N.W.; Mermin, N.D. *Solid State Physics*; Holt, Rinehart and Winston: New York, NY, USA, 1976; ISBN 0-03-083993-9.
26. Herbert, A.W.; Jackson, C.P.; Lever, D.A. Coupled groundwater flow and solute transport with fluid density strongly dependent upon concentration. *Water Resour. Res.* **1988**, *24*, 1781–1795. [CrossRef]
27. El-Amin, M.F. Double dispersion effects on natural convection heat and mass transfer in non-Darcy porous medium. *Appl. Math Comput.* **2004**, *156*, 1–17. [CrossRef]

Article

Experimental Study of Particle Deposition on Surface at Different Mainstream Velocity and Temperature

Fei Zhang **, Zhenxia Liu *, Zhengang Liu and Yanan Liu**

School of Power and Energy, Northwestern Polytechnical University, Xi'an 710129, China;
zhangfei089@mail.nwpu.edu.cn (F.Z.); zgliu@nwpu.edu.cn (Z.L.); liuyanan1993@mail.nwpu.edu.cn (Y.L.)
* Correspondence: zxliu@mail.nwpu.edu.cn; Tel.: +86-029-8843-1118

Received: 23 January 2019; Accepted: 20 February 2019; Published: 24 February 2019

Abstract: The effect of mainstream velocity and mainstream temperature on the behavior of deposition on a flat plate surface has been investigated experimentally. Molten wax particles were injected to generate particle deposition in a two-phase flow wind tunnel. Tests indicated that deposition occurs mainly at the leading edge and the middle and backward portions of the windward side. The mass of deposition at the leading edge was far more than that on the windward and lee sides. For the windward and lee sides, deposition mass increased as the mainstream velocity was increased for a given particle concentration. Capture efficiency was found to increase initially until the mainstream velocity reaches a certain value, where it begins to drop with mainstream velocity increasing. For the leading edge, capture efficiency followed a similar trend due to deposition spallation and detachment induced by aerodynamic shear at high velocity. Deposition formation was also strongly affected by the mainstream temperature due to its control of particle phase (solid or liquid). Capture efficiency initially increased with increasing mainstream temperature until a certain threshold temperature (near the wax melting point). Subsequently, it began to decrease, for wax detaches from the model surface when subjected to the aerodynamic force at the surface temperature above the wax melting point.

Keywords: gas turbine engine; particle deposition; capture efficiency; multiphase flow

---

1. Introduction

Aero-engines would encounter particle laden air flows when operating in environments with a high concentration of airborne particles during extended service. Sand, fly ash, debris and other external particles may flow into the engine combustion, once introduced into the inlet airflow. A majority of particles flow through the combustion along the main channel flow, and subsequently attack the hot component surface through deposition. The temperature is in excess of 2400 K in the primary combustion zone, far above the particle melting temperature (T_{melt}). Particles may flow through the engine without any effect, or impact on the surface by means of rebounding, spreading, spattering or adhering [1–3], possibly experiencing phase transition and deposition afterwards. Deposition is a complex physical and chemical process. Deposition on the turbine blade would dramatically increase the surface roughness and block film cooling holes in certain conditions, which could lead to a loss in aerodynamic and cooling efficiency [4,5]. For land-based gas turbines, trace amounts of foreign matter in fuels and carbon microparticles generated from the burning of raw energy sources can be injected into the turbine along with the main flow and subsequently deposited on the blade surface, which will affect its heat transfer characteristics and aerodynamic performance. In addition, a growing number of impurities are found in the urban environment, indicating more particles flow into the combustion along the turbine cascade passage once ingested into the engine. This could shorten part life and increase the risk of operation failure [6,7].

To date, the deposition of sand and coal ash on gas turbine engines has been studied by multiple researchers. Kim et al. [8] conducted an experimental study of volcanic ash particle deposition. They found that blockage of film cooling holes could cause damage to vanes at the normal inlet temperature. Koenig et al. [9] explored the negative effect of gas-particle two phase flow on gas turbine blade surfaces by experimental investigation. They found that if particle-laden gas flowed over the blade surface at high temperatures, molten particles would be deposited in the proximity of the blade cooling holes, causing a degradation in turbine blade cooling performance.

However, deposition study on a turbine is neither cost nor time efficient. To improve efficiency and enable cost savings, Jensen et al. [10] designed the Turbine Accelerated Deposition Facility (TADF) to generate deposition in a 4 h test that could simulate 10,000 h of turbine operation. They conducted validation tests to simulate the ingestion of foreign particles typically found in an urban environment by seeding a combustion with large concentrations of airborne particles. Crosby et al. [11] studied the effects of particle size, inlet temperature, and impingement cooling on deposition. They noted that deposition growth increased as particle mass mean diameter, inlet temperature, and target coupon temperature increased, based on a series of 4-h deposition tests. Furthermore, deposition thickness became more uniform at a lower test model surface temperature for a given inlet temperature. Wammack et al. [12] studied the behaviors of deposition on a flat plate with three kinds of surface treatments. They observed extensive spallation of the thermal barrier coated coupons caused by successive deposition. Ai et al. [13] utilized the TADF to generate particle deposition on a test model with film cooling holes. Hot gas flow impinged on the film-cooled target surface at 45 deg. The coupon surface temperature was measured by a red, green, bule (RGB) camera, allowing measurement of spatial temperature distribution to evaluate the influence of deposition thickness on surface heat transfer and film cooling performance. Bonilla et al. [14] studied how particle size affected ash deposition on nozzle guide vanes (NGVs) from a CFM56-5B aero engine. Lundgreen et al. [15] constructed a new turbine cascade to study the deposition on an actual blade surface at turbine mainstream temperatures of 1350 °C, 1265 °C, and 1090 °C. They found that deposition mass on the pressure surface increased as the mainstream temperature increased. The largest amount of deposition occurred on the leading edge. Whitaker et al. [16] noted that the mass of deposition on the blade surface was dependent on particle size distribution, indicating that microparticles with larger sizes were easier to deposit on the surface. They concluded that a reduction of inflow turbulence intensity could slow down the deposition buildup and extend the service life of the turbine. Laycock et al. [17] investigated the independent effects of mainstream temperature and surface temperature on fly ash deposition at elevated operating temperatures. They found capture efficiency increased as the mainstream temperature increased. It showed an increase until a certain threshold temperature and then a decrease, with increasing initial deposition surface temperature.

Recently, a considerable effort has been devoted to developing computational models for the prediction of particle deposition behavior. Numerical models of deposition mainly include two types of models, the critical velocity model and the critical viscosity model. As proposed by Brach and Dunn [18] in the former model, a particle will adhere to a surface when the incoming velocity of the particle (critical velocity) is larger than the particle normal impact velocity. Based on the former model, Zhou and Zhang [19] studied the characteristics of particle deposition on the film-cooled wall numerically and the effect of particle size and blowing ratio on deposition efficiency of shaped holes. Bons et al. [20] subsequently improved the critical velocity model and simulated the particle deposition on a turbine vane surface [21]. They found that capture efficiency increased initially with Stokes number (S_t), and then decreased rapidly when $S_t > 1.5$. In the latter model, the particle viscosity at the softening temperature is considered as the reference viscosity and the sticking probability value is dependent on the relationship between particle viscosity and the reference value. When the particle viscosity is less than or equal to the critical value, the sticking probability value is assumed to be 100%, whereas for higher particle viscosity, capture efficiency decreased with increasing particle viscosity. Based on the latter model, Sreedharan and Tafti [22] investigated the deposition of ash particles impinging on a flat of 45° wedge-shape geometry numerically. They found that the majority of the ash deposition occurred near the stagnation region and capture efficiency increased with increase in jet temperature. Yang and Zhu [23] investigated the particle deposition in the vane passage numerically according to the model. They found that the mainstream temperature and capture efficiency were positively related, but deposition distribution was less sensitive to temperature. Considering the coupled effect that deposition has on the flow geometry, Forsyth et al. [24,25] took advantage of the dynamic mesh morphing technique to make an accurate prediction of particle deposition and found that varied surface topology caused by deposition could be the reason for marked change in fluid and particle velocity. With a similar method, Liu et al. [26] modelled the deposition of particles of different sizes numerically. They investigated the independent effect of particle size on the deposition mass and found that the deposition mass increased with the increasing particle size.

Model-size scaling and test condition scaling are required, due to the stringent test conditions and limitations of necessary measurements at the engine-representative temperature. Lawson et al. [27] constructed a low-speed open loop wind tunnel facility to simulate wax deposition on an endwall with film cooling holes. They modeled the sand and coal ash with atomized molten wax droplets. Particle Stokes number and thermal scaling parameter (TSP) were matched based on the similarity laws. Albert et al. [28,29] adopted a similar method to investigate the wax microparticle deposition on a blade leading edge and pressure side in a closed loop wind tunnel at low operating temperatures. The effect of particle deposition on surface heat transfer characteristics has been thoroughly investigated. However, few studies have been performed to shed light on deposition distribution under different inlet operating conditions.

A deeper understanding of the particle deposition mechanism and characteristics would provide a more accurate quantification of deposition effect on the efficiency and performance of engines. The objectives of the study were to investigate wax particle deposition behavior experimentally in an open loop wind tunnel facility and explore the individual effects of mainstream velocity and temperature on spatial distribution of particle deposition on a flat plate.

2. Experimental Facilities and Procedures

2.1. Experimental Facilities and Model

Tests for this study were carried out in an open loop wind tunnel. The wind tunnel inlet mainstream is driven by two axial fans, after which the flow passes through an air heater, turbulence grids, and a wax sprayer before entering the test section (Figure 1). Mainstream flow is warmed in the air heater to simulate gas from engine combustors. Turbulence grids are situated behind the air heater and comprised of a series of vertically-oriented bars. They help to generate turbulence, thus simulating the actual operating conditions for engines. A wax sprayer is positioned at the center of the wind tunnel behind the turbulence grids. Wax particles could be seeded across the span of test section from the sprayer head. Atomized molten wax particles are delivered to the test section by the wind tunnel mainstream flow. They then impact the model surface to simulate the deposition phenomenon that occurs for sand particles in an actual gas turbine flow-path. The surface temperature is measured by an IR camera and calibrated in situ by a thermocouple (error ≤ 1 °C). The mass of deposition on the test model is measured by an analytical balance with a high precision degree of 0.005 mg (METTLER TOLEDO XPE206DR).

The test section is 0.3 m in width, 0.3 m in height, and 2.5 m in length (Figure 2). The model could be viewed through glass windows in the test section walls. The mainstream approach velocity could be maintained in a range of 5–40 m/s. The mainstream flow temperature is measured by a thermal resistor (uncertainty: ± 0.15 °C). The thermal resistor sends temperature feedback signals to a temperature controller, which adjusts power output of the air heater based on the input current. With this method, mainstream flow could attain a stabilized temperature during testing. The mainstream temperature could be increased by 70 °C at most. Both operating parameter range and sensor accuracy are summarized and listed in Table 1.

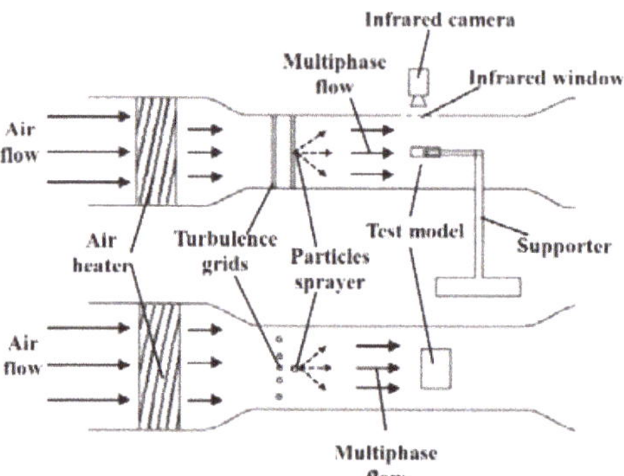

Figure 1. Experimental system schematic.

Figure 2. Photograph of the test section.

Table 1. Operating parameter range and sensor accuracy.

Mainstream Velocity (m/s)	Mainstream Temperature (°C)	Mainstream Velocity Measurement Error (%)	Mainstream Temperature Measurement Precision (°C)	Deposition Mass Measurement Precision (mg)	Surface Temperature Measurement Error (°C)
5–40	25–95	2	±0.15	0.005	±0.5

The wax spray system schematic is shown in Figure 3. The system serves to heat the wax particles above their melting point temperature and spray liquid wax particles into the test section mainstream flow. Wax mass flow rate was in the range of 5–80 g/min with a relative uncertainty of 1.6%. Calibration experiments were carried out based on the relationship between the wax mass flow rate and gear pump speed to acquire calibrated data. A laser particle analyzer is used to measure particle size with a precision degree of 3% (OMEC Instrument CO. DP-02). Wax particle size and flow rate can be controlled by adjusting the air supply pressure and gear pump speed, respectively. The adjustable range for mean particle size is 8–100 μm.

A photograph of the spray system in the wind tunnel is shown in Figure 4. The spray nozzle is located in a hole at the middle of a hollow column (Spray Co. 1/8JJAUCO). The wax heating system is composed of a molten wax reservoir, water-bath vessel, resistive heater and control system. The molten wax reservoir is made up of stainless steel. It was heated uniformly and stably in the water-bath vessel. The control system is used to send temperature feedback signals and minimize fluctuations in the water temperature. The precision uncertainty for the wax temperature is 1 °C. The wax can be heated to 80 °C by the heating system. The purposes of compressed air are to atomize the molten wax and provide the start-stop control of the wax spray system. The wax has a nominal solidification temperature around 41 °C, above the ambient temperature. Consequently, air heating is required to avoid the solidification of wax after its mixing with the atomized air in the spray nozzle. The air temperature can reach 100 °C by using the air heater. All lines are wrapped with electrical heating belts and foaming polyurethane materials as an outer layer to keep the liquid wax above the molten point and achieve thermal insulation and energy saving. A thermocouple is applied to measure the pipeline surface temperature for temperature control purposes. The working conditions of the wax spray system are listed in Table 2.

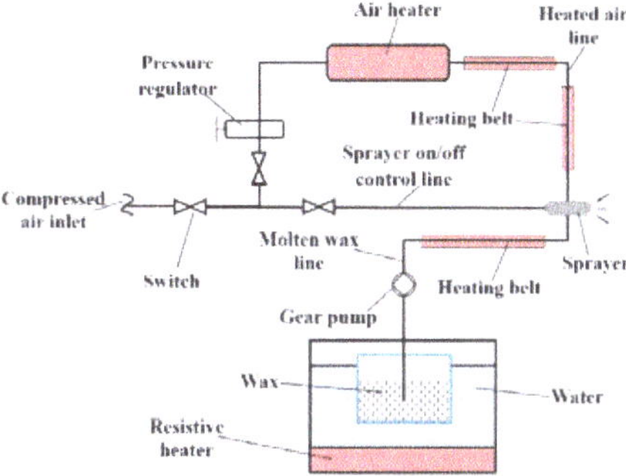

Figure 3. Schematic of the wax spray system.

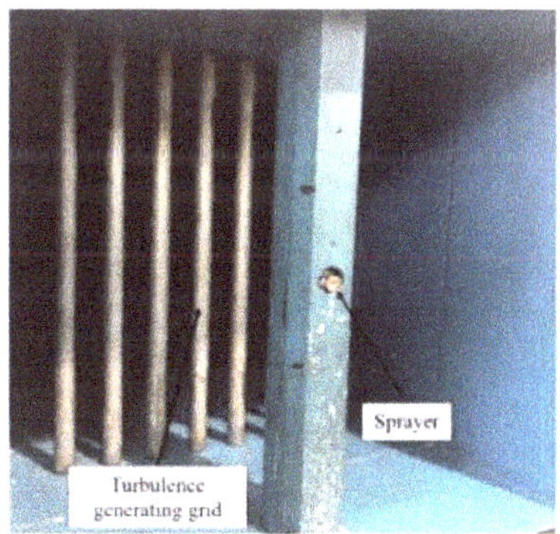

Figure 4. Picture of the wax sprayer and turbulence generating.

Table 2. Wax spray system working conditions.

Particle Size (μm)	Wax Mass Flow Rate (g/min)	Wax Heating Temperature (°C)	Atomized Air Temperature °C)	Atomized Air Pressure (MPa)
8–100	5–80	30–80	25–100	0.1–0.8

The test model and its fixing holder are shown in Figure 5. The test model is a flat plate, which is 150 mm in length, 150 mm in width, and 13 mm in thickness. The test model is fixed by a holding device. Flow fields around the test model might be affected, in that the holding device is thicker than the test model. To mitigate the possible effect, the holding device adopt streamlined outer surface and three air guide holes. Both the pitch angle and rotating angle of the test model can be adjusted. Schematic of the particle-laden flow and angle of attack (AOA) is shown in Figure 6. The adjustable ranges for AOA, pitch angle and rotating angle are $-15°{\sim}+15°$, $-20°{\sim}+20°$, $-90°{\sim}+90°$, respectively. The height of the test model can be adjusted by a mechanical screw lift.

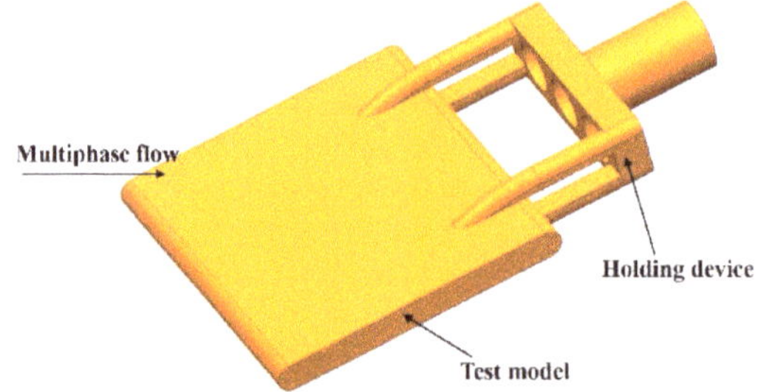

Figure 5. Three-dimensional maps of the test model and its fixing holder.

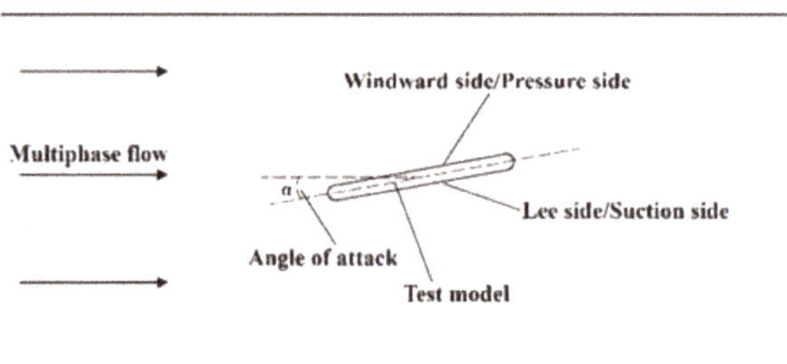

Figure 6. Schematic of multiphase flow and angle of attack.

2.2. Experimental Parameters

The experimental parameters in this paper are shown in Table 3. The mainstream velocity range was 10–30 m/s and the temperature range was 32–48 °C. Wax with a molten point of 41 °C was selected as the particle for injection to observe its deposition behavior. This enables an experimental simulation of particle deposition in a wind tunnel facility. The wax particle size distribution was measured by a laser particle analyzer, as shown in Figure 7. Its main range was 10–20 µm. According to the study conducted by Dring et al. [30], the trajectory of the particles is dominated by the Stokes number. The Stokes number of particles in air flow can be defined as:

$$St_k = \frac{\rho_p d_p^2 U_p}{18\mu_\infty L} \tag{1}$$

where ρ_p is the wax particle density, d_p is the wax particle diameter, U_p is the wax particle velocity, L is the characteristic length and μ_∞ is the gas dynamic viscosity. The Stokes number characterizes the behavior of particles suspended in a fluid flow, which provides an indicator of how well the particles follow the mean fluid streamlines. A particle with a $St_k < 1$ follows fluid streamlines closely, indicating a perfect convection-diffusion. On the contrary, a particle with a $St_k > 1$ is determined by the inertia force and it is likely to continue along its initial trajectory.

The density of molten wax used in this paper was approximately 800 kg/m³. The corresponding St_k, in the range of 0.04–0.12, could be estimated based on the length of the test model. The distance between the sprayer and the test model was chosen as about 10 times length of the test model and the corresponding St_k is 0.004–0.012, ensuring that the atomized wax particles could diffuse in mainstream flow as uniformly as possible.

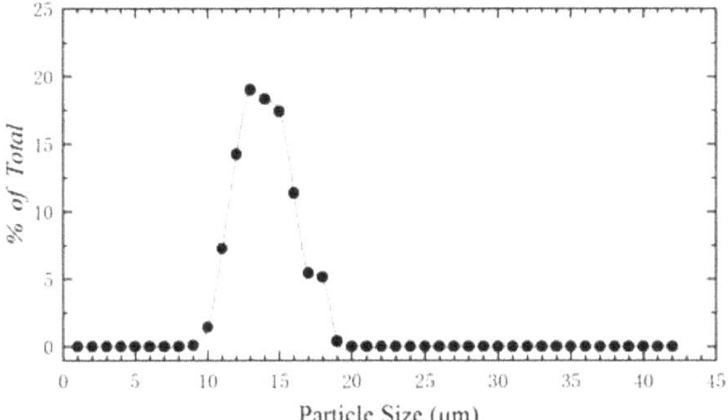

Figure 7. Wax particle size distribution.

Table 3. Test parameters for deposition testing.

Case No.	Mainstream Velocity U_∞ (m/s)	Mainstream Temperature T_∞ (°C)	Test Time t (min)	Wax Volume Concentration c	AOA α (deg)
1	10	40			
2	15	40			
3	20	40			
4	25	40			
5	30	40	5	4.21×10^{-7}	-5
6	10	32			
7	10	37			
8	10	44			
9	10	48			

2.3. Experimental Procedures

The test procedure was as follows: (1) For a typical experiment, the test model should be fixed by the holder at first. The status of all test devices was checked before testing, including the wax spray system, air heater, and AOA of the flat plate. Full test preparations are required. (2) The wax heating system was held at the desired temperature to keep wax in a molten state. The gear pump for starting the wax spray system was closed at the time. (3) The axial fan was opened at the front of the wind tunnel. The air speed was brought to the required value and then held stable. (4) Then the air heater for the wind tunnel was opened and adjusted to a certain temperature so that the test model surface and air can reach a thermal steady state. (5) The gear pump for starting the wax spray system was opened and adjusted according to the required wax mass flow rate in order to achieve different test conditions. The surface temperature of the test model was detected precisely with an infrared radiation thermometer. The data were recorded every second. (6) At the end of an experiment, the wax spray system, air heater for the wind tunnel and axial fan need to be turned off successively. (7) The test model was moved away from the wind tunnel to be photographed. The wax depositions from the entire leading edge, windward side, and lee side were collected to measure the wax deposition mass, respectively. Finally, the remaining wax on the test model was removed in preparation for the next experiment.

2.4. Uncertainty Analysis

An uncertainty analysis serves to validate the measurement technique effectiveness and repeatability of test results. The test Case 1 in Table 3 was performed five times for uncertainty analysis.

Deposition mass standard deviation σ and uncertainty $\Delta\overline{m}_{dep}$ were calculated by Equations (2) and (3), respectively.

$$\sigma = \sqrt{\frac{\sum\limits_{i=1}^{n}\left(m_i - \overline{m}_{dep}\right)^2}{n-1}}, \tag{2}$$

$$\Delta\overline{m}_{dep} = \frac{t_\alpha \sigma}{\sqrt{n}}, \tag{3}$$

where m_i is an observed value for deposition mass, $\overline{m}_{dep}$ is the mean deposition mass, n is the sample size, and t_α is the upper quantile of t-distribution. When the confidence interval is 0.95, t_α is 2.766.

The standard deviation σ of the deposition mass was calculated as 0.046 with a sample size of five (Table 4). Uncertainty in the deposition mass was calculated as 0.056 g (relative uncertainty 3.3%). The resulting deposition mass showed sufficient repeatability between experiments to support the discussions and conclusions.

Table 4. Test results of the deposition mass.

i	m_{dep} (g)	$\overline{m}_{dep}$ (g)	σ (g)	$\Delta\overline{m}_{dep}$ (g)	$\frac{\Delta\overline{m}_{dep}}{\overline{m}_{dep}}$ (%)
1	1.71				
2	1.67				
3	1.66	1.70	0.046	0.056	3.3
4	1.77				
5	1.67				

3. Results and Discussion

3.1. The Effect of Mainstream Velocity

Mainstream velocity through the wind tunnel was set to investigate its effect on the particle deposition mass and efficiency. Mainstream temperature and particle concentration were held constant in tests.

Deposition distributions created on the model leading edge, windward side and lee side are presented in Figures 8–10, respectively. The particle deposition reached the maximum at the leading edge and they became sparser on the windward and lee sides. The forward portions of the windward and lee sides were not visible, with very thin and sparse depositions, while the middle and backward portions of both the windward and lee sides were visible with increasing amounts of deposition towards the trailing edge. Significant depositions at the leading edge region were attributed to particle inertia force. When the particle-laden flow approached the stagnation line, change to the motion direction occurred and it was more evident for the air phase. However, particles cannot follow fluid well or flow around the semi-circular surface due to their inertia. They impacted and then deposited on the leading edge surface. Small particles with a low Stokes numbers tended to follow the flow readily along the model surface and they were less prone to deposition under the drag force. This resulted in a much larger amount of deposition at the leading edge than that on the windward and lee sides. Sreedharan and Tafti [22] conducted numerical simulations of ash deposition on a flat plate oriented at 45 deg to an impinging ash laden jet. The obtained results were validated by experiments from Crosby et al. [11]. Sreedharan and Tafti presented that most particle deposition occurred near the stagnant region. Herein, the stagnant region is located at the leading edge of the test model where maximum deposition mass occurred. Liu et al. [26] conducted numerical simulations of particle deposition on a flat plate under operating conditions similar to those of this study. They noted that deposition mass at the leading edge was far larger than that on the windward side. The result is consistent with that obtained in this study. Deposition buildup at the leading edge affected its geometry structure, serving as a shield from the mainstream, which resulted in less deposition on the forward portions of the windward and lee sides. On the other hand, a deposition layer was generated first at the leading edge with a faster deposition rate. The model surface became rough gradually, which increased the deposition at the leading edge. Sparsely distributed spotted particle deposition was observed on the windward and lee sides. Compared with the leading edge, surface roughness changed less for the windward and lee sides, which caused a lower deposition growth rate. Wax deposition was almost symmetric about the stream-wise direction for both sides. It was less at each end than that in the middle portion along the span-wise direction, due to the flow field edge effect.

Deposition behavior at the leading edge changed evidently with the increasing mainstream velocity (Figure 8). At low velocity, particle deposition was less at the stagnant line. Deposition thickness increased as an evident central depression belt extended to the ends of the windward and lee surfaces. Deposition flaking occurred at the boundary lines between the leading edge and windward side or lee side. The deposition had an irregular shape with a slightly serrated and rough border. With velocity increasing, the mass of deposition adjacent to the stagnant line increased fast and the depressed belt was filled up gradually. Also, deposition on the boundary lines became smooth. Deposition flaking in long strips was observed at the same position in Figure 8c,d. The shedding region expanded with the increasing velocity. Inspection of Figure 8e revealed the spallation of deposition that occurred at the maximum velocity, which caused an irregular physical appearance of the deposition and changed the streamlined shape of the leading edge. At low velocity, deposition was sparsely distributed in spots at the windward and lee sides. The deposition pattern became uniform and the deposition range was almost constant with increasing velocity.

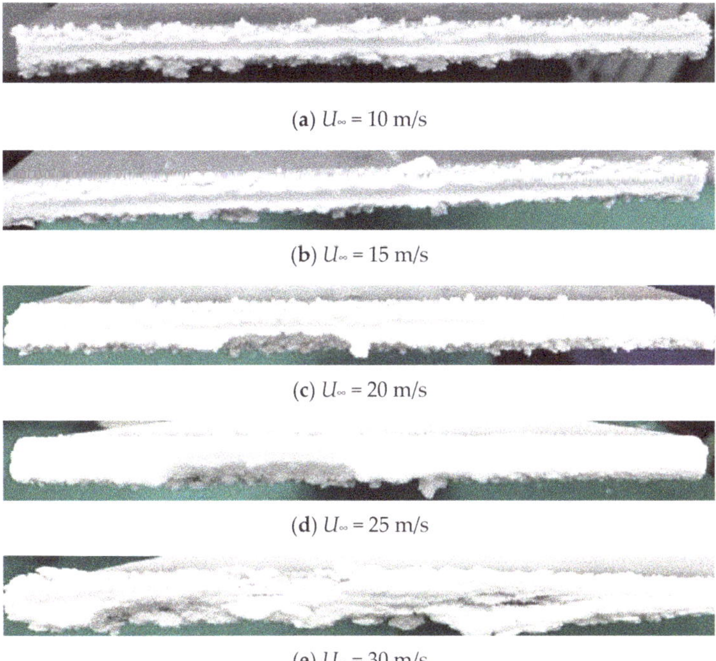

(a) U_∞ = 10 m/s

(b) U_∞ = 15 m/s

(c) U_∞ = 20 m/s

(d) U_∞ = 25 m/s

(e) U_∞ = 30 m/s

Figure 8. Photographs of wax deposition on leading edge with different mainstream velocities.

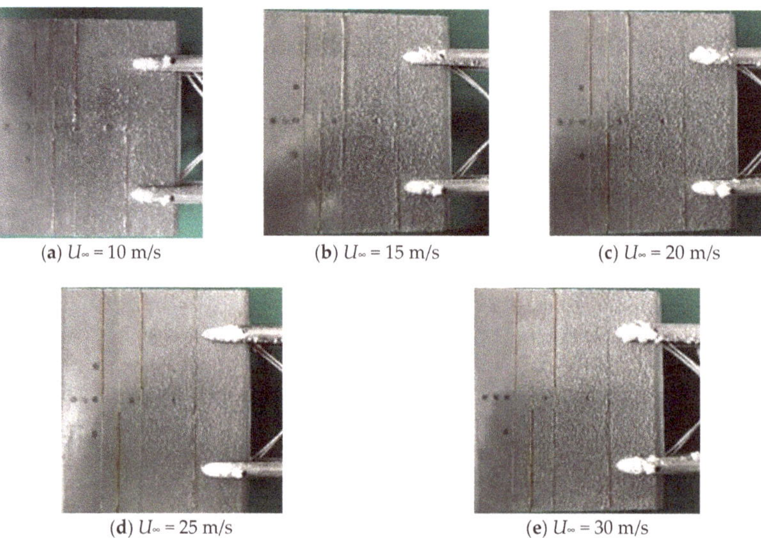

(**a**) $U_\infty = 10$ m/s   (**b**) $U_\infty = 15$ m/s   (**c**) $U_\infty = 20$ m/s

(**d**) $U_\infty = 25$ m/s   (**e**) $U_\infty = 30$ m/s

Figure 9. Photographs of wax deposition on windward surface with different mainstream velocities.

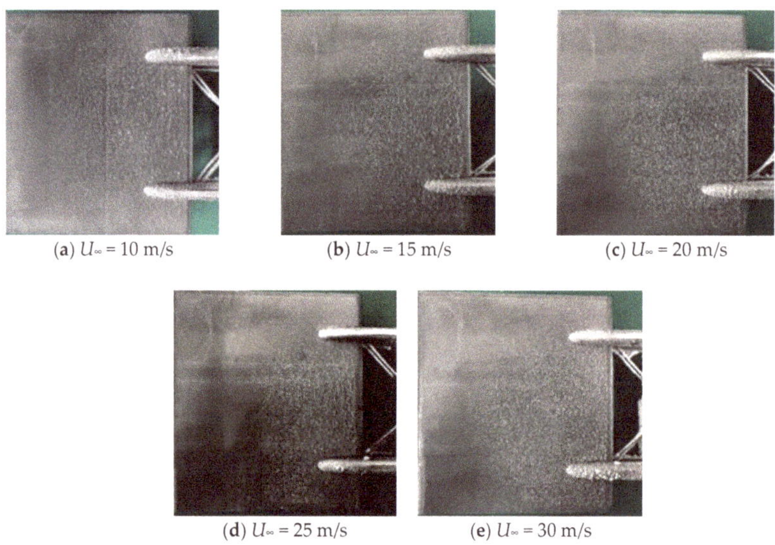

(**a**) $U_\infty = 10$ m/s   (**b**) $U_\infty = 15$ m/s   (**c**) $U_\infty = 20$ m/s

(**d**) $U_\infty = 25$ m/s   (**e**) $U_\infty = 30$ m/s

Figure 10. Photographs of wax deposition on lee surface with different mainstream velocities.

The deposition mass at different mainstream velocities is presented in Figure 11. Deposition mass increased gradually with increasing velocity. This was explained by the increased mainstream velocity, which indicated an increased number of impinging particles in unit time for a constant particle concentration. Deposition began to decrease just above a threshold velocity of 25 m/s. This was caused by the deposition spallation occurring at the maximum velocity (Figure 8e). A large number of particles detached from the deposition at the leading edge under the shear force, which resulted in a loss of deposition mass. The windward and lee sides of the test model experienced a decreasing deposition mass difference as the velocity increased.

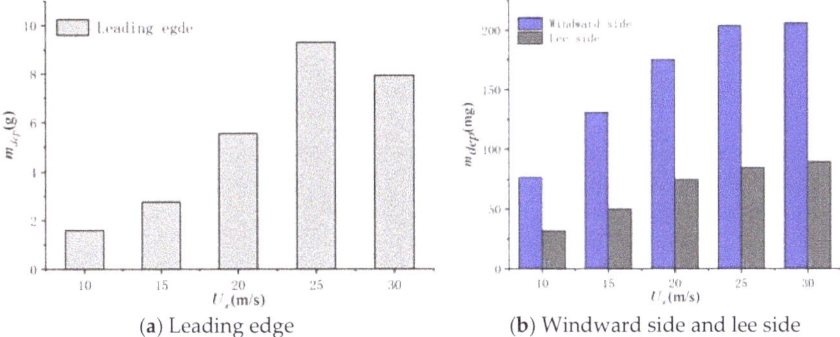

(**a**) Leading edge (**b**) Windward side and lee side

Figure 11. Deposition mass with different mainstream velocities.

Capture efficiencies were measured and calculated in numerous literatures related to effects on deposition to characterize the deposition behavior. The capture efficiencies can be defined as:

$$\eta_{cap} = \frac{m_{dep}}{m_{in}} = \frac{m_{dep}}{q_v \times c \times t},$$ (4)

where m_{dep} is the mass of particle deposition, m_{in} is the total mass of particle injected, q_v is the mainstream volume flow rate, c is the particle volume concentration and t is the test time. Figure 12 shows the effects of the mainstream velocity on deposition capture efficiency. For the leading edge, capture efficiency increased over the first four tests and then decreased as the velocity increased. The dominant force for wax particles to impinge on the test model is inertial force. The increasing capture efficiency could be attributed to increased particle Stokes number, indicating a relative increase in the inertial force. Deposition mass increased with increasing velocity. The increase in deposition mass was related to deposition thickness with a given exposed coupon surface area for 5 min of exposure time. As the deposition thickness increased, the resulting vane contour was no longer in the streamline shape, which enhanced the capture efficiency. Capture efficiency decreased at a velocity of 30 m/s. The main reason for this is that the inlet air mass flow rate increased with increasing mainstream velocity, resulting in increased deposition gravity. The leading edge could have been 'saturated' with wax particles. Meanwhile, the increase in velocity causes the aerodynamic shear force to increase. When the sum of the gravity and aerodynamic force is greater than the bonding force of the deposition, splitting and detachment may occur. For the windward and lee sides, the capture efficiency increases first and then decreases with increasing mainstream velocity. The capture efficiency is at maximum at a mainstream velocity of 20 m/s. Deposition was easier to generate with higher mainstream Reynolds number and particle Stokes number, which resulted in the enhanced capture efficiency. The effect of particle impingement on the wall surface could not be neglected [20]. Due to increased mainstream Reynolds number, particles impacted the test model at the AOA of 5 deg. with more normal impact energy and momentum transfer, and subsequently rebounded more easily. This could lead to the loss in capture efficiency. The change trend of the capture efficiency on the windward and lee sides could be the result of both particle tracing fidelity to the mainstream and rebounding against the wall surface.

The total capture efficiency for a test model was consistent with the capture efficiency for the leading edge (Figure 12c). This could be attributed to the fact that the mass of the deposition at the leading edge occupied a large proportion of the total deposition mass. Variations in mainstream velocity would account for change in the Stokes number. Apart from the Stokes number, particle collision and rebound characteristics combined with the deposition splitting and detachment due to aerodynamic force could account for the change of the capture efficiency. For a flat plate, total capture efficiency was dominated by the mass of deposition at the leading edge. Figure 13 shows the capture efficiency as a function of mainstream velocity and its comparisons with that reported by Taltavull [31] and Bowen [32]. The horizontal coordinate is the particle kinetic energy (E_{kin}), a function of velocity. The expression for E_{kin} is shown in Equation (5):

$$E_{kin} = \frac{1}{2} m_p U_p^2,$$ (5)

where m_p is the particle mass, U_p is the particle velocity, which is equal to U_∞. The variation trend of capture efficiencies with E_{kin} is consistent to that reported by Taltavull and Bowen, but the value of capture efficiency is not consistent, mainly due to the different geometric structure. Taltavull et al. and Bowen et al. studied the behavior of deposition on flat plates oriented at 30 deg, 90 deg to an impinging gas flow, respectively. The area of windward sides is bigger than that herein, accounting for the difference in the capture efficiency.

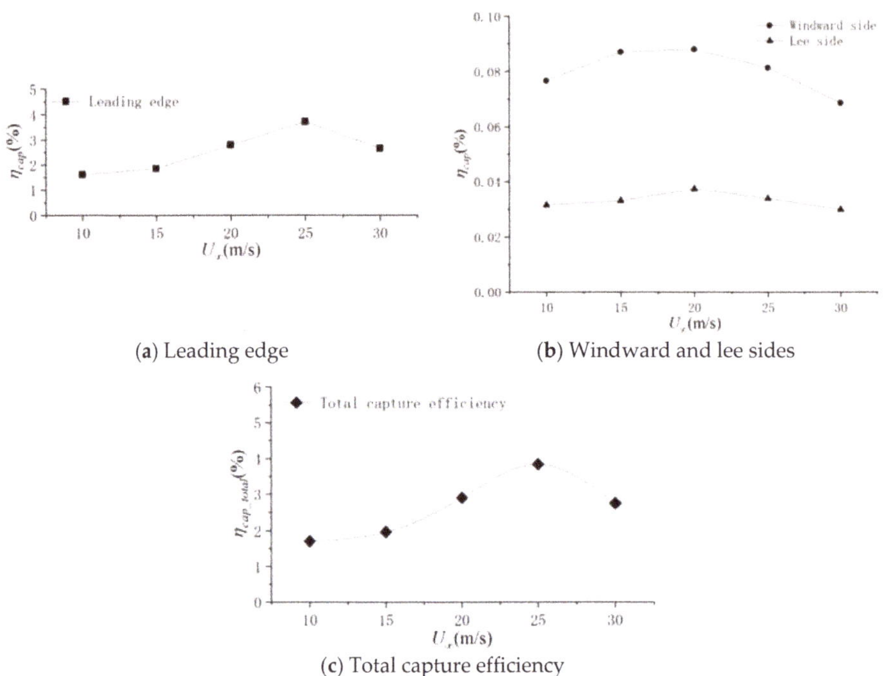

(a) Leading edge

(b) Windward and lee sides

(c) Total capture efficiency

Figure 12. Capture efficiency at different mainstream velocities.

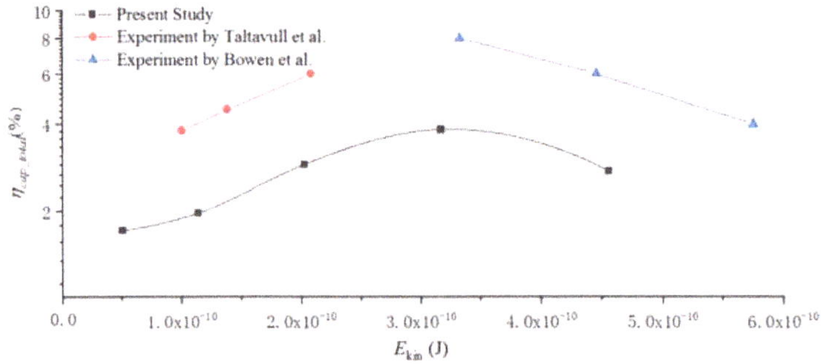

Figure 13. Total capture efficiency in this study compared with data from Taltavull [31] and Bowen [32].

3.2. The Effect of Mainstream Temperature

Five series of tests were set up to investigate the individual effects of mainstream temperature (T_∞) on particle deposition for a given mainstream Reynold number and particle concentration. The tests were run with a mainstream velocity of 10 m/s and particle concentration of 4.21×10^{-7}. The temperature was increased from 32 °C up to a maximum of 48 °C.

Wax deposition distribution at the leading edge, windward and lee sides are presented in Figures 14–16, respectively. The maximum deposition amount was observed at the leading edge, while the amount of deposition was less in the forward portions of the windward and lee sides than that in the middle and backward portions. At the T_∞ of 32 °C, far below the wax melting temperature, deposition near the stagnation line reached maximum and it became gradually less towards the two ends, in a V shaped pattern. The deposition patterns at the two ends were uniform and no flaking was observed. As T_∞ increased, a nonuniform deposition structure occurred, indicating softened particle behavior and the effect of the mainstream flow along the surface. When the test was conducted at the T_∞ of 37 °C, particle deposition was less than that at 32 °C in the vicinity of the stagnation line. An evident depressed belt at the leading edge could be observed from Figure 14b. Particle deposition increased gradually from the middle to the two ends. It was also observed during tests that larger bulks of wax detached from the deposition at the bottom side of the leading edge. At the T_∞ of 40 °C, close to the wax melting temperature, the depressed belt appeared to be more evident. The deposition pattern at both ends was nonuniform, for particles there were blown off from the deposition. This could mean that the deposition stickiness increased with more trapping particles at the leading edge. Deposition decreased significantly as T_∞ increased from 40 °C to 43 °C, above the wax melting temperature. A transparent deposition layer was observed at the leading edge at 43 °C, while no deposition was observed on the surface of the leading edge at 48 °C. When T_∞ was below the wax melting point, sparse individual deposition was observed at 32 °C on the windward and lee sides, whereas more deposition could be discerned at the higher T_∞. However, deposition decreased significantly when the temperature increased to 43 °C. Sparse deposition was observed at the backward portions of the windward and lee sides at the T_∞ of 43 °C, slightly above the melting point. At the T_∞ of 48 °C, a transparent wax deposition layer was observed only at the trailing edge.

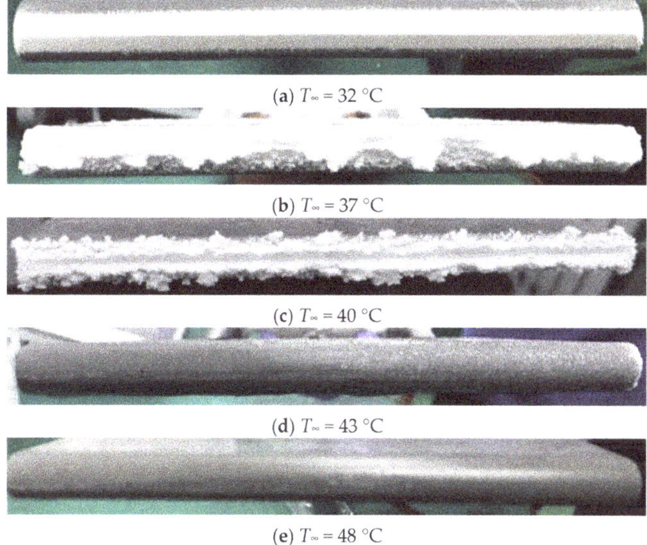

Figure 14. Photographs of wax deposition on leading edge with different mainstream temperature.

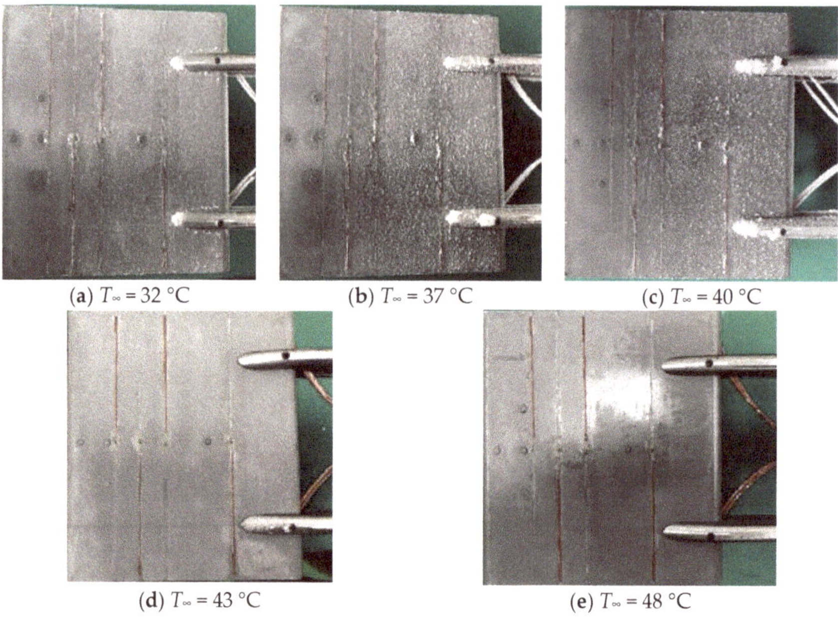

(**a**) $T_\infty = 32\ ^\circ$C   (**b**) $T_\infty = 37\ ^\circ$C   (**c**) $T_\infty = 40\ ^\circ$C

(**d**) $T_\infty = 43\ ^\circ$C   (**e**) $T_\infty = 48\ ^\circ$C

Figure 15. Photographs of wax deposition on windward surface with different mainstream temperatures.

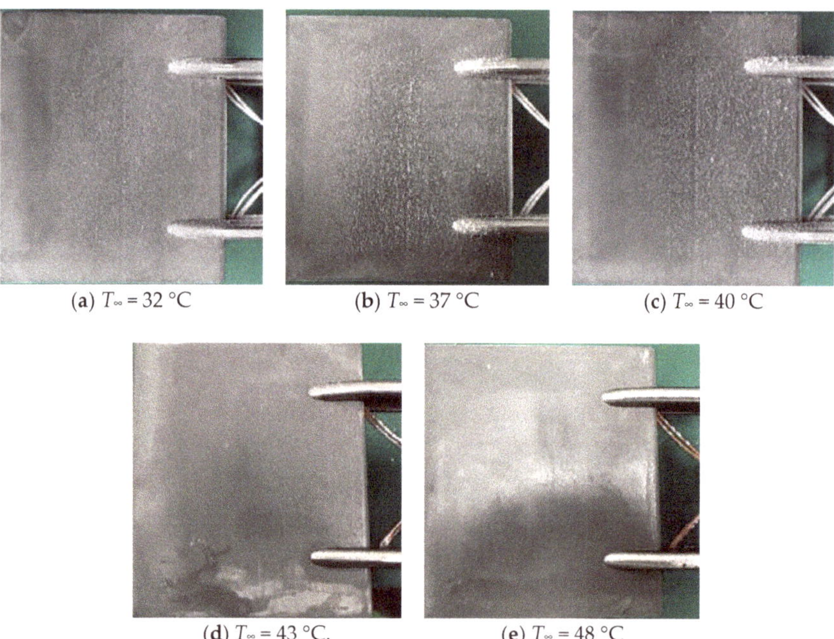

(**a**) $T_\infty = 32\ ^\circ$C   (**b**) $T_\infty = 37\ ^\circ$C   (**c**) $T_\infty = 40\ ^\circ$C

(**d**) $T_\infty = 43\ ^\circ$C.   (**e**) $T_\infty = 48\ ^\circ$C

Figure 16. Photographs of wax deposition on lee surface with different mainstream temperatures.

The deposition mass variation with T_∞ was shown in Figure 17. Deposition mass initially increased and then began to decrease at a mainstream temperature of 40 $^\circ$C for the leading edge, windward sides and lee sides. The threshold temperature was found to be approximately equal to the wax melting point. Once a molten wax particle was injected from the nozzle head at the T_∞ of 32 °C, it was very likely to quickly cool down and solidify just before it impacted the surface. Lower T_∞ increased the probability of a particle rebounding upon impact and not sticking to the surface. As T_∞ approached the wax melting point, particles attached to the test model in a molten, or partially molten state, making them less susceptible to rebounding or detachment. Instead, deposition appeared to spread and solidify. This increased deposition mass was attributed to the larger number of molten particles in the larger temperature mainstream. Liquid wax spatter could occur due to the model surface temperature above the wax melting point. The uncoagulated particles flew along the surface in the form of a liquid film. The aerodynamic stress caused the liquid particles to detach from the surface of the test model, resulting in less deposition.

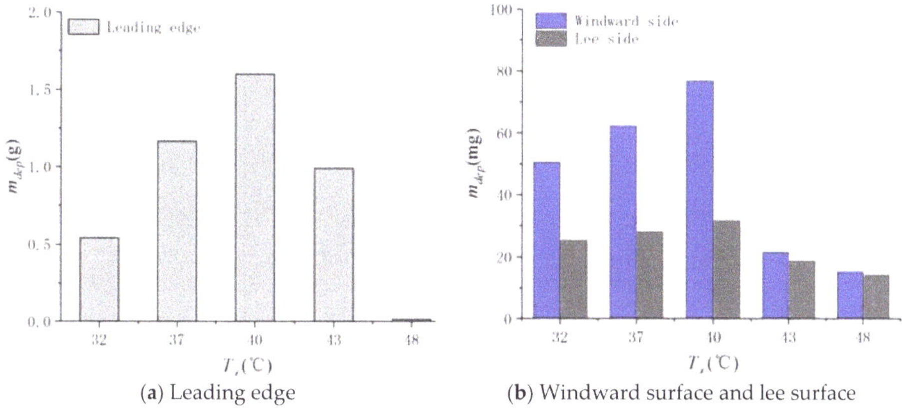

(a) Leading edge   (b) Windward surface and lee surface

Figure 17. Deposition mass with different mainstream temperature.

Capture efficiency at different T_∞ was shown in Figure 18. Capture efficiency increased to a certain threshold temperature and then decreased with increasing T_∞. It reached the maximum when T_∞ was close to the melting point. When T_∞ was below the melting temperature, the trend of capture efficiency in this paper was consistent with that reported by Lundgreen and Crosby [11,15]. They noted that capture efficiency increased with the mainstream temperature increasing. However, for T_∞ above the wax melting temperature, the capture efficiency decreased with T_∞ increasing. This behavior was also observed by Laycock et al. [17] They concluded that capture efficiency increased until a certain temperature, where capture efficiency began to decrease as surface temperature increased. This was attributed to the increased number of particles in molten form as they encountered the flat plate and increased probability of liquid wax spatter at increased T_∞. Since the test model was non-cooled, its surface temperature was highly dependent on T_∞. Increasing T_∞ could elevate the model surface temperature. When the surface temperature of the test model exceeded the wax melting point, liquid wax film would detach from the structure surface under the aerodynamic force and thus it would not solidify upon reaching it. Figure 19 shows total capture efficiency reported by this paper, Crosby and Laycock. The variation trend of the capture efficiency reported in the paper is basically consistent with the data from Crosby and Laycock. The result is more closely matched to the data from Crosby, mainly for Stokes number of wax particles adopted herein is close to that of sand particles adopted by Crosby. For the particles of larger sizes used by Laycock, capture efficiency is higher due to a larger Stokes number.

Temperature increases were found to decrease particle capture efficiency at inlet temperatures above the particle melting point. However, excessively high inlet temperatures may cause damage to the component surface [12,33]. Therefore, it is not reasonable to achieve low capture efficiency and normal engine performance only by increasing gas temperature. Further investigation is still required to decrease the negative effects of deposition on plate surface heat transfer characteristics and aerodynamic performance.

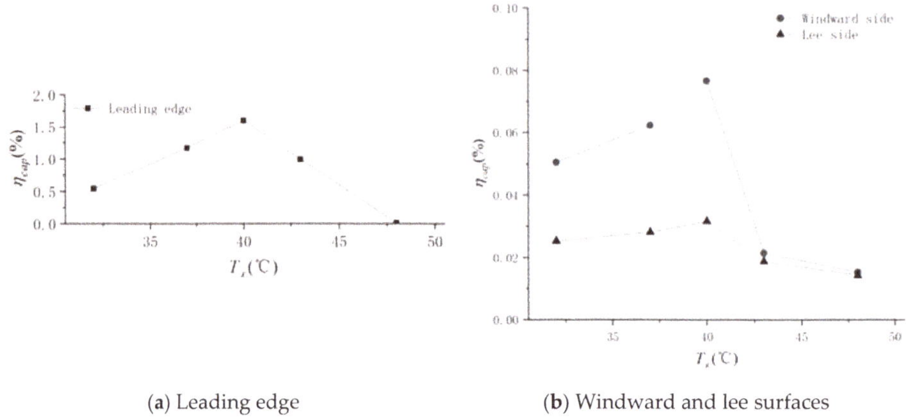

(**a**) Leading edge   (**b**) Windward and lee surfaces

Figure 18. Capture efficiency with different mainstream temperature.

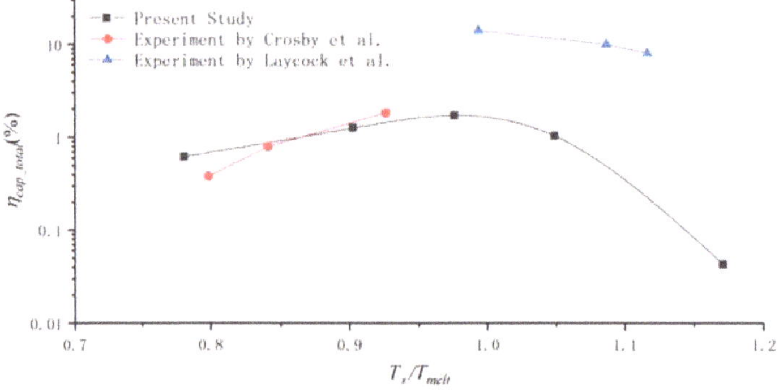

Figure 19. Total capture efficiency herein compared to data from Crosby [11] and Laycock [17].

4. Conclusions

Nine series of tests were conducted in a two-phase flow wind tunnel to investigate the independent effects of mainstream Reynolds number and mainstream temperature on particle deposition mass and distribution. Based on the results presented in this paper, the following conclusions can be drawn:

(1) Maximum wax deposition occurred at the leading edge, which increased levels of surface roughness and affected the vane contour. For the windward and lee sides, deposition was distributed in sparse spots, mainly in the middle and backward portions. No position was observed in the forward portion.

(2) Increased mainstream velocity denoted more particles reaching the surface in unit time for a given particle concentration. For the windward and lee sides, deposition mass increased with increasing mainstream velocity. Capture efficiency increased initially to a certain value and then decreased as mainstream velocity increased, for particles were easier to rebound with more kinetic energy. For the leading edge, capture efficiency increased and then began to decrease at a mainstream velocity of 25 m/s.

(3) T_∞ was adjusted to change the phase of particles impacting the model surface. When T_∞ was below the particle melting point, deposition mass and capture efficiency increased with increase in T_∞

for both the leading edge and windward and lee sides. The mainstream temperature, higher than the particle melting point, would result in lower deposition mass and capture efficiency.

The study focused on particle deposition on a flat plate. Further work should consider the effect of geometric structure on the particle deposition and the effect of particle deposition on surface heat transfer characteristics.

Author Contributions: Investigation, Writing-Original Draft Preparation, F.Z.; Resources, Z.L. (Zhenxia Liu); Supervision, Z.L. (Zhengang Liu); Data Curation, Y.L.

Funding: This study was funded by the National Natural Science Foundation of China, grant number 51606155.

Conflicts of Interest: The authors declare no conflict of interest.

References

1. Dean, J.; Taltavull, C.; Clyne, T. Influence of the composition and viscosity of volcanic ashes on their adhesion within gas turbine aeroengines. *Acta Mater.* **2016**, *109*, 8–16. [CrossRef]
2. Shin, D.; Hamed, A. *Advanced High Temperature Erosion Tunnel for Testing TBC and New Turbine Blade Materials*; ASME Paper 2016, No. 2016-57922; ASME: New York, NY, USA, 2016. [CrossRef]
3. Tabakoff, W. Measurements of particles rebound characteristics on materials used in gas turbines. *J. Propuls. Power* **2015**, *7*, 805–813. [CrossRef]
4. Chambers, J.C. The 1982 Encounter of British Airways 747 with the Mt. Galuggung Eruption Cloud. In Proceedings of the 23rd Aerospace Science Meeting, Reno, NV, USA, 14–17 January1985.
5. Tabakoff, W. Review—Turbomachinery performance deterioration exposed to solid particulates environment. *J. Fluids Eng.* **1984**, *106*, 125–134. [CrossRef]
6. Dunn, M.G.; Moller, J.C.; Moller, J.E.; Adams, R.M. Performance deterioration of a turbofan and a turbojet engine upon exposure to a dust environment. *J. Eng. Gas Turbines Power* **1987**, *109*, 336–343. [CrossRef]
7. Boulanger, A.; Patel, H.; Hutchinson, J.; DeShong, W.; Xu, W.; Ng, W.; Ekkad, S. Preliminary experimental investigation of initial onset of sand deposition in the turbine section of gas turbines. In Proceedings of the ASME Turbo Expo 2016, Seoul, Korea, 13–17 June 2016. [CrossRef]
8. Kim, J.; Dunn, M.C.; Baran, A.J. Deposition of volcanic materials in the hot sections of two gas turbine engines. *J. Eng. Gas Turbines Power* **1993**, *115*, 641–651. [CrossRef]
9. Koenig, P.; Miller, T.; Rossmann, A. Damage of High Temperature Components by Dust-Laden Air. In Proceedings of the Agard Conference, Rotterdam, The Netherlands, 25–28 April 1994; p. 25.
10. Jensen, J.W.; Squire, S.W.; Bons, J.P.; Fletcher, T.H. Simulated Land-based Turbine Deposits Generated in an Accelerated Deposition Facility. *J. Turbomach.* **2005**, *127*, 462–470. [CrossRef]
11. Crosby, J.M.; Lewis, S.; Bons, J.P.; Ai, W.; Fletcher, T.H. Effects of temperature and particle size on deposition in land based turbines. *J. Eng. Gas Turbines Power* **2008**, *130*, 819–825. [CrossRef]
12. Wammack James, E.; Crosby, J.; Fletcher, D.; Bons, J.P.; Fletcher, T.H. Evolution of surface deposits on a high-pressure turbine blade-Part I: Physical Characteristics. *J. Turbomach.* **2008**, *130*, 021020. [CrossRef]
13. Ai, W.; Murray, N.; Fletcher, T.H.; Harding, S.; Lewis, S.; Bons, J.P. Deposition near film cooling holes on a high pressure turbine vane. *J. Turbomach.* **2012**, *134*, 041013. [CrossRef]
14. Bonilla, C.; Webb, J.; Clum, C.; Casaday, B.; Brewer, E.; Bons, J.P. The effect of particle size and film cooling on nozzle guide vane deposition. *J. Eng. Gas Turbines Power* **2012**, *134*, 101901. [CrossRef]
15. Lundgreen, R.; Sacco, C.; Prenter, R.; Bons, J.P. Temperature Effects on Nozzle Guide Vane Deposition in a New Turbine Cascade Rig. In Proceedings of the ASME Turbo Expo 2016, Seoul, Korea, 13–17 June 2016. [CrossRef]
16. Whitaker, S.M.; Prenter, R.; Bons, J.P. The effect of freestream turbulence on deposition for nozzle guide vanes. *J. Turbomach.* **2015**, *137*, 121001. [CrossRef]
17. Laycock, R.; Fletcher, T.H. Independent effects of surface and gas temperature on coal fly ash deposition in gas turbines at temperatures up to 1400 °C. *J. Eng. Gas Turbines Power* **2016**, *138*, 021402. [CrossRef]
18. Brach, R.; Dunn, P. A mathematical model of the impact and adhesion of microspheres. *Aerosol Sci. Technol.* **1992**, *16*, 1–14. [CrossRef]
19. Zhou, J.H.; Zhang, J.Z. Numerical investigation of particle deposition on converging slot-hole film-cooled wall. *J. Cent. South Univ.* **2017**, *24*, 2819–2828. [CrossRef]

20. Bons, J.P.; Prenter, R.; Whitaker, S. A Simple Physics-Based Model for Particle Rebound and Deposition in Turbomachinery. *J. Turbomach.* **2017**, *139*, 081009. [CrossRef]

21. Prenter, R.; Ameri, A.; Bons, J.P. Computational simulation of deposition in a cooled high-pressure turbine stage with hot streaks. *J. Turbomach.* **2017**, *139*, 091005. [CrossRef]

22. Sreedharan, S.S.; Tafti, D.K. Composition dependent model for the prediction of syngas ash deposition in turbine gas hotpath. *Int. J. Heat Fluid Flow* **2011**, *32*, 201–211. [CrossRef]

23. Yang, X.J.; Zhu, J.X. Numerical simulation of particle deposition process inside turbine cascade. *Acta Aeronaut. Astronaut. Sin.* **2017**, *38*, 120530. [CrossRef]

24. Forsyth, P. High Temperature Particle Deposition with Gas Turbine Applications. Ph.D. Dissertation, University of Oxford, Oxford, UK, 2017.

25. Connolly, J.; Forsyth, P.; McGilvray, M.; Gillespie, D. The Use of Fluid-Solid Cell Transformation to Model Volcanic Ash Deposition within a Gas Turbine Hot Component. In Proceedings of the ASME Turbo Expo 2018, Oslo, Norway, 11–15 June 2018. [CrossRef]

26. Liu, Z.; Zhang, F.; Liu, Z. A Numerical Model for Simulating Liquid Particles Deposition on Surface. In Proceedings of the ASME Turbo Expo 2018, Oslo, Norway, 11–15 June 2018. [CrossRef]

27. Lawson, S.A.; Thole, K.A. Simulations of Multiphase Particle Deposition on Endwall Film-Cooling. *J. Turbomach.* **2012**, *134*, 011003. [CrossRef]

28. Albert, J.E.; Keefe, K.J.; Bogard, D.G. Experimental Simulation of Contaminant Deposition on a Film Cooled Turbine Airfoil Leading Edge. In Proceedings of the ASME Turbo Expo 2009, Lake Buena Vista, FL, USA, 13–19 November 2009. [CrossRef]

29. Albert, J.E.; Bogard, D.G. Measurements of Adiabatic Film and Overall Cooling Effectiveness on a Turbine Vane Pressure Side with a Trench. In Proceedings of the ASME Turbo Expo 2011, Vancouver, BC, Canada, 6–10 June 2011. [CrossRef]

30. Dring, R.P.; Caspar, J.R.; Suo, M. Particle trajectories in turbine cascades. *J. Energy* **1979**, *3*, 161–166. [CrossRef]

31. Taltavull, C.; Dean, J.; Clyne, T.W. Adhesion of volcanic ash particles under controlled conditions and implications for their deposition in gas turbines. *Adv. Eng. Mater.* **2016**, *18*, 803–813. [CrossRef]

32. Bowen, C.P.; Libertowski, N.D.; Mortazavi, M.; Bons, J.P. Modeling Deposition in Turbine Cooling Passages with Temperature-Dependent Adhesion and Mesh Morphing. *J. Eng. Gas Turbines Power* **2019**, *141*, 071010. [CrossRef]

33. Davidson, F.T.; Kistenmacher, D.A.; Bogard, D.G. A Study of Deposition on a Turbine Vane with a Thermal Barrier Coating and Various Film Cooling Geometries. *J. Turbomach.* **2014**, *136*, 1769–1780. [CrossRef]

Article

Development and Assessment of Two-Stage Thermoacoustic Electricity Generator

Ahmed Hamood [1], Artur J. Jaworski [1,*] and Xiaoan Mao [2]

1   School of Computing and Engineering, University of Huddersfield, Queensgate, Huddersfield HD1 3DH, UK; A.Hamood@hud.ac.uk
2   Faculty of Engineering, University of Leeds, Woodhouse Lane, Leeds LS2 9JT, UK; X.Mao@leeds.ac.uk
*   Correspondence: a.jaworski@hud.ac.uk; Tel.: +44-148-447-2965

Received: 16 April 2019; Accepted: 7 May 2019; Published: 11 May 2019

Abstract: This paper presents the development and assessment of a two-stage thermoacoustic electricity generator that aims to mimic the conversion of waste heat from the internal combustion engine exhaust gases into useful electricity. The one wavelength configuration consists of two identical stages which allow coupling a linear alternator in a "push-pull" mode because of the 180° out of phase acoustic excitation on two sides of the piston. This type of coupling is a possible solution for the low acoustic impedance of looped-tube traveling-wave thermoacoustic engines. The experimental set-up is 16.1 m long and runs at 54.7 Hz. The working medium is helium at maximum pressure of 28 bar. In practice, the maximum generated electric power was 73.3 W at 5.64% thermal-to-electric efficiency. The working parameters, namely load resistance, mean pressure and heating power, were investigated. System debugging illustrates the effect of local acoustic impedance of the regenerator on the start-up process of the thermoacoustic engine. The additional modelling showed that the feedback loop length can be reduced by using a combination of acoustic inertance and compliance components.

Keywords: thermoacoustic electricity generator; multi-stage; traveling-wave heat engine; push-pull; inertance-compliance; acoustic streaming

---

1. Introduction

Due to the widespread utilisation of high-grade heat sources in industry and transportation, there has been an increase in waste heat rejected to the environment. Therefore, many technologies have been developed for waste heat recovery applications for a range of scales and heat-grades. Thermoacoustic traveling-wave engines (TATWE) have drawn attention because of their advantages which include no mechanical moving parts, longevity and the use of environmentally friendly gases as working media [1]. Thermoacoustic engines are capable of converting heat into acoustic power. The acoustic power generated by the TATWE can be used to generate electricity by driving an electromechanical linear alternator or generate cooling by driving a thermoacoustic refrigerator [2] or a pulse tube cryocooler [3,4]. In general, TATWE take the form of an acoustic resonator filled with a gas and containing a thermoacoustic core consisting of a porous medium (stack or regenerator) with heat source and heat sink (i.e., heat exchangers) adjacent to it. The gas in the vicinity of the solid surface of the porous medium undergoes a thermodynamic cycle somewhat similar to the Stirling cycle.

The first TATWE in a looped-tube configuration was presented by Yazaki et al. [5]. It can be likened in certain respects to a standing-wave thermoacoustic engine because of the two-wavelength loop containing the thermoacoustic core at a specific location. Yazaki et al. [5] designed and built their air-filled engine to study the spontaneous gas oscillations in a travelling wave setting. The experimental results showed that the travelling wave outperformed the standing wave engines at the same frequency and wavelength. The low efficiency of this engine was discovered to be caused by low acoustic impedance (acoustic impedance is the ratio of oscillating pressure to volume flow rate). Looped-tube TATWE can have more than one thermoacoustic core in the same engine. These so-called multi-stage TATWEs are a solution to a low acoustic impedance and work at low onset temperature [6,7]. The multi-stage TATWE can be built either having identical stages each having a power extraction point or a number of geometrically non-identical stages in series and a single alternator in the loop. Although the cross section of the thermoacoustic cores in series increases in the wave propagation direction to reduce acoustic loses [8,9], identical stages run at lower acoustic losses as the staged power extraction points allow the device to run without high acoustic power "spots".

de Blok [8,10] built four engines demonstrating the possibility of cascading stages in looped tube TATWE. The engines consisted of four identical stages each having a power extraction point, and they were generating electricity in a range of 18 W to 1.64 kW. Zhang and Chang [7] studied the onset temperature, mean pressure, type of working medium (gas), hydraulic radius as well as the number of stages in a four-stage engine configuration similar to de Blok's [8]. They also investigated replacing one of the engine stages with a refrigerator stage [11]. The results showed that the device can reach a relative Carnot COP (coefficient of performance) of 28.5% at a cooling temperature of 5 °C. Yang et al. [12] adopted the four-stage configuration and adapted it by replacing one stage with acoustic compliance acting as acoustic impedance control. The engine design aimed at low-grade temperature. A thermal-to-electric efficiency of 1.51% was achieved at the hot temperature of 120 °C. Yang et al. [13] developed the engine by adding five acoustic loads. A maximum thermal efficiency achieved was 9.6% at the hot temperature of 195 °C. Li et al. [14] suggested an upgrade for the system with four identical stages by installing a branch containing a refrigerator stage and a dual linear alternator.

Another multi-stage configuration proposed by Li et al. [15] was built as three identical stages with linear alternators connected in-line to shift the phase difference. At a mean pressure of 40 bar of helium, each stage of the engine generated 1080 W of acoustic power at 36% total efficiency. Wu et al. [16] developed the system to generate useful electricity. At 50 bar mean pressure and 650 °C heating temperature, the engine generated 1.57 kW of electricity at thermal-to-electric efficiency of 16.8%. Bi et al. [17] optimized the acoustic network of the engines and ran it at a higher mean pressure of 60 bar; the engine generated 4.69 kW at 15.6% of thermal-to-electric efficiency.

Linear alternators convert the acoustic power into electricity at high transduction efficiency. The alternators used in thermoacoustic technology must have flexure bearings and gas clearance seals, which require no lubrication, to make them compatible with the low-maintenance requirements of thermoacoustic devices. The main disadvantage is that they are expensive. Hamood et al. [18] showed that the linear alternator performance improves with increasing the acoustic impedance. It is then advantageous to utilize multi-stage TATWE where high acoustic impedance is favourable for the linear alternator to generate higher electrical power. However, for "self-matched" identical stages they require many expensive linear alternators to run at high acoustic impedance and relatively low acoustic losses—hence finding a way to limit the number of alternators is important. The aim of the current research is to demonstrate a configuration that enables coupling one linear alternator to two power extraction points of two stages that work in pressure antiphase, thus reducing the number of expensive alternators in the system. Here, the effective acoustic field driving the linear alternator is the algebraic sum of the two fields belonging to the two stages.

2. Experimental Apparatus

The experimental apparatus consists of two identical engine stages each having a power extraction point, and a linear alternator connected to these two power extraction points. The conceptual design of the electricity generator is shown in Figure 1a, while Figure 1b shows a photograph of the actual device. The identical stages generate acoustic waves having similar pressure and volume flow rate amplitudes, but which are out of phase by 180° between the two stages. When these out of phase acoustic fields act upon the alternator, one is "pushing" while the other is "pulling" the piston. Hence, the active acoustic impedance running the alternator is the sum of the two push-pull acoustic fields, and this will increase the power output at a specific acoustic impedance [18].

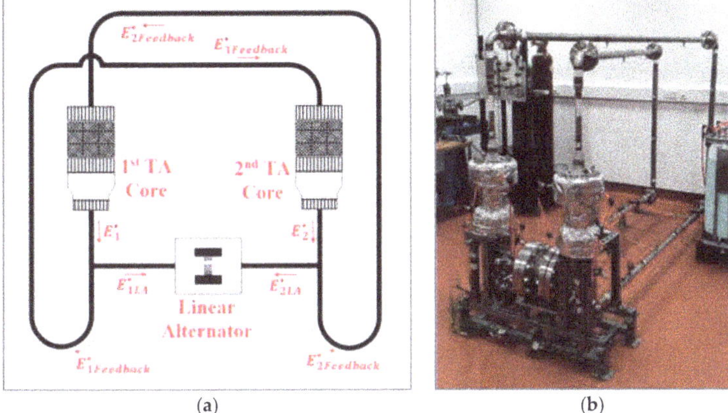

Figure 1. (**a**) Conceptual drawing of the engine; (**b**) Photo of the thermoacoustic engine reported here.

The DeltaEC (design environment for low-amplitude thermoacoustic energy conversion) simulation tool was used to simulate the acoustic field and optimize the dimensions of the device components. DeltaEC is a design tool for thermoacoustic applications developed based on the linear thermoacoustic theory [19]. It enables a continuous optimization process to investigate the dimensions that offer the best performance. After a complex trade-off optimization process which considers the performance and parts' availability, a successful model was generated.

In addition to the system optimisation, DeltaEC was also used to study the favourable acoustic field for the linear alternator to generate high electrical power out of the acoustic input with high efficiency. Specifications of the linear alternator (Q-Drive 1S132M) were used as input values for the model. Figure 2 shows the generated power (black curves) and efficiency (blue curves) at a frequency of 56 Hz. They are plotted as a function of phase difference between velocity and pressure. The local acoustic impedance (which is a parameter here) is shown in the legend. The simulation results confirm that the linear alternator generates higher electrical power and runs more efficiently at higher acoustic impedance. In addition, the phase difference of around 15° is favourable for the linear alternator to generate the highest power, while between 40–60° to run at highest possible efficiency.

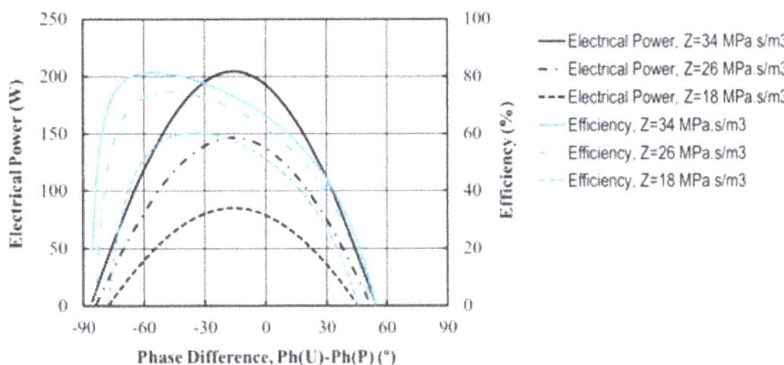

Figure 2. LA electrical power generated and efficiency at push-pull coupling for different acoustic impedance and phase difference.

This paper describes the experimental rig only in brief, since the full details can be found in [18]. The thermoacoustic generator is a 16.1 m long, looped tube two-stage thermoacoustic engine, and uses pressurized helium at 28 bar as the working gas. It runs at a frequency of 54.7 Hz. The simulation results for the one-wavelength acoustic field in the engine are presented in Figure 3. The figure shows that the regenerators are located near maximum pressure and minimum volumetric velocity amplitudes to be near the highest acoustic impedance and minimize the viscus dissipation. The two branches leading to the linear alternator sides are placed near the regenerator to ensure the highest possible acoustic impedance at the linear alternator branches. The acoustic power distribution along the engine shows that the acoustic power is generated in regenerators and mainly dissipated through the linear alternator branches.

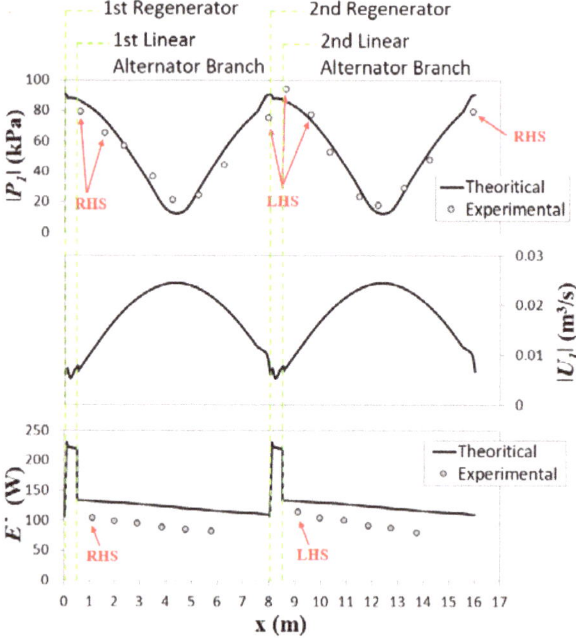

Figure 3. Distribution of pressure amplitude, volumetric velocity and acoustic power along the engine.

Figure 4 shows a cross-section of the thermoacoustic core. The regenerator holder, hot heat exchanger and thermal buffer tube have been manufactured as one piece to eliminate potential gas leakage problems which might appear at elevated temperatures at the seals of hot parts. The ambient heat exchanger is a cross flow heat exchanger having staggered fins at both water and helium flow directions. It is made out of a block of copper. The diameter of the heat exchanger on the helium side is 101.75 mm, and its thickness is 30 mm. The fins are 0.5 mm in width leaving 1 mm channels; on the helium side the fins are 8 mm long, while they are 5 mm long on the water side. At the design drive ratio, the peak-to-peak displacement is roughly half of the heat exchanger length. The porosity of the ambient heat exchanger is 31.2% on helium side.

The regenerator is made of 445 stainless steel mesh screen layers, piled up inside the regenerator holder. Regenerator length is 73 mm, and its diameter is 102 mm. The wire diameter in the mesh screen is 65 μm while the wire-to-wire aperture is 180 μm. On each end of the regenerator, there are coarse diamond mesh screens of 1.2 mm thickness, which act as spacers. The spacers allow the gas leaving the heat exchangers to mix and spread over the entire regenerator cross section. The regenerator hydraulic radius and the volume porosity have been calculated using the wire diameter, aperture and the amount of packed mesh per unit volume. The hydraulic radius is 60.5 μm and the volumetric porosity is 78.9%.

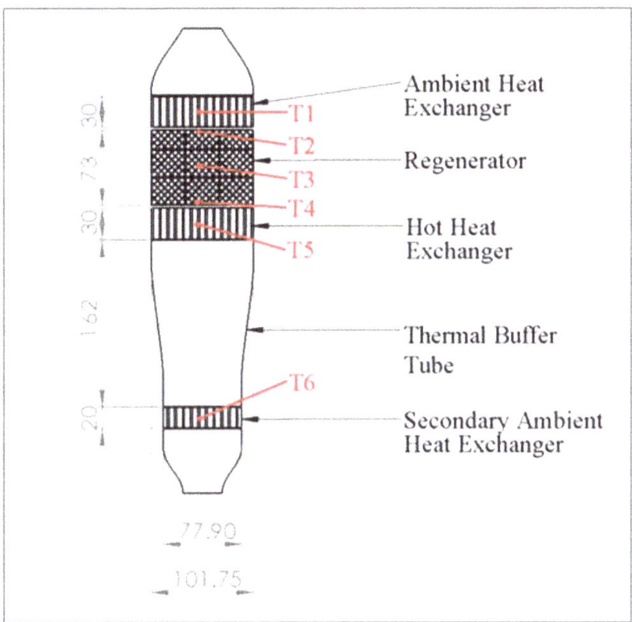

Figure 4. Cross-section of the thermoacoustic core consisting of ambient heat exchanger, regenerator packing, hot heat exchanger, thermal buffer tube and secondary ambient heat exchanger.

The hot heat exchanger has been manufactured from a low carbon steel. The choice of the material is based on a trade-off between the thermal conductivity and mechanical strength at elevated temperature. It has the face diameter of 102.2 mm (4 inch) and length of 40 mm along the flow direction. It is equipped with pairs of 100 W cartridge heaters. On the helium side, the comb-like structure creates channels of 1 mm width and fins that are 7 mm long and 0.5 mm thick. The porosity of the hot heat exchanger on the helium side is 34.4%. At the design amplitude, the peak-to-peak displacement is roughly one third of the heat exchanger length.

Below the hot heat exchanger is the thermal buffer tube providing thermal buffer between the hot and secondary ambient heat exchangers. It is 162 mm long having a conical middle section which reduces the internal diameter from 102.2 mm to 77.9 mm. The conical section is expected to reduce the Rayleigh streaming in the thermal buffer tube, as recommended by Swift [1].

The last part of the thermoacoustic core is the secondary ambient heat exchanger. The purpose of this part is to prevent heat from flowing beyond the core into the resonator. It is similar to the main ambient heat exchanger but with smaller dimensions. It is made of copper and has a porosity of 38%. The diameter of the heat exchanger on helium side is 77.5 mm, and its thickness is 20 mm. The fins are 0.5 mm in width; the fins are 9 mm long on the helium side and 5 mm long on the water side. At the design amplitude, the peak-to-peak displacement is roughly equal to the heat exchanger thickness.

The acoustic network delivers the acoustic power generated in the thermoacoustic core to the linear alternator branch and the rest is fed to the other thermoacoustic core. The network comprises of a straight standard $1\frac{1}{2}$ inch tube. The last 275 mm of the feedback loop is a standard 1-inch tube to adjust the phase difference at the linear alternator for a better performance. The linear alternator used in the rig is Q-Drive 1S132M. This alternator is asymmetric in that on one side the piston is exposed to the gas while the other side is connected to a shaft forming part of the electromagnetic armature.

Subsequently, the gas flow on two sides of the piston is not symmetrical—this feature is being corrected to some extent by introducing bespoke PVC inserts on the armature side, cf. [18].

3. Experimental Results

The experiment preparation starts with charging the engine with helium to 28 bar and then turning the cooling and heating systems on. The regenerators start generating weak oscillations when helium at the hot side of the regenerator reaches a temperature of 230 °C at a temperature difference of 185 °C. Normally, the engine does not amplify the weak acoustic oscillations (even at much higher temperature differences) to a level intense enough to drive the linear alternator. However, it has been found that in practice the intense acoustic wave can be excited by driving the linear alternator as an acoustic driver at a specific frequency. For instance, a few cycles of the piston excitation using a function generator and an amplifier at a frequency of 50.8 Hz was enough to excite the intense oscillation. This allows delivering an acoustic power to the cold side of the regenerator at a favourable acoustic phasing. An electrical control circuit was designed to protect the alternator and facilitate starting the engine. It switches the linear alternator connection in three ways based on the piston displacement measured by the laser displacement sensor: namely to function generator/amplifier, load resistance and a short circuit. At no oscillations present, the circuit connects the linear alternator to the function generator/amplifier which excites the piston for a few cycles at about 1.5 mm peak displacement. Once the engine amplifies the acoustic power and drives the piston over 2 mm peak displacement threshold, the circuit connects the linear alternator to the load resistance to dissipate the generated electricity and control the piston displacement. In case the engine drives the linear alternator close to its maximum stroke of 6 mm, the circuit switches the connection of the linear alternator to a short circuit to protect the alternator by stopping the piston oscillation.

At no oscillation condition, there is a high heat loss of about 450 W per stage from the hot heat exchanger (this value is deducted in performance calculation in this paper). As the hot heat exchanger is manufactured as one piece with the thermal buffer tube and the regenerator holder, the hot heat exchanger cannot be insulated from these two pieces. A possible way to reduce the conduction heat loss from the regenerator holder to the ambient heat exchanger is to place a low heat conductivity gasket between them. A gasket made out of thermiculite 715, Flexitallic model number SCRC04003T71515, was used. This gasket material has a low thermal conductivity of 0.3 W/m.K. The minimum available gasket thickness of 1.5 mm was selected. At this thickness, the gasket can seal up to 140 bar at a temperature of up to 540 °C.

The experiments showed that the insulating gasket improved the regenerator temperature difference and the performance of the engine. For example, the regenerator temperature difference increased from 297 °C to 308 °C and the generated electricity from 48.6 W (cf. previous work [18]) to 62.2 W at 900 W heating power, 28 bar mean pressure and 30.8 Ω load resistance.

There is an acceptable agreement between the measurements and the calculated results. The circular symbols in Figure 3 indicate the measured pressure amplitude and acoustic power (calculated using a two-microphone method, [20]), while the continuous line shows the calculated values along the engine. The measured values of pressure amplitude showed small differences between corresponding points for the two stages. This is caused by the construction of the asymmetrical linear alternator. All the left-hand side (LHS) points which are facing the armature of the linear alternator have slightly higher amplitudes than the right-hand side (RHS) points which are facing a flat side of the piston.

3.1. Effect of Load Resistance

In the experiments, a resistive load was connected to the linear alternator to measure and dissipate the generated electricity. The load value varied from 26.3 Ω to 92.5 Ω. Any value lower than 26.3 Ω damps the oscillations and the performance decreases at loads higher than 92.5 Ω, at similar heating power and mean pressure. For the linear alternator acting alone, at the nominal operating frequency, an increase in the load resistance will normally lead to a decreased acoustic load imposed by the linear alternator upon the oscillatory flow into the branch. As a result, the acoustic pressure in the branch, as well as the acoustic pressure difference across the alternator, increase in amplitude. The piston displacement also increases slightly, and so the linear alternator with the branch act more like a standing wave resonator which will draw less power from the engine loop. When such an alternator is coupled to the engine loop (as is the case here), the resulting acoustic pressure in the engine loop is of a higher amplitude.

The linear alternator piston applies an acoustic load to the acoustic field at each of the linear alternator branches. The value of the load resistance dominates the acoustic load which dominates the acoustic field and performance of the engine. Figure 5a shows the experimental results for the acoustic power generated by one engine stage, the acoustic power delivered to one side of the linear alternator and the piston displacement at different load resistances. Increasing the load resistance will decrease the linear alternator acoustic load which allows the piston to oscillate at higher displacement. Figure 5b shows the electricity output measured using the load resistance connected to the linear alternator against the predicted values obtained from the DeltaEC model. The experimental values are indicated by the symbols, while the continuous line shows the model prediction. The experimental values represent the average of four experimental readings, while the error bars correspond to their standard deviation. In experiments, the device generated 62.2 W of electricity at load resistance of 30.8 Ω (the best performance of the engine will be presented later in Section 4.1). The load resistance, the amplitude of acoustic pressure at the linear alternator and the temperature differential across the regenerator taken from experiments were applied as the boundary conditions for the DeltaEC model. A maximum electrical power of 85.02 W was predicted when the load resistance is 30.8 Ω.

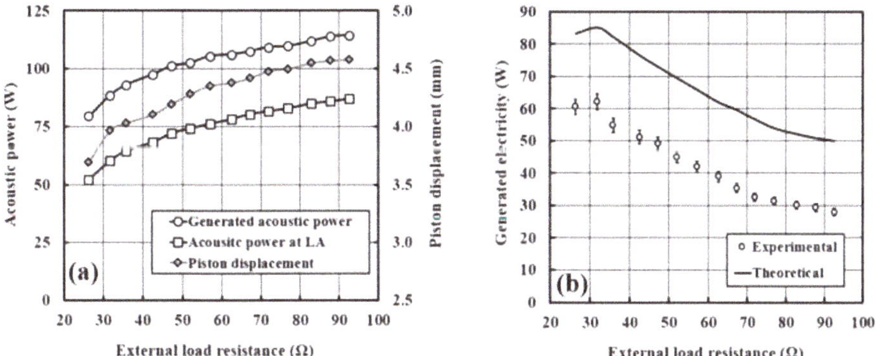

Figure 5. (**a**) Acoustic power generated by one engine stage, acoustic power delivered to one side of the linear alternator and piston displacement, (**b**) Electricity generated as a function of load resistance on the linear alternator. Heating power is 900 W.

The experimental and simulated electricity output profiles are comparable at all magnitudes of load resistance. However, significant discrepancies are observed. The main reason is that the phase difference between the volumetric flow rate and pressure at the linear alternator in the experiment is not the linear alternator's favourable acoustic condition set during modelling. For instance, the phase difference in the simulation is $-30°$, while in the experiment it is $10.5°$. Figure 2 shows that the linear alternator does not favour the experimental phase difference value. Unfortunately, when fitting the DeltaEC model to the experimental results one can only take care of a limited number of the most important parameters, for instance the pressure amplitudes and temperature data will take precedence over phase relationships. However, there are additional reasons for discrepancies between modelling and experiments. For example, acoustic streaming which occurs in the experiment (explained in Section 4.2) and which is responsible for transferring heat from the hot to ambient heat exchanger is not included in the model. Similarly, the acoustic power dissipation through major and minor losses was calculated in the simulations using steady flow loss correlations for oscillating flow. In addition, DeltaEC performs calculations by integrating the one-dimensional wave and heat transfer equations, while the actual flow and heat transfer physics is three-dimensional in experiments. The accuracy of DeltaEC simulation results in predicting turbulence phenomena remains questionable, which may also be the underlying problem.

The power output of the electricity generator is a product of the acoustic power delivered to the alternator and its acoustic-to-electric transduction efficiency. The transduction efficiency should reach the maximum when the load resistance is equal to the coil resistance of the alternator [21], i.e., 2 Ω. The electrical power produced is also proportional to the piston displacement to the power of two. This increases continuously as seen in Figure 5a. Figure 6a shows that the acoustic-to-electric efficiency falls from 62.7% to nearly 28.4% by increasing the load resistance from 26.3 Ω to 92.5 Ω. The thermal-to-electric efficiency reaches the maximum of 6.91% at the highest electrical output when applying a load resistance of 30.8 Ω. Figure 6b shows the temperature difference measured across the regenerator (T2 and T4 shown in Figure 4) at various load resistances. At the same heating power, the temperature differential across the regenerator reduces gradually vs. the load resistance. Clearly, the heat transfer between hot and ambient heat exchangers increases due to a high-volume flow rate, but unfortunately this is not coupled with the increase in electricity production. This is because, while the acoustic power increases (cf. Figure 5a) the phasing between pressure and velocity (cf. Figure 2) becomes less favourable, and so the electrical power extraction decreases.

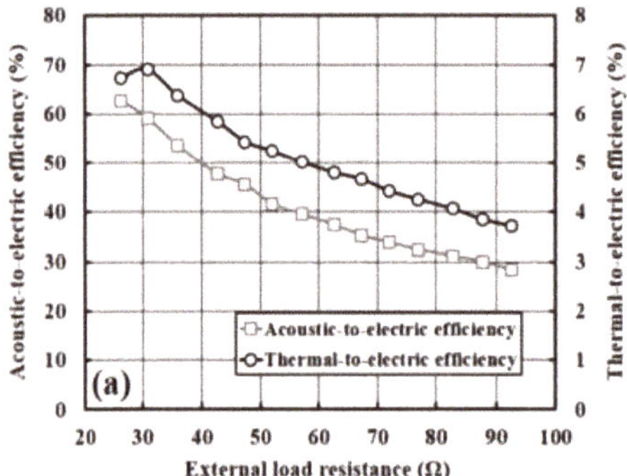

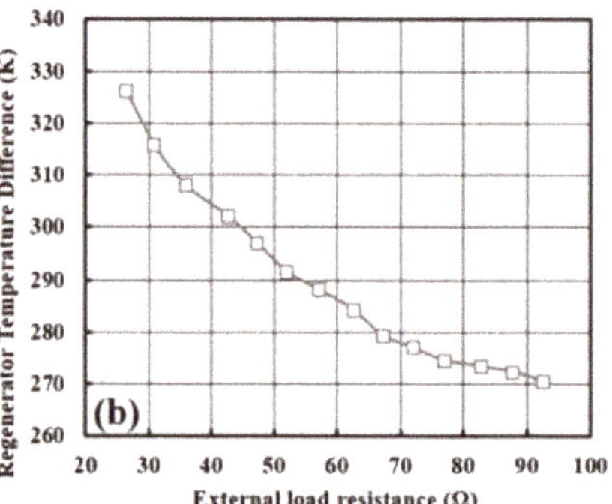

Figure 6. (**a**) Acoustic-to-electric and thermal-to-electric efficiencies vs. load resistance; (**b**) Temperature differential across the regenerator vs. load resistance. Heating power is 900 W.

3.2. Effect of Mean Pressure

The values of the mean pressure will affect both the power density of the acoustic field and the thermodynamic properties of the working gas. Swift et al. [1] determined the power density factor to be $p_m a A$, where a is the speed of sound, p_m is the mean pressure and A is the cross-sectional area. Higher power density will enable the thermoacoustic engine to run at higher acoustic impedance which in turn will allow higher acoustic to electric conversion at the linear alternator [18]. Varying the mean pressure changes the thermodynamic properties of the gas, including density and thermal and viscous penetration depths. These influence the processes of energy conversion in the thermoacoustic system. The mean pressure was varied in the range of 14–28 bar, at a load resistance of 30.8 Ω and heating power of 900 W. Any mean pressure less than 14 bar led to a non-harmonic oscillation which failed to maintain itself and was quickly damped.

Figure 7a shows the experimental values of the net acoustic power generated in one engine stage, the acoustic power delivered to one side of the linear alternator and the piston displacement vs. load resistance applied. It indicates that the engine performs better at higher mean pressure, as it provides higher power density and favourable phase difference to the linear alternator. Figure 7b shows the measured electrical power and the values predicted by the DeltaEC model. Symbols denote the experimental results, while the line shows the model prediction. In experiments, the engine generated 62.2 W of electricity at mean pressure of 28 bar (the best performance of the engine will be presented later in Section 4.1). The mean pressure, load resistance, the amplitude of acoustic pressure at the linear alternator and the measured temperature differential across the regenerator were applied as DeltaEC boundary conditions. A maximum electrical power of 85.02 W was predicted when the mean pressure is 28 bar. There is a clear trend of decreasing the generated electrical power with the decreasing mean pressure. The experimental and simulated electricity output profiles are comparable at all magnitudes of mean pressure. However, significant discrepancies are observed, which were explained in Section 3.1 in some detail and these explanations are applicable here too. Additional Figure 8 shows the effect of mean gas pressure on the acoustic-to-electric efficiency and thermal-to-electric efficiency (cf. Figure 8a) and the measured drive ratio (cf. Figure 8b).

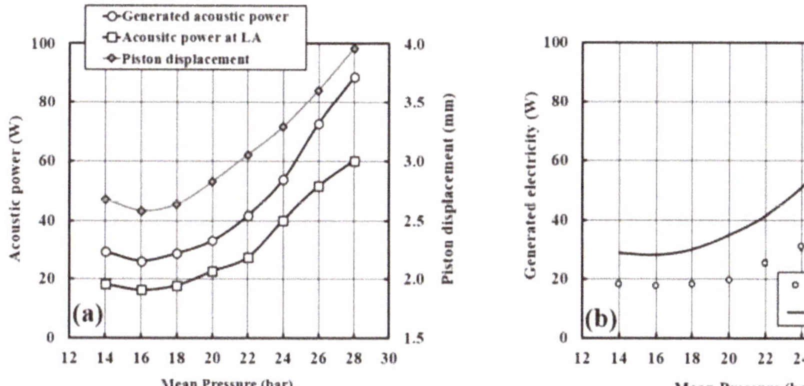

Figure 7. (**a**) Acoustic power produced in one engine stage, acoustic power on one side of the linear alternator and piston displacement vs. mean pressure; (**b**) Electricity generated by the device vs. mean pressure. Heating power is 900 W.

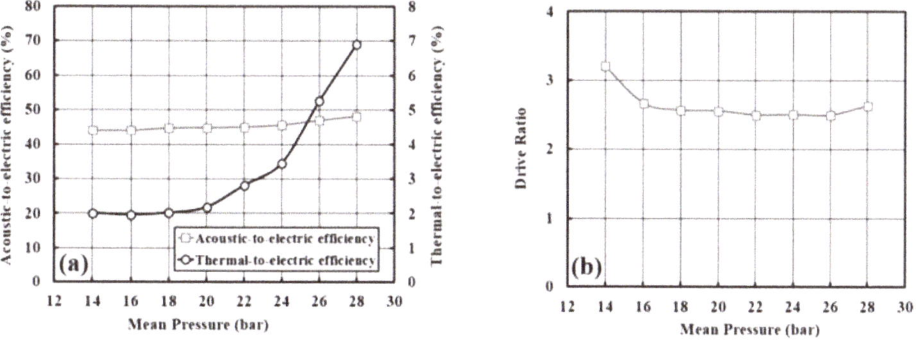

Figure 8. (**a**) Acoustic-to-electric and thermal-to-electric efficiencies vs. mean pressure; (**b**) Drive ratio vs. mean pressure. Heating power is 900 W.

Focusing on the low mean pressure range, it is not clear why the generated electrical power, generated acoustic power, drive ratio and piston displacement increase slightly when the mean pressure drops from 16 to 14 bar. Theoretically, these values should decrease based on the argument of power density being proportional to mean pressure. Most likely, this counterintuitive performance enhancement might be attributed to a phase difference at the regenerator being closer to the traveling wave for 14 bar (compared to 16 bar), which could lead to generating a higher acoustic power. Swift [1] pointed out that a resonator channel acts as an acoustic inertance and compliance simultaneously. Both contribute to the behaviour of the wave propagation in the channel. However, reducing the mean pressure increases the acoustic compliance effect which shifts the volumetric flow rate phase, while decreasing the acoustic inertance effect of a resonator which shifts the pressure phase. Unfortunately, the current setup does not allow the detailed measurements to validate this point. However, it is possible to inspect the DeltaEC modelling results in terms of the phase angle between oscillating pressure and volumetric velocity. When the mean pressure reduces from 28 to 16 bar the phase angle increases from 58.9° to 68.4°, i.e., the wave becomes "less travelling" and "more standing". On the other hand, a further decrease of mean pressure from 16 to 14 bar causes the phase angle to decrease from 68.4° to 68.0° to make the wave slightly "more travelling" again, which explains the apparent improvement of generator performance.

3.3. Effect of Heating Power

Heating power and oscillation intensity are the two parameters determining the regenerator hot side temperature. However, heating power is the dominant parameter determining the ability to maintain a high temperature difference across the regenerator during the oscillation. In this section, the value of the heating power represents the summation of the equal heating power of the two stages. At no oscillation, there is a high heat loss of about 450 W per stage from the hot heat exchanger which is deducted in performance calculations in this paper. The heating power was varied from the minimum power of 500 W capable of maintaining oscillations to a maximum of 1700 W, at 28 bar mean pressure and load resistance of 30.8 Ω. Figure 9a shows the generated electrical power at different heating power for both experiments and simulation. For both, the maximum is reached at a heating power of 1300 W. In the experiments, a maximum electrical power output of 72.5 W was obtained at 5.58% of thermal-to-electric efficiency, while the maximum efficiency of 7.3% was obtained at heating power of 700 W generating 51.1 W of electricity, as shown in Figure 10a. The thermal-to-electric efficiency decreases between 700 W and 1700 W.

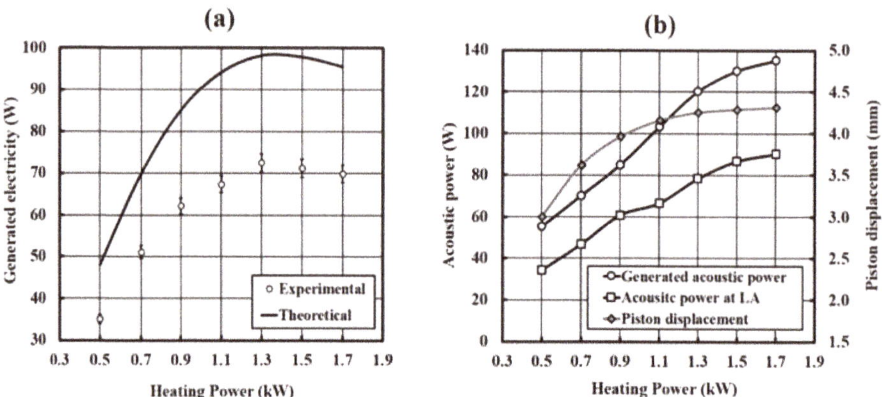

Figure 9. (**a**) Electricity generated by the device vs. heating power; (**b**) Acoustic power generated in one engine stage, acoustic power on one side of the alternator and piston displacement vs. heating power.

The existence of maximum generated electricity for heat input of 1300 W can be explained as a combination of two effects: On the one hand, increasing the heating power leads to the increase of the regenerator temperature difference (Figure 10b), generated acoustic power, acoustic power at the linear alternator and the piston displacement (Figure 9b). At the same time, the measured difference between volumetric flow phase and pressure phase increases towards the unfavourable values for the linear alternator which leads to the decrease in the alternator acoustic-to-electric efficiency, as shown in

Figure 10a. In experiments, the phase difference at 900 W heating power is 10.5° and it increases up to 34° at 1700 heating power. Figure 2 shows how shifting the phase difference affects the generated power and efficiency of the linear alternator. These competing effects lead to a maximum electricity production at 1300 W heating power rather than the highest heating power.

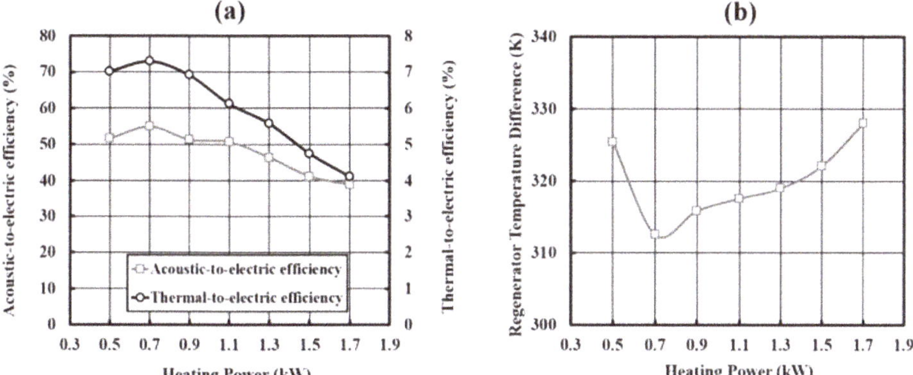

Figure 10. (**a**) Acoustic-to-electric and thermal-to-electric efficiencies vs. heating power; (**b**) Regenerator temperature difference vs. heating power.

The regenerator acts as an acoustic power amplifier. However, the flow resistance inside the regenerator plays a vital role in the power amplification as reported by Yu and Jaworski [22]. At a certain acoustic impedance, the flow resistance will dissipate most of the acoustic power fed through the regenerator cold end and this will decrease the acoustic power generation. Under such circumstances, the externally set temperature gradient will not have a significant effect. In fact, low conversion of heat into sound will lead to heating up of the regenerator hot side as shown at 500 W heating power in Figure 10b.

4. System Debugging

The aim of the debugging and optimization process was to solve and/or eliminate two problems: self-starting and streaming.

4.1. Start-Up Improvement

As mentioned in Section 3, the engine in its baseline configuration could not self-start and required "kick-starting" where a few cycles of initial excitation came from externally exciting the linear alternator. The successful solution to this problem turned out to be a slight reduction in the flow resistance. Yu and Jaworski [22] highlighted the relation between the flow resistance and local acoustic impedance and their effect on the net acoustic power and acoustic power input. It was concluded that the flow resistance plays a key role in determining the regenerator impedance as it determines the volumetric flow rate at a specific pressure amplitude. At a given pressure amplitude, the higher flow resistance increases the acoustic impedance by decreasing the volumetric flow rate.

Reducing the flow resistance in the regenerator was a possible solution to reduce the acoustic impedance, and hence increase the acoustic power leaving the regenerator, at specific acoustic power entering it, by reducing the acoustic power dissipated at the regenerator. The flow resistance could be reduced by decreasing the length of the regenerator or increasing the cross-sectional area. In the current research, the regenerator holder was welded to the hot heat exchanger and a heavy flange, therefore its length and diameter are fixed. The only way to reduce the regenerator length is to replace some of the regenerator mesh screens with coarse mesh (same as used for the spacers, cf. Section 2). The coarse mesh screens were applied on the cold side of the regenerator, for ease of replacement.

The effect of the regenerator length was investigated experimentally at 30.8 Ω load resistance, 1300 W heating power and 28 bar mean pressure. The regenerator length was increased once and reduced twice by a 1.2 mm step, which is the thickness of a single coarse mesh.

The engine self-starts at a regenerator length of 71.8 mm and 70.6 mm. The oscillation starts at a regenerator temperature difference of 280 °C. The reduction of the flow resistance was found to enhance the performance by a very small fraction. Figure 11 shows the effect of regenerator length on the generated electricity and thermal-to-electric efficiency. The new maximum generated electricity is 73.3 W at 5.64% thermal-to-electric efficiency. The relative Carnot efficiency is 11.3%, drive ratio is 3.4% at a regenerator temperature difference of 288.8 °C.

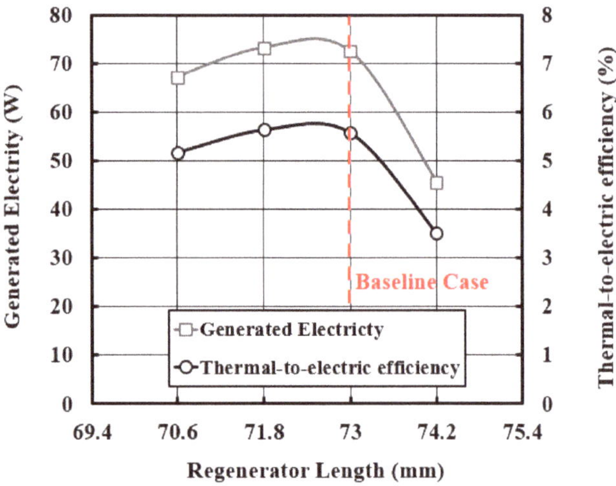

Figure 11. Generated electricity and thermal-to-electric efficiency versus the regenerator length.

4.2. Efforts Towards Supressing Streaming

Gedeon streaming exists in looped-tube or toroidal devices only. The reason is that a closed loop topology encourages a steady flow to circulate along such resonators. Gedeon [23] explained it as mass flow in the Stirling engines and pulse tube cryocoolers with a closed loop which leads to time-averaged convection enthalpy flux from the hot to the cold side. This phenomenon wastes heat in a thermoacoustic engine by removing heat from the hot side to the ambient of the regenerator without generating acoustic power. The devices suffering from a non-zero mass flow through the porous medium will show a non-linear temperature distribution within the porous medium. All the experimental tests showed a non-uniform temperature distribution, an example being shown in Figure 12. Many researchers [1,24,25] summarized that this kind of streaming can be suppressed either by placing a latex membrane or applying a non-symmetric flow resistance such as jet pump. The latex or elastic membrane will be transparent to the acoustic power while forming a barrier to the streaming flow, hence eliminating it.

The elastic membrane needs to be placed close to the minimum volumetric flow rate to suppress this streaming at the lowest possible acoustic power loss. Figure 3 shows that the best location is near the main ambient heat exchanger. Unfortunately, this location in the experimental rig was used to feed through the thermocouples, and hence, the membrane could not be placed there. Potential locations are between two flanges at three locations, as shown in Figure 13a. Figure 13b shows the locations of the membrane with reference to the theoretical volumetric flow rate.

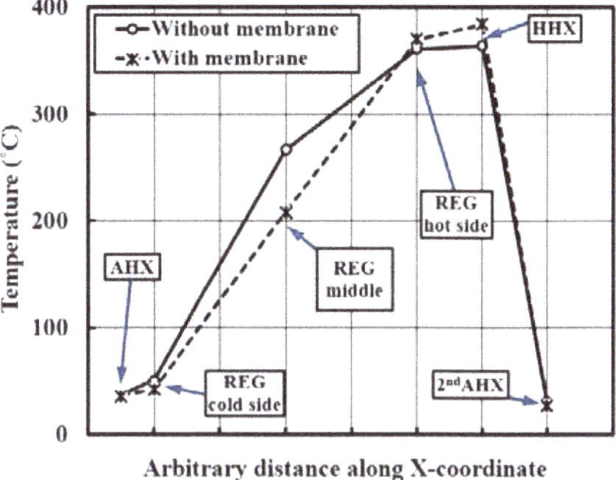

Figure 12. The effect of the presence of a concave membrane on the temperature profiles in the thermoacoustic core.

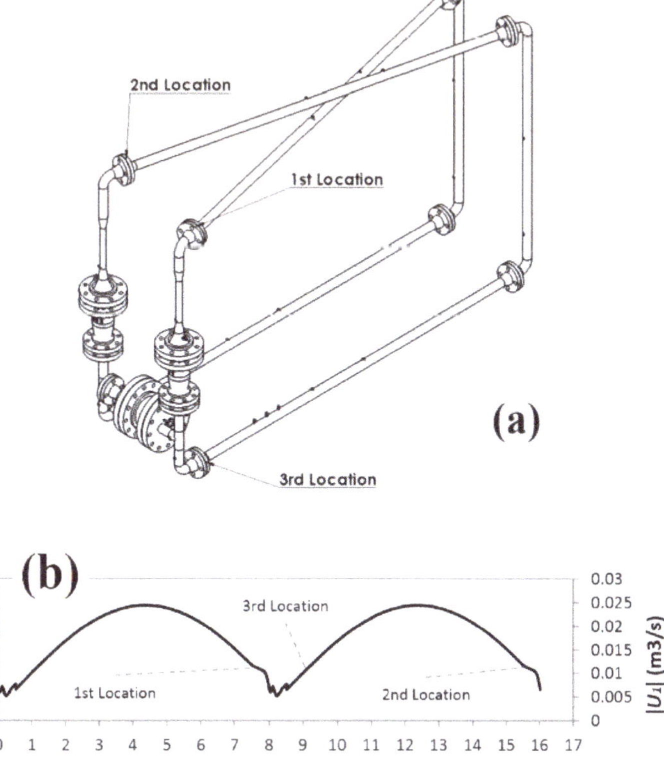

Figure 13. (**a**) Locations of the membrane along the engine loop; (**b**) locations of the membrane on the volumetric flow rate graph.

The membrane was selected based on its material elastic properties and thickness. A sheet of 100% genuine latex of 0.25 mm thickness, was used. Figure 14b–d show the three profiles of the membrane that were tested: flat, concave and loose. Figure 14a shows an example of an assembled membrane.

All three profiles were used in single and double locations. They were used on their own at the 1st and 3rd location, as shown in Figure 13a, and together at the 1st and 2nd location. The concave profile was made by continuous stretching and heat treatment.

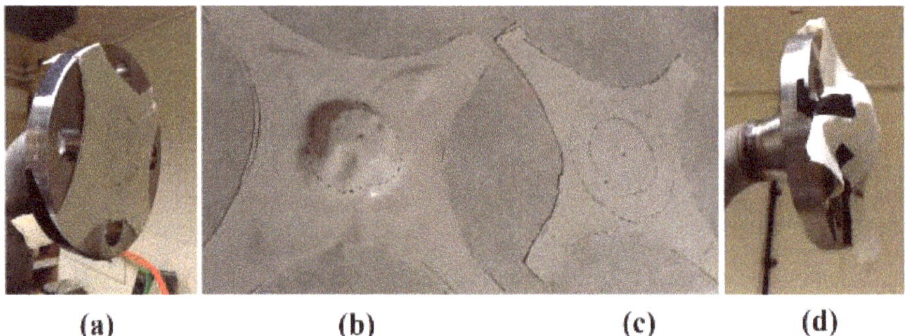

Figure 14. (**a**) A membrane assembled on a flange; (**b**) concave membrane; (**c**) flat membrane; (**d**) loose membrane.

The experimental results showed that a single membrane placed at any location or a double membrane can suppress the Gedeon streaming and generate a uniform temperature distribution along the regenerator, as shown in Figure 12. Unfortunately, the membranes also act as flow resistance and dissipate the acoustic power. The generated electricity for the tested membrane locations and profiles varied from 0.5 W to 4.2 W. The highest performance was achieved by using one concave membrane at the 1st location.

5. Feedback Loop Optimization

This section presents a DeltaEC study to propose a modified design of the experimental apparatus to reduce its size. In particular, the new model considers shortening the feedback loop while keeping the current thermoacoustic cores and alternator holder unchanged. The current engine is 16.1 m long, of which approximately 15 m is a constant diameter feedback loop.

The function of the feedback loop is to deliver acoustic power to the regenerator at a favourable acoustic phasing. The current uniform section feedback loop shifts the pressure phase by 175° and volumetric flow rate phase by 50°. This phase shift could be achieved within a much shorter length by using a variable cross-section feedback loop. A wide cross-section pipe shifts the phase of the volumetric velocity since it acts as an acoustic compliance, while a narrow pipe shifts the pressure phase since it acts as an acoustic inertance [1]. The combination of compliance-inertance loop shifts the acoustic phasing in much shorter length than the constant diameter loop.

Many configurations of feedback loop combining inertances and compliances were studied, however, the paper will present the one that provides the shortest length without dissipating a high share of engine's generated acoustic power. Firstly, DeltaEC was used to simulate the acoustic field in the feedback loop only. The model considered the acoustic wave characteristics at the beginning and the end of the thermoacoustic core as boundary conditions of the compliance-inertance feedback loop. The new feedback loop reduced the engine length from 16.1 to 7.5 m. Subsequently, it was tested numerically on a full model and showed the same performance.

For a pipe of a certain length and diameter, the phase shifting capabilities strongly depend on the acoustic wave characteristics at the inlet. In this study, the phase shifting results for the local acoustic wave at the engine feedback inlet are shown in Figure 15. The selection of the pipe dimensions to act as an acoustic inertance is based on the pressure phase shifting and acoustic power dissipation. Figure 15 shows an example of the pressure phase shifting and acoustic power dissipation for different sizes of pipes at an acoustic impedance of 5.1 M Pa·s/m^3, 55° phase difference and 56.6 Hz frequency (which are the values at the inlet of the feedback loop to be replaced). A small diameter pipe can shift the pressure phase at a shorter length than larger diameter, however, it will dissipate higher acoustic power. For every pipe diameter there is a length range that appears to be very sensitive to the pressure phase shifting. This region needs to be avoided.

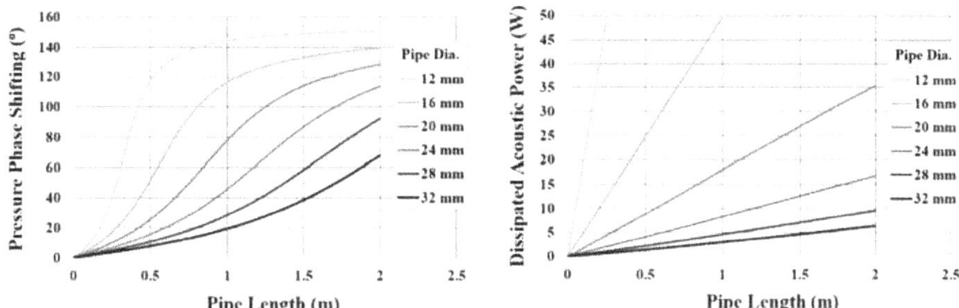

Figure 15. Effect of pipe length and diameter of the inertance pipe on (**a**) pressure phase shifting; (**b**) dissipated acoustic power.

Large diameter pipe shifts the velocity phase at low acoustic power dissipation. The phase shifting capabilities strongly depend on the acoustic wave characteristics at the inlet. The selection of the pipe diameter and length is based on the required phase shifting, at the acoustic wave characteristics at the inlet. Figure 16 shows an example of the effect of pipe length and diameter on the velocity phase shifting at an acoustic impedance of 3.8 M Pa·s/m^3, 39° phase difference and 56.6 Hz frequency (which are the values in the middle of the feedback loop where the compliance will be placed). Similar to the selection of inertance, the steep change regions of the phase shifting need to be avoided for better solution stability.

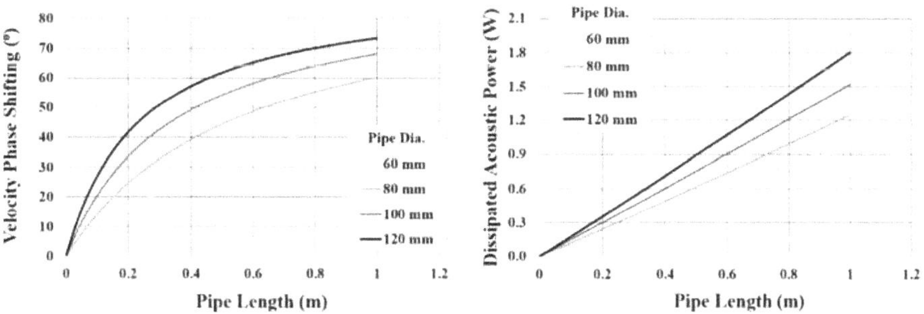

Figure 16. Effect of pipe length and diameter of the compliance pipe on (**a**) velocity phase shifting; (**b**) dissipated acoustic power.

The best feedback loop configuration studied consisted of inertance-compliance-inertance. The inertance was split into two parts with a compliance sandwiched between them so that the inertance after the compliance will be in lower acoustic impedance region. This will allow the use of a thinner pipe in the second inertance (after the compliance) which will lead to a shorter feedback loop without creating a high acoustic power loss. After continuous optimisation process, the ideal diameters of the inertances and compliance were replaced by those available for the commercially available pipes. This generated some discrepancy which was actually found to be advantageous as the new configuration allowed a reduction of the phase difference near the middle of the regenerator closer to the traveling wave phase difference (namely from 26° to 9°). This helped to increase the generated acoustic power, at a similar regenerator temperature difference, from 123.5 W to 159.5 W. However, the extra generated acoustic power is dissipated in the feedback loop. The original feedback loop dissipates 23.8 W while the new loop dissipates 59.5 W, as shown in Figure 17c. Figure 18 compares the engine configuration for both old and new feedback loops.

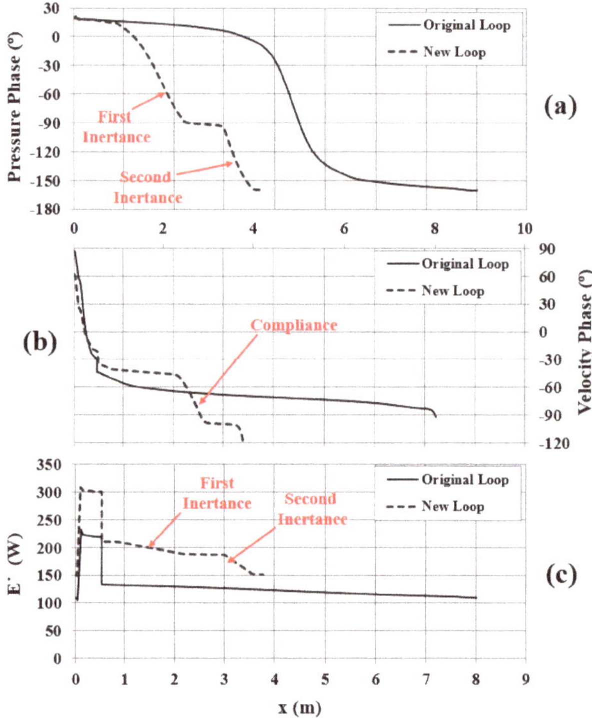

Figure 17. One-stage simulation results comparing the original and new feedback loop: (a) pressure phase, (b) velocity phase and (c) acoustic power flow along the engine.

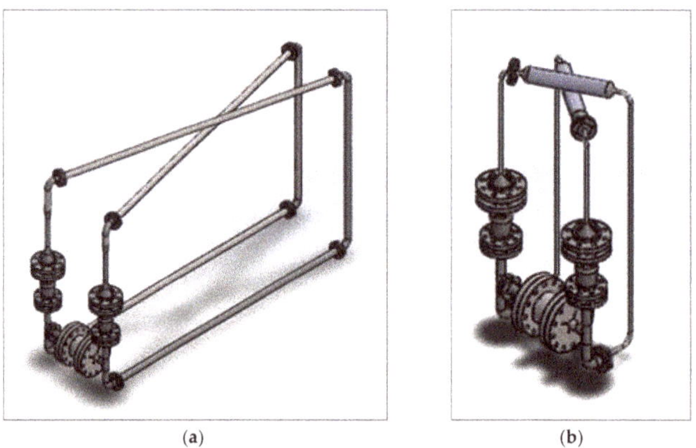

Figure 18. Electricity generator with: (**a**) original feedback loop and (**b**) improved feedback loop.

Figure 17 compares the pressure phase, velocity phase and acoustic power in one stage for the original and new feedback loop. The first section of the feedback loop is a $1\frac{1}{2}$ inch diameter pipe with 300 mm length, which is part of the previous configuration. This is followed by a standard cone leading to the first inertance which is 1313 mm long and has a $\frac{3}{4}$ inch (20.9 mm) diameter. This shifts the pressure phase by approximately 100°, as shown in Figure 17a. The acoustic compliance is 420 mm long and 3 inches (77.9 mm) in diameter, which shifts the volumetric flow velocity phase by approximately 40°, as shown in Figure 17b. The second inertance is 584 mm long with dimeter of $\frac{1}{2}$ (12.2 mm) inch, and shifts the pressure phase by approximately 62°, as shown in Figure 17a. Both reducers connecting the compliance to the two inertances are non-standard, and of 50 mm length. The length and diameter of the two inertances and compliance were carefully optimized aiming to achieve the acoustic conditions at the shortest length possible.

6. Conclusions

Current work is focused on detailed studies and potential further improvements of a two-stage traveling-wave thermoacoustic engine. Here, the configuration of two identical half-wavelength stages allows the coupling of the linear alternator to two points with out-of-phase acoustic field, i.e., in so-called "push-pull" mode in an attempt to improve the impedance matching of the alternator to the engine, as well as reduce the ultimate cost by requiring only one alternator for two power extraction points. Modelling approaches are combined with experimental work in order to improve the overall performance of the prototype as well as improve the design to achieve more compact size.

In particular, the work presented here deals with system debugging, for instance improvements in electrical power output through limiting the axial heat leaks, and investigating the effects of regenerator length (i.e., regenerator impedance) on the start-up conditions in order to allow the engine self-excite without external power input and application of elastic membrane to eliminate Gedeon streaming. In addition, the paper presents a detailed account of the characterisation of the electricity generator system from the point of view of the mean pressure (range 14–28 bar) and heating power (500–1700 W) and load resistance (26.3 Ω to 92.5 Ω). It was found that the maximum electricity generated can reach 73.3 W at the heat input of 1300 W, load resistance of 30.8 Ω and mean pressure of 28 bar, with the overall thermal-to-electric efficiency of 5.64%. The maximum thermal-to-electric efficiency of 7.3% was obtained at heat input power of 700 W, while generating 51.1 W of electricity.

Finally, a design exercise was carried out aiming at reducing the size of the device while maintaining the same levels of performance. DeltaEC simulations have shown that introducing an inertance-compliance-inertance coupling instead of the constant diameter feedback pipe can reduce the resonator length from 16.1 m to 7.5 m, leading to a much smaller volume of the device.

Author Contributions: A.J.J. and X.M. guided the research. A.H. carried out the research by conceptualising the design, DeltaEC modelling, experimental data collection and producing the first draft. All authors critically evaluated the data and contributed to improving the manuscript through a number of iterations, in particular interpretation and discussion.

Funding: This research was funded by Royal Society Industry Fellowship scheme (Ref. IF110094, 2012-2015) and EPSRC UK under HARP[2] programme (Ref. EP/R023328/1, 2018-2021).

Acknowledgments: Artur J. Jaworski acknowledges funding from the Royal Society Industry Fellowship scheme (Ref. IF110094, 2012-2015) and EPSRC UK under HARP[2] programme (Ref. EP/R023328/1, 2018-2021). All authors acknowledge Innovate UK for financial support under TITAN project (ref no. 131497).

Conflicts of Interest: The authors declare no conflict of interest.

References

1. Swift, G.W. *Thermoacoustics: A Unifying Perspective for Some Engines and Refrigerators*; Acoustical Society of America (through the American Institute of Physics): Melville, NY, USA, 2002.
2. Luo, E.; Dai, W.; Zhang, Y.; Ling, H. Thermoacoustically driven refrigerator with double thermoacoustic-Stirling cycles. *Appl. Phys. Lett.* **2006**, *88*, 074102. [CrossRef]
3. Dai, W.; Luo, E.; Hu, J.; Ling, H. A heat-driven thermoacoustic cooler capable of reaching liquid nitrogen temperature. *Appl. Phys. Lett.* **2005**, *86*, 224103. [CrossRef]

4.   Dai, W.; Yu, G.; Zhu, S.; Luo, E. 300 Hz thermoacoustically driven pulse tube cooler for temperature below 100 K. *Appl. Phys. Lett.* **2007**, *90*, 024104. [CrossRef]

5.   Yazaki, T.; Iwata, A.; Maekawa, T.; Tominaga, A. Traveling wave thermoacoustic engine in a looped tube. *Phys. Rev. Lett.* **1998**, *81*, 3128. [CrossRef]

6.   Chen, B.; Yousif, A.A.; Riley, P.H.; Hann, D.B. Development and assessment of thermoacoustic generators operating by waste heat from cooking stove. *Engineering* **2012**, *4*, 25859. [CrossRef]

7.   Zhang, X.; Chang, J. Onset and steady-operation features of low temperature differential multi-stage travelling wave thermoacoustic engines for low grade energy utilization. *Energy Convers. Manag.* **2015**, *105*, 810–816. [CrossRef]

8.   de Blok, K. Novel 4-stage traveling wave thermoacoustic power generator. In Proceedings of the ASME 2010 3rd Joint US-European Fluids Engineering Summer Meeting Collocated with 8th International Conference on Nanochannels, Microchannels, and Minichannels, Montreal, QC, Canada, 1–5 August 2010; pp. 73–79.

9.   Hu, J.Y.; Luo, E.C.; Zhang, L.M.; Chen, Y.Y.; Wu, Z.H.; Gao, B. Analysis of a displacer-coupled multi-stage thermoacoustic-Stirling engine. *Energy* **2018**, *145*, 507–514. [CrossRef]

10.   de Blok, K. Multi-stage traveling wave thermoacoustics in practice. In Proceedings of the 19th International Congress on Sound and Vibration, Vilnius, Lithuania, 8–12 July 2012.

11.   Zhang, X.; Chang, J.; Cai, S.; Hu, J. A multi-stage travelling wave thermoacoustic engine driven refrigerator and operation features for utilizing low grade energy. *Energy Convers. Manag.* **2016**, *114*, 224–233. [CrossRef]

12.   Yang, R.; Wang, Y.; Jin, T.; Feng, Y.; Tang, K. Development of a three-stage looped thermoacoustic electric generator capable of utilizing heat source below 120 C. *Energy Convers. Manag.* **2018**, *155*, 161–168. [CrossRef]

13.   Yang, R.; Wang, Y.; Feng, Y.; Jin, T.; Tang, K. Performance of a looped thermoacoustic engine with multiple loads capable of utilizing heat source below 200 °C. *Appl. Therm. Eng.* **2019**, *148*, 516–523. [CrossRef]

14.   Li, L.; Wu, Z.; Hu, J.; Yu, G.; Luo, E.; Dai, W. A novel heat-driven thermoacoustic natural gas liquefaction system. Part I: Coupling between refrigerator and linear motor. *Energy* **2016**, *117*, 523–529. [CrossRef]

15.   Li, D.; Wu, Z.; Luo, E.; Zhang, L. Experimental Investigation on the Conversion between Heat and Power of the kW-class Thermoacoustic Engine. *Energy Procedia* **2014**, *61*, 1058–1062.

16.   Wu, Z.; Yu, G.; Zhang, L.; Dai, W.; Luo, E. Development of a 3kW double-acting thermoacoustic Stirling electric generator. *Appl. Energy* **2014**, *136*, 866–872. [CrossRef]

17.   Bi, T.; Wu, Z.; Zhang, L.; Yu, G.; Luo, E.; Dai, W. Development of a 5kW traveling-wave thermoacoustic electric generator. *Appl. Energy* **2017**, *185*, 1355–1361. [CrossRef]

18.   Hamood, A.; Jaworski, A.J.; Mao, X.; Simpson, K. Design and construction of a two-stage thermoacoustic electricity generator with push-pull linear alternator. *Energy* **2018**, *144*, 61–72. [CrossRef]

19.   Ward, B.; Clark, J.; Swift, G. *Design Environment for Low (Amplitude Thermoacoustic Energy Conversion DeltaEC Version 6.3 b11 Users Guide*; Los Alamos National laboratory: Santa Fe, NM, USA, 2012.

20.   Fusco, A.M.; Ward, W.C.; Swift, G.W. Two-sensor power measurements in lossy ducts. *J. Acoust. Soc. Am.* **1992**, *91*, 2229–2235. [CrossRef] [PubMed]

21.   Yu, Z.; Backhaus, S.; Jaworski, A.J. Design and Testing of a Travelling-Wave Looped-Tube Engine for Low-Cost Electricity Generators in Remote and Rural Areas. In Proceedings of the 7th International Energy Conversion Engineering Conference, Denver, CO, USA, 2–5 August 2009. [CrossRef]

22.   Yu, Z.; Jaworski, A.J. Impact of acoustic impedance and flow resistance on the power output capacity of the regenerators in travelling-wave thermoacoustic engines. *Energy Convers. Manag.* **2010**, *51*, 350–359. [CrossRef]

23.   Gedeon, D. DC gas flows in Stirling and pulse tube cryocoolers. In *Cryocoolers 9*; Springer: New York, NY, USA, 1997; pp. 385–392.

24.   Wilcox, D.; Spoor, P. Gedeon streaming suppression in a small scale thermoacoustic-Stirling engine-generator. *J. Acoust. Soc. Am.* **2014**, *135*, 2408. [CrossRef]

25.   Tijani, M.E.H.; Spoelstra, S. High temperature thermoacoustic heat pump. In Proceedings of the 19th International Congress on Sound and Vibration, Vilnius, Lithuania, 8–12 July 2012; pp. 8–12.

Article

Investigation of the Concepts to Increase the Dew Point Temperature for Thermal Energy Recovery from Flue Gas, Using Aspen®

Nataliia Fedorova [1,2], Pegah Aziziyanesfahani [1], Vojislav Jovicic [1,2,*], Ana Zbogar-Rasic [1], Muhammad Jehanzaib Khan [1] and Antonio Delgado [1,2]

[1] Institute of Fluid Mechanics (LSTM), Friedrich-Alexander University (FAU), 91058 Erlangen, Germany; nataliia.fedorova@fau.de (N.F.); pegah.aziziyan@gmail.com (P.A.); ana.zbogar-rasic@fau.de (A.Z.-R.); muhammad.j.khan@fau.de (M.J.K.); antonio.delgado@fau.de (A.D.)

[2] Erlangen Graduate School in Advanced Optical Technologies (SAOT), 91054 Erlangen, Germany

* Correspondence: vojislav.jovicic@fau.de; Tel.: +49-9131-85-29492

Received: 31 March 2019; Accepted: 23 April 2019; Published: 26 April 2019

Abstract: Thermal energy of flue gases (FG) dissipating from industrial facilities into the environment, constitute around 20% of the total dissipated thermal energy. Being part of the FG, water vapour carries thermal energy out of the system in the form of the latent heat, which can be recovered by condensation, thus increasing the overall efficiency of an industrial process. The limiting factor in this case is the low dew point temperature (usually 40–60 °C) of the water vapour in the FG. The increase of the dew point temperature can be achieved by increasing the water content or pressure. Taking these measures as a basis, the presented work investigated the following concepts for increasing the dew point temperature: humidification of the flue gas using water, humidification using steam, compression of the FG and usage of the steam ejector. Modelling of these concepts was performed using the commercial software Aspen®. The humidification of the FG using water resulted in the negligible increase in the dew point (3 °C). Using steam humidification the temperatures of up to 92 °C were reached, while the use of steam ejector led to few degrees higher dew point temperatures. However, both concepts proved to be energy demanding, due to the energy requirements for the steam generation. The FG compression enabled the achievement of a 97 °C dew point temperature, being both energy-efficient and exhibiting the lowest energy cost.

Keywords: thermal energy recovery; flue gas; dew point temperature; condensation; Aspen®

---

1. Introduction

The dissipation of thermal energy from industrial facilities during production processes is and has been a challenging issue worldwide. Flue gases from industrial processes constitute around 20% of the total dissipated energy [1–3]. Water vapour, as a part of the flue gas, carries latent heat, which can be recovered. Depending on the industry, flue gas temperatures vary from 120 to 200 °C [4–6], whereas water vapour content can be up to 90 %vol. For example, as presented in Table 1, in the flue gas of natural gas-fired boilers 20 %vol accounts for the water vapour [3,6], while in the potato crisps manufacturing its content is around 40 %vol [5]. Unlike the flue gas from the drying processes, which can contain up to 90% of the water vapour [7], most of the flue gases from the other processes have a much lower water vapour content. Baking, textile, pulp and paper industries have a potential for the recovery of thermal energy from the flue gas, due to the large quantity of water vapour in their flue gases and massive production rates [8–10].

Table 1. Composition of different industrial flue gases.

Industry	H_2O [%]	CO_2 [%]	N_2 [%]	O_2 [%]
Potato crisps manufacturing [5]	41.1	5	50.6	3.3
Natural gas-fired boiler [11]	18–20	8–10	67–72	2–3
Coal-fired boiler [11]	8–10	12–14	72–77	3–5

The difference in the concentration levels of water vapour in flue gases of different processes is in general less influenced by the type of the fired fuel and more by the amount of the water vapour originating from the process itself. In many of the processes, especially in the food industry, a single fan at the outlet of the system evacuates all the gases from the system. In this way, the flue gas originating from the burners is commonly mixed with the gases originating from the processed product and also often with the excess air, sucked into the system from the surrounding. In the processes like drying or baking, the raw material (paper, potato, dough, etc.) can release significant amounts of the water vapour leading to the increase of the overall water vapour concentration in the flue gas at the outlet of the system.

The higher the amount of water vapour in the flue gas, the more latent heat can be recovered from it. Therefore, this issue is essentially relevant for industries with water vapour-rich flue gases. Recovering sensible and latent heat by water vapour condensation from the flue gas has been reported by different sources as a promising way to improve the total energy efficiency by around 10% [3,7,12]. Herewith come economic and environmental benefits.

A traditional unit for thermal energy recovery from the flue gas is a gas-to-liquid condensing heat exchanger. During the heat exchange with the cooling liquid, the flue gas temperature is reduced below the dew point, so that the water vapour is condensed and the release of the latent and sensible heat occurs. Porous and non-porous gas separation membranes have been actively developed for the simultaneous heat and water recovery from the flue gas [7,13]. The growing interest for the thermal energy recovery application is connected to the Organic Rankine Cycle (ORC) technology for generation of electricity and absorption refrigerator (AR) technology for driving cooling processes [5,14–16].

One of the factors limiting the application of the aforementioned recovery systems is the low dew point temperature of the water vapour in the flue gas, usually in the range of 40–60 °C, i.e., the temperature increase of the working liquid is not sufficient to be used as process heat within the production cycle. Increasing the dew point temperature can help to upgrade the temperature level of working fluids and to reach higher energy efficiencies. For example, in AR the increase of the thermal source temperature from 80 °C to 86 °C, increases the coefficient of performance by 7% [17,18].

On the example of the condensing heat exchangers (HE) as a recovery technology, the increase of the dew point temperature of the water vapour contained in the flue gas leads to obtaining the condensate and the cooling water at the higher temperature level. In this way, these otherwise waste streams can be further used within an industry as process heat for a wide range of applications, e.g., space heating, heating, ventilation and air-conditioning (HVAC) of office areas, as sanitary water, for washing and cleaning in the production areas, for preparation of dough in the case of a baking industry, for preheating of fuel and combustion air in the case of a power generation industry [19–21], etc. Other industry-dependent in-plant demands for recovered thermal energy should be generally determined by conducting a full energy audit.

Process modelling is one of the ways to analyse the thermal energy recovery from industrial processes. The licensed software Aspen Plus® and Aspen® HYSYS® find applications in modelling of chemical, biological and physical processes. It enables to model complex processes using simple models with built in unit operation models (e.g., heat exchangers, columns, reactors, mixers, splitters, etc.) and property methods.

The Aspen® software is a widely used tool for analysing system performances. Jana et al. [22] used it to model the utilisation of the waste heat by means of a condensing HE for the post-combustion CO_2 capture. Luyben [23] simulated HEs with phase changes, namely a condenser and an evaporator, for low-level energy recovery with n-hexane as a working fluid. Duan et al. [24] analysed coal gasification, which included a gasifier and a boiler to recover steam and blast furnace slag. Ishaq et al. [25] modelled a trigeneration system for electricity, hydrogen and fresh water production from the flue gas of a glass melting furnace. Mazzoni et al. [26] proposed ORC plant arrangements, based on the turbo-expander pumping system and internal regeneration processes for low grade waste thermal energy recovery, aiming to improve the plant efficiency and to reduce the cooling load on the condenser.

The motivation behind the presented research was to investigate the concepts, leading to the increase of the dew point temperature of the water vapour contained in the flue gas (preferably above 80 °C), in order to recover its latent and sensible heat. It is generally known, that the increase of the dew point temperature can be achieved through the increase of the water vapour share in the flue gas or through the increase of the flue gas pressure. Taking these measures as a basis, the presented work investigated the following concepts for increasing the dew point temperature: humidification of the flue gas using water, humidification of the flue gas using steam, compression of the flue gas and the usage of the steam ejector.

The investigated models were developed in Aspen Plus® V8.8 and Aspen® HYSYS® V8.8. The condensing shell-and-tube HE was chosen as a thermal energy recovery technology. Under examination was the influence of the dew point temperature of the water vapour in the flue gas and the flow rate of the cooling water in the HE on the amount of the recovered thermal energy, the temperature of the cooling water at the HE outlet and the flow rate of the condensate. In order to analyse each concept from the economic point of view, the energy cost was estimated.

2. Model Description and Methodology

Within the presented work, four concepts leading to the increase of the dew point temperature of the water vapour in the flue gas were investigated, by means of the aspenONE® Engineering Suite:

1. Humidification of the flue gas using water (Aspen Plus® V8.8)
2. Humidification of the flue gas using steam (Aspen Plus® V8.8)
3. Compression of the flue gas (Aspen Plus® V8.8)
4. Usage of the steam ejector (Aspen® HYSYS® V8.8)

The steady state process modelling based on the energy balance was performed for each concept. The Peng-Robinson equation of state was chosen as a physical property method, commonly used in the gas processing industry. The investigated models were established under the following assumptions: the flue gas follows the ideal gas behaviour, the flue gas is composed of noncondensable (dry air) and condensable gas (water vapour), which is the approach followed in many studies [3,6,27].

In the tested concepts, the parameters of the flue gas at the outlet of the industrial baking oven were measured and used for calculations. The flue gas at the exit of an industrial baking oven had the temperature of 120 °C and the pressure of 1.01 bar, with the dew point temperature of 67 °C. The flue gas mass flow rate was 276.2 kg/h, with the air mass flow rate of 229.4 kg/h (83 %mas) and 46.8 kg/h for water vapour (17 %mas).

2.1. Concept 1: Humidification of the Flue Gas Using Water

Modelling of the concept with humidification of the flue gas using water, shown in Figure 1, is realised by adding WATER (95 °C and 1.01 bar) into the flue gas stream EXGAS. During perfect mixing (MIXER unit model) water evaporates using the thermal energy of the flue gas. Evaporation of the sprayed water increases the concentration of the water vapour in the flue gas stream and consequently its partial pressure and the GASIN dew point temperature.

The humidified flue gas stream GASIN leaves the mixer and is cooled in the counter-current heat exchanger HE (HEATX model) by the flow of WATERIN, supplied at a temperature of 60 °C and pressure of 1.01 bar. During the heat exchange between the two streams, the water vapour from the humidified flue gas stream condenses at constant pressure, and the condensate CONDENS occurs. The heated cooling water WATEROUT and the cooled dried flue gas GASOUT leave the HE.

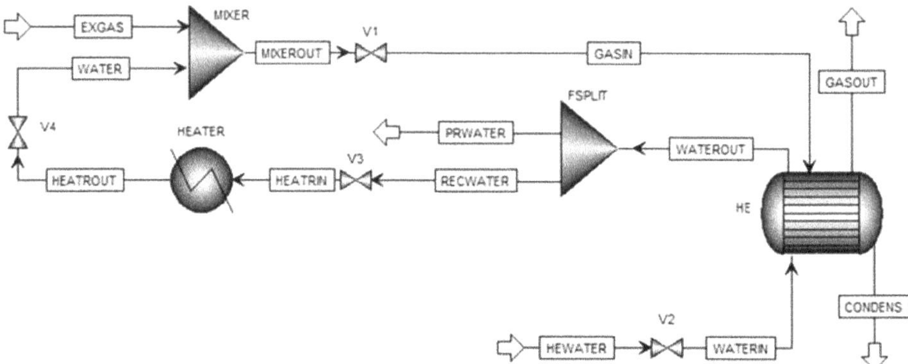

Figure 1. Flowsheet for the humidification of the flue gas using water.

The heated cooling water is then separated in the stream splitter (FSPLIT model), depending on the required amount of water for the GASIN dew point temperature increase. The first part, the water for further recycling RECWATER, is heated in the HE (HEATER model), in order to be sprayed into the MIXER. The second part, the process water PRWATER, can be used in other applications required within the industry, and is considered as the useful process stream, obtained from the thermal energy recovery cycle of the flue gas.

2.2. Concept 2: Humidification of the Flue Gas Using Steam

In the process model of the second investigated concept, the steam generator SG (HEATER model) is provided for the production of STEAM (120 °C and 1.01 bar). This steam is further used for humidification of the flue gas stream EXGAS, as presented in Figure 2. The recycling water in the concept RECWATER passes through the SG, in order to be evaporated. The other components are the same as in the previous model.

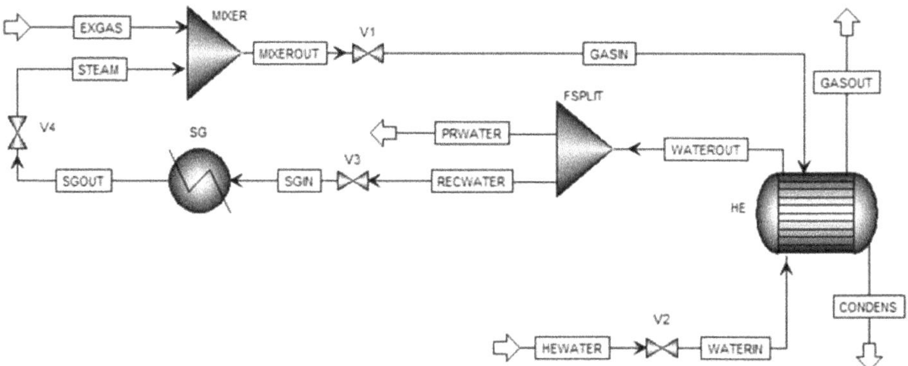

Figure 2. Flowsheet for the humidification of the flue gas using steam.

It is expected that in the case of the steam humidification, the higher GASIN dew point temperature can be reached, since there is no need to use the thermal energy of the flue gas for the evaporation of water, as in concept 1.

2.3. Concept 3: Compression of the Flue Gas

The third investigated concept is the compression of the flue gas, demonstrated in Figure 3. The flue gas stream EXGAS is directed to the compressor (COMPR model), where during the isentropic compression its temperature and pressure increase. The compressed flue gas stream GASIN is then led to the heat exchanger HE, as in the previously described cases. The heated cooling water WATEROUT at the outlet of the HE is not needed further in the recovery process, thus it can be fully utilized as process heat for different applications within the industry.

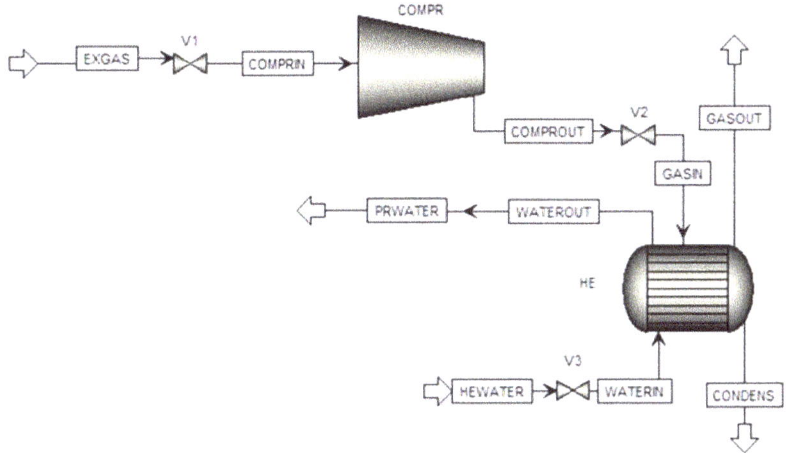

Figure 3. Flowsheet for the concept with the compression of the flue gas.

2.4. Concept 4: Usage of the Steam Ejector

Using the steam ejector leads to the GASIN dew point temperature increase, due to the simultaneous increase in the water content of the flue gas and its pressure, resulting from the geometry of the device. The flowsheet for concept 4 is given in Figure 4. The modelling was performed in the Aspen® HYSYS® software, since Aspen Plus® does not incorporate the steam ejector model. The other units are the same as in the previously described concepts.

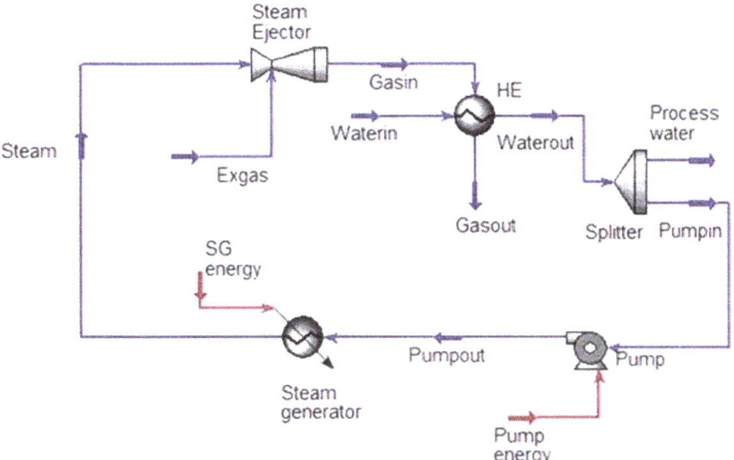

Figure 4. Flowsheet for the concept with the steam ejector.

The flue gas stream EXGAS is introduced from the low pressure side of the STEAM EJECTOR (Ejector model), whereas STEAM as a motive fluid is supplied from the high-pressure side. The discharged stream GASIN is directed to the HE (Heat Exchanger model), where the water vapour condenses from the flue gas during water cooling by WATERIN. The heated cooling water WATEROUT is separated in the SPLITTER (Tee model) into two parts: (1) PUMPIN, recirculated to generate the high pressure steam, and (2) PROCESS WATER, utilized for the needs of the industry. Pressure of the water PUMPIN is increased in the PUMP (Pump model), and then the high pressure water PUMPOUT is directed to the STEAM GENERATOR (Heater model) to be evaporated at the constant pressure. The generated high pressure STEAM is led to the STEAM EJECTOR to humidify the flue gas.

2.5. Heat Exchanger Design

The shell-and-tube heat exchanger for extracting the thermal energy from the flue gas was designed using the Aspen Shell & Tube Exchanger™. The following inlet parameters for the HE were used: physical properties of the inlet streams (T, p, mass flow and composition), temperature change, pressure drops in the shell and the tube side and the fouling factors for each inlet fluid. The possible HE design, according to the given inlet parameters, coupled with the failure analysis was provided by the software. The suggested design was improved by trial-and-error calculations following Kern's method [28]. The HE with the key design parameters, shown in Table 2, was used for modelling of all the previously described concepts.

Table 2. Design parameters of the HE.

Shell OD (mm)	Tube OD (mm)	Tube Length (mm)	Tube Pattern	Number of Tubes	Baffle Spacing (mm)	Number of Baffles
219	19	2700	Triangular	49	0.15	15

The effectiveness of the heat exchanger was calculated using the Number of Transfer Units (NTU) method. The typical values for the reference case with the dew point temperature of 67 °C lay in the range of 83–93% for the flow rate of cooling water of 400–2000 kg/h, respectively.

The allowable pressure drops were determined, based on the standard values given by Aspen®, as 0.1 bar and 0.2 bar for the shell and the tube side, respectively. The fouling factor of 0.0003 m²·°C/W for the cooling water and 0.0005 m²·°C/W for the flue gas were taken from the literature [29].

2.6. Theoretical Calculations

All the calculations and considerations in this work are related to one specific industrial facility, namely the natural gas fired, tunnel baking oven. The limitations related to the investigated facility (composition, temperature and flow rate of the flue gas, fluid parameters at the inlet/outlet of the process components, etc.) influence most of the obtained results. Therefore, it is imperative to perform the similar analysis for each specific system of interest with its unique properties.

Based on the parameters of the flue gas, previously listed in the preface to chapter two of this work, the inlet values for process modelling were determined by thermodynamic calculations. Few major ones are presented below with Equations (1)–(4). The used terminology can be found in Nomenclature.

Humidity ratio of an air-vapour mixture:

$$x = \frac{m_w}{m_{air}} \tag{1}$$

Partial pressure and density of water vapour in the flue gas:

$$P_w = \frac{x \cdot P}{0.622 + x} \tag{2}$$

$$\rho_w = \frac{P_w}{R_w \cdot T} \tag{3}$$

Dalton's law for air-vapour mixtures:

$$P = y_{air}P + y_{vap}P \tag{4}$$

The dew point of the water vapour was taken from the thermodynamic tables for the saturated vapour pressure values (P_i^s), calculated using Raoult's law:

$$y_i P = x_i P_i^s \tag{5}$$

The amount of the recovered thermal energy, representing the enthalpy difference between the flue gas at the HE inlet (GASIN stream) and at HE outlet (GASOUT stream), was calculated with Equations (6)–(8):

$$Q_{RE} = m_{air} \cdot (h_{in} - h_{out}) \tag{6}$$

$$h_{in} = \left(C_{p,air} \cdot t_{gas,in}\right) + x_{in}\left(C_{p,vap} \cdot t_{gas,in} + \Delta h_{vap}\right) \tag{7}$$

$$h_{out} = \left(C_{p,air} \cdot t_{gas,out}\right) + x_{out}\left(C_{p,vap} \cdot t_{gas,out} + \Delta h_{vap}\right) \tag{8}$$

For each investigated concept described above (Figures 1–4), the energy gain and the energy demand were estimated. The energy dissipation in the pipelines between the system components was neglected. As energy gain (Equations (9)–(12)) was considered the case, when the temperature of the cooling water, of the flue gas and of the condensate at the HE outlet exceeded $t_{washing}$. The temperature $t_{washing}$ was set to be 60 °C, based on the average temperature level useful for washing purposes within the production.

$$Q_{gain} = Q_{w,out} + Q_{gas,out} + Q_{condens} \tag{9}$$

$$Q_{w,out} = m_{prwater} \cdot C_{p,w} \cdot \left(t_{w,out} - t_{washing}\right) \tag{10}$$

$$Q_{gas,out} = m_{gas,out} \cdot C_{p,gas} \cdot \left(t_{gas,out} - t_{washing}\right) \tag{11}$$

$$Q_{condens} = m_{condens} \cdot C_{p,condens} \cdot \left(t_{condens} - t_{washing}\right) \tag{12}$$

The energy demand for increasing the dew point temperature in each concept was calculated as the difference between the thermal energy of a stream at the outlet and the inlet of a device, taking the corresponding device's efficiency into account (90% for the heater, 90% for the steam generator, 85% for the compressor and 90% for the pump [30,31]):

$$Q_{dem} = \frac{Q_{d,out} - Q_{d,in}}{\eta} \tag{13}$$

The energy cost was estimated for each tested concept, assuming that: (1) the flue gas compressor (concept 3) is driven by the electro motor, while (2) the heater and the steam generator (concepts 1, 2 and 4) use natural gas as fuel. The average electricity and the natural gas prices for German industries in 2017 were 0.127 €/kWh and 0.026 €/kWh, respectively [32].

3. Results and Discussion

The effects of the dew point temperature of the water vapour in the flue gas (Θ) and the flow rate of the cooling water in the HE ($m_{w,in}$) on the recovered thermal energy (Q_{RE}), the water temperature at the HE outlet ($t_{w,out}$) and the condensate flow rate ($m_{condens}$) were investigated for each concept described above. The fitting curves were created using the Origin® 2019 by polynomial fit of the second order with $R^2 \in [0.9882–1]$.

3.1. Concept 1: Humidification of the Flue Gas Using Water

In this concept, water (t = 95 °C and p = 1.01 bar) for humidification of the flue gas is sprayed directly into the flue gas stream. The computational results of the calculations, obtained for concept 1, are presented in Figure 5.

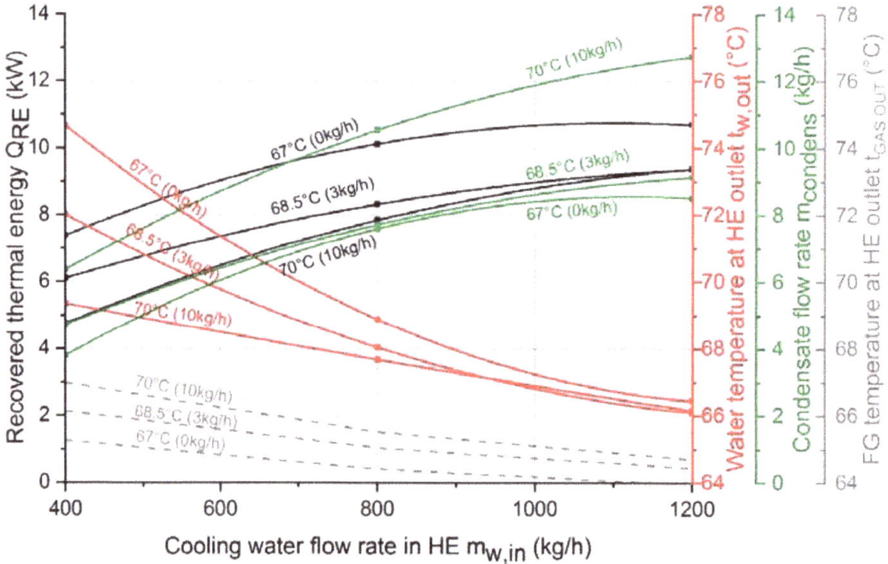

Figure 5. Influence of the cooling water flow rate in the heat exchanger (HE) and the dew point temperature on the recovered thermal energy (black lines), the water temperature at the HE outlet (red lines), the condensate flow rate (green lines) and the flue gas temperature at the HE outlet (gray lines) for concept 1.

Increasing the flow rate of sprayed water (0, 3, 10 kg/h) leads to an increase of the dew point temperature from 67 °C (no water added) to 70 °C (water addition of 10 kg/h). The relatively low temperature increase of $\Delta T = 3$ °C can be attributed to the negligible rise in the water content of the flue gas. On the other hand, a further increase of the sprayed water amount above 10 kg/h is limited by the ability of the flue gas to evaporate the sprayed water.

The increase of the dew point temperature from 67 °C to 70 °C results in the 8% reduction of the water temperature at the outlet of the HE ($t_{w,out}$) for 400 kg/h flow rate of cooling water. This is due to the fact that in case of the low flow rate of cooling water, the transferred heat is mostly sensible and, therefore, higher for the flue gas with the lower dew point temperature. By increasing the flow rate of cooling water more condensate is generated, meaning that the transferred heat is both sensible and latent. The same explanation is applicable to the thermal energy recovery trend line, which increases with the increase of the cooling water flow rate.

Nevertheless, due to the low amount of the recovered thermal energy (maximum of 10.5 kW) and unattainability of the high dew point temperature, this concept does not meet the goals of the research and is not recommended for the practical use in the investigated facility.

3.2. Concept 2: Humidification of the Flue Gas Using Steam

The computational results of the calculations, obtained for concept 2, are presented in Figure 6. Steam for the humidification purpose is supplied in this case at the temperature of 120 °C and pressure of 1.01 bar.

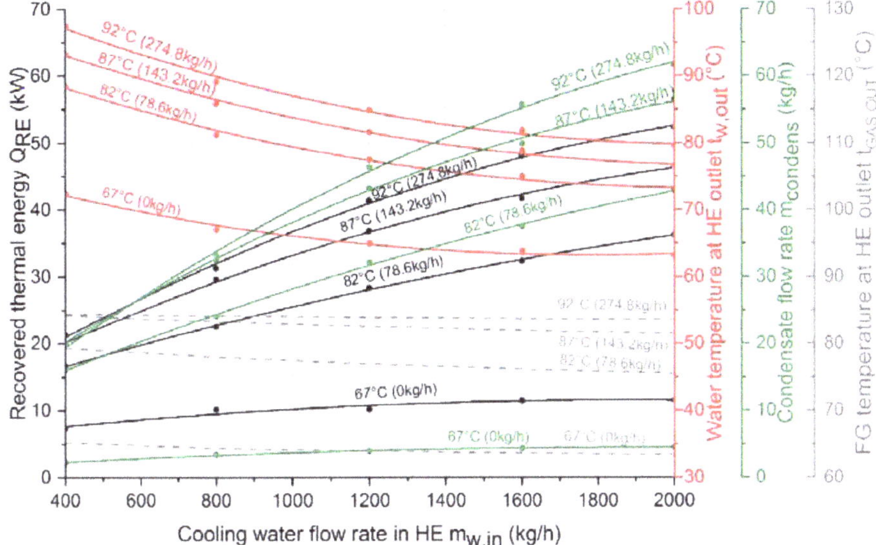

Figure 6. Influence of the cooling water flow rate in the HE and the dew point temperature on the recovered thermal energy (black lines), the water temperature at the HE outlet (red lines), the condensate flow rate (green lines) and the flue gas temperature at the HE outlet (gray lines) for concept 2.

The increase of the dew point temperature (67, 82, 87 and 92 °C) is achieved by increasing the flow rate of the introduced steam (0, 78.6, 143.2 and 274.8 kg/h, respectively). As expected, this results in the increase of all investigated parameters. For the flow rate of cooling water of 2000 kg/h, the recovered thermal energy is increased by 80% (from 10 to 52 kW), due to the enthalpy increase of the flue gas at the inlet of the HE caused by the steam humidification. The flow rate of condensate is increased by around 75% and reaches the maximum absolute value of 62 kg/h for the case with 92 °C dew point and 2000 kg/h flow rate of cooling water. Since by steam introduction, the enthalpy of the flue gas at the HE inlet increases, water which is passing through the HE is increasingly heated up. The temperature of water at the HE outlet increases on average by around 25%, independent of the cooling water flow rate at the HE, with the increase of the dew point temperature.

The flow rate of the cooling water in the HE has the similar influence on investigated parameters compared to the previous concept with water humidification. Supplying more cooling water to the HE results in the release of latent heat from the flue gas, so the condensate flow rate and the recovered thermal energy show the intensive growth. On the other hand, the higher flow rate of the cooling water results in the decrease of the water temperature at the HE outlet. This follows from the fact that the same amount of the flue gas at the same conditions has to warm up the significantly (five times) increased amount of the cooling water in the HE.

3.3. Concept 3: Compression of the Flue Gas

In the third investigated concept the increase of the flue gas pressure (1.01, 1.92, 2.33, 2.83 and 3.41 bar) results in the increase of the dew point temperature (67, 82, 87, 92 and 97 °C, respectively), whereas the water vapour content of the flue gas remains constant (17 %mas). The computational results, obtained for concept 3 are shown in Figure 7.

The dependences of Q_{RE}, $t_{w,out}$ and $m_{condens}$ are qualitatively similar to the ones obtained in the concept with the steam humidification. However, the achieved values are higher in the case of steam humidification. For instance, the recovered thermal energy displays the rise while increasing the dew point temperature, but the maximum absolute value is 30 kW in the case of the flue gas compression, which is ca. 20 kW less than in the case of the steam humidification. This is because the flue gas contains four times less water vapour: 46.8 kg/h in the case of the flue gas compression concept and (46.8 + 143.2) kg/h in the case of the steam based humidification.

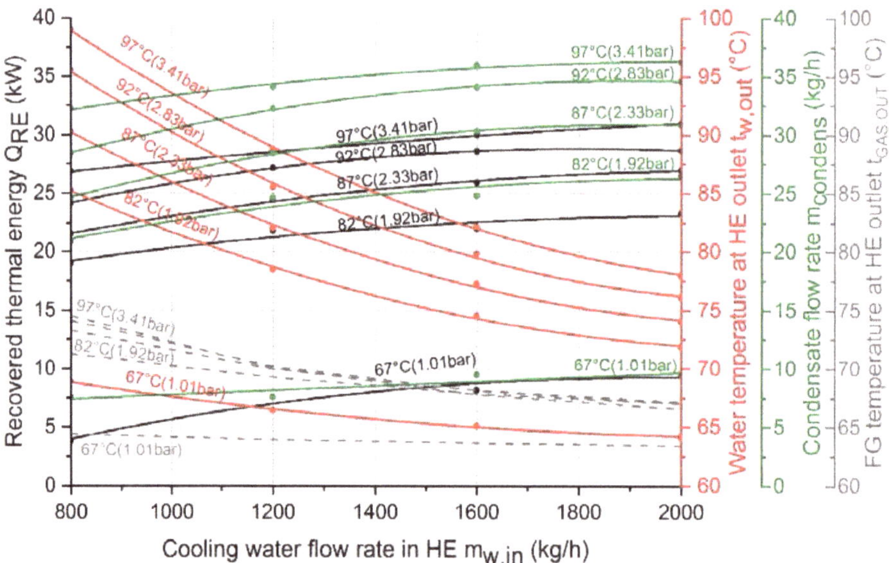

Figure 7. Influence of the cooling water flow rate in the HE and the dew point temperature on the recovered thermal energy (black lines), the water temperature at the HE outlet (red lines), the condensate flow rate (green lines) and the flue gas temperature at the HE outlet (gray lines) for concept 3.

Nevertheless, looking at the condensation rate, the pressure rise of the flue gas of around 2 bar (from 1.01 to 3.41 bar) leads to the 70% increase in the condensate flow rate.

The temperature of the flue gas is significantly increased during the isentropic compression and the high amount of the sensible heat is transferred to the water supplied to the HE. For this reason the water temperature at the outlet of the HE rises with the increase of the dew point temperature by around 30% for the low flow rate of the cooling water (800 kg/h) and by around 17% for the high flow rate (2000 kg/h).

3.4. Concept 4: Usage of the Steam Ejector

The modelling results for concept 4 (steam supply via the steam ejector) in comparison to concept 2 (steam supply via the steam generator) are presented in Figure 8. The comparison is eligible, due to the same flow rate of steam introduced in both cases, which amounts to 143.2 kg/h. While the humidification via the steam generator was performed by adding steam at 120 °C and 1.01 bar, the steam ejector was supplied with steam at 160 °C and 4 bar. In concept 2, the dew point temperature was 87 °C, whereas in concept 4 the higher dew point temperature was reached, namely 91 °C.

The process tendencies, corresponding to concepts 2 and 4, are qualitatively similar, but the target temperatures (dew point and cooling water outlet temperature) are higher in the case of the steam ejector. This is attributed to the fact that the usage of the steam ejector leads to the increase in both the water content and the pressure of the flue gas.

The condensate flow rate and the recovered thermal energy are almost the same in the case of the low flow rate of the cooling water (400–1200 kg/h) for both concepts. In this case the condensate flow rate dependencies are similar and therefore, the recovered thermal energy is mainly influenced by the sensible heat of the flue gas.

Increasing the cooling water flow rate further demonstrates the more noticeable difference between the two concepts. For 2000 kg/h the recovered thermal energy and the condensate flow rate are around 8% and 12% higher in the case of the steam ejector usage, because steam is introduced at higher thermodynamic parameters.

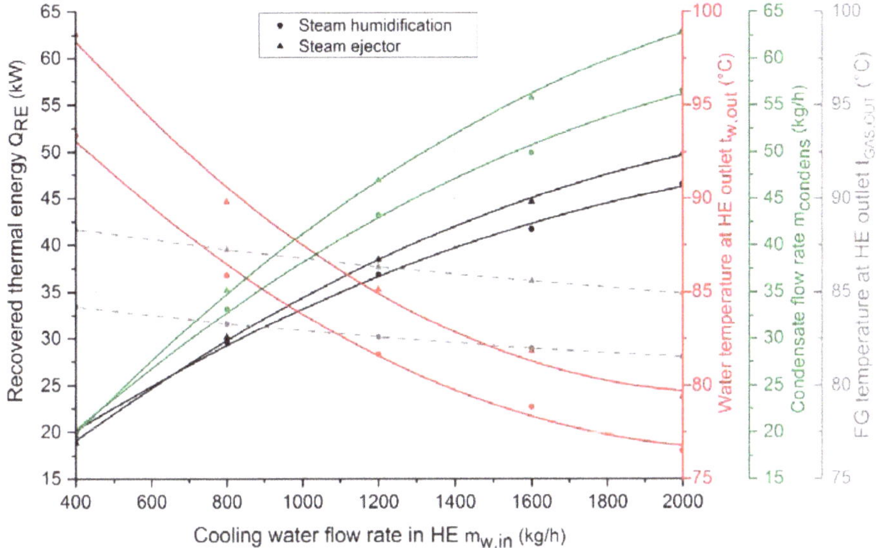

Figure 8. Comparison between the results calculated for concepts 2 and 4. Influence of the cooling water flow rate in the HE on the recovered thermal energy (black lines), the water temperature at the HE outlet (red lines), the condensate flow rate (green lines) and the flue gas temperature at the HE outlet (gray lines).

The water temperature at the outlet of the HE by using the steam ejector is 10 °C higher compared to the steam humidification case in the whole range of the cooling water flow rate. This follows from the fact that the flue gas at the inlet of the HE has higher temperature in the case of the steam ejector concept, therefore the flue gas transfers more sensible heat to the cooling water passing through the HE.

3.5. Comparison of the Results of the Investigated Concepts

The energy balance with the gain and the demand trend lines for concepts 2, 3 and 4, at the cooling water flow rate of 1200 kg/h, are shown in Figure 9. Concept 1 (humidification of the flue gas using water) is not considered, due to the relatively low increase in the dew point temperature.

During the humidification of the flue gas using steam (concept 2), additional energy input is required for the steam generator. In the concept with the compression of the flue gas (concept 3), additional energy input is required to run the flue gas compressor. In the case of the steam ejector usage (concept 4), additional energy is required by the pump, for the pressure rise of water, and by the steam generator, for the production of steam.

The analysis indicates that the humidification of the flue gas using steam (concept 2) is the most energy demanding process for increasing the dew point temperature. The energy demand of the steam generator rises with the dew point temperature increase, since the greater mass flow rate of water has to be evaporated in order to get the required amount of steam for injection into the flue gas. The concept with the steam ejector (concept 4) has the lower energy demand than concept 2, because the desirable dew point temperature is reached using the lower amount of steam but at the higher pressure. Moreover, the pump has a relatively high efficiency and is considered as the low-energy consuming equipment for increasing the pressure. In case of the third investigated concept, where the flue gas is compressed, the energy demand is the lowest of all three concepts. The reason is that the mass flow rate of the flue gas to the compressor remains constant for all of the investigated dew point temperatures. The maximum considered pressure increase by the compressor is about 2 bar (for increasing the dew point temperature from 67 °C to 97 °C), which requires a relatively low isentropic work.

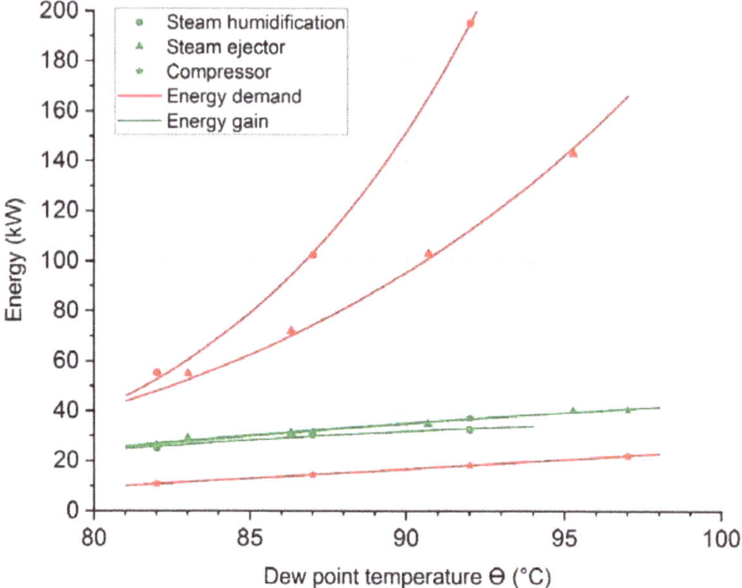

Figure 9. The energy demand and the energy gain for increasing the dew point temperature (Θ) for the investigated concepts.

Further analysis of the obtained results is related to the ratio of the recovered and additionally introduced energy. Since the energy gain stays almost the same for all investigated concepts, the conclusion is that only the third investigated concept, based on the flue gas compression, is of practical interest. In this case the energy gain exceeds the energy demand, therefore, the concept with compression is considered as the most promising for the energy recovery in the systems with the dew point temperature increase.

Cost analysis of the investigated concepts was conducted based on the current energy costs and additional energy demands of different concepts (Figure 10). The prices of different energy sources were introduced in chapter 2.6. The required investment costs including the prices of equipment were not taken into consideration.

As it was expected, the energy cost follows the trend of the energy demand, and grows as the dew point temperature increases. The growth of the energy cost per temperature degree for the steam humidification concept is especially steep, since the generation of the high amount of steam is coupled with the high energy demand. The energy cost can be reduced up to 35% (for the dew point temperature of 92 °C) by applying the steam ejector, which uses the lower amount of steam, but at higher pressure.

The energy cost in the case of the flue gas compression concept also increases, since the greater pressure is required for increasing the dew point temperature. Compared to two other concepts, concept 3 exhibits the lowest energy cost (20,000 €/year), which is half of the costs of concept 2 in the case of the dew point temperature of 92 °C. Nevertheless, the installation, the maintenance and the repair costs of the compressor are normally higher than the equipment used in the other proposed concepts and should be taken into account in the future work.

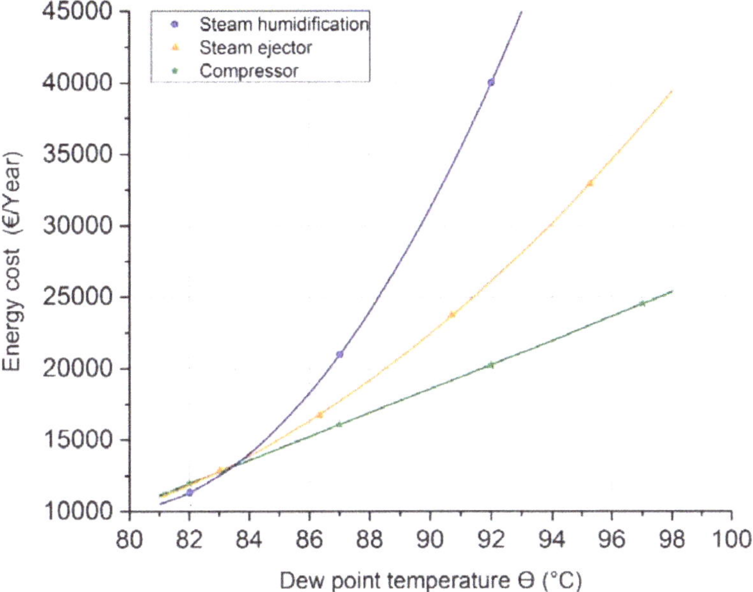

Figure 10. Estimated energy costs for the investigated concepts.

The potential practical application of the theoretical concepts investigated in this work is also related to several major challenges. The first one is an appropriate selection of the optimal method based on the actual size, requirements and properties of each specific facility. For example, all the calculations and considerations in this work are related to one specific industrial facility, namely the natural gas fired, tunnel baking oven. The limitations related to the investigated facility (composition, temperature and flow rate of the flue gas, fluid parameters at the inlet/outlet of the process components, etc.) influence most of the obtained results. Therefore, it is imperative to perform the similar analysis for each specific system of interest with its unique properties. The next technical challenge is to compensate for the eventual dynamic changes within the process. The presented work is based on the continuous industrial process with time independent process parameters which is often not the case. Although compensation of the dynamic process behavior is out of the scope of this work, the use of cognitive algorithms for the process management in connection to the dew point temperature increase, could result in the optimal use of the available potentials for the thermal energy recovery. Nevertheless, one of the major challenges for the practical implementation of the investigated techniques would probably be of the financial nature and related to the investment costs for the additional equipment. Taking into account the investment costs, the additional maintenance costs and the price of the low-temperature thermal energy in Germany, some of the simplified financial calculations, not presented in this work, indicate that most of the investigated processes are financially justified only for the larger facilities and pay-off periods of 20–30 years. Therefore, a full economic analysis for the considered concepts will be part of a future detailed investigation.

4. Conclusions

Recovering both the sensible and latent heat is part of the energy efficiency optimization of every process that has thermal energy dissipating with the flue gas. Essential relevance is related to the industrial processes where flue gases are rich in water vapour (e.g., baking, textile, pulp and paper industries). Cooling the flue gas below the water vapour dew point temperature in a shell-and-tube heat exchanger, leads to the condensation of the water vapour and to the release of the sensible and latent heat. In order to use this energy as process heat, the increase of the usually low dew point temperatures (40–60 °C) to the higher levels (80–95 °C) is of special interest.

The increase of the dew point temperature of the water vapour contained in the flue gas, leads to the increase of the temperature level of the cooling fluid and the condensate. In this way, these otherwise waste streams can be further used within an industrial process, thus improving its energy efficiency.

The present study was focused on the investigation of four concepts for the increase of the dew point temperature of the industrial flue gas: humidification of the flue gas using water, humidification of the flue gas using steam, compression of the flue gas and the usage of the steam ejector. All the calculations and considerations in this work are related to one specific industrial facility, namely the natural gas fired, tunnel baking oven. The process modelling was performed using the commercial software Aspen®. For each considered concept the effects of the dew point temperature and the cooling water flow rate in the HE on the recovered thermal energy, the water temperature at the HE outlet and the condensate flow rate were investigated.

The major conclusions for the considered concepts are summarized as follows:

- The increase of the dew point temperature above 90 °C is possible by the steam humidification, compression of the flue gas and using the steam ejector.
- The humidification of flue gas using water is not recommended, due to the relatively low thermal energy recovery level (maximum of 10 kW is achieved) and the negligible increase in the dew point temperature (up to 70 °C) in comparison to other tested concepts.
- Although the steam humidification shows the highest potential for the thermal energy recovery, it is also the most energy demanding and, consequently, the most expensive process, in the case when the steam has to be specially produced for the dew point temperature increase. When this low parameter steam is available as a waste product, the investigated concept gains on its importance.
- In the steam humidification concept the maximum of 52 kW of the recovered thermal energy is achieved and the flow rate of the condensate is increased by app. 75% at $m_{w,in}$ = 2000 kg/h. The maximal calculated dew point temperature is 92 °C. The energy demand of the steam generator increases with the dew point temperature increase, since the greater water mass flow rate has to be heated and evaporated, in order to get the required amount of steam.
- Using the steam ejector leads to the increase in both the water content and pressure of the flue gas, due to the geometry of the device. Therefore, the recovered thermal energy and the condensate flow rate are around 8% and 12% higher compared to the steam humidification concept. Yet, the energy demand exceeds the energy gain, making this concept not suitable for the use in a thermal energy recovery cycle, in the case when there is no waste steam available in the facility.
- Both from the energetic and economic point of view, the compression of the flue gas has the highest potential, as the energy gain exceeds the energy demand and the energy cost is the lowest out of all the investigated methods. Although the low required pressure increase of app. 2 bar corresponds to the relatively low isentropic work by the compressor, a detailed economic analysis should be performed, taking into account the equipment price, the installation and the maintenance costs for each concrete industrial facility.
- The concept with the flue gas compression will be the subject of further investigations, in combination with the Organic Rankine Cycle and the absorption refrigerator.

Author Contributions: Conceptualization, N.F., V.J., A.Z.-R. and A.D.; Methodology, N.F., V.J. and P.A.; Validation, P.A.; Investigation, P.A., N.F.; Resources, A.D.; Data Curation, P.A., N.F., V.J., A.Z.-R. and M.J.K.; Writing—Original Draft Preparation, N.F., P.A., A.Z.-R. and V.J.; Writing—Review and Editing, N.F., V.J., A.Z.-R., M.J.K. and P.A.; Software, P.A. and M.J.K.; Visualization, P.A. and N.F.; Supervision, N.F., V.J.; Funding acquisition, N.F., V.J., A.D.

Funding: This research received no external funding.

Acknowledgments: The authors gratefully acknowledge the financial support of the German Academic Exchange Service (DAAD), and the funding of the Erlangen Graduate School in Advanced Optical Technologies (SAOT) by the German Research Foundation (DFG) in the framework of the German excellence initiative. We also acknowledge the support by Deutsche Forschungsgemeinschaft and Friedrich-Alexander-Universität Erlangen-Nürnberg (FAU) within the funding programme Open Access Publishing.

Conflicts of Interest: The authors declare no conflict of interest. The funders had no role in the design of the study; in the collection, analysis, or interpretation of data; in the writing of the manuscript, or in the decision to publish the results.

Nomenclature

C_p	Specific heat capacity (kJ kg^{-1} K^{-1})
h	Specific enthalpy (kJ kg^{-1})
Δh_{vap}	Vaporization enthalpy of water (kJ kg^{-1})
m	Mass flow rate (kg h^{-1})
P	Total pressure (bar)
P_i^s	Saturation vapour pressure (bar)
Q	Heat (kW)
R	Gas constant (kJ kg^{-1} K^{-1})
t	Temperature (°C)
x	Humidity ratio of an air-vapour mixture (kg$_w$ kg$_{air}$$^{-1}$)
x_i	Mole fraction of a component in liquid phase (-)
y	Mole fraction of a component (-)
y_i	Mole fraction of a component in gas phase (-)

Greek symbols

ρ	Density (kg m^{-3})
Θ	Dew point temperature (°C)
η	Efficiency (-)

Abbreviations

AIR	Dry air
COMPR	Compressor
CONDENS	Condensate
D	Device
DEM	Energy demand
EXGAS	Flue gas at the exit of an industrial flue before a concept modification
FG	Flue gas
FSPLIT	Splitter
GAIN	Energy gain
GAS	Flue gas after a concept modification
HE	Heat exchanger
HEWATER	Cooling water for a heat exchanger
IN	Inlet
OD	Outer diameter
OUT	Outlet
PRWATER	Useful water for processes in a company
RE	Recovered thermal energy
RECWATER	Recycling water to be used in a concept
SG	Steam generator
VAP	Vapour
W	Water

References

1. Hu, Y.; Gao, Y.; Lv, H.; Xu, G.; Dong, S. A New Integration System for Natural Gas Combined Cycle Power Plants with CO_2 Capture and Heat Supply. *Energies* **2018**, *11*, 3055.
2. Schwabe, K.; Walsdorf-Maul, M.; Schaudienst, F.; Vogdt, F.U. Using Waste Heat for Sustainable Manufacturing Based on the Example of a Conventional Industrial Bakery. *Int. J. Mater. Mech. Manuf. (IJMMM)* **2013**, 274–277. [CrossRef]
3. Terhan, M.; Comakli, K. Design and economic analysis of a flue gas condenser to recover latent heat from exhaust flue gas. *Appl. Therm. Eng.* **2016**, *100*, 1007–1015. [CrossRef]
4. Xu, G.; Huang, S.; Yang, Y.; Wu, Y.; Zhang, K.; Xu, C. Techno-economic analysis and optimization of the heat recovery of utility boiler flue gas. *Appl. Energy* **2013**, *112*, 907–917. [CrossRef]
5. Aneke, M.; Agnew, B.; Underwood, C.; Wu, H.; Masheiti, S. Power generation from waste heat in a food processing application. *Appl. Therm. Eng.* **2012**, *36*, 171–180. [CrossRef]
6. Che, D.; Da, Y.; Zhuang, Z. Heat and mass transfer characteristics of simulated high moisture flue gases. *Heat Mass Transf.* **2005**, *41*, 250–256. [CrossRef]
7. Wang, D.; Bao, A.; Kunc, W.; Liss, W. Coal power plant flue gas waste heat and water recovery. *Appl. Energy* **2012**, *91*, 341–348. [CrossRef]
8. Bajpai, P. Pulp and Paper Production Processes and Energy Overview. In *Pulp and Paper Industry*; Elsevier: Amsterdam, The Netherlands, 2016.
9. Hasanbeigi, A.; Price, L. A review of energy use and energy efficiency technologies for the textile industry. *Renew. Sustain. Energy Rev.* **2012**, *16*, 3648–3665. [CrossRef]
10. Jank, R.; Schulte, S. Verfahren zur Energiegewinnung aus Wasserdampf enthaltenden Schwaden und Vorrichtung zur Durchführung dieses Verfahrens. European Patent No. WO2017064036, 20 April 2017.
11. Song, C.; Pan, W.; Srimat, S.T.; Zheng, J.; Li, Y.; Wang, Y.-H.; Xu, B.-Q.; Zhu, Q.-M. Tri-reforming of Methane over Ni Catalysts for CO_2 Conversion to Syngas With Desired H_2/CO Ratios Using Flue Gas of Power Plants Without CO_2 Separation. In *Carbon Dioxide Utilization for Global Sustainability, Proceedings of the 7th the International Conference on Carbon Dioxide Utilization, Seoul, Korea, 12–16 October 2003*; Elsevier: Amsterdam, The Netherlands, 2004; pp. 315–322.
12. Osakabe, M. Heat exchanger for latent heat recovery. *Mech. Eng. Rev.-Bull. JSME* **2015**, *2*, 1–24. [CrossRef]
13. Zhao, S.; Yan, S.; Wang, D.K.; Wei, Y.; Qi, H.; Wu, T.; Feron, P.H.M. Simultaneous heat and water recovery from flue gas by membrane condensation: Experimental investigation. *Appl. Therm. Eng.* **2017**, *113*, 843–850. [CrossRef]
14. Law, R.; Harvey, A.; Reay, D. Opportunities for low-grade heat recovery in the UK food processing industry. *Appl. Therm. Eng.* **2013**, *53*, 188–196. [CrossRef]
15. Hung, T.C.; Shai, T.Y.; Wang, S.K. A review of organic rankine cycles (ORCs) for the recovery of low-grade waste heat. *Energy* **1997**, *22*, 661–667. [CrossRef]
16. Tchanche, B.F.; Lambrinos, G.; Frangoudakis, A.; Papadakis, G. Low-grade heat conversion into power using organic Rankine cycles—A review of various applications. *Renew. Sustain. Energy Rev.* **2011**, *15*, 3963–3979. [CrossRef]
17. Ziegler, F. EnEff Wärme: Absorptionskältetechnik für Niedertemperaturantrieb—Grundlagen und Entwicklung von Absorptionskältemaschinen für die fernwärme-und solarbasierte Kälteversorgung. In *Abschlussbericht*; Technische Universität Berlin Fakultät III-Prozesswissenschaften: Berlin, Germany, 2013.
18. Offizielles Stadtportal für die Hansestadt Hamburg. Solares Kühlen für Büro- und Dienstleistungsgebäude, Österreichisches Forschungs-und Prüfzentrum Arsenal Ges.m.b.H. Available online: https://www.hamburg.de/contentblob/1356374/8760da9f3d940e22e7f926723591eae7/data/solare-kuehlung.pdf (accessed on 25 March 2019).
19. U.S. Department of Energy. Waste Heat Reduction and Recovery for Improving Furnace Efficiency, Productivity and Emissions Performance. Available online: https://www.energy.gov/sites/prod/files/2014/05/f15/35876.pdf (accessed on 25 March 2019).
20. Mukherjee, S.; Asthana, A.; Howarth, M.; Mcniell, R. Waste heat recovery from industrial baking ovens. *Energy Procedia* **2017**, *123*, 321–328. [CrossRef]

21. EcoStep. Energieeffizienz in Bäckereien-Energieeinsparungen in Backstube und Filialen, ttz Bremerhaven. Available online: http://www.ecostep-online.de/cms_uploads/files/eneff_baeckerei_-_leitfaden_-_juli_2014.pdf (accessed on 25 March 2019).

22. Jana, K.; De, S. Utilizing waste heat of the flue gas for post-combustion CO_2 capture—A comparative study for different process layouts. *Energy Sources Part A Recovery Util. Environ. Eff.* **2016**, *38*, 960–966. [CrossRef]

23. Luyben, W.L. Heat exchanger simulations involving phase changes. *Comput. Chem. Eng.* **2014**, *67*, 133–136. [CrossRef]

24. Duan, W.; Yu, Q.; Wang, K.; Qin, Q.; Hou, L.; Yao, X.; Wu, T. ASPEN Plus simulation of coal integrated gasification combined blast furnace slag waste heat recovery system. *Energy Convers. Manag.* **2015**, *100*, 30–36. [CrossRef]

25. Ishaq, H.; Dincer, I.; Naterer, G.F. New trigeneration system integrated with desalination and industrial waste heat recovery for hydrogen production. *Appl. Therm. Eng.* **2018**, *142*, 767–778. [CrossRef]

26. Mazzoni, S.; Arreola, M.J.; Romangoli, A. Innovative Organic Rankine arrangements for Water Savings in Waste Heat Recovery Applications. *Energy Procedia* **2017**, *143*, 361–366. [CrossRef]

27. Hu, H.W.; Tang, G.H.; Niu, D. Experimental investigation of convective condensation heat transfer on tube bundles with different surface wettability at large amount of noncondensable gas. *Appl. Therm. Eng.* **2016**, *100*, 699–707. [CrossRef]

28. Towler, G.; Sinnott, R. *Chemical Engineering Design: Principles, Practice, and Economics of Plant and Process Design*, 2nd ed.; Butterworth-Heinemann: Waltham, MA, USA, 2013.

29. Incropera, F.P.; Dewitt, D.P.; Bergman, T.L.; Lavine, A.S. *Principles of Heat and Mass Transfer*, 7th ed.; John Wiley & Sons: Hoboken, NJ, USA, 2011.

30. Baehr, H.D.; Kabelac, S. *Thermodynamik*; Springer Vieweg: Berlin, Germany, 2012.

31. Spirax Sarco. Grundlagen der Dampf-und Kondensattechnologie. Available online: http://www.spiraxsarco.com/global/de/Resources/Documents/Grundlagen-der-Dampf-und-Kondensattechnologie.pdf (accessed on 25 March 2019).

32. Statistisches Bundesamt (Destatis). Data on Energy Price Trends, Long-Time Series from January 2000 to March 2018. Available online: https://www.destatis.de (accessed on 25 March 2019).

Article

SExperimental Study on the Fire-Spreading Characteristics and Heat Release Rates of Burning Vehicles Using a Large-Scale Calorimeter

Younggi Park, Jaiyoung Ryu * and **Hong Sun Ryou ***

School of Mechanical Engineering, Chung-Ang University, 84, Heukseok-ro, Dongjak-gu, Seoul 06974, Korea; pyg0511@cau.ac.kr
* Correspondence: jairyu@cau.ac.kr (J.R.); cfdmec@cau.ac.kr (H.S.R.);
  Tel.: +82-2-820-5279 (J.R.); +82-2-820-5280 (H.S.R.)

Received: 8 March 2019; Accepted: 12 April 2019; Published: 17 April 2019

Abstract: In this article, large-scale experimental studies were conducted to figure out the fire characteristics, such as fire-spreading, toxic gases, and heat release rates, using large-scale calorimeter for one- and two-vehicle fires. The initial ignition position was the passenger seat, and thermocouples were attached to each compartment in the vehicles to determine the temperature distribution as a function of time. For the analysis, the time was divided into sections for the various fire-spreading periods and major changes, e.g., the fire spreading from the first vehicle to the second vehicle. The maximum temperature of 1400 °C occurred in the seats because they contained combustible materials. The maximum heat release rates were 3.5 MW and 6 MW for one and two vehicles, respectively. Since the time to reach 1 MW was about 240 s (4 min) before and after, the beginning of the car fire appears to be a medium-fast growth type. It shows the effect on the human body depending on the concentration of toxic substances such as carbon monoxide or carbon dioxide.

Keywords: fire-spreading characteristics; real vehicle experiments; toxic gases; temperature distributions; unsteady heat release rate

---

1. Introduction

Many vehicle fire studies are being carried out since vehicle accidents result in catastrophe due to various reasons such as drivers' negligence, electrical faults, deliberately lit and arson in parking lots [1], buildings, and tunnels, and so forth. To prevent vehicle disasters and suggest a guideline for evacuees in case of emergency, figuring out the fire-spreading mechanism and measuring other key parameters such as heat release rates and temperatures are necessary to install sprinklers, smoke control systems, and setting up fire extinguishers.

In fire researches, calorimeters, which can estimate a heat release rate, smoke generation, carbon dioxide, carbon monoxide, and so on, have been widely used for experimental study [2–5]. In the case of heat release rate, it has been regarded as one of the most important factors during experimental studied since it can be used to calculate the size of the fire and be more easily used in performance-based design (PBD). Large-scale tests are considered the most accurate way to secure fire-spreading effect and heat release rate, however, there is limited research, as it cost too much [6–8]. Katsuhiro et al. [6] found that the temperature distribution and maximum HRR reached 3 MW by changing the initial ignition location of the fire. In addition, they found that the fire spread radically after the windows were broken. Throughout related large-scale experiments, a single vehicle represents the heat release rate of 2.5–5 MW. However, most of the experimental studies were conducted only on one vehicle. Vehicle fires normally occur between two vehicles because of accidents. Li et al. [9] conducted large-scale experiments using two sedan vehicles to determine the change of the temperature distribution; however, in this study, the heat release rate was not considered, although it is an imperative part not only for analyzing the fire phenomena, but also suggesting guidelines for rescue as a function of time. Therefore, we investigated fire-spreading phenomena considering the effect of two vehicles fire scenario to figure out an important parameter such as heat release rates, fire-spreading time, and so forth.

A secondary factor that causes catastrophe in vehicle fires is the influence of toxic gases, e.g., carbon monoxide (CO) and carbon dioxide (CO_2), on human breathing and other functions [10–12]. Truchot et al. [13] investigated toxic-gas emissions from vehicle fires in tunnels. They did experimental studies using large-scale facilities, where the air flow in the tunnel could be controlled. They found and measured the toxic substances through a flame ionization detector/analyzer. It showed the toxic substances emitted in the tunnel shaft and analyzed the effects of various heat release rates. Smoke and toxic gases that are produced in the open are released into the air, due to their buoyancy. However, when a vehicle burns in an enclosed space, e.g., a tunnel or an indoor parking lot, the gas cannot escape to the atmosphere. Therefore, it is one of the main factors that should be considered in fire research to understand the concentration of the toxic substances generated when a vehicle fire occurs. However, there is limited information regarding the toxic gases that appear in the event of the fire.

Furthermore, many experimental studies are being actively carried out with pool-fire tests to represent fire accidents in tunnels or high-rise buildings [14–23]. Beak et al. [23] conducted tunnel-fire experiments using small-scale tunnels with pool fires, because it is challenging to experiment in real tunnels with real vehicles. He provided detailed information on the fire's heat release rate and its hole-plugging effect on tunnels; however, the results were obtained from pool-fire experiments, not real vehicles. It is clear that the results obtained from real products and pool-fire experiments will be different.

Therefore, it is necessary to analyze the fire-spreading characteristics of an actual vehicle and analyze the heat release rate and temperature distribution in order to understand the fire phenomena and its applications. Figure 1 represents a schematic diagram of the basic concept for this study. In this study, the fire-spreading characteristics, unsteady heat release rate, and the toxic substances in vehicle fires are measured using a large-scale calorimeter. All information, such as the change of heat release rate, temperature distributions and toxic gases as a function of time, gained throughout this research can be applied to other fire-related researches regarding vehicles and tunnels, parking lots, and so forth.

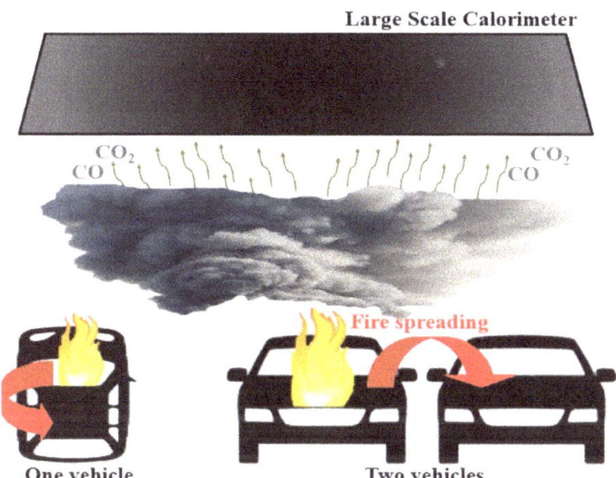

Figure 1. Schematic diagram of the basic concept for this study.

2. Experimental Setup and Conditions

Experiments were conducted using a large-scale calorimeter (LSC) from the Korea Institute of Construction Technology (KICT), applicable up to 10 MW, to determine the unsteady heat release rate and toxic gases, e.g., CO and CO_2, generated by the fire. The schematic diagram of LSC is represented in Figure 2 to figure out the values regarding change of the heat release rate and toxic substances as a function of time, and specifications of the experimental apparatus are represented in Table 1.

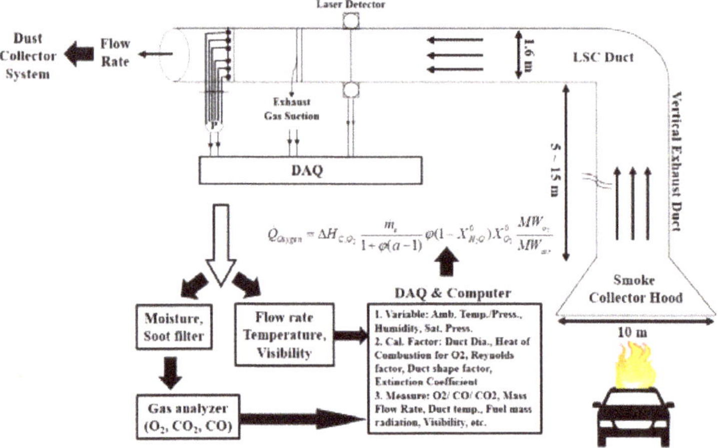

Figure 2. Schematic diagram of large-scale calorimeter (LSC) used to detect the changes in the heat release rate and toxic substances as a function of time.

Table 1. Specification of experimental apparatus.

Measurement	Specification
Duct pressure difference	Output: 4–20 mA, Range: 0–1245 Pa Model: PADT-D1000 Pa
Duct Temperature	K-type wire, Range: −200–1000 °C
Gas Analyzer	Output: 4–20 mA, Range: O_2 20.95%, CO_2 8%, CO 0.8% Model: Servomax 4100
Laser	Output: 0–8.4 mV, Range: 0–100% Model: 25-LHP-213-249
Load Cell	Output: 4–20 mA, Range: 0–3000 kg Manufacturer: Sartorious
Heat Flux	Plate Type, Range: −200–1000 °C Model: GTW-10-32-485A
Mass Flow	Output: 4–20 mA, Range: 0–2500 L/min Model: DPE-S

To investigate the fire-spreading characteristics and temperature distributions for one and two vehicles, K-type thermocouples (OMEGA, measuring range: −200–1260 °C) were attached to the engine room, bumpers, seats, and fuel tank. Temperature data were transmitted to a data-acquisition unit (DAQ) on a PC every second, and the experiments were recorded by a video recorder. The initial ignition location was assumed to be the passenger seat. Figure 3 represents the experimental setup and conditions. In the two-vehicle experiment, the distance between vehicles was set to 50 cm, and the thermocouples were attached in the same positions. The fuel was almost eliminated to prevent explosions during the experimental study.

Figure 3. Experimental setup and measurement tools.

Four-door sedan vehicles were used in the experiments. The vehicle size was 4.7 m (length) × 1.8 m (width) × 1.4 m (height), and was Hyundai's EF sonata released in 1998. These vehicles have not been used for about 10 years, and the experiment was carried out after the pressure of tires were removed with opening full front windows. The location and number of thermocouples in each compartment are represented in Figure 4. Thermocouples were attached to the engine room, bumpers, seats, and fuel tank. Sixteen thermocouples were attached to the engine room, four each at the top and bottom, and two on each side. Ten thermocouples were attached to the bumpers; five each on the front and rear. In addition, 14 thermocouples were attached to each seat in the vehicle interior, representing the head, waist, legs, and feet. Twelve thermocouples were attached to the fuel tank, one on each side, and four each on the top and bottom. Thus, 42 thermocouples were used to figure out the change of the temperature distribution as a function of time.

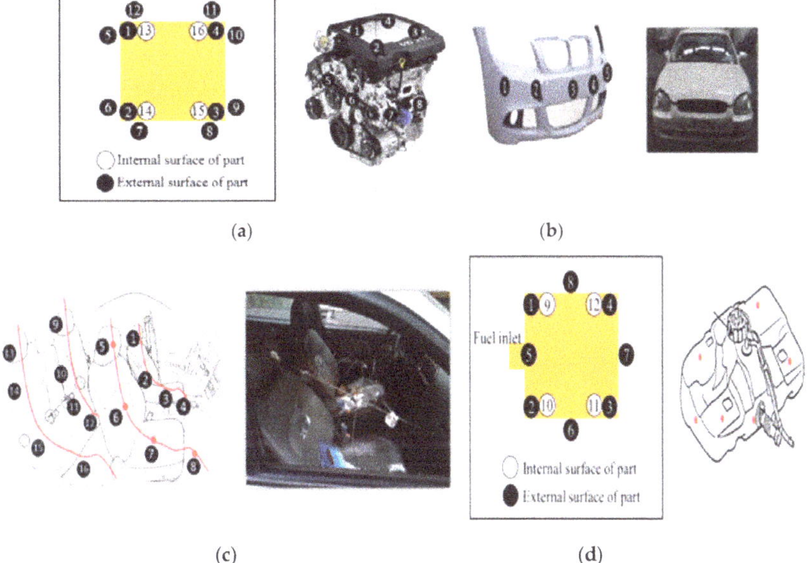

Figure 4. Location and number of thermocouples: (**a**) Engine room, (**b**) bumpers, (**c**) vehicle interior, and (**d**) fuel tank.

3. Results and Discussion

3.1. Fire-Spreading Characteristics

Fire-spreading over one and two vehicles as a function of time while on experiments are presented in Figure 5. Also, a flow-chart about fire-spreading inside a vehicle and two vehicles as a function of time are represented in Figure 6. These represent the major fire-spreading points as a function of time. As represented in the Figure 5 and the flow-chart, in the one-vehicle experiments, the fire spread to the driver seat and rear seat almost simultaneously, followed by the fuel tank, engine room, and finally the front and back bumpers. In the two-vehicle experiments, the fire followed the same sequence until it spread to the next vehicle, due to radiation. It then spread simultaneously to the rear seat and driver seat of the second vehicle, then the engine room, fuel tank, and bumpers in a regular sequence. Detailed information on the fire-spreading characteristics and special events as a function of time are represented in Table 2.

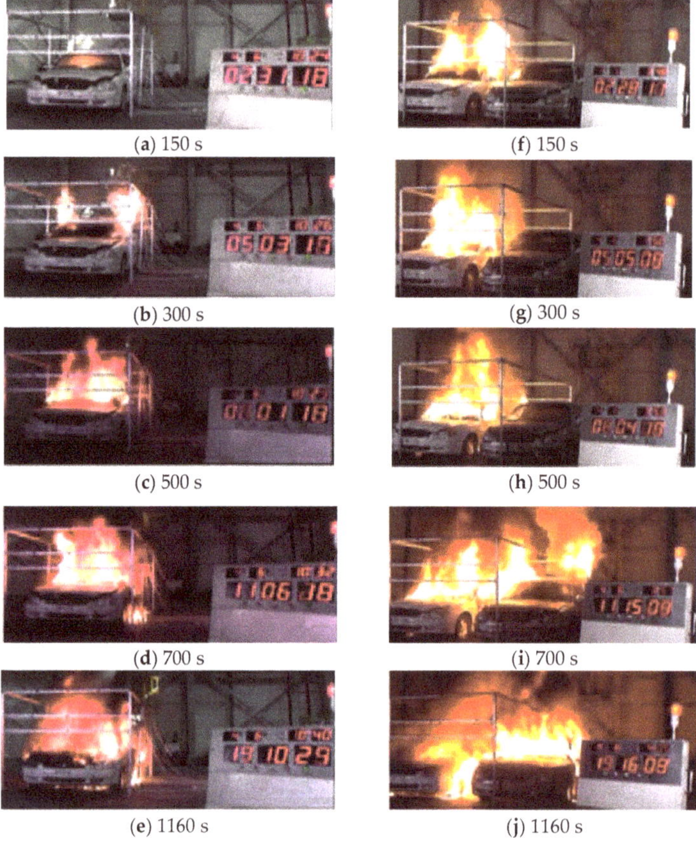

(a) 150 s   (f) 150 s

(b) 300 s   (g) 300 s

(c) 500 s   (h) 500 s

(d) 700 s   (i) 700 s

(e) 1160 s   (j) 1160 s

Figure 5. Fire spreading as a function of time for one and two vehicles; (**a**) 150 s after fires in one vehicle, (**b**) 300 s after fires in one vehicle, (**c**) 500 s after fires in one vehicle, (**d**) 700 s after fires in one vehicle, (**e**) 1160 s after fires in one vehicle, (**f**) 150 s after fires in two vehicles, (**g**) 300 s after fires in two vehicles, (**h**) 500 s after fires in two vehicles, (**i**) 700 s after fires in two vehicles, and (**j**) 1160 s after fires in two vehicles.

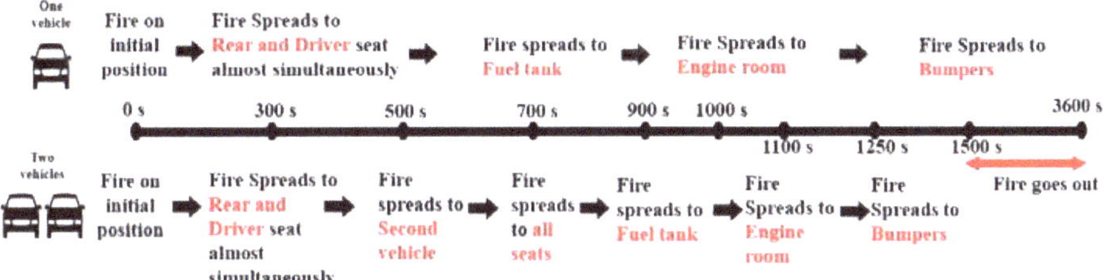

Figure 6. Major fire-spreading flow-chart.

Table 2. Detailed information on the fire-spreading characteristics and special events as a function of time.

Time (s) After Ignition	Fire Reaches Maximum Temperature and Goes Out	
	First Vehicle	Second Vehicle
0	Ignition in passenger seat	
150	Fires active in passenger seat	
300	Fire spreads to driver and rear seats almost simultaneously	
500	-	Fire spreads to the second vehicle
700	Fire spreads to fuel tank	Fire dramatically spreads to seats
900	-	Fire spreads to fuel tank
1000	Fire spreads to engine room	-
1100	-	Fire spreads to engine room
1250	-	Fire spreads to bumpers
1500	Fire spreads to bumpers	-
1500–3600	Fire goes out	Fire goes out

Therefore, the main temperature changes in the seats, engine room, fuel tank, and bumpers were represented. The temperature distribution on seats are represented in Figure 7. The passenger seat actively caught fire around 300 s (5 min) after ignition, and the highest temperature was observed in the interior, due to its combustible materials compared with the other compartments.

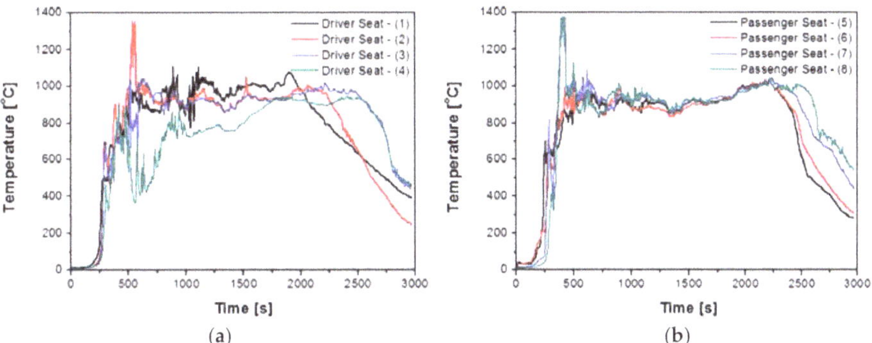

Figure 7. Temperature distribution on seats of the first vehicle; (**a**) temperature of driver seat in first vehicle and (**b**) temperature of passenger seat in first vehicle.

The fire spread to the driver and rear seats simultaneously; however, the temperature differed slightly, as can be seen in Figure 8, which represents the temperature distribution of the driver and rear seats for the two sets of vehicles. The temperature increase in the driver seat and the rear seat can be observed almost simultaneously. The beginning of the temperature increase and the highest temperature are represented as the green and red lines, respectively. This means that the fire in the driver seat was larger than the rear seat, even though the fire-spreading time was similar because the driver seat was located closer to the fire than the rear seat.

To investigate the effect of temperature propagation on the number of vehicles, the driver seat and the rear seat inside the first vehicle are represented in Figure 8a,b. In the case of Figure 8c,d, it showed the driver seat and the rear seat inside the first vehicle when the two-vehicle experiment. After 500 s (8 min 20 s), the fire spread to the next vehicle, and the interior's temperature rapidly increased around 700 s (11 min 40 s), as shown in Figure 8c,d. Since the second passenger seat was located next to the first vehicle, the fire reached the thermocouple's temperature limit of 1370 °C.

The analysis was begun by dividing the time into sections for the various fire-spreading periods and major changes. Section 1 represented the period of the fire spreading from the passenger seat to the driver and rear seats after 300 s (5 min). Section 2 was 500 s (8 min 20 s) after the ignition, in which the fire spread to the next vehicle. Section 3 was when both the passenger seat and driver seat of the second vehicle caught fire. Section 4 was the period when the HRR reached maximum in the case of two-vehicle fire. These same sections were used for representing the heat release rates and the toxic substances in Sections 3.2 and 3.3, respectively.

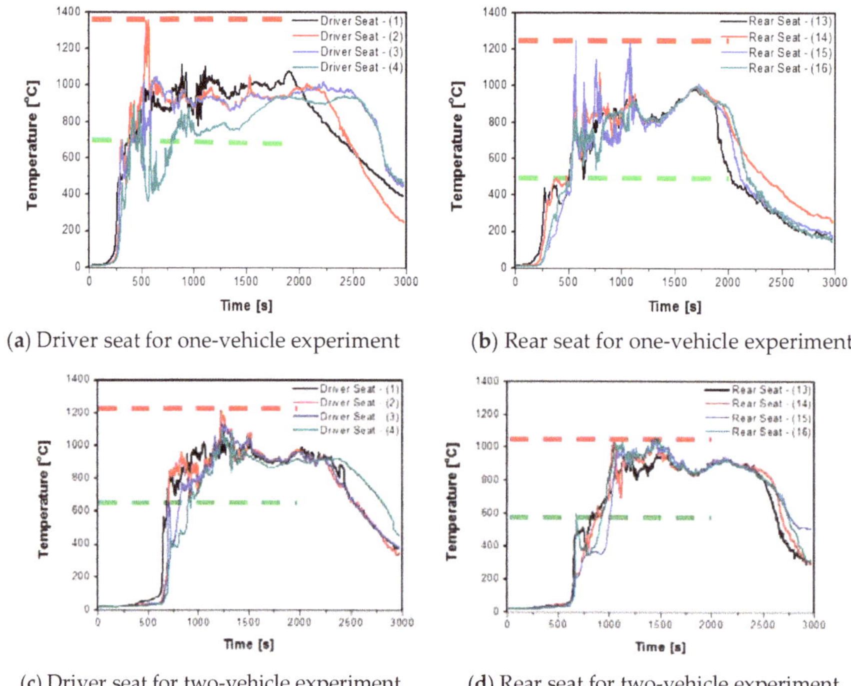

(**a**) Driver seat for one-vehicle experiment

(**b**) Rear seat for one-vehicle experiment

(**c**) Driver seat for two-vehicle experiment

(**d**) Rear seat for two-vehicle experiment

Figure 8. Temperature distribution results inside the driver and rear seats for one and two vehicles: (**a**) Temperature of driver seat inside the first vehicle for the one-vehicle experiment, (**b**) temperature of rear seat inside the first vehicle for the one-vehicle experiment, (**c**) temperature of driver seat inside the first vehicle for the two-vehicle experiment, and (**d**) temperature of rear seat inside the first vehicle for the two-vehicle experiment.

The temperature distribution in engine room and fuel tank were represented in Figure 9. The engine room was comprehensively burning within about 1200 s (20 min) of the fire ignition. In addition, the fire in the initial vehicle did not propagate directly to the engine room, but first spread to the seats. After about 1300 s (21 min 40 s), the fire spread to the adjacent vehicle. The fire went out 2500 s (41 min 20 s) after reaching the maximum temperature.

In the fuel tank, the temperature change was relatively small compared to the other combustibles. This was because most of the fuel was removed; however, in a real vehicle fire, it is the most vulnerable and dangerous compartment for the fire to reach.

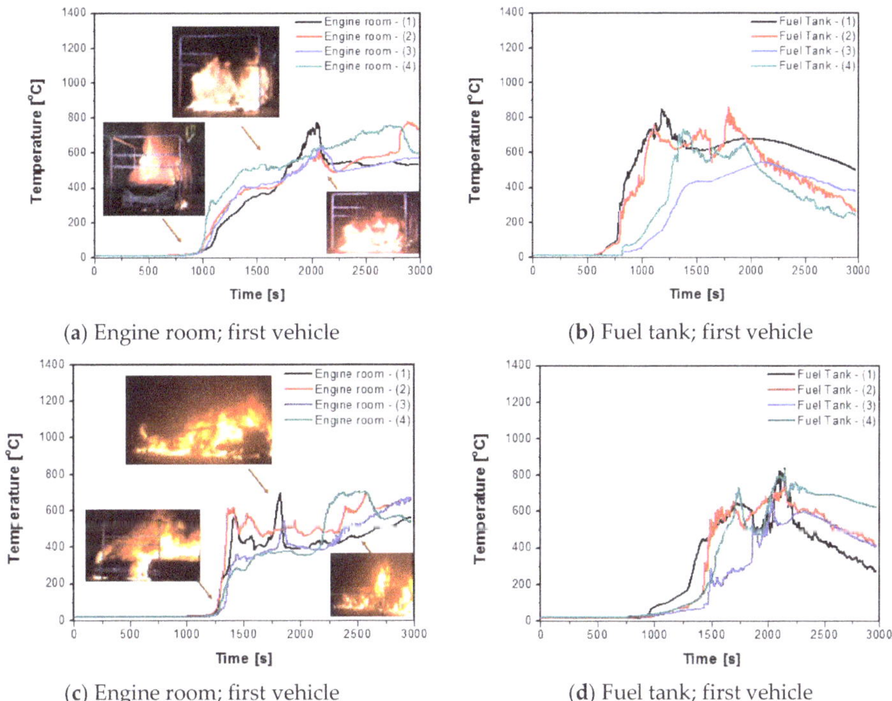

(**a**) Engine room; first vehicle

(**b**) Fuel tank; first vehicle

(**c**) Engine room; first vehicle

(**d**) Fuel tank; first vehicle

Figure 9. Temperature distribution in the engine room and fuel tank for one and two vehicles: (**a**) Temperature of engine room in first vehicle in the one-vehicle experiment, (**b**) temperature of fuel tank in first vehicle in the one-vehicle experiment, (**c**) temperature of engine room in first vehicle in the two-vehicle experiment, and (**d**) temperature of fuel tank in first vehicle in the two-vehicle experiment.

The temperature distributions at the bumpers are represented in Figure 10. For the front bumper, the temperature rose from the place nearest the initial fire vehicle. After about 1400 s (23 min 30 s), the temperature rose in the next vehicle. The temperature rose later because it was located farthest from the first vehicle. In addition, in the rear bumper, the fire spread more quickly than the front bumper because it was relatively close to the seat.

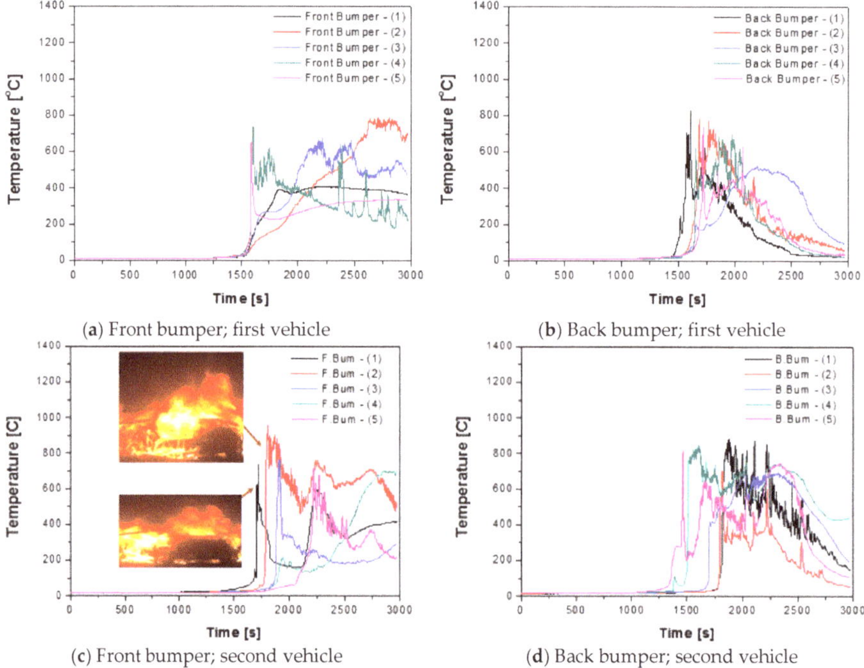

(**a**) Front bumper; first vehicle   (**b**) Back bumper; first vehicle

(**c**) Front bumper; second vehicle   (**d**) Back bumper; second vehicle

Figure 10. Temperature distribution at bumpers for one and two vehicles: (**a**) Temperature of front bumper in first vehicle, (**b**) temperature of back bumper in first vehicle, (**c**) temperature of front bumper in second vehicle, and (**d**) temperature of back bumper in second vehicle.

3.2. Heat Release Rate

The heat release rates considering the unsteady fire phenomenon are represented in Figure 11. As can be seen in Figure 6, the heat release rate increased rapidly around 180 s (3 min) after ignition because the fire spread into the vehicle interior. Further, when the windshield broke around 300 s (5 min), the heat release rate increased sharply to about 2.3 MW. However, it continued to increase because the fire spread from the passenger seat to the driver seat. The passenger-seat cushion spontaneously ignited and then temporarily decreased. Because the cushion was composed of a composite material, the heat release rate increased and decreased repeatedly until it reached its maximum.

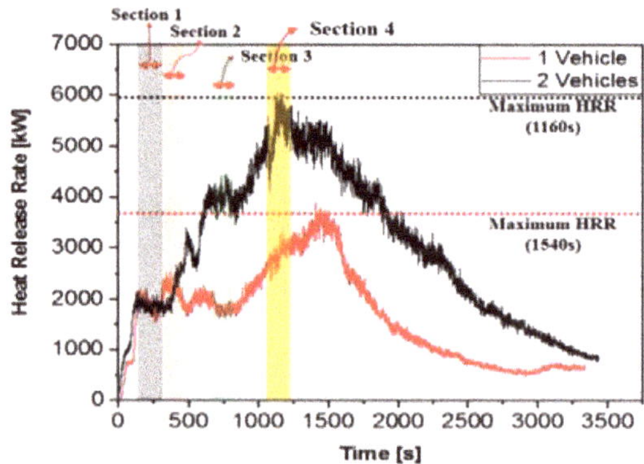

Figure 11. Heat release rate for one and two vehicles as a function of time.

For the engine room and the fuel tank, the increasing and decreasing heat release rate was not apparent because there was little fuel in the engine room and the engine oil was low. After reaching the maximum heat release rate of 3.5 MW, the fire slowly went out.

In the case of the two-vehicle experiments, the heat release rate followed a similar pattern until the initial window breakage. After that, the fire spread to the entire passenger seat and then the next vehicle at about 600 s (10 min), due to radiation. After the ignition, the fire spread rapidly through the open passenger window. The spacing between the vehicles was about 50 cm, and there was a 2 MW fire around the vehicle. When both vehicles were burning, the fire spread dramatically to all of the seats. The maximum heat release rate was about 6 MW, and the fire went out gradually. Since the time to reach 1 MW was about 240 s (4 min), the fire growth in a vehicle is considered to be a medium-fast fire.

The heat release rate is one of the most important parameters to calculate the flame height [24,25]. McCaffrey [24] presented several formulas under various conditions, but, one of the functions under the open condition can be calculated the flame height using heat release rate. McCaffrey's formula is as follows:

$$Zc = 0.08\,Q^{2/5} \tag{1}$$

where Zc (m), Q (kW) are flame height and the heat release rates, respectively. Throughout this formula, the maximum flame heights were calculated as 2.1 m and 2.6 m for one and two vehicles, respectively.

Furthermore, the height of the flame in normal atmospheric conditions was indicated by non-dimensional analysis [25], and it is as follows:

$$L_f = 0.235 \cdot Q^{2/5} - 1.02D \tag{2}$$

where L_f(m), Q (kW), D (m) are flame height, the heat release rates, and diameter of pool, respectively. In case of the function suggested by Heskastad [25], the flame height can be calculated using the heat release rate and pool diameter. Throughout this formula, the maximum flame heights were calculated 4.3 m and 7.6 m, 1.35 m and 2.83 m, and 3.49 m and 4.97 m for one and two vehicles for varying diameters, respectively. Because the width and length of the vehicle have different diameters, the flame heights were calculated with the diameters at 1.8 m width, 4.7 m length, and 2.6 m hydraulic diameter of the vehicle. Flame heights are represented as a function of time and maximum flame heights in Figure 12 and Table 3, respectively. However, measuring the change of flame heights was affected significantly under various conditions.

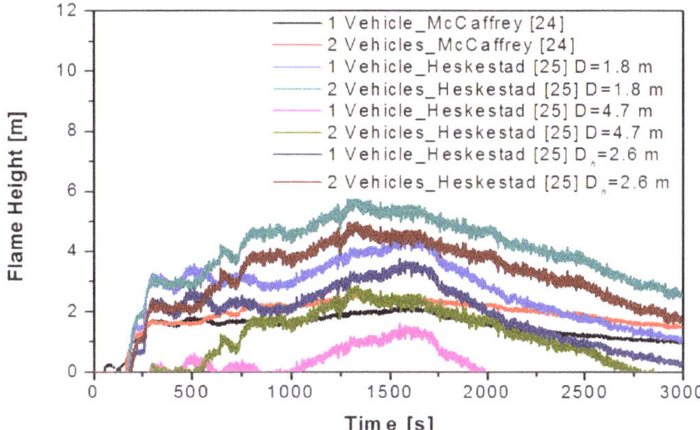

Figure 12. Flame height calculated by Equation 1 (McCaffrey [24]) and Equation 2 (Heskestad [25]) for varying diameter as a function of time.

Table 3. Comparison of flame heights under the open condition.

Number of Vehicle	Maximum Flame Height (m)			
	McCaffrey [24]		Heskestad [25]	
	-	$D = 1.8$	$D = 4.7$	$D_h = 2.6$
One	2.1	4.3	1.35	3.49
Two	2.6	7.6	2.83	4.97

The fire growth can be evaluated using the following generic fire-growth curve:

$$Q_{\max} = a \cdot (t - t_i)^2 \tag{3}$$

where a is the fire-growth coefficient (kW/s^2), t means time (s), and t_i means the time of ignition (s). In this study, the time of ignition can be taken as zero. The fire-growth coefficients for vehicle fires up to 1 MW are represented in the Appendix A. Reducing the flame height and fire growth were decisive parameters for suppressing a fire; a fire extinguisher should be considered, especially in enclosed spaces, e.g., tunnels and indoor parking lots.

It is generally well acknowledged that fire experiments are really difficult to do repeated experiments, due to time consumption and expensive time. Thus, various fire articles did not contain a detailed uncertainty analysis. Melcher et al. [26] suggested the impact of random deviations that may occur in a single experiment. Mass loss rates and heat release rates for one and two vehicles were represented with error range in Figure 13. Yellow and lavender indicate the error ranges that occurred in one and two vehicles, respectively.

Applying the actual heat release rates obtained from this study to the numerical analysis study, Park et al. published a study on the effect of a vehicle accident on the evacuation in various tunnel aspect ratio [27].

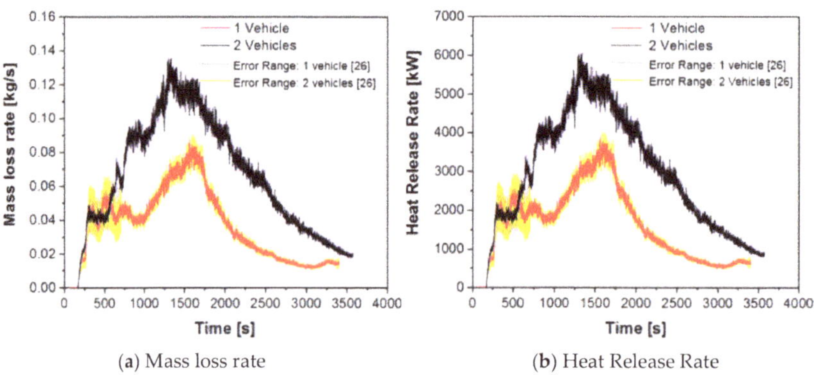

(a) Mass loss rate   (b) Heat Release Rate

Figure 13. Mass loss rate and heat release rate for one and two vehicles as a function of time, error range estimated following Melcher et al. [26]: (a) Mass loss rate and (b) heat release rate.

The maximum heat release values obtained by other studies are shown in Table 4 and compared with the results obtained through this experiment. As a rule of thumb, the maximum heat release rates were normally represented 2.5 MW to 5 MW. Okamoto et al. [8] found the heat release rates as a function of time in real-scale experiments, and the vehicles used in this experiment were similar to those used in this study. The experimental conditions were quite similar, such as conducting the vehicle fire test in the absence of fuel, but only with different experimental measuring equipment. In addition, the maximum heat release rate was 3.5 MW, the same as this study. However, in the case of research done by Shipp et al. [28], the maximum heat release rate was significantly higher than other studies since the vehicle tests involved fuel spill from the petrol tank. Furthermore, Ingason [29] conducted the vehicle fire experiments in a tunnel so that the maximum heat release rates were different because of ceiling temperatures, ventilation system on maximum heat release rate, and fire growth rates.

Table 4. Comparison of maximum heat release rates throughout large-scale experiments.

Type of Vehicles	Maximum Heat Release Rate (MW)			
	Okamoto et al. [6]	Shipp et al. [28]	Ingason [29]	Park et al.
Small passenger car	3.5	8	2.5	3.5
Large passenger car	4.2	-	<5	-
2 passenger cars	-	-	3.5–10	6

3.3. Toxic Substances

The changes in the amounts of carbon dioxide and carbon monoxide as a function of time are represented in Figure 14. These values represent the toxic gases obtained through the LSC from the vehicle as it burned. The main changes on carbon monoxide and carbon dioxide as a function of time are represented in Tables 5 and 6, respectively. Furthermore, it showed similar trends to the distribution of change of heat release rate. However, judging from the influence on the human body, it is obvious that these values might be not correct, because of the many differing conditions in fire phenomena, e.g., enclosed or open spaces, scale of space, ventilation systems, sprinklers, and so on.

With fires in open spaces, the smoke and toxic gases will be released into the air. In enclosed spaces, e.g., tunnels or indoor parking lots, the smoke will accumulate continuously, and the concentration will greatly increase. However, the effects on the human body can be analyzed based on concentrations of CO and CO_2 from the vehicle itself [30].

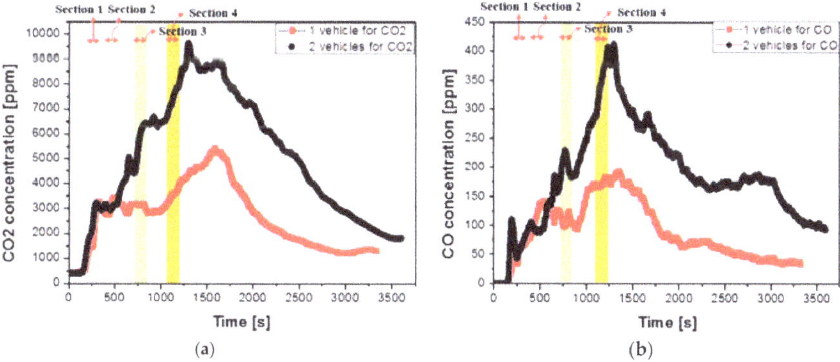

Figure 14. Toxic-substance concentration results: (**a**) Carbon dioxide and (**b**) carbon monoxide.

Table 5. Carbon-dioxide concentration at the major change periods, and influences on the body based on value.

Section	CO_2 Concentration (ppm)		Effect on Health [30]	
	One Car	Two Cars	One Car	Two Cars
Section 1	1000–3000		General condition and mild headaches	
Section 2	3000		Poor air condition and headaches	
Section 3	3000	4500–6200	Poor air condition and headaches	
Section 4	3000–3500	6500–8000	Headaches	Respiratory, circulatory, and cerebral impairment

Table 6. Carbon-monoxide concentration at the major change periods, and influences on the body based on value.

Section	CO Concentration (ppm)		Effect on Health [30]	
	One Car	Two Cars	One Car	Two Cars
Section 1	0–50	0–100	Slight headaches	
Section 2	75–150		Headaches	
Section 3	100–125	150–225	Headaches	Dizziness, nausea, fatigue, headaches
Section 4	150–175	300–425	Dizziness, nausea, fatigue, headaches	Headache and nausea; life threatening in 3 h

Based on the values obtained from the LSC, the CO_2 generated from one vehicle will give a person a slight headache. However, in a two-vehicle fire, the slight influence occurs before 1000 s (16 min 40 s); however, after that, it will affect the respiratory system and nervous system, and cause cerebral impairment. Furthermore, only a slight headache can be felt about 500 s (8 min 20 s) after ignition. However, after that, the area should be evacuated. In the worst case, life is threatened, and evacuation may be difficult because of the high concentration of CO over 1200 s (20 min). The higher the heat release rate is, the higher the concentration of CO and CO_2. Therefore, the larger the fire is, the faster the area should be evacuated. Based on the CO and CO_2 concentrations absorbed through the LSC, the impact on human health applies when people were directly exposed. It is apparent that the values obtained through the LSC will vary depending on the location and conditions of the fire inside of buildings, but it can provide indirect guidance on ventilation and evacuation in compartment space while presenting the toxic concentration occurring in the vehicle itself.

4. Conclusions

The aim of this study was to investigate how fire spreads through a vehicle using an actual vehicle-fire test, and to analyze the heat release rate for two vehicles using a large-scale calorimeter. In addition, the influence of the toxic gases generated from the vehicle fires was analyzed. An analysis of the experimental results provided the following conclusions.

(1) In actual vehicle tests, the fire spread from the initial ignition location to the rear seat, engine room, fuel tank, and bumper in regular sequence. In the two-vehicle situation, a similar tendency was observed, and the fire spread to the next vehicle after about 500 s (8 min 20 s).

(2) The fire rose sharply after 200 s (3 min). The maximum heat release rates of one and two vehicles were represented as 3.5 MW and 6 MW, reached at about 1540 s (25 min 40 s) and 1160 s (19 min 20 s), respectively. Since the time to reach 1 MW was about 240 s (4 min) before and after, the fire growth in a vehicle fire is considered to be a medium-fast fire phenomenon.

(3) The evacuation should be totally completed within 20 min after the fire starts, because of the high concentrations of carbon dioxide and carbon monoxide in an enclosed space. This may vary, depending on the size of the location where the fire occurs.

(4) In this study, uncertainty analysis cannot be included because of financial problems like other related studies [6–8], however, we place emphasis on fire-spreading characteristics inside of a vehicle and toxic gases.

Author Contributions: Conceptualization, methodology, investigation, formal analysis, Y.P. and J.R.; writing—original draft preparation, Y.P.; writing—review and editing, J.R.; project administration, supervision, H.S.R.

Funding: This research was supported by the Research and Development to Enhance Firefighting Response Ability founded by National Fire Agency (2018-NFA002-007-01010002-2018) and Chung-Ang University Research Grants in 2017.

Conflicts of Interest: The authors declare no conflict of interest.

Appendix A. Fire-Growth Coefficients for Vehicle Fires

Table A1. Fire-Growth Coefficients for one and two vehicle fires.

Time	One Vehicle	Two Vehicles	Time	One Vehicle	Two Vehicles
55	0.000122845	0	111	−0.00003228	0
56	0.000110589	0	112	−0.000100544	0
57	0.000111826	0	113	−0.000102829	0
58	0.000119181	0	114	−0.000125789	0
59	0.000151472	0	115	−0.000112598	0
60	0.000143496	0	116	−0.000120963	0
61	0.000141763	0	117	−0.000118931	0
62	0.000142229	0	118	−0.000123675	0
63	0.00014096	0	119	−0.000136618	0
64	0.00013857	0	120	−0.000132197	0
65	0.000130619	0	121	−0.000123733	0
66	0.000143594	0	122	0.000045496	0
67	0.000153154	0	123	5.5872E-06	0
68	0.000143394	0	124	4.16544E-05	0
69	0.000115387	0	125	3059376E-05	0
70	0.000326219	0	126	3.63424E-05	0
71	0.000223675	0	127	3.67568E-05	0
72	0.000277621	0	128	2.67056E-05	0
73	0.000272314	0	129	−1.11936E-05	0
74	0.000274293	0	130	1.77728E-05	0
75	0.000288267	0	131	9.4848E-06	0
76	0.000282707	0	132	1.01104E-05	0
77	0.000298621	0	133	−3.34768E-05	0
78	0.000276741	0	134	0.00000116	0
79	0.000281992	0	135	0.000144683	0
80	0.000264557	0	136	0.000136411	0
81	0.000300158	0	137	0.000158899	0
82	0.000287592	0	138	0.000142838	0
83	0.000257986	0	139	0.000129723	0
84	0.000279203	0	140	0.000139946	0
85	0.000259619	0	141	0.000131667	0
86	0.000282691	0	142	0.000127328	0
87	0.000290432	0	143	0.000127696	0
88	0.000129774	0	144	0.000236862	0
89	0.000130891	0	145	0.000222347	0
90	0.000116182	0	146	0.000211238	0
91	0.000112029	0	147	0.000220477	0
92	0.000117714	0	148	0.000215518	0
93	0.000104933	0	149	0.00025139	0
94	0.00010421	0	150	0.00021375	0
95	0.000104573	0	151	0.000351069	0
96	0.000099011	0	152	0.000364856	0
97	0.0000978	0	153	0.000371894	0
98	0.00009746	0	154	0.00035011	0
99	0.0000985872	0	155	0.000354869	0
100	0.000088056	0	156	0.000278704	0
101	0.000102904	0	157	0.000365112	0
102	0.0000951808	0	158	0.00045428	0
103	0.000102624	0	159	0.00038917	0
104	−1.50448E-05	0	160	0.000341139	0
105	−9.12816E-05	0	161	0.000341352	0
106	−0.000100579	0	162	0.000358957	0
107	−9.69392E-05	0	163	0.000454278	0
108	−9.98464E-05	0	164	0.000407691	0
109	−2.65664E-05	0	165	0.000914008	0
110	−3.36192E-05	0	166	0.000862808	0

Table A1. *Cont.*

Time	One Vehicle	Two Vehicles	Time	One Vehicle	Two Vehicles
167	0.001054307	0.000220694	203	0.0100165	0.0120997
168	0.001063165	0.000611402	204	0.0110054	0.0114384
169	0.00104391	0.00068039	205	0.0110963	0.0124061
170	0.001228091	0.000880261	206	0.0101954	010114076
171	0.001635272	0.000812598	207	0.0104325	0.0122057
172	0.001834352	0.001029643	208	0.0106026	0.0121818
173	0.001916245	0.001240034	209	0.0111004	0.0124914
174	0.00199748	0.002309933	210	0.0111064	0.0135382
175	0.001945301	0.002533862	211	0.0126027	0.0129667
176	0.002285653	0.002699115	212	0.0107972	0.01428483
177	0.002352554	0.002952645	213	0.011416	0.0135615
178	0.002293226	0.002970237	214	0.0117953	0.0138855
179	0.003488533	0.003338856	215	0.0112167	0.0142959
180	0.003595838	0.003579187	216	0.0117955	0.0135957
181	0.003715666	0.00413451	217	0.0123651	0.0139167
182	0.004403666	0.0041032	218	0.011948	0.0153101
183	0.004578416	0.004038008	219	0.0121137	0.013121
184	0.00505969	0.005178826	220	0.0118276	0.0146671
185	0.004863582	0.006157565	221	0.0121345	0.0160241
186	0.005264023	0.007279477	222	0.0121715	0.0140538
187	0.005925605	0.007346544	223	0.0121353	0.0150714
188	0.005557421	0.006510378	224	0.0125515	0.0146725
189	0.006212605	0.007681355	225	0.0120098	0.0150361
190	0.006590218	0.007854469	226	0.0126213	0.0147933
191	0.005883032	0.007422917	227	0.0113123	0.0146584
192	0.003584963	0.007919211	228	0.0113451	0.0153618
193	0.0066938	0.0087581	229	0.0112896	0.0158492
194	0.0082796	0.0090616	230	0.0117224	0.0150186
195	0.008573	0.0095076	231	0.0121374	0.0151767
196	0.0081337	0.010333	232	0.0122526	0.0159671
197	0.0084333	0.0105719	233	0.0122324	0.0157199
198	0.0092459	0.0096946	234	0.0120766	0.0158552
199	0.0098028	0.0101741	235	0.0121153	0.0149351
200	0.0101057	0.0114833	236	0.0119933	0.0169582
201	0.009347	0.011416	237	0.0112681	0.0161582
202	0.0100612	0.0113177	238	0.0123996	0.016821

References

1. Li, Y.; Spearpoint, M. Analysis of vehicle fire statistics in New Zealand parking buildings. *Fire Technol.* **2007**, *43*, 93–106. [CrossRef]
2. Babrauskas, V. The Cone Calorimeter. In *SFPE Handbook of Fire Protection Engineering*; Springer: New York, NY, USA, 2016; pp. 952–980.
3. Wang, Z.; Hu, X.; Jia, F.; Galea, E.R. A two-step method for predicting time to flashover in room corner test fires using cone calorimeter data. *Fire Mater.* **2012**, *37*, 457–473. [CrossRef]
4. Lai, C.M.; Ho, M.C.; Lin, T.H. Experimental Investigations of Fire Spread and Flashover Time in Office Fires. *J. Fire Sci.* **2010**, *28*, 279–302. [CrossRef]
5. Hakkarainen, T. Rate of Heat Release and Ignitability Indices in Predicting SBI Test Results. *J. Fire Sci.* **2001**, *19*, 284–305. [CrossRef]
6. Katsuhiro, O.; Norimichi, W.; Yasuaki, H.; Tadaomi, C.; Ryoji, M.; Hitoshi, M.; Satoshi, O.; Hideki, S.; Yohsuke, T.; Kimio, H.; et al. Burning behavior of sedan passenger cars. *Fire Saf. J.* **2009**, *44*, 301–310. [CrossRef]
7. Mangs, J.; Keski-Rahkonen, O. Characterization of the Fire Behaviour of a Burning Passenger Car. Part1: Car Fire Experiments. *Fire Saf. J.* **1994**, *23*, 17–35. [CrossRef]

8.   Katsuhiro, O.; Takuma, O.; Hiroki, M.; Masakatsu, H.; Norimichi, W. Burning behavior of minivan passenger cars. *Fire Saf. J.* **2013**, *62*, 272–280. [CrossRef]

9.   Li, D.; Zhu, G.; Zhu, H.; Yu, Z.; Gao, Y.; Jiang, X. Flame spread and smoke temperature of full-scale fire test of car fire. *Case Stud. Therm. Eng.* **2017**, *10*, 315–324. [CrossRef]

10. Loladze, I. Rising atmospheric CO_2 and human nutrition: Toward globally imbalanced plant stoichiometry? *Trends Ecol. Evol.* **2002**, *17*, 457–461. [CrossRef]

11. Bailey, J.E.; Argyropoulos, S.V.; Kendrick, A.H.; Nutt, D.J. Behavior and Cardiovascular Effects of 7.5% CO_2 in Human Volunteers. *Med. J. Aust.* **2005**, *21*, 18–25. [CrossRef]

12. Mayr, F.B.; Spiel, A.; Leitner, J.; Marsik, C.; Germann, P.; Ullrich, R.; Wagner, O.; Jilma, B. Effects of Carbon Monoxide inhalation during Experimental Endotoxemia in Humans. *Am. J. Respir. Crit. Care Med.* **2005**, *171*, 354–360. [CrossRef] [PubMed]

13. Truchot, B.; Fouillen, F.; Collet, S. An experimental evaluation of toxic gas emissions from vehicle fires. *Fire Saf. J.* **2018**, *97*, 111–118. [CrossRef]

14. Deckers, X.; Haga, S.; Sette, B.; Merci, B. Smoke control in case of fire in a large car park: Full-scale experiments. *Fire Saf. J.* **2013**, *57*, 11–21. [CrossRef]

15. Yan, Z.; Guo, Q.; Zhu, H. Full-scale experiments on fire characteristics of road tunnel at high altitude. *Tunn. Undergr. Space Technol.* **2017**, *66*, 134–146. [CrossRef]

16. Blanchard, E.; Boulet, P.; Desanghere, S.; Cesmat, E.; Meyrand, R.; Garo, J.P.; Vantelon, J.P. Experimental and numerical study of fire in a midscale test tunnel. *Fire Saf. J.* **2012**, *47*, 18–31. [CrossRef]

17. Lonnermark, A.; Ingason, H. Fire Spread and Flame Length in Large-Scale Tunnel Fires. *Fire Technol. J.* **2006**, *42*, 283–302. [CrossRef]

18. Horvath, I.; Beeck, J.V.; Merci, B. Full-scale and reduced-scale tests on smoke movement in case of car park fire. *Fire Saf. J.* **2013**, *57*, 35–43. [CrossRef]

19. Ingason, H. Model scale railcar fire tests. *Fire Saf. J.* **2007**, *42*, 271–282. [CrossRef]

20. Lee, S.R.; Ryou, H.S. An experimental study of the effect of the aspect ratio on the critical velocity in longitudinal ventilation tunnel fires. *Fire Sci. J.* **2005**, *23*, 119–138. [CrossRef]

21. Lee, S.R.; Ryou, H.S. A numerical study on smoke movement in longitudinal ventilation tunnel fires for different aspect ratio. *Build. Environ. J.* **2006**, *41*, 719–725. [CrossRef]

22. Baek, D.; Bae, S.; Ryou, H.S. A numerical study on the effect of the hydraulic diameter of tunnels on the plug-holing phenomena in shallow underground tunnels. *J. Mech. Sci. Technol.* **2017**, *31*, 2331–2338. [CrossRef]

23. Baek, D.; Sung, K.H.; Ryou, H.S. Experimental study on the effect of heat release rate and aspect ratio of tunnel on the plug-holing phenomena in shallow underground tunnels. *Int. J. Heat Mass Transf.* **2017**, *113*, 1135–1141. [CrossRef]

24. McCaffrey, B. Flame Height. In *SFPE Handbook of Fire Protection Engineering*, 2nd ed.; National Fire Protection Assoc.: Quincy, MA, USA, 1995.

25. Heskestad, G. Fire Plumes. In *SFPE Handbook of Fire Protection Engineering*, 2nd ed.; National Fire Protection Assoc.: Quincy, MA, USA, 1995.

26. Melcher, T.; Zonke, R.; Trott, M.; Krause, U. Experimental investigations on the repeatability of real scale fire test. *Fire Saf. J.* **2016**, *82*, 101–114. [CrossRef]

27. Park, Y.; Lee, Y.; Na, J.; Ryou, H.S. Numerical Study on the Effect of Tunnel Aspect Ratio on Evacuation with Unsteady Heat Release Rate Due to Fire in the Case of Two Vehicles. *Energies* **2019**, *12*, 133. [CrossRef]

28. Shipp, M.; Spearpoint, M. Measurements of the severity of fires involving private motor vehicles. *Fire Mater.* **1995**, *19*, 143–151. [CrossRef]

29. Ingason, H. *An Overview of Vehicle Fires in Tunnels*; SP Swedish National Testing and Research Institute: Boras, Sweden, 2001.

30. Agency for Toxic Substances and Disease Registry (ATSDR). Available online: www.atsdr.cdc.gov/2012 (accessed on 27 February 2019).

Article

Macro and Meso Characteristics of In-Situ Oil Shale Pyrolysis Using Superheated Steam

Lei Wang [1,2], Dong Yang [2], Xiang Li [2], Jing Zhao [1,2], Guoying Wang [2] and Yangsheng Zhao [1,2,*]

[1] College of Mining Engineering, Taiyuan University of Technology, Taiyuan 030024, China; leiwang0327@163.com (L.W.); zhaojing19860207@163.com (J.Z.)

[2] Key Laboratory of In-situ Property Improving Mining of Ministry of Education, Taiyuan University of Technology, Taiyuan 030024, China; ydscience@hotmail.com (D.Y.); ivanobstinate@163.com (X.L.); wangguoyingscience@gamil.com (G.W.)

* Correspondence: y-s-zhao@263.net; Tel.: +86-138-3415-1069

Received: 1 August 2018; Accepted: 30 August 2018; Published: 31 August 2018

Abstract: The efficiency of oil shale pyrolysis is directly related to the feasibility of in-situ mining technology. Taiyuan University of Technology (China) proposed the technology of in-situ convective heating of oil shale, which uses superheated steam as the heat carrier to heat the oil shale's ore-body and transport the pyrolysis products. Based on the simulated experiments of in-situ oil shale pyrolysis using superheated steam, the changes in fracture characteristics, pyrolysis characteristics and mesoscopic characteristics of the oil shale during the pyrolysis have been systematically studied in this work. The Xinjiang oil shale's pyrolysis temperature ranged within 400–510 °C. When the temperature is 447 °C, the rate of pyrolysis of kerogen is the fastest. During the pyrolysis process, the pressure of superheated steam changes within the range of 0.1–11.1 MPa. With the continuous thermal decomposition, the horizontal stress difference shows a tendency to first increase and then, decrease. The rate of weight loss of oil shale residue at various locations after the pyrolysis is found to be within the range of 0.17–2.31%, which is much lower than the original value of 10.8%, indicating that the pyrolysis is more adequate. Finally, the number of microcracks (<50 μm) in the oil shale after pyrolysis is found to be lie within the range of 25–56 and the average length lies within the range of 53.9636–62.3816 μm. The connectivity of the internal pore groups is satisfactory, while the seepage channel is found to be smooth. These results fully reflect the high efficiency and feasibility of in-situ oil shale pyrolysis using superheated steam.

Keywords: superheated steam; triaxial stress; thermogravimetry; X-ray microtomography; thermal cracking

1. Introduction

As an unconventional oil and gas resource, oil shale is a fine-grained sedimentary rock, which is rich in solid organic matter (kerogen) and has fine bedding [1,2]. Oil shale can generate shale oil through retorting. After shale oil's hydrocracking, refined oil, such as gasoline, kerosene and diesel oil can be obtained, which is of great significance to alleviate the current oil shortage. The reserves of oil shale resources in China are huge and can be converted into 47.6 billion tons of shale oil [3,4].

At present, most countries around the world use in-situ retorting to exploit oil shale [5,6]. In-situ mining of oil shale only needs to pass the heat-injection well to the ore-body and directly heat the ore-body. After the oil shale ore-body is fully pyrolyzed, the organic matter is also pyrolyzed to generate oil and gas, whereas the hydrocarbon is discharged to ground through production well [7]. According to different forms of heating, the in-situ mining of oil shale can be divided into three classifications, namely heat conduction, convection heating and radiation [8,9]. Furthermore, in-situ conversion process [10] uses high temperature of electrode to heat the ore-body, whereas the heater temperature can reach up to 1000 °C. However, the thermal conductivity of the oil shale is extremely poor, and the heating efficiency is low. Therefore, it takes a long time for the oil shale to reach the effective pyrolysis temperature. Han et al. [11] found that it takes about 10 years or more to reach the initial pyrolysis temperature of oil shale in the area of 400 m^2. Kyung et al. [12] simulated the effect of electric heating on the behavior of pyrolysis products and concluded that, after pyrolysis, almost 60% of the shale oil was trapped in mineral matrix because of poor fluidity. Raytheon's radio frequency/critical flow (RF/CF) technology [13] uses (RF) transmitters to heat the oil shale's ore-body, and then, extract oil and gas, which is produced from pyrolysis, using supercritical carbon dioxide. Yang et al. [14] proposed an in-situ oil shale recovery method, which combines microwave heating and hydraulic fracturing, and then, simulated the thermal decomposition of oil shale under microwave irradiation. Compared with the conventional heating method, microwave heating requires shorter time, and has lower energy consumption, higher oil production and quality. Radiation produces strong heat penetration and faster heating, though the technology is not yet mature enough [15,16]. The in-situ fracturing and heating technology using nitrogen injection [17,18] uses high-temperature and high-purity nitrogen to pyrolyze the oil shale's ore-body. Therefore, high-temperature nitrogen can play an important role in the recovery of oil and gas. Allawzi et al. [19,20] found that the solubility of organic matter increased due to the interaction of supercritical carbon dioxide and cosolvent. Zhang et al. [21] studied the changes in shale oil composition and yield after bioleaching the oil shale, and found that, after bioleaching, the yield of shale oil increased by 15.38%, whereas the contents of high molecular weight and low molecular weight hydrocarbons in shale oil also increased.

It can be said that, regardless of the mining technology used, the most important thing is to find a way to maximize the efficiency of pyrolysis, whereas the pyrolysis temperature and the development of pores and fractures directly determine the efficiency of pyrolysis. At present, studies have focused on describing the evolution of pore construction in the pyrolysis of oil shale [22–26]. Geng et al. [27] have systematically analyzed the evolution of pores and the structure of fractures in oil shale under high temperature and high pressure using a combination of X-ray microtomography (μCT) and mercury intrusion porosimetry. It is considered that 300–500 °C is the stage, where the porosity and the number and aperture of fractures increase significantly. Bai et al. [28–30] studied the evolution characteristics of pore structure during the pyrolysis of Huadian oil shale at the temperature of 100–800 °C and found that the permeability of oil shale significantly increases within the temperature range of 350–450 °C. Kang et al. [9,31] calculated the percolation probability of true three dimensional (3D) digital CT cores of oil shale specimens under different temperatures. The results showed that, when the porosity is higher than 12%, the connectivity of pore-connected clusters is very good and the connection of seepage channels is smooth, which is favorable for oil and gas production and high temperature fluid injection. Saif et al. [32–34] studied the evolution of pores and fractures during pyrolysis of Green River oil shale and found that the critical temperature for a sharp increase in the porosity of oil shale lies within the range of 390–400 °C. After the critical temperature, the porosity of oil shale rapidly increased to 22–25%. Liu et al. [35] analyzed the evolution of pore structure of Fushun oil shale under pressure and temperature conditions and found that the lithostatic pressure would significantly inhibit the development of pores. Pan et al. [36,37] reported that the mineral matters have an insignificant effect on the pyrolysis reactions of kerogen in Jimsar oil shale. Barshefsky et al. [38] reported that the isolated kerogen of Russian oil shale was completely pyrolyzed at 420 °C, while the raw oil shale decomposition was only 65% complete.

In short, many experts and scholars have done a lot of research on the relationship between temperature and pyrolysis characteristics; however, there is little research on the study of pyrolysis characteristics of oil shale under stress constraints. In 2005, the in-situ convection heating of oil shale was put forward by Zhao Yangsheng's team at the Taiyuan University of Technology, China [39]. The technology used superheated steam as the heat carrier to heat the oil shale ore-body, while the produced oil and gas were transported using steam. During the in-situ mining of oil shale using superheated steam, the internal pores and fractures of oil shale are not only a channel for the migration of steam and kerogen pyrolysis products but also a location for heat exchange and transfer in the rock mass, which is directly related to the efficiency of pyrolysis. In this work, based on the simulated experiments of in-situ oil shale pyrolysis using superheated steam, the cracking and pyrolysis characteristics of oil shale samples are thoroughly studied during the pyrolysis and after the pyrolysis, the evolutionary characteristics of pores and fractures inside the oil shale are carefully discussed. The study provides a necessary prerequisite for the application and commercialization of in-situ oil shale mining technology using superheated steam.

2. Experimental

2.1. Thermogravimetric Experiments

The experimental sample was taken from Jimsar County, Xinjiang, China. The oil shale was crushed and sieved to a particle size of $\leq$180 μm for pyrolysis experiments. Table 1 summarizes the results of oil shale industrial analysis and low-temperature carbonization.

Table 1. Proximate and Fischer assay analyses of the Xinjiang oil shale.

Analysis	Composition
Proximate analysis (wt %, ad)	
Moisture	0.56
Ash	77.89
Volatile matter	17.78
Fixed carbon	3.77
Fischer assay analysis (wt %, ad)	
Oil yield	9.08
Water yield	1.50
residue	86.48
Gas + loss	2.94

The pyrolysis weight loss experiment of oil shale was conducted using DTU-2B thermogravimetric analyzer. The device has a temperature measuring accuracy of 0.1 °C and a sensitivity of less than 1 μg. The ground oil shale samples were evenly spread in the crucible. The cooling water was turned on and high-purity nitrogen was slowly passed into the crucible. The temperature was increased from 70 °C to 900 °C at the rate of 3.5 °C/min. The thermogravimetric (TG) and differential thermogravimetric (DTG) curves of the oil shale were obtained using the thermogravimetric experiments. The TG curve reflects the change in sample's mass with temperature, while the DTG curve reflects the relationship between the rate of change of sample's mass and the temperature.

2.2. Simulated Experiments for In-Situ Oil Shale Pyrolysis Using Superheated Steam

For the simulated experiments of in-situ oil shale pyrolysis using superheated steam, the process of preparing the samples is shown in Figure 1. Large oil shale samples were cast through concrete, making the fabricated specimen a cube with the dimensions of 300 mm × 300 mm × 300 mm. After the sample was fully dried, a well-shaped diversion trough was ground on the surface of the specimen using the polisher, which facilitates the outflow of oil shale pyrolysis products. Meanwhile, core drilling was carried out in the middle of the specimen. The diameter of the drill hole was 32 mm, whereas the depth was 200 mm. The drill hole was used as the location for the insertion of heat injection tube. The heat injection tube was mainly composed of a lower-end flower tube and an upper-end sleeve tube. The flower tube was used as the channel of oil shale pyrolysis using the superheated steam and the sleeve tube played the role of sealing and insulation.

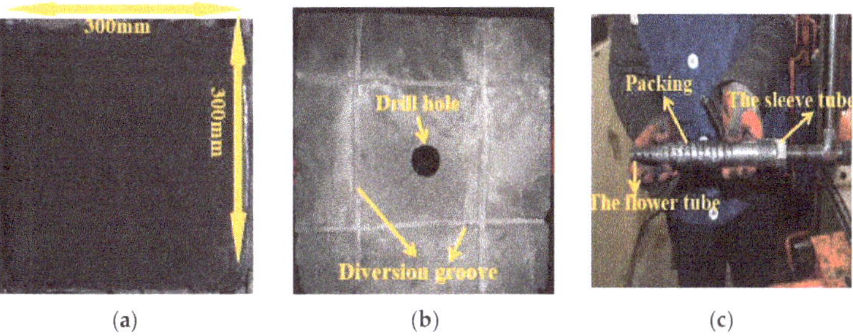

Figure 1. The process of specimen preparation: (**a**) Casting large the oil shale sample with concrete; (**b**) Drill hole and diversion groove of the processed sample; (**c**) Structure of the heat injection tube.

The vertical stress of 3 MPa and the horizontal stress of 4 MPa (Figure 2) were applied to the specimen using a large-size true triaxial press (Figure 3) to simulate the geo-stress environment where the oil shale was located. The press was mainly composed of test loading frame, axial and lateral hydraulic cylinder loading system, numerically controlled hydraulic instrument and other auxiliary devices. The superheated steam, generated by the steam generator, pyrolyzed the specimen under the condition of stress constraint using the heat injection tube. The numerically controlled hydraulic instrument can monitor the stress characteristics of the specimen in real time during the pyrolysis process.

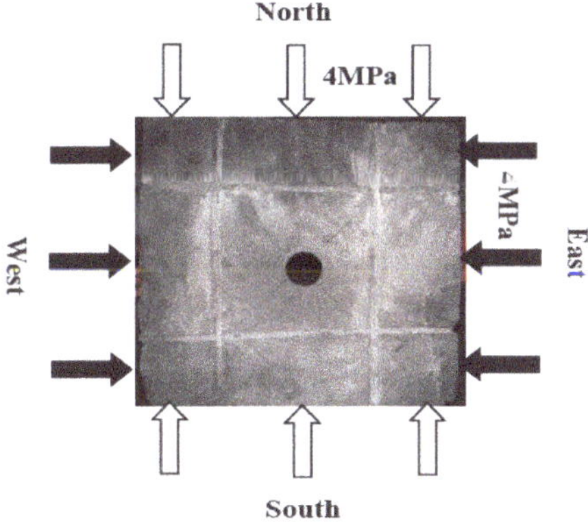

Figure 2. Schematic of the applied horizontal stress.

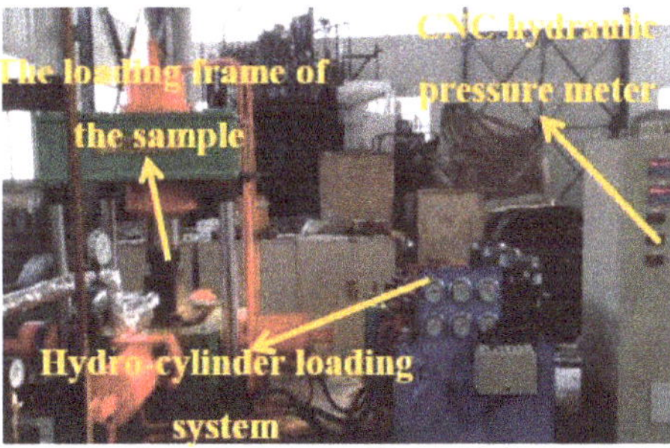

Figure 3. Large-size true three-axes press.

2.3. Micro-CT Scan and the Analysis of Pyrolysis Effect

According to the difference of vertical distance between the outlet of heat injection tube and the pyrolysis oil shale, the oil shale residue was divided into three locations. The thermogravimetric analyses of the oil shale residues at different locations were carried out to analyze the pyrolysis of oil shale. The oil shale residues at different locations were processed into cylindrical samples having the length and diameter of 10 mm and 5 mm, respectively. The internal structure of the oil shale residue was scanned using a μCT225kVFCB high-precision CT analysis system (Figure 4) and the mesoscopic characteristics of the oil shale after pyrolysis were obtained. In this experiment, the scanning voltage was 90 kV and the electric current was 70 μA. The scans were obtained in with 400 frames, the superimposed frame rate was 2 fps and a plane image was generated after the reconstruction. There were 1500 scanned layers and the size of the scanning cell was 2.66 μm × 2.66 μm.

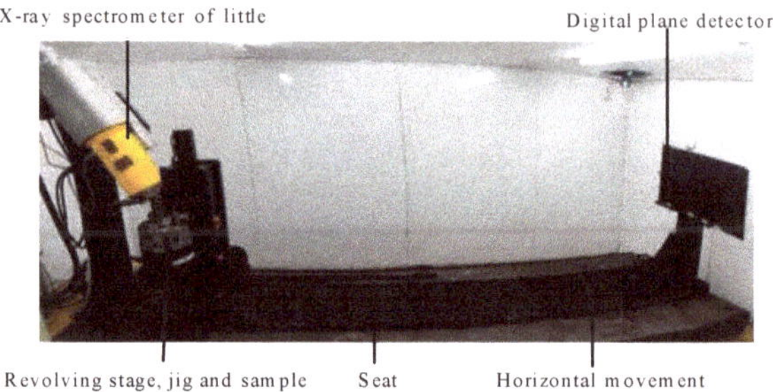

Figure 4. X-ray microtomography experimental system.

3. Results and Discussion

3.1. Thermogravimetric Analysis of Xinjiang Oil Shale

Figure 5 shows the TG and DTG curves of Xinjiang oil shale obtained using thermogravimetric analysis. As can be seen from Figure 5, the major stage of weight loss of Xinjiang oil shale occurs in a relatively small temperature range of 400–510 °C, while the rate of weight loss reaches 10.8%. During this stage, the pyrolysis of oil shale kerogen is closer to completion and the rate of pyrolysis is faster. The pyrolysis of oil shale is considered to be a process in which oil shale is pyrolyzed into oil, gas and semi-coking products in two steps. Firstly, the oil shale was pyrolyzed into tar, which then, was further pyrolyzed to obtain the final products. When the temperature is 447 °C, the rate of mass loss of oil shale is the highest and the pyrolysis rate of kerogen is the fastest.

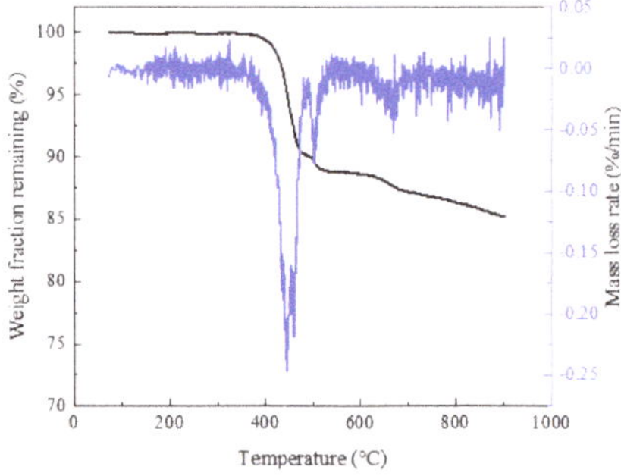

Figure 5. Thermogravimetric (TG) and differential thermogravimetric (DTG) curves of Xinjiang oil shale.

3.2. Stress Characteristics of the Specimen during Pyrolysis

In the process of in-situ oil shale pyrolysis using superheated steam, the weak cementing surface inside the oil shale will break under the action of high-temperature and high-pressure steam, due to which, the heat exchange area inside the rock mass increases. The superheated steam will heat the rock mass along the fracture surface. After the decomposition of organic matter, more pores and fractures are formed inside the oil shale and the hydrocarbon generated by the pyrolysis will further widen the pores and fractures in the migration process, which forms a huge seepage channel. At the same time, due to the continuous development of internal fractures inside oil shale and the continuous injection of superheated steam, the cohesive force of the molecular bonds of oil shale is reduced, which reduces the tensile strength of oil shale and makes the oil shale more prone to tensile fracture. The variation in steam pressure with pyrolysis time during the pyrolysis process is shown in Figure 6.

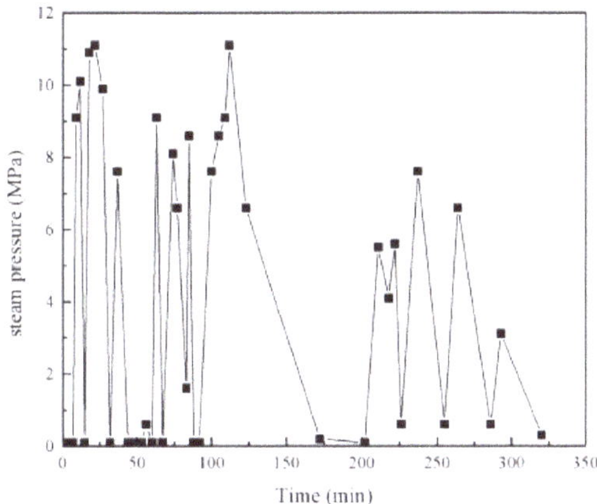

Figure 6. Variation in steam pressure with pyrolysis time.

It can be seen from Figure 6 that the superheated steam pressure varies between 0.1–11.1 MPa as the oil shale pyrolysis proceeds. This is due to the reason that the thermal cracking of oil shale is a process of gradual expansion. With the continuous injection of superheated steam, the stress at the tip of the fracture gradually increases. When the stress value reaches the threshold point of crack initiation, the fracture expands. The expansion process is also the process of energy release, which shows the decrease in stress. Once the stress is lower than the threshold point of the crack, the fracture stops expanding. Meanwhile, the stress concentration is produced again in the tip of the fracture and the fracture continues to expand. Therefore, the fracture expansion process inside the oil shale is characterized by the continuous cycle of stress concentration—fracture expansion—stress reduction and stress re-centralization.

The critical state of tensile fracture of oil shale is:

$$p - \sigma_v \geq T_0 \tag{1}$$

where T_0 is the tensile strength of oil shale (*MPa*), p is the superheated steam pressure (MPa) and σ_v is the vertical stress (MPa).

When the tensile fracture of the specimen occurs, the fracture direction is perpendicular to the vertical principal stress direction, thereby forming multiple horizontal fractures around the heat injection tube, as shown in Figure 7.

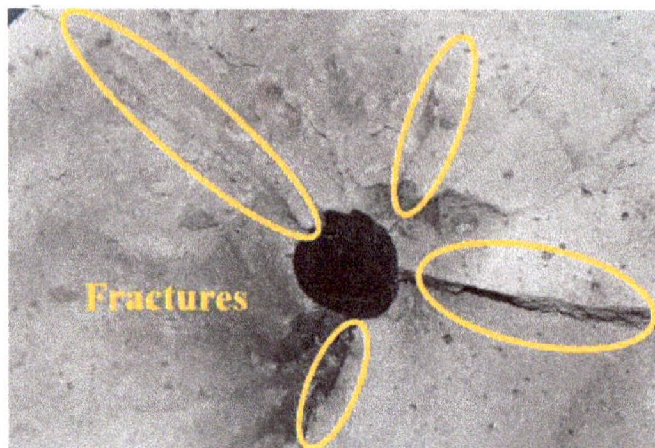

Figure 7. Characteristics of the oil shale fracture after pyrolysis.

The anisotropy of the oil shale is obvious. In the process of pyrolysis, the thermal expansion coefficient of particles in different positions inside the oil shale is different, which leads to the change in stress state in the horizontal direction.

Figure 8 shows the variation in horizontal stress difference with time during the pyrolysis process.

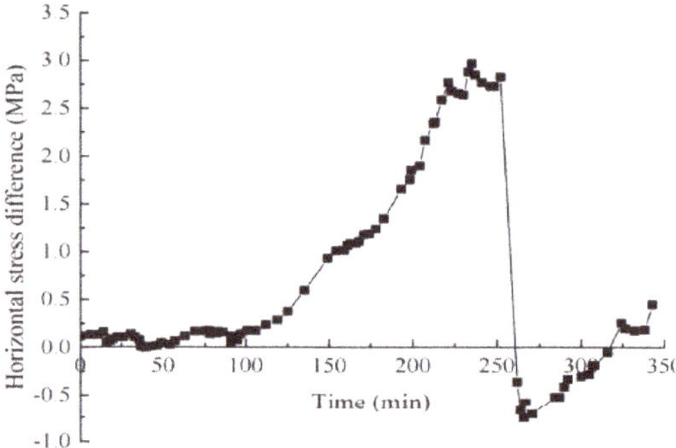

Figure 8. Variation in horizontal stress difference with time in pyrolysis process.

In Figure 8, the horizontal stress difference is the difference between the horizontal stress in the north-south direction and the horizontal stress in the east-west direction. When the pyrolysis time is within the range of 0–252 min, the horizontal stress difference increases with the increase in pyrolysis time and the maximum horizontal stress difference is found to be 2.95 MPa. During this period of time, the north-south direction is the direction of the maximum principal stress and the fracture will expand perpendicular to the north-south direction. Additionally, the degree of cracking will become more obvious during this time period. This is due to the reason that, in the experiment, the direction of oil shale bedding is perpendicular to the north-south direction, whereas the bedding plane undergoes tensile brittle fractures, which appear to be open perpendicular to the north-south direction, resulting in an increase in the stress in the north-south direction. When the pyrolysis time changes from 256 min to 316 min, the maximum horizontal stress difference is 0.75 MPa. During this time period, the rock mass between the oil shale bedding surface in in-situ condition undergoes shear failure under the action of high-temperature and high-pressure steam, which results in larger horizontal stress in the east-west direction than that in the north-south direction. Under these conditions, the east-west direction is the direction of the maximum principal stress. Overall, the extent of shear failure of rock mass between the bedding planes is lower than that of the brittle fractures of the bedding plane. From the macroscopic point of view, a large number of cracks formed by thermal fracturing and pyrolysis of oil shale will cause the injected heat-carrying fluid to seep into the ore-body from the injection well, which continues to pyrolyze the oil shale and continuously extract oil and gas. Therefore, the results of this study provide a scientific basis and technical support for the implementation of in-situ retorting of oil shale.

3.3. Microscopic Characteristics of the Specimen after Pyrolysis

The pyrolyzed sample is shown in Figure 9a. The oil shale around the heat injection tube has been broken into many small pieces, indicating that the oil shale has been fully pyrolyzed and its color has changed from yellow brown to black. Figure 9b shows the sampling position of the oil shale after pyrolysis. The distance between Location A and the outlet of the heat injection tube is the smallest, followed by Location B, whereas the distance from Location C to the outlet of the heat injection tube is the largest.

(a)   (b)

Figure 9. Pyrolized sample: (**a**) Morphology of the oil shale after pyrolysis; (**b**) Sampling locations.

The TG curves of oil shale at different locations after pyrolysis are shown in Figure 10. At the temperature of 510 °C, the rates of weight loss of oil shale at Locations A, B and C are 0.17%, 0.72% and 2.31%, respectively, while that of the original oil shale is 10.8%, indicating that the oil shale at each location has been fully pyrolyzed. At the same time, the pyrolysis effect decreases with the increase in distance from the outlet of the heat injection tube. This is because the process of oil shale pyrolysis using superheated steam is an energy consuming process. Farther from the outlet of the heat injection tube, lower is the temperature and worse is the pyrolysis effect.

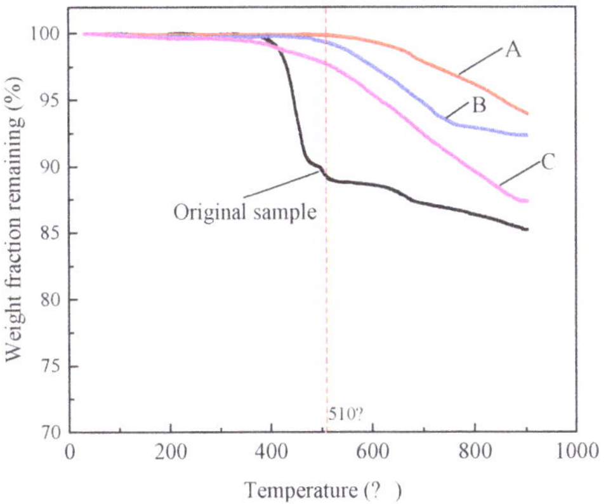

Figure 10. TG curves of oil shale after pyrolysis.

Xue et al. [40,41] used thermogravimetry to analyze the oil shale residue, pyrolyzed using low-temperature dry distillation technology and found that the rate of weight loss of oil shale residue was 6.52%.

The thermogravimetric results of oil shale at different locations after pyrolysis show that the rates of weight loss of oil shale at Locations A, B and C are 0.63%, 2.49% and 4.61%, respectively, which are lower than that of the oil shale residue after low-temperature dry distillation. Therefore, it can be said that the method of oil shale pyrolysis using superheated steam can achieve a high degree of pyrolysis of organic matter in oil shale. In order to further study the pyrolysis properties of oil shale and its residue, the Coats-Redfern method [42] was used to analyze the pyrolysis kinetics of oil shale and its residue. The expression of the Coats-Redfern method is given by Equation (2).

$$In\left[-\frac{In(1-\alpha)}{T^2}\right] = In\frac{AR}{\beta E} - \frac{E}{R} \cdot \frac{1}{T} \tag{2}$$

The equation represents a straight line with as $In\frac{AR}{\beta E}$. the intercept and $-\frac{E}{R}$ as the slope. The activation energy E can be obtained by fitting the least square method. The activation energy of oil shale and its residue in the main weight loss stage (400–510 °C) is presented in Table 2.

Table 2. Analysis of the activation energy of oil shale and its residue.

	Activation Energy (kJ/mol)	
Temperature Range	**400 °C–450 °C**	**450 °C–510 °C**
Original sample	24.804	25.396
Sample A	10.582	6.294
Sample B	9.532	4.151
Sample C	5.935	7.177

CT scanning technology uses the principle that X-rays have different penetration capabilities for different density materials and therefore, the density is reflected in voxel of different gray levels. In the grayscale image of CT, greater the brightness, higher is the density of the material. As the density of the pores and fractures is the lowest, it appears black in the CT image [43,44]. Figure 11 shows a micro-CT reconstructed image of the internal structure of the 500th and 1000th layers of the cross-sections of oil shale at Locations A, B and C.

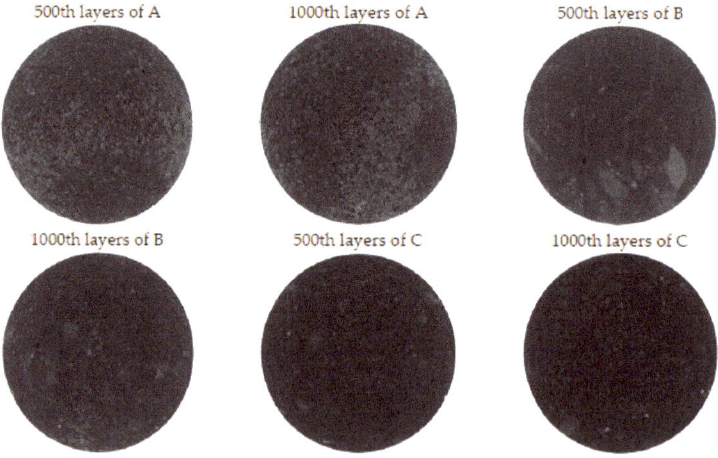

Figure 11. CT-scan grayscale imaging of oil shale after pyrolysis.

In Figure 11, the scattered white areas are the undecomposed minerals during pyrolysis. There are less pores and fractures inside the oil shale. This is because the pores and fractures, formed by the pyrolysis of oil shale under triaxial stress constraint, are constrained in the process of expansion and the fracture surface may be closed. Due to these reasons, the thermal cracking of oil shale under in-situ condition is the result of the combined effect of thermal stress of superheated steam and triaxial stress. In order to visually obtain the distribution of pores and fractures in oil shale after pyrolysis, the CT grayscale images of Figure 11 need to be subjected to "image segmentation" (binarization processing). To conduct segmentation (separate the image into pore and solid phases), 288 the maximum entropy method proposed by Kapur et al [45] was adopted. The Kapur et al [46] 289 method quantitatively considers the gray values of all pixels of an image, and assigns a unique 290 threshold to each image (Figure 12). Because there are a lot of noises in the binary image, the cracks less than 3 voxel are cleared by MATLAB software. Figure 13 shows a micro-CT image of a cross-section of the sample processed using binarization, where the white areas represent the pores and fractures (the length greater than 7.98 μm) and the black areas represent the oil shale matrix.

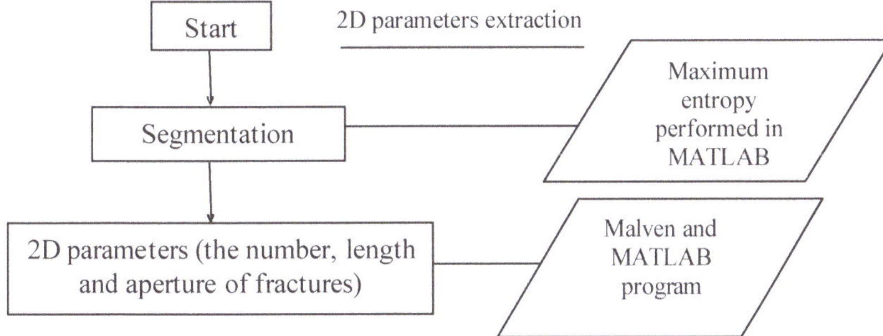

Figure 12. The workflow for processing images from CT scan.

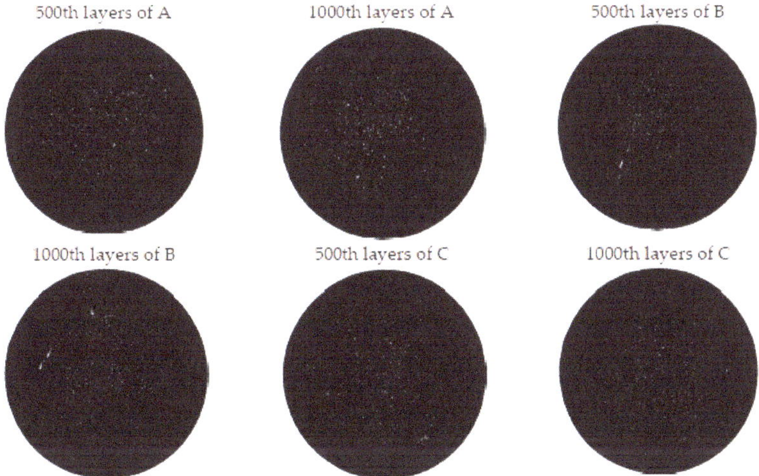

Figure 13. Oil shale CT-scan imaging after the binarization processing.

In order to quantitatively evaluate the development characteristics of pores and fractures of the oil shale after pyrolysis, the number, average length and average aperture of fractures with the length greater than 50 μm are determined. The statistical results are presented in Table 3.

Table 3. Parameters for the fracture of oil shale after pyrolysis.

Parameters of Fractures / Sample Number		Number	Average Length (μm)	Average Aperture (μm)
A	500th layers	48	54.4334	24.6902
	1000th layers	64	53.2752	24.4329
B	500th layers	30	59.4005	21.7064
	1000th layers	28	65.3627	27.0685
C	500th layers	25	57.5110	23.8858
	1000th layers	25	50.4162	23.0557

As can be seen from Table 3, the number of fractures in the 500th layer of oil shale at Location C is the lowest (only 25), while the number of fractures of oil shale at Location A is the highest (up to 56). The average length of fractures in oil shale at different locations lies within the range of 53.9636–62.3816 μm. Among them, the average length of fractures in the 1000th layer of oil shale at Location C is the smallest, whereas that at Location B is the largest. The average aperture of fractures in the 500th layer of oil shale at Location C is only 23.4708 μm and is the minimum, whereas that in the 1000th layer of oil shale at Location A is 24.5616 μm and is the largest. The fracture parameters of the 500th and 1000th layers of oil shale at different locations are averaged to obtain the variation characteristics of oil shale fractures, as shown in Figure 14.

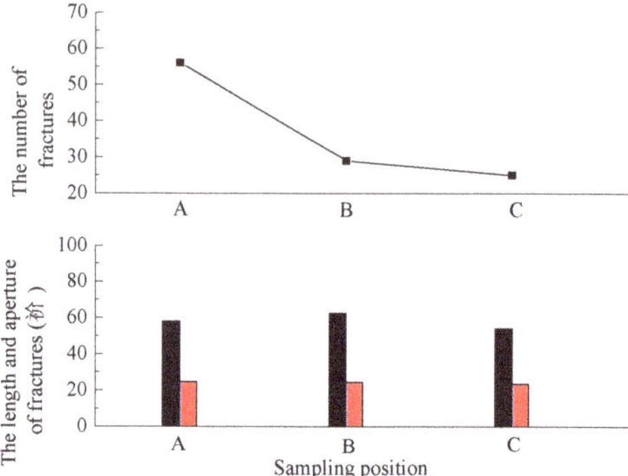

Figure 14. Variation characteristics of fracture parameters in oil shale.

In general, as the distance from the outlet of heat injection tube increases, the number of fractures continuously decreases, however the average length and the aperture of the fractures do not change significantly. This is due to the reason that, greater the distance from the outlet of the heat injection tube, lower is the temperature of superheated steam, and worse is the effect of thermal cracking of oil shale, which result in fewer number of fractures. The oil and gas generated by the pyrolysis of oil shale will both extend and expand the formed fractures in the process of migration, thus forming a more developed fracture channel. Furthermore, there will be little change in the length and aperture of fractures inside the oil shale at different locations. Generally speaking, oil shale is an anisotropic and heterogeneous rock, and the effect of thermal expansion of internal particles is different under the action of temperature. During the in-situ pyrolysis, many micron-scale cracks can be formed in oil shale. In this type of cracks, fluids, such as shale oil and shale gas, can migrate freely and obey the law of hydrostatic mechanics.

After CT scanning of the samples, a series of two-dimensional grayscale images are obtained, which can reflect the density distribution of different layers inside the samples. The 700th to 900th layers are selected and imported into the AVIZO 9.0 software. The threshold segmentation of these grayscale images is done through appropriate thresholds and a binary image characterizing fractures inside the oil shale is obtained [47]. Then, all the binary images obtained in the foregoing operation are successively stacked in the vertical direction, so that the reconstruction of the three-dimensional fracture structure can be achieved. In the process of three-dimensional reconstruction, a three-dimensional digital model of 200 × 200 × 200 voxel is obtained. In order to fully reflect the connectivity and distribution of fractures inside the oil shale, the computing load of the computer in the three-dimensional reconstruction is considered. Figure 15 shows a three-dimensional image of the distribution of fractures inside the oil shale at Locations A, B and C (the color of the fracture space is blue, while that of the matrix is gray), whereas the image is an 8-bit undefined grayscale image. The color range is 0–255.

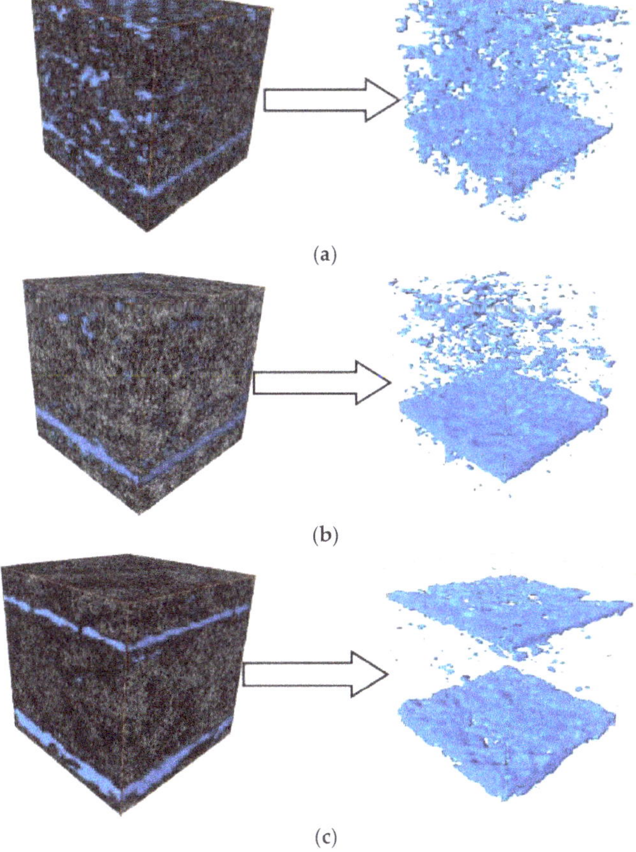

(a)

(b)

(c)

Figure 15. Distribution characteristics of internal holes and fractures in oil shale: (**a**) pore and fracture groups of oil shale at location A; (**b**). pore and fracture groups of oil shale at location B; (**c**) pore and fracture groups of oil shale at location C.

From Figure 15, it can be seen that the distribution of pore and fracture groups inside the oil shale is denser at Location A and the number of pores and fractures is large, which form a large permeate channel, which connects the two relative surfaces. The distribution of oil shale pore and fracture groups at Locations B and C is more scattered in the three-dimensional space and its size and number of oil shale fractures are smaller than that of oil shale at Location A. However, the seepage channel is still formed between the two relative surfaces, which is beneficial to the extraction of oil and gas. In general, the pyrolysis reaction of kerogen occurs obviously and a large number of pores and cracks are formed in the solid skeleton of oil shale, which constitute the entrance and exit channels of pyrolysis fluids and products in the pyrolysis process of oil shale. After pyrolysis using superheated steam, the oil shale can be regarded as a porous medium with high permeability. The results have shown that the in-situ oil shale pyrolysis technology using superheated steam is an efficient and feasible method for oil and gas production from shale oil.

4. Conclusions

The process of oil shale pyrolysis using superheated steam is a multi-field coupling process. The pyrolysis process involves the crack initiation, the decomposition of organic matter and the migration of the products. During the pyrolysis, the temperature distribution inside the oil shale will have serious inhomogeneity, which leads to the different pyrolysis effects of oil shale at different positions. On the basis of simulated experiments of in-situ oil shale pyrolysis using superheated steam, the variations in steam pressure and constrained stress during the pyrolysis are obtained. At the same time, the pyrolysis effect of the oil shale and the evolution of pores and fractures are studied after pyrolysis, which provide a certain level of guidance for the technological design of in-situ oil shale pyrolysis using superheated steam.

With the continuous development of pyrolysis, two forms of rupture occur inside the oil shale. These are the brittle fractures of the bedding plane and the shear failure of the rock mass between the bedding planes. After the pyrolysis, the rate of weight loss of oil shale residue was much lower than that of the original sample, indicating that the pyrolysis of the oil shale was more complete. After the pyrolysis, the pores and fractures inside the oil shale are widely distributed, whereas the oil shale can be regarded as a porous medium with high permeability. The feasibility of in-situ mining of oil shale using superheated steam is verified based upon an efficient pyrolysis process.

Author Contributions: All authors contributed to the research in the paper. Y.Z., D.Y. and L.W. conceived and designed the experiments; X.L. and G.W. performed the experiments; J.Z. analyzed the data; L.W. wrote the paper.

Acknowledgments: This work was supported by the National Natural Science Foundation of China (grant NOs. 11772213, U1261102), the National Youth Science Foundation of China (grant NOs. 51704206).

Conflicts of Interest: The authors declare no conflict of interest.

References

1. Dyni, J.R. Geology and resources of some world oil-shale deposits. *Oil Shale* **2003**, *20*, 193–252.
2. Yu, X.; Luo, Z.; Li, H.; Gan, D. Effect of vibration on the separation efficiency of oil shale in a compound dry separator. *Fuel* **2018**, *214*, 242–253. [CrossRef]
3. Hao, Y.; Gao, X.Q.; Xiong, F.S.; Zhang, J.L.; Li, Y.J. Temperature distribution simulation and optimization design of electric heater for in-situ oil shale heating. *Oil Shale* **2014**, *31*, 105–120. [CrossRef]
4. Wu, Y.; Lin, C.; Ren, L.; Yan, W.; An, S.; Chen, B. Reconstruction of 3D porous media using multiple-point statistics based on a 3D training image. *J. Nat. Gas Sci. Eng.* **2018**, *51*, 129–140. [CrossRef]
5. Rabbani, A.; Baychev, T.G.; Ayatollahi, S.; Jivkov, A.P. Evolution of Pore-Scale Morphology of Oil Shale during Pyrolysis: A Quantitative Analysis. *Transp. Porous Media* **2017**, *4*, 1–20. [CrossRef]
6. Eseme, E.; Urai, J.L.; Krooss, B.M.; Littke, R. Review of mechanical properties of oil shales: implications for exploitation and basin modelling. *Oil Shale* **2007**, *24*, 159–174.
7. Crawford, P.M.; Killen, J.C. New Challenges and Directions in Oil Shale Development Technologies. *Oil Shale A Solut. Liq. Fuel Dilemma* **2010**, *1032*, 21–60.

8. Lin, L.; Lai, D.; Guo, E.; Zhang, C.; Xu, G. Oil shale pyrolysis in indirectly heated fixed bed with metallic plates of heating enhancement. *Fuel* **2016**, *163*, 48–55. [CrossRef]

9. Kang, Z.; Yang, D.; Zhao, Y.; Hu, Y. Thermal cracking and corresponding permeability of Fushun oil shale. *Oil Shale* **2011**, *28*, 273–283. [CrossRef]

10. Rangel-German, E.R.; Schembre, J.; Sandberg, C.; Kovscek, A.R. Electrical-heating-assisted recovery for heavy oil. *J. Petroleum Sci. Eng.* **2004**, *45*, 213–231. [CrossRef]

11. Han, H.; Zhong, N.N.; Huang, C.X.; Liu, Y.; Luo, Q.Y.; Dai, N.; Huang, X.Y. Numerical simulation of in-situ conversion of continental oil shale in Northeast China. *Oil Shale* **2016**, *33*, 45–57.

12. Lee, K.J.; Moridis, G.J.; Ehlig-Economides, C.A. Compositional simulation of hydrocarbon recovery from oil shale reservoirs with diverse initial saturations of fluid phases by various thermal processes. *Energy Explor. Exploit.* **2017**, *35*, 172–193. [CrossRef]

13. Anonymous. Raytheon Technology Shows Promise in Extracting Oil from Shale. *Microw. J.* **2006**, *49*, 40.

14. Yang, Z.; Zhu, J.; Li, X.; Luo, D.; Qi, S.; Jia, M. Experimental Investigation of the Transformation of Oil Shale with Fracturing Fluids under Microwave Heating in the Presence of Nanoparticles. *Energy Fuel* **2017**, *31*, 10348–10357. [CrossRef]

15. Gerasimov, G.; Khaskhachikh, V.; Potapov, O. Experimental study of kukersite oil shale pyrolysis by solid heat carrier. *Fuel Process. Technol.* **2017**, *158*, 123–129. [CrossRef]

16. Chen, C.; Gao, S.; Sun, Y.; Guo, W.; Li, Q. Research on Underground Dynamic Fluid Pressure Balance in the Process of Oil Shale In-Situ Fracturing-Nitrogen Injection Exploitation. *J. Energy Resour. Technol.* **2017**, *139*. [CrossRef]

17. Yang, Y.; Liu, S.C.; Li, Q.; Sun, Y.H. Research on experiment of in-situ pyrolysisof oil shaleand production analysis. *J. Chem. Pharm. Res.* **2013**, *5*, 763–767.

18. Yang, Y.; Sun, Y.H.; Li, Q. Experiment and simulation of oil shale pyrolysis. *Int. J. Earth Sci. Eng.* **2013**, *6*, 1311–1317.

19. Allawzi, M.; Al-Otoom, A.; Allaboun, H.; Ajlouni, A.; Al Nseirat, F. CO_2 supercritical fluid extraction of jordanian oil shale utilizing different co-solvents. *Fuel Process. Technol.* **2011**, *92*, 2016–2023. [CrossRef]

20. Lu, J.; Hawthorne, S.; Sorensen, J.; Pekot, L.; Kurz, B.; Smith, S. Advancing CO 2 enhanced oil recovery and storage in unconventional oil play-Experimental studies on Bakken shales. *Appl. Energy* **2017**, *208*, 171–183.

21. Zhang, X.Q.; Li, Y.S. Changes in shale oil composition and yield after bioleaching by bacillus mucilaginosus and thiobacillus ferrooxidans. *Oil Shale* **2017**, *34*, 146. [CrossRef]

22. Tong, J.; Han, X.; Wang, S.; Jiang, X. Evaluation of Structural Characteristics of Huadian Oil Shale Kerogen Using Direct Techniques (Solid-State13C NMR, XPS, FT-IR, and XRD). *Energy Fuel* **2011**, *25*, 4006–4013. [CrossRef]

23. Mao, J.; Fang, X.; Lan, Y.; Schimmelmann, A.; Mastalerz, M.; Xu, L.; Schmidt-Rohr, K. Chemical and nanometer-scale structure of kerogen and its change during thermal maturation investigated by advanced solid-state 13C NMR spectroscopy. *Geochim. Cosmochim. Acta* **2010**, *74*, 2110–2127. [CrossRef]

24. Sun, Y.; He, L.; Kang, S.; Guo, W.; Li, Q.; Deng, S. Pore Evolution of Oil Shale during Sub-Critical Water Extraction. *Energies* **2018**, *11*, 842. [CrossRef]

25. Modica, C.J.; Lapierre, S.G. Estimation of kerogen porosity in source rocks as a function of thermal transformation: Example from the Mowry Shale in the Powder River Basin of Wyoming. *AAPG Bull.* **2012**, *96*, 87–108. [CrossRef]

26. Chen, J.; Xiao, X. Evolution of nanoporosity in organic-rich shales during thermal maturation. *Fuel* **2014**, *129*, 173–181. [CrossRef]

27. Geng, Y.; Liang, W.; Liu, J.; Cao, M.; Kang, Z. Evolution of pore and fracture structure of oil shale under high temperature and high pressure. *Energy Fuel* **2017**, *31*, 10404–10413. [CrossRef]

28. Bai, F.; Sun, Y.; Liu, Y.; Li, Q.; Guo, M. Thermal and kinetic characteristics of pyrolysis and combustion of three oil shales. *Energy Convers. Manag.* **2015**, *97*, 374–381. [CrossRef]

29. Sun, Y.; Bai, F.; Liu, B.; Liu, Y.; Guo, M.; Guo, W. Characterization of the oil shale products derived via topochemical reaction method. *Fuel* **2014**, *115*, 338–346. [CrossRef]

30. Bai, F.; Sun, Y.; Liu, Y.; Guo, M. Evaluation of the porous structure of huadian oil shale during pyrolysis using multiple approaches. *Fuel* **2017**, *187*, 1–8. [CrossRef]

31. Kang, Z.; Zhao, J.; Yang, D.; Zhao, Y.; Hu, Y. Study of the evolution of micron-scale pore structure in oil shale at different temperatures. *Oil Shale* **2017**, *34*, 42–45. [CrossRef]

32. Saif, T.; Lin, Q.; Bijeljic, B.; Blunt, M.J. Microstructural imaging and characterization of oil shale before and after pyrolysis. *Fuel* **2017**, *197*, 562–574. [CrossRef]

33. Saif, T.; Lin, Q.; Singh, K.; Bijeljic, B.; Blunt, M.J. Dynamic imaging of oil shale pyrolysis using synchrotron x-ray microtomography. *Geophys. Res. Lett.* **2016**, *43*, 6799–6807. [CrossRef]

34. Saif, T.; Lin, Q.; Butcher, A.R.; Bijeljic, B.; Blunt, M.J. Multi-scale multi-dimensional microstructure imaging of oil shale pyrolysis using X-ray micro-tomography, automated ultra-high resolution SEM, MAPS Mineralogy and FIB-SEM. *Appl. Energy* **2017**, *202*, 628–647. [CrossRef]

35. Liu, Z.; Yang, D.; Hu, Y.; Zhang, J.; Shao, J.; Song, S. Influence of In Situ Pyrolysis on the Evolution of Pore Structure of Oil Shale. *Energies* **2018**, *11*, 755. [CrossRef]

36. Pan, L.; Dai, F.; Huang, J.; Liu, S.; Li, G. Study of the effect of mineral matters on the thermal decomposition of Jimsar oil shale using TG-MS. *Thermochim. Acta* **2016**, *627*, 31–38. [CrossRef]

37. Pan, L.; Dai, F.; Huang, J.; Liu, S.; Zhang, F. Investigation of the gas flowdistribution and pressure drop in Xinjiang oil shale retort. *Oil Shale* **2015**, *32*, 172–185. [CrossRef]

38. Aboulkas, A.; Harfi, K.E.; Bouadili, A.E. Thermal degradation behaviors of polyethylene and polypropylene. Part I: Pyrolysis kinetics and mechanisms. *Energy Convers. Manag.* **2010**, *51*, 1363–1369. [CrossRef]

39. Zhao, Y.S.; Feng, Z.C.; Yang, D. the Method for Mining Oil & Gas from Oil Shale by Convection Heating. China Invent Patent CN200,510,012,473, 20 April 2005.

40. Liu, Y.; Xue, X.; He, Y. Solvent Swelling Behavior of Fushun Pyrolysis and Demineralized Oil Shale. *J. Comput. Theor. Nanosci.* **2011**, *4*, 1838–1841. [CrossRef]

41. Li, Y.; Feng, Z.; Xue, X.; He, Y.; Qiao, G. Ecological utilization of oil shale by preparing silica and alumina. *J. Chem. Ind. Eng.* **2008**, *59*, 1051–1057.

42. Coats, A.W.; Redfern, J.P. Kinetic parameters from thermogravimetric data. *Nature* **1964**, *201*, 68–69.

43. Moine, E.C.; Groune, K.; Hamidi, A.E.; Khachani, M.; Halim, M.; Arsalane, S. Multistep process kinetics of the non-isothermal pyrolysis of Moroccan Rif oil shale. *Energy* **2016**, *115*, 931–941. [CrossRef]

44. Zhang, Y.; Lebedev, M.; Al-Yaseri, A.; Yu, H.; Xu, X.; Iglauer, S. Characterization of nanoscale rockmechanical properties and microstructures of a Chinese sub-bituminous coal. *J. Nat. Gas Sci. Eng.* **2018**, *52*, 106–116. [CrossRef]

45. Luo, L.F.; Lin, H.; Li, S.C.; Lin, H.; Flu-hler, H.; Otten, W. Quantification of 3-d soil macropore networks in different soil types and land uses using computed tomo- graphy. *J. Hydrol.* **2010**, *393*, 53–64. [CrossRef]

46. Kapur, J.N.; Sahoo, P.K.; Wong, A.K.C. A new method for gray-level picture thresholding using the entropy of the histogram. *Comput. Vis. Graph. Image Process* **1985**, *29*, 273–285. [CrossRef]

47. Ni, X.; Miao, J.; Lv, R.; Lin, X. Quantitative 3D spatial characterization and flow simulation of coal macropores based on μCT technology. *Fuel* **2017**, *200*, 199–207. [CrossRef]

Article

A Numerical Study on Influence of Temperature on Lubricant Film Characteristics of the Piston/Cylinder Interface in Axial Piston Pumps

Yueheng Song [1], Jiming Ma [2] and Shengkui Zeng [1,3,*] 

[1] School of Reliability and System Engineering, Beihang University, Beijing 100191, China; songyh125@buaa.edu.cn
[2] Sino-French Engineering School, Beihang University, Beijing 100191, China; jiming.ma@buaa.edu.cn
[3] Science and Technology on Reliability and Environmental Engineering Laboratory, Beihang University, Beijing 100191, China
* Correspondence: zengshengkui@buaa.edu.cn; Tel.: +86-108-231-6369

Received: 13 June 2018; Accepted: 12 July 2018; Published: 13 July 2018

Abstract: The loss of kinetic energy of moving parts due to viscous friction of lubricant causes the reduction of piston pump efficiency. The viscosity of lubricant film is mainly affected by the thermal effect. In order to improve energy efficiency of piston pump, this research presents a numerical method to analyze the lubricant film characteristics in axial piston pumps, considering the thermal effect by the coupled multi-disciplinary model, which includes the fluid flow field expressed by Reynolds equation, temperature field expressed by energy equation and heat transfer equation, kinematics expressed by the motion equation. The velocity and temperature distributions of the gap flow of piston/cylinder interface in steady state are firstly numerically computed. Then the distributions are validated by the experiment. Finally, by changing the thermal boundary condition, the influence of thermal effect on the lubricant film, the eccentricity and the contact time between the piston and cylinder are analyzed. Results show that with the increase of temperature, the contact time increases in the form of a hyperbolic tangent function, which will reduce the efficiency of the axial piston pump. There is a critical temperature beyond which the contact time will increase rapidly, thus this temperature is the considered as a key point for the temperature design.

Keywords: thermodynamic; numerical simulation; thermal effect; axial piston pumps

1. Introduction

As sliding pairs of axial piston pump move, both the metal parts and lubricants heat up, which causes temperature to increase significantly, decreases oil viscosity, reduces the bearing capacity of the oil, and intensifies the radial movement of the parts simultaneously [1]. The shape of the oil film changes, as does the lubrication mode, and the trajectory of piston could vary with the changing of oil film. The temperature impacts on oil film characteristics are important for analysis of the efficiency of the sliding pair.

Practical experience also indicates that the medium temperature significantly affects life. Cai et al. [2] tested the life of steel materials with a lubricating medium in the range of 23–175 °C and verified that the wear degree was directly related to the temperature. Specifically, the life of steel materials was closely related to the oil temperature distribution in the gap, and the distribution condition had a distinct influence on the oil characteristics. Unfortunately, the general test method could take temperature as the only variables; thus the results obtained by this way could only indicate the existence of a fuzzy relationship between temperature and life.

To indicate the relationship between the temperature and life more precisely, the behavior of lubricant film between the moving pair needs to be studied. Because the moving process is affected by fluid, solid, and thermal conditions together, it is a typical multiple-domain coupled problem. Many researchers have studied this problem around the bearing. McCallion et al. [3] solved the Reynolds and energy equations separately and neglected the effects of pressures during the calculation of temperature distribution. Ferron et al. [4] thoroughly studied the thermohydrodynamic (THD) performance of a plain journal bearing and determined its thermal characteristics. Rohde et al. [5] studied the elastic and thermal deformations in a slider bearing, noted that the variations in fluid viscosity with temperature were much more influential than solid deformation due to thermal and elastic effects. Researchers are still interested in the journal bearing where oil or gas is used as lubricants [6,7]. Lu et al. [8] conducted a thermal-fluid coupling study on characteristics of air–oil two phase flow and heat transfer in a micro unmanned aerial vehicle (UAV) bearing chamber. Some researchers took the cavitation into account [9] and others developed an efficient numerical method [10].

Axial piston pumps with constant pressure are the main type of engine-driven pumps (EDP), whose lifetime is affected by three sliding pairs, specifically the swash plate/slipper pair, cylinder/valve plate pair, and piston/cylinder pair, as shown in Figure 1.

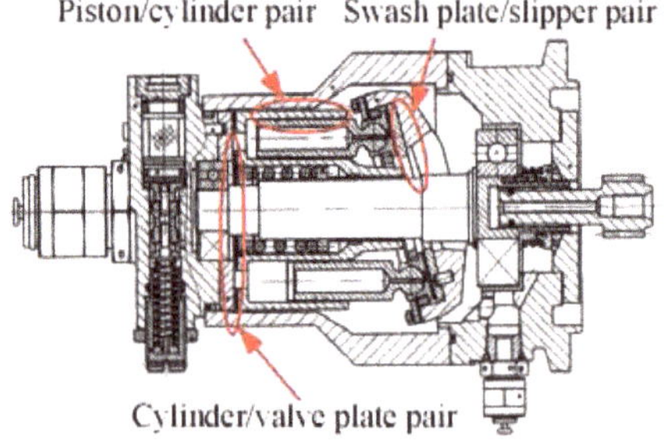

Figure 1. Sliding pairs in axial piston pump.

Several Elasto Hydrodynamic Lubrication (EHL) researches have focused on the piston pump. Gels et al. developed a simulation tool based on the Reynolds equation, which allowed to vary the geometry of the sliding parts. With the help of this tool, Gels have found an optimal compromise of piston/cylinder interface which reduces the losses of energy [11]. Ma et al. presented a method on the basis of the EHL model to analyze the behavior of the swash plate/slipper pair [12]. Olems presented an analytical model that integrates the multiphysics characteristic equations including Reynolds Equation and energy equation [13]. Given inlet medium temperature and operating conditions, the temperature distribution in the clearance of piston/cylinder interface was determined using the model. A series of studies of MAHA research center were carried out based on Olem's results. Ivantysynova et al. established a complete model of main sliding pairs for thermal analysis of oil film gap [14]. Kazama established a non-isothermal model of slipper/swash plate pair based on thermos-hydraulic dynamic theory [15]. Pelosi, Ivantysynova [16], and Zecchi [17] built an EHL model for the piston/cylinder pair and the swash plate/slipper pair. However, as shown in all these studies, few studies have modelled the interaction relationship among elastic, thermal and hydraulic behaviors only use numerical method without establishing the computer-aided design (CAD) model or using commercial software.

This study was designed to investigate the temperature distribution for the further accurate analysis on the thermal effect on the piston/cylinder interface. The flow velocity distribution in the piston/cylinder interface at a steady operating state of the piston pump (the output pressure of pump is steady) was calculated using the Reynolds equation. Then, the velocity distribution was imported into the model, considering the comprehensive effects of flow and temperature distributions in the gap, and the movement of moving components.

Only the numerical method is used during the whole simulation process. So, a parametric simulation model without the construction of a three-dimensional (3D) digital model was established, which is convenient for further application to the simulation of sliding pairs with different structures.

2. Description of Model

Under the assumption that the gap clearance (radial direction) was much smaller than the other two dimensions (circumferential and axial directions), an unwrapped Cartesian reference system was used for the description of the computation domain. The coordinates in this reference system are defined as Equation (1) and presented in Figure 2.

$$\begin{cases} x = \theta \\ y = r \\ z = z \end{cases} \tag{1}$$

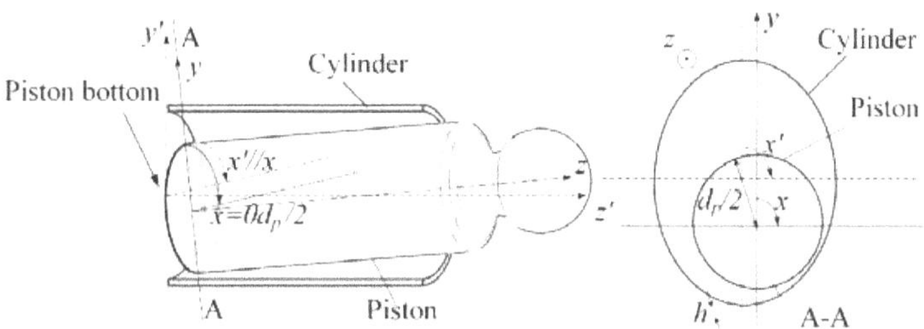

Figure 2. Definitions of coordinates and gap height.

2.1. Mathematical Model

The axial piston pump has an odd number of pistons arranged in a circular array within a housing that is commonly referred to as a cylinder block. This cylinder block is driven to rotate about its axis of symmetry by an integral shaft that is aligned with the pumping pistons. A piston/cylinder interface is an assembly with one piston, one cylinder, and a film of lubricant oil. When the piston pump is operating, the piston moves in a linear reciprocating pattern within the cylinder block, suctioning or discharging oil.

Analysis of thermodynamic behavior of piston/cylinder interface in piston pumps is a fluid-structure interaction problem. The motion for fluid flow, heat transfer, and the motion of the piston are expected to be determined simultaneously. Because it is a time-dependent problem, one shaft revolution is divided into a number of discrete time steps. For each time step, the problem was considered time-independent.

The momentum transport for the fluid flow in the gap is described by the Reynolds equation [18]:

$$\frac{\partial}{\partial x}\left(\frac{\partial p}{\partial x}\frac{h^3}{\mu}\right) + \frac{\partial}{\partial z}\left(\frac{\partial p}{\partial z}\frac{h^3}{\mu}\right) = 6\left((U_h - U_0)\frac{\partial h}{\partial x} + (V_h - V_0)\frac{\partial h}{\partial z} + 2\frac{\partial h}{\partial t}\right) \tag{2}$$

where h is the film thickness, μ is the oil viscosity, U_h is fluid velocity component along the circumferential direction on the boundary $y = h$, U_0 is fluid velocity component along the circumferential direction on the boundary $y = 0$, V_h is fluid velocity component along the axial direction on the boundary $y = h$, V_0 is fluid velocity component along the axial direction on boundary $y = 0$. $y = h$ indicates the interface between the cylinder and oil, and $y = 0$ indicates the interface between the piston and oil. We assumed that the velocity of oil at oil/solid interface was the same as the solid. So U_h, U_0, V_h, and V_0 were constant.

The temperature field of the fluid is governed by the convection-conduction equation [18]:

$$c_o \frac{\partial(\rho_o T_o)}{\partial t} + \rho_o c_o \text{div}(\mathbf{V} \cdot T_o) = k_o \text{div}(\text{grad} T_o) + \phi_o \tag{3}$$

where, T_o is oil temperature, c_o is the specific heat capacity at constant pressure of oil, ρ_o is the density of oil, k_o is the thermal conductivity of oil, and ϕ_o is the viscous dissipation term.

The vector $\mathbf{V}$ in Equation (3) is the velocity of fluid composed by u and v which are determined by followed equations:

$$u = \frac{1}{2\mu} \frac{\partial p}{\partial x} (y^2 - yh) + (U_h - U_0)\frac{y}{h} + U_0$$
$$v = \frac{1}{2\mu} \frac{\partial p}{\partial z} (y^2 - yh) + (V_h - V_0)\frac{y}{h} + V_0 \tag{4}$$

The temperature depends on the fluid velocity, so the last term in Equation (3) describes the energy dissipation:

$$\phi_o = \mu \left[\left(\frac{\partial u}{\partial y}\right)^2 + \left(\frac{\partial v}{\partial y}\right)^2 \right] \tag{5}$$

Then the local viscosity of grid point in the gap can be calculated using the temperature distribution with the Reynolds viscosity temperature equation [18]:

$$\mu = \mu_{\text{ref}} e^{-\sigma(T_o - T_{\text{ref}})} \tag{6}$$

where T_{ref} is the reference temperature, μ_{ref} is the viscosity of oil at reference temperature, and σ is the viscosity-temperature index.

The local viscosity is then used as new value in the following iteration cycle of algorithm.

For the temperature distribution inside the piston, the temperature distribution was determined using the heat conduction equation, which is written as Equation (7) [18]:

$$c_p \frac{\partial(\rho_p T_p)}{\partial t} = k_p \text{div}(\text{grad} T_p) + \phi_p \tag{7}$$

where T_p is the piston temperature, c_p is the specific heat capacity at constant pressure of piston, ρ_p is the density of piston, k_p is the thermal conductivity of piston, and ϕ_p is the viscous dissipation term.

Because the cylinder is geometrically much larger than the gap and the piston, the temperature is here considered a constant, T_c (case temperature).

2.2. Boundary Conditions

To simplify the model, only one piston/cylinder interface was considered. This piston/cylinder formed the system boundary of the thermal calculation, as shown in Figure 3. Because the nine piston/cylinder interfaces were symmetrical around the distribution circle, the behavior of these interfaces could be expected to be almost the same. For this reason, it was here considered reasonable to choose one interface as a representative. The numbers here represent the corresponding boundaries as labeled in Figure 3.

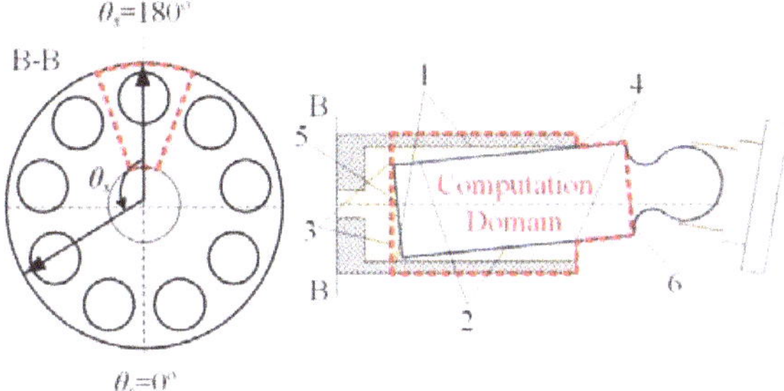

Figure 3. Computation domain and boundary.

The boundary conditions of fluid mechanics equations (Reynolds equation) are shown in Equation (8):

$$\text{Interface 3}: \ P|_{\text{bottom edge}} = \begin{cases} P_{\text{suc}}(\theta_s = 0° - 180°) \\ P_{\text{dis}}(\theta_s = 180° - 360°) \end{cases} \tag{8}$$

$$\text{Interface 4}: \ P|_{\text{top edge}} = P_{\text{case}}$$

where P_{suc} is the output pressure of the pump when the piston is in the oil suction area ($\theta_s = 0°$–$180°$), P_{dis} is the output pressure of the pump when the piston is in the oil discharge area ($\theta_s = 180°$–$360°$), and P_{case} is the case pressure of the pump.

The boundary conditions of thermal analysis are the environmental temperature and temperatures of the medium at the inlet and outlet temperatures. For one piston/cylinder interface, solving the energy equation of fluid and heat conduction equation of solid covers the changes in heat in one phase and between two phases. For each specific boundary, the following boundary conditions are defined. The numbers here represent the corresponding boundaries as labeled in Figure 3.

(1)   On the film-cylinder interface:

$$\left.\frac{\partial T_o(x,z)}{\partial y}\right|_{y=\frac{d_c}{2}} = -\frac{k_c}{k_o d_p} h(x,z) \left(\frac{T_o(x,z) - T_a}{\Delta y}\right) \tag{9}$$

(2)   On the film-piston interface:

$$\left.\frac{\partial T_o(x,z)}{\partial y}\right|_{y=\frac{d_p}{2}} = -\frac{k_p}{k_o d_p} h(x,z) \left.\frac{\partial T_p(x,z)}{\partial y}\right|_{\text{piston}-\text{film}} \tag{10}$$

(3)   Oil on the bottom edge of the film:

$$T_o|_{\text{bottom edge}} = T_{\text{mix}} \tag{11}$$

where T_{mix} is the mixing temperature, which is determined based on the energy balance [19].

(4)   Oil on the top edge of the film:

$$T_o|_{\text{top edge}} = T_{\text{in}} \tag{12}$$

where T_{in} is inlet temperature of oil.

(5)   On the bottom surfaces of the piston:

$$\left.\frac{\partial T_{\mathrm{p}}(x,y)}{\partial z}\right|_{z=0} = -\frac{\eta d_{\mathrm{p}}}{2k_{\mathrm{p}}}\left(\left.T_{\mathrm{p}}(x,y)\right|_{z=0} - T_{\mathrm{mix}}\right) \tag{13}$$

where η is piston convection heat transfer coefficient.

(6)   On the top of the piston:

$$\left.\frac{\partial T_{\mathrm{p}}(x,y)}{\partial z}\right|_{z=l_t} = -\frac{\eta d_{\mathrm{p}}}{2k_{\mathrm{p}}}\left(\left.T_{\mathrm{p}}(x,y)\right|_{z=l_t} - T_{\mathrm{in}}\right) \tag{14}$$

Once the oil film temperature distribution was determined, mechanical analysis could progress.

2.3. Deformation Equation

For an accurate description of thermodynamic behavior of piston/cylinder interface, it is not sufficient to just solve the energy and Reynolds equation, but one also needs to consider deformation due to temperature and pressure. In this paper, the deformation matrix method was used to calculate the elastic deformation and thermal deformation of metal parts. The deformation equations are shown in Equation (15) [20]:

$$\begin{aligned}
h_{\mathrm{E}}(i,j) &= \sum_f \sum_g \left[p_{f,g} + \left.p_c\right|_{f,g}\right] \left.C_p\right|_{f,g}^{i,j} \Delta A \\
h_{\mathrm{T}}(i,j) &= \sum_f \sum_g \left[\left.T_P\right|_{f,g} - T_{\mathrm{ref}}\right] \left.C_T\right|_{f,g}^{i,j}
\end{aligned} \tag{15}$$

where h_{E} is the elastic deformation, $p_{f,g}$ is the oil pressure at node (f,g), $\left.p_c\right|_{f,g}$ is the contact pressure, which is zero in this case, because the piston bears no force in the initial free case, $\left.C_p\right|_{f,g}^{i,j}$ is the elastic deformation matrix, which represents the displacement of node (i,j) caused by the unit load acting on the node (f,g).

Similarly, h_{T} is the thermal deformation, $\left.C_T\right|_{f,g}^{i,j}$ is the thermal deformation matrix, $\left.T_P\right|_{f,g}$ is the temperature at node (f,g), and T_{ref} is the reference temperature and set to 25 °C.

2.4. Force Analysis

The force must be analyzed to solve motion equations. To clarify the description of force analysis, some definitions are presented in Figure 4.

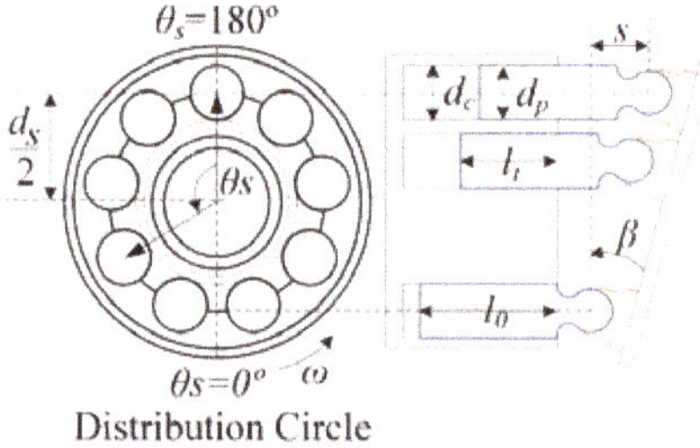

Figure 4. Definitions used for mechanical analysis.

When the pump is operating, the cylinder block slides over the valve plate. With the change in position within the distribution circle (θ_s), the displacement chambers of each piston are connected to the low-pressure port ($\theta_s = 0°$–$180°$) or the high-pressure port ($\theta_s = 180°$–$360°$). The pistons were joined to the slippers that glided over the swash plate. By changing the swash plate angle β, the piston stroke s could be adjusted. θ_s, l_t, and $\mathbf{s}$ were subject to Equation (16).

$$\theta_s = \omega t$$
$$l_t = \frac{d_s}{2}(1 - \cos\theta_s)\tan\beta$$
$$\mathbf{s} = l_0 - l_t = l_0 - \frac{d_s}{2}(1 - \cos\theta_s)\tan\beta \cdot \mathbf{k} \tag{16}$$

The loads acted on the piston in the cavity of piston/cylinder interface in the specific coordinates system is shown in Figure 5.

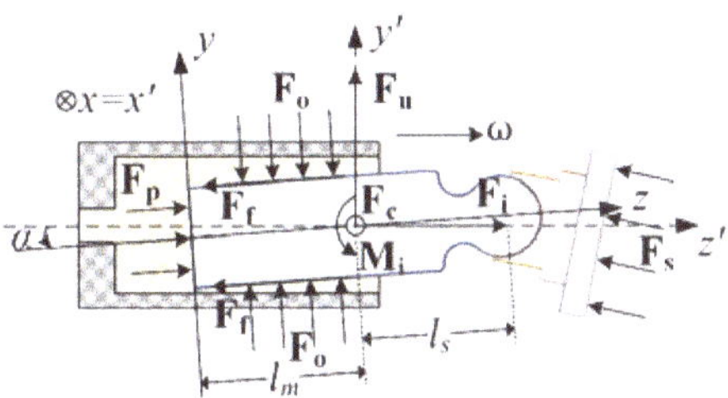

Figure 5. Graph of mechanical analysis.

Because of the overturning angle α, the velocity of piston $\dot{s}$ had a component normal to the ω, so the Coriolis force $\mathbf{F_c}$ took place. With the movement of cylinder block around the shaft, the centrifugal force $\mathbf{F_u}$ acted on the center of gravity of the piston. Because the piston movement took place within six degrees of freedom, there was also the inertia force in z-direction $\mathbf{F_i}$ and the inertia moment $\mathbf{M_i}$. All these forces acted on the center of gravity and were defined using Equation (17):

$$\mathbf{F_c} = -2m\omega \times \dot{s}$$
$$\mathbf{F_u} = m\omega^2 \frac{d_s}{2}\mathbf{j'}$$
$$\begin{cases} \mathbf{F_i} = -m\ddot{s} \\ \mathbf{M_i} = -J\dot{\omega} \end{cases} \tag{17}$$

The swash plate was removed for the calculation of acting force of slipper $\mathbf{F_s}$. $\mathbf{F_s}$ was resolved through the force equilibrium equation in the x-direction. $\mathbf{F_s}$ was decomposed into the y-direction and z-direction as presented in Equation (18):

$$\mathbf{F_{sz}} = F_s \cos(\alpha + \beta) \cdot \mathbf{k}$$
$$\mathbf{F_{sy}} = F_s \sin(\alpha + \beta) \cdot \mathbf{j} \tag{18}$$

The pressure force $\mathbf{F_p}$ was calculated using the pressure of displacement chamber P_c, which differed between the low-pressure port ($\theta_s = 0°$–$180°$) and the high-pressure port ($\theta_s = 180°$–$360°$):

$$\mathbf{F_p} = \frac{\pi d_p^2}{4} P_c \cdot \mathbf{k} \tag{19}$$

The acting force of oil was normal to the piston surface and is defined as follows:

$$\mathbf{F_o} = \int_0^{2\pi} \int_0^{l_t} p(x,z) \cdot \mathbf{j} \cdot \cos x \frac{d_p}{2} dz dx \tag{20}$$

Due to the movement of piston, the friction of oil $\mathbf{F_f}$ acted on the piston. $\mathbf{F_f}$ was calculated using Equation (21):

$$\mathbf{F_f} = \left(\int_0^{2\pi} \int_0^{l_t} \left(-h(x,z) \left(\frac{\partial p(x,z)}{\frac{d_p}{2} \partial x} + \frac{\partial p(x,z)}{\partial z} \right) + \frac{\mu \dot{s}}{h(x,z)} \right) \frac{d_p}{2} dz dx \right) \cdot \mathbf{k} \tag{21}$$

Because the forces defined by Equations (17) and (19) acted through the center of gravity of the piston, their resultant moments were all zero. This left the moments produced by $\mathbf{F_s}$, $\mathbf{F_o}$, and $\mathbf{F_f}$, which can overturn the piston. These moments were calculated using Equations (22)–(24):

$$\mathbf{M_{F_s}} = \mathbf{F_s} \cdot l_s \sin(\alpha + \beta) \tag{22}$$

$$\mathbf{M_{F_o}} = \int_0^{2\pi} \int_0^{l_t} p(x,z) \cdot \mathbf{j} \cdot \cos x (z - l_m) \frac{d_p}{2} dz dx \tag{23}$$

$$\mathbf{M_{F_f}} = \left(\int_0^{2\pi} \int_0^{l_t} \left(-h(x,z) \left(\frac{\partial p(x,z)}{\frac{d_p}{2} \partial x} + \frac{\partial p(x,z)}{\partial z} \right) + \frac{\mu \dot{s}}{h(x,z)} \right) \cos x \frac{d_p^2}{2} dz dx \right) \cdot \mathbf{k} \tag{24}$$

where $p(x,z)$ in Equations (20), (21), (23) and (24) is the pressure distribution obtained by Section 2.1. $h(x,z)$ will be determined in Section 2.6.

Assuming the vector sum of forces and moments are $\mathbf{F_{sum}}$ and $\mathbf{M_{sum}}$.

$$\begin{aligned}
\mathbf{F_{sum}} &= \mathbf{F_c} + \mathbf{F_u} + \mathbf{F_i} + \mathbf{F_s} + \mathbf{F_p} + \mathbf{F_o} + \mathbf{F_f} \\
\mathbf{M_{sum}} &= \mathbf{M_i} + \mathbf{M_{F_s}} + \mathbf{M_{F_o}} + \mathbf{M_{F_f}}
\end{aligned} \tag{25}$$

2.5. Movement Analysis

The movement of the piston was analyzed during force analysis. According to Newton's second law, several types of force were identified. They are shown in Equation (26).

$$\begin{aligned}
\mathbf{v_e}|_{t+dt} &= \mathbf{v_e}|_t + \frac{\mathbf{F_{sum}}}{m} dt \\
\mathbf{e}|_{t+dt} &= \mathbf{e}|_t + \mathbf{v_e}|_t dt \\
\mathbf{v_\alpha}|_{t+dt} &= \mathbf{v_\alpha}|_t + \frac{\mathbf{M_{sum}}}{J} dt \\
\mathbf{\alpha}|_{t+dt} &= \mathbf{\alpha}|_t + \mathbf{v_\alpha}|_t dt
\end{aligned} \tag{26}$$

2.6. Calculation of the Thickness of the Oil Film

The thickness of the oil film $h(x,z)$ was determined as described below.

As shown in Figure 6, supposing that the space between the piston and cavity was filled with an oil film, the thickness of that oil film distribution could be determined according to the geometry characteristics of piston and cavity. In section C-C, O_1 was the center of the cavity, O_1' was the projection of O_1 on the y axis and O_2 was the center of piston, assuming that the piston overturning angle was α and projection of eccentric distance at bottom of piston e on y axis was e_0. Piston shaft and cavity shaft intersected in the position $(0, 0, l_1)|_{(x,y,z)}$. The following geometric relationships were established:

$$e_0 = l_1 \tan \alpha \tag{27}$$

For an arbitrary cross section C-C of piston, O_1 and O_2 were used to define a straight line that crossed the surface of the piston at P_1 and crossed the cavity surface at P_2. Any position on surface of piston P_1 could correspond to the circumferential angle θ. $|P_1P_2|$ was the thickness of the oil film, whose value is here written as $\bar{h}(\theta, z)$. In order to simplify the calculation, there was a transformation of system of coordinates. As shown in Figure 6b, O_2 was located at $(0, 0, z)$ and O_1 at $(-e_2, -e_1, z)$. e_1 and e_2 were subject to Equation (28):

$$e_1 = |O_1'O_2| = (z - l_1)\tan\alpha$$
$$e_2 = \frac{e_1}{\tan\theta} \tag{28}$$

In this way, the algebraic expression of straight line determined by O_1O_2 was as follows:

$$y = x\tan\theta \tag{29}$$

On the cross section C-C, cavity was an ellipse, and the piston was a circle. Their algebraic expression was as shown in Equation (30):

$$\frac{4(x+e_2)^2}{d_c^2} + \frac{4(y+e_1)^2}{d_c^2\sec^2\alpha} = 1$$
$$x^2 + y^2 = \frac{d_p^2}{4} \tag{30}$$

The coordinates of P_1 and P_2 were determined by solving Equations (29) and (30) simultaneously. The results are presented in Equations (31) and (32):

$$\begin{cases} x_{P_1} = \frac{d_p}{2}\cos\theta \\ y_{P_1} = \frac{d_p}{2}\sin\theta \end{cases} \tag{31}$$

$$\begin{cases} x_{P_2} = \sqrt{\dfrac{d_c^2\sec^2\alpha}{4\left(\sec^2\alpha+\tan^2\theta\right)}} - e_2 \\ y_{P_2} = \sqrt{\dfrac{d_c^2\sec^2\alpha}{4\left(\sec^2\alpha+\tan^2\theta\right)}}\tan\theta - e_1 \end{cases} \tag{32}$$

Oil film thickness $\bar{h}(\theta, z)$ under specific overturning angle α was determined using the coordinates of P_1 and P_2:

$$\bar{h}(x, z) = \sqrt{\left(x_{P_1} - x_{P_2}\right)^2 + \left(y_{P_1} - y_{P_2}\right)^2} \tag{33}$$

Along with the elastic and thermal deformations, oil film thickness should be written as:

$$h(x, z) = \bar{h}(x, z) + h_E(x, z) + h_T(x, z) \tag{34}$$

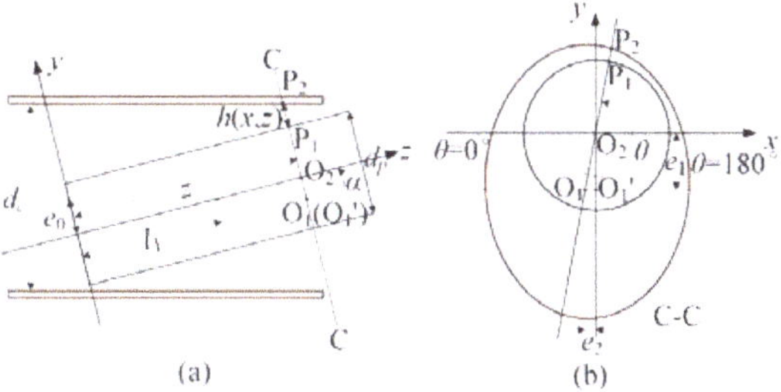

Figure 6. Graph of calculation of oil film thickness: (**a**) Axial section view; (**b**) Bottom view.

3. Simulation Algorithm and Conditions

Simulation of the thermodynamic behavior of the piston requires knowledge of the eccentric position, velocity of the piston, and all forces acting on it. Among these, eccentric position and forces are profoundly influenced by temperature distribution within the gap. An iterative method for the simultaneous calculation of all the requirements was here developed. Figure 7 shows a diagram of these calculations.

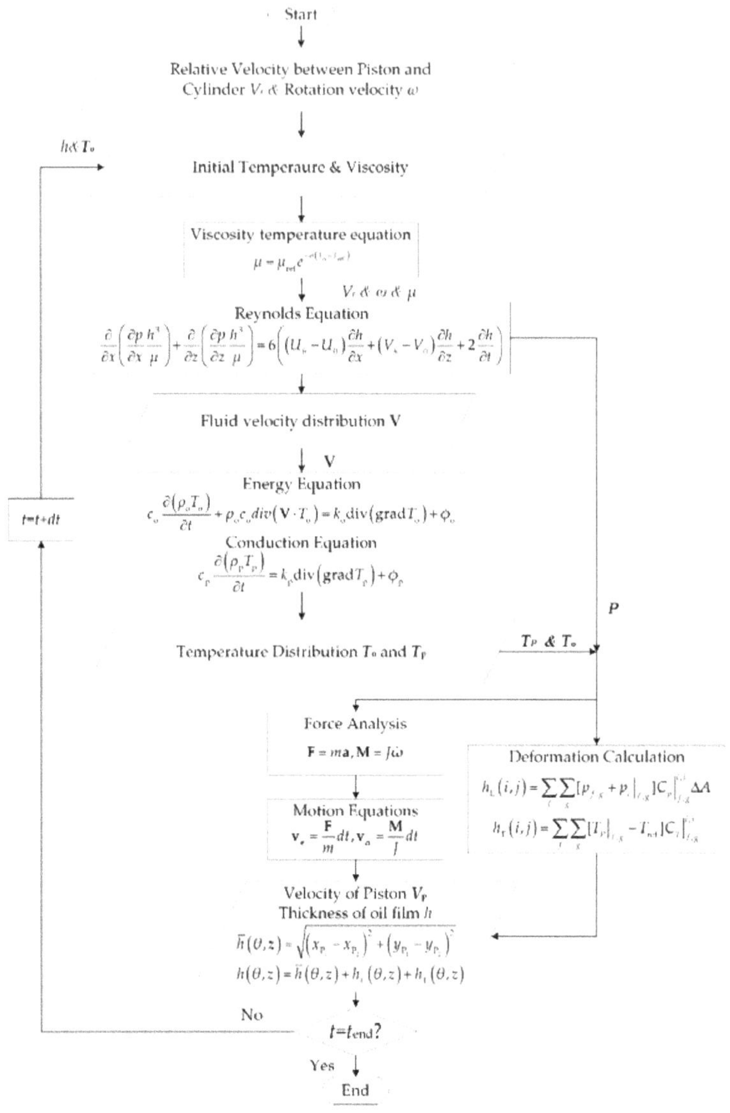

Figure 7. Flow chart of calculation.

With constant inlet and outlet pressure, the velocity distribution is considered constant and a basic condition of the Reynolds equation. When the velocity results were considered as the input of energy equation, the temperature distribution indicated by resolving of energy equation was the temperature distribution of oil film in steady state. The mechanical analysis was completed considering hydrodynamic forces and temperature distribution as inputs of motion equations. Using the overrelaxation iterative formula of the finite difference method, the Reynolds equation and the energy equation are solved by programming in MATLAB. If the absolute error of the results on all interior points obtained by two adjacent iterations is not greater than that of the prescribed error, then the iterative solution is considered to be convergent.

An axial piston pump with nine pistons was here selected for study. To start the simulation, some fundamental conditions were defined, which were the structure and working parameters of the pump, the material properties and thermal coefficients of the piston/cylinder interface, and the pressure of displacement chamber.

The structure and working parameters of a specific type of pump are presented in Table 1.

Table 1. Structure and working parameters.

Parameter	Value
Number of pistons	9
Rated working pressure	225 bar
Rated rotation speed	1000 rpm
Distribution circle diameter	80 mm
Piston diameter	30.14 mm
Cylinder cavity diameter	31.2 mm

The material properties and thermal coefficients of the pump are shown in Table 2 [21].

Table 2. Material properties.

	Density (kg/m^3)	Young Modulus (Gpa)	Specific Heat Capacity (J/(kg·K))	Heat Conductivity (W/(m·K))	Viscosity (20°C) (Pa·s)	Heat Convection Coefficient (W/(m^2·K))
Cylinder	8100	130	390	41.9	/	/
Piston	7850	210	490	50.28	/	50
Oil	862.1	/	1880	0.143	0.15	/

4. Results and Discussion

A simulation of the chosen piston pump was conducted with the model and numerical algorithm introduced in this paper.

In order to validate the developed model in this article, the simulated velocity and temperature distribution were compared with the experiment performed by Pelosi [22]. Constant inlet and outlet pressures on both sides of the piston when the pump was in normal operation were 16 bar ($\theta_s = 0°$–$180°$) and 225 bar ($\theta_s = 180°$–$360°$) (boundary $z = 0$ in Figure 8) and 5 bar (boundary $z = 1$ in Figure 8) respectively. For the temperature simulation, the input parameters were same as Pelosi [22], which are 57.5 °C for the case temperature, 48 °C for temperature at high pressure port, and 43 °C for temperature at low pressure port.

We selected the state where $l_t = l_0$, where the piston was in the position of $\theta_s = 0°$ on the distribution circle, and the whole piston was in the cylinder, The corresponding $\alpha = 0.0067°$. Figure 8 shows the velocity distribution of oil film at the interface of the oil and the piston. Figure 9a shows the result of thermal simulation, and Figure 9b shows the measurement result of the oil film at the interface of the oil and the piston carried by Pelosi [22]. Thus, we can conclude that the model could describe the oil temperature distribution, with reference to previous research.

Results indicated that the velocity distribution of oil film in piston/cylinder interface was affected not only by the relative velocity between piston and cylinder in the form of sinusoidal wave (z direction) but also the rotation of piston (x direction). The motion of the piston would lead the oil to follow its motion, and the change of the velocity direction of the oil would have hysteresis because of the viscous effect, so the vortex would be produced.

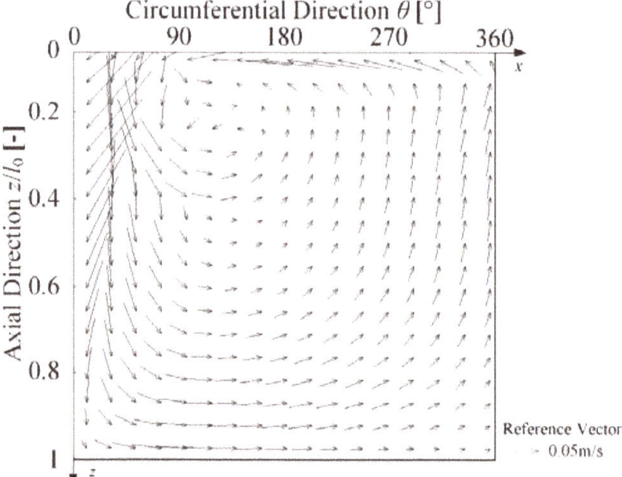

Figure 8. Velocity distribution.

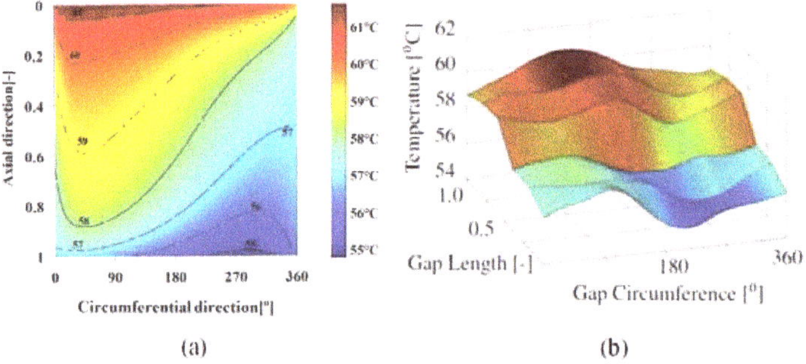

(a)   (b)

Figure 9. (**a**) Simulated temperature distribution; (**b**) Measured temperature distribution by experiment [22].

Analysis of the Figure 8 also showed that the velocity of flow near the boundaries ($z = 0$ and $z = 1$) was faster than in the middle.

Comprehensive analysis of the contents of Figures 8 and 9 indicated that heat transfer was faster where the velocity was faster. Temperature diffused along the velocity, whether the temperature was high or low. This indicated that the accurate calculation of the velocity distribution was very important to the calculation of temperature distribution, considering the velocity has a highly significant effect on temperature diffusion.

The higher temperature region corresponded to a region where the fluid film was thinner, where more heat was generated. In these areas, the oil film tended to become thinner in next time step because high temperature reduced the carrying capacity of oil under the condition of lower viscosity. The eccentricity may be increased, but it still depended on the structure and the pressure in the chamber.

5. Analysis of Thermal Effect

With the thermal results, it was possible to calculate the fluid pressure and piston motion in the piston/cylinder interface. In order to investigate the effect of inlet oil temperature on the lubricating characteristics of oil film, we used the model proposed in this paper to simulate the movement of the bottom of piston in the chamber at different inlet temperatures. To facilitate the simulation, we assumed that the temperatures at high pressure port and low pressure port were the same as the inlet temperature T_{in}, and the T_{in} of the six simulations was 20 °C, 40 °C, 60 °C, 80 °C, 100 °C, and 140 °C respectively, and the case, temperature T_c was always 50 °C.

Figure 10 presents simulated trajectory of the piston bottom in the cavity over one shaft revolution with different inlet temperatures. The following conclusions can be summarized:

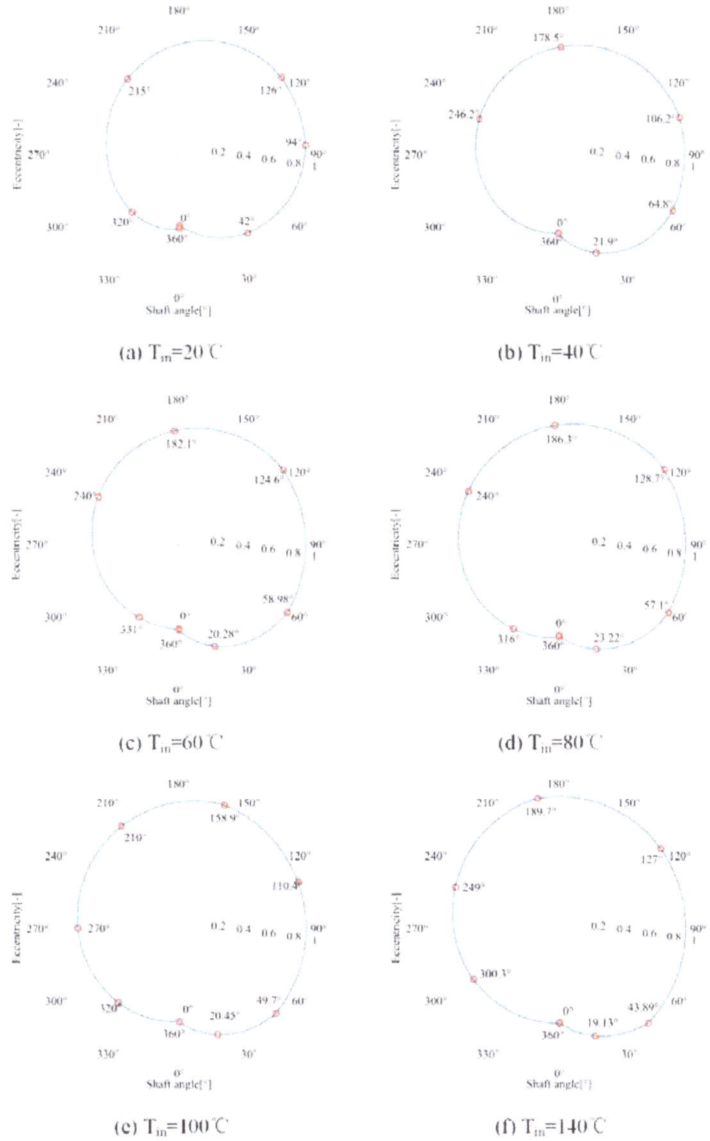

Figure 10. Simulated trajectory of the piston bottom in the cavity over one shaft revolution with different inlet temperatures.

(1). When the piston rotates within the suction zone ($\theta_s = 0°–180°$), the oil pressure of the cylinder cavity is smaller, therefore the oil film supporting force is smaller, and the radial movement of the piston intensifies under the same inertial force compared with the condition when the pistons rotates into the outlet zone ($\theta_s = 180°–360°$). The reduction of the eccentricity during the discharge area is caused by the very high pressure difference between the cylinder and the ambient pressure.

(2). With the increase of oil inlet temperature, the contact time between piston and cavity becomes longer and longer.

These results show that when the pump is operating, the eccentricity appears to be up and down with the changes of piston position in the distribution circle. The bottom of piston makes physical contact with the cylinder during a large part of the operating time, there will be more energy consumption.

According to the six simulation results, the curve of contact time with the inlet oil temperature is fitted out (see Figure 11). In general, there should be a positive correlation between contact time and solid friction. Therefore, the contact time is regarded as a simple measurement of solid friction, the longer the contact time, and the more the energy consumption. From our analysis, it is concluded that this relationship between contact time and inlet oil temperature should be expressed by Equation (35). The values A, B, C and D are constants, which are determined by the material properties, physics and operating conditions of the sliding pair and lubricating oil and can be obtained by the simulation method presented in this paper. In this case, $A = 0.15$, $B = 0.04$, $C = -84.37$, and $D = 0.25$. The R-square for the fitting, analyzed by MATLAB, is 0.97.

$$\frac{t_c}{t_{\text{total}}} = A \cdot \tanh(B(T_{\text{in}} + C)) + D \tag{35}$$

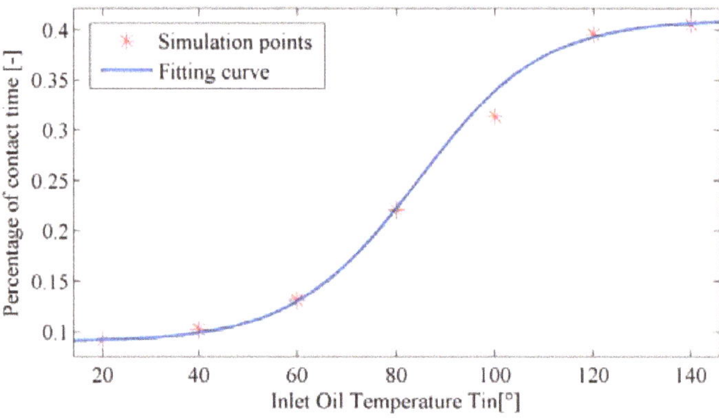

Figure 11. Curve of contact time with oil inlet temperature.

Figure 11 shows that the contact time increased with the increase of inlet oil temperature, and the growth rate increased first and then decreased, but would not increase all the time. From 20°C to 60°C, the contact time increased slowly and the temperature did not have much effect on the movement of the piston. From 60 °C to 100 °C, the contact time increased by a large margin, which may be due to the decrease in viscosity as the temperature continues to increase, resulting in a significant decrease in the bearing capacity of the lubricant film. The piston is in continuous contact with the cylinder under the action of centrifugal force. After 100 °C, the contact time between the piston and the cylinder will finally reach a balance due to the constraints of the structure of piston/cylinder interface and the working conditions (pressure).

6. Conclusions

A numerical multi-discipline modeling method suitable for describing the lubricant oil film characteristics in piston pump has been developed. The results of velocity field and temperature field are accurate compared with the experiment. Considering the thermal effect, the quantitative relationship between oil inlet temperature and contact time between piston and cylinder is obtained in this paper, and it is shown that the relationship is a hyperbolic tangent. Thus, it can be predicted when and where the contact will appear under different temperature conditions. This method can be helpful when it refers to the design and analysis of a piston pump.

In this paper, we consider the thermal effect on the oil film characteristics. In the future, we would include the oil pressure, axis rotate speed and the cavitation in the model and analyze all these factors' effects on the pump operation performance, such as oil leakage and pressure fluctuation. We will also establish a test rig to verify the simulation results.

Author Contributions: Conceptualization, Yueheng Song and Jiming Ma; Methodology, Formal Analysis, Investigation and Writing-Original Draft Preparation, Yueheng Song; Writing-Review & Editing, Supervision and Project Administration, Shengkui Zeng.

Funding: This research received no external funding.

Conflicts of Interest: The authors declare no conflict of interest.

References

1. Yang, H.Y.; Ding, F.; Yang, X.P.; Lu, Q. Hydraulic power systems for trunk line aircrafts. *China Mech. Eng.* **2009**, *18*, 2152–2159.
2. Cai, Z.B.; Zhou, Y.; Qu, J. Effect of oil temperature on tribological behavior of a lubricated steel–steel contact. *Wear* **2015**, *332–333*, 1158–1163. [CrossRef]
3. McCallon, H.; Yousif, F.; Lloyd, T. The analysis of thermal effects in a full journal bearing. *J. Lubr. Technol.* **1970**, *92*, 578–587. [CrossRef]
4. Ferron, J.; Frene, J.; Boncompain, R.A. Study of the thermohydrodynamic performance of a plain journal bearing comparison between theory and experiments. *J. Lubr. Technol.* **1983**, *105*, 422–428. [CrossRef]
5. Rohde, S.M.; Oh, K.P. A thermoelastohydrodynamic analysis of a finite slider bearing. *J. Lubr. Technol.* **1975**, *97*, 450–460. [CrossRef]
6. Mi, J.T.; Meng, Y.G. THD analysis of rolling piston and journal bearings in rotary compressors. *Tribol. Trans.* **2016**, *59*, 195–207. [CrossRef]
7. Cheng, F.; Ji, W. A velocity-slip model for analysis of the fluid film in the cavitation region of a journal bearing. *Tribol. Trans.* **2016**, *97*, 163–172. [CrossRef]
8. Lu, P.; Zheng, X.; Li, W.; Yang, P.; Huang, H.; Yu, Y. A thermal-fluid coupling numerical study on the characteristics of air-oil two phase flow and heat transfer in a micro UAV bearing chamber. *RSC Adv.* **2017**, *7*, 44598–44604. [CrossRef]
9. Braun, M.J.; Hannon, W.M. Cavitation formation and modelling for fluid film bearings: A review. *J. Eng. Tribol.* **2016**, *224*, 839–863. [CrossRef]
10. Chan, C.W.; Han, Y.F.; Wang, Z.; Wang, J.; Shi, F.; Wang, N.; Wang, Q.J. Exploration on a fast ehl computing technology for analyzing journal bearings with engineered surface textures. *Tribol. Trans.* **2014**, *57*, 206–215. [CrossRef]
11. Gels, S.; Murrenhoff, H. Simulation of the lubricating film between contoured piston and cylinder. *Int. J. Fluid Power* **2010**, *11*, 15–24. [CrossRef]
12. Ma, J.M.; Chen, J.; Li, J.; Li, Q.L.; Ren, C.Y. Wear analysis of swash plate/slipper pair of axis piston hydraulic pump. *Tribolo. Int.* **2015**, *90*, 467–472. [CrossRef]
13. Olems, L. Investigations of the temperature behaviour of the piston cylinder assembly in axial piston pumps. *Int. J. Fluid Power* **2000**, *1*, 27–39. [CrossRef]
14. Ivantysynova, M.; Huang, C.C.; Andreas, J.J. Determination of Gap Surface Temperature Distribution in Axial Piston Machines. In Proceedings of the ASME International Mechanical Engineering Congress and Exposition, Chicago, IL, USA, 5–10 November 2006.

15. Kazama, T. Thermohydrodynamic lubrication model applicable to a slipper of swashplate type axial piston pumps and motors (Effects of operating conditions). *Tribol. Online* **2010**, *5*, 250–254. [CrossRef]

16. Pelosi, M.; Ivantysynova, M. The impact of axial piston machines mechanical parts constraint conditions on the thermo-elastohydrodynamic lubrication analysis of the fluid film interfaces. *Int. J. Fluid Power* **2013**, *14*, 35–51. [CrossRef]

17. Zecchi, M. A Novel Fluid Structure Interaction and Thermal Model to Predict the Cylinder block—Valve Plate Interface Performance in Swash Plate Type Axial Piston Machines. Ph.D. Thesis, Purdue University, West Lafayette, IN, USA, December 2013.

18. Wen, S.Z.; Huang, P. *Principle of Tribology*, 4th ed.; Tsinghua University Press: Beijing, China, 2012.

19. Khonsari, M.M.; Wang, S.H. On the fluid-solid interaction in reference to thermoelastohydrodynamic analysis of journal bearings. *J. Tribol.* **1991**, *113*, 398–404. [CrossRef]

20. Zhang, Z. Analysis of the Thermohydrodynamic Performance of Dynamically Loaded Bearings in Consideration of Non-Netonian, Derformation and Surface Topography. Ph.D Thesis, Shanghai Jiao Tong University, Shanghai, China, 2014.

21. Yang, S.M.; Tao, W.S. *Heat Transfer*, 4th ed.; Higher Education Press: Beijing, China, 2006.

22. Pelosi, M.; Ivantysynova, M. A Novel Thermal Model for the Piston/Cylinder Interface of Piston Machines. In Proceedings of the ASME Dynamic Systems and Control Conference, Hollywood, CA, USA, 12–14 October 2009.

Article

Investigation on Water Hammer Control of Centrifugal Pumps in Water Supply Pipeline Systems

Wuyi Wan *, **Boran Zhang and Xiaoyi Chen**

Department of Hydraulic Engineering, College of Civil Engineering and Architecture, Zhejiang University, Hangzhou 310058, China; zhangboran@zju.edu.cn (B.Z.); zjuchenxy@zju.edu.cn (X.C.)
* Correspondence: wanwuyi@zju.edu.cn; Tel.: +86-571-8795-1346

Received: 28 November 2018; Accepted: 29 December 2018; Published: 29 December 2018

Abstract: Water hammer control in water supply pipeline systems is significant for protecting pipelines from damage. The goal of this research is to investigate the effects of pumps moment of inertia design on pipeline water hammer control. Based on the method of characteristics (MOC), a numerical model is established and plenty of simulations are conducted. Through numerical analysis, it is found that increasing the pumps moment of inertia has positive effects both on water hammer control as well as preventing pumps rapid runaway speed. Considering the extra cost of space, starting energy, and materials, an evaluation methodology of efficiency on the increasing moment of inertia is proposed. It can be regarded as a reference for engineers to design the moment of inertia of pumps in water supply pipeline systems. Combined with the optimized operations of the valve behind the pumps, the pipeline systems can be better protected from accident events.

Keywords: transient analysis; pumps; moment of inertia; water hammer; pipe flow

1. Introduction

Due to the non-uniformity on spatial and temporal distribution of water resources, long distance water supply systems are required to deliver water from humid regions to arid regions, and from the wilderness to the cities [1]. Water hammer is an undesired hydraulic phenomenon in this kind of engineering, especially in pressured pipelines. It is a pressure wave caused when the running condition of the system is changed, for example, a sudden valve closure at the end of a pipeline system, causing a pressure wave to propagate in the pipe. Extreme water hammer can cause pipe and device damage by high overpressure or low negative pressure [2–5]. Operations such as rapid opening and closing of the valves, starting and shutting down the pumps, and other sudden changes on devices along the pipeline will cause water hammers. To reduce harmful extreme water hammers, designers usually use pressure control devices such as surge tanks [6–8], air valves [9,10], and pressure vessels [11–13], as well as changing the operation time of valves [14–16], to keep the extreme pressure in a safe range. Usually, it is the water hammer occurring place that will load the most extreme pressure. For example, the valve at the end of a pipeline will bear the highest pressure in the whole pipeline in a water hammer after a rapid closing of it, and the outlet tube of a pump will suffer the lowest negative pressure in a water hammer caused by a pump stop event [17–19]. The magnitude of extreme pressure of the worst case is related to the wave speed of water hammer and the time for valve closing according to the Joukowsky formula [20], and it can be significantly reduced by expanding the closing time of the valves. The pressure control devices can protect the part of the pipeline on the other side of it, opposite to the pressure wave source [21]. Thus, the pressure control devices are normally recommended to be set somewhere most likely to produce a water hammer, if the geographical conditions permit this [22]. Nevertheless, pressure control devices set at other special positions may have a great positive influence too. For instance, in-line valves are selected to isolate parts of the system from a high-pressure source to a low-pressure water hammer event [14], in this way greatly reducing or eliminating any return surge.

Sometimes, the pressure control devices are limited by the geographical conditions. For instance, surge tanks provide a free surface to reflect the pressure wave and store energy, reducing the reflection time of the system and limits the influence of the pressure wave, which helps to protect the upstream section from higher pressure. Meanwhile, there is a problem with the limited heights, caused by mass oscillations between the reservoir and the surge tanks, which can be limited by throttles in the surge tanks [23–25]. On the other hand, when the pipeline is buried deep underground, the air valve is difficult to install. Except for adding pressure control devices in the pipeline systems, there are other methods to reduce the extreme pressure. Li et al. [26] proposed an improved bypass pipe which overcomes the limitations of conventional bypass pipes and can significantly help to reduce the maximum extreme pressure in a pumped water supply system. Triki [27,28] investigated other water hammer control strategies using an in-line polymeric short-section, and a branched polymeric penstock, respectively, and found both have a significant effect on mitigating the pressure increase and decrease induced by water hammer waves. Zeng et al. [29] analyzed different guide vane closing schemes on reducing water hammer intensity. Considering controlling the pulsating pressure and runaway speed during the transient processes, a three-phase valve closing scheme showed an obvious advantage, and is validated by correspondingly conducted experiments. Ballun [30] investigated advantages of different kinds of valves and proposed a methodology for designers to predict and ultimately prevent the valve slam. Hur et al. [31] used a frequency response analysis to study the homologous relationships between the rotational speed of a pump and its head or discharge, and derived the transfer functions for the head and discharge between the upstream and downstream pumps. Wan et al. [32], on the other hand, presented an optimal collaborative scheme to prevent inverse rotation and overpressure during the startup process.

In this research, another strategy is proposed to control the transient processes by adjusting the pumps moment of inertia in pumped water supply systems. Compared to other strategies, it can be accessed directly by changing the size of the flywheel of a pump and does not require further artificial operations. Through 1D pressured water supply pipeline model establishment and plenty of numerical simulations, the principles of adding a moment of inertia are analyzed. It is found that this strategy has a positive effect on both extreme pressure and pumps runaway speed control. Combining the operations on valves, more cases are studied, and a kind of evaluation methodology is proposed to judge the improvement. The results show that an optimized strategy of staging the valve closing period, and an appropriate increase of pumps moment of inertia can significantly reduce the water hammer intensity and protect the devices.

2. Materials and Methods

2.1. Method of Characteristics

In this study, the simulation was conducted by programming in FORTRAN based on MOC and the boundary conditions of the involved devices. To simulate a transient process, there are a lot of numerical methods, including the Finite Difference Method (FDM) [33] and the Finite Element Method (FEM) [34,35]. The FDM method of characteristics (MOC) is one of the most widely used methods recently, because of its convenience and accuracy. In water supply pipeline systems, using the method of characteristics [36], the complete dynamic governing equations can be expressed as Equations (1) and (2). There are some assumptions made in the derivation of the equations. The flow in a pipe is one-dimensional and the control volume is fixed relative to the pipe. The velocity and pressure are uniform at the cross sections of the pipe, and the fluid density is constant. The pipe walls and fluid are linearly elastic, that is, stress is proportional to the strain. And the temperature is assumed to be constant [36]:

$$g\frac{\partial h(x,t)}{\partial x} + v(x,t)\frac{\partial v(x,t)}{\partial x} + \frac{\partial v(x,t)}{\partial t} + \tau_w = 0 \tag{1}$$

$$v(x,t)\frac{\partial h(x,t)}{\partial x} + \frac{\partial h(x,t)}{\partial t} - v(x,t)\sin\alpha_{\text{Pipe-Plane}} + \frac{a^2}{g}\frac{\partial v(x,t)}{\partial x} = 0 \tag{2}$$

where x and t are independent variables denoting the location and the time, respectively, and correspondingly, the h and v are dependent variables denoting, respectively, the hydraulic pressure and the flow velocity. $\tau_w = \tau_{ws} + \tau_{wu}$ is the instantaneous wall shear stress, which can be divided into two parts [37]. One is $\tau_{ws} = f\frac{v(x,t)|v(x,t)|}{2D}$, representing the quasi-steady component, where f denotes the Darcy friction factor, which is valued as 0.029 in the simulation of this study, D denotes the diameter of the pipeline, and is the dominant friction factor in a steady flow. τ_{wu}, the unsteady component, has a significant influence on the water hammer intensity reducing process. However, it affects little on the value of the extreme pressure, which is usually the first strike when the water hammer occurs. To simulate the unsteady component, various numerical methods were proposed. Instantaneous acceleration-based (IAB) models and weighting function-based (WFB) are the two popular models [37]. $\alpha_{\text{Pipe-Plane}}$ represents the angle between the pipeline and the horizontal plane. However, in this research, it is the extreme pressure that we care about, the unsteady friction factor was therefore neglected. Controlling the characteristic relation between wave speed and meshes, which is the essential assumption of MOC, is shown as Equation (3),

$$\frac{dx}{dt} = \pm a \tag{3}$$

the variables at the next time step on the middle node can be inferred from the parameters at the last time step on those three adjacent nodes, expressed as Equation (4) derived from Equation (1)–(3):

$$\begin{cases} C^+ : h(i,t+\Delta t) = h(i-1,t) - Rq(i-1,t)|q(i-1,t)| + Bq(i-1,t) - Bq(i,t+\Delta t) \\[2mm] C^- : h(i,t+\Delta t) = h(i+1,t) + Rq(i+1,t)|q(i+1,t)| - Bq(i+1,t) + Bq(i,t+\Delta t) \end{cases} \tag{4}$$

in which $B = \frac{a}{gA}$, $R = \frac{f\Delta x}{2gDA^2}$, $h(i,t+\Delta t)$ denotes the water head at section i at time $t+\Delta t$; $q(i,t+\Delta t)$ denotes the discharge at section i at time $t+\Delta t$; $h(i-1,t)$ denotes the water head at section $i-1$ at time t; $q(i-1,t)$ denotes the discharge at section $i-1$ at time t; $h(i+1,t)$ denotes the water head at section $i+1$ at time t, and $q(i+1,t)$ denotes the discharge at section $i+1$ at time t. The access of the variables at the next time step can be expressed as two characteristic lines, shown in Figure 1. The positive characteristic line represents the controlling relationship between the variables of the node at section i at time $t+\Delta t$, and the hydraulic head $h(i-1,t)$ and discharge $q(i-1,t)$ of the node at section $i-1$ at time t. While the negative characteristic line represents the controlling relationship between the variables of the node at section i at time $t+\Delta t$, and the variables of the node at the downstream adjacent section at time t. As the variables of the adjacent nodes at time t are already known, the middle node at the next time step can be solved by combining the two characteristic lines. In this way, the variables can be solved in every inner section except the boundary sections.

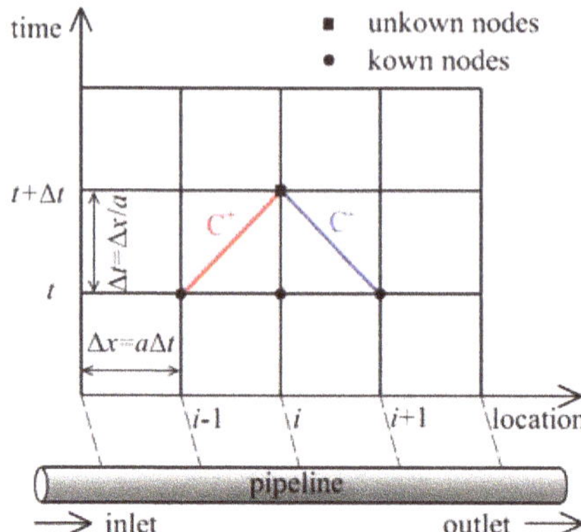

Figure 1. Meshing of the method of characteristics (MOC) in the pipeline.

Furthermore, in a water supply system, there are always many kinds of devices set along the pipelines, such as valves, tanks, pumps, etc. Therefore, to simulate the transient processes in water supply systems, those boundary conditions need to be solved first. In previous research, boundary conditions have been a significant topic in numerical simulations [8,38–41]. Till now, researchers have investigated a series of reliable and effective accesses in numerical simulations of transients in water supply systems.

2.2. Boundary Conditions of Pumps

Pumps are frequently used in hydraulic systems. To simulate a transient process of a hydraulic system with a pump station, the boundary conditions of pumps are necessary to be solved correctly. The running characteristics of a pump depend on several parameters. Four quantities are involved in the characteristics, including the dynamic head H, the flow discharge Q, the shaft torque T, as well as the rotational speed N.

In addition, every single pump has its own complete characteristics curve, which determines the relationship among these four quantities. According to the rated dynamic head H_R, the rated flow discharge Q_R, the rated shaft torque T_R, and the rated rotational speed N_R, the complete characteristics can be expressed in the following form as Equation (5):

$$\begin{cases} WH(x_p) = \dfrac{H/H_R}{(N/N_R)^2+(Q/Q_R)^2} \\[2ex] WB(x_p) = \dfrac{T/T_R}{(N/N_R)^2+(Q/Q_R)^2} \end{cases} \tag{5}$$

where $WH(x_p)$ and $WB(x_p)$ are two defined dimensionless quantities representing the features of the pumps, the curve of which consists the relationships among the four dominant variables mentioned above. Once two of the variables are determined, the others can be derived using the characteristic curves represented by $WH(x_p)$ and $WB(x_p)$. Figure 2 shows the complete characteristic curve of the pump used in this simulation model. It is an imaginary turbine with the parameters changed according to the example described in reference [2]. The abscissa $x_p = \pi + \tan^{-1}\left(\frac{QN_R}{NQ_R}\right)$ represents the instantaneous position of pump operation. That means only two quantities among them are independent. Along with the complete characteristic curve, a certain operating condition can be concretely expressed numerically.

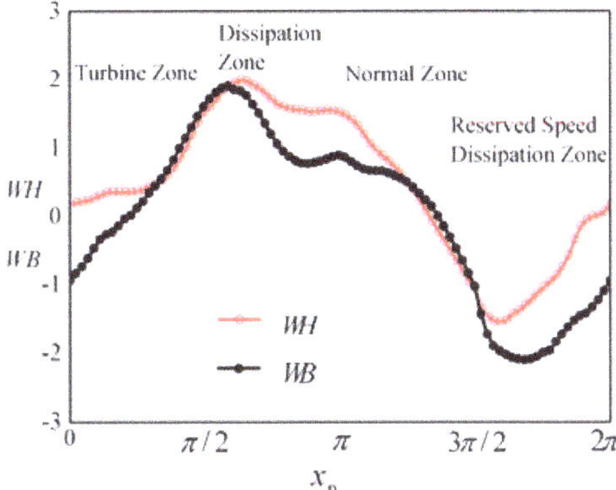

Figure 2. Complete characteristic curve of the pumps. [2].

The characteristic lines of a pump are shown in Figure 3. A working pump provides a water head rise. The water head and discharge on both sides of a pump can be solved using the following set of equations [8], which combine the complete characteristic curve in Figure 3:

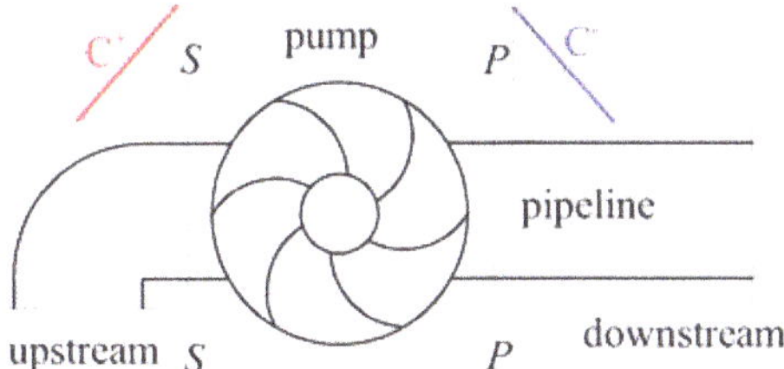

Figure 3. Characteristic lines of a pump.

$$\begin{cases} h(i_{s-s}, t + \Delta t) + \mathrm{tdh} = h(i_{p-p}, t + \Delta t) \\ \mathrm{tdh} = H_R \left(\dfrac{N}{N_R}^2 + \dfrac{Q}{Q_R}^2 \right) WH(x_\mathrm{p}) \\ \left(\dfrac{N}{N_R}^2 + \dfrac{Q}{Q_R}^2 \right) WB(x_\mathrm{p}) + \dfrac{T_0}{T_R} - \dfrac{WR_g^2}{g} \dfrac{N_R}{T_R} \dfrac{\pi}{30\Delta t} \dfrac{N_0 - N}{N_R} = 0 \\ q(i_{s-s}, t + \Delta t) = q(i_{p-p}, t + \Delta t) \\ h(i_{s-s}, t + \Delta t) = C_P - Bq(i_{s-s}, t + \Delta t) \\ h(i_{p-p}, t + \Delta t) = C_M + Bq(i_{p-p}, t + \Delta t) \end{cases} \quad (6)$$

The solutions have been described in detail in previous research [2,32], which is complex and lengthy. Since it is not the main focus of this study, the details of these solutions are partly omitted here.

3. Model Establishment

3.1. Pipeline Meshing

In a long water supply system, there are various kinds of pipelines, including steel pipes, PCCP pipes and tunnels etc. that divides a water supply system into several zones. Due to the differences in the diameter, roughness, and elasticity, the wave velocity and rubbing effect in the zones are not the same. At the connecting profile of two continuous pipeline sections, usually, it may not be divided just exactly at the abrupt change profile. To deal with the meshing problem, the wave speed or pipe length must be modified in the numerical simulation model, and that inevitably leads to a system error. The best way to limit the simulation system error is to compact the meshing. The closer the adjacent sections are set, the more accurate the simulation will be. However, while increasing the accuracy by using dense mesh, the calculating time and internal storage required in the computer are enhanced at the same time.

In the simplified model shown in Figure 4, there are two reservoirs upstream and downstream, whose boundary conditions are assumed to regard the water level of them kept constant during the transient processes. A pump station is located at the upstream side to provide hydraulic head, and the pipeline is a long tunnel through a mountain connecting the two reservoirs. The meshing is uniformly distributed according to Equation (3), which should satisfy the Courant condition [36]. The maximum value of time step mainly depends on the rate at which the transient process occurs. There is no compulsory restriction on the minimum value of time step, in theory, however, the less the time step is set, the longer the running time of the program, and the higher the cost. From previous investigations [42], it is found that when the time step is in the order of 0.01 s, the detailed reflection of device operations, and the simulation accuracy can be ensured. Considering the running time and accuracy of the calculation, the time step is chosen to be 0.01 s in this numerical model empirically. According to the basic assumption of MOC, Equation (3), which requires the location step to equal the time step multiplying the wave speed. Therefore the mesh density is set every 10 m, as the wave speed is considered as 1000 m/s in the model.

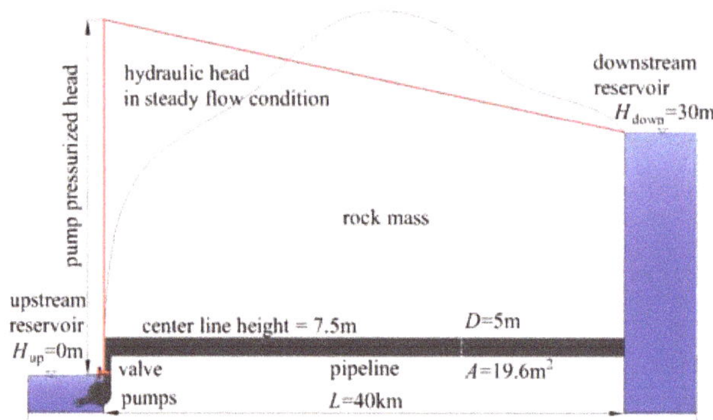

Figure 4. The physical model of a simplified case established for simulation.

3.2. Pumps Moment of Inertia

In a water supply system, pumps are the energy providing devices, pressurizing the delivering liquids meters to tens of meters' water head. Both starting and stopping the pumps can cause water hammer in the pipelines. Starting a pump can cause a positive pressure rising wave delivering downstream, while stopping a pump may lead to a sudden pressure decrease. In a pressured pipeline system, extreme negative pressure is always an undesired phenomenon that may cause the stability failure of pipelines. Determined by the material, the pipelines can tolerate a certain range of negative pressure, which is the available static pressure, similar to its loading capacity of positive pressure. When the intensity of negative pressure is within the tolerable range, it is not severe and can be resisted by the pipeline. However, when the intensity of the negative pressure is beyond the tolerable range of the pipelines, it will be critical.

The moment of inertia is a kind of inherent parameter that determines the rotate speed changing processes, with a dominant influence on the restrictive relationship among the four variables illustrated above, connected with Equation (5). In Equation (7), T is the unbalanced torque, $I = WR_g^2/g$ is the moment of inertia, W is the weight of the rotating part plus entrained liquid, R_g is the radius of gyration of the rotating mass, ω is the angular velocity in radians, and $\frac{d\omega}{dt}$ represents the angular acceleration. It shows that the angular acceleration is inversely proportional to the moment of inertia.

$$T = -I\frac{d\omega}{dt} \tag{7}$$

Figure 5 shows the negative pressure wave producing process in a controlled volume behind the pump after pump stopping. After the pump power off, while the discharge q_{outlet} through section j remains as large as the steady condition, the discharge q_{inlet} through section i decreases with the rotate speed n of the pump. In this way, the water supplied from upstream cannot fulfill the room in the control volume anymore. The faster the angular velocity ω decreases, the worse the negative water hammer will be. Therefore, we prefer the angular acceleration to be less, which demands the moment of inertia of the pump to raise. When the moment of inertia of the pump increases, the acceleration processes of pump start and stop last longer. It should have a positive effect on water hammer control by expanding the transient responding time after a sudden operation.

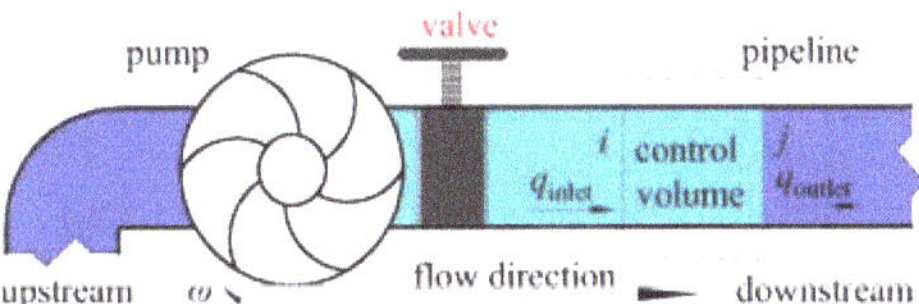

Figure 5. Physical analysis of negative pressure occurring in a pump stopping progress.

4. Simulation and Analysis

4.1. Pump Load Process and Extreme Pressure Along Pipeline

The parameters of the simulated pipeline and pumps are listed in Table 1. The simulated conditions are shown in Table 2, which are changed at the moment of inertia or valve operations to investigate different aspects as follows. t_{V0} is the start time of valve operation, Δt_{V1} is the time period of the first stage of valve operation, Δt_{V2} is the time period of the first stage of valve operation, $\Delta \tau_1$ is the closed ratio of the first stage, and $\Delta \tau_2$ is the closed ratio of the second stage. To investigate the influence of changing pumps moment of inertia on water hammer protection, the first five cases shown in Table 2, which use different moments of inertia without valve operations, are simulated respectively for comparison. In all the simulated cases, the pump failure times are set as 1000 s uniformly. Actually, the model does not need that long to achieve stability, however, it is set at this value to imitate a sudden failure during a normal running condition. As the length of the pipeline is 40 km, and the pressured wave speed is 1000 m/s, the reflection time is therefore $\frac{2L}{a} = 80$ s after the pump's failure. It is the time of the first water hammer response transferring back after the event occurrence. The upstream water level is 0 m, while the downstream water level is 30 m higher, which needs the pumps to provide the head difference and extra energy to overcome the friction along the pipeline. Therefore, after pumps failure, the water will flow backward and push the pumps to inverse if the valve is not closed. In this part of the analysis, the valve is kept open, thus, the backflow would cause the pumps inversion. The operations of the valve will be analyzed in the following section. Increasing the pumps moment of inertia can be easily accessed by adding a flywheel onto the pump's spindles, as shown in Figure 6.

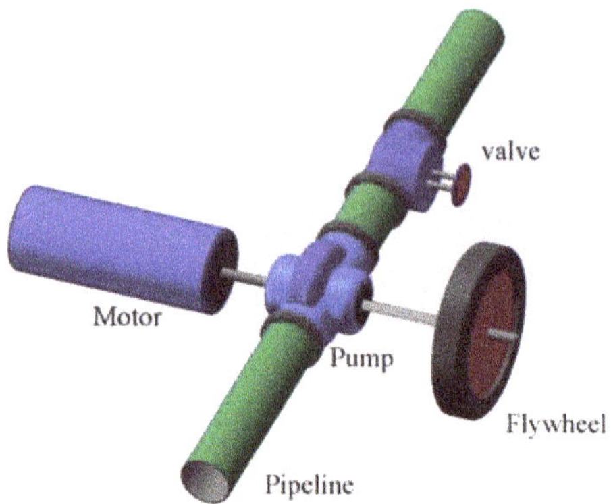

Figure 6. Physical model of a pump with an additional flywheel to increase the moment of inertia.

Table 1. Parameters of the simulated pipeline.

Pumps				Pipeline			
H_R (m)	Q_R (m³/s)	N_R (r/min)	T_R (N · m)	L (km)	D (m)	a (m/s)	A (m²)
35.57	8.60	375	84,583	40	5	1000	19.6

Table 2. Parameters of the simulated conditions.

Cases	I (kg · m²)	Valve Closing				
		t_{V0}	Δt_{V1}	$\Delta\tau_1$	Δt_{V2}	$\Delta\tau_2$
i_C# ($i_C = 1,2, \ldots ,7$)	$i_C \cdot 10^5$	/	/	/	/	/
8#	10^5	1000	600 s	100%	0 s	0%
9#	10^5	1100	600 s	100%	0 s	0%
10#	10^5	1000	60 s	80%	540 s	20%

After a sudden power off, the pumps lose electric force resource to sustain the water supply function, turning gradually slower due to the inertia. In this way, the water discharge drawn through the pumps decreases, while the outlet of the control volume in Figure 5 remains at a high velocity. Therefore, the negative pressure wave occurs as the momentum of the water in the control volume decreases. Figure 7 shows the complete pressure processes behind the pump of the whole simulation under a series of conditions including the pumps' start, normal running condition, pumps' failure, and backflow steady condition. Under different conditions of additional moment of inertia, the start period, normal running, and backflow steady conditions differ little, while the difference in the pump failure period is obvious. For a clear comparison, a detailed view of the pressure changing process in the pump failure period at the location right behind the pumps is shown in Figure 8. After the power off, the hydraulic head behind the pumps decreases dramatically, which produces a negative water hammer. Using pumps with a high moment of inertia, the head dropping process shows a relatively smooth trend compared to other conditions. That leads to a protection effect for the water supply system by reducing the negative water hammer intensity. However, there is an interesting inversion of those pressure curves under different conditions. In this set of curves, the inversion zone appears at about 150 s after the pumps power off. It reveals that the pressure process under a low moment of inertia condition not only decreases rapidly but also has a quicker rise back than the other conditions. Obviously, it can be regarded as a sign of a stronger intensity of the produced negative water hammer.

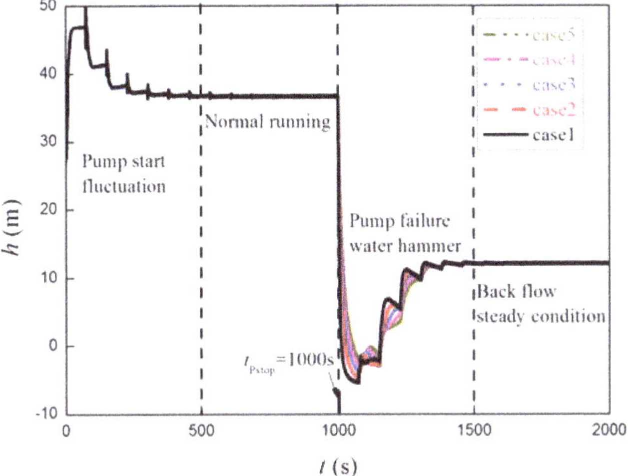

Figure 7. Pressure processes behind the pump of complete time series under different moment of inertia conditions.

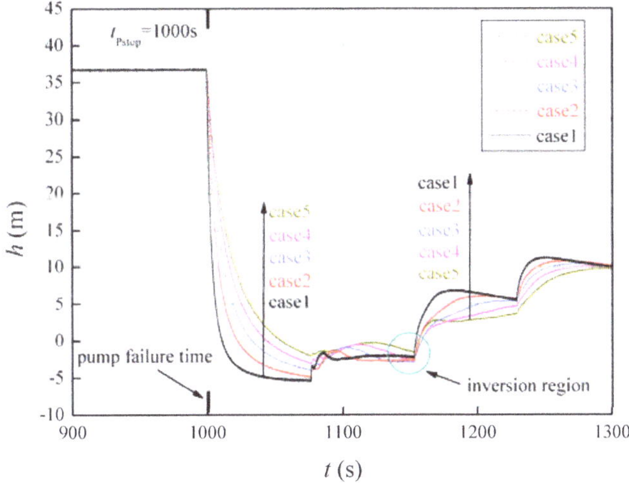

Figure 8. Detailed comparison of pressure processes behind the pump under different moment of inertia conditions.

Figure 9 shows the non-dimensional rotate speed of the pumps and the non-dimensional velocity of the water flowing through the pumps. The non-dimensional rotate speed is using the actual rotate speed divided by the rated rotate speed of the pump, and the non-dimensional velocity is using the actual discharge divided by the rated discharge of the pump. During the period of the normal running condition, the flow discharge is 3.74 m³/s, the cross area of the pipeline is 19.6 m², and the flow velocity is 0.19 m/s. After about 500 s since the pump failure with no operations on the valve, the system reaches another steady condition, where the backflow discharge is −6.1 m³/s, and the corresponding velocity is −0.31 m/s. The hydraulic head h analyzed in this study, sometimes called the piezometer head, does not include the velocity head. The velocity head is as large as $\frac{v^2}{2g}$, and the velocity $v = \frac{Q}{A}$. Then, the velocity head of the normal condition and backflow steady condition are, respectively, 0.0018 m and 0.0049 m. Compared to the changes on the hydraulic head, the velocity head affects little. As for the pumps' behavior, in normal conditions, the rotate speed is 375 r/min, equal to the rated rotate speed. In the backflow steady condition after the pumps' failure without valve operations, the pumps will be forced to inverse by the backflow force provided by the water level difference between the upstream and downstream reservoirs. The non-dimensional rotate speed of inversion will reach −0.87, finally, which is about −326 r/min. The inversion needs to be controlled, because the extended rapid runaway speed may cause damage to the pumps as well. In comparison, it can be seen that the normal running condition and backflow steady condition do not differ as the moment of inertia changes. The moment of inertia only affects the unsteady transient processes. Apparently, regardless of the rotate speed or fluid velocity, pumps of higher moment of inertia show better stability and produce slighter water hammer than those of low moment of inertia. Thus, it is clear that using pumps with a high moment of inertia is definitely a more secure choice for water hammer control in engineering.

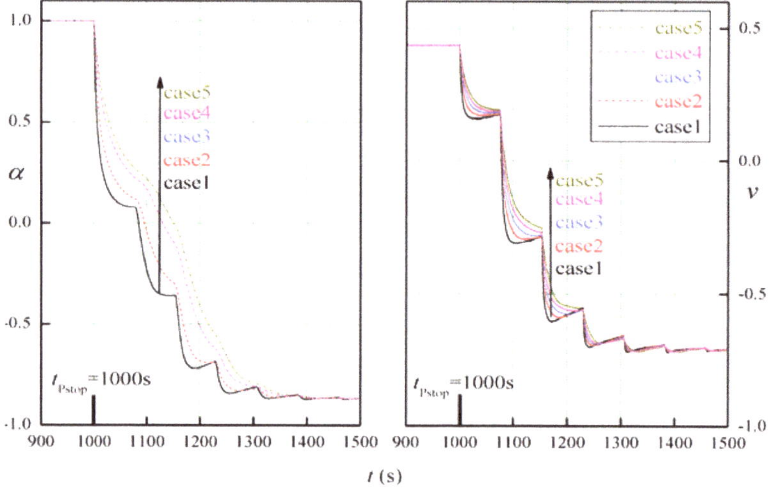

Figure 9. Comparisons of dimensionless rotate speed and dimensionless velocity changing progress under different moment of inertia conditions.

In reality, we are concerned more about the extreme pressure along the pipeline rather than the water hammer process at a certain place. That is because the center pipeline altitude can be changed limited by the topographic conditions. For instance, the pipeline should be deeper underground when crossing a river, and it may rise along with a hillside. Thus, the long-distance water supply pipelines are usually not horizontal. Figure 10 shows the comparison of the extreme pressure curve along the pipeline. The simulation results show that increasing the pumps moment of inertia can not only protect the most severe negative pressure zone but also reduce the water hammer intensity in the whole pipeline system.

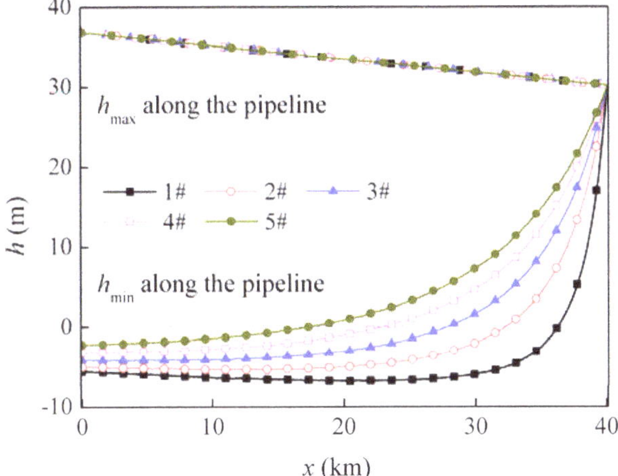

Figure 10. Comparison of the extreme dynamic head along the pipeline under different moment of inertia conditions.

4.2. Quantity of Additional Moment of Inertia

The moment of inertia can be accessed by changing the mechanic structure of the pumps. Except for improving the working accuracy by reducing the water hammer intensity, it will accelerate the cost and the occupied space at the same time. Except for adding an additional moment of inertia of pumps, there are still plenty of ways to reduce the water hammer intensity, such as surge tanks, air valves, optimization of operations, etc. Therefore, there is no need to increase the pumps moment of inertia with no limit. That will cost unnecessary economical and space resources. To numerically compare the value of positive effect, two quantities are defined. $\Delta h_{p\ n1,n2}$ represents the difference of the extreme low pressure at the side of pipe right next to the pumps, where the negative pressure wave first occurs under the conditions of case n1# and case n2#, where n1 and n2 are two integers, representing the order number of two different cases selected from five different cases shown in Table 1. In most conditions, that is the location of the worst zone in a water hammer caused by a pump power off event. Therefore, it is regarded as a typical quantity to measure the protecting effect. However, in the conditions with a low moment of inertia, the rotate speed of the pumps would reduce quickly. The filled water inlet may only relieve the negative pressure of the upstream part but cannot access the downstream part to relieve the negative pressure wave delivered there. Therefore it is possible that the minimum pressure will occur somewhere else in the pipeline. The hydraulic head h, sometimes called piezometer head, equals the pipe center line height adding the relative pressure inside the pipe. When the hydraulic head curve is lower than the altitude of the pipe center line, the pipe will suffer negative pressure. In normal conditions, there is no negative pressure in the pipeline. However, in a water hammer process, the negative pressure wave may bring about a dramatic head decrease, and when the head decrease is greater than the original positive pressure at that point, negative pressure occurs. Therefore, other than additional devices and optimization of operations, reducing the pipeline height can significantly help to prevent negative pressure occurrence in the pipeline, while increasing the positive pressure. Take case 1# for example, the negative pressure wave influence is shown in Figure 11. There are three locations along the pipeline picked out to show the negative pressure delivering process, which are, respectively, $x = 1$ km, $x = 20$ km, and $x = 39$ km. The other one is ΔW, the whole area surrounded by two extreme pressure curves of different conditions, which represents the effect along the whole water supply system. Figure 12 shows the definition of Δh_p and ΔW.

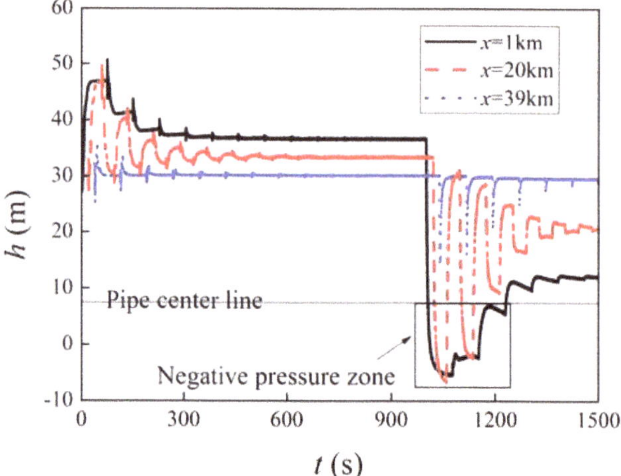

Figure 11. The negative pressure wave delivering process in case 1#.

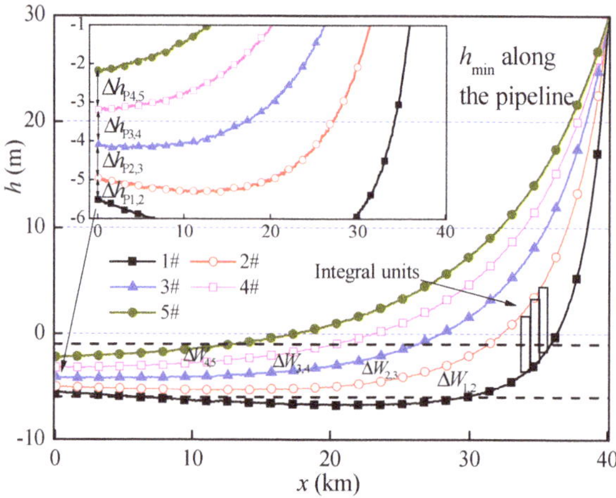

Figure 12. Description of the defined quantities Δh_P and ΔW.

To find a rule between the additional moment and the caused positive effect on pressure control, more conditions are simulated, since the original five cases were not adequate to reflect the trend. Seven of them were selected to show the trend of the two defined quantities. Figure 13 shows the relationship between the times of additional moment and the caused positive effect represented by those two quantities of adjacent cases, using the first seven cases listed in Table 2.

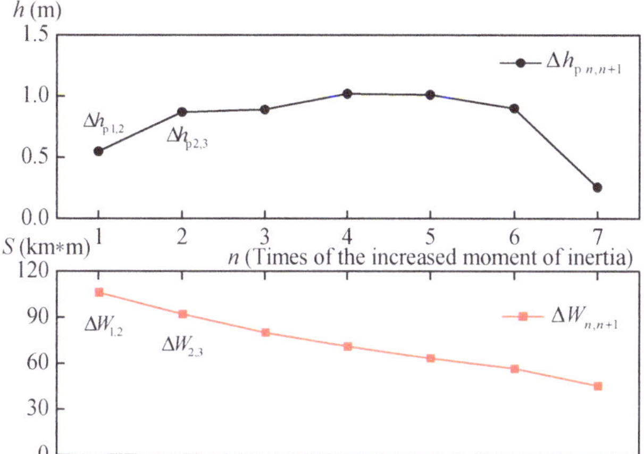

Figure 13. The trend of Δh_p and ΔW changing with increased moment of inertia.

Obviously, as the increase of the additional moment of inertia, the efficiency of the improvement on extreme pressure control effect will not be maintained at a high level. Especially ΔW, which represents the sight of the whole pipeline system, which keeps decreasing while the moment grows. Concerning the most severe zone in the pipeline, which is the point right behind the pump station, the helpful quantity of additional moment is limited below 6 times of the original moment of inertia. However, that can only be a qualitative conclusion, because the concrete numerical value is only suitable for the model established in this study, thus there is a lack of universality.

4.3. Considering the Valve

The simulations conducted above are without considering operations of valves. Normally, to ensure the pumps are not damaged from rapid runaway speed, there is always a non-returning valve set behind the rotors, as shown in Figure 5. Once the pump starts to inverse, the valve should be closed to protect the pump. However, operation of the valve can also bring undesired consequences to the system. If the valve is closed too fast or not at a suitable time, it will cause another water hammer event even more severe than the one caused by pump failure. In order to figure out the effect of putting off the valve closing time, as well as staging the valve operation, three more cases were added to make a comparison. The valve operations' corresponding connection with the system is reflected in Figure 14. Figures 15 and 16 show, respectively, the pressure processes behind the pump and the dimensionless rotate speed, and velocity changing progresses in the conditions considering valve closing listed as cases 8#, 9#, and 10# in Table 2.

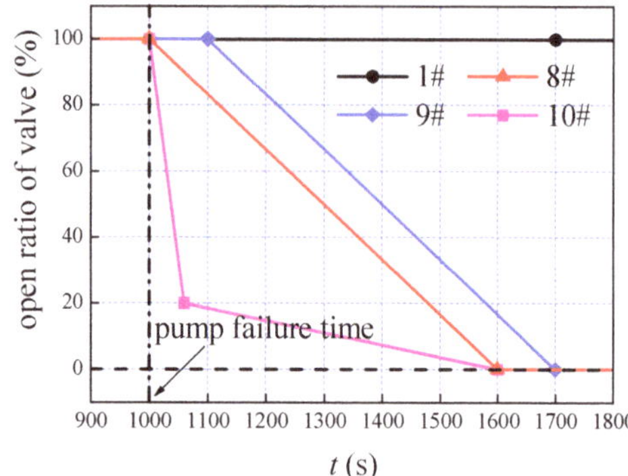

Figure 14. Parameters of the simulated operating conditions considering valve operations.

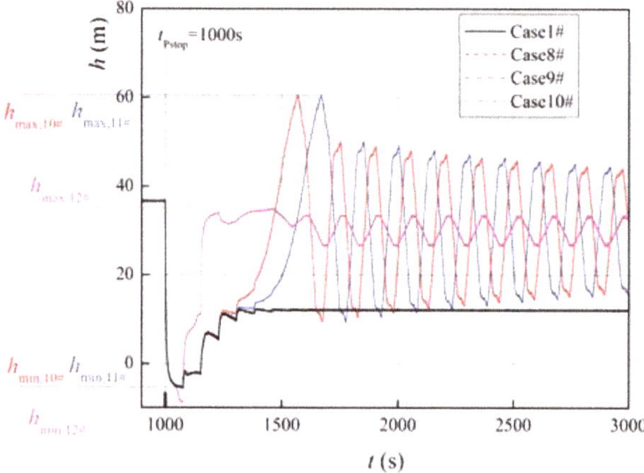

Figure 15. Comparison of pressure processes behind the pump under different valve closing operations.

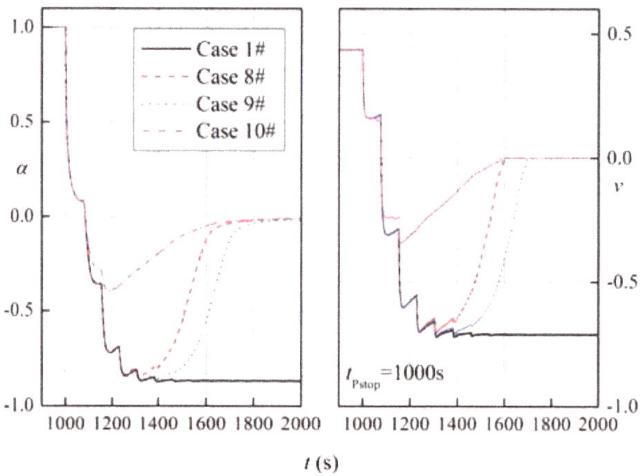

Figure 16. Comparison of dimensionless rotate speed and dimensionless velocity changing progresses under different valve closing operations.

Comparisons among case 8#, case 9#, and case 10# represents two kinds of ways to adjust the operation of the valve, putting off the valve operation time and staging the closing operation. Obviously, putting off the valve closing operation time has a slight influence on pressure control, and a negative effect on controlling the runaway speed. That means the backflow discharge should be controlled by a quick reduction of the valve open ratio before it increases to cause a rapid runaway speed. Though case 10# shows a significant improvement on both runaway speed and pressure control, it magnifies the extreme negative pressure behind the pump at the same time. That is because the rapid close stage I limited the discharge supplement into the negative pressure zone.

Adding pumps moment of inertia, and staging valve closing operation, plenty of more cases are investigated. The remaining other parameters were unchanged, only adjusting the moment of inertia and the time of valve closing in stage I. Ranging moment of inertia from 100,000 to 600,000, and time of valve closing stage I from 60 s to 120 s, the lowest pressure behind the pumps and the maximum runaway speed non-dimensional rotate speed are shown in Figure 17.

As shown, increasing pumps moment of inertia benefits on both aspects. Prolongation of valve closing time can help reduce extreme negative pressure but enlarge runaway speed at the same time. The key region is the period that the flow direction remains forward, which is within about 90 s shown in Figure 9. The best operation is to let the flow go on running to fill the negative pressure zone and close the valve to prevent runaway speed when backflow returns. Staging the time of valve closing is for accessing two aims at the same time. The first is stage I, rapid closing till 20% opening ratio, that is for preventing runaway speed. The other is stage II, slow closing till entirely closed, to avoid another water hammer caused by valve closing.

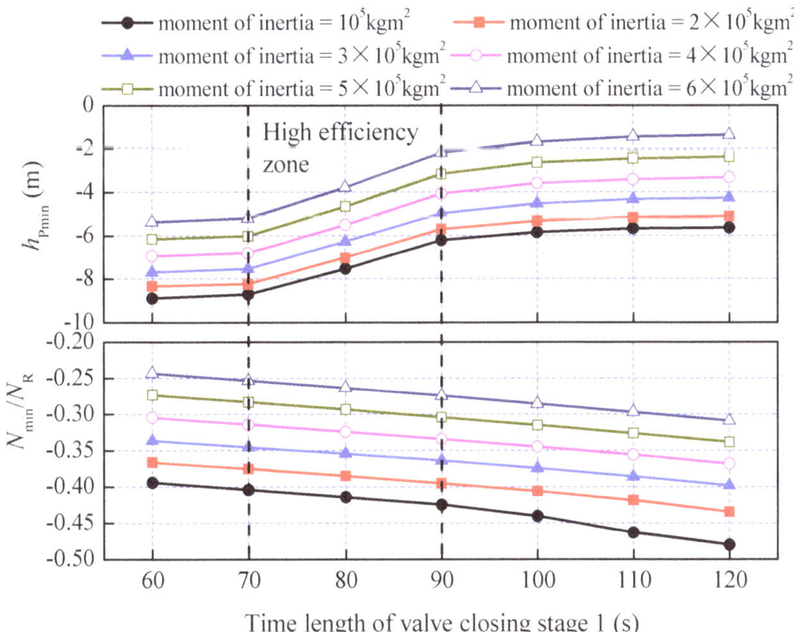

Figure 17. Comparisons of pressure processes behind the pump and dimensionless rotate speed under different additional moment of inertia when combining different valve closing operations.

5. Discussion

Surge tanks, air valves, and pressure vessels are widely used to the control intensity of water hammer. They are usually of good value to improve the system running security. However, the pipeline arrangement of a long-distance water supply system is subject to the geographical conditions. For instance, to cross a river, the pipeline is usually put under the riverbed, or above the water surface like a bridge, and surge tanks are usually limited by the height of the covered thickness above the pipeline. Indeed, the height of the surge tanks can be increased by building a tower upon the land but it leads to much labor and financial cost at the same time. Therefore, reasonable adjustment on devices is a relatively convenient and effective way to control extreme pressure in water hammers. Especially in the projects when the altitude of the pipeline is higher than the pumps. The rising zone of the pump's outlet pipe cannot be protected by surge tanks or other additional devices. The pressure control ability primarily depends on the characteristics of pumps and the operations of valves. Optimizing the operations processes and adding pumps moment of inertia can also help to control the extreme pressure along the pipeline and protect the pumps from fast runaway speed. The comparison among case 1# to 7# indicates that additional moment of inertia has a positive effect on negative pressure control and there is a most effective quantity of it. Furthermore, comparison among case 1#, and 8# to 10#, show the best cooperation of the valve to control the pump's runaway speed is staging time and to avoid the forward flow period. The simulated cases in this research are under specific limitations on pipeline design and pump parameters, the conditions of actual water supply engineering may have complex situations on the devices and pipeline design, including faster flow velocity and larger water hammer intensity. Nevertheless, the advantages of the additional moment of inertia remain the same.

To claim the assumptions used in this research, expect for the physical model simplification, the numerical model used for simulation has some assumptions. The first is the basic assumption of MOC, which requires the relationship between meshing grid and pressure wave speed, which is illustrated in Equation (3). The second is the neglect of unsteady friction factor, which actually has little influence on the prediction of extreme pressure value and does not affect the investigation results either. Besides, the flow in the pipe is one-dimensional and the control volume is fixed relative to the pipe. The velocity and pressure are uniform at the cross sections of the pipe, and the fluid density is constant. The pipe walls and fluid are linearly elastic, that is, stress is proportional to strain. And the temperature is assumed to be constant. Increasing the pumps moment of inertia can be easily accessed by adding a flywheel to the pump's spindles. It makes the response of pumps to accident events slower and slighter, which is normally a helpful change for device protection. It has to be pointed out that the efficiency of improvement will decline as the additional moment increases. For different engineering, there are corresponding valued limits of the additional moment of inertia. In addition, an added moment of inertia requires extra materials, space, and energy consumption. However, it does not need extra power to keep running in steady running conditions. Considering the security improvement on accident events in water supply systems, it is a recommendation to design the pumps with a reasonably high moment of inertia. Though, it has to be clarified that there are still some limitations, the effect of additional pump moment of inertia has just a relieving function for negative pressure when pumps fail. It has little benefit for water hammer control caused by downstream valve rapid closing or other devices operations inside the pipeline. In further research, the investigation on other parameters of pumps and pipeline design will be conducted, such as the complete characteristic curves of the pumps and draft tube design of the pump station.

6. Conclusions

In this study, a 1D numerical model containing a pump station, valves, reservoirs, and pipeline was established. After investigations on the effect of changing the pumps moment of inertia and operations of valves behind the pumps after a power off event, the following conclusions were drawn:

In water supply pipe systems, except for normal pressure control devices, additional pumps moment of inertia can help prevent extreme negative pressure occurring behind the pumps in accident events such as electric power off. Moreover, the extreme pressure curve along the pipeline can also be reduced. Considering the construction cost and the efficiency, there is always an efficient zone of the additional moment of inertia according to different engineering.

The pumps design and optimized valve operations are significant to protect pipe and pump from damage when accident events occur. Through various cases of numerical simulations based on the method of characteristics, it was found that increasing the moment of inertia benefits both negative pressure control after pumps power off and preventing pumps runaway speed. Closing the valve is for protecting the pumps from rapid runaway speed.

Staging the valve closing time can significantly reduce the water hammer caused by valve closing. To limit the negative pressure intensity, the dominant part of valve closing should begin after the period that the pipe flow direction remains forward. As when the flow direction is still forward, the filling water can help reduce the negative pressure caused by pump failure. The best operation is to let the flow go on running to fill the negative pressure zone and close the valve to prevent runaway speed when backflow returns.

Author Contributions: Conceptualization, W.W., and B.Z.; methodology, B.Z.; software, B.Z. and X.C.; validation, B.Z.; formal analysis, B.Z.; investigation, B.Z.; resources, W.W.; data curation, B.Z. and X.C.; writing-original draft preparation, B.Z.; writing-review and editing, W.W.; visualization, W.W.; supervision, W.W.; project administration, W.W.; funding acquisition, W.W.

Funding: This work was supported by the National Key R&D Program of China (2018YFC0407203), the National Natural Science Foundation of China (Nos. 51779216, 51279175) and the Zhejiang Provincial Natural Science Foundation of China (Grant No. LZ16E090001).

Conflicts of Interest: The authors declare no conflict of interest.

Acronyms

MOC	method of characteristics

Nomenclature

g	Acceleration of gravity (m/s^2)
h	Pressure head (m)
x	Distance along pipe from the inlet (m)
t	Time, as subscript to denote time (s)
v	Flow velocity (m/s)
τ_w	Instantaneous wall shear stress
$\alpha_{Pipe-Plane}$	The angle between pipe and the horizontal plane.
τ_{ws}	The quasi-steady component
τ_{wu}	The unsteady component
f	Darcy–Weisbach friction factor
D	Main pipe diameter (m)
a	Wave speed of water hammer (m/s)
Δt	Time step (s)
B	A defined constant parameter of a pipeline
R	A defined constant parameter of a pipeline
i	Serial number of nodes (s)
q	Discharge (m^3/s)
A	Area of section (m^2)
Δx	length of element, space interval step (m)
H	Head of the pump (m)
Q	Discharge through the pump (m^3/s)
T	Torque of the pump (Nm)
N	Rotate speed of the pump
WH	A defined dimensionless variable representing a pump's characteristic
WB	A defined dimensionless variable representing a pump's characteristic
x_p	Instantaneous position of pump operation

H_R	Rated head of the pump (m)
N_R	Rated speed of the pump
Q_R	Rated discharge of the pump (m^3/s)
T_R	Rated torque of the pump (Nm)
π	Constant
I	The moment of inertia (kgm^2)
W	The weight of rotating parts plus entrained liquid (kg)
R_g	The radius of gyration of the rotating mass (m)
ω	The angular velocity in radians
n	Non-dimensional rotate speed of the pump
j	Section number
L	The length of pipeline (m)
t_{Pstop}	Time of the pump powering off (m^2)
h_{max}	The maximum hydraulic head (m)
h_{min}	The minimum hydraulic head (m)
Δh	The difference of hydraulic head (m)
i_C	The number of cases
ΔW	The whole area surrounded by two extreme pressure curves
λ	Non-dimensional coefficient
$h_{Psteady,2}$	The hydraulic head at pump in the second steady condition (m)
h_{Pmin}	The minimum hydraulic head at pump in a transient process (m)
$h_{Psteady,1}$	The hydraulic head at pump in the first steady condition (m)

References

1. Kummu, M.; de Moel, H.; Ward, P.J.; Varis, O. How Close Do We Live to Water? A Global Analysis of Population Distance to Freshwater Bodies. *PLoS ONE* **2011**, *6*. [CrossRef] [PubMed]
2. Wylie, E.B.; Streeter, V.L. *Fluid Transients*; McGraw-Hill International Book Co.: New York, NY, USA, 1978.
3. Kwan, E.S.K.; Heilman, C.B.; Shucart, W.A.; Klucznik, R.P. Enlargement of basilar artery aneurysms following balloon occlusion—"water-hammer effect" Report of 2 cases. *J. Neurosurg.* **1991**, *75*, 963–968. [CrossRef] [PubMed]
4. Brunone, B.; Karney, B.W.; Mecarelli, M.; Ferrante, M. Velocity profiles and unsteady pipe friction in transient flow. *J. Water Resour. Plann. Manag.-Asce* **2000**, *126*, 236–244. [CrossRef]
5. Bergant, A.; Simpson, A.R.; Tijsseling, A.S. Water hammer with column separation: A historical review. *J. Fluids Struct.* **2006**, *22*, 135–171. [CrossRef]
6. Xu, B.; Chen, D.; Zhang, H.; Wang, F. Modeling and stability analysis of a fractional-order Francis hydro-turbine governing system. *Chaos Solitons Fractals* **2015**, *75*, 50–61. [CrossRef]
7. Fang, H.; Chen, L.; Dlakavu, N.; Shen, Z. Basic Modeling and simulation tool for analysis of hydraulic transients in hydroelectric power plants. *IEEE Trans. Energy Convers.* **2008**, *23*, 834–841. [CrossRef]
8. Wan, W.; Zhang, B. Investigation of Water Hammer Protection in Water Supply Pipeline Systems Using an Intelligent Self-Controlled Surge Tank. *Energies* **2018**, *11*, 16. [CrossRef]
9. Stephenson, D. Effects of air valves and pipework on water hammer pressures. *J. Transp. Eng.-ASCE* **1997**, *123*, 101–106. [CrossRef]
10. Bergant, A.; Kruisbrink, A.; Arregui, F. Dynamic Behaviour of Air Valves in a Large-Scale Pipeline Apparatus. *Strojniski Vestn.-J. Mech. Eng.* **2012**, *58*, 225–237. [CrossRef]
11. Wan, W.; Huang, W.; Li, C. Sensitivity Analysis for the Resistance on the Performance of a Pressure Vessel for Water Hammer Protection. *J. Press. Vessel Technol. Trans. Asme* **2014**, *136*. [CrossRef]
12. Stephenson, D. Simple guide for design of air vessels for water hammer protection of pumping lines. *J. Hydraul. Eng. Asce* **2002**, *128*, 792–797. [CrossRef]
13. De Martino, G.; Fontana, N. Simplified Approach for the Optimal Sizing of Throttled Air Chambers. *J. Hydraul. Eng. Asce* **2012**, *138*, 1101–1109. [CrossRef]
14. Karney, B.W.; Simpson, A.R. In-line check valves for water hammer control. *J. Hydraul. Res.* **2007**, *45*, 547–554. [CrossRef]

15. Tian, W.; Su, G.; Wang, G.; Qiu, S.; Xia, Z. Numerical simulation and optimization on valve-induced water hammer characteristics for parallel pump feedwater system. *Annals Nucl. Energy* **2008**, *35*, 2280–2287. [CrossRef]
16. Bazargan-Lari, M.R.; Kerachian, R.; Afshar, H.; Bashi-Azghadi, S.N. Developing an optimal valve closing rule curve for real-time pressure control in pipes. *J. Mech. Sci. Technol.* **2013**, *27*, 215–225. [CrossRef]
17. Zhou, J.; Xu, Y.; Zheng, Y.; Zhang, Y. Optimization of Guide Vane Closing Schemes of Pumped Storage Hydro Unit Using an Enhanced Multi-Objective Gravitational Search Algorithm. *Energies* **2017**, *10*, 911. [CrossRef]
18. Zhou, D.; Chen, H.; Zhang, L. Investigation of Pumped Storage Hydropower Power-Off Transient Process Using 3D Numerical Simulation Based on SP-VOF Hybrid Model. *Energies* **2018**, *11*, 1020. [CrossRef]
19. Nagode, K.; Skrjanc, I. Modelling and Internal Fuzzy Model Power Control of a Francis Water Turbine. *Energies* **2014**, *7*, 874–889. [CrossRef]
20. Carratelli, E.P.; Viccione, G.; Bovolin, V. Free surface flow impact on a vertical wall: A numerical assessment. *Theor. Comput. Fluid Dyn.* **2016**, *30*, 403–414. [CrossRef]
21. Moghaddas, S.M.J.; Samani, H.M.V.; Haghighi, A. Transient protection optimization of pipelines using air-chamber and air-inlet valves. *Ksce J. Civ. Eng.* **2017**, *21*, 1991–1997. [CrossRef]
22. Kim, H.; Hur, J.; Kim, S. The Optimization of Design Parameters for Surge Relief Valve for Pipeline Systems. In Proceedings of the Applied Mathematics and Computer Science Conference, Rome, Italy, 27–29 January 2017.
23. Richter, W.; Zenz, G.; Schneider, J.; Knoblauch, H. Surge tanks for high head hydropower plants—Hydraulic layout—New developments/Wasserschlösser für Hochdruck-Wasserkraftanlagen—Hydraulische Auslegung—Neue Entwicklungen. *Geomech. Tunn.* **2015**, *8*, 60–73. [CrossRef]
24. Gabl, R.; Righetti, M. Design criteria for a type of asymmetric orifice in a surge tank using CFD. *Eng. Appl. Comp. Fluid Mech.* **2018**, *12*, 397–410. [CrossRef]
25. Adam, N.J.; De Cesare, G.; Nicolet, C.; Billeter, P.; Angermayr, A.; Valluy, B.; Schleiss, A.J. Design of a Throttled Surge Tank for Refurbishment by Increase of Installed Capacity at a High-Head Power Plant. *J. Hydraul. Eng.* **2018**, *144*, 10. [CrossRef]
26. Li, X.; Zhu, M.; Xie, J. Numerical Simulation of Transient Pressure Control in a Pumped Water Supply System Using an Improved Bypass Pipe. *Strojniski Vestn.-J. Mech. Eng.* **2016**, *62*, 614–622. [CrossRef]
27. Triki, A. Water-Hammer Control in Pressurized-Pipe Flow Using a Branched Polymeric Penstock. *J. Pipeline Syst. Eng. Pract.* **2017**, *8*. [CrossRef]
28. Triki, A. Water-hammer control in pressurized-pipe flow using an in-line polymeric short-section. *Acta Mech.* **2016**, *227*, 777–793. [CrossRef]
29. Zeng, W.; Yang, J.; Hu, J.; Yang, J. Guide-Vane Closing Schemes for Pump-Turbines Based on Transient Characteristics in S-shaped Region. *J. Fluids Eng.-Trans. Asme* **2016**, *138*. [CrossRef]
30. Ballun, J.V. A methodology for predicting check valve slam. *J. Am. Water Work Assoc.* **2007**, *99*. [CrossRef]
31. Hur, J.; Kim, S.; Kim, H. Water hammer analysis that uses the impulse response method for a reservoir-pump pipeline system. *J. Mech. Sci. Technol.* **2017**, *31*, 4833–4840. [CrossRef]
32. Wan, W.; Li, F. Sensitivity Analysis of Operational Time Differences for a Pump-Valve System on a Water Hammer Response. *J. Press. Vessel Technol.-Trans. Asme* **2016**, *138*. [CrossRef]
33. Chaudhry, M.; Hussaini, M. Second-order accurate explicit finite-difference schemes for waterhammer analysis. *J. Fluids Eng.* **1985**, *107*, 523–529. [CrossRef]
34. Kontzialis, K.; Moditis, K.; Paidoussis, M.P. Transient Simulations of the Fluid-Structure Interaction Response of a Partially Confined Pipe Under Axial Flows in Opposite Directions. *J. Press. Vessel Technol.-Trans. Asme* **2017**, *139*. [CrossRef]
35. Kochupillai, J.; Ganesan, N.; Padmanabhan, C. A new finite element formulation based on the velocity of flow for water hammer problems. *Int. J. Press. Vessels Pip.* **2005**, *82*, 1–14. [CrossRef]
36. Wylie, E.B.; Streeter, V.L.; Suo, L. *Fluid Transients in Systems*; Prentice Hall Englewood Cliffs: Englewood, NJ, USA, 1993.
37. Meniconi, S.; Duan, H.F.; Brunone, B.; Ghidaoui, M.S.; Lee, P.J.; Ferrante, M. Further Developments in Rapidly Decelerating Turbulent Pipe Flow Modeling. *J. Hydraul. Eng.* **2014**, *140*. [CrossRef]
38. Karney, B.W.; McInnis, D. Efficient calculation of transient flow in simple pipe networks. *J. Hydraul. Eng.-Asce* **1992**, *118*, 1014–1030. [CrossRef]

39. Wan, W.; Huang, W. Investigation on complete characteristics and hydraulic transient of centrifugal pump. *J. Mech. Sci. Technol.* **2011**, *25*, 2583–2590. [CrossRef]
40. Wan, W.; Huang, W. Investigation of Fluid Transients in Centrifugal Pump Integrated System With Multi-Channel Pressure Vessel. *J. Press. Vessel Technol.-Trans. Asme* **2013**, *135*. [CrossRef]
41. Hou, Q.; Zhang, L.; Tijsseling, A.S.; Kruisbrink, A.C.H. Rapid filling of pipelines with the SPH particle method. In Proceedings of the International Conference on Advances in Computational Modeling and Simulation, Kunming, China, 14–16 December 2011.
42. Wan, W.; Huang, W. Water hammer simulation of a series pipe system using the MacCormack time marching scheme. *Acta Mech.* **2018**, *229*, 3143–3160. [CrossRef]

Article

One-Log Call Iterative Solution of the Colebrook Equation for Flow Friction Based on Padé Polynomials

Pavel Praks [1,2,*]  and Dejan Brkić [1,*]

1   European Commission, DG Joint Research Centre (JRC), Directorate C: Energy, Transport and Climate, Unit C3: Energy Security, Distribution and Markets, Via Enrico Fermi 2749, 21027 Ispra (VA), Italy
2   IT4Innovations National Supercomputing Center, VŠB—Technical University Ostrava, 17. listopadu 2172/15, 708 00 Ostrava, Czech Republic
*   Correspondence: Pavel.Praks@ec.europa.eu or Pavel.Praks@vsb.cz (P.P.); dejanbrkic0611@gmail.com (D.B.)

Received: 19 June 2018; Accepted: 11 July 2018; Published: 12 July 2018

Abstract: The 80 year-old empirical Colebrook function ζ, widely used as an informal standard for hydraulic resistance, relates implicitly the unknown flow friction factor λ, with the known Reynolds number Re and the known relative roughness of a pipe inner surface ε^*; $\lambda = \zeta(Re, \varepsilon^*, \lambda)$. It is based on logarithmic law in the form that captures the unknown flow friction factor λ in a way that it cannot be extracted analytically. As an alternative to the explicit approximations or to the iterative procedures that require at least a few evaluations of computationally expensive logarithmic function or non-integer powers, this paper offers an accurate and computationally cheap iterative algorithm based on Padé polynomials with only one *log*-call in total for the whole procedure (expensive *log*-calls are substituted with Padé polynomials in each iteration with the exception of the first). The proposed modification is computationally less demanding compared with the standard approaches of engineering practice, but does not influence the accuracy or the number of iterations required to reach the final balanced solution.

Keywords: Colebrook equation; Colebrook-White; flow friction; iterative procedure; logarithms; Padé polynomials; hydraulic resistances; turbulent flow; pipes; computational burden

1. Introduction

The empirical Colebrook equation [1,2] implicitly relates the unknown flow friction factor λ with the known Reynolds number Re and the know relative roughness of inner pipe surface, ε^*; $\lambda = \zeta(Re, \varepsilon^*, \lambda)$, where ζ is functional symbol, Equation (1).

$$\frac{1}{\sqrt{\lambda}} = -2 \cdot \log_{10}\left(\frac{2.51}{Re} \cdot \frac{1}{\sqrt{\lambda}} + \frac{\varepsilon^*}{3.71} \right) \tag{1}$$

In Equation (1) Re is Reynolds number; $4000 < Re < 10^8$, ε^* is relative roughness of inner pipe surface; $0 < \varepsilon^* < 0.05$, and λ is Darcy flow friction factor; $0.0064 < \lambda < 0.077$ (all three quantities are dimensionless). All values are in correlation with the diagram of Moody [3–5].

The Colebrook equation is based on experiments performed by Colebrook and White in 1937 with the flow of air through a set of artificially roughened pipes [2]. The accuracy of this 80 year-old equation is disputed many times [6–8] but it is still accepted in engineering practice as an informal standard for hydraulic resistance. Therefore, to repeat results and for comparisons, it is required to solve the Colebrook equation accurately. Numerous evaluations of flow friction factor such as in the case of complex networks of pipes pose extensive burden for computers, so not only an accurate but also a simplified solution is required. Calculation of complex water or gas distribution networks [9] which requires few evaluations of logarithmic function for each pipe, presents a significant and extensive burden which available computer resources hardly can easily manage [10–14].

The Colebrook equation is based on logarithmic law where the unknown flow friction factor λ is given implicitly, i.e., it appears on both sides of Equation (1) in form $\lambda = \xi(Re, \varepsilon^*, \lambda)$, from which it cannot be extracted analytically (an exception is through the Lambert W-function [15–17]). The common way to solve it is to guess an initial value λ_0 for friction factor and then to try to balance it using the iterative algorithm [18] which needs to be terminated after the certain number of iterations when the final balanced value λ_n is reached. As an alternative to the iterative procedure, numerous approximate formulas are available [19–22]. Usually, more complex approximations are more accurate, but also more computationally expensive because they contain at least a few logarithmic expressions and/or terms with non-integer powers which require use of demanding algorithms (non-integer exponential or natural logarithm) to be evaluated in processor units of computers and to be stored in registers [10–16]. The most accurate explicit approximations up to date are by Buzzelli [23], Zigrang and Sylvester [24], Serghides [25], Romeo et al. [26], and Vatankhah and Kouchakzadeh [27,28]. They introduce the relative error of up to 0.14% [20] and they at least require evaluation of two or more computationally expensive functions [10,11].

The presented scheme for solving the Colebrook equation requires only one single call of the logarithmic function in respect to the whole iterative procedure. It is equally accurate as a standard iterative approach and does not require additional iterations to reach the same accuracy. Instead of the computationally expensive logarithmic function, its Padé polynomial equivalent [29] is used in all iterations, exception the first. The Padé approximant is the approximation of a function by a rational algebraic fraction where both the numerator and the denominator are polynomials [29]. Because these rational functions only use the elementary arithmetic operations, they can be evaluated numerically very easily. In the computer environment, they required less basic floating-point operations compared with the logarithmic function [30–32].

The presented simplified iterative method can be profitable for future computing software in terms of having a high level of accuracy and speed with a decreased computational burden.

2. Evaluation of Logarithmic Function through Padé Polynomials

Basic floating-point operations such as addition and multiplication are carried out directly in the Central Processor Unit (CPU) while logarithmic functions, exponents or square roots require expensive operations based on more complex algorithms [30–32]. In addition to logarithms and non-integer powers, Biberg [33] adds also division in the group of more costly functions for evaluation while addition, subtraction and multiplication has treated as low-cost operations according to Biberg [33]. Winning and Coole [34] report average time for 100 million operation in seconds and relative effort, respectively as follows: addition 23.40 s and 1, subtraction 27.50 s and 1.18, division 31.70 s and 1.35, multiplication 36.20 s and 1.55, squared 51.10 s and 2.18, square root 53.70 s and 2.29, cubed 55.58 s and 2.38, natural log 63.00 s and 2.69, cubed root 63.40 s and 2.71, fractional exponential 77.60 s and 3.32, and log to 10-base 78.80 s and 3.37.

To illustrate the complexity of computing in modern computers it should be noted that even such a relatively simple equation such as Colebrook's can make a numerical problem in computer registers due to overflow error. Its transformed version in term of the Lambert W-function can give such large numbers for some pairs of the Reynolds number Re and the relative roughness of inner pipe surface ε^* which are from the practical domain of applicability in engineering practice and which cannot be stored in 32- or 64-bit registers of modern computers [15,16].

In order to simplify the common iterative procedure from engineering practice for solving the Colebrook equation, the logarithmic function is replaced with its relevant Padé polynomial equivalent in all iterations with exception to the first. The Padé polynomials can accurately approximate logarithmic function only in a limited domain. For example, knowing that $log_{10}(100) = 2$, value of $log_{10}(90)$

can be obtained from $log_{10}(100/90) = log_{10}(100) - log_{10}(90) \rightarrow log_{10}(90) = log_{10}(100) - log_{10}(100/90)$ using the fact that $100/90 \approx 1.111$ is near 1. Logarithmic function can be replaced by its Padé polynomial equivalent very accurately in a limited domain, instead of $log_{10}(1.111)$, already calculated $log_{10}(100) = 2$ and Padé polynomial which is accurate around 1 for argument $z = 1.1111$ can be used to calculate $log_{10}(90)$.

Because of linearization of the unknown parameter λ, a more suitable form of the Colebrook equation for computation is $x = -2 \cdot log_{10}\left(\frac{2.51 \cdot x}{Re} + \frac{\varepsilon^*}{3.71}\right)$, where $x = \frac{1}{\sqrt{\lambda}}$. The argument of logarithmic function in the Colebrook equation is $y = \frac{2.51 \cdot x}{Re} + \frac{\varepsilon^*}{3.71}$ where only evaluation through its native logarithmic form $log_{10}(y)$ need go only in the first iteration where further evaluation can go through the appropriate Padé polynomial which is accurate for its argument z around 1, knowing that $z_{01} = \frac{y_0}{y_1}$, $z_{02} = \frac{y_0}{y_2}$, $z_{03} = \frac{y_0}{y_3}$, etc. or $z_{01} = \frac{y_0}{y_1}$, $z_{12} = \frac{y_1}{y_2}$, $z_{23} = \frac{y_2}{y_3}$, etc. in the case of the Colebrook equation it is always near 1; $z \approx 1$. Evaluation of 10-base logarithmic function in many computing languages goes through natural logarithm where $log_{10}(z) = \frac{ln(z)}{ln(10)}$ and where $ln(10) \approx 2.302585093$ is constant, and therefore the Padé polynomials that approximate accurately $ln(z)$ for $z \approx 1$ are shown; Equations (2)–(7). Padé polynomials of different orders can be used for approximation of $ln(z)$, here all accurate for arguments z close to 1; $z \approx 1$. As the expansion point $z_0 = 1$ is a root of $ln(z)$, the accuracy of the Padé approximant decreases. Setting the OrderMode option in Matlab Padé command to relative compensates for the decreased accuracy. Thus, here, the Pade approximant of (m,n) order uses the form $ln(z) \approx \frac{(z-z_0) \cdot (\alpha_0 + \alpha_1 \cdot (z-z_0) + ... + \alpha_m (z-z_0)^m)}{1 + \beta_1 \cdot (z-z_0) + ... + \beta_n (z-z_0)^n}$, where α and β are coefficients (the coefficients of the polynomials need not be rational numbers).

Horner algorithm transforms polynomials into a computationally efficient form and therefore, Horner nested polynomial representations of the Padé polynomials of different orders for $ln(z)$ where $z \rightarrow 1$ are shown here; Equations (2)–(7). Relative error introduced by them; Equations (2)–(5) compared with $ln(z)$ is shown in Figure 1 and for Equation (6) in Table 1. Higher order of Padé approximants are more accurate, but more complex. For example, Padé polynomial of order (2,3) is with polynomial of order 2 in numerator and of order 3 in denominator; Equation (6). Of course, low order formulas are simpler, but they have larger errors than high order formulas and vice versa.

Order (1,1), Equation (2):

$$ln(z) \approx \frac{z \cdot (z + 4) - 5}{4 \cdot z + 2} \tag{2}$$

Order (1,2), Equation (3):

$$ln(z) \approx \frac{3 \cdot (z - 1) \cdot (z + 1)}{z \cdot (z + 4) + 1} \tag{3}$$

Order (2,1), Equation (4):

$$ln(z) \approx \frac{-z \cdot (z \cdot (z - 9) - 9) - 17}{18 \cdot z + 6} \tag{4}$$

Order (2,2), Equation (5):

$$ln(z) \approx \frac{z \cdot (z \cdot (z + 18) - 9) - 10}{z \cdot (9 \cdot z + 18) + 3} \tag{5}$$

Order (2,3), Equation (6):

$$ln(z) \approx \frac{(z - 1) \cdot (11 \cdot z^2 + 38 \cdot z + 11)}{3 \cdot (z^3 + 9 \cdot z^2 + 9 \cdot z + 1)} \tag{6}$$

Order (3,2), Equation (7):

$$ln(z) \approx \frac{z \cdot (z \cdot (11 \cdot z + 27) - 27) - 11}{z \cdot (z \cdot (3 \cdot z + 27) + 27) + 3} \tag{7}$$

In Equations (2)–(7), z is from $z_{01} = \frac{y_0}{y_1}$, $z_{02} = \frac{y_0}{y_2}$, $z_{03} = \frac{y_0}{y_3}$, etc., or $z_{01} = \frac{y_0}{y_1}$, $z_{12} = \frac{y_1}{y_2}$, $z_{23} = \frac{y_2}{y_3}$, etc.; and $y = \frac{2.51 \cdot x}{Re} + \frac{\varepsilon^*}{3.71}$.

Relative error of Padé approximants (2,2) for $z \approx 1$ of $ln(z)$ is negligible for $0.8 < z < 1.2$. Thus, relative error of the used Padé approximants (2,3) of $ln(z)$ in the proposed iterative procedure is even more negligible and therefore it is not presented in Figure 1, but it is available in Table 1. As can be seen from Figure 1, even the very simple form of Padé polynomials (1,2) and (2,1) are of high accuracy in respect of domain of interest for solving the Colebrook equation which is $z \approx 1$; $z \in [0.9, 1.1]$.

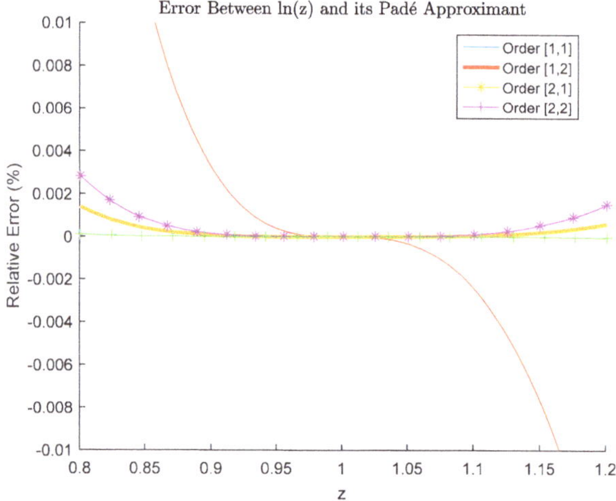

Figure 1. Relative error between $ln(z)$ and its Padé approximants accurate for $z \approx 1$.

Table 1. Relative error in % of Padé approximant (2,3) for z in interval [0.6; 1,6].

z	$log_{10}(z) = \frac{ln(z)}{ln(10)}$	Padé Approximants (2,3)	Relative Error %
0.6	-0.22184875	-0.221847398	$6.1 \times 10^{-4}\%$
0.65	-0.187086643	-0.187086228	$2.2 \times 10^{-4}\%$
0.7	-0.15490196	-0.154901848	$7.2 \times 10^{-5}\%$
0.75	-0.124938737	-0.124938712	$2.0 \times 10^{-5}\%$
0.8	-0.096910013	-0.096910009	$4.4 \times 10^{-6}\%$
0.85	-0.070581074	-0.070581074	$6.6 \times 10^{-7}\%$
0.9	-0.045757491	-0.045757491	$4.9 \times 10^{-8}\%$
0.95	-0.022276395	-0.022276395	$6.5 \times 10^{-10}\%$
1	0	0	0%
1.05	0.021189299	0.021189299	$4.8 \times 10^{-10}\%$
1.1	0.041392685	0.041392685	$2.7 \times 10^{-8}\%$
1.15	0.06069784	0.06069784	$2.7 \times 10^{-7}\%$
1.2	0.079181246	0.079181245	$1.3 \times 10^{-6}\%$
1.25	0.096910013	0.096910009	$4.4 \times 10^{-6}\%$
1.3	0.113943352	0.113943339	$1.2 \times 10^{-5}\%$
1.35	0.130333768	0.130333735	$2.6 \times 10^{-5}\%$
1.4	0.146128036	0.146127961	$5.1 \times 10^{-5}\%$
1.45	0.161368002	0.161367854	$9.2 \times 10^{-5}\%$
1.5	0.176091259	0.176090987	$1.5 \times 10^{-4}\%$
1.55	0.190331698	0.190331231	$2.5 \times 10^{-4}\%$
1.6	0.204119983	0.204119223	$3.7 \times 10^{-4}\%$

3. Initial Starting Point for the Proposed Iterative Method

In the case of the Colebrook equation, practical experience shows that trying to get a good initial starting point x_0 has limited value until it is chosen in the domain of applicability of the equation which is $3.68 < x < 12.47$. Every initial starting point x_0 chosen from the domain of applicability of the Colebrook equation will lead to the final accurate solution surely, with the only difference that in some cases more additional iterations would be needed. Usually, with the initial guess x_0 that is close to the exact solution, the iterative procedure converges to it in five or fewer iterations. To date, cases which lead to divergence, fluctuation, or convergence to a possible far away solution outside of the practical domain of applicability of the Colebrook equation are not known. In the proposed approach, a good starting point should be chosen within the domain of applicability of the Colebrook equation and should not contain any logarithmic function and/or non-integer power term.

A number of options to choose an optimal starting point x_0 are considered: (1) special case of the Colebrook equation when $Re \to \infty$, (2) integration of the Colebrook equation, (3) explicit approximations of the Colebrook equation, and (4) fixed value.

1. The common approach is to choose an initial starting point from the zone of fully developed turbulent rough hydraulic flow $x_0 = -2 \cdot log_{10}\left(\frac{\varepsilon^*}{3.71}\right)$, because in this special case of the Colebrook equation where $Re \to \infty$, the equation is in explicit form with respect to x; $x_0 = \xi(\varepsilon^*)$, where ξ is functional symbol [18,22]. Here the goal is to avoid use of logarithmic functions and therefore, this starting point is not suitable.

2. An efficient procedure for finding a sufficiently good initial starting point x_0 is proposed by Yun [35] in the integral form; Equation (8):

$$x_0 = \frac{1}{2} \cdot \left\{ a + b + sgn(F(a)) \cdot \int_a^b \tanh(F(x))dx \right\} \tag{8}$$

In Equation (8), $F = x - \xi(x) = 0$, ξ represents the Colebrook equation, a is the lower while b is the upper limit from which an initial starting point x_0 should be chosen; $a = 3.68$ and $b = 12.47$ because the domain of applicability of the Colebrook equation that is between 3.68 and 12.47 in respect to x, sgn is signum function: if $F(a) > 0 \to sgn(F(a)) = 1$, $F(a) = 0 \to sgn(F(a)) = 0$, and $F(a) < 0 \to sgn(F(a)) = -1$, while $tanh$ is hyperbolic tangent which is defined through the exponential function e^x with non-integer power x the use of which is as computationally expensive as the use of the logarithmic function and which therefore cannot be recommended.

3. Every explicit approximation of the Colebrook equation [19–28]; $x \approx \varsigma(Re, \varepsilon^*)$, where ς is the functional symbol, can be used to choose an initial starting point x_0. On the other hand, almost all available approximations contain logarithmic or/and terms with non-integer powers, which makes them unsuitable for use in the developed approach. On the other hand, having previous experience with training Artificial Neural Networks (ANN) to simulate the Colebrook equation [36], i.e., to use ability of artificial intelligence to simulate the Colebrook equation not knowing its logarithmic nature but only knowing raw input and corresponding output datasets $\{Re, \varepsilon^*\} \to \{x\}$, a computationally cheap explicit approximation of the Colebrook equation is developed through genetic programming [21,37–40]. The developed approximation is computationally efficient because of its polynomial structure; Equation (9):

$$x_0 = 5.05 - 30.73 \cdot \varepsilon^* + \frac{3.4 \cdot Re + \frac{Re^2}{4696477}}{46137.9 + Re + \frac{Re^2}{3250657.6} + \frac{\varepsilon^* \cdot Re^2}{515.25}} \tag{9}$$

Eureqa [computer software] by Nutonian, Inc., Boston, MA, USA. [41,42] is used to generate Equation (9). The Eureqa-polynomial approximation; Equation (9) has up to 40% relative error, but it is very cheap and sufficiently accurate to serve as an initial starting point x_0.

4. Extensive tests over the domain of applicability of the Colebrook equation shows that one fixed value can also be used as the initial starting point x_0 for the iterative procedure in all cases.

Results indicate that the proposed Padé approach works in all cases, as the argument z of $ln(z)$ is always close to one. When Equation (9) is used, values of z are within the range 0.91–1.05. Moreover, for the most pairs of the Reynolds number Re and the relative roughness of inner pipe surfaces ε^* which are in the domain of applicability, the initial starting point $x_0 = 7.273124147$ requires the least number of iterations.

To avoid using a computationally expensive logarithmic function in the initial stage of the iterative procedure, the recommendation is to start calculation with fixed-value starting point $x_0 = 7.273124147$ or to use a polynomial expression; Equation (9). Power-law formulas from Russian practice which does not contain logarithmic function can also be used as an alternative although they usually contain integer power in fractional form [43–45].

4. Proposed Iterative Method

The Colebrook equation is usually solved iteratively using the Newton-Raphson method [46] or even more using a simplified Newton-Raphson method known as the fixed-point method [18]. Recently, hybrid three-point methods have been proposed [47,48].

Here is presented an adjusted very accurate, fast and computationally cheap version of the Newton-Raphson method suitable for the Colebrook equation in which the logarithmic function is replaced after the first iteration with the Padé approximant in polynomial form [29].

Knowing that the Colebrook equation is based on logarithmic law [1,2], the achievement with this simplified approach is more significant. Numerical examples are shown in Section 5 of this paper.

Iteration 0:

In order to avoid use of computationally expensive logarithmic functions or functions with non-integer powers, a required initial starting point x_0 should be chosen using recommendations from Section 3 of this paper; points 3 or 4.

Iteration 1:

Having provided an initial starting point x_0, new value x_1 can be calculated using Equation (10):

$$x_1 = x_0 - \frac{F(x_0)}{F'(x_0)} \tag{10}$$

In Equation (10), $F(x)$ represents the Colebrook equation $x = \xi(x)$ which needs to be in suitable form, $F = x - \xi(x) = 0$; Equation (11):

$$F(x_0) = x_0 + 2 \cdot \log_{10}(y_0) = 0 \tag{11}$$

In Equation (11), $y_0 = \frac{2.51 \cdot x_0}{Re} + \frac{\varepsilon^*}{3.71}$ which will also be used in the next iteration (in an additional variant of the proposed method y_0 is used in all subsequent iterations), while in Equation (10), the first derivative of F in respect to x; $F'(x)$ is from Equation (12):

$$F'(x_0) = 1 + \frac{2 \cdot 2.51}{2.302585093 \cdot Re \cdot \left(\frac{100 \cdot \varepsilon^*}{371} + \frac{2.51 \cdot x_0}{Re} \right)} \tag{12}$$

In Equation (12), $ln(10) \approx 2.302585093$ is with constant value, and therefore only $\log_{10}(y_0)$ from Equation (11) requires evaluation of the logarithmic function.

In many programming languages, evaluation of logarithmic function of any base is processed by natural logarithm [14]. Change of 10-base logarithm from the Colebrook equation to e-based natural logarithm where $e \approx 2.718$ and where $ln(10) \approx 2.302585093$ is implemented as $\log_{10}(z) = \frac{ln(z)}{ln(10)}$.

Iteration 2:

New value x_2 should be calculated using Equation (13):

$$x_2 = x_1 - \frac{F(x_1)}{F'(x_1)} \tag{13}$$

In Equation (13), $F(x_1)$ is not calculated by $log_{10}(y_1)$, where $y_1 = \frac{2.51 \cdot x_1}{Re} + \frac{\varepsilon^*}{3.71}$, but using Padé polynomial replacement for logarithmic function which is accurate for $z \to 1$ and using the already calculated value of $log_{10}(y_0)$ from the previous iteration; Equation (14):

$$F(x_1) = x_1 + 2 \cdot log_{10}(y_0) - \frac{(z_{01} - 1) \cdot (11 \cdot z_{01}^2 + 38 \cdot z_{01} + 11)}{2.302585093 \cdot (3 \cdot z_{01}^3 + 9 \cdot z_{01}^2 + 9 \cdot z_{01} + 1)} \tag{14}$$

In Equation (14), $log_{10}(y_0) - \frac{(z_{01}-1) \cdot (11 \cdot z_{01}^2 + 38 \cdot z_{01} + 11)}{2.302585093 \cdot (3 \cdot z_{01}^3 + 9 \cdot z_{01}^2 + 9 \cdot z_{01} + 1)} \equiv log_{10}(y_1)$, $2.302585093 \approx ln(10)$, and $z_{01} = \frac{y_0}{y_1}$. In the first iteration, $log_{10}(y_0)$ is already known; Equation (11). The Padé polynomial used in Equation (14) is of order (2,3) which means that the polynomial in the numerator is of the order of 2 while in the denominator of order 3. The Padé polynomials are also known as Padé approximants and here the maximal relative error of the polynomial expression term in Equation (14) in domain $z\epsilon[0.6, 1.6]$; $z \to 1$ is minor as shown in Table 1. Value of z for the procedure shown in practice is $z\epsilon[0.9, 1.1]$ and therefore the error of the used Padé approximant can be neglected in the case shown.

The first derivative $F'(x_1)$ does not contain any logarithmic functions and should be evaluated using Equation (12), where x_0 should be replaced with the new value x_1 or knowing that the value of the derivative does not change significantly between two iterations, $F'(x_0)$ can be reused in all subsequent iterative cycles. Even knowing that the value of the first derivate in the procedure shown is always near 1; for rough calculations it can be assumed that $F'(x) \approx 1$ which gives the fixed-point method as a special case of the Newton-Raphson scheme.

Iteration 3:

New value x_3 is again evaluated in the same way using Equation (15):

$$x_3 = x_2 - \frac{F(x_2)}{F'(x_2)} \tag{15}$$

In Equation (15), $F'(x_2)$ can be calculated or $F'(x_1)$ or $F'(x_0)$ can be reused. In additon, $F(x_2)$ can be calculated using $z_{02} = \frac{y_0}{y_2}$, where $y_2 = \frac{2.51 \cdot r_2}{Re} + \frac{\varepsilon^*}{3.71}$. Input parameter for Padé polynomial z_{02} here refers to y_0 from the first iteration; Equation (16):

$$F(x_2) = x_2 + 2 \cdot log_{10}(y_0) - \frac{(z_{02} - 1) \cdot (11 \cdot z_{02}^2 + 38 \cdot z_{02} + 11)}{2.302585093 \cdot (3 \cdot z_{02}^3 + 9 \cdot z_{02}^2 + 9 \cdot z_{02} + 1)} \tag{16}$$

In Equation (16), $log_{10}(y_0) - \frac{(z_{02}-1) \cdot (11 \cdot z_{02}^2 + 38 \cdot z_{02} + 11)}{2.302585093 \cdot (3 \cdot z_{02}^3 + 9 \cdot z_{02}^2 + 9 \cdot z_{02} + 1)} \equiv log_{10}(y_2)$.

The Padé polynomial is a very accurate approximation of logarithmic function, so knowing that y_0 is evaluated directly through the logarithmic function, while y_1, y_2, y_3, etc. is based on its Padé polynomial equivalent, it is obvious that the sequence $z_{01} = \frac{y_0}{y_1}$, $z_{02} = \frac{y_0}{y_2}$, $z_{03} = \frac{y_0}{y_3}$, etc. is slightly more accurate compared with the sequence $z_{01} = \frac{y_0}{y_1}$, $z_{12} = \frac{y_1}{y_2}$, $z_{23} = \frac{y_2}{y_3}$, etc. which accumulates error introduced with Padé approximations. Anyway, the introduced error is so small that it can practically be neglected. The second sequence $z_{01} = \frac{y_0}{y_1}$, $z_{12} = \frac{y_1}{y_2}$, $z_{23} = \frac{y_2}{y_3}$, etc. yields Equation (17):

$$F(x_2) = x_2 + 2 \cdot log_{10}(y_1) - \frac{(z_{12} - 1) \cdot (11 \cdot z_{12}^2 + 38 \cdot z_{12} + 11)}{2.302585093 \cdot (3 \cdot z_{12}^3 + 9 \cdot z_{12}^2 + 9 \cdot z_{12} + 1)} = x_2 + 2 \cdot$$

$$log_{10}(y_0) - \frac{(z_{01} - 1) \cdot (11 \cdot z_{01}^2 + 38 \cdot z_{01} + 11)}{2.302585093 \cdot (3 \cdot z_{01}^3 + 9 \cdot z_{01}^2 + 9 \cdot z_{01} + 1)} - \frac{(z_{12} - 1) \cdot (11 \cdot z_{12}^2 + 38 \cdot z_{12} + 11)}{2.302585093 \cdot (3 \cdot z_{12}^3 + 9 \cdot z_{12}^2 + 9 \cdot z_{12} + 1)} \tag{17}$$

In Equation (17), $log_{10}(y_1) - \frac{(z_{12}-1) \cdot (11 \cdot z_{12}^2 + 38 \cdot z_{12} + 11)}{2.302585093 \cdot (3 \cdot z_{12}^3 + 9 \cdot z_{12}^2 + 9 \cdot z_{12} + 1)} \equiv log_{10}(y_2)$ and, $log_{10}(y_0) - \frac{(z_{01}-1) \cdot (11 \cdot z_{01}^2 + 38 \cdot z_{01} + 11)}{2.302585093 \cdot (3 \cdot z_{01}^3 + 9 \cdot z_{01}^2 + 9 \cdot z_{01} + 1)} \equiv log_{10}(y_1)$

Iteration i:

All indexes i in respect the third iteration should be updated as $i = i + 1$ with exemption of index 0 in Equation (16). The calculation is finished when $x_{i+1} \approx x_i$.

The algorithm for the proposed improved procedure is given in Figure 2.

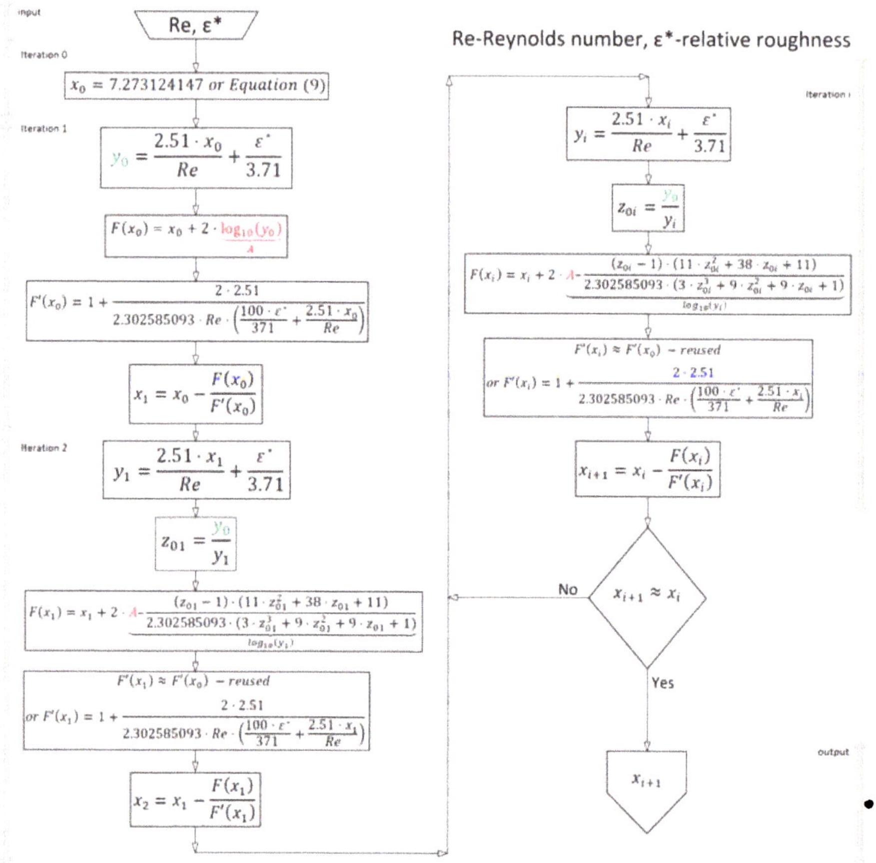

Figure 2. Algorithm for the proposed one *log*-call improved procedure.

Only a one-off evaluation of the logarithmic function is needed in the proposed algorithm from Figure 2, which is clearly marked in red; $A = log_{10}(y_0)$. On the other hand, y_0 calculated in iteration 1 is reused in all next steps and it is marked in green in Figure 2.

The proposed procedure can be simplified assuming that $F'(x_i) = 1$, which gives the simple fixed-point procedure [18] instead of the Newton-Raphson.

5. Numerical Examples

Here are two numerical examples:

Example 1:	Example 2:	
$Re = 8.31 \cdot 10^3,\ \varepsilon^* = 0.024$	$Re = 2.5 \cdot 10^6,\ \varepsilon^* = 4 \cdot 10^{-4}$	
Iteration 0		
$x_0 = 6.279860788$	$x_0 = 7.401979091$	(9)
Iteration 1		
$y_0 = 0.008365808$	$y_0 = 0.000115248$	
$\log_{10}(y_0) = -2.077492116$	$\log_{10}(y_0) = -3.938365477$	
$F(x_0) = 2.124876556$	$F(x_0) = -0.474751864$	(11)
$F'(x_0) = 1.001337518$	$F'(x_0) = 1.001024781$	(12)
$x_1 = 4.157822498$	$x_1 = 7.876244936$	(10)
Iteration 2		
$y_1 = 0.007724855$	$y_1 = 0.000115724$	
$z_{01} = \frac{y_0}{y_1} = 1.082972765$	$z_{01} = \frac{y_0}{y_1} = 0.995885374$	
0.034617535	-0.001790646	Padé approximant (6)
$F(x_1) = -0.066396805$	$F(x_1) = 0.003095273$	(14)
$FF'(x_1) = 1.001986711$	$F'(x_1) = 1.000970478$	
$x_2 = 4.224087653$	$x_2 = 7.873152664$	(13)
Iteration 3		
$y_2 = 0.00774487$	$y_2 = 0.000115721$	
$z_{02} = \frac{y_0}{y_2} = 1.080174034$	$z_{02} = \frac{y_0}{y_2} = 0.995912092$	
0.033493733	-0.001778995	Padé approximant (6)
$F(x_2) = 0.002115955$	$F(x_2) = -2.03017 \cdot 10^{-5}$	(16)
$F'(x_2) = 1.001957048$	$F'(x_2) = 1.000970813$	
$x_3 = 4.221975832$	$x_3 = 7.873172946$	(15)
Final value:		
$x = 4.22204103$	$x = 7.873172814$	
$\lambda = \frac{1}{x^2} = 0.056098998$	$\lambda = \frac{1}{x^2} = 0.016132454$	

6. Conclusions

An efficient algorithm for the iterative calculation of the Colebrook equation by both an accurate and computationally efficient Padé approximation is presented in this paper. It requires only one evaluation of the logarithmic function in respect to the whole iterative procedure and more specifically only in the first iteration, while the common procedures from current engineering practice require at least one evaluation of logarithmic function for every single iteration. The logarithmic function in the proposed procedure is replaced in all iterations (except the first), with simple, accurate and efficient Padé polynomials [29]. In this way the same accuracy is reached through the proposed less demanding procedure, after the same number of iterations as in the standard algorithm which uses *log*-call in each iterative step. This is a good achievement, knowing that the nature of the Colebrook equation is logarithmic. For their evaluation in the Central Processor Unit (CPU), Padé polynomials require a lower number of floating-point operations to be executed compared with the logarithmic function [10–14,30–34,44].

The here presented iterative approach only introduces a computationally cheaper alternative to the standard iterative procedure. It does not reduce the number of required iterations to reach the final desired accuracy nor provide more accurate results. The proposed method reduces the burden for the Central Processor Unit (CPU) as less floating-point operations need to be executed. In that way, the presented approach also increases speed of computation. On the other hand, many explicit non-iterative approximations to the Colebrook equation are available in literature [19–28,49] which initially appear simple for computation, but are not. They are widely used, but although some of them are very accurate, they contain relatively complex internal iterative steps and also a number of computationally demanding functions. For example, the widely used Haaland approximation [49,50] introduces relative error up to 1.5%, but with the cost of evaluation of one logarithmic expression and one non-integer power. In addition, the approximation by Romeo et al. [26,39] reaches extremely high accuracy with the relative error of up to 0.14%, but with a cost of evaluation of even three logarithmic expressions and two non-integer powers. Regarding alternative iterative procedures, Clamond [10] provides an accurate iterative approach using Ω function, but this algorithm requires at least two *log*-calls; one for initialization and one in the first iteration, which is more expensive compared with the here presented approach.

The procedure proposed in this paper can significantly reduce the computational burden for evaluating complex distribution networks with various applications (water, gas) [9,50–55]. For example, a probabilistic approach using time dependent modeling of distribution or transmission networks requires many millions of evaluations of Colebrook's equation. For such kinds of computations it is always good to have a cheaper but still accurate approach to speed up the process.

Author Contributions: The paper is a product of joint efforts of the authors who worked together on models of natural gas distribution networks. P.P. has scientific background in computer science and applied mathematics while D.B. in petroleum and mechanical engineering. This multidisciplinary approach yields the simplification in the proposed iterative calculation. Both authors tested the proposed method and controlled the results independently. D.B. wrote the draft of this paper based on software implementation provided by P.P.

Funding: The European Commission covers the Article Processing Charges to make this paper available to all interested parties through the gold open access model.

Acknowledgments: We thank Adrian O'Connell who as a native speaker kindly checked the correctness of English expression through the paper.

Conflicts of Interest: The authors declare no conflict of interest. The views expressed are those of the authors and may not in any circumstances be regarded as stating an official position of the European Commission or of the Technical University Ostrava.

References

1. Colebrook, C.F. Turbulent flow in pipes with particular reference to the transition region between the smooth and rough pipe laws. *J. Inst. Civ. Eng. (Lond.)* **1939**, *11*, 133–156. [CrossRef]
2. Colebrook, C.; White, C. Experiments with fluid friction in roughened pipes. *Proc. R. Soc. Lond. Ser. A Math. Phys. Sci.* **1937**, *161*, 367–381. [CrossRef]
3. Moody, L.F. Friction factors for pipe flow. *Trans. ASME* **1944**, *66*, 671–684.
4. LaViolette, M. On the history, science, and technology included in the Moody diagram. *J. Fluids Eng.* **2017**, *139*, 030801. [CrossRef]
5. Flack, K.A. Moving beyond Moody. *J. Fluid Mech.* **2018**, *842*, 1–4. [CrossRef]
6. Allen, J.J.; Shockling, M.A.; Kunkel, G.J.; Smits, A.J. Turbulent flow in smooth and rough pipes. *Philos. Trans. R. Soc. Lond. A Math. Phys. Eng. Sci.* **2007**, *365*, 699–714. [CrossRef] [PubMed]
7. Langelandsvik, L.I.; Kunkel, G.J.; Smits, A.J. Flow in a commercial steel pipe. *J. Fluid Mech.* **2008**, *595*, 323–339. [CrossRef]
8. McKeon, B.J.; Zagarola, M.V.; Smits, A.J. A new friction factor relationship for fully developed pipe flow. *J. Fluid Mech.* **2005**, *538*, 429–443. [CrossRef]
9. Praks, P.; Kopustinskas, V.; Masera, M. Monte-Carlo-based reliability and vulnerability assessment of a natural gas transmission system due to random network component failures. *Sustain. Resil. Infrastruct.* **2017**, *2*, 97–107. [CrossRef]
10. Clamond, D. Efficient resolution of the Colebrook equation. *Ind. Eng. Chem. Res.* **2009**, *48*, 3665–3671. [CrossRef]
11. Giustolisi, O.; Berardi, L.; Walski, T.M. Some explicit formulations of Colebrook–White friction factor considering accuracy vs. computational speed. *J. Hydroinform.* **2011**, *13*, 401–418. [CrossRef]
12. Danish, M.; Kumar, S.; Kumar, S. Approximate explicit analytical expressions of friction factor for flow of Bingham fluids in smooth pipes using Adomian decomposition method. *Commun. Nonlinear Sci. Numer. Simul.* **2011**, *16*, 239–251. [CrossRef]

13. Winning, H.K.; Coole, T. Explicit friction factor accuracy and computational efficiency for turbulent flow in pipes. *Flow Turbul. Combust.* **2013**, *90*, 1–27. [CrossRef]

14. Vatankhah, A.R. Approximate analytical solutions for the Colebrook equation. *J. Hydraul. Eng.* **2018**, *144*, 06018007. [CrossRef]

15. Sonnad, J.R.; Goudar, C.T. Constraints for using Lambert W function-based explicit Colebrook–White equation. *J. Hydraul. Eng.* **2004**, *130*, 929–931. [CrossRef]

16. Brkić, D. Comparison of the Lambert W-function based solutions to the Colebrook equation. *Eng. Comput.* **2012**, *29*, 617–630. [CrossRef]

17. Brkić, D. W solutions of the CW equation for flow friction. *Appl. Math. Lett.* **2011**, *24*, 1379–1383. [CrossRef]

18. Brkić, D. Solution of the implicit Colebrook equation for flow friction using Excel. *Spreadsheets Educ. (eJSiE)* **2017**, *10*, 2. Available online: http://epublications.bond.edu.au/ejsie/vol10/iss2/2 (accessed on 11 July 2018).

19. Gregory, G.A.; Fogarasi, M. Alternate to standard friction factor equation. *Oil Gas J.* **1985**, *83*, 120–127.

20. Brkić, D. Review of explicit approximations to the Colebrook relation for flow friction. *J. Pet. Sci. Eng.* **2011**, *77*, 34–48. [CrossRef]

21. Brkić, D.; Ćojbašić, Ž. Evolutionary optimization of Colebrook's turbulent flow friction approximations. *Fluids* **2017**, *2*, 15. [CrossRef]

22. Brkić, D. Determining friction factors in turbulent pipe flow. *Chem. Eng.* **2012**, *119*, 34–39.

23. Buzzelli, D. Calculating friction in one step. *Mach. Des.* **2008**, *80*, 54–55.

24. Zigrang, D.J.; Sylvester, N.D. Explicit approximations to the solution of Colebrook's friction factor equation. *AIChE J.* **1982**, *28*, 514–515. [CrossRef]

25. Serghides, T.K. Estimate friction factor accurately. *Chem. Eng.* **1984**, *91*, 63–64.

26. Romeo, E.; Royo, C.; Monzón, A. Improved explicit equations for estimation of the friction factor in rough and smooth pipes. *Chem. Eng. J.* **2002**, *86*, 369–374. [CrossRef]

27. Vatankhah, A.R.; Kouchakzadeh, S. Discussion of "Turbulent flow friction factor calculation using a mathematically exact alternative to the Colebrook–White equation" by Jagadeesh R. Sonnad and Chetan T. Goudar. *J. Hydraul. Eng.* **2008**, *134*, 1187. [CrossRef]

28. Vatankhah, A.R.; Kouchakzadeh, S. Discussion of "Exact equations for pipe-flow problems". *J. Hydraul. Res.* **2009**, *47*, 537–538. [CrossRef]

29. Baker, G.A.; Graves-Morris, P. Padé approximants. In *Encyclopedia of Mathematics and Its Applications*; Cambridge University Press: Cambridge, UK, 1996. [CrossRef]

30. Kropa, J.C. Calculator algorithms. *Math. Mag.* **1978**, *51*, 106–109. [CrossRef]

31. Rising, G.R. *Inside Your Calculator: From Simple Programs to Significant Insights*; John Wiley & Sons: Hoboken, NJ, USA, 2007. [CrossRef]

32. Pineiro, J.A.; Ercegovac, M.D.; Bruguera, J.D. Algorithm and architecture for logarithm, exponential, and powering computation. *IEEE Trans. Comput.* **2004**, *53*, 1085–1096. [CrossRef]

33. Biberg, D. Fast and accurate approximations for the Colebrook equation. *J. Fluids Eng.* **2017**, *139*, 031401. [CrossRef]

34. Winning, H.K.; Coole, T. Improved method of determining friction factor in pipes. *Int. J. Numer. Methods Heat Fluid Flow* **2015**, *25*, 941–949. [CrossRef]

35. Yun, B.I. A non-iterative method for solving non-linear equations. *Appl. Math. Comput.* **2008**, *198*, 691–699. [CrossRef]

36. Brkić, D.; Ćojbašić, Ž. Intelligent flow friction estimation. *Comput. Intell. Neurosci.* **2016**, 5242596. [CrossRef] [PubMed]

37. Giustolisi, O.; Savić, D.A. A symbolic data-driven technique based on evolutionary polynomial regression. *J. Hydroinform.* **2006**, *8*, 207–222. [CrossRef]

38. Petković, D.; Gocić, M.; Shamshirband, S. Adaptive Neuro-Fuzzy computing technique for precipitation estimation. *Facta Univ. Ser. Mech. Eng.* **2016**, *14*, 209–218. [CrossRef]

39. Ćojbašić, Ž.; Brkić, D. Very accurate explicit approximations for calculation of the Colebrook friction factor. *Int. J. Mech. Sci.* **2013**, *67*, 10–13. [CrossRef]

40. Mitrev, R.; Tudjarov, B.; Todorov, T. Cloud-based expert system for synthesis and evolutionary optimization of planar linkages. *Facta Univ. Ser. Mech. Eng.* **2018**. [CrossRef]

41. Schmidt, M.; Lipson, H. Distilling free-form natural laws from experimental data. *Science* **2009**, *324*, 81–85. [CrossRef] [PubMed]

42. Dubčáková, R. Eureqa: Software review. *Genet. Program. Evol. Mach.* **2011**, *12*, 173–178. [CrossRef]
43. Альтшуль, А.Д. Гидравлические Сопротивления; Недра: Moscow, Russia, 1982. (In Russian)
44. Lipovka, A.Y.; Lipovka, Y.L. Determining hydraulic friction factor for pipeline systems. Journal of Siberian Federal University. *Eng. Technol.* **2014**, *7*, 62–82. Available online: http://elib.sfu-kras.ru/handle/2311/10293 (accessed on 11 July 2018).
45. Черникин, В.А.; Черникин, А.В. Обобщенная формула для расчета коэффициента гидравлического сопротивления магистральных трубопроводов для светлых нефтепродуктов и маловязких нефтей. Наука и Технологии Трубопроводного Транспорта Нефти и Нефтепродуктов **2012**, *4*, 64–66. (In Russian)
46. Abbasbandy, S. Improving Newton–Raphson method for nonlinear equations by modified Adomian decomposition method. *Appl. Math. Comput.* **2003**, *145*, 887–893. [CrossRef]
47. Brkić, D.; Praks, P. Discussion of "Approximate analytical solutions for the Colebrook equation". *J. Hydraul. Eng.* **2019**, *144*, 06018007.
48. Džunić, J.; Petković, M.S.; Petković, L.D. A family of optimal three-point methods for solving nonlinear equations using two parametric functions. *Appl. Math. Comput.* **2011**, *217*, 7612–7619. [CrossRef]
49. Haaland, S.E. Simple and explicit formulas for the friction factor in turbulent pipe flow. *J. Fluids Eng.* **1983**, *105*, 89–90. [CrossRef]
50. Wood, D.J.; Haaland, S.E. Discussion and closure: "Simple and explicit formulas for the friction factor in turbulent pipe flow" (Haaland, S.E., 1983, ASME J. Fluids Eng., 105, pp. 89–90). *J. Fluids Eng.* **1983**, *105*, 242–243. [CrossRef]
51. Praks, P.; Kopustinskas, V.; Masera, M. Probabilistic modelling of security of supply in gas networks and evaluation of new infrastructure. *Reliab. Eng. Syst. Saf.* **2015**, *144*, 254–264. [CrossRef]
52. Brkić, D. Spreadsheet-based pipe networks analysis for teaching and learning purpose. *Spreadsheets Educ. (eJSiE)* **2016**, *9*, 4. Available online: https://epublications.bond.edu.au/ejsie/vol9/iss2/4/ (accessed on 11 July 2018).
53. Brkić, D. Discussion of "Economics and statistical evaluations of using Microsoft Excel solver in pipe network analysis" by I.A. Oke, A. Ismail, S. Lukman, S.O. Ojo, O.O. Adeosun, and M.O. Nwude. *J. Pipeline Syst. Eng. Pract.* **2018**, *9*, 07018002. [CrossRef]
54. Brkić, D. Iterative methods for looped network pipeline calculation. *Water Resour. Manag.* **2011**, *25*, 2951–2987. [CrossRef]
55. Brkić, D. A gas distribution network hydraulic problem from practice. *Petrol. Sci. Technol.* **2011**, *29*, 366–377. [CrossRef]

MDPI

St. Alban-Anlage 66

4052 Basel

Switzerland

Tel. +41 61 683 77 34

Fax +41 61 302 89 18

www.mdpi.com

Energies Editorial Office

E-mail: energies@mdpi.com

www.mdpi.com/journal/energies